Anthropogeomorphology of Bhagirathi-Hooghly River System in India

Anthropogeomorphology of Bhagirathi-Hooghly River System in India

Edited by
Balai Chandra Das, Sandipan Ghosh, Aznarul Islam
and Suvendu Roy

CRC Press
Taylor & Francis Group
Boca Raton London New York

CRC Press is an imprint of the
Taylor & Francis Group, an **informa** business

First edition published 2021
by CRC Press
6000 Broken Sound Parkway NW, Suite 300, Boca Raton, FL 33487-2742

and by CRC Press
2 Park Square, Milton Park, Abingdon, Oxon, OX14 4RN

ISBN: 978-0-367-86102-5 (hbk)
ISBN: 978-1-003-03237-3 (ebk)

Typeset in Times
by codeMantra

To our parents and teachers

Contents

Foreword

I was very pleased to be invited by Dr. Suvendu Roy on behalf of the editorial board to write a Foreword to this edited book entitled *Anthropogeomorphology of Bhagirathi-Hooghly River System in India*.

Anthropogeomorphology was proposed more than 50 years ago (Golomb and Eder, 1964; Fels, 1965) as a discipline devoted to the study of humans as geomorphological agents. It now seems strange to reflect that human impact was not significantly considered by geomorphologists in relation to environmental impact until about 1960 (Gregory and Lewin, 2014, Chapter 16). In Chapter 3, phases of contributions were distinguished: the first from 1864 to 1976 providing the foundation, the second from 1960 to 1999 including a brilliant paper (Wolman, 1967) which changed ways of geomorphological thinking (Gregory, 2011) and a third group from 2000 to 2010 which included a series of contributions documenting present processes. With the Anthropocene now officially proposed as a subdivision in the geological timescale, recognising the extent to which human agency has affected the earth's land surface, it must be appropriate to recognise a fourth phase since 2010 in which the significance of results of anthropogenic investigations is documented and their importance for future understanding and sustainable management of environment demonstrated and utilised.

It is in this context that it is a great pleasure to congratulate the editors for their vision in producing this timely book on the Bhagirathi-Hooghly River System, a major system which has an essential role including irrigation and water supply, for densely populated West Bengal. There have not been enough publications focused on major complex systems so this is a very welcome addition to the research literature. The way in which the chapters allow a multidisciplinary focus, reconstruct stages and character of anthropogenic impact and embrace approaches at several scales is very appropriate. Including products of spatial survey often employing remote sensing with results including a human impact map, floodplain inundation maps, geodiversity index as well as considering hazards, risks and implications of future climate change is very warmly commended. I hope that the results of each investigation will provide the enhanced understanding and appreciation necessary to inform future hazard management.

Previous emphasis in anthropogeomorphology has often been on small-scale problems, complex situations have been avoided and not enough applications have been drawn out. Therefore, I am sure that this book will make a significant contribution and provide a foundation for future integrated management approaches and sustainable development.

Kenneth J Gregory
CBE BSc, PhD, DSc, DSc (Hon), DUniv, CGeog., FRGS, FGCL, FBSG
Visiting Professor University of Southampton, Emeritus Professor University of London

REFERENCES

Fels, E. 1965. Nochmals: Anthropogen Geomorphologie. *Petermanns Geographische Mitteilungen* 109, 9–15.
Golomb, B. and Eder, H.M. 1964. Landforms made by man. *Landscape* 14, 4–7.
Gregory, K.J. 2011. Wolman MG (1967) A cycle of sedimentation and erosion in urban river channels. *Geografiska Annaler* 49A: 385–395. *Progress in Physical Geography* 35, 831–41.
Gregory, K.J. and Lewin, J. 2014. *The Basics of Geomorphology: Key Concepts.* Sage, London.
Wolman, M.G. 1967. A cycle of sedimentation and erosion in urban river channels. *Geografiska Annaler* 49A, 385–95.

Foreword

This book is both timely and prescient. It is founded on an objective, scientific synthesis of the outcomes of decades of research on fluvial forms, processes and functions in a river network that is of both national and global significance. The chapters are authored by those responsible for the primary research, and they have been edited by four leading academics, each of whom is an expert in his field, and each of whom is of national and international repute. Indeed, as an admirer of the editors and their original contributions to knowledge, I was honoured to be offered the opportunity to write the Foreword to this book.

This book uses a new lens through which to examine the history, current status and future prospects of rivers comprising the Bhagirathi-Hooghly drainage system. That lens is *Anthropogeomorphology*: an approach that applies the principles of *geomorphology* (*geo*—the earth, *morph*—the form, *ology*—the study of) to reveal the role of *anthro* (i.e. humans) in affecting the physical, chemical and biological processes responsible for river forms, functions and changes therein. The influence of human impacts on the environment is now known to be pervasive and, perhaps, dominant, giving rise to the generally accepted proposal that we are now in a new geological era—the Anthropocene. However, the date on which this era began is still to be agreed, with arguments made for a start date as early as the agricultural revolution (12,000–15,000 years BP) and as late as the *great acceleration* that began after World War II.

As chronicled in this book, human impacts on rivers in the Bhagirathi-Hooghly system can be traced back at least 4,500 years—to the ancient Bengal civilisation, though the most serious consequences of catchment land-use intensification, extractive industries, development on floodplains and river management/regulation for land drainage, waste disposal, water supply and flood control may be attributed to the 20th century and, especially, the period since the 1970s. The worrying fact is that environmental degradation of rivers, wetlands and floodplains of the types forensically reported in this book is not unique to that region: on the contrary, every large, deltaic river in the world is in trouble, from the Mississippi to the Mekong.

Rivers in the Bhagirathi-Hooghly system are exploited by humankind in multiple ways and detailed accounts provided in this book demonstrate how rivers and riverine landscapes are used across a range of geographical scales to support primary industries (agriculture, forestry, fishing, sand and mineral mining), accommodate transport networks (road, rail and water), supply fresh water (irrigation, municipal and conjunctive uses) and dispose of industrial pollutants and urban contaminants. Numerous dams, barrages, channelisation projects and embankments all serve to enhance the capacity of river functions highly valued by society but at the cost of reductions in the capacity of the river to support less-valued functions such as wildlife and biodiversity. Inevitably, as pressure on the natural system has increased, more and more of the less-valued functions have been sacrificed to fuel economic growth. The clear message that emerges from the multiple case examples related in this book is that, while the development history of each stream and river differs, in every case over-exploitation and careless development eventually threatens even the highest-valued river functions. In short, it is the path to environmental, economic and societal catastrophe.

The factual evidence included in this book is compelling and unequivocal: both practitioners within the relevant river and water management authorities and representatives of the communities served by those authorities should read it and heed the messages it delivers. This is necessary if the parlous state of river in the Bhagirathi-Hooghly system is to be addressed successfully and sustainably. The good news is that this is still possible; the future prospects of these rivers may seem bleak, but the great thing about the future is that we can change it.

To achieve this, difficult choices will have to be made, particularly in prioritising some rivers for restorative action above others. In prioritising projects, authorities should note that while economic development and environmental conservation are worthy goals, whether they are meeting the needs of the people remains of paramount importance. In this context, sustainable river management is a matter of social science just as much as it is an issue in economics, engineering and environmental science. It follows that solutions to the problems faced by regional rivers and their managers should strive to improve the lives of citizens living in the vulnerable communities so often involved in unsustainable encroachment into the river's domain: this opens the way for solutions that deliver and distribute societal benefits in ways that are socially equitable.

In closing, I will add only this thought; while the economic, environmental and social benefits provided by healthy rivers are of inestimable value, rivers do not exist to serve the needs of humankind. In the words of the great German philosopher Immanuel Kant, in the world of ends, "everything has either a *price* or a *dignity*. Anything that has a *price* can be replaced by something else as its equivalent, but anything that is above all *price*, and therefore admits of no equivalent, has a *dignity*".[1] To paraphrase Kant in the context of the Bhagirathi-Hooghly system, 'rivers are not the means to an end (i.e. supplying society's needs), but are an end in themselves, which means they have not merely a relative worth, that is, a *price*, but an inner worth, that is a *dignity*'. But what is this thing, *dignity*? To answer that, consider the root, the etymology of the word:

dignity (n.)
from Latin *dignitas* "worthy, proper, fitting, honorable",
Etymological definitions from: http://www.etymonline.com

---

[1] From *Groundwork of the Metaphysics of Morals*, translated and edited by Mary Gregor (Cambridge: Cambridge, 1998), pp. 42–43.

Modern definitions build on the concept of 'worth' to define *dignity* as the "innate right to be **valued, respected**, and to **receive ethical treatment**. In the modern context, **dignity** can function as an extension of the Enlightenment-era concept that living entities have **inherent, inalienable rights**". https://en.wikipedia.org/wiki/Dignity.

Hence, restoring the dignity of a river starts by admitting that its worth does not stem only from the benefits it provides to society, the values we put on those benefits or the price society is willing to pay for them. For many readers, this may be a new and novel view of the river, but in 2017, the High Court of Uttarakhand declared that the Ganges and Yamuna rivers are living entities and granted them the same legal rights as a person.[2] This enshrined protection of these great rivers in a way that manifestly recognises their innate right to be valued, respected and gives their representatives (named by the HC as the director general of Namami Gange project and the Uttarakhand chief secretary and advocate general) recourse to legal remedies against any person or organisation that would do them harm. In this context, time is especially short for the Yamuna, which has been described as being "close to death"[3].

My sincere hope is that, notwithstanding current, erstwhile restoration efforts, Uttarakhand HC's move (and similar moves in New Zealand[4], Colombia[5] and Bangladesh[6]) will bolster efforts to restore the utility, health and dignity of not only the Ganges and Yamuna themselves but also thousands of kilometres of degraded and disrespected rivers across the world, including, particularly, those in the Bhagirathi-Hooghly river system.

Colin R. Thorne
Professor and Chair of Physical Geography
University of Nottingham, UK

---

[2] https://www.hindustantimes.com/india-news/uttarakhand-hc-says-ganga-is-india-s-first-living-entity-grants-it-rights-equal-to-humans/story-VoI6DOG71fyMDihg5BuGCL.html.

[3] https://www.hindustantimes.com/delhi/yamuna-a-dead-river-says-report-even-as-focus-on-clean-ganga/story-4R6VXEcjNOILSelnREqrxN.html.

[4] https://www.theguardian.com/world/2017/mar/16/new-zealand-river-granted-same-legal-rights-as-human-being.

[5] https://www.earthlawcenter.org/blog-entries/2018/7/rights-for-the-anchicay-river-in-colombia.

[6] https://asiancorrespondent.com/2019/07/bangladesh-declares-its-rivers-legal-persons/.

Preface

Although, as of June 2019, neither the International Commission on Stratigraphy (ICS) nor the International Union of Geological Sciences (IUGS) has officially approved the term 'Anthropocene' as a recognised subdivision of geologic time (https://en.wikipedia.org/wiki/Anthropocene), yet the fact is that during the era of Great Acceleration, man plays an increasingly significant role in changing the face of the inhabited earth. The earth scientists all over the world are keenly concerned about it and many scientific papers are coming each year illuminating anthropogeomorphology of different corners of the world. Man as a geomorphic agent is shaping and reshaping the physical landscapes incessantly either intentionally or unintentionally by depositing made ground, excavating worked ground, accelerating processes of erosion through agricultural practices, interfering flow processes of the river channel. The basin area of the Bhagirathi-Hooghly River system is the densest in the densely populated India, and obviously, the geomorphology of the region is largely interfered by man. Unsurprisingly, the role of man in shaping and reshaping the landforms of the Bhagirathi-Hooghly river system is increasing at an accelerated rate. Humans are not only shaping or reshaping the earth forms but also interfering and controlling the processes responsible for shaping those landforms. With intensive agricultural practices, man flattened the sloping ground and made irrigation canals translocating a million tonnes of earth materials, induced more and more silt into the river and made dams and barrages not only to reshape the earth forms but also to control river regime and its processes of erosion, transportation and deposition. Farakka barrage not only changed the channel dynamics of the river Bhagirathi-Hooghly but also put long-lasting signature on the river Padma and its branches such as Bhairab, Jalangi and Mathabhanga-Churni. Channel and basin geomorphology of tributaries coming from the right bank of the Bhagirathi-Hooghly system has intensely been reshaped by the processes of construction of dams and barrages, sand and stone mining, terracing for agriculture, excavation of irrigation canals, making embankments, etc. Tidal creeks of active delta have lost their spill areas and are confined between embankments which is used for premature reclamation of lands. Bheries of hundreds of square kilometres have reshaped the earth forms extensively.

We four simultaneously felt the urgency of collecting valuable chapters on anthropogeomorphology of different sub-basins of the Bhagirathi-Hooghly system. Dr. Islam took the leading role, as he is used to doing, and invited chapters from expert authors of respective sub-basins. After a long process of 1.5 years, we could manage 16 chapters on anthropogeomorphology of the Bhagirathi-Hooghly system within a single pair of cover. Hope, this will be a 'mall' at which, reader's thirst for anthropogeomorphology of any basin or sub-basin of Bhagirathi-Hooghly system will achieve contentment.

Frequently asked question faced from students preparing dissertation papers on geomorphology and anthropogeomorphology of Bhagirathi-Hooghly river system is that they are facing a serious problem of absence of reference literature and how they will prepare their dissertation? This book, we think at least to some extent, will meet their need. Academicians from a wide group of disciplines such as earth science, geology, geography, geomorphology, river science and river engineering, urban and rural planning and management, policymakers and NGOs who are very much interested in knowing about what, where, why and how anthropogenic signatures on landforms and processes are involved in the region will gain from the volume. Hope they will also meet their want, to an extent, in this book.

Chapter 1, in general, illuminates the theme—concept, ideas and issues of anthropogeomorphology of the Bhagirathi-Hooghly river system. Chapter 2 deals mainly with the execution of Farakka barrage and its effect on channel geomorphology of the Bhagirathi river. Chapter 3 focuses on anthropogeomorphology of the Hooghly basin and Lower Deltaic West Bengal. In this chapter, river 'Hooghly' is spelled as *Hugli*. Chapter 4 deals with the ecological and geomorphic impacts of

Farakka barrage project on the Bagmari–Bansloi–Pagla river system. Anthropogenic involvement in geomorphic processes through the construction of dams and barrages, sand mining, embankments, etc. in the Mayurakshi River Basin is illustrated in Chapter 5. Chapter 6 depicts the anthropogeomorphology of the Ajay River Basin. Damodar was the *river of sorrow*, and to change it into the 'river of joyfulness' several dams were constructed on it and flow was channelised through hundreds of kilometres of desired canals changing basin and channel geomorphology which is presented in the Chapter 7. Chapter 8 deals with human imprints on the forms and processes of the Kangsabati River Basin. Chapter 9 includes the anthropogenic impact of urbanisation, embankment, sand mining and other activities on the geomorphology of the Dwarkeswar River Basin. Chapter 10 focuses on anthropogeomorphology of the Silabati River Basin. Chapter 11 focuses on anthropogenic imprints on the hydro-geomorphology Rasulpur River Basin. Chapter 12 deals with the made and worked earth forms and mourns the killing of a dying river Jalangi. Anthropo-footprints on forms and processes of the Churni River Basin were measured in Chapter 13. Anthropogeomorphological features and their associated terrain attributes of Saraswati and Jamuna rivers were deciphered in the Chapter 14. Chapter 15 focuses on the systematic exhumation of the facets of human encroachment within the palaeo-fluvial regime of the Kana–Ghia–Kunti system of Damodar Fan Delta. At last, the Chapter 16 explains the anthropogeomorphology of an abandoned channel, Anjana river, dealing with the human role on the evolution from a river to a canal.

We acknowledge our heartiest gratitude to authors who give us their best for preparing 16 chapters of this book. We are thankful to them. Kenneth J. Gregory, the Visiting Professor, University of Southampton, and Emeritus Professor, University of London, and Colin R. Thorne, Professor and Chair of Physical Geography, University of Nottingham, UK, have written forewords for this book. We acknowledge our deepest gratitude to them. Special thanks to our colleagues and friends for their sincere support. We are thankful to all our students who helped in the field and in the lab for successful completion of this book. Finally, we acknowledge our sincere gratitude to the CRC Press, Taylor & Francis Group especially Irma Shagla Britton, Senior Editor, Environmental Sciences, GIS & Remote Sensing, CRC Press and her team for their interest in working with us.

Balai Chandra Das
Sandipan Ghosh
Aznarul Islam
Suvendu Roy
January 31, 2020

Editors

Balai Chandra Das is an Associate Professor of Geography, Krishnagar Government College, Nadia, West Bengal. He earned a postgraduate degree in geography from the University of Burdwan and a PhD in geography from the University of Calcutta. He has published more than 30 research articles in reputed national and international journals, proceedings and edited volumes. Dr. Das has served as an editorial board member for two international journals and as a reviewer for five more. He is the main editor of two books (1) *Neo-thinking on Ganges–Brahmaputra Basin Geomorphology* (ISBN 978-3-319-26442-4) and (2) *Quaternary Geomorphology in India: Case Studies from the Lower Ganga Basin* (ISBN 978-3-319-90426-9) published by Springer International Publishing, Switzerland. He is one of the members of the *Scientific Committee* of *IWC-2016* and *WRAA-2020 Oman*. His current research interest is on the fundamental geomorphology of rivers and lakes.

Sandipan Ghosh is an Applied Geographer with a postgraduate, MPhil and PhD in geography from The University of Burdwan. He has published more than forty book chapters, international and national research articles in various renowned journals of geography and geo-sciences. He is the author of two books, *Flood Hydrology and Risk Assessment: Flood Study in a Dam-Controlled River of India* (ISBN 978-3-659-50098-5) and *Laterites of the Bengal Basin: Characterization, Geochronology and Evolution* (ISBN 978-3-030-22937-5). He is also one of editors of two books: (1) *Neo-Thinking on Ganges – Brahmaputra Basin Geomorphology* (ISBN 978-3-319-26442-4) and (2) *Quaternary Geomorphology in India: Case Studies from the Lower Ganga Basin* (ISBN 978-3-319-90426-9). He has performed as one of editors in the Asian Journal of Spatial Science and Journal of Geography and Cartography. Alongside he has worked as a reviewer in many international geo-science journals of Taylor & Francis (*Geo-Carto International*), Springer (*Environment Development and Sustainability, Arabian Journal of Geosciences and Geosciences, Spatial Information Research, Journal of Earth System Sciences* and *Journal of Geological Society of India*), the International Water Association (*Water International*), Journal of Nicolaus Copernicus University, Torun (*Bulletin of Geography Physical Geography Series*) and Indian Academy of Sciences (*Current Science*). Dr. Ghosh is a lifetime member of The International Association of Hydrological Sciences (IAHS), Eastern Geographical Society (EGS) and Indian Geographical Foundation (IGF). His principal research fields are various dimensions of fluvial geomorphology, flood geomorphology, quaternary geology, soil erosion and laterite study. Currently he has worked on (a) the gully morphology and soil erosion on the lateritic terrain of West Bengal and (b) quaternary geomorphology and active tectonics in the Bengal Basin, West Bengal. At present he is an Assistant Professor at the Department of Geography, Chandrapur College, Purba Barddhaman, West Bengal.

Aznarul Islam is an Applied Fluvial Geomorphologist with an MSc in geography from Kalyani University, West Bengal, and an MPhil and PhD in geography from the University of Burdwan, West Bengal. He is currently an Assistant Professor and Head (Officiating) in the Department of Geography, Aliah University, Kolkata. Previously, he was engaged in teaching and research at the Department of Geography, Barasat Government College, West Bengal. He has already published more than 25 research papers in different journals of national and international repute including *Natural Hazards,* Springer*; Environmental Earth Sciences,* Springer; *Physical Geography,* Taylor & Francis; *Environmental Monitoring and Assessment*, Springer; *Arabian Journal of Geoscience*, Springer; *Environment, Development and Sustainability*, Springer; *Chinese Geographical Science*, Springer; *International Journal of River Basin Management*, Taylor & Francis; *SN Applied Sciences*, Springer, *River Behaviour and Control*, India. He has contributed seven book chapters in edited volumes and one conference proceedings. He is an editor of *Neo-Thinking on Ganges–Brahmaputra Basin Geomorphology*, Springer International Publishing, Switzerland and *Quaternary Geomorphology in India: Case Studies from the Lower Ganga Basin*, Springer International Publishing AG, part of Springer Nature. He has presented papers in more than 20 national and international seminar and conferences. He has completed one Major Research Project funded by the Indian Council of Social Science Research (ICSSR), Ministry of Human Resource Development, Government of India. He has delivered several invited lectures and special lectures in different national and regional programmes. He has been performing the role of a reviewer in different international journals including *Scientific Reports,* Nature; *Trees, Forests and People*, Elsevier; *Environment, Development and Sustainability*, Springer; *Modeling Earth Systems and Environment*, Springer; *Spatial Information Research*, Springer. He is a life member of Foundation of Practicing Geographers (FPG), Kolkata, and National Association of Geographers, India (NAGI), New Delhi. He was an Assistant Convenor of the two-day national seminar: Geography of Habitat (26–27 February 2016), organised by FPG. He has to date successfully supervised more than 35 dissertations on various topics of geomorphology at the master's level. Currently he is supervising four PhD students in geomorphology and related fields. His principal area of research includes hydro-geomorphological issues of the Bengal Delta, including channel shifting riverbank erosion, flood, ecological stress of the riverine tract and environmental flow.

Suvendu Roy focuses his research interest on the interface between anthropogenic activities and changing channel geomorphology, especially on headwater streams of the tropical region. This includes field-based studies to identify micro-geomorphological processes and landforms. He is also interested in the application of GIS and remote sensing to develop new ways to identify the impact of human activities on landscapes. He earned a BA in geography at Burdwan Raj College, The University of Burdwan, and an MA in geography (specialised in advanced geomorphology) at The University of Burdwan (India). He earned a PhD in geography (anthropogeomorphology) at the University of Kalyani (India). He has more than 20 research publications in various journals of international and national repute. Since 2013, he is deeply involved in his research: human-induced changes in river systems.

He is the life member of International Association of Hydrological Sciences (IAHS); Foundation of Practicing Geographer (Kolkata, India). Indian Institute of Geomorphologists (Allahabad, India) He is an invited reviewer of different Scopus and SCI indexed journals of Springer, Elsevier, Taylor & Francis, Willy, Science Domain groups of publication. His main areas of innovative research include Forest River Geomorphology, Anthropogeomorphology and Archaeogeomorphology. Dr. Roy is an Assistant Professor in the Department of Geography, Kalipada Ghosh Tarai Mahavidyalaya, Bagdogra, Siliguri, West Bengal, India.

Contributors

Ananta Gope
Department of Geography
Vivekananda Mahavidyalaya
Burdwan, India

Arghyadip Sen
Department of Geography
Banaras Hindu University
Varanasi, India

Arijit Majumder
Department of Geography
Jadavpur University
Kolkata, India

Avijit Kar
Department of Zoology
Vidyasagar University
Midnapore, India

Aznarul Islam
Department of Geography
Aliah University
Kolkata, India

Balai Chandra Das
Department of Geography
Krishnagar Government College
Krishnanagar, India

Bidhan Chandra Patra
Centre for Aquaculture Research, Extension
and Livelihood
Department of Aquaculture Management and
Technology
Vidyasagar University
Midnapore, India

Biplab Sarkar
Department of Geography
Aliah University
Kolkata, India

Debabrata Das
Department of Geography
Krishnagar Government College
Krishnanagar, India

Debabrata Mondal
Department of Geography
S.C. Bose Centenary College
Murshidabad, India

Gouri Sankar Bhunia
Aarvee Associates Architects, Engineers and
Consultants Pvt Ltd
Hyderabad, India

Kashif Imdad
Department of Geography
Pandit Prithi Nath (PG) College (affiliated with
CSJM University)
Kanpur, India

Mainul Islam
Department of Geography
Aliah University
Kolkata, India

Manojit Bhattacharya
Department of Zoology
Fakir Mohan University
Balasore, India

Md. Mofizul Hoque
Department of Geography
Aliah University
Kolkata, India

Mehebub Sahana
School of Environment, Education and
Development
University of Manchester
Manchester, United Kingdom

Mohd Rihan
Department of Geography
Jamia Millia Islamia
New Delhi, India

Nabendu Sekhar Kar
Department of Geography
Shahid Matangini Hazra Government College
for Women
Purba Medinipur, India

Nilanjana Biswas
Department of Geography
Vivekananda College for Women
Kolkata, India

Pravat Kumar Shit
Department of Geography
Raja Narendra Lal Khan Women's College
Midnapore, India

Priyank Pravin Patel
Department of Geography
Presidency University
Kolkata, India

Rahaman Ashique Ilahi
District Spatial Data Centre
Cooch Behar District, India

Rishikesh Prasad
Department of Education
Government College of Education (CTE)
Habra, India

Sadhan Malik
Department of Geography
University of Burdwan
Burdwan, India

Samrat Deb
Indira Gandhi Conservation Monitoring Centre
WWF-India
New Delhi, India

Sanat Kumar Guchhait
Department of Geography
University of Burdwan
Burdwan, India

Sandipan Ghosh
Department of Geography
Chandrapur College
Bardhaman, India

Sayantan Das
Department of Geography
Dum Dum Motijheel College
Kolkata, India

Sayoni Mondal
Department of Geography
Presidency University
Kolkata, India

Shambhu Nath Sing Mura
Assistant Professor of Geography
Vivekananda Mahavidyalaya
Burdwan, India

Soma Bhattacharya
Post Graduate Department of Geography
Vivekananda College for Women
Kolkata, India

Subodh Chandra Pal
Department of Geography
University of Burdwan
Burdwan, India

Suman Deb Barman
Department of Geography
University of Burdwan
Burdwan, India

Sunando Bandyopadhyay
Department of Geography
University of Calcutta
Kolkata, India

Susmita Ghosh
Department of Geography
Aliah University
Kolkata, India

Suvendu Roy
Department of Geography
Kalipada Ghosh Tarai Mahavidyalaya
Bagdogra, Darjeeling, India

Ujwal Deep Saha
Department of Geography
Vivekananda College for Women
Kolkata, India

Wani Suhail Ahmad
Department of Geography
Aligarh Muslim University
Aligarh, India

1 An Appraisal to Anthropogeomorphology of the Bhagirathi-Hooghly River System
Concepts, Ideas and Issues

Balai Chandra Das, Sandipan Ghosh,
Aznarul Islam and Suvendu Roy

CONTENTS

1.1 INTRODUCTION

Rivers are a much-cherished feature of the natural world, performing countless vital functions in both social and ecosystem terms. In many parts of the world, human-induced degradation has profoundly altered the natural functions of river systems (Graf, 2006; Meybeck and Vorosmarty, 2004; Brierley and Fryirs, 2005; Szabo, 2010; James et al., 2013; Jain et al., 2016; Tarolli et al., 2019). It is very true that humans have been bequeathed a legacy of ageing river engineering projects whose objectives were simply designed, whose effectiveness are uncertain and which were planned in ignorance of their long-term physical and environmental impacts on the river system (Williams, 2001). Human disturbance has introduced a source of change that is foreign to the geomorphic and biotic conditions of river systems. Human disturbance has modified the nature and rate of river adjustments, altering the spatial and temporal distribution of river forms and processes (Brierley and Fryirs, 2005). According to Tarolli et al. (2019), human societies have been reshaping the geomorphology of landscapes for thousands of years, producing anthropogenic geomorphic features ranging from earthworks and reservoirs to settlements, roads, canals, ditches and plough furrows that have distinct characteristics compared with landforms produced by natural processes.

The Anthropocene represents the time since human impacts have become one of the major external forcing on natural processes. Humans have interacted with rivers from the time of ancient civilisations. The Indian sub-continent, which hosts many large and perennial rivers with significant hydrological and geomorphic diversity, is also home to an ancient civilisation and is currently one of the most populated regions on the globe (Sinha et al., 2005; Jain et al., 2016). The Ganga River Basin of India is characterised by significant variability in human population and nature of disturbances, which poses a number of challenges in the sustainable management of the river. For example, 61,948 million litres of urban sewage is generated on a daily basis in India and more than 38,000 million litres of wastewater goes into the major rivers of India (Sengupta, 2018). Estimated polluted riverine length (mainly the rivers of West Bengal, Maharashtra, Assam, Madhya Pradesh and Gujarat) is 12,363 km, having a BOD (biological oxygen demand) range of 10–30 mg l^{-1}, i.e. severely polluted (Sengupta, 2018). On the other side, India has by now about 4500 reservoirs (created by dams on rivers) which have now lost their storage capacity and functionality due to excessive siltation rate (0.475–9.44 mm yr^{-1}) (CWC, 2019). The anthropogenic disturbances have caused significant decrease in forest cover from 89 million ha to 63 million ha and an increase in agricultural area from 92 million ha to 140 million ha in India (Jain et al., 2016). Large dams have caused more pronounced disconnectivity on the sediment fluxes, and as a result, the sediment supply from rivers to oceans has decreased around 70%–80% in most of the Indian river basins (Gupta et al., 2012).

The human dimension of geomorphology is a vital area of research that has been largely neglected beyond the local and sub-regional scale (James et al., 2013). Much less has been written about the effectiveness of humans as geomorphic agents in the tropical densely populated countries, like India. Despite this situation of Indian rivers, the impact of anthropogenic forcing on natural geomorphic systems has not been analysed in detail. It is very essential that the multidisciplinary river studies at modern and historical timescales be pursued vigorously for securing the health and futures of the Indian rivers (Jain et al., 2016). One of the major research concerns is the development of hydrology–morphology–ecology relationship in the river system and the assessment of the anthropogenic disturbances on this or part of this relationship (Sinha, 2009;

Jain et al., 2016). Such geomorphic investigations will lead to a new understanding about the present status of the river systems and will help to project the future behaviour and forms of rivers in the scenario of uncertainties associated with climate change and the ever-growing impacts of anthropogenic activities. Before going into deep discussion on the emerging issues of fluvial responses to human disturbance, it is of utmost necessity to apprehend the key concepts and imperative ideas about anthropogenic geomorphology. Basically, this chapter examines and summarises how humans have modified fluvial landforms and the processes that formed them during the Anthropocene.

1.2 KEY CONCEPTS AND IDEAS

1.2.1 ANTHROPOCENE

Unless there is a global catastrophe—a meteorite impact, a world war or a pandemic—mankind will remain a major environmental force for many millennia (Crutzen, 2002). Because of these anthropogenic emissions of carbon dioxide, global climate may depart significantly from natural behaviour for many millennia to come. Recent global environmental changes suggest that Earth may have entered a new human-dominated geological epoch (Lewis and Maslin, 2015). Destruction of tropical rainforests, agricultural expansion, fossil fuel burning, dam building and river diversion have become common place. It seems appropriate to assign the term 'Anthropocene' to the present, geological epoch, supplementing the Late Holocene—the warm period of the past 10–12 millennia (Crutzen, 2002). Here, the fluvial response to human disturbance has been studied in the timeframe of Anthropocene. So, it is necessary to clarify the term first.

The 'Anthropocene' is itself a contested concept, both in terms of whether or not it exists and when it began (Goudie and Viles, 2016). Those who propose that the Anthropocene should become formally established as part of the geological timescale do so on the grounds that human activities now dominate the Earth System and have led to a marked shift in its state (Goudie and Viles, 2016). The term Anthropocene was introduced by Crutzen (2002) as a name for a new epoch in Earth's history—an epoch when human activities have become so profound and pervasive that they rival or exceed the great forces of nature in influencing the functioning of the Earth System (Goudie and Viles, 2016). The Anthropocene could be said to have started in the latter part of the 18th century, when analyses of air trapped in polar ice showed the beginning of growing global concentrations of carbon dioxide and methane (Crutzen, 2002). This date also happens to coincide with James Watt's design of the steam engine in 1784 (Crutzen, 2002).

1.2.1.1 A Human Golden Spike

Defining the beginning of the Anthropocene as formal geologic unit of time requires the location of a global marker of an event in stratigraphic material, known as a Global Stratotype Section and Point (GSSP), which is renowned as 'golden spikes'—the preferred boundary markers (Lewis and Maslin, 2015). Each 'golden spike' is a single physical manifestation of a change recorded in a stratigraphic section, often reflecting a global change phenomenon. Several approaches have been put forward to define when the Anthropocene began, including those focusing on the impact of fire, pre-industrial farming, sociometabolism and industrial technologies (Table 1.1).

In the last 300 years, as Steffen et al. (2007) suggest, we have moved from the Holocene into the Anthropocene. They have identified three stages in the Anthropocene: (1) Stage 1, which lasted from 1800 to 1945, 'The Industrial Era'; (2) Stage 2, extending from 1945 to 2015, 'The Great Acceleration'; and (3) Stage 3, which may perhaps now be starting, is a stage when people have become aware of the extent of the human impact and then human wants stewardship of the Earth System (Goudie and Viles, 2016).

Primarily, the beginning of the Industrial Revolution (1760–1880) has often been suggested as the beginning of the Anthropocene. Since the 1950s, the 'Great Acceleration' is marked by a major

TABLE 1.1
Potential Start Dates for a Formal Anthropocene Epoch

Event	Date	Geographical Extent	Primary Stratigraphic Marker	Potential GSSP Date
Extensive farming	~8000 yr BP to present	Eurasian event, global impact	CO_2 inflection in glacier ice	None, inflection too diffuse
Rice production	6500 yr BP to present	Southeast Asian event, global impact	CH_4 inflection in glacier ice	5020 yr BP CH4 minima
Industrial revolution	1760 to present	Northwest Europe event, local impact, becoming global	Fly ash from coal burning	~1900 diachronous over ~200 yr
Nuclear weapon detonation	1945 to present	Local events, global impact	Radionuclides (^{14}C) in tree rings	1964 ^{14}C peak
Persistent industrial chemicals	~1950 to present	Local vents, global impacts	For example, SF_6 peak in glacier ice	Peaks often very recent so difficult to accurately date

Source: Lewis and Maslin (2015).

expansion in human population, large changes in natural processes and the development of novel materials from minerals to plastics (Lewis and Maslin, 2015). The earliest potential GSSP primary marker Lewis and Maslin (2015) identify is the inflection of atmospheric methane at 5020 years BP. Many scientists find that only two other events, namely, the Orbit Spike dip in CO_2 with a minimum at 1610 and the bomb spike 1964 peak in ^{14}C, appear to fulfil the criteria for a GSSP to define the inception of the Anthropocene (Lewis and Maslin, 2015).

1.2.2 POPULATION, TECHNOLOGY AND ENVIRONMENTAL IMPACTS

Exponential growth in the global human population is often cited as a factor influencing a rapid rate of Anthropocene environmental change. In addition to accelerating geomorphic change, the growing population also increases the vulnerability of society to natural hazards, like floods (Meybeck and Vorosmarty, 2004). Urbanisation of lands that are susceptible to floods, tsunamis, volcanic eruptions and the like puts large numbers of people at risk reducing the resiliency of society to catastrophic events. A series of theorems presented by Ehrlich and Holdren (1971) ultimately led to a simple, iconic conceptual relationship, IPAT, that is commonly used to express environmental impacts in terms of population, affluence and technology (Holdern and Ehrlick, 1974):

$$I = PAT$$

where I is the environmental impact, P is the population, A is the consumption or affluence and T is the technology.

The nature, extent and timing of human impacts on geomorphic systems are not only merely academic questions but may have serious physical and social implications. For example, soil loss is a geomorphic consequence of human activities (e.g. deforestation and slope modification) that is highly relevant to feeding global populations (Ehrlich and Holdren, 1971). Human intervention with river and consequent deterioration of its channel leads to permanent occupational shift of the society (Das 2015a, 2015b). Interrelationships among human activities, climate and natural hazards may take many forms, of which direct responses of landforms to human activities are but one aspect (James et al., 2013). Human activities have indirect influences on landforms through cascades, biological systems and natural hazards (Figure 1.1). Cascades in the movement and storage of mass

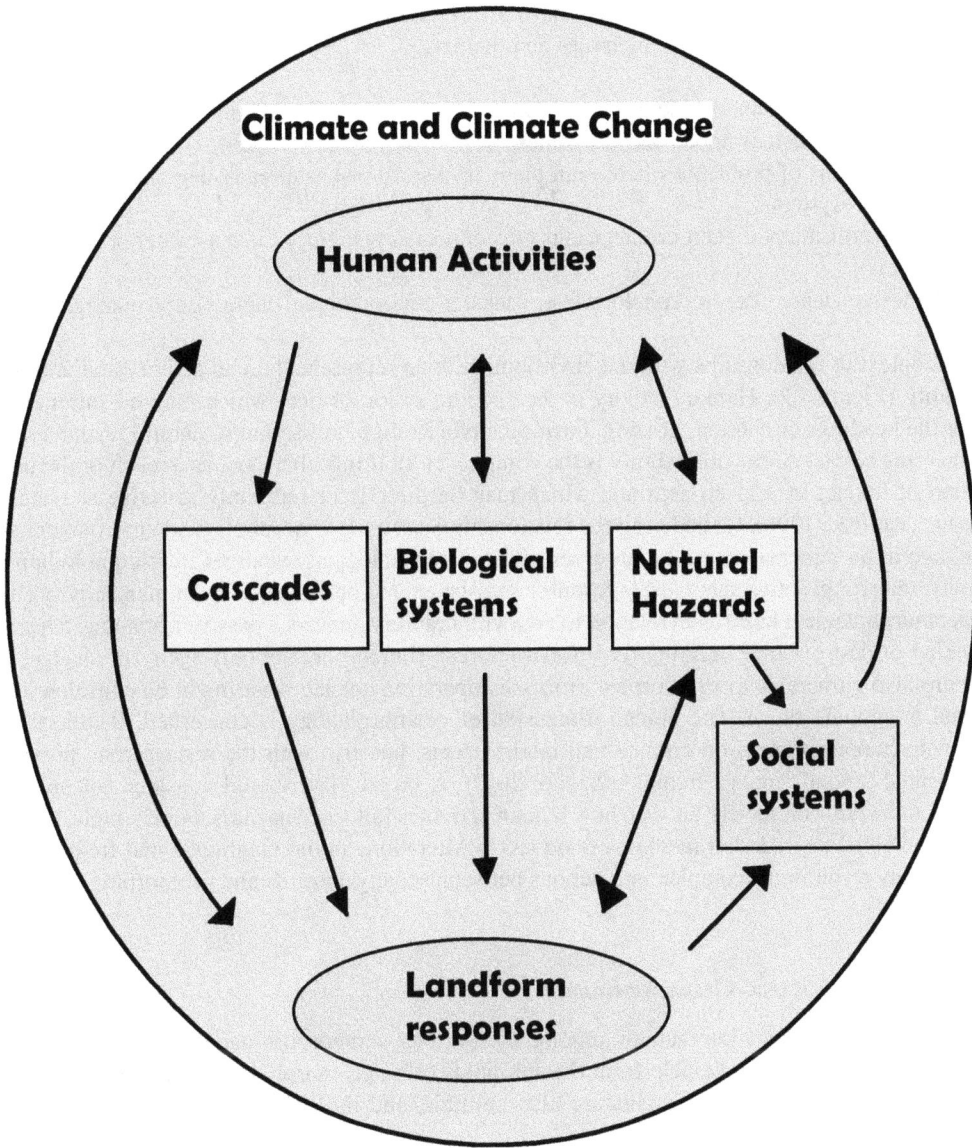

FIGURE 1.1 System diagram for interrelationships between human activities and landform responses. (Modified from James et al. (2013).)

or energy commonly link human activities to landform responses, so that geomorphic responses may be propagated indirectly and may be delayed, mitigated or extended in time and space (James et al., 2013). Rates and intensities of human activities are influenced by cultural factors of the social system, especially the level of technology employed and population dynamics, such as population growth and mitigations.

1.2.3 ENVIRONMENTAL GEOMORPHOLOGY

To understand the human impact on landforms with a correct and satisfactory procedure, environmental geomorphology, a sub-field of geomorphology, was introduced by Coates in 1971. It is defined that environmental geomorphology is the practical use of geomorphology for the solution of problems where man wishes to transform or to use and change surficial processes (Panizza, 1996). This sub-field involves the following issues and themes:

1. The study of geomorphic processes and terrain that affects man, including hazard phenomena such as floods and landslides
2. The analysis of problems where man plans to disturb or has already degraded the land–water ecosystem
3. Man's utilisation of geomorphic agents or products as resources, such as water or sand and gravel
4. How the science of geomorphology can be used in environmental planning and management

In the context of relationships with the environment, man represents human activity and area vulnerability (Figure 1.2). Human activity is the specific action of man which may be summarised under the headings of hunting, grazing, farming, deforestation, utilisation of natural resources and engineering works. Area vulnerability is the complex of all things that exist as a result of the intervention of human in a given area and which may be directly or indirectly sensitive to material damage (Panizza, 1996). Considering the relationship between geomorphological environment and man, two main dimensions can be observed: (1) geomorphological resources in relation to human activity, man as an active agent (e.g. a resource may be altered or destroyed by human activity), and (2) geomorphological hazards in relation to area vulnerability, man as a passive agent (e.g. a hazard may alter or destroy some buildings or infrastructures). Humans are not only agents of change, but they are also vulnerable to geomorphic processes operating outside what might be considered the normal magnitude range. The human dimension of geomorphology is concerned, therefore, not only with human impacts on climate and land systems, but also with the reverse role, in which geomorphic systems impact humans (Szabo, 2010). A broad view should consider not only the impacts of hazards on society but also how human activities influence hazards. For example, anthropogenic channel aggradation increases flood risks. Alterations in the magnitude and frequency of hazards may result in the complex interactions between society, hazards and geomorphic processes (Szabo, 2010).

1.2.4 Anthropogenic Geomorphology

To better understand all interactions among the various geomorphic agents shaping the landscape, geomorphologists take help from the sub-fields of biogeomorphology and zoogeomorphology (Hupy, 2017). Arguably, humans are also animals, and the impacts they have rendered on

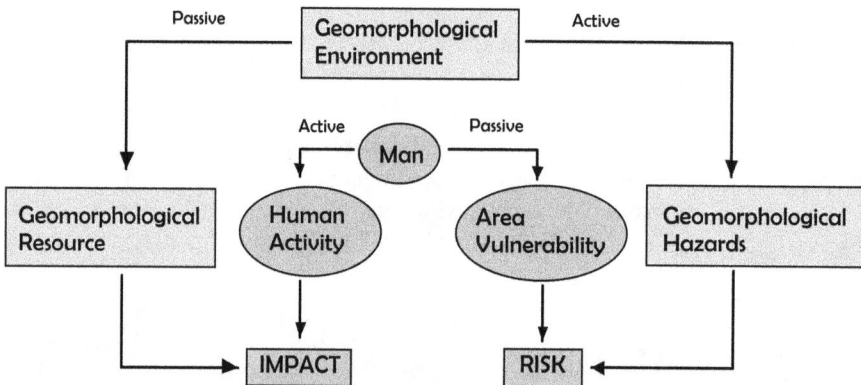

FIGURE 1.2 Interrelationships between geomorphological environment and man within the domain of environmental geomorphology. (Modified from Panizza (1996).)

the Earth's surface supersede any past landscape transformation, denudation or deposition, by other species in the animal kingdom. Therefore, the anthropogenic element or anthropogenic geomorphology is yet another sub-field within the discipline of geomorphology that is increasing in appearance within scientific literature (Hupy, 2017). Shaping the landscape more than any other natural process, it is a topic that has received scant attention in India and nowhere near the amount of research deserved.

Anthropogeomorphology, a term invented by Golomb and Eder (1964), is the study of the human role in creating landforms and modifying the operation of geomorphological processes (Goudie and Viles, 2016). Anthropogenic geomorphology is an emerging systematic field that overlaps with climate change and natural hazards research. Collectively these three topics form a human dimension of geomorphology that should gain increasing prominence in the 21st century with mounting concerns over the ability to reconcile population growth, dwindling recourses, global environmental change, climate warming, public safety and sustainability with the stability of geomorphic systems (James et al., 2013). Human creates landforms directly and indirectly by altering geomorphic process rates and landscape sensitivities (Table 1.2).

Interest and focus on how humans change geomorphic systems accelerated in the late 20th century as concerns mounted about global environmental change and growing population pressures (Holdren and Ehrlich, 1974). Anthropogenic geomorphology focuses on many key aspects of geomorphological processes within the Anthropocene. Most of the classic textbooks of geomorphology, including those from the past few decades, however, ignore it totally (Goudie and Viles, 2016). The geomorphic works of Marsh (1864), Gilbert (1917), Bennett (1938), Jacks and Whyte (1939), Thomas (1956), Brown (1970), Nir (1983) and Szabo (2010) cover the extent of human influence on the physical environment and landforms (Table 1.3). The nature, extent and timing of human impacts on geomorphic systems are not merely academic questions but may have serious physical and social implications. For example, soil loss is a geomorphic consequence of human activities that is highly relevant to feeding global population (most of the global food supply is derived from the land). Study of the effects of human activities on geomorphic systems is an emerging branch of geomorphology that focuses on humans as agents of change (Holdren and Ehrlich, 1974).

TABLE 1.2
Some Examples of Human-Created or Anthropogenic Landforms

Feature	Cause
Pits and ponds	Mining, marling
Broads	Peat extraction
Spoil heaps	Mining
Terracing, lynchets	Agriculture
Ridge and furrow	Agriculture
Cuttings	Transport
Embankments	Transport; river and coast management
Dikes	River and coast management
Mounds	Defence, memorials
Craters	War
City mounds	Human occupation
Canals	Transport, irrigation
Reservoirs	Water management
Subsidence depressions	Mineral and water extraction
Moats	Defence

TABLE 1.3
Some Milestones in Anthropogenic Geomorphology

Research Work/Main Theme	Authors
Man and Nature—The pioneer work on the human impact on the environment	Marsh (1864)
Hydraulic-mining debris in the Sierra Nevada—An in-depth geomorphic study on the consequences of gold mining inland from San Francisco	Gilbert (1917)
The Rape of the Earth: A World Survey of Soil Erosion—A popular global survey and polemic on the global menace of soil erosion	Jacks and Whyte (1939)
Man's role in Changing the face of the Earth—An edited volume based on ground-breaking symposium on the human impact	Thomas (1956)
Man makes the Earth—A thoughtful and largely neglected study on anthropogeomorphology	Brown (1970)
Man, a Geomorphological Agent: An Introduction to Anthropic Geomorphology—A thorough review of knowledge to focus on the man's role in changing face of earth	Nir (1983)
Anthropogenic Geomorphology—A largely Hungarian review that is especially strong on constructed and excavated landforms	Szabo, David and Loczy (2010)
Geomorphology of Human Disturbances. Climate Change and Hazards—An edited volume to depict the current contributions on anthropogenic geomorphology	James, Harden and Clague (2013)
Geomorphology in the Anthropocene—An outstanding contribution to understand the human role in different aspects of geomorphology	Gouide and Viles (2016)
From features to fingerprints: a general diagnostic framework for anthropogenic geomorphology—Advances towards empirical and theoretical framework to integrate the natural and socio-cultural forces which shapes the earth	Tarolli, Cao, Sofia, Evans and Ellis (2019)

1.2.4.1   Man as Geomorphological Agent

Man is a geomorphological agent who was originally multizonal but has progressively become azonal up to the present day. Unlike other agents, such as water, ice, wind and glacier, human is not limited or localised but, on the contrary, is less and less conditioned by environmental variables (Panizza, 1996). Man has a great capability of movement and adaptation: his activities on the environment area are result of his technological development and are guided by economic, social and cultural needs (Nir, 1983). Man transforms, corrects and modifies natural processes by increasing or decreasing their rate of action and by causing the rupture of certain equilibrium which nature will try to reconstitute in different ways (Panizza, 1996; Goudie and Viles, 2016). With the passing of time, this modifying action has assumed increasingly widespread and intense patterns in different geomorphic systems (Table 1.4).

TABLE 1.4
Some Examples of Anthropogeomorphic Changes on Geomorphic Systems

System	Anthropogenic Changes	Geomorphic Responses
Rivers	Damming; bank and bed armouring; river bed mining	Sedimentation, deposition, flow frequencies, roughness changes
Coasts	Dredging; armouring; sea-level rise	Beach erosion, sediment redistribution
Hillslopes	Deforestation, ploughing, road cuts	Rills, gullies
Groundwater	Mining	Subsidence
Glacial	Global warming	Ice margin retreat
Per glacial	Road embankments, global warming	Permafrost degradation
Arid and aeolian	Vegetation change, vehicular traffic	Deflation and deposition

Source: Szabo (2010).

It is not easy to distinguish and separate 'natural' forms from 'artificial' ones. There is a law governing reciprocal actions and the unity of geographical actions according to which it is not possible to consider facts and phenomena separately (Panizza, 1996). The various components of Earth's surface are to be considered not according to unilateral relations but as reciprocal and complex. There is an action of the physical environment on man and a reaction of man towards the environment (Figure 1.3). In some cases, man's settlements and activities can cause the worsening of erosional processes in an unstable environment, such as mountainous areas of Himalayas, where the extremes of the climate are accompanied by high-slope gradients. It has been found that man's interventions take place in three ways (Panizza, 1996):

1. Artificial forms, directly modelled by man's activities
2. Works aiming to divert, correct or upgrade natural processes
3. Modifications of natural phenomena, indirectly resulting from man's activities

The intensity and extension of these activities depend upon four factors (Panizza, 1996):

1. Demographic: that is the numerical consistency of the individuals practicing any activity which is reflected on natural dynamics
2. Historical: length of time of human presence and activity
3. Technological and cultural: capacity to exploit the environment according to one's own purposes
4. Socio-economic: demand for better living conditions and consumer goods

The historical order of the effects of man's intervention on the physical world can be schematised as follows (Nir, 1983): (1) *Hunting* (limited impact—deforestation of some areas), (2) *Animal Farming* (changes soil profile and enhances soil erosion), (3) *Agriculture* (expansion of agriculture ruins forest and enhances soil erosion; degradation of arable lands through introduction of fertilisers and pesticides; modifies slope), (4) *Resource Exploitation* (mining, increased population density, industrialisation and expansion of transport intensify pollution and modify landforms; technological development is a synonym of environmental catastrophism) and (5) *Engineering Works* (construction of roads, railways, bridges, dams, buildings, other hydraulic works, harbours and coastal structures and urban growth, etc. have direct impact on the proliferation of flood hazards, river planform changes, landslides, slope instability, reduction of soil permeability and different types of pollution) (Panizza, 1996).

Human impacts can be indirect and direct. 'Indirect human impacts' on rivers result from activities that occur outside of the river network and do not directly alter channel form or process. Human-induced changes in atmospheric circulation, temperature and precipitation patterns are one category of indirect impacts that have received increasing attention within the past two decades (Goudie and Viles, 2016). Changes in land cover have been occurring for thousands of years and recognition of the effects on the river networks goes back more than a century (Petts, 1979; Wohl, 2014). 'Direct human impacts' result from activities within the river network that directly alter channel form and process. Flow regulation, and specifically the construction of dams, has received the most scientific and public scrutiny. Other typically smaller-scale alternations of channel form and connectivity—associated with flow regulation, construction of levees, dredging and channelisation, bank stabilisation and in-stream mining—can have cumulative effects equally substantial to those associated with large dams (Schumm, 1969; Brandt, 2000; Goudie and Viles, 2016). Construction of piers within channel and subsequent intervention with the flow process lead to substantial shifting of thalweg and changes in channel's cross-sectional forms (Das 2019).

1.2.4.2 Classification of Anthropogenic Impacts

It is now understood that today human agent is equal in importance to other geomorphic factors. Although the energy released by human society is insignificant compared to the endogenic forces

FIGURE 1.3 Google images (2019) show (a) modification of Himalayan hillslopes into settlements and roads at Namchi Bazar, Sikkim, India; (b) converting basaltic hill to stone mines and quarries at Tildanga, Jharkhand, India; and (c) modifying Damodar River into reservoir through dam at Panchet, Jharkhand, India.

TABLE 1.5

Classification of Anthropogenic Landforming Processes

1. Direct anthropogenic processes	2. Indirect anthropogenic processes
1.1 *Constructional*	2.1 *Acceleration of erosion and sedimentation*
Tipping: loose, compacted, molten	Agricultural activity and clearances of vegetation
Graded: moulded, ploughed	Engineering, especially road construction and
Terraced	urbanisation
1.2 *Excavational*	Incidental modification of hydrological regime
Digging, cutting, mining, blasting of cohesive or	2.2 *Subsidence: collapse, settling*
non-cohesive materials	Mining
Cratered	Hydraulic
Trampled, churned	Thermokarst
1.3 *Hydrological interference*	2.3 *Slope failure: landslide, flow*
Flooding, damming, canal	Accelerated creep
Construction, dredging, channel modification	Loading undercutting
Draining	Shaking
Coastal protection	Lubrication
	2.4 *Earthquake generation*
	Loading (reservoirs)
	Lubrication (fault plane)

Source: Haigh (1978).

of the Earth, human impact is not only commeasurable to the influence of exogenic processes but even surpasses their efficiency (Szabo, 2010). Humans interfere with the complex system, including geomorphological ones, from outside and, thus, necessarily disturb the natural order (dynamic equilibrium) of the processes, which has evolved over time spans of various lengths (Table 1.5). An obvious aspect is to classify anthropogenic impacts according to the character of the human activity. Over recent decades, the following classification of anthropogenic impacts has been identified (Szabo, 2010):

1. Mining—the processes involved and the resulting landforms are usually called *montanogenic*
2. Industrial impact is reflected in *industrogenic* landforms
3. Settlement (urban) expansion exerts a major influence on the landscape over ever-increasing areas. The impacts are called *urbanogenic*
4. Traffic also has rather characteristic impacts on the surface
5. As the first civilisations developed, highly advanced farming relied on rivers; water management (river channelisation, drainage) occupies a special position in anthropogenic geomorphology
6. Agriculture is another social activity causing changes on the surface. *Agrogenic* impacts also include transformation due to forestry
7. Although warfare is not a productive activity, it has long-established surface impacts
8. In contrast, the impacts of tourism and sports activities are rather new fields of study in anthropogenic geomorphology

1.3 METHODOLOGICAL OUTLOOK

Geomorphology and anthropogeomorphology alike are best understood knowing that changes rendered upon the surface of the Earth come about from a collective body of geomorphic agents (James et al., 2013). A variety of forces are at play, both eroding and depositing materials in a constant ongoing process. Anthropogeomorphologists study these agents as well as connecting the direct and indirect activities of human as a geomorphic agent (Hupy, 2017). For the anthropogeomorphologist, finding changes on the Earth's surface is not difficult, and indeed, what may prove more challenging is finding what geomorphic agents today are not related to some type of human activity (Hupy, 2017). There are a number of challenges in the sustainable management of Indian rivers in this time of climate change. It is found that fluvial geomorphological studies in India have mostly focused on the river response to climate and tectonic forcing at quaternary timescale (Jain et al., 2016). It is very essential that the multi-disciplinary river studies at modern and historical timescales need to be purchased vigorously for securing the health and future of Indian rivers. Now, one of the major research concerns is the development of hydrology–morphology–ecology relationship in the river system and the assessment of the anthropogenic disturbances on this or a part of this relationship (Jain et al., 2016). The approaches to this study and methodology outlook to deal with anthropogenic geomorphology are mainly found in the valuable writings of Nir (1983), Brandt (2000), Brierley and Fryirs (2005), Graf (2006), Szabo (2010), Wohl (2014), and Goudie and Viles (2016). In this book, we have focused only on the methodology of fluvial geomorphology which reflects variable fluvial responses to different anthropogenic activities. The scope of this book does include not only the study of man-made landforms but also the investigation of man-induced surface changes, the prediction of corollaries of upset natural equilibrium as well as the formulation of proposal in order to preclude harmful impacts.

Human impacts on river character can only be reliably interpreted if longer-term controls on river evolution are understood. Hence, appraisals of system response to human disturbance must be made in context of inferred adjustments that would have occurred under natural disturbance regimes. Whatever the form of disturbance and whether a site-specific direct impact such as dam constriction or indirect disturbance such as forest clearance, the effects can be transmitted long distances from their source (Brierley and Fryirs, 2005; Szabo, 2010). Ultimately, however, management must consider cumulative responses to disturbance, interpreting how these adjustments will shape likely future patterns and rates of river changes (Fryirs and Brierly, 2013). The analysis of anthropogenic activities cannot take into account only *what* man has done but also *how* and for *how long*, since human action is discontinuous in space, in time and in intensity (Panizza, 1996). A correct assessment of man-induced morphological effects becomes possible only by means of a systematic study carried out according to the following procedure (Nir, 1983):

1. Historical approach, in order to identify man's interventions on landforms
2. Geomorphological approach, in order to assess the amount and extension of the geomorphological processes observed
3. Socio-economic approach, in order to investigate the dynamicity of human activities by considering economic and social-structural parameters
4. Planning approach, in order to unite and integrate the various points of view

1.3.1 Analysing River Change and Complex Response

River change is defined as adjustments to the assemblage of geomorphic units along a reach that record a marked shift in river charterer and behaviour (Brierley and Fryirs, 2005). To understand the ongoing impacts of human activities, we have to identify and analyse the changes in river morphology and hydrology in Anthropocene. Analysis of river change at the reach scale, viewed in context of changes to catchment-scale linkages, provides a critical basis to develop proactive river management programs (Brierley and Fryirs, 2005). To the anthropogeomorphologists, it is very important to determine what components of a river system are likely to change over any given timeframe and what the consequences of those changes are likely to be. Instinctively, human attention is drawn to landscapes that are subject to change. Observation of bank erosion, river responses to flood events, anecdotal records of river adjustments or analyses of historical maps and satellite images provide compelling evidence of the nature and rate of river changes (Panizza, 1996; Brierley and Fryirs, 2005; Wohl, 2014).

River character and behaviour in any given reach reflect cumulative responses to a range of disturbance events. Evolutionary investigations must consider long time periods in order to identify the critical processes involved in river change. Spatial and temporal controls on river adjustment and change must be meaningfully integrated to provide insights into system vulnerability (Brierley and Fryirs, 2005). The notion of complex response indicates that there may be pronounced variability in the nature and rate of system responses to the same external stimuli, resulting in stark differences in the pattern of spatial and temporal lags (Schumm, 1969; Brierley and Fryirs, 2005). Responses to impacts in one place may dampen or buffer effects elsewhere. Alternatively, small anthropogenic disturbances in one part of a system may set up chain reactions that breach threshold condition elsewhere, resulting in chaotic responses. In Anthropocene, it is very essential to realise the complex response of a river system and its trend at reach scale. Among the continuum of responses that may be observed are the following situations which should be analysed in field (Brierley and Fryirs, 2005):

1. No response may be detected, as system absorbs the impacts of disturbance
2. Effects may be short-lived or intransitive. Humid alluvial channels often show little change flowing disturbance events or they have the capacity to recover rapidly
3. Part of progressive change—river may respond rapidly at first after disruption but at a steadily declining rate thereafter
4. Change may be instantaneous, as breaching of threshold conditions prompts the adoption of a new state or even a new type of river
5. Change may be lagged. Off-site impacts of major disturbances may induce lagged responses in downstream reaches
6. Equivalent responses may occur via differing pathways of adjustment

1.3.2   FRAMING A RIVER STYLE FRAMEWORK

In the River Style Framework (RSF), contemporary attributes of rivers are related to the capacity for change under current conditions, whether that represents an irreversibly altered human-induced set of conditions or otherwise field (Brierley and Fryirs, 2005). Assessment of whether river response to human disturbance is reversible or permanent is appraised in terms of the assemblage of river forms and processes along the reach (i.e. assemblage of geomorphic units). The RSF provides a catchment-based physical platform with which to guide management activities that respect the inherent diversity and ever-changing nature of river systems. River styles record the character and behaviour of rivers throughout a catchment, providing a geomorphic appraisal of what a river system looks like, how it behaves and how it has adjusted over time (Brierley and Fryirs, 2005). This spatially and temporally integrative frame appraises contemporary river morphology and formative process in light of river change, thereby providing critical insights with which to interpret geomorphic river condition. This forms a basis to predict river futures and the potential for geomorphic river recovery (Brierley and Fryirs, 2005; Fryirs and Brierley, 2013). The RSF does not present a quantitative summary of river character, behaviour, condition or recovery potential for differing river types. Rather, it provides the guiding principles upon which quantitative assessments can be made on a catchment. The key attributes of RSF are focused on the river behaviour, catchment-framed survey, landscape evolution and contemporary geomorphic condition (Table 1.6).

The nested hierarchical basis of the framework is structured into five scales: catchments, landscape units, reaches, geomorphic units and hydraulic units. River forms and processes are interpreted using a building block, 'constructivist' approach to analyse reach-scale assemblages of geomorphic units (Fryirs and Brierley, 2013). The RSF provides a coherent package of baseline data upon which an array of additional information can be applied, providing a consistent platform

TABLE 1.6

Key Attributes of River Style Framework

Attributes	Description
River behaviour	RSF evaluates river behaviour, indicating how a river adjusts within its valley setting. This is achieved through appraisal of the form–process associations of geomorphic units that make up each river style.
Catchment frame	It provides a catchment-framed baseline survey of river character and behaviour throughout a catchment. Downstream patterns and connections among reaches are examined, demonstrating how disturbance impacts in one part of a catchment are manifested elsewhere over differing timeframes.
Landscape evolution	It evaluates recent river changes in context of longer-term landscape evolution, framing river responses to human disturbance in context of the capacity for adjustment of each river style.
Geomorphic condition	The contemporary geomorphic condition of the river is assessed with understanding of downstream patterns. It provides key insights with which to determine geomorphic river recovery potential.

Source: Fryirs and Brierley (2013).

for decision-making for a range of management activities. The structure of nested hierarchy allows for top-down explanations of controls on river character and behaviour and a bottom-up constructivist approach to interpretation of river character and behaviour (Figure 1.4). The RSF comprises four stages: (1) identification, interpretation and mapping of river styles throughout a catchment; (2) system evolution; (3) predicting pathway of future river adjustment and (4) river management applications and implications (Figure 1.5).

1.3.3 ASSESSING VULNERABILITY, SUSCEPTIBILITY AND RIVER SENSITIVITY

In Anthropocene, direct and indirect human impacts on the fluvial forms and processes have introduced key physical phenomena, such as vulnerability, susceptibility and river sensitivity, into a river system. 'Vulnerability' refers to the potential of a reach to experience a shift in state within its natural capacity for adjustment or to be transformed to a different type of river. Vulnerability can result from the breaching of either an intrinsic or an extrinsic threshold. Under circumstances of incipient instability that arise as threshold conditions are approached, minor disturbance events may bring about changes to a new state (Begin and Schumm, 1984). In contrast, 'susceptibility' refers to the ease with which a system or reach is able to adjust within its natural capacity for adjustment (Schumm, 1985). Susceptible landscapes retain a characteristic identity as they form and reform under a given process regime (Fryirs and Brierley, 2013). At first glance, continual adjustment

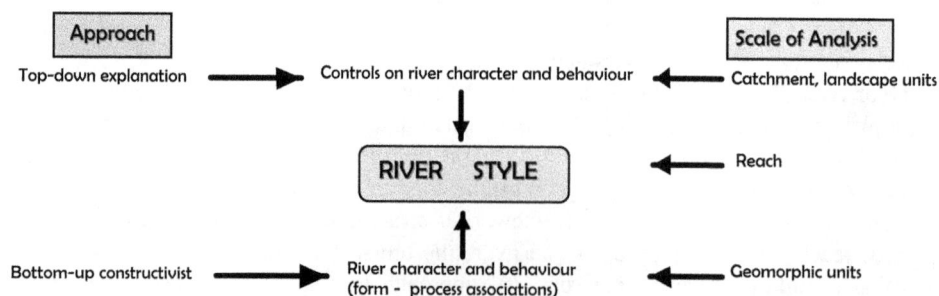

FIGURE 1.4 Approaches and scales of analysis adopted in the RSF.

STAGE ONE : Catchment-wide baseline survey of river character and behaviour

STAGE TWO : Catchment-framed assessment of river evolution and geomorphic river condition

STAGE THREE : Assessment of the future trajectory of change and geomorphc river recovery potential

STAGE FOUR: River management applications and implications: Catchment-based vision building, identification of target conditions and prioritization of management efforts

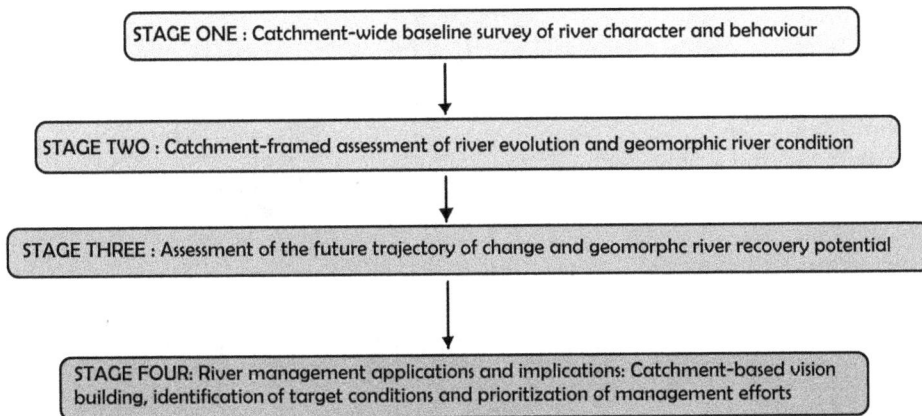

FIGURE 1.5   Stages of the RSF.

may be perceived as a form of instability, but this is not always the case, as a susceptible river is not necessarily vulnerable to adjustment or change. Measures of 'river sensitivity' to change are considered to reflect the sum of susceptibility and vulnerability for any given river type (Brierley and Fryirs, 2005). This reflects the ease with which adjustment can take place (i.e. the way in which the reach has adjusted its form to resist change) and the proximity to threshold conditions. The sensitive rivers are readily able to adjust to perturbations but are prone to dramatic adjustment or change. Assessment of river sensitivity can be grounded in a stepwise procedure, as summarised in Table 1.7.

Key principles form a series of filters of information upon which the four stages of the RSF are built. Analyses of geomorphic river character and behaviour, viewed from cross-sectional and

TABLE 1.7
Guiding Principles for Interrupting System Sensitivity and Vulnerability

Stage and Principle	Action
Identify the type of river and its valley setting	Assess the potential range of variability and contemporary capacity for adjustment.
Appraise the balance of erosional and depositional landforms along the reach	Interpret the ease with which various features are likely to be reworked, based on the composition (e.g., texture, cohesively) and position along the reach (e.g., proximity to the thalweg).
Interpret reach evolution as a basis to interpret how the balance of impelling and resisting forces along a reach adjusts over time	Analyse the formative factors that drive geomorphic change, such as the dominant discharge, the role extreme events and the history of events. Interpret the type of events that have shaped the contemporary geomorphic state of the reach. Appraisal of geomorphic effectiveness and how sensitive a river is to change.
Based on analysis of the trajectory of change, interpret what system is evolving towards. Identify limiting factors and pressures that may modify the prevailing balance.	River change can be driven by an increase in impelling forces, a decrease in resisting forces or a combination of these factors. Interpretation of system responses to past events provides guidance into these relationships.
Identify threshold conditions that guide interpretations of indicators of change	Thresholds of probable concern are likely to be specific to a particular type of river. In some instances, direct tools can be applied, whether empirically based or theoretically based to aid in threshold spotting.

Source: Brierley and Fryirs (2005).

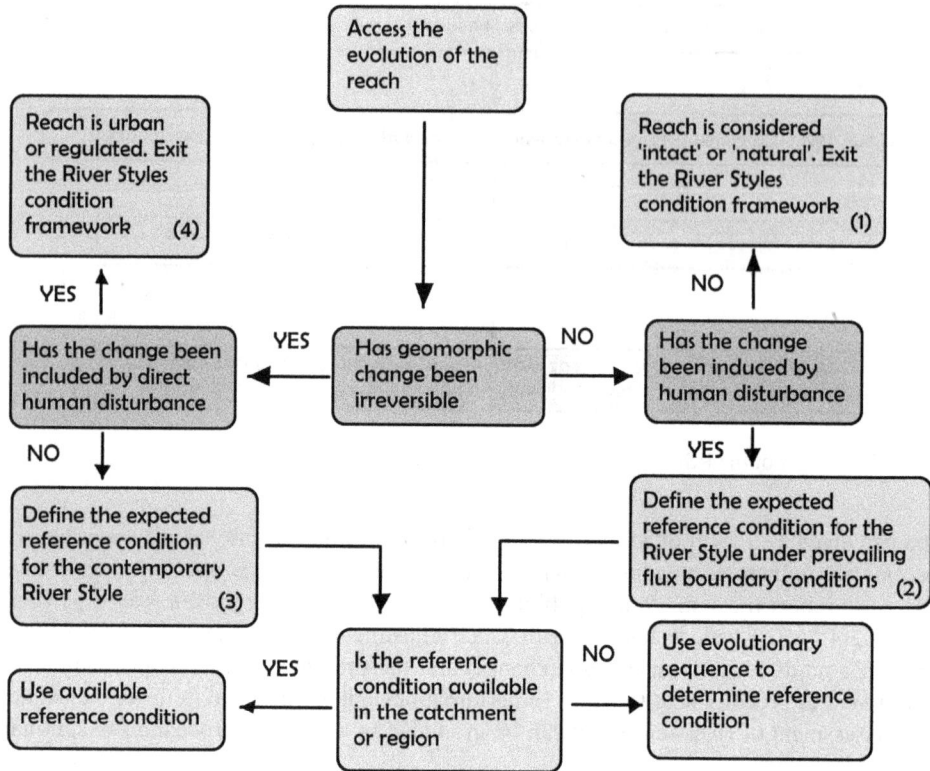

FIGURE 1.6 Decision tree to identify reference condition for the study of anthropogenic geomorphology. (Modified from Brierley and Fryirs (2005).)

planform perspectives, provide a platform upon which separate sets of procedures are used to appraise geomorphic river condition. Any assessment of river condition must be framed relative to some benchmark or reference reach, thereby providing a determination of the extent to which human-induced changes to river character and behaviour fall outside the long-term pattern (Brierley and Fryirs, 2005; Fryirs and Brierley, 2013). A decision-making tree is applied to identify which reference condition should be used for any given situation. This provides a generic tool to assess river responses to disturbance (Figure 1.6). In field, four types of reference condition can be differentiated and adopted to perform river sensitivity analysis (Brierley and Fryirs, 2005):

1. Remnant reaches that have been minimally disturbed by humans, such that geomorphic changes to river character and behaviour remain reversible
2. Reaches where human disturbance has occurred, but geomorphic changes to river character and behaviour remain reversible
3. Reaches where change has been induced by indirect human disturbance and irreversible change has resulted
4. Reaches where change has been induced by direct human disturbance and irreversible geomorphic change has resulted

1.4 BHAGIRATHI-HOOGHLY RIVER BASIN—A GEOGRAPHICAL ACCOUNT

The Bhagirathi-Hooghly is one such river basin, perhaps the only one of its kind, that poses so many problems directly to its 15 million inhabitants because of the progressive silting up of the Bhagirathi reach of the river and decrease of water even in the tidal Hooghly reach, and it reduces the functionality of the Kolkata Port (Chatterjee, 1972). The future of Kolkata City, the greatest metropolis, port and commercial-cum-industrial complex of India, is linked up with the great river. The Ganga and Brahmaputra Rivers together with their numerous branches and adjuncts intersect the country of Ganga–Brahmaputra–Meghna (GBM) delta in such a variety of directions as to form the most complete and easy navigation that can be conceived (Bose, 1972; Chatterjee, 1972). In this book, the different river systems of the Bhagirathi-Hooghly River Basin (e.g. Pagla, Bansloi, Mayurakshi, Ajay, Damodar, Dwarkeswar, Silai, Kangsabati, Jalangi, Rasulpur, Saraswati and Anjana rivers) are considered as key study areas to assess the anthropogenic impact on the river morphology and hydrology within the premise of the Bengal Basin and Peninsular Shield.

Beyond the outfall of the Kosi River, the River Ganga turns into the plains of West Bengal round the outcrops of the Rajmahal Hills receiving only a few local drainage channels on the left bank. A few kilometres below the Farakka Barrage, the river starts throwing off distributaries that join the right-hand channel Bhagirathi which formed once the main arm of the Ganga (Bose, 1972). The left arm of Ganga, Padma River is joined by the Brahmaputra and later on debouches into the Bay of Bengal as the Meghna River (Bose, 1972). The delta between the two arms Bhagirathi-Hooghly and Padma–Meghna is now known as the Ganga–Brahmaputra–Meghna (GBM) delta, having average area of 58,752 km² (Figure 1.7). The Bhagirathi-Hooghly River is the western branch of the Ganga

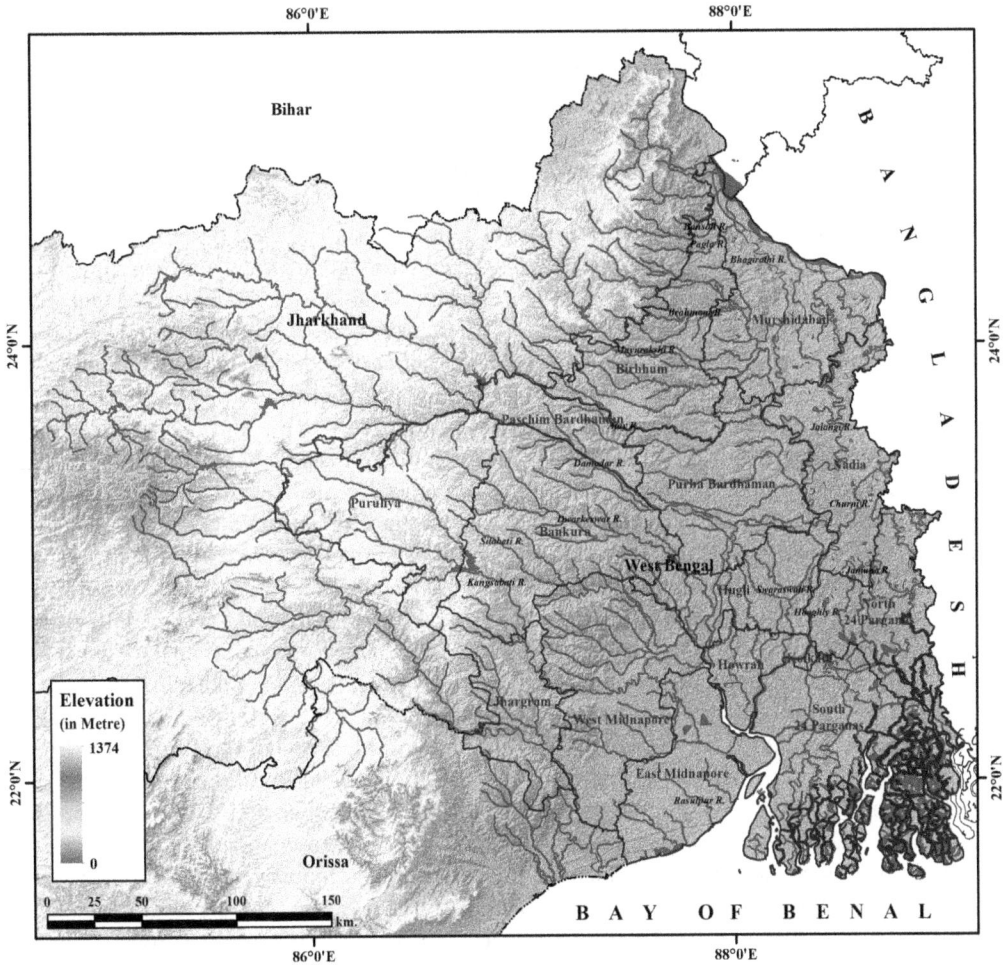

FIGURE 1.7 The elevation character and drainage network of the Bhagirathi-Hooghly River Basin (elevation derived from SRTM 90m).

and flows more than 500 km through West Bengal. The upper non-tidal part traversing 230 km from Jangipur to Nabadwip is known as the Bhagirathi River while the lower tidal stretch of 288 km extending from Nabadwip to its confluence at the Bay of Bengal is known as the Hooghly River (Parua, 1992; Islam and Guchhait, 2017a). The Bhagirathi remains delinked from the Ganga for about 9 months of the year and receives water from 38 km long feeder canal originating from the Farakka Barrage (Rudra, 2018).

All arrays of ancient and modern centres of art, culture and industries indicate that right arm of the GBM delta was the main channel of Ganga River before the left arm Padma–Meghna started driving more and more discharges towards the end of the 18th century (Bose, 1972: Chatterjee, 1972). The Bhagirathi being practically cut off from the main Ganga by a sand bar, now carries only the local rainfall except in the monsoonal months when the level of the river water rises above the sand bar. It was mentioned by Sir William Willcocks that this river was really an 'overflow irrigation canal', brought into being by the greatest irrigation expert of ancient times, *Bhagirath* (Chatterjee, 1972). But according to Chatterjee (1972), it is more likely that the name 'Bhagirathi' as applied to the principal Himalayan headwater of the Ganga was given to its lower reach as well. The name 'Hooghly' was given to the lower reach of the river by English sailors only in 17th century after the great emporium, Hooghly of that century (Chatterjee, 1972).

1.4.1 Geology and Stratigraphy

The Bengal Basin is occupied dominantly by the fluvial system of GBM delta, consisting the largest fluvio-deltaic to shallow marine sedimentary basin of the world (Hossain et al., 2019; Ghosh and Guchhait, 2019). This is one of the thickest sedimentary basins of the world consisting of ~21 km thick Early Cretaceous–Holocene sedimentary succession. Geographically, the Bengal Basin lies approximately between 20° 34′ to 26° 40′ N and 87° 00′ to 92° 45′ E with its major portion being constituted by Bangladesh (Hossain et al., 2019). The Bhagirathi-Hooghly Basin of GBM delta is a part of north-west Stable Shelf which consists of an easterly dipping shelf that is separated from the Precambrian Indian Shield to the west by a prominent Basin Margin Fault Zone (i.e. more than 350 km in length) with dislocation and cataclasis (Hossain et al., 2019). The GBM delta took the present shape in 3000 years BP and the delta started to grow since 7060±120 years BP (Rudra, 2018; Ghosh and Guchhait, 2019). This part constitutes the eastern continuation of the Indo-Gangetic Alluvial Plain (IGAP) which separates the Extra Peninsular India to the north and the Peninsular India to the south. The Rajmahal Hills of West Bengal and Bihar is a fault-bounded small volcanic and tectonic element situated in the western edge of the Stable Shelf of the Bhagirathi-Hooghly Basin. The delta is expected to grow southwards due to estuarine deposition but such had not been the case along the western part where the sea has encroached inland (Rudra, 2018). Bandyopadhaya (2007) proposed a wider extension of the GBM delta and included '*para-deltas*' formed by the western tributaries to the Bhagirathi-Hooghly within its geographical area. These undulating parts of laterites along with the adjoining Late Quaternary alluvial plains lying between the Chotanagpur Plateau and the Bhagirathi-Hooghly River are described as *Rarh Bengal* (Rudra, 2018; Ghosh and Guchhait, 2019).

An important tectonic feature of BHB, lies between the Indian Shield to west and the Eocene Hinge Zone (EHZ) to the east, is a prominent fault zone with the dislocation and cataclasis running along the N15°E–S15°W marked by the crowding of gravity contours (Alam et al., 2003; Hossain et al., 2019). EHZ is also known as shelf break or palaeo-continental slope or trace of the Eocene shelf edge. Sengupta (1996) named this NNE–SSW running narrow 25–100 km Hinge Zone as the 'Calcutta–Mymensingh gravity high' (−30 to −15 mGal and sloping towards southeast). In the Bolpur, Galsi and West Ghatal areas, bordering the eastern margin of the zone of shallow, buried basement ridges, seismic surveys have detected a series of normal, down-to-the-basin strike faults arranged in *en echelon* pattern (Sengupta, 1966). According to Nath et al. (2014), the Stable Shelf of the Bengal Basin can be divided structurally into four tectonic elements from northwest to southwest: (1) the North Bengal Foreland, (2) the Basin Margin Fault Zone/western carp zone, (3) the Central Stable Shelf and (4) the EHZ. This part of the Bengal Basin is intersected by several faults and lineaments, e.g. Rajmahal Fault, Ganga–Padma Fault, Chotanagpur Foothill Fault, Damodar Fault, Saithia–Bahmani Fault, Purulia Shear Zone, etc. (Hossain et al., 2019; Ghosh and Guchhait, 2019). The EHZ is also identified as a seismically active tectonic element. This hinge has reportedly triggered two earthquakes of magnitude M_w 7.3 and 6.2 in 1842 and 1935 (Nath et al., 2014).

In the beginning of the Late Eocene, thick Tertiary clastic sediments accumulated in the basin with depiction accelerating with the arrival of clearly orogenic sediments in the earliest Miocene. Sediments attain a thickness of 3–4.5 and ~3.5 km to the northern and southern edge of the Stable Shelf, respectively (Hossian et al., 2019). The basement grabens are filled up by the Gondwana sediments comprising sandstone, shale and coal. The Rajmahal Basalt Traps lies above the Gondwana sediments, followed by the fossilferous Eocene Sylhet Limestone and then the post-Eocene sediments comprising sands, gravels and clay (Sengupta, 1966; Hossian et al., 2019). Just west of the Durgapur–Baripada line, the great Indian Shield disappears under a thick blanket of river-borne alluvium. The basement under this alluvial cover is marked by buried ridges, and some of these buried ridges presumably were vents for the eruption of a part of the Rajmahal Basalt flows which underlie the Cretaceous–Tertiary sediments of West Bengal (Sengupta, 1966). Depositional conditions in post-Eocene times were controlled chiefly by movements on the EHZ. Only freshwater and estuarine marine deposits accumulated over most of the Stable Shelf region during that time. With the regression of the sea since Pliocene, the estuarine and fluvatile conditions of deposition prevailed in most of the Bhagirathi-Hooghly River Basin (Sengupta, 1972). During Early Pleistocene time, shallow marine conditions prevailed only in the deeper parts of the Bengal Basin. Since Late Pleistocene, the sea finally receded from the Bengal Basin area, and finally, older alluvium and laterite deposits were covered completely by a thick mantle of floodplain Holocene alluvium (Sengupta, 1972; Mahata and Maiti, 2019).

1.4.2 HYDRO-GEOMORPHOLOGY

The hydrologic and geomorphic investigation of Bhagirathi and delta growth was extensively studied by Willcocks (1930), Majumdar (1942), Bagchi (1944), Morgan and McIntire (1959), Chatterjee (1972), Bandyopadhyay (2007), Parua (2009), Islam (2011), Sinha and Ghosh (2012), Bandyopadhyay et al. (2014), Guchhait et al. (2016), Islam and Guchhait (2017a, 2017b) and Rudra (2014, 2018). The Bhagirathi-Hooghly River has an undulating catchment area of 66,000 km² along right bank drained by eight major tributaries which are Bansloi, Pagla, Mayurakshi, Ajay, Damodar, Rupnarayan, Khari, Dwarkeswar, Kangsabati and Haldi which carry eroded sediments from the Peninsular Shield. These tributaries together contribute about 48,410 million m³ of water annually into the Bhagirathi-Hooghly River (Rudra, 2018). The eastern part of the catchment covering 5971 km² is drained by two other distributaries of the Ganga, namely, Bhairab–Jalangi and the Mathabhanga–Churni which contribute 2922 million m³ of the water annually (Rudra, 2018). The monsoon rain supplies substantial water from the catchment, and the groundwater contribution during lean months is also important. The lower reach of the river is replenished regularly by tidal water. The mean monthly freshwater flow of this river at Nabadwip and Gangasagar (i.e. outfall) is estimated taking into account rainfall, evapotranspiration, infiltration and storage in the basin (Figure 1.8). The mean flow of lean months varies from 1822 to 4111 million m³, and the flow of monsoon varies from 5021 to 13,074 million m³ in the Bhagirathi reach (at Nabadwip) (Rudra, 2018). The Farakka project has connected the Ganga with the Bhagirathi through feeder canal and the Bhagirathi between Jangipur and Nabadwip which as fordable before 1133 m³s⁻¹ of water was induced in 1975 (Rudra, 2014). The water level of the Bhagirathi was raised due to induced water during the monsoon months, and four tributaries, namely, Bansloi, Pagla, Mayurakshi and Ajay in the west bank and the Jalangi and Churni in the east bank, faced drainage congestion at their respective outfalls, promoting recurrent floods in the basin (Parua, 2009, Rudra, 2018).

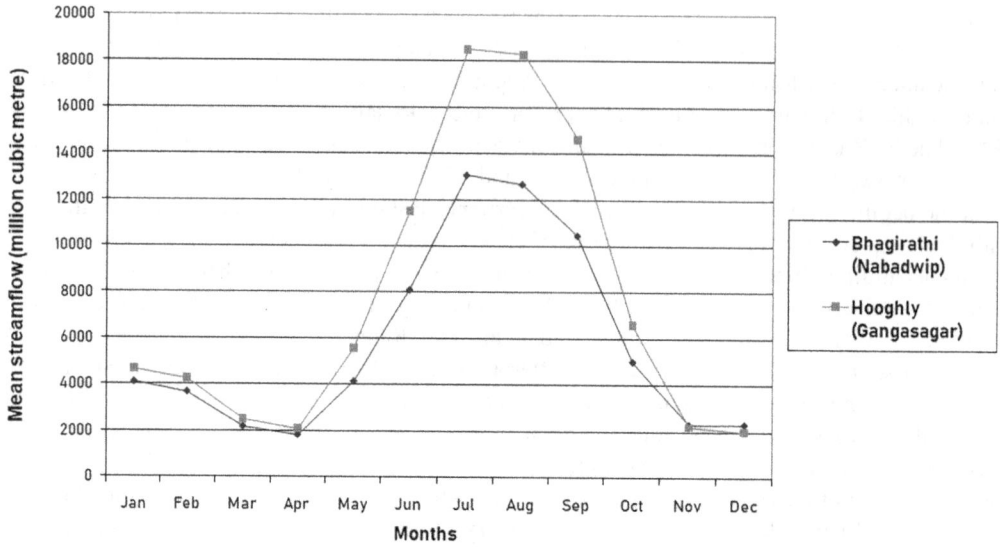

FIGURE 1.8 Mean annual freshwater flowing through the Bhagirathi-Hooghly River Basin (million m³).

The Bhagirathi-Hooghly meanders all through over its floodplain from its head near Ramkantapur to its mouth at Gangasagar and shows all the typical features of a meandering river (Chatterjee, 1972). Bagchi (1944) suggested three types of river in the GBM delta: (1) defunct channels like Adi Ganga, (2) dissected and interrupted channels due to human interference and (3) primitive channels. The following features were identified in the Bhagirathi-Hooghly River Basin by Chatterjee (1972):

1. Ox-bow lakes representing the cut-off portion of meander bends, especially between Katwa and Kalna
2. Meander scrolls, depressions and rises on the convex side of bends formed as the channel migrated laterally down valley and towards the concave bank. This is very much pronounced on all the convex bends between Agradwip and Purbasthali
3. Sloughs, areas of stagnant water, formed both in meander-scroll depressions and along the valley walls as flood flows move directly down valley, scouring adjacent to the valley walls. These features can best be seen at Dainhat and Patuli
4. Natural levees, raised bermes or narrow shelfs above the floodplain surface adjacent to the channel, usually containing coarser materials deposited as flood flows over the top of the channel banks. These are most frequently found on both the banks of the Bhagirathi between Katwa and Dainhat on the west and between Nakashipara and Muragachha on the east

The rivers of the Bhagirathi-Hooghly River Basin, especially the distributaries, in the deltaic tract form meander and change their courses mainly in monsoon peak discharge (Sinha and Ghosh, 2012; Rudra, 2018). Migration of meander loops and formation of ox-bow lakes are governed by two main factors, namely, annual variability of peak discharge and stratigraphy of floodplain. The Bhagirathi is very active between Jangipur and Nabadwip, but the southern tidal regime has fewer tendencies to erode its banks due to formation of clay-slit resist banks (Rudra, 2018). Another important physical feature is the Sundarbans which occupies the southern portion of the Bhagirathi-Hooghly Basin bordering the sea. Sundarbans (about 3–5 m from mean sea level) is intersected by a network of big tidal rivers and islands (now 128 islands in India), the estuaries of which penetrate far inland (Chatterjee, 1972). The Hooghly estuary is a funnel-shaped mouth which allows tidal water to invade the river, allowing huge sediment-laden water to penetrate and choke the channel. The report said that 2.96×10^6 tons of sediment from the channel and 23.2×10^6 tons of the bank forming materials were flushed downstream during the period 1976–1991 (Rudra, 2018). The coastal tract of the Sundarbans continues to grow by the deposition of silt mostly pushed back from the estuary by the tidal waves. The rate of accretion may be as high as $12 \, cm \, yr^{-1}$, as observed in Prentice Island (Rudra, 2018). This active delta spreads over $9630 \, km^2$ in India and $9610 \, km^2$ in Bangladesh, having world's most dense mangrove forest and noted for the UNESCO World Heritage Site and swamp Bengal Tiger (Parua, 2009; Rudra, 2018).

1.4.3 River System

The rivers of the Bhagirathi-Hooghly River Basin can be divided into three major groups: (1) the rivers of *Rarh Bengal* (western tributaries), (2) eastern distributaries and (3) tidal creeks. These three groups have their own distinct topographic, morphologic and hydrologic characteristics, forming variable landforms and morpho-stratigraphic lithofacies in the floodplains. Alongside, these three groups are differently influenced by the anthropogenic activities, like engineering structures, mining, agricultural pattern, urbanisation and other developmental activities, having prominent regional variation.

1.4.3.1 The Rivers of *Rarh Bengal*

The western tributaries have eroded the Chotanagpur Plateau, Gondwana sediments and lateritic *Rarh plain*, forming dissected interfluves and vast fertile floodplains before meeting with the Bhagirathi-Hooghly River. Taken from the north, Bansloi is the first tributary of the Bhagirathi, which has its source at Ramgarh near Rajmahal Hills and outfall at Jangipur. This river has a basin area of $1176 \, km^2$ and its total length is $112 \, km$. The discharge that runs off the basin is highest during July when about 394 million m^3 water flows (Rudra, 2018). Another River Pagla, having a basin area of $628 \, km^2$, originates at Masina of Jharkhand and it has mean peak monsoon discharge of 138 million m^3 during July (Rudra, 2018). Both the rivers meet in a waterlogged area of $50 \, km^2$, named *Ahiron* Wetland which was formed due to drainage congestion of Bhagirathi River. Further downstream, the following tributaries flow into the river Bhagirathi, namely, Dwarka, Brahmani, Kopai and mainly Mayurakshi rivers. Mayurakshi River originates from Trikut Hills in Deoghar district of Jharkhand and finally discharges into Bhagirathi near Kalyanpur (Rudra, 2018). The river has its length of $250 \, km$ and its basin area is about $6400 \, km^2$. At upstream, a 47 m high and 660 m long Massanjore Dam was built in 1955 near Dumka, Jharkhand. Then, at Siuri, Birbhum (32 km downstream of Massanjore Dam), the 309 m long Tilpara barrage was built to regulate flood flow and irrigation flow. All tributaries of Mayurakshi draining into the *Hijal bill*, finally bifurcated into two outlets, namely, Uttarasan and Babla towards the Bhagirathi River (Rudra, 2018). Maximum discharge flows (about 2031 million m^3) during July, and this river is notorious for recurrent flood at downstream (Islam and Baraman, 2020).

The Ajay River originates from the plateau of Hazaribagh and flows about $276 \, km$ before joins Bhagirathi at Katwa, Purba Barddhaman. It has a basin area of $6074 \, km^2$, and the maximum runoff of 1352 million m^3 is generated in July. The annual suspended load of Ajay was estimated to be 0.59 million ton (Bandyopadhyay et al., 2014; Rudra, 2018). There have been more than 14 major flood episodes in the 20th century, and recently, the devastating flood of September 2000 opened many branches in the old embankment, and vast agricultural lands became non-productive due to sand splays. The Hinglo Dam (developed on the Hinglo River, a tributary of Ajay) has a capacity of $17,102,000 \, m^3$ which provides irrigation in areas between the Ajay and the Kopai. The next is the most important tributary of the Bhagirathi-Hooghly River, and it is the most notorious river due to occurrence of annual devastating floods in the southern West Bengal. The Damodar River originates from the Khamarpat Hill of Palamu in Chotanagpur Plateau and flows for a length of $540 \, km$, before discharging into Bhagirathi-Hooghly River through two distributaries (viz. Mundeswari and Damodar) being bifurcated near Jamalpur, Purba Barddhaman (Rudra, 2018). The total basin area of Damodar is 20,874 km^2 and about 6042 km^2 area lies in Jharkhand. It has a number of tributaries, such as Barakar, Konar, Bokaro, Haharo, Jamunia, Khadia and Sali. It is found that highest discharge of 5366 million m^3 flows in July, but occasionally, the monsoon peak discharge suddenly goes above 10,000 m^{3}s^{-1} (experienced 18,406 m^{3}s^{-1} in the floods of 1913 and 1935), recorded at Rhondia Weir, Purba Barddhaman (Rudra, 2018). To manage flash floods and other developments of the region, the five major dams (Tenughat, Tilaiya, Konar, Maithon and Panchet) and the Durgapur Barrage were built in Jharkhand and West Bengal under the Damodar Valley Corporation (DVC) which is India's first multipurpose river valley project.

Rivers Silabati (basin area of 1449 km^2) and Dwarkeswar (basin area of 4866 km^2) meet at Bandar, and the combined flow is named as Rupnarayan River which joins with Hooghly River at Geonkhali covering distance of 78 km. Alongside the Rupnarayan receives a portion of water from the Damodar through the Mundeswari. Since these rivers have common outlet with the Rupnarayan, the Khanakul, Arambag, Goghat and Ghatal areas of Hooghly and Purba Medinipur districts suffered severe drainage congestion during rainy season (Bandyopadhyay et al., 2014; Rudra, 2018). An important river, Kangsabati (or Kasai), having its origin in the uplands of Jhalda, Purulia district, flows eastward and ultimately discharges through two outlets, one into the Rupnarayan and other through the Haldi. It has the basin area of 9527 km^2, having a high monsoonal runoff of 1990 million m^3 in the month of August. The tributaries of Kangsabati have created drainage congestion in Moyna basin and the downstream is notoriously flood prone. A dam 38 m high and 10 km long was built at Mukutmanipur, Bankura district, in 1956 to fulfil the Kansai Irrigation Project. Rasulpur and Pichhabani are two major tributaries which drain the coastal plains of Purba Medinipur and discharge into the Hooghly River. These two rivers together constitute a catchment area of 2074 km^2 (Rudra, 2018).

1.4.3.2 Eastern Distributaries

The Jalangi used to take off from Padma near a settlement 'Jalangi' (where it got its name from the river) in the district of Murshidabad. At present, it is a beheaded moribund channel and receives supply during 1 or 2 months of rains through Bhairab offtake at Akhriganj. Upper reach from offtake to Bhairab confluence is now a series of linear pools. Several ponds and brick-fields have encroached the river badly (Das, 2014). Jalangi meets the river Bhagirathi at Swarupganj, opposite to Nabadwip. Mathabhanga–Churni, another eastern distributary-cum-tributary, is a river of Indo-Bangladesh importance. It also takes off from the River Padma near Jalangi offtake and falls into the River Bhagirathi at Sibpur near Ranaghat. Pollution stress deteriorated water quality for fish community (Sarkar and Islam, 2020) and for irrigation (Sarkar and Islam, 2019). The Jamuna is a moribund channel which connected the Bhagirathi-Hooghly River and Ichhamati and was presumably opened since the desiccation of the Bidyadhari (Bandyopadhyay et al., 2014; Rudra, 2018). The tides advancing northwards through formerly active Saraswati and the Bidyadhari channels would meet at Tribeni and accelerate the process of sedimentation leading to formation of a mid-channel bar. The Jamuna is presently having no link with the Hooghly River, but its extremely sinuous course is traceable for about 66 km down to Charghat where it joins the Ichhamati (Rudra, 2018). Another distributary, the Bidyadhari, has its source near the Mathura Bil which appears as an abandoned meander cut-off of the Bhagirathi located very close to Kalyani, Nadia district. This was an important navigational route for Indo-Roman coastal trade during 300 BC–500 AD and the business was conducted from the port of Chandraketugarh and stood on bank of Bidyadhari (Rudra, 2018). The Saraswati is a moribund distributary (77 km length) of Hooghly River having its source at Tribeni and outfall into the same river at Sankrail (Bandyopadhyay et al., 2014; Rudra, 2018). It is important to note that the Saraswati in its lower reach flowed through a different course till the 7th century AD and discharged through the Rupnarayan estuary. At present, this river is linear pool of stagnant swamp having feeble flow at both ends and the river bed is encroached by the brick kilns and horticulture.

1.4.3.3   Tidal Creeks

The littoral tract of GBM delta is a unique area with 13 major estuaries and many interlacing channels with intervening islands. The region facilitates the largest mangrove ecosystem, entitled 'Sundarbans' which has been declared as world heritage site. The creeks lying between the Hooghly estuary in the west and Padma–Jamuna–Meghna estuary in the east are mostly beheaded having no upstream flow (Rudra, 2018). Water flowing in creeks is induced from the sea, and the flow is governed by the tide-velocity asymmetry and thus has two-way flow. Most channels in Sundarbans can be broadly identified as tidal creeks with few exceptions such as Ichhamati–Haribhanga, Gorai–Madhumati and Arial Khan which receive feeble supply of freshwater during peak of the monsoon (Rudra, 2018). The major channels and creeks are Saptamukhi, Thakuran, Matla, Piyali, Bidya, Gosaba, Bidyadhari, Kalindi and Raimangal, etc. The silt-laden water travels up to the northern tidal limit of these creeks. This process of accretion operates to build up the floodplain, and the deposition of silt gradually blocks the flow of the river and the old creek splits up around obstructions, thus forming numerous channels, tidal creeks and distributaries (Rudra, 2018). In these creeks and the Hooghly estuary, the fluvial sediments undergo metamorphosis in the estuarine environments of Bay of Bengal and many diverse geomorphic features are formed in this active delta of GBM.

1.4.4   Summarising Human Impacts on Rivers

The Ganga has been oscillating within wide limits and also discharged through any distributaries to facilitate water and sediment dispersal. The old courses of the Ganga are now left as moribund channels. In an uncontrolled situation, the Bhagirathi enjoyed the opportunity of free swing, and the limit of swing was determined by sedimentary layers of different resistance. Since the construction of Farakka Barrage which impounds 87 million m³ of water, the river has tried to adjust with the new hydraulic regime. Major structural interventions include constructions of embankments, protection of bank with the boulders and constructions of spurs which may deflect impinging current. The mighty river continues to impinge its banks with immense power during the monsoon and causes damage to arable land and human settlements of Malda, Murshidabad and Nadia districts. Alongside, the key factors are considered to be responsible for the rapid deterioration of the navigability of the Hooghly River (Bose, 1972):

1. The duration and intensity of upland supply from the main stream Ganga into the Bhagirathi-Hooghly channel was limited to the months of mid-June to mid-November, and the volume of supply diminished from about 75 to 80 thousand cusecs to nil
2. As against this, the arm Bhagirathi-Hooghly receives a string tidal inflow from the Bay of Bengal throughout the year

Since many times before, the rivers of West Bengal always have tendency to exceed the critical limit of overflow during the monsoon months and the vast area of floodplains are submerged. The people of this region have been living with floods from time immemorial. The inhabitants learnt the art of utilising the sediment-laden floodwaters during the floods and agricultural prosperity was very much linked to the ecological service of the rivers. But since the 9th century, the British rulers in collaboration with landlords implemented the jacketing of flood-prone rivers, such as Ajay, Damodar and Mayurakshi, through building of wide and solid embankments. Now these embankments ensured protection from low-intensity floods but trapped the sediment load in riverbed (less sediment input to floodplain), causing decay of drainage systems. Alongside, the palaeochannels are abandoned and decayed due to encroachment of agricultural land, delinking from main rivers. Next, to command and control over hydraulic system of West Bengal, the dams were built on the western tributaries of Bhagirathi-Hooghly River, following the model of Tennessee Valley Authority. The DVC developed its multipurpose river valley projects in the 1950s to enrich the agro-economical prosperity of West Bengal, but the experiences of DVC suggest a gap between target work and implemented work and reservoir's capacity is reducing due to high siltation rate. Nowadays, the monsoon floods are inevitable in the southern West Bengal due to sudden release of gigantic water into the Damodar River which is now decaying its carrying capacity. Additionally, the extensive deforestation in the upper catchments, multiple cropping in the lower reaches especially in the floodplains and indiscriminate extraction of groundwater leading to diminution of the base flow towards the rivers during lean months had been major causes of the decay. Now, the rivers of *Rarh Bengal* have faced the crisis of water even in monsoon months.

1.5   FLUVIAL RESPONSES TO ANTHROPOGENIC ACTIVITIES—KEY ISSUES

A legacy of ageing river engineering projects is now observed and the effectiveness is uncertain. The projects were planned in ignorance of their long-term physical and environmental impacts on the river system (Williams, 2001). It is necessary to understand that human disturbance has introduced a source of change that is foreign to the geomorphic and biotic conditions of river systems (Figure 1.9) (Brierley and Fryirs, 2005). At world scale and also regional scale, it is evident that human disturbance or activity has modified the nature and rate of river adjustments, altering the spatial and temporal distribution of river forms and processes. Over the last 100–500 years, human modifications have been the dominant form of disturbance to the fluvial environment of India, exerting a greater influence than adjustments caused by climate change (Brierley and Fryirs, 2005). Though India has dense network of rivers and monsoon climatic regime, the country now suffers

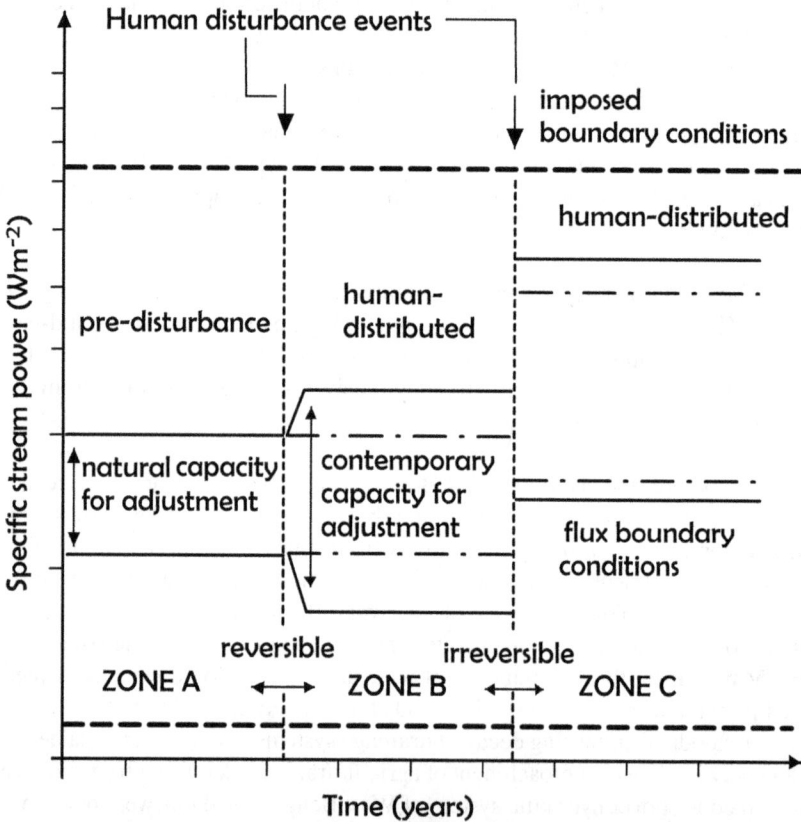

FIGURE 1.9 Fluvial adjustments to human disturbances in respect of stream power with time. (Modified from Brierley and Fryirs (2005).)

TABLE 1.8

Forms of Human Disturbance to River Courses

Direct Channel Changes	Indirect Catchment Changes
River regulation	*Land-use changes*
• Water storage by reservoirs and water diversion schemes	• Changes to ground cover, including forest clearance, afforestation and changes in agricultural practice
Channel modifications	
• River engineering, channelisation programs include flood control works, bed/bank stabilisation structures and channel realignment—sand/gravel extraction and dredging programs	• Urbanisation and building/infrastructure construction, including storm water systems
• Clearance of riparian vegetation and removal of woody debris	• Mining activity

from severe water crisis due to the ever-increasing demand of water for unscientific river basin projects and rigorous urbanization processes (Islam, 2012).

Hydraulic civilisation developed along some of the world's great river, marking a turning point in social organisations. Profound variability in population and resource pressures, tied to the contemporary nature and extent of river regulation and management activities, shapes opportunities for ongoing and future developments (Brierley and Fryirs, 2005; Wohl, 2014; Hupy, 2017). In most instances, human endeavours have sought to control and stabilise rivers. While responses to human disturbance may be reversible or at least can be stopped or reduced in some instances, elsewhere impacts are irreversible, marking a change in the direction of long-term evolutionary trends (Brierley and Fryirs, 2005). Human modifications to biophysical attributes of river systems can be direct or indirect (Brierley and Fryirs, 2005). While most direct modifications are intended, indirect modifications are inadvertent (Table 1.8).

Schemes that set out to stabilise and regulate river systems typically endeavour to fix a channel with a given size and configuration in a given position. Change is natural. Innately, the river will adjust. Once river is fixed in place, the cost of keeping it there rises inexorably (Brierley and Fryirs, 2005). Protection of societal, economical and institutional infrastructure has ensured that this course of action must be maintained. Emphasis is placed on dams and reservoirs, barrages, connectivity in river system, channelisation programs and sand and gravel extractions, and the resultant impacts are flood hazard, bank failure and land loss (Guchhait et al., 2016), urban flood risk, water crisis in river, channel migration and changes in channel morphology.

1.5.1 IMPACT OF DAMS ON RIVERS

The number of publications on the effects of dam construction on the environment around the world has increased as dam construction has increased. In 1900, there were 427 large dams, i.e. higher than 15 m, around the world, while in 1950 and 1986, there were 5268 and about 39,000, respectively (Brandt, 2000). At present, China has the highest number of dams (23,842) followed by the United States of America (9261). Nearly 5254 major/medium dams and barrages had been constructed in India by the year 2017 (NRLD, 2019). It is found that the storage capacity created by these large infrastructures is 253 billion m^3, and another 51 billion m^3 storage is under construction stages. Very large dams on rivers have large, statistically significant effects on downstream hydrology and geomorphology, summarised in the specific construct in Figure 1.10. These hydrologic and geomorphic effects translate themselves into far-reaching adjustments to riparian ecosystems and are likely to be part of the explanation for the decline of some threatened or endangered riparian wildlife, particularly riparian obligate avian species (Graf, 2006). Konar, Maithon, Panchet, Tenughat, Tilaiya, Khandoli dams and Durgapur Barrage on Damodar River System, Massanjore Dam and Tilpara Barrage on Mayurakshi River System, Mukutmanipur Dam on Kangsabati River System and Farakka Barrage

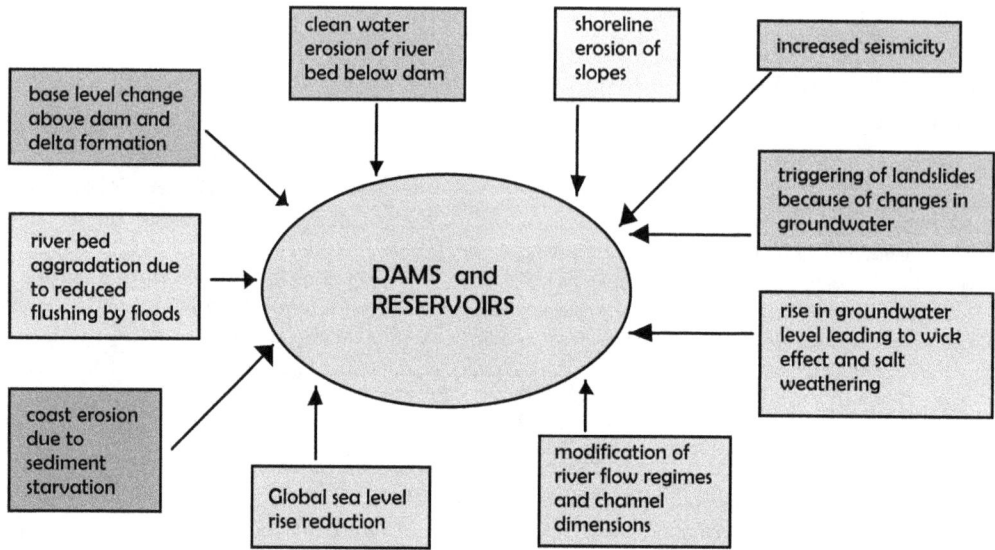

FIGURE 1.10 The multiple geomorphic effects of dams and reservoirs in the landscapes. (Modified from Goudie and Viles (2016).)

on Ganga River are the important dams and barrages which have profound impact on the physical and socio-economic environment of the Bhagirathi-Hooghly River Basin.

The dominant form of direct human-induced disturbance to river courses reflects sachems that have endeavoured to control and regulate their flow and associated concerns for water supply, whether for agricultural (irrigation), commercial/industrial or residential purposes (Brierley and Fryirs, 2005; Wohl, 2014). Enormous effects have been undertaken to make dry lands wetter and wet lands drier, ensuring that water is available for human purposes. The extent of these programs is staggering. The global volume of freshwater trapped in reservoirs now exceeds the volume of flow along rivers (Brierley and Fryirs, 2005). Surely, dam construction has played a major part in pivotal societal changes, such as the development of hydraulic civilisations. Although dams have been constructed for more than 5000 years, the pace of construction quickened dramatically after the Second World War, and each year more than 200 large dams are completed. Reservoirs, dams and diversions are vital for securing water supplies and controlling floods, but the changes in fluvial processes they impose on downstream rivers, dominated by reduced sediment loads and lower flood magnitudes, can be dramatic (Petts and Gurnell, 2013).

1.5.1.1 Collective Impact of Dams

The function of dams may vary markedly, ranging from water supply facilities to flood control impoundments. A common basis for all water supply programs, however, is the disruption they place upon patterns and rates of flow. Dams not only disrupt the longitudinal continuity of flow along rivers; they also act as major barriers to sediment transfer. Collectively, disruption to water and sediment transfer mechanisms impacts directly on river structure and functions both upstream and downstream of the control structure (Brierley and Fryirs, 2005) (Figure 1.11). Dam construction traps sediment in a delta, creating on accumulation zone at the entrance to the reservoir (point A). Suspended sediment load drapes the former channel at this point, which now lies beneath the reservoir. At point B, immediately downstream of the dam, reduced bed load and increased erosive potential of the 'hungry river' have induced bed incision following dam closure. The original floodplain is increasingly decoupled from the channel because of changes to the flow regime and morphological adjustments. Sediments released from this zone have accumulated downstream at

(a) Catchment-scale adjustments

headcut retreat
up tributaries

A

B ▬ dam

bars

C

sediment
slug

Channel cross-sections

A

B

C

post-dam channel profile
pre-dam channel profile
direction of sediment
accumulation and erosion

coastal erosion and
shoreline recession

(b) Longitudinal profile adjustments

Hungry
water

reservior

DAM

sediment
slug
formation

Material
deposition

Original channel bed

incision

New channel bed

Bed can become
armoured

Bed incision declining
in intensity downstream

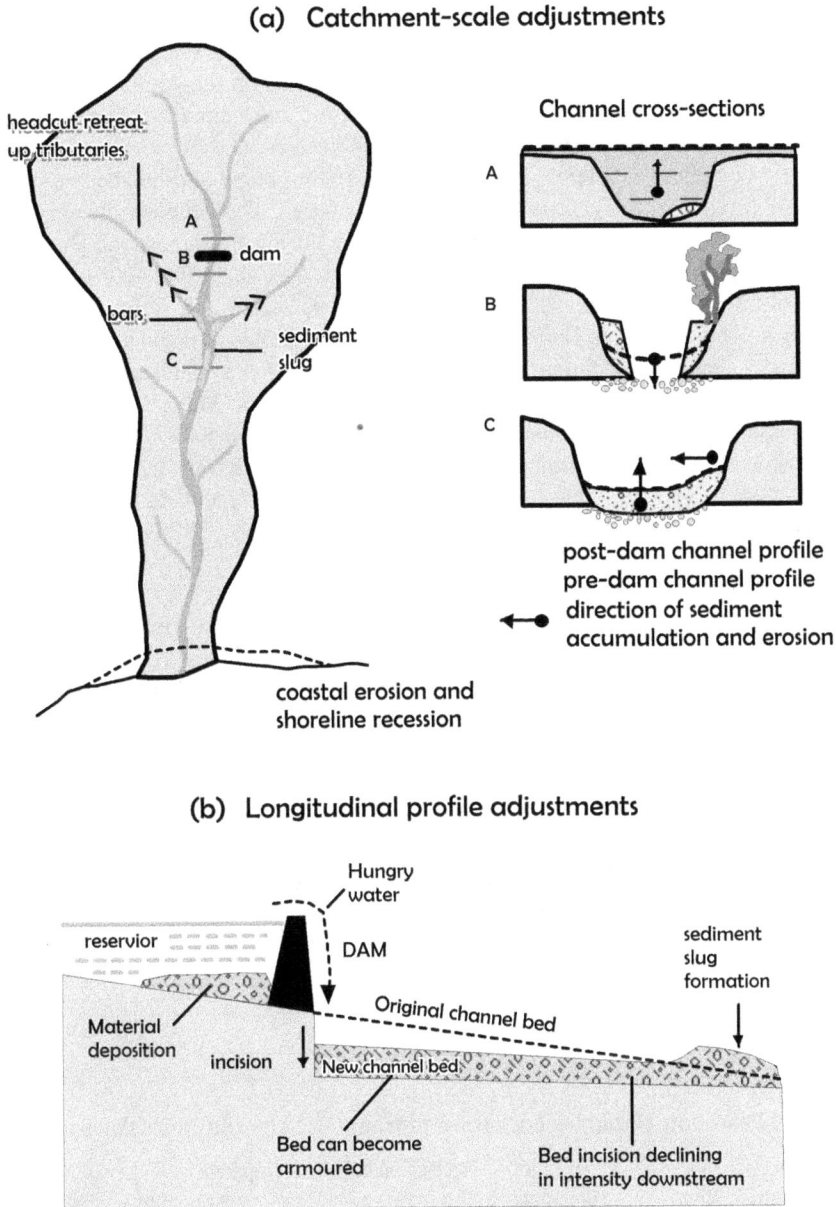

FIGURE 1.11 Geomorphic impacts of dam construction on river character and behaviour. (Modified from Brierley and Fryirs (2005).)

point C, where the channel has contracted through the formation of lateral bars. Off-site impacts of dam construction may include incision of tributary streams and altered morphodynamics at the coastline.

1.5.1.2 Upstream Effect

The reduction in channel gradient following elevation of base-level upstream of dams reduces the transporting capacity of flow as it enters the reservoir. This promotes delta development at the backwater limit, reducing the water storage capacity of the reservoir (Graf, 2006). Reservoirs make excellent sediment traps, commonly retaining more than 90% of the total load and the entire coarse fraction (i.e. all bedload sediment and all or part of the suspended load) (Brandt, 2000; Brierley and Fryirs, 2005). Rapidly flowing rivers from the high lands bring fine soil particles called silt through the process of erosion. The silt gets laterally deposited on the banks of the rivers. When the river is dammed, its flow velocity suddenly gets lessened and stagnant or semi-stagnant conditions result. The buoyancy of the silt particles then gets lowered as they slowly or gradually settle on the reservoir bed or bottom, thereby filling the same. In the course of time, the live storage of the reservoir gets reduced, and this phenomenon is called siltation. The siltation of reservoirs reduces the carrying capacity of dam and flood storage capacity during the monsoon. The reservoirs of India are losing about 1.3 billion m^3 of storage capacity (it cost about Rs. 1448 crores) each year. In India, the overall picture indicates that reservoir sedimentation is a serious national problem which requires immediate action. Heavily silted reservoirs of India are listed as, Matatila on Betwa River (38% gross capacity loss), Gumti on Gumti River (20.4% live storage capacity loss), Maithon on Barakar River (25.29% live storage capacity loss) and Kadna on Mahi River (12.85% live storage capacity loss) (Table 1.9). It is found that some of the dams have shown high siltation rate, viz. Panchet on Damodar River (1.05 mm yr^{-1}), Maithon on Barakar River (1.282 mm yr^{-1}), Pong on Beas River (2.785 mm yr^{-1}), Koyna on Koyna River (1.52 mm yr^{-1}) and Ramganga on Ramganga River (2.294 mm yr^{-1}) (Table 1.10).

1.5.1.3   Downstream Effect

Downstream impacts of dam construction reflect lowered flood peak magnitudes and marked reductions in sediment load. Worldwide, dams have reduced the mean annual flow by up to 80% as a result of abstraction for human uses, increased evaporation and seepage (Garf, 2000). The mean annual flow (2.33 year recurrence interval) is typically reduced by 25%–30%, and the influence of reservoir lag and flood storage declines for higher floods (Graf, 2000). So, rarer events may be altered less than more frequent ones (Table 1.11).

1.5.1.4   The Hungry Water

The downstream reach below the dam has observed an unexpectable hydrogeomorphic change which was not at all predictable to the ground extent by the river engineers. The released floodwater has the kinetic energy to move sediment from the dam closure, but little or no sediment load is available to them due to rise in base level and resultant sedimentation. This 'hungry water'

TABLE 1.9

A Database on Reservoir Capacity Loss (Live Storage) in Selected Reservoirs of India

Name of Reservoir	River, State	Year of Impounding	Original Live Storage (MCM)	Capacity Loss (MCM) up to 2006	Loss (%)	Siltation Rate (MCM yr^{-1})
Bhadar	Narmada, Gujarat	1964	223.703	35.913	16.05	0.998
Gumti	Gumti, Tripura	1984	249.07	63.83	20.4	3.36
Isapur	Penganga, Maharashtra	1983	928.262	28.633	3.08	1.43
Kadana	Mahi, Gujarat	1983	1712	220.29	12.87	2.03
Maithon	Barakar, Jharkhand	1955	607.268	153.578	25.29	3.34
Matatila	Betwa, Uttar Pradesh	1956	1132.7	430.37	38	10.01
Palitana	Shetrunji, Gujarat	1959	374.832	127.544	18.84	1.908
Sondur	Sondur, Chhattisgarh	1988	179.611	44.823	24.95	2.99
Srisailam	Krishna, Andhra Pradesh	1984	7165.83	2013.33	28.1	1.87
Ramsagar	Godavari, Telangana	1951	29.397	4.734	16.1	0.05

Source: Thakkar and Bhattacharyya (2006).
Notes: MCM, Million cubic metre.

TABLE 1.10
Actual and Design Siltation Rate in Selected Reservoirs of India

Name of Reservoir	River, State	Design Rate (mm yr^{-1})	Actual Rate (mm yr^{-1})
Gumti	Gumti, Tripura	0.362	9.940
Kyrdemkulai	Umtru, Meghalaya	0.138	0.144
Halali	Betwa, Madhya Pradesh	0.476	2.032
Matatila	Betwa, Uttar Pradesh	0.132	0.370
Ramsagar	Bamni, Rajasthan	0.157	0.524
Kadna	Mahi, Gujarat	0.357	0.475
Massanjore	Mayurakshi, Jharkhand	0.364	0.696
Maithon	Barakar, Jharkhand	0.905	1.282
Sondur	Mahanadi, Chhattisgarh	0.357	5.768
Kallada	Kallada, Kerala	1.450	4.780

Source: Thakkar and Bhattacharyya (2006).

TABLE 1.11
Examples of Hydrological Changes to below Dams

Hydrological Change	Region, River, Dam	Source
Fifty years flood reduced by 20%	Central Europe	Lauterbach and Leder (1969)
Reduced the 2-year flood by 5%	Utah, Green River, Flaming Gorge Dam	Grams and Schmidt (2002)
Compressed flow range from 150–9000 to 500–2000 m^{3}s^{-1}	Canada, Peace River, WAC Bennett Dam	Kellerhals and Gill (1973)
Flood stages reduced for 1200 km downstream	Zambezi, Lake Kariba	Guy (1981)
Increased minimum flow from 503 m^{3}s^{-1} in 1966 to 890 m^{3}s^{-1} to meet hydropower demand		
Transformed the seasonal river into a perennial one	India, Damodar River Reservoirs	Jain et al. (1973)

Source: Graf (2000).

(Brierley and Fryirs, 2005) is able to expend its energy on erosion of the channel bed and banks. Typically, the channel incises and decreases in bankfull cross-sectional area of over 50% are not uncommon, whereas there is no sediment supply immediately downstream of a dam and bed materials are relatively fine-grained and bed degradation may be experienced (Brierley and Fryirs, 2005). In general terms, downstream channel contraction and degradation continue until development of bed armour or reduction in the energy slope stabilises the channel. An initial decrease in channel width following dam closure may be followed by a widening phase as the bed becomes armoured and relatively more resistant. This may be accompanied by a change in channel planform, such as an increase in channel sinuosity or a decrease in channel multiplicity. If vegetation encroachment occurs, the channel may become increasingly stable (Brierley and Fryirs, 2005). Bed incision and channel narrowing inevitably entail a time lag, as materials are removed and redeposited. It is found from many researches that the changes to sediment transfer regimes following dam closure are manifest over timeframes ranging from 10 to over 500 years. Since many dams were constructed in the 20th century, it may be a century or more before the river adjustment process is fully realised, especially in downstream reaches.

1.5.2 CHANNELISATION PROGRAMMES

As many settlements are established along valley floors, concerns for flood control and hazard reduction to support infrastructure development have prompted calls for the training of river courses. Most streamlines in urban and peri-urban areas have been channelised via concrete lining and piping of flow. Swampy areas have been extensively drained for agricultural purposes. In contrast to dam construction, this essentially represents a point disturbance with off-site impacts. Channelisation activities are applied over varying lengths of river.

Levee construction deepens flows, potentially increasing rates of bed erosion. Since channelisation involves manipulation of one or more of the dependent hydraulic variables of slope, depth, width and roughness, feedback effects promote adjustments towards a new characteristic state. The agricultural prosperity of West Bengal was linked with the ecological services of the rivers, and the people of Bengal have been living with floods from time immemorial. Since the 19 century, the British rulers in collaboration with landlords planned to achieve freedom from flood and built earthen embankments along the banks of rivers which had the tendency to spill over during the monsoon (Rudra, 2018). The main objective of jacketing rivers (mainly Kansai, Silai, Dwarkeswar, Ajay, Damodar and Mayurakshi rivers) was to protect the interest of landlords who uses to pay the treasury to the British East India Company (Rudra, 2018). That human disturbance has great impact mainly on the Lower Damodar River Basin, as decline in agricultural prosperity and groundwater level. The embankment ensured protection from low-intensity floods but trapped the sediment load in river bed, causing the decay of drainage systems and consequent crisis (Rudra, 2018). With the passage of time, the palaeochannels are abandoned and decayed, disconnecting the rivers from distant floodplains. Since the railways and highways in the Bhagirathi-Hooghly Basin were transverse to the flow direction of rivers and were built with narrow culverts, there were drainage congestions, expansion of flood contours and decline in agricultural productivity (Rudra, 2018). So, the embankments, roads and railways were described as satanic chains to the floodplains.

Channelisation may include instability not only in the improved reach but also upstream and/ or downstream (Brierley and Fryirs, 2005). Impacts are particularly pronounced in response to channel slope modifications pronounced in response to channel slope modifications or straightening programs that increase local bed steepens and hence erosive potential (Table 1.12). Greater concentration of flow within the channel may accelerate the transmission of flood waves and accentuate flood peaks, relative to the period prior to channelisation. Transfer of excess load to reach downstream of the area of maximum disturbance may result in accelerated channel aggradation and/or bank accretion, reducing channel capacity (Brierley and Fryirs, 2005).

1.5.3 FLOOD HAZARDS AND HUMAN ROLE

Flood is one of the most severe natural hazards which causes misery to a large section of human population and severe damages to agricultural livelihood. Floods could be broadly categorised into overbank flooding, channel shifting and outburst flooding (Jain et al., 2016). The Himalayan Rivers are some of the worst affected rivers in recent times, and this has been attributed to high monsoonal discharges and very high silt load of these rivers. Due to dam construction and upper catchment erosion, extensive siltation is responsible for reduction in the channel capacity which creates potential conditions for channel avulsion and flooding (Jain et al., 2016). The floods of the Bhagirathi-Hooghly Basin are interpreted as hazards not because of the intrinsic attributes of natural phenomena but because of the failure of the socio-technical system to cope up with the situation (Rudra, 2018). Now, total freedom from the flood through structural measures has proved futile. The flood control and post-flood management programmes have not been able to provide the expected relief in the most flood-prone areas. Human-induced flood may occur due to the sudden release of excess water from the dam, failure of hydraulic structures such as embankments or reservoirs and the impacts may be disastrous when the downstream regions are densely populated (Parua, 2009;

TABLE 1.12
Geomorphic Impacts of Channelisation Procedures

Methods	Purpose	Description	Impacts
Straightening (realignment)	Flood protection, infrastructure development	River is shortened artificially by cut-offs	Steepened gradient, increase in flow velocity and transport capacity, degradation of channel, headcut irrigation, increase in sediment load downstream, promoting aggradation
Levee and floodwall construction	Increased conveyance capacity to reduce overbank flooding	Widening and/or deepening of the channel	Reduction in velocity and unit power stream, lowering sediment transport capacity, promoting bench deposition
Channel stabilisation and bank protection works	Control bank erosion	Use of structures such as paving, gabions, steel piles and dikes	Alteration in channel width and roughness components, promoting sedimentation adjacent to the bank, potentially increasing flooding if channel capacity is reduced
Dredging	Maintain navigable channels	Sediment removal from the bed to deepen the channel	May promote degradation through thalweg, enabling knickpoints to migrate upstream, thereby contributing additional sediment to the degraded reach

Source: Knighton (1998).

Rudra, 2018). The unusually concentrated rainfall due to cloudburst is the major cause of flood in West Bengal. The peak discharge in case of the rivers of the Bhagirathi-Hooghly Basin is generally observed in late August or September. It is observed that the rain during the first part of monsoon recharges the groundwater table and rejuvenates both rivers and wetlands. There is more runoff and less infiltration in the second half of the rainy season when a cloudburst and cyclonic rainfall may cause disaster. In addition, the reservoirs have already lost their pervious flood storage capacity, and the extensions of railways and highways across the channels of floodplains have aggravated the area and time period of flood with inadequate passage for floodwater (Parua, 2009; Rudra, 2018).

It is found that with increasing population density, the flood-prone areas of the Bhagirathi-Hooghly Basin have expanded continuously from the 17 m contour line in the 1950s to 20 m in the 1960s and to 26 m in the 1978 floods (Rudra, 2018). The flooded area crossed 20,000 km² in four different years, and the flood of medium magnitude, i.e. 2000–10,000 km², occurred on ten occasions in between 1960 and 2010 (Rudra, 2018). In late September 1978, the lower basins of Mayurakshi, Ajay and Damodar overflowed and the floodwater submerged about 30,000 km² of southern West Bengal rendering 15 million people homeless. In 2000, the flood caused distress to 20 million people of Birbhum, Murshidabad, Nadia, Barddhaman and North 24-Parganas districts. When DVC-induced water synchronises with rainwater of lower catchment and tidal backflow from the Hooghly River creates a hydraulic dam intercepting free flow of floodwater, the situation becomes grim. For those reasons, the vast region of Khanakul-I, Khanakul-II, Amta, Pursura, Jangipara, Arambagh and Tarakeswar (parts of the Lower Damodar Basin) is annually flooded. The flood issue is more pertinent nowadays because the highest monetary damage (including crops, houses and public utilities) due to annual flood amount to Rs. 25,353.270 crore in 2015. Some of the statistics show the maximum impact of flood (1953–2015) on the human population and socio-economic structures of West Bengal—(1) maximum area affected (3.08 million ha) in the 1978 flood (Figure 1.12), (2) maximum population (21.8 million) affected in the 2000 flood, (3) maximum damage to cropped area (2.49 million ha) in the 2007 flood, (4) maximum damage to crops in respect of value (Rs. 11,433.680 crore) in the 2015 flood, (5) maximum damage to houses in respect

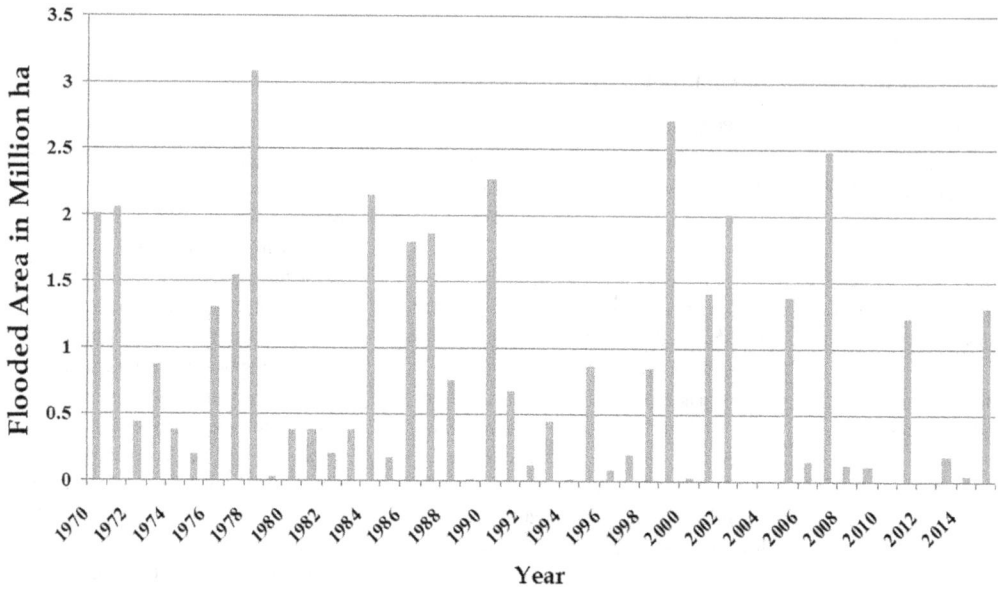

FIGURE 1.12   Post-dam flood affected area of West Bengal from 1970 to 2015 (IWD, 2015).

of value (Rs. 7895.630 crore) in the 2015 flood, (6) maximum cattle lost (no. 221,826) due to flood in 1978, (7) maximum loss of human lives (no. 2730) in 1978 and (8) maximum damage to public utilities in respect of value (Rs. 6023.960 crore) in 2015 (IWD, 2015).

1.5.3.1   Urban Flood Risk

Recent flood-situations, in New Delhi (27 September 1988), Bharauch (3 August 2004), Mumbai (26 July 2005), Chennai (November–December 2015), Kolkata (30 June 2007), Noida (9 September 2010) and Srinagar (September 2014), have led to disruptions in civic life (Narain, 2016; Ray et al., 2019). Flood is defined as an overflow of a large body of water over areas not usually inundated. Thus, flooding in urban areas is primarily caused by intense and/or prolonged rainfall, which overwhelms the capacity of the drainage system (Figure 1.13). Urban flooding is significantly different from rural flooding. Urbanisation increases flood risk by up to three times, increased peak flow results in flooding very quickly; further, it affects a large number of people due to high production density in urban areas. A major concern is blocking of natural drainage pathways through construction activity and encroachment on catchment areas, riverbeds and lakebeds. Some of the major hydrological effects of urbanisation are as follows (Zevenbergen et al., 2010; Narain, 2016):

1. Increased water demand, often exceeding the available natural resources
2. Increased waste water, polluting rivers and lakes and endangering the ecology
3. Destruction of lakes and wetlands is a major issue in cities. Lakes and wetlands can store the excess water and regulate the flow of water, but converting them into development purposes has increased urban flood risk
4. Reduced infiltration due to paving of surfaces which decreases ground absorption and increases surface runoff
5. Reduced groundwater recharge and diminishing base flow of streams
6. Sewer flooding associated with blocked sewer overflows to larger rivers or surcharging through manhole covers
7. Degradation of channel inside the cities and loss of carrying capacity during increased peak flow

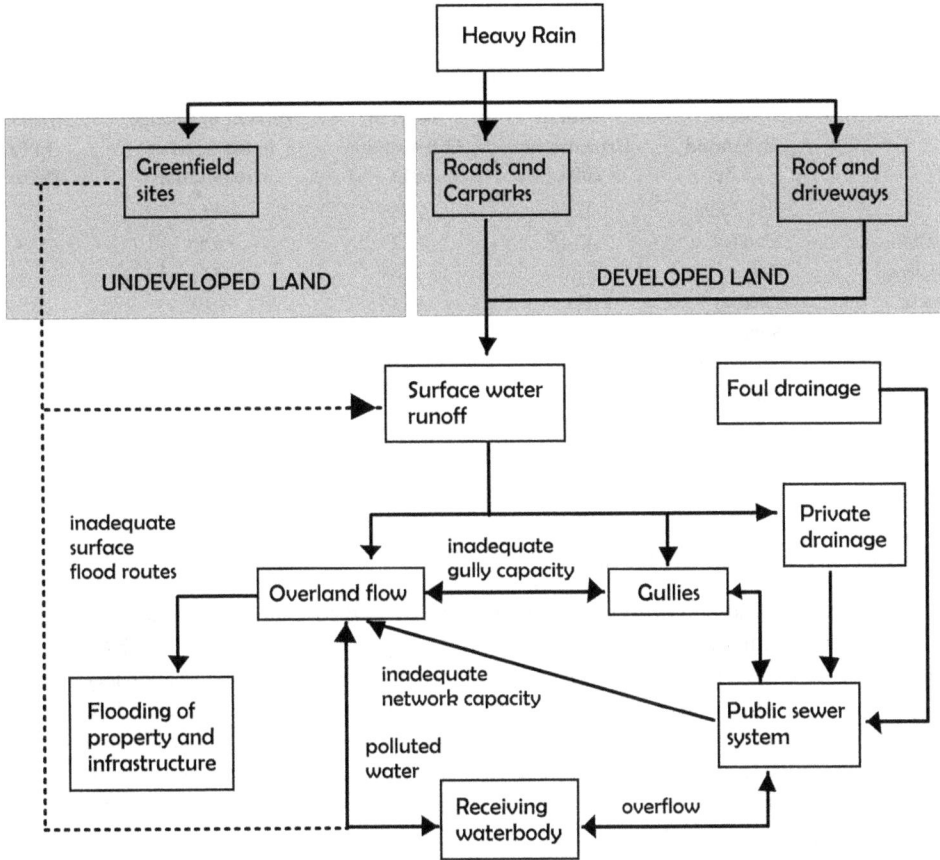

FIGURE 1.13   A systematic diagram showing the processes of urban flooding. (Modified from Rafiq et al. (2016).)

In Asian cities, both urbanisation and industrialisation processes provide a base for socio-economic gain. Such processes considerably increase the potential risk of damage from environmental pollution, ecological disruptions, hydrometeorological disasters and climate change exacerbations (Table 1.13) (Ray et al., 2019). In urban environments, farmland, vegetation cover and bare soil have been converted into built-up areas. As a result, water runs off through the concrete structures, sometimes known as pluvial flooding or urban flooding. In the outskirts of main cities, human encroachments onto the active flood channel, poor flood management strategies, lack of flood early warning systems and disposal of solid waste in drainage lines are the major causes of urban flooding (Ray et al., 2019). Similarly, increasing urbanisation is consistently linked to increasing urban problems, which indicates that urbanisation and urban risks have a positive correlation. In the urban areas, it is very true that floods are a natural phenomenon, and it is a human construct to label flooding as being acceptable or not (Zevenbergen et al., 2010). Taking lessons from the past flood episodes, the Municipal Corporations have taken few steps to manage urban floods, viz. installation of proper functional drainage system, better forecasting rainfall events, installing many automatic weather stations, maintenance of existing drainage channels and proving alternative drainage path. Few cities have accepted the concept of 'sponge cities' which already functions to tackle urban wastewater issues in China (Zevenbergen et al., 2010). Sponge city envisages replacing concrete pavements with porous pavements to ensure better infiltration, and it further aims to restore wetlands, develop rain garden and rooftop garden for better absorption of rainwater.

TABLE 1.13

Increased Urbanisation Prompting Increased Flood Risk in India

State	Most Urbanised City	Level of Urbanisation in 2001 (%)	Level of Urbanisation in 2011 (%)	Percent of Change in the Level of Urbanisation	Urban Flood in Past Decades
Delhi	New Delhi	93.18	97.50	4.64	3
Maharashtra	Mumbai	42.43	45.23	6.56	9
West Bengal	Kolkata	27.97	31.89	14.01	5
Tamilnadu	Chennai	44.04	48.45	10.01	7
Jammu & Kashmir	Srinagar	24.81	27.21	9.60	2
Gujarat	Ahmedabad	37.36	42.58	13.97	7
Karnataka	Bengaluru	33.99	38.57	13.47	4

Source: Ray et al. (2019).

1.5.4 GRAVEL AND SAND EXTRACTION

Instream mining for sand and gravel used in construction and for placer deposits of precious metals has occurred for millennia in some regions. By altering substrate grain-size distribution and abundance, all forms of instream mining disrupt sediment dynamics and cause a variety of channel adjustments (Mossa and James, 2013; Wohl, 2014). In the western tributaries of the Bhagirathi-Hooghly Basin, the sand mining has now appeared as a growing economy, and with increasing level of urbanisation, the demand for sands is increasing. Gravel and sand extraction can take the form of instream (wet mining) where sediment is extracted from instream bar and bed surfaces or open floodplain pits through innovative machines. Sand mining may involve extensive clearing, diversion of flow, stockpiling of sediment and excavation of deep pits. Eventually, the new channel morphology is developed and that can be regarded as anthropogenic landform, developed due to sand mining activities. By removing sediment from the channel, the pre-existing balance between sediment supply and transport capacity is disrupted (Figure 1.14) (Brierley and Fryirs, 2005). Typical responses include lowering of the streambed, local increase in slope and flow velocity upon entering the pit and adjustments to channel geometry. Once bed armour is destroyed, enhanced bed scour may generate headcuts in oversteepened reaches and hungry water erodes the bed downstream (Brierley and Fryirs, 2005). Mined reaches can widen from material extraction along the channel perimeter or by capturing the anthropogenic form of a pit through migration or avulsion (Mossa and James, 2013). An avulsion into one or more floodplain pits can have the same effect as instream mining, because these along-channel pits can trap much of the bedload transport for decades (Mossa and James, 2013).

As geomorphologically effective sediment transporting events are infrequent in many gravel-bed rivers, instream mining activities may operate for several years without obvious effects upstream or downstream. Headcuts may propagate upstream for kilometres on the main river and tributaries potentially undermining bridges and weirs. Undercutting of banks promotes channel expansion and frequent shifts of channel thalweg are observed in the river (Mossa and James, 2013). Planform changes may ensure, typically entailing the adoption of a low sinuosity, single-channelled river. Bed and bank erosion downstream from mining can be exacerbated by reduced sediment supply. Disruption of a coarse surface layer can increase sediment mobility and downstream turbidity, aggradation, overbank deposition and lateral channel mobility (Wohl, 2014). Enhanced rates of downstream sediment delivery may promote channel aggradation and instability. These adjustments alter the availability and viability of aquatic habitat, groundwater levels and riparian vegetation associations (Brierley and Fryirs, 2005). If managed effectively, with clear separation of

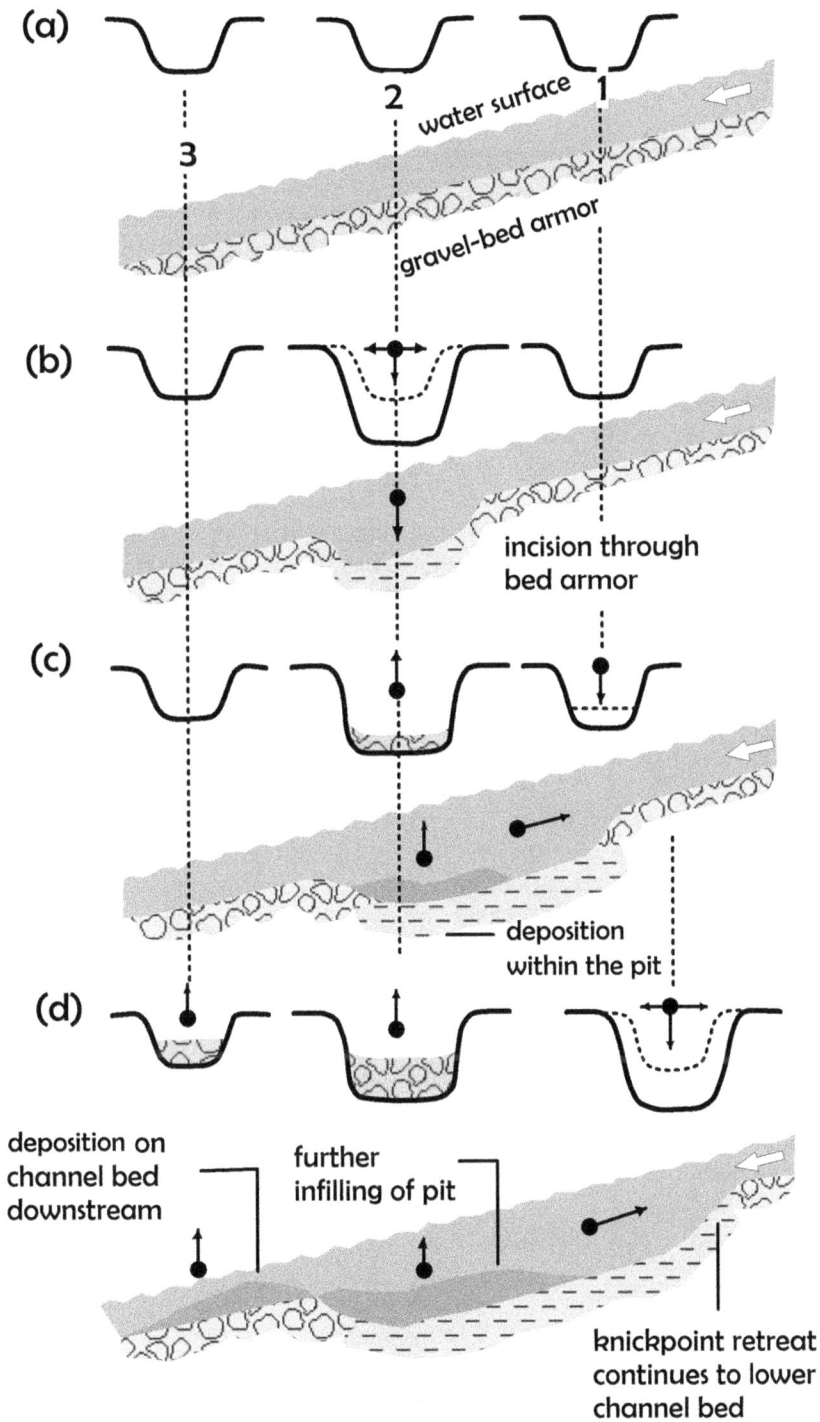

FIGURE 1.14 Geomorphic impacts of gravel and sand mining in the river bed: (a) river's sediment load and force available to transport sediment are continuous through the reach, (b) excavation of an instream pit breaks the bed armour and instigates a headcut at upstream, (c) headward extension of headcut acts to maintain bed slope and hungry water erodes downstream and (d) sediments released following the upstream progression of headcut and associated channel expansion partially infill the incised and expanded reach of downstream in the form of bars and benches. (Modified from Brierley and Fryirs (2005) and Mossa and James (2013).)

geomorphic activity from the channel zone, pits may be stabilised and left open once mining activities are complete, creating habitat for local flora and fauna, in large open-water ponds (Brierley and Fryirs, 2005).

1.6 SUMMARY OF CONTRIBUTING CHAPTERS

Given the complexity of biophysical linkages and the cumulative nature of disturbance impacts, along with profound variability in the inherent sensitivity of individual river systems to adjustment, it is often difficult to isolate cause–effect relationships that directly link changes in river morphology to discrete underlying factors (Brierley and Fryirs, 2005; Szabo, 2010; Wohl, 2014). Within any given catchment of the Bhagirathi-Hooghly Basin, not all landforms may have responded to the last external influence in the same way, resulting in considerable complexity in patterns and/ or rates of landscape responses to disturbance events. Fluvial responses to human disturbance vary markedly across the region, affected by factors such as environmental setting, population pressure (today and in the past) and level of economic and industrial development. Different reaches within a river basin are typically at differing stages of adjustment to differing forms of human and natural disturbance. In the Bhagirathi-Hooghly Basin contemporary river forms and processes have recorded system adjustments to the totality of these disturbance impacts and their inter-connected consequences. Thus, it is often very difficult to isolate the consequences of any one form of disturbance from others.

It is important to consider that a river is not a closed geomorphic system. River processes are governed by flux and energy transfers, and it maintains equilibrium with the inherent climatic, geological and landscape settings (Jain et al., 2016). Unless human impact on modern rivers is documented in such complex framework, threshold dynamics and sensitivity of the various components of large river basin cannot be understood. Anthropogenic disturbance and associated physical changes are ongoing in the Bhagirathi-Hooghly Basin. Progressive adjustments to natural events and cumulative human impacts ensure that river morphology and hydrogeomorphic processes are in a permanent state of adjustment, with marked variability in flow and sediment fluxes over a range of spatial and temporal scales. Understanding the trajectory and rate of change and the capacity for ongoing and future adjustments is key consideration for the floodplain management in the Bhagirathi-Hooghly Basin. In this book, the following questions are addressed, focusing on the prime theme of anthropogeomorphology:

1. Will the present trajectory and rate of change continue? If so, how long?
2. When will the available sediment stores be depleted?
3. What state is the system moving towards?
4. What lagged and off-site impacts are involved?
5. Is change irreversible?
6. To what extent will the inhabitants be affected by risk and hazard?

Understanding of contemporary river forms and processes, tied to interpretations of longer-term natural variability derived from studies of river evolution, provides a basis to assess river responses to differing forms of disturbance. From this analysis, predictions of likely future character, behaviour and condition can be made in the context of changes to within-catchment linkages of river processes. The human dimension of geomorphology is a vital area of research that has been largely neglected beyond the local and sub-regional scale. Much less has been written about the effectiveness of humans as geomorphic agents in the tropical and densely populated countries, like India. In this context of increasing concern over global change and human impacts on the environment, a volume of anthropogeomorphology (with special emphasis on fluvial geomorphology of Bhagirathi-Hooghly River Basin) is much demanded which will be dedicated to addressing the human disturbances, change and vulnerability in the alluvial river basin, focusing on human dimensions of fluvial geomorphology in India. The central goal of this volume is to pull together several syntheses and reviews from the writings on these topics and show how geomorphology can benefit from an anthropogeomorphic viewpoint.

In this volume, totally 16 chapters are included, and each chapter provides a wide and in-depth geomorphic understanding of fluvial responses on human disturbances in the important sub-basins of the Bhagirathi-Hooghly River Basin. The present chapter generally focuses on the scientific procedures for understanding the human impact on fluvial dynamics and to integrate catchment-scale geomorphic perceptive of river forms, processes and linkages. Chapter 2 deals mainly with the post-barrage channel behaviour of Bhagirathi River through understanding the changes in width/depth ratio, cubic capacity, bank erosion, channel oscillation and frequent inundation. Chapter 3 mainly focuses on the impacts of anthropogenic changes (the location-specific studies of montanogenic, industrogenic, urbanogenic, hydrologic, traffic, agrogenic and recreational interventions) on the landforms of Lower Deltaic West Bengal. Chapter 4 deals with the ecological and geomorphic impacts of construction of various engineering works of the Farakka Barrage project on the floodplains of the Bagmari–Bansloi–Pagla River Basin. Chapter 5 includes a detailed anthropogeomorphic study of the Mayurakshi River Basin, which mainly focuses on the impacts of dams and barrages, sand mining, embankment, land-use land-cover change, deforestation, stone quarrying, filling of inland water bodies, intensive cultivation on channel and extra channel areas, landfilling near riverbank, road crossing and contamination of water through industrialisation. Chapter 6 depicts the spatio-temporal anthropogeomorphic changes in the Ajay River Basin, showing the resultant effects of deforestation, embankments, sand mining and agricultural expansion. Chapter 7 focuses on how extensive and significant human impacts on fluvial geomorphology have been and how these impacts are likely to increase in future in the Damodar River Basin. Chapter 8 deals with human imprints on the Kangsabati River Basin through a historical overview and a cartographic and quantitative analysis. Chapter 9 analyses the geomorphic impacts of urbanisation, embankment, sand mining and other types of river regulation activities on the Dwarkeswar River Basin using remote sensing and geographic information system. Chapter 10 provides an environmental geomorphic outlook on the Silabati River Basin, with a main focus on the geodiversity assessment and the examination of the human linkages with the different physiographic units. Chapter 11 deals with the hydro-geomorphology of a tidal river, Rasulpur River Basin, with a special emphasis on the human disturbances in Anthropocene. Chapter 12 mainly depicts the anthropogeomorphology of Jalangi River, showing the historical and contemporary evolution of basin and channel geomorphology of the river in respect of human disturbances. Chapter 13 deals with the origin, evolution and anthropogenic signatures of a dying river Churni and the impact of human on the changing landforms. Chapter 14 attempts to identify and demarcate the palaeochannels of Saraswati–Jamuna system based on the existing anthropogeomorphological features and their associated terrain attributes using geo-spatial techniques. Chapter 15 generally focuses on systematic exhumation of the facets of human encroachment within the palaeo-fluvial regime of *Kana*–Ghia–Kunti system which is in the lower part of Damodar Fan Delta. Finally, Chapter 16 explains the palaeogeography of an abandoned channel, Anjana River, and also deals with the human role in the evolution of geomorphic changes.

REFERENCES

Bagchi, K. 1944. *The Ganges Delta*. Calcutta: University of Calcutta.
Bandyopadhyay, S. 2017. Evolution of Ganga – Brahmaputra delta: a review. *Geological Review of India* 69 (3): 235–268.
Bandyopadhyay, S., N.S. Kar, S. Das and J. Sen. 2014. River systems and water resources of West Bengal: a review. *Journal of the Geological Society of India* 3: 63–84.
Begin, Z.A. and S.A. Schumm. 1984. Gradational thresholds and landforms singularity: significance for Quaternary studies. *Quaternary Research* 21: 264–274.
Bennett, H.H. 1938. *Soil Conservation*. New York: McGraw-Hill.

Bose, N.K. 1972. The Bhagirathi-Hooghly Basin – a few remarks. In *The Bhagirathi – Hooghly Basin*, ed. Bagchi, K.G., xii–xvii, Calcutta: Proceedings of the Interdisciplinary Symposium University of Caluctta.

Brandt, S.A. 2000. Classification of geomorphological effects downstream of dams. *Catena* 40: 375–401.

Brierley, G.J. and K.A. Fryirs. 2005. *Geomorphology and River Management*. Maiden: Blackwell Publishing.

Brown, E.H. 1970. Man shapes the earth. *Geography Journal* 136: 74–85.

Central Water Commission (CWC). 2019. *National Register of Large Dams*. New Delhi: Ministry of Water Resources, Govt. of India, CWC.

Chatterjee, S.P. 1972. The Bhagirathi-Hooghly Basin. In *The Bhagirathi – Hooghly Basin*, ed. Bagchi, K.G., xix–xxiv, Calcutta: Proceedings of the Interdisciplinary Symposium University of Calcutta.

Crutzen, P.J. 2002. Geology of mankind. *Nature* 419: 23.

Das, B.C. 2014. Impact of in-bed and on-bank soil cutting by brick fields on moribund deltaic rivers: a study of Nadia river in West Bengal. *The NEHU Journal* XII (2): 101–111.

Das, B.C. 2015a. Socio-economic impact of a decaying river on fishermen: a case study of Taranipur Village, West Bengal. *International Journal of Research in Management, Science & Technology* 3 (4), E-ISSN: 2321-3264, December 2015. Available at www.ijrmst.org.

Das, B.C. 2015b. Crying with the river: a study on a dying river and her famished fishermen. In *Life and Living Through Newer Spectrum of Geography*, eds. Ismail, M. and A. Alam, 3–22. Mohit Publications, New Delhi.

Das, B.C. 2019. A study on the impact of bridge construction on channel dynamics of the river Jalangi, West Bengal, India. *Scientific Journal of K F U (Humanities and Management Sciences)* 20 (1): 265–279.

Ehrlich, P.R. and J.P. Holdren. 1971. Impact of population growth. *Science* 171 (12): 12–17.

Fryirs, K.A. and G.J. Brierly 2013. *Geomorphic Analysis of River System: An Approach to Reading the Landscape*. Wiley-Blackwell, New York.

Ghosh, S. and Guchhait, S.K. 2019. *Laterites of Bengal Basin – Characterization, Geochronology and Evolution*. Cham: Springer Nature.

Gilbert, G.K. 1917. *Hydraulic-Mining Debris in the Sierra Nevada*. Washington: United States Geological Survey, Professional Paper 105.

Golomb, B. and H.M. Eder. 1964. Landforms made by man. *Landscape* 14: 4–7.

Goudie, A.S. and H.A. Viles. 2016. *Geomorphology in the Anthropocene*. Cambridge: Cambridge University Press.

Graf, W.L. 2006. Downstream hydrologic and geomorphic effects of large dams on American Rivers. *Geomorphology* 79: 336–360.

Guchhait, S.K., A. Islam, S. Ghosh, B.C. Das and N.K. Maji (2016). Role of hydrological regime and floodplain sediments in channel instability of the Bhagirathi River, Ganga-Brahmaputra Delta, India. *Physical Geography* 37 (6): 476–510. DOI: 10.1080/02723646.2016.1230986.

Gupta, H., S. Kao and M. Dai. 2012. The role of mega dam in reducing sediment fluxes: a case study of large Asian Rivers. *Journal of Hydrology* 464–465: 447–458.

Haigh, M.J. 1978. *Evolution of Slopes on Artificial Landforms*. Blaenavon, UK: University of Chicago Depart of Geography, Research Paper 183.

Holdren, J.P. and P.R. Ehrlich. 1974. Human population and the global environment. *American Scientist* 62: 282–292.

Hossain, M.S., M.S.H. Khand, K.R. Chowdhury and R. Abdullah. 2019. Synthesis of the tectonics and structural elements of the Bengal Basin and its surroundings. In Tectonics and Structural Geology: Indian Context, ed. Mukherjee, S., 135–218. Cham: Springer Geology.

Hupy, J.P. 2017. Anthropogeomophology. In *The International Encyclopedia of Geography*, eds. Richardson, D., N. Castree, M.F. Goodchild, A. Kobayashi, W. Liu and R.A. Martson, 1–6, London: John Wiley & Sons Ltd.

Irrigation and Waterways Directorate (IWD). 2015. *Annual Flood Report for the Year 2015 West Bengal*. Kolkata: IWD, Govt. of West Bengal.

Islam, A. 2011. Variability of stream discharge and bank erosion—A case study on the river Bhagirathi. *Journal of River Research Institute River Behaviour and Control* 31(1): 55–66.

Islam, A. 2012. Water scarcity in the north eastern states of India: Mechanisms and mitigations. *Indian Streams Research Journal* 2(11): 1–7.

Islam, A. and S.K. Guchhait. 2017a. Search for social justice for the victims of erosion hazard along the banks of river Bhagirathi by hydraulic control: A case study of West Bengal, India. *Environment, Development and Sustainability* 19(2): 433–459.

Islam, A. and S.K. Guchhait. 2017b. Analysing the influence of Farakka Barrage Project on channel dynamics and meander geometry of Bhagirathi river of West Bengal, India. *Arabian Journal of Geosciences* 10(11): 245.

Islam, A. and S.D. Barman. 2020. Drainage basin morphometry and evaluating its role on flood-inducing capacity of tributary basins of Mayurakshi River, India. *SN Applied Sciences*. https://link.springer.com/article/10.1007/s42452-020-2839-4

Jacks, G.V. and R.O. Whyte. 1939. *The Rape of the Earth: A World Survey of Soil Erosion.* London: Faber and Faber.

Jain, V., R. Sinha, L.P. Singh and S.K. Tandon. 2016. River systems in India: the anthropocene context. *Proceedings of the Indian national Science Academy* 82 (3): 747–761.

James, L.A., C.P. Harden and J.J. Claugue. 2013. Geomorphology of human disturbances, climate change and hazards. In *Treatise on Geomorphology* (vol. 13), ed. Shroder, J. 1–13. San Diego: Elsevier.

Knighton, D. 1998. *Fluvial Forms and Processes.* London: Arnold.

Lewis, S.L. and M.A. Maslin. 2015. Defining the anthropocene. *Nature* 519: 171–180.

Mahata, H.K. and R. Maiti. 2019. Evolution of Damodar Fan-Delta in the Western Bengal Basin, West Bengal. *Journal of the Geological Society of India* 93 (6): 645–656.

Marsh, G.P. 1864. *Man and Nature.* New York: Scribner.

Meybeck, M. and C.J. Vorosmarty. 2004. The integrity of river and drainage basin systems: challenges from environmental change. In *Vegetation, Water, Humans and Climate: A New Perspective*, eds. Kabat, C., M. Claussen, P.A. Dirymeyer and J.H.C. Gash, 297–463, Berlin: Springer-Verlag.

Morgan, J.P. and W.G. McIntire. 1959. Quaternary geology of the Bengal Basin, East Pakistan and India. *Geological Society of America Bulletin* 70 (3): 319–342.

Mossa, J and L.A. James. 2013. Impacts of mining on geomorphic systems. In *Geomorphology of Human Disturbances, Climate Change and Hazards*, eds. James, L.A., C.P. Harden and J.J. Claugue, 74–95. San Diego: Elsevier.

Narain, S. 2016. An urban nightmare. In *Why Urban India Floods*, ed. Narain, S., 4–10. New Delhi: Down to Earth, Centre for Science and Environment.

Nath, S.K., M.D. Adkikari, S.K. Maiti, N. Devaraj, N. Srivastava and L.D. Mohapatra. 2014. Earthquake scenario in West Bengal with emphasis in seismic hazard microzonation of the city of Kolkata, India. *Natural Hazards and Earth System Sciences* 14: 2549–2575.

Nir, D. 1983. *Man, a Geomorphological Agent: An Introduction to Anthropic Geomorphology.* Jerusalem: Keter.

Panizza, M. 1996. *Environmental Geomorphology.* Amsterdam: Elsevier.

Parua, P.K. 2009. *The Ganga: Water Use in the Indian-Subcontinent.* Dordrecht: Springer.

Petts, G. and A. Gurnell. 2013. Hydrogeomorphic effects of reservoirs, dams and diversions. In *Geomorphology of Human Disturbances, Climate Change and Hazards*, eds. James, L.A., C.P. Harden and J.J. Claugue, 96–114. San Diego: Elsevier.

Petts, G.E. 1979. Complex response of river channel morphology subsequent to reservoir construction. *Progress in Physical Geography* 3: 329–362.

Rafiq, F, S. Ahmed, S. Ahmad and A.A. Khan. 2016. Urban floods in India. *International Journal of Scientific & Engineering Research* 7 (1): 721–734.

Ray, K., P. Pandey, C. Pandey, A.P. Dimri and K. Kishore. 2019. On the recent floods in India. *Current Science* 117 (2): 204–218.

Rudra, K. 2014. Changing river courses in the western part of the Ganga-Brahmaputra delta. *Geomorphology* 227: 87–100.

Rudra, K. 2018. *Rivers of the Ganga – Brahmaputra – Meghna Delta: A Fluvial Account of Bengal.* Cham: Springer Nature.

Sarkar, B. and A. Islam. 2019. Assessing the suitability of water for irrigation using major physical parameters and ion chemistry: a study of the Churni River, India. *Arabian Journal of Geosciences* 12 (20): 637.

Sarkar, B. and A. Islam. 2020. Drivers of water pollution and evaluating its ecological stress with special reference to macrovertebrates (fish community structure): a case of Churni River, India. *Environmental Monitoring and Assessment* 192 (1): 45.

Schumm, S.A. 1969. River metamorphosis. *Journal of the Hydraulics Division ASCE HYI* 7: 255–273.

Sengupta, S. 1966. Geological and geophysical studies in western part of Bengal Basin, India. *Bulletin of the American Association of Petroleum Geologists* 50 (5): 1001–1017.

Sengupta, S. 1972. Geological framework of Bhagirathi – Hooghly Basin. In *The Bhagirathi – Hooghly Basin*, ed. Bagchi, K.G., 3–8, Calcutta: Proceedings of the Interdisciplinary Symposium University of Caluctta.

Sengupta, S. 2018. Cleaning India's polluted rivers. In *State of India's Environment 2018: A Down to Earth Manual*, eds. Narain, S., C. Bhusan, R. Mahapatra, A. Misra and S. Das, 52–63. New Delhi: Centre for Science and Environment.

Sinha, R. 2009. The great avulsion of Kosi on 18 August 2008. *Current Science* 97: 429–433.

Sinha, R. and S. Ghosh. 2012. Understanding dynamics of large rivers aided by satellite remote sensing: a case study from Lower Ganga Plains, India. *Geocarto International* 27: 207–219.

Sinha, R., V. Jain, G. Prasad Babu and S. Ghosh. 2005. Geomorphic characterization and diversity of the fluvial systems of the Gangetic plains. *Geomorphology* 70: 207–225.

Szabo, J. 2010. Anthropogenic Geomorphology: subject and system. In *Anthropogenic Geomorphology: A Guide to Man-Made Landforms*, eds. Szabo J, L. David and D. Loczy. 3–10. Dordrecht, Springer.

Tarolli, P., W. Cao, G. Sofia, D. Evans and E.C. Ellis. 2019. From features to fingerprints: a general diagnostic framework for anthropogenic geomorphology. *Progress in Physical Geography* 43 (1): 95–128.

Thakkar, H. and S. Bhattacharyya. 2006. Reservoir siltation in India: latest studies. *Dams, Rivers and People* 4 (7–8): 1–15.

Thomas, W.F. 1956. *Man's Role in Changing the Face of the Earth*. Chicago: University of Chicago Press.

Willcocks, W. 1930. *Ancient System of Irrigation in Bengal*. Calcutta: University of Calcutta.

Williams, P.B. 2001. River engineering versus river restoration. *ASCE Wetlands Engineering & River Restoration Conference*, Reno, Nevada.

Wohl, E. 2014. *Rivers in the Landscape: Science and Management*. Chichester: Wiley-Blackwell.

Zevenbergen, C., A. Cashman, N. Evelpidou, E. Pasche, S. Garvin and R. Ashley. 2010. *Urban Flood Management*. Boca Raton: CRC Press.

2 Anthropogeomorphology of the Bhagirathi River

Aznarul Islam, Sanat Kumar Guchhait and Md. Mofizul Hoque

CONTENTS

2.1 INTRODUCTION

Evolution of Ganga bears the marks of myth and controversy. Origin and evolution of the Ganga are difficult to predict in an accurate spatio-temporal sequence as scientific record of this river is very meagre as well as limited in broad temporal framework. Various scholars have attempted to unfold the dynamics in the light of riddle of Ganga evolution in different time periods on the basis of previous literatures. Review of previous literature helps in understating as well as confusing the nature of evolution of Ganga. Thus, there are some contesting ideas regarding origin of Bhagirathi River in West Bengal. As far as the previous records are concerned, it is common illustration that the Bhagirathi is the oldest and the earliest to attain its superiority (Oldham, 1870; Sherwill, 1858; Reaks, 1919; Mukherjee, 1938; Majumder, 1942; Bagchi, 1944; Umitsu, 1987). In support of Bhagirathi's antiquity, there have been three different strands of evidence: (1) textual, (2) travelogue and (3) deltaic posture (Rudra, 2010).

The textual evidences include description and interpretation regarding Bhagirathi in the epics and religious texts such as the Ramayana, the Mahabharata, Matsya Purana and Baiyu Purana. These texts interpret Ganga as Bhagirathi. Early settlements were developed in the western portion of the delta than the eastern. Tamralipta, Karnasubarna, Chandraketugarh, etc. attest the fact in the interpretation of Puranas. Bhagirathi was the ancient channel of Ganga. These settlements were developed in the western part of the delta as ancient river centric civilisation. But their descriptions are influenced by personal emotion and experiences rather than scientific judgement. That's why these textual evidences appear dubious as the elements of the history (Rudra, 2010). Besides, travel accounts of the foreign emissaries like that of Megasthenes (300 BC), Ptolemy (150 BC), Fa-Hien (300 AD) and Hsuan Tsang (639 AD) clearly noted about the river Bhagirathi and port Tamralipta. During 300 BC to 700 AD, there was ongoing foreign trade from Tamralipta to Southeast Asia and Rome. This description bears the mark of antiquity of Bhagirathi. But trade on waterways depends not only on navigability of the river but on other factors like hinterland, socio-economic advancement of the area too. Besides this, there is sky-high difference between historic time period and geological time scale. Furthermore, as an evidence of antiquity of Bhagirathi than Padma, the shape and posture of the Bengal delta is often used. The western part of the delta is extended more in southward direction than its eastern counterpart. As delta is prograding towards south, the higher southward extension bears a mark of early delta building process. This supports the view of the flow of ancient Ganga through Bhagirathi.

The above evidences point towards Bhagirathi's antiquity. Most of the previous scientists such as Oldham (1870), Sherwill (1858), Reaks (1919), Mukherjee (1938), Majumder (1942), Bagchi (1944) and Umitsu (1987) have supported the paradigm of Bhagirathi's antiquity. They believed that Bhagirathi is the natural and initial path of Ganga. However, there are two distinct groups of scientists that contradict each other regarding the time when Padma became supreme at the expense of Bhagirathi. The first group believes in antiquity of Padma's prominence, whereas the other in relative recency of Padma's supremacy. According to Bagchi (1944), one of the stalwarts of the first group, the river Bhagirathi was unimportant even during 300 BC though originating earlier and superior in sanctity. Birth of Padma took place before 300 BC, and during the historic period from 300 BC to present, it had uninterruptedly been the principal course of Ganga. Bhagirathi was the oldest course and earliest to attain prominence. But it began silting up as early as 2,000 years ago. Padma was the latter to attain its present prominence, evidently at the expense of Bhagirathi. This indicates that delta being formed in its embryonic stage near the apex was gradually extended eastwards and southwards. Mukherjee (1938), one of the supporters of the second group, opines that Padma took its supremacy not earlier than the 15th century. They believe in the tilting of Bengal basin towards the east in the 16th century for this topsy-turvy. Therefore, before the 15th century, Bhagirathi was the main channel of Ganga and Padma was the minor one. These two philosophical stand points are contrasting with each other with their own arguments, but neither of these provides a complete understating of evolutions; rather, both are fragmentary in argument. The antithesis of Bhagirathi's antiquity was propounded by Willcocks (1930). He thought that Bhagirathi along with the other channels of South Bengal are artificial canals for the purpose of irrigation excavated by then Hindu kings to mitigate severe drought. Rudra (2010) comments that 'Bhagirathi as older path of Ganga' is more emotional statement than geographical logic. Rudra (2010) has supported Padma's supremacy since its inception and provided three alternative reasons. First is the logic of Para delta formation and canal construction which dictates that in past Ajay, Mayurakshi, Damodar, Rupnarayan and Haldi built their small delta along the western bank of the present Bhagirathi. These small deltas are called para deltas. These rivers had fallen in the estuary. In the dry period, estuary became saline for dearth of fresh water from these rain-fed rivers. Consequently, canal in the name of Bhagirath was constructed to make the environment saline free. The second argument is related to channel width and belt width asymmetry which states that in Murshidabad district, the width of the river Bhagirathi and Padma is 300 m and 2 km, respectively. There is no such evidence of floodplain of Bhagirathi which goes against the huge discharge of Ganga passing through Bhagirathi. And the third relates to flood impulse that propounds that presently 18–27 lac cusec discharge goes down the Ganga–Padma channel (Rudra, 2010). Surplus water in the peak period overflows the bank. Rudra (2010) thinks that Bhagirathi probably took its birth by a past high-magnitude flood event.

Looking at the close perusal of the previous arguments and counterarguments regarding the evolution of Bhagirathi, it can be mentioned that till now all these arguments are lacking soundness of scientific logic as well as geology because river research in this area is at inception phase. Since 1980, a group of scholars viz. Allison et al. (2003), Goodbred Jr. and Kuehl (1998, 2000a, 2000b), have extensively conducted geological exploration with the help of modern dating techniques to reveal some mysteries for a scientific understanding of the evolutionary history of Ganga–Brahmaputra (GB) delta on quantitative basis. They have scientifically proved the stages of delta evolution, lobe migration (eastward stepping of Ganga delta), impact of climate change on delta evolution, stratigraphic sequence of delta. In spite of those works, historical reconstruction of the channel evolution is still incomplete. The above discussion points up that origin and evolution of the river Bhagirathi are still shrouded with mysteries. But undeniably, it is true that eastward tilting by neo-tectonic movements is not at all new. Actually it has been occurring at the boundary of Indo-Burmese plate since Eocene. But there was a latest episode of neo-tectonic movement during the 16th–18th centuries (Sen, 2010) which has accelerated the decay of Bhagirathi relatively and supremacy of Padma by eastward tilting of Bengal basin. Therefore, to save the then Calcutta port (Syama Prasad Mookerjee Port) and provide saline-free water to the urban and industrial set-up in and around Kolkata, the Farakka Barrage was constructed in 1975. Thus, the geomorphic evolution of the Bhagirathi canal (doubtful) since its inception or the evolution of the Bhagirathi River (doubtful) since the construction of the Farakka Barrage is largely a response to human interventions. This work would systematically address the major anthropogenic signatures over the river and its hydro-geomorphic consequences.

2.2 THE RIVER AND ITS BASIN

2.2.1 LOCATION AND AREA

The Bhagirathi River Basin (BRB) is a part of Ganga–Brahmaputra–Meghna delta. It is a distributary of the Ganga flowing through the state of West Bengal traversing a length of about 230 km up to Nabadwip where it has a confluence with the river Jalangi. Thus, the BRB extends from 23°23′ 8″ N to 24°49′ 35″ N and 87°54′ 15″ E to 88°22′ 2″ E. Unlike an ideal drainage basin characterised by production, transfer and deposition zones, the BRB is the lower part of the Ganga basin. This poses a problem of delineating the basin boundary explicitly. Irrigation and Waterways Department, in its annual flood report, 2016, has delineated the Bhagirathi-Hooghly drainage area as 5,452 km² of which 1,687.45 km² is accounted for by the Bhagirathi River (Figure 2.1). For the present study to trace out the impact of the anthropogenic interventions, a buffer has been delineated taking a width of 1.381 km on either side of the river. The buffer width is justified on the consideration of the channel oscillation in the past, and present and future i.e. factor of safety (Ness, 2004).

2.2.2 DRAINAGE SYSTEM

The river Bhagirathi is a one-way channel i.e. non-tidal, while Hooghly is a two-way channel influenced by the tidal upsurge. Fluvial dynamics of the river Bhagirathi is to some extent controlled by its tributaries. Major right bank tributaries of Bhagirathi are Bansloi, Pagla, Mayurakshi and Ajay, while Jalangi River is the only left bank tributary (Figure 2.1). For detailed hydro-geomorphic study, the river Bhagirathi was conveniently subdivided by Kolkata Port Trust (presently Syama Prasad Mookerjee Port Trust) into three reaches: head reach (upstream), middle reach (mid-stream) and tail reach (downstream). Head reach extends from Feeder Canal outfall at Ahiron to Nasipur, middle reach from Nasipur to Sitahati and tail from Sitahati to Nabadwip (Guchhait et al., 2016). Similarly, head reach is constituted by three stretches or sectors (1–3), middle reach by four stretches (4–7) and tail reach by another four stretches (8–11) totalling 11 stretches or the sector of the river. Currently there are four major gauge stations on river Bhagirathi viz. Feeder Canal (head reach), Berhampore (middle reach), Katwa and Nabadwip (tail reach) (Figure 2.1). From the perspective of the anthropogeomorphology, the BRB is unique i.e. the basin is profusely altered by the human interventions mainly by Farakka Barrage (Figure 2.2). Moreover, the dams and barrages in the tributary basins like Tilpara Barrage and Massanjore dam over the Mayurakshi River and Sikatia Barrage over the Ajay River have also had a substantial control on the sediment flux and the hydrologic regime of the BRB.

2.2.3 Geology

Stratigraphic provinces in and around the Bengal basin is of diverse origin bearing the marks of antiquity. However, major formation of Bengal basin is the alluvium of either recent origin or Pleistocene period. Virtually, this covers the whole GB delta. Barind tract and Madhupur terrace contain formation of Pleistocene upland. Assam and Myanmar-I are mainly composed of tertiary upland formation. Formation of Eocene, Miocene and Oligocene can be found in the areas of Garo hill and Shillong massif. The oldest Achaean formation is found along the margin of Bengal basin mainly in the Chotonagpur plateau fringe (Figure 2.3).

Eocene hinge zone across the Bengal basin is a remarkable geologic characteristic. Dauki fault which separates Meghalaya plateau and Sylhet basin is an important thrust fault in the Seismotectonics map of the Bengal basin and surroundings (Figure 2.3).

2.2.4 Physiography

South Bengal is physiographically diversified. Alluvial plain is built by the deposition of the lower Ganga system; coastal plain is formed along the Digha–Kanthi coast and Sundarban, while plateau fringe is lying along the western portion of the Bengal basin. It is worthy to mention that there is a clear eastward gradation of relief. Therefore, gradual fall in relief has been observed from the west to east (Figure 2.4a and b).

The fall in elevation is greatly controlled by three main faults viz. Chhotanagpur Foot-hill Fault (CFF), Medinipur Farakka Fault (MFF) and Damodar Fault (DF). Along these fault lines, there has been step-like fall in the elevation (Singh et al., 1998; Islam, 2016).

2.2.5 Climate and Soil

Indian climate is basically influenced by south-western and north-eastern monsoon. According to Koeppen's Scheme of climatic classification, BRB lies in monsoon type with dry winters (Cwg) with 100–200 cm annual rainfall and tropical savanna type (Aw) with 75 cm annual rainfall. The western part of the Bhagirathi River is drier than that of the eastern counterpart. For example, Karimpur located on the eastern part of the Bhagirathi basin records 1,434 mm of rainfall while Purbasthali located on the western part records 1,313 mm. Similarly, rainfall has a southward increasing trend in the basin. For example, Farakka at the north of the basin records 1,326 mm while Krishnanagar at the south records 1,353 mm. Furthermore, annual average temperature is around 25°C–26°C with average minimum temperature in January (11°C) and average maximum temperature in May (34°C).

The Bhagirathi River drains on almost flat alluvial plain surface and flows from north to south. The eastern part of this river is low-lying with heavy fertile alluvial soil. The western part is slightly

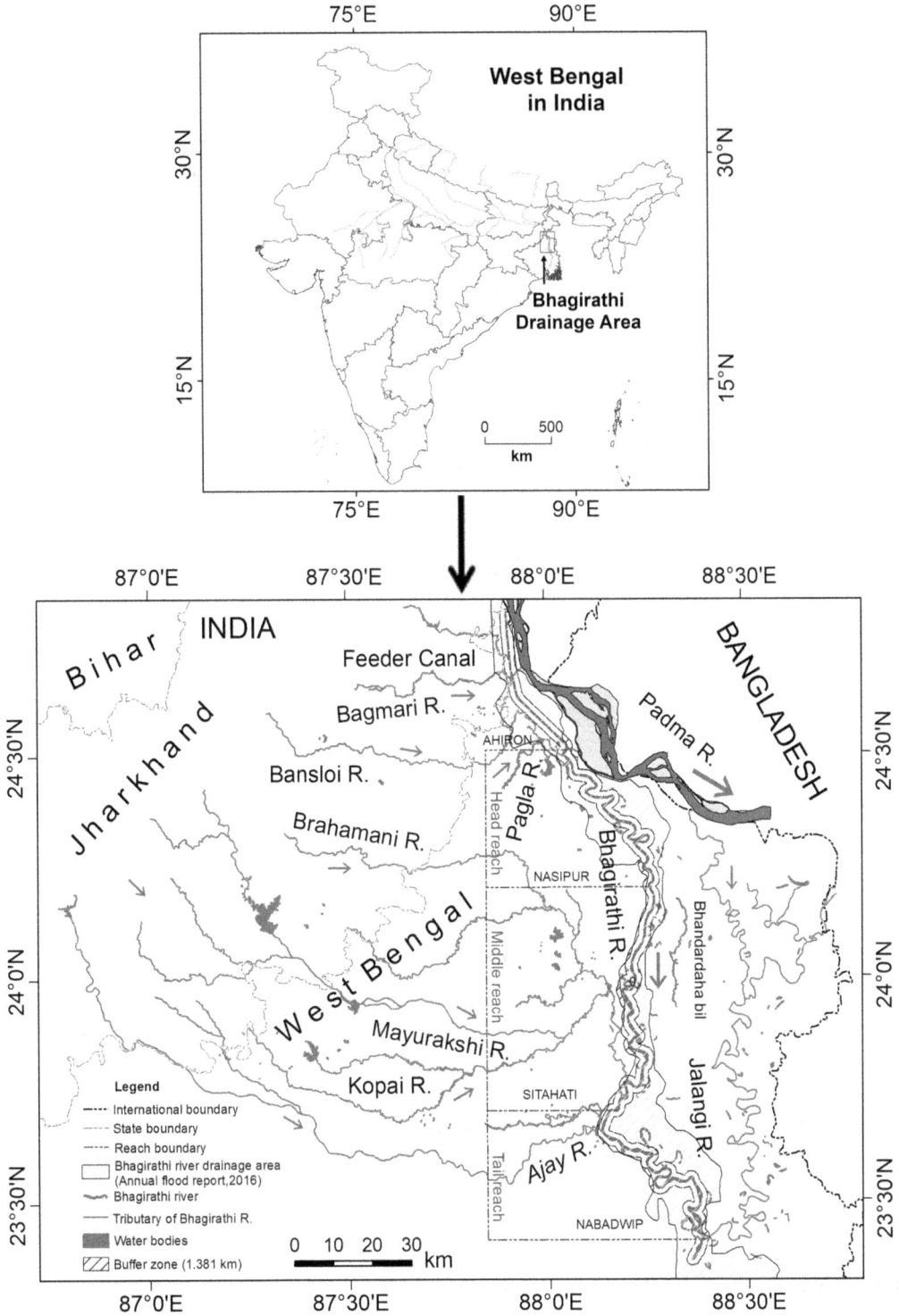

FIGURE 2.1 Location of the Bhagirathi River and its tributaries. (*Source*: Drainage lines are extracted from Google Earth dated 25 October 2019 and Landsat OLI-path 139, row 43 dated 6 February 2019 and path 138, row 44 dated 15 February 2019.)

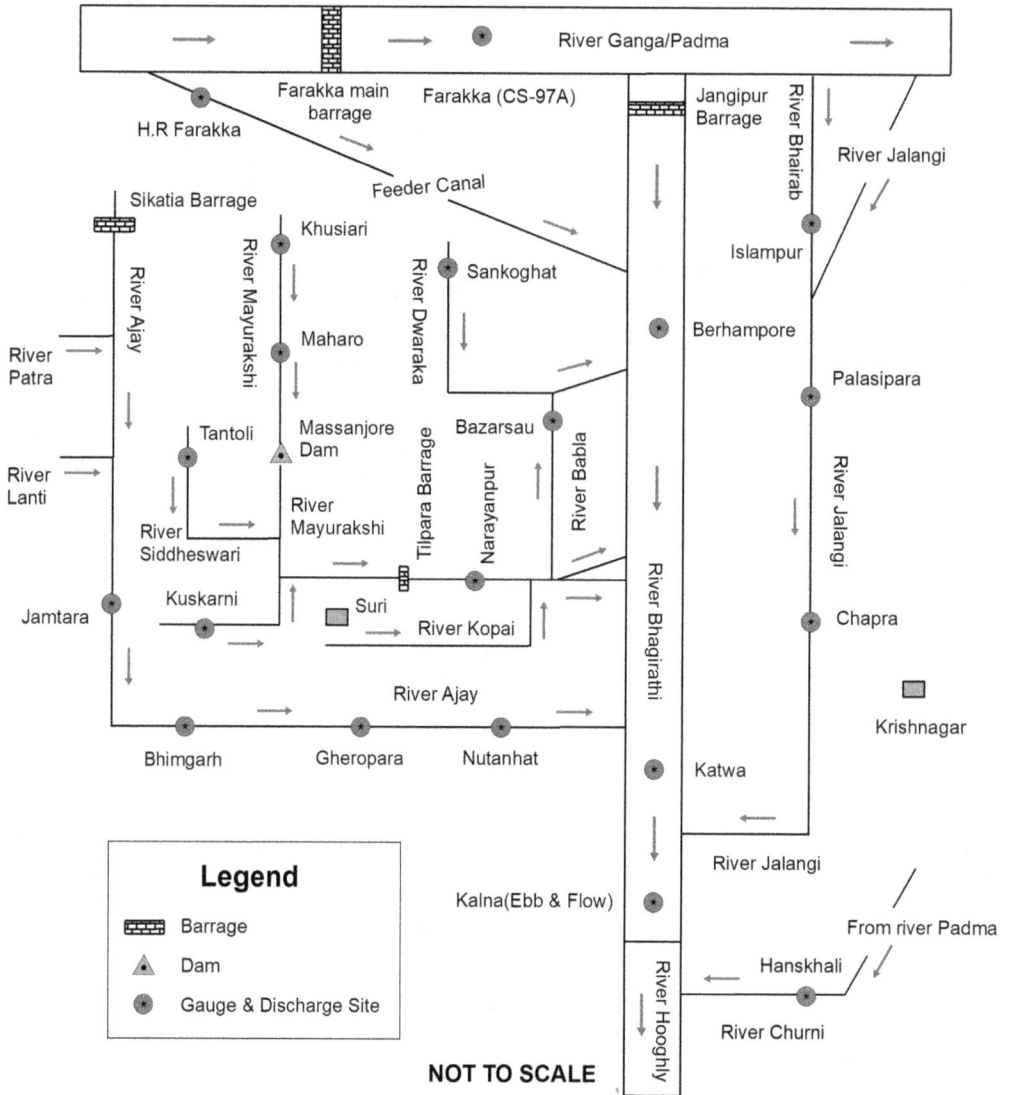

FIGURE 2.2   Schematic diagram of the Bhagirathi system. (*Source*: CWC (2015–2016).)

undulating surface and high relief. BRB is a young deltaic plain with similar sedimentary structure. The floodplain surface along with highly fertile soil is common on the river side with loose and unconsolidated sediment structure. The soils on the eastern part of this river is composed of sands and clays brought by rivers and its tributary while the soils on the western part is composed of older alluvium and lateritic hard clay (Guchhait et al., 2016).

2.3   ANTHROPOGENIC SIGNATURES ON FORMS AND PROCESSES

Each system is governed by some principles of nature that keeps it in a state of equilibrium. Natural control theory believes that any perturbations appeared naturally would be corrected with the passage of time because nature knows the limit. However, alternation of any system behaviour due to human activities takes the system away from its initial condition to a farther point akin to positive

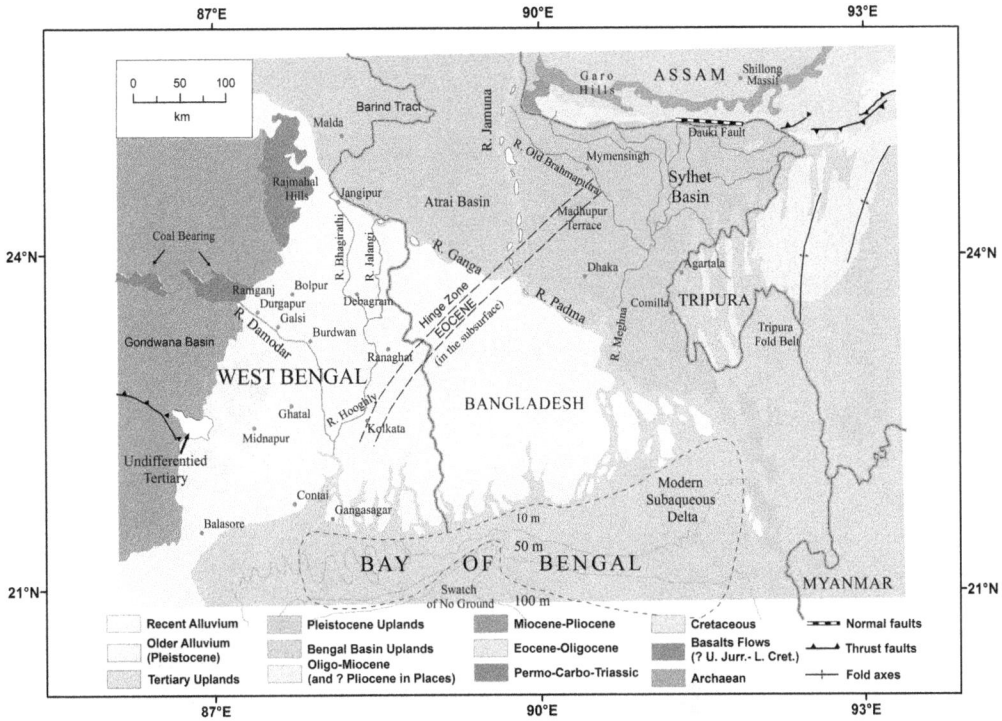

FIGURE 2.3 Stratigraphic and geologic provinces in and around Bengal basin. (Based on Chakraborty (1970).)

feedback mechanism. This ultimately pushes it to such an extent that it cannot recover naturally. In the BRB, Farakka Barrage Project (FBP) is such a massive intervention that fluvio-hydrologic behaviour of the Ganga River has radically changed. Besides, how the other small-scale interventions like brick field and road crossing, guide bank with bamboo fencing and frequent ship movements have altered the fluvial system of the Bhagirathi River has been addressed in the following sections.

2.3.1 FARAKKA BARRAGE PROJECT: A MEGA-SCALE INTERVENTION

2.3.1.1 Location and Characteristics

The FBP is a massive engineering project constructed by the Government of India over the Ganga River in Murshidabad and Malda districts of West Bengal with the aims of (1) maintaining navigability of the river Bhagirathi especially in the lean months through flushing out sediments deposited in Hooghly River by inducing 40,000 cusecs of water through Feeder Canal (2) saving the Kolkata Port and (3) providing saline-free water to the inhabitants of Kolkata (Parua, 2009; Rudra, 2011; Islam and Guchhait, 2017a). To execute the plan of the FBP, an authority was set-up in 1961 that materialised the plan in 1975. This project comprised the three major components: the Farakka Barrage (2,245 m long), Jangipur Barrage (213 m long) and 38 km long Feeder Canal (Figure 2.5). Furthermore, the FBP complex is constituted by the 112 Gates including 108 main Gates and 4 Fish Lock Gates along with 11 Head Regulator Gates for diversion of about 40,000 cusecs of discharge into the Feeder Canal (Parua, 2009). The Jangipur Barrage was constructed with an intention to steady the flow from the Feeder Canal outfall at Ahiron to Bhagirathi by retraining it from returning the flow into the Padma through Bhagirathi.

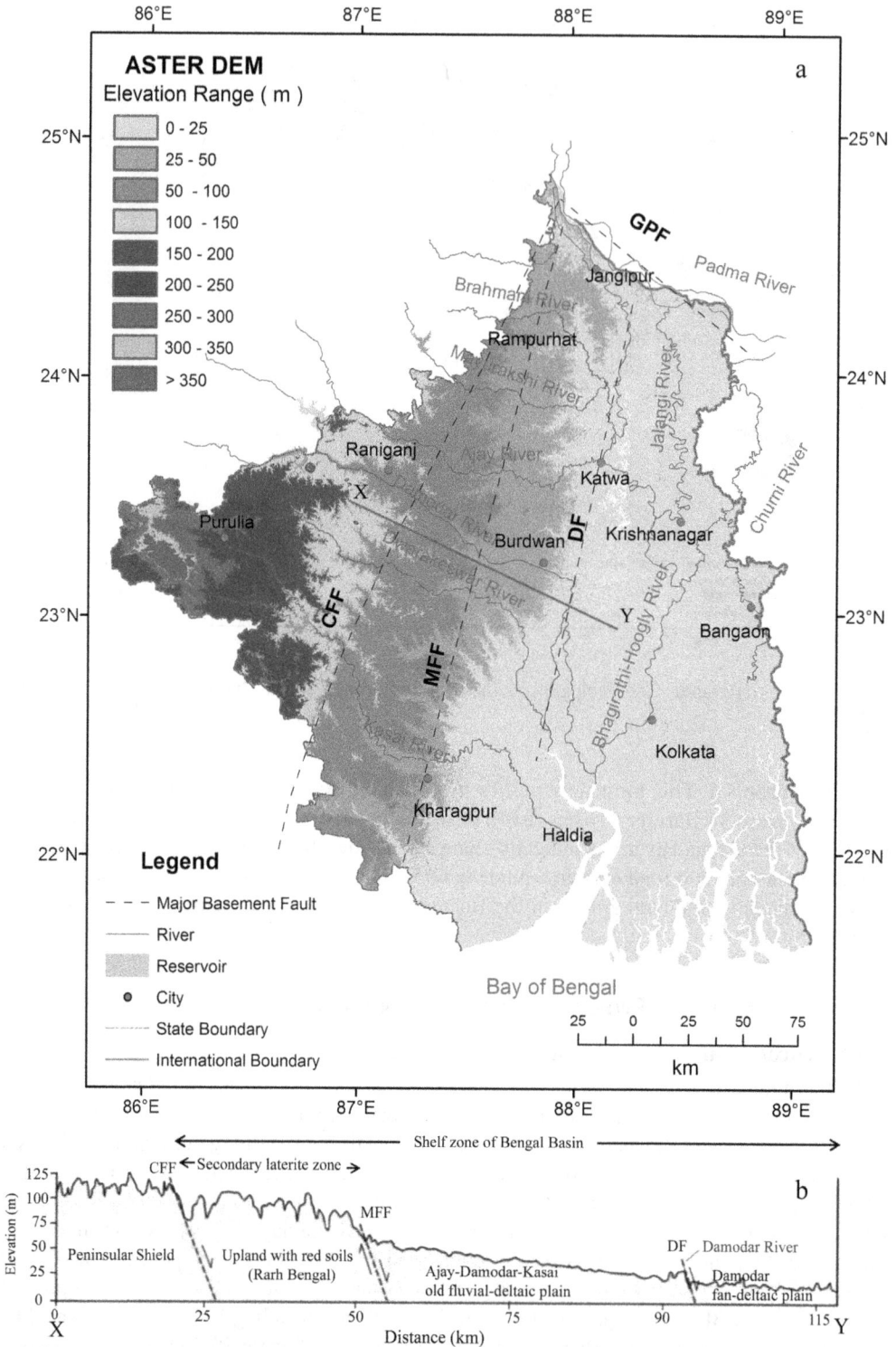

FIGURE 2.4 (a) Elevation map (prepared from ASTER DEM (2011)) and identified major basement faults (modified from Singh et al. (1998)) in the western part of the Bengal basin; (b) west–east cross profile (X-Y) with emplacement of faults. (N.B.: CFF, Chhotanagpur Foot-hill Fault; MFF, Medinipur–Farraka Fault; DF = Damodar Fault; GPF = Ganga–Padma Fault.)

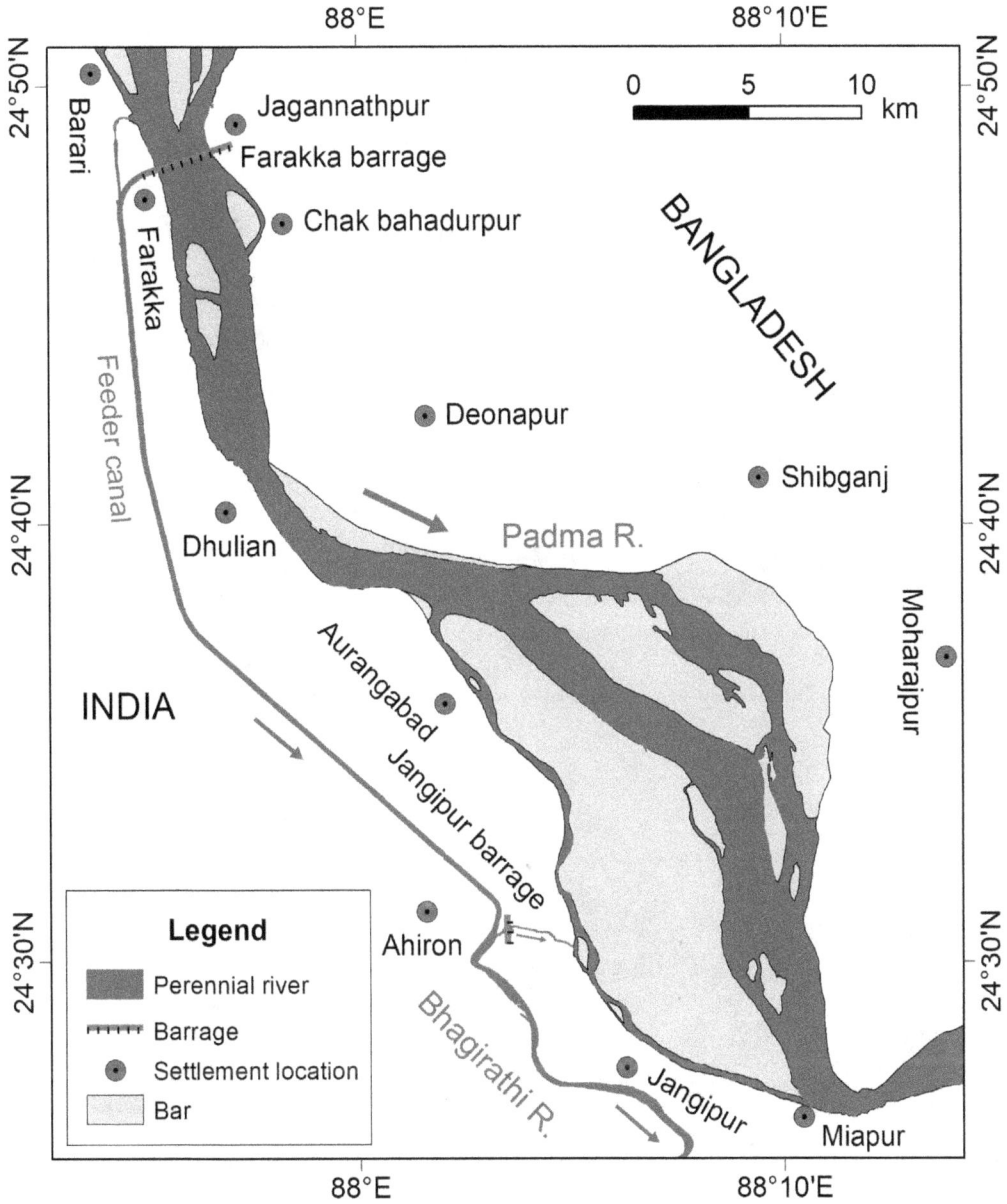

FIGURE 2.5 Index plan of Farakka Barrage Project (prepared from the Google Earth image dated 9 November 2019 and OLI-path 139, row 43 dated 6 February 2019).

Besides, the road and rail bridge (Figure 2.6a and b) across the river Ganga at Farakka connected north Bengal with the south Bengal. Moreover, national waterway number 1 i.e. Haldia–Allahabad Inland Waterway is maintained through this Feeder Canal. The Feeder Canal also supplies water to 2,100 MW Farakka Super Thermal Power Project (FSTPP) of NTPC Ltd. at Farakka (Parua, 2009).

The FBP has profound impacts on the channel morphology and hydraulic behaviour of the Ganga River system both on the upstream and the downstream of the barrage (Parua, 2009; Mirza, 2004). The problem of inundation and river bank erosion are common fluvial hazards associated with this project in Malda, Murshidabad and Nadia districts of West Bengal (Rudra, 2011; Islam, 2016). It can aptly be mentioned that the fluvial regime of the Lower Ganga system is controlled by the

FIGURE 2.6 Photographs (a) and (b) showing rail-cum-road bridge across the river Ganga. (*Source*: Todaytimesnews.com.)

anthropogenic modification of the hydraulic behaviour by the construction of barrage since 1975. In the following sections, how FBP has controlled the channel dynamics of Bhagirathi in West Bengal especially meander geometry of the river has been analysed.

2.3.1.2 Farakka Barrage Project and Hydro-Geomorphology

I. Changes in hydrologic regime

In pre-Farakka period, variability in discharge was mainly dependent on the variability of monsoon climatic regime. Variability of river discharge was seasonal in nature and less frequent during that period. However, after the commissioning of the FBP in 1975 river regime has thus been artificially controlled on the consideration of arithmetic hydrology. After the construction of Farakka Barrage on Ganga, Indo-Bangladesh water dispute emerged immediately. This resulted in water sharing treaties in 1977 and 1996 which

TABLE 2.1

Comparative Statement Showing Water Availability (in Cusecs) under 1977[a] and 1996[b] Agreement

	Bangladesh			India			
Period	1977	1996	Incr. / Decr.	1977	1996	Incr. / Decr.	Remarks
1–10 January	58,500	67,516	9,016	40,000	40,000	0	[a]The flow was calculated on the
11–20 January	51,250	57,673	6,423	38,500	40,000	1,500	basis of 75% dependable flow
21–31 January	47,500	50,154	2,654	35,000	40,000	5,000	during 1948–1973 i.e. prior to
1–10 February	46,250	46,323	73	33,000	40,000	7,000	the operation of Farakka Barrage.
11–20 February	42,500	42,859	−359	31,500	40,000	8,500	[b]The flow was calculated on the
21–28/29 February	39,250	39,106	−144	30,750	40,000	9,250	basis of average flow (50%
1–10 March	38,500	35,000	−3,500	26,750	39,419	12,669	dependable) during the period
11–20 March	38,000	35,000	−3,000	25,500	33,931	8,431	from 1948 to 1988 i.e. the period
21–31 March	36,000	29,688	−6,312	25,000	35,000	10,000	of 25 years. (1948–1973) during
1–10 April	35,000	35,000	0	24,000	28,180	4,180	pre-Farakka operation and
11–20 April	34,750	27,633	−7,117	20,750	35,000	14,250	15 years (1974–1988) during
21–30 April	34,500	35,000	500	20,500	25,992	5,492	post-Farakka operation under
1–10 May	35,000	32,351	−2,649	21,500	35,000	13,500	mutual agreement. The period of
11–20 May	35,250	35,000	−250	24,000	38,590	14,590	1988–1996 i.e. 8 years of
21–31 May	38,750	41,854	3,104	26,750	40,000	13,250	unilateral withdrawal was not
							considered.

initiated variable discharge for both the countries on a 10-day scale in the lean months from January to May (Table 2.1).

It is clear from Table 2.1 that river regime, as expected, is fluctuating frequently especially in the months from March to May as per rule of water sharing treaty in 1996. It is true that discharge is more variable in the lean months especially from March to May and relatively less variable for rest of the year as evident from the discharge hydrograph of the river at Feeder Canal (Farakka) and Berhampore gauge stations (Figure 2.7a and b).

From the discharge hydrograph, peak and valley configuration appears in the treaty period (January–May) for the gauge stations at Feeder Canal (Farakka), Berhampore and Katwa, but in the normal period (June–December), hydrograph appears normal at the Feeder Canal and Berhampore but oscillates at Katwa (Islam and Guchhait, 2017b). Moving average on ten-day scale shows the pattern of peak and valley configuration as well as smooth curve.

II. Morphological changes

Channel morphology responds to hydraulic changes in a quick succession. Due to the hydrologic alternation by the FBP, channel morphology has undergone a radical change during the post-Farakka stage. The morphological changes have been detected through four major variables: channel width, cross-sectional area, hydraulic mean radius (HMR) and cubic capacity. The analysis has been carried for the entire 11 stretches of the river from the cross section (CS) 1 to CS 372a as measured by the Kolkata Port Trust (Figure 2.8).

a. Channel width

Channel width is perhaps the most sensitive indicator to hydrologic alteration (Islam, 2011). The width has been measured at datum level. It has been observed that out of the 11 stretches of the river, ten stretches portray a positive increase in the channel width (Figure 2.9).

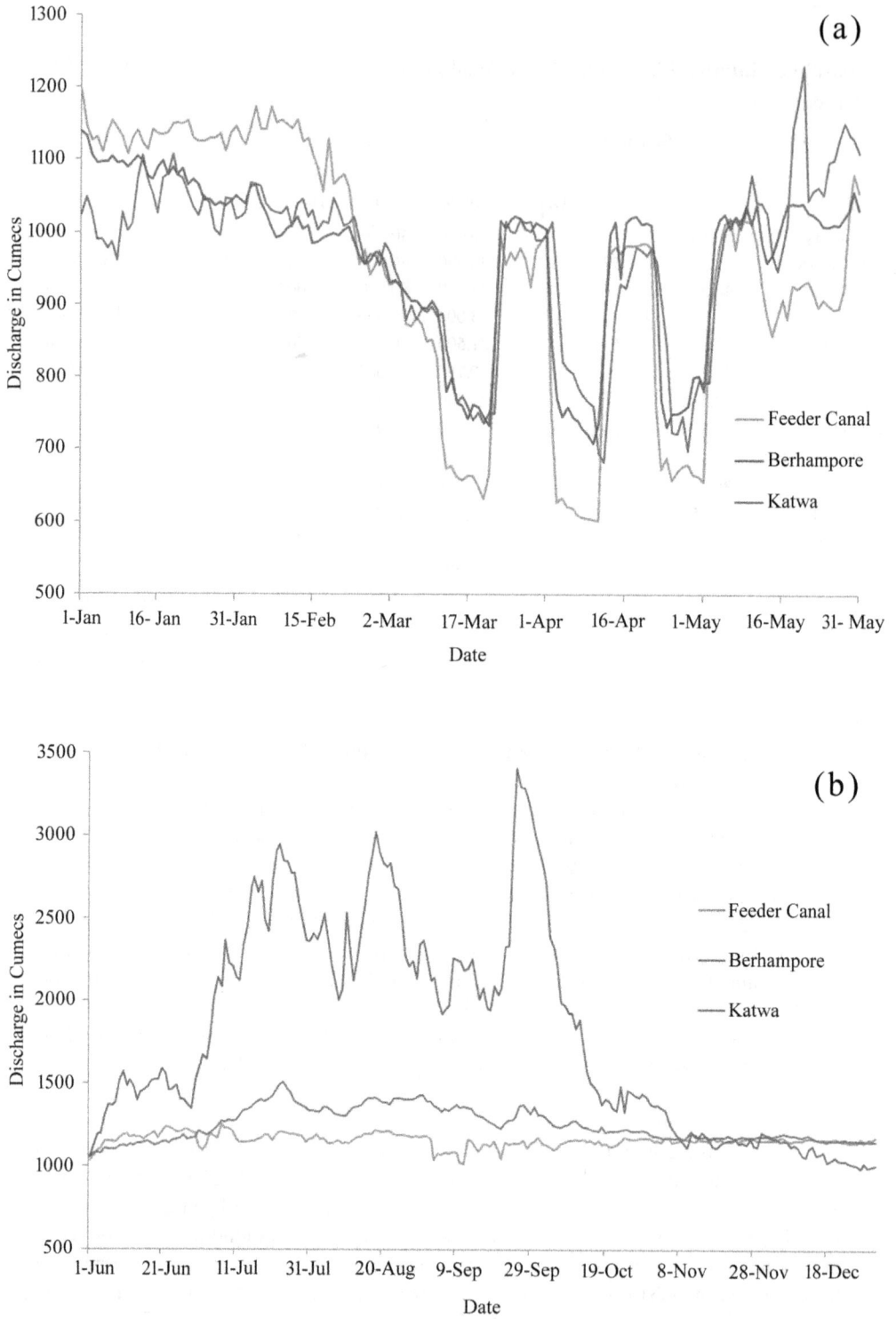

FIGURE 2.7 Discharge hydrograph of the river Bhagirathi (a) treaty period and (b) normal period.

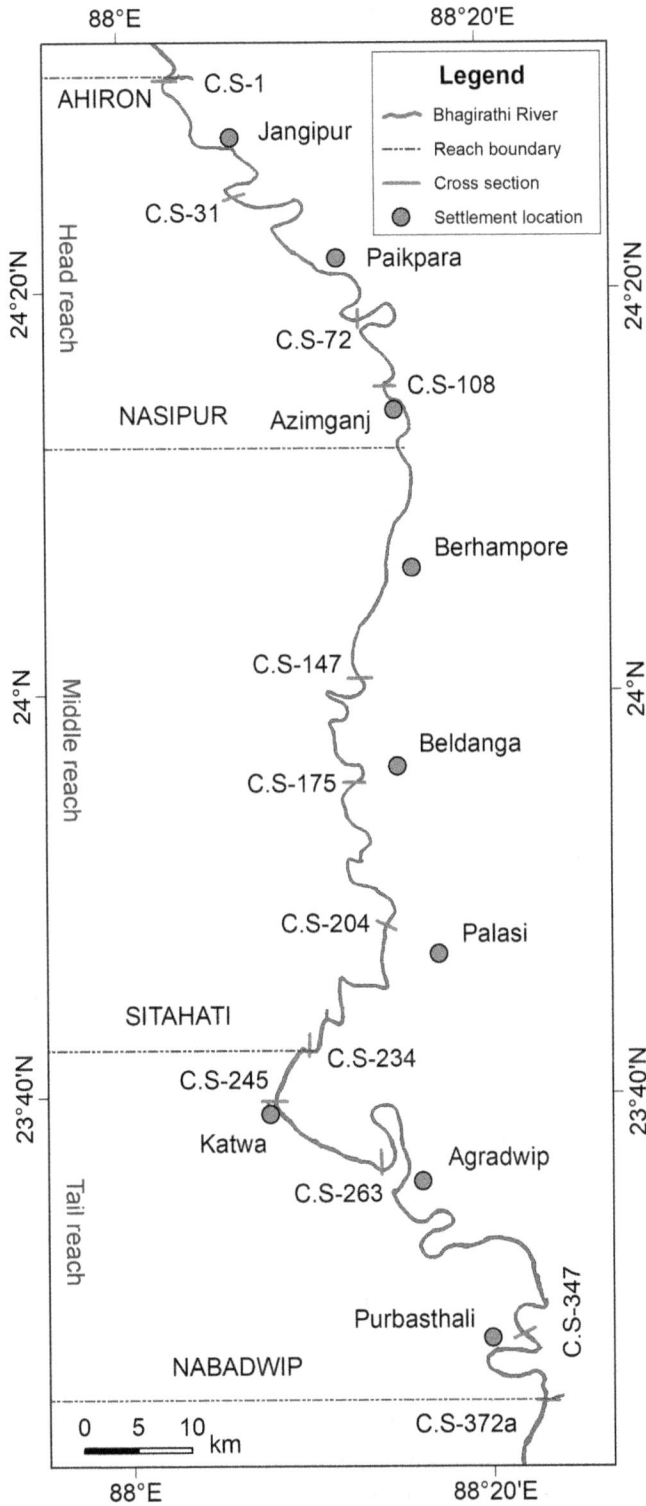

FIGURE 2.8   Location of the cross sections on the river Bhagirathi. (Based on Ghosh (2000).)

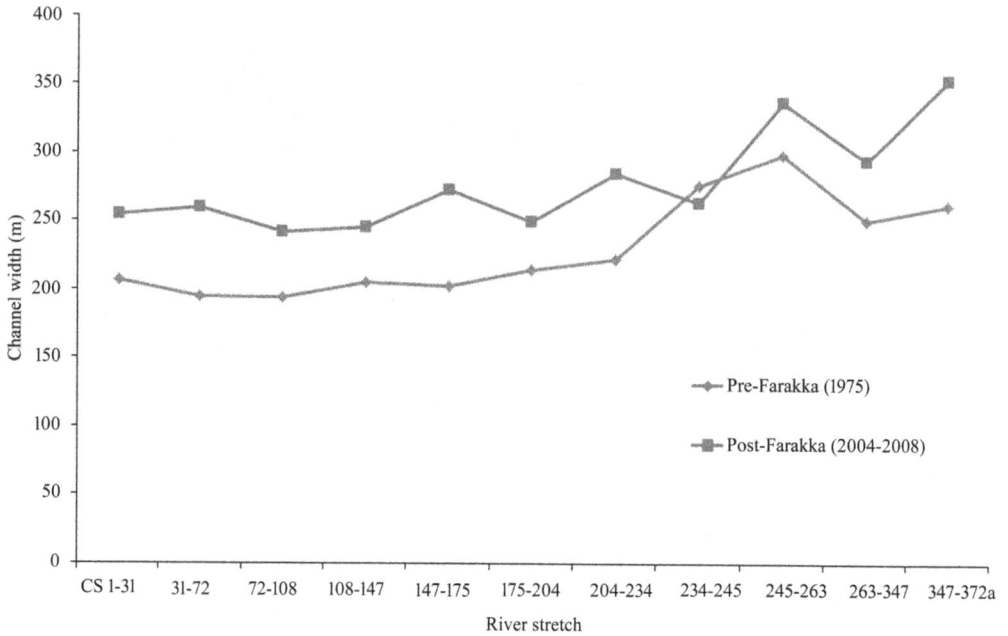

FIGURE 2.9   Changing channel width in the post-Farakka period (computed from the annual reports on the river Bhagirathi (2004–2005 to 2008–2009 prepared by the KoPT).

The highest increase of about 36% has been found for the stretch extending from CS 347 to CS 372, while the stretch extending from CS 234 to 245 has recorded a negative change in the growth index of width (about −5%) during the measurement period. However, in general, rising trend is observed for the river in the post-Farakka period. This increase is nothing but due to the increasing hydraulic impulse from the FBP in the lean months (January–May).

b. Cross-sectional area

Cross-sectional area on an average has stepped out from a pre-Farakka minimum of about 723 m^2 to the height of 941 m^2 with an increase of 23%. This increasing trend has reach-wise variations. In the upper reach where the effect of the hydraulic impulse is the maximum due to proximity effect, areal increase is also maximum. For example, an increase of about 46% during the period of the observation has been recorded by the stretch 1 (Figure 2.10).

The middle reach has also substantial increase of the area of about 10%–25%. However, the tail reach where the effect of the major tributaries like Ajay and Mayurakshi is notable has registered lesser increase in the area (~3%–24%). The stretch 8 extending from CS 234 to CS 245 shows this minimum increase (~3%) in the CS area. This may be due to the distance decay effect of the hydrologic input from the Farakka. Here the channel is controlled by the FBP only on the lean months, while the Ajay and Mayurakshi have had a great control on the lower stretch during the monsoon period.

c. Hydraulic mean radius

The HMR for each segment obtained by dividing the respective weighted average of the cross-sectional area by the weighted average of width has increased in the recent decade compared to the 1975 episode. The pattern of increase follows the trend of the CS area. During the period of the observation, the stretch's average increase is about 8%. However, there is notable spatial trend i.e. declining from the Feeder Canal outfall at Ahiron to the Jalangi Bhagirathi confluence at Nabadwip (Figure 2.11).

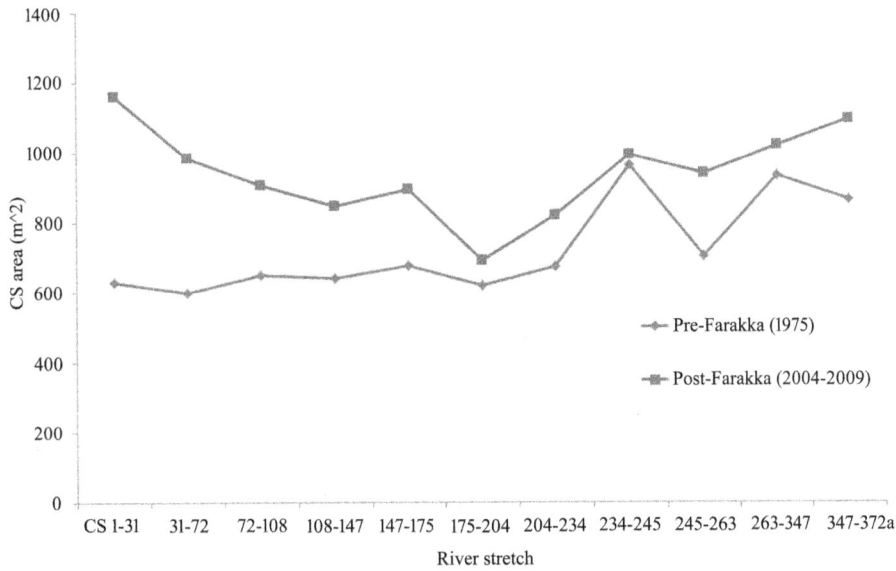

FIGURE 2.10 Changing cross-sectional area in the post-Farakka period (computed from the annual reports on the river Bhagirathi (2004–2005 to 2008–2009 prepared by the KoPT).

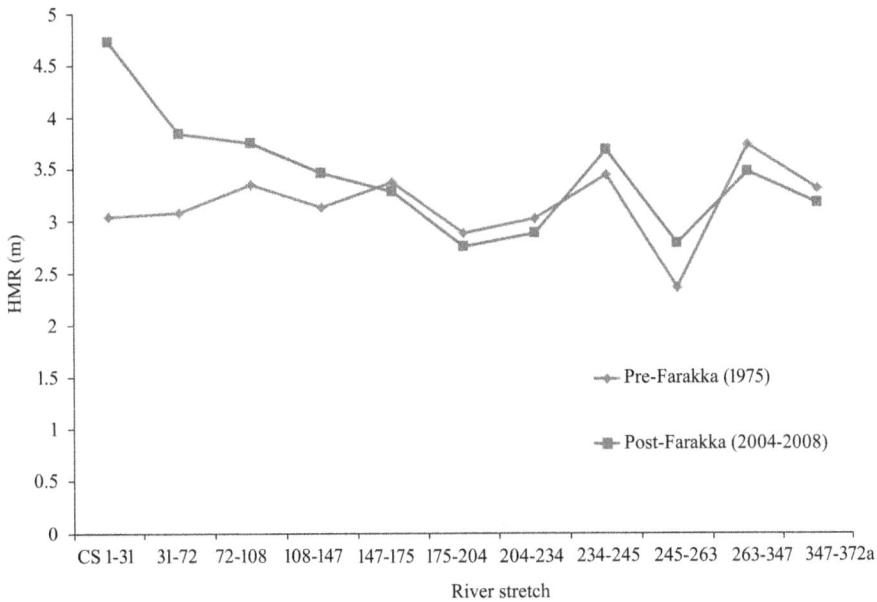

FIGURE 2.11 Changing HMR in the post-Farakka period (computed from the annual reports on the river Bhagirathi (2004–2005 to 2008–2009 prepared by the KoPT).

The head reach has registered the increase of about 10%–37%. However, the middle reach portrays a great diversity ranging from −4.7% to 9.6%. Similarly, the tail reach also shows most fluctuating results ranging from −7.4% to 15%. It implies that channel efficiency has decreased in the middle and lower reaches compared to the upper stretch.

d. Cubic capacity

Cubic capacity for each segment measured at datum level by multiplying the average of the cross-sectional area by the length of the segment implies the capacity to carry loads in the channel. Before the construction of the Farakka Barrage, the absence of flow in the lean period especially in the upper stretch of the river was remarkable (Figure 2.12).

Therefore, the hydrologic input via FBP augmented a high flow in the lean months that actually increased the cubic capacity of the head reach (Islam and Guchhait, 2018). The results show that head reach registered maximum increases (41%–51%), while the gradual fall has been observed downwards. In the middle reach, it ranges from 21% to 49%. Similarly, in the tail reach, some areas have even portrayed a decrease in the cubic capacity. For example, stretch 11 (CS 347–372a) shows a decrease of about −57%. This is due to the bed scouring in the upper stretch and deposition in the lower reaches of the river.

III. Planform changes

Planform changes are analysed for the period from 1973 to 2019 using Landsat satellite images. During this period of observation, channel oscillation, stream meandering, sinuosity and braiding have been detected mainly as a consequence of hydrologic changes after 1975.

a. Channel oscillation and stream avulsion

The course of river Bhagirathi-Hooghly has undergone changes during last few centuries (Chatterjee, 1989). Descriptions by various scholars viz. Rennell (1788), Colebrooke (1801), Sherwill (1858), Hirst (1915), Mukherjee (1938), Bandyopadhyay (1996), Ray (1999), Bhattacharya (2000) and Rudra (2010) have outlined frequent shifting nature of Ganga. Hence, channel shifting of Ganga or the Bhagirathi-Hooghly system has drawn attention from time immemorial. In the recent decades, especially after the construction of the FBP in 1975, channel changes of the Bhagirathi have achieved a momentum. The entire course of the Bhagirathi became unstable immediately after the barrage was constructed (Figures 2.13–2.15).

However, upper stretch has become relatively adjustable to the morphogenic variables, but the lower middle and the lower stretch still oscillate at a considerable rate.

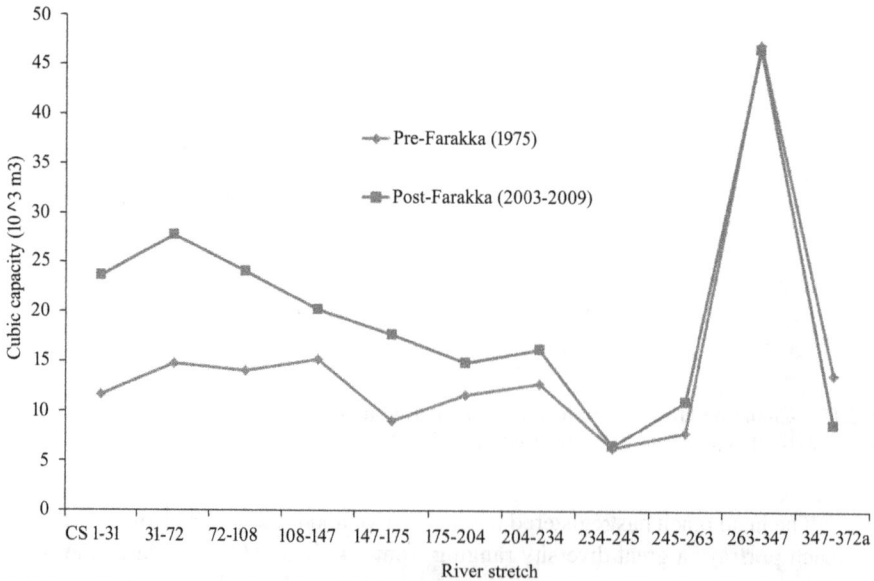

FIGURE 2.12   Changing cubic capacity in the post-Farakka period (computed from the annual reports on the river Bhagirathi (2004–2005 to 2008–2009 prepared by the KoPT).

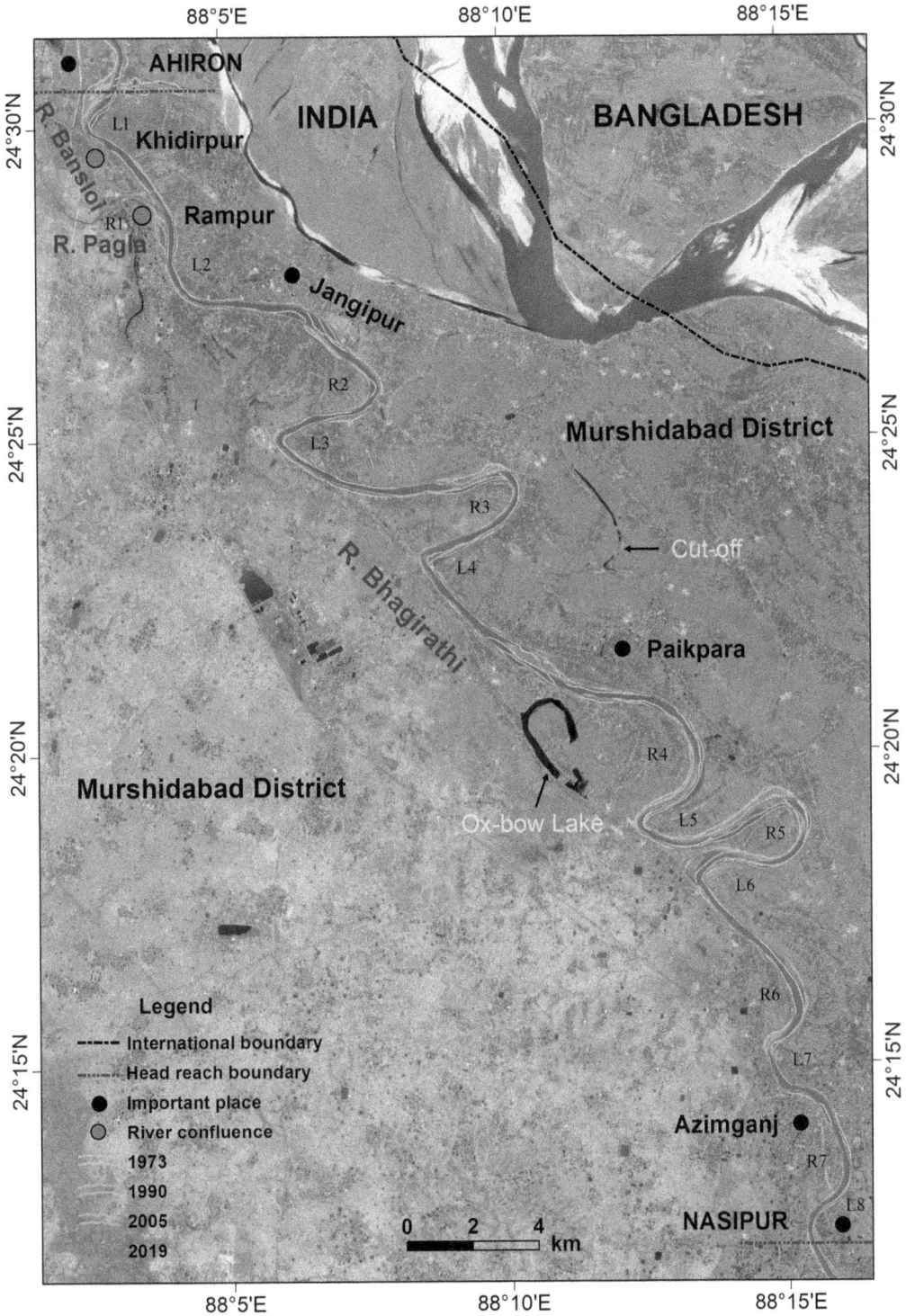

FIGURE 2.13 Channel oscillation in the head reach during 1973–2019 (*Note*: L1–L8 and R1–R7 indicate the number of the left bank and right bank meander loops, respectively, for the years 1973, 1990, 2005 and 2019.) (*Source*: Multispectral Scanner or MSS-path 149, row 43, dated 17 January 1973; Thematic Mapper or TM-path 139, row 43 dated 6 February 1990, 15 February 2005; Operational Land Imager or OLI-path 139, row 43 dated 6 February 2019 and Sentinel 2A T45RXH and T45QXG dated 5 January 2019.)

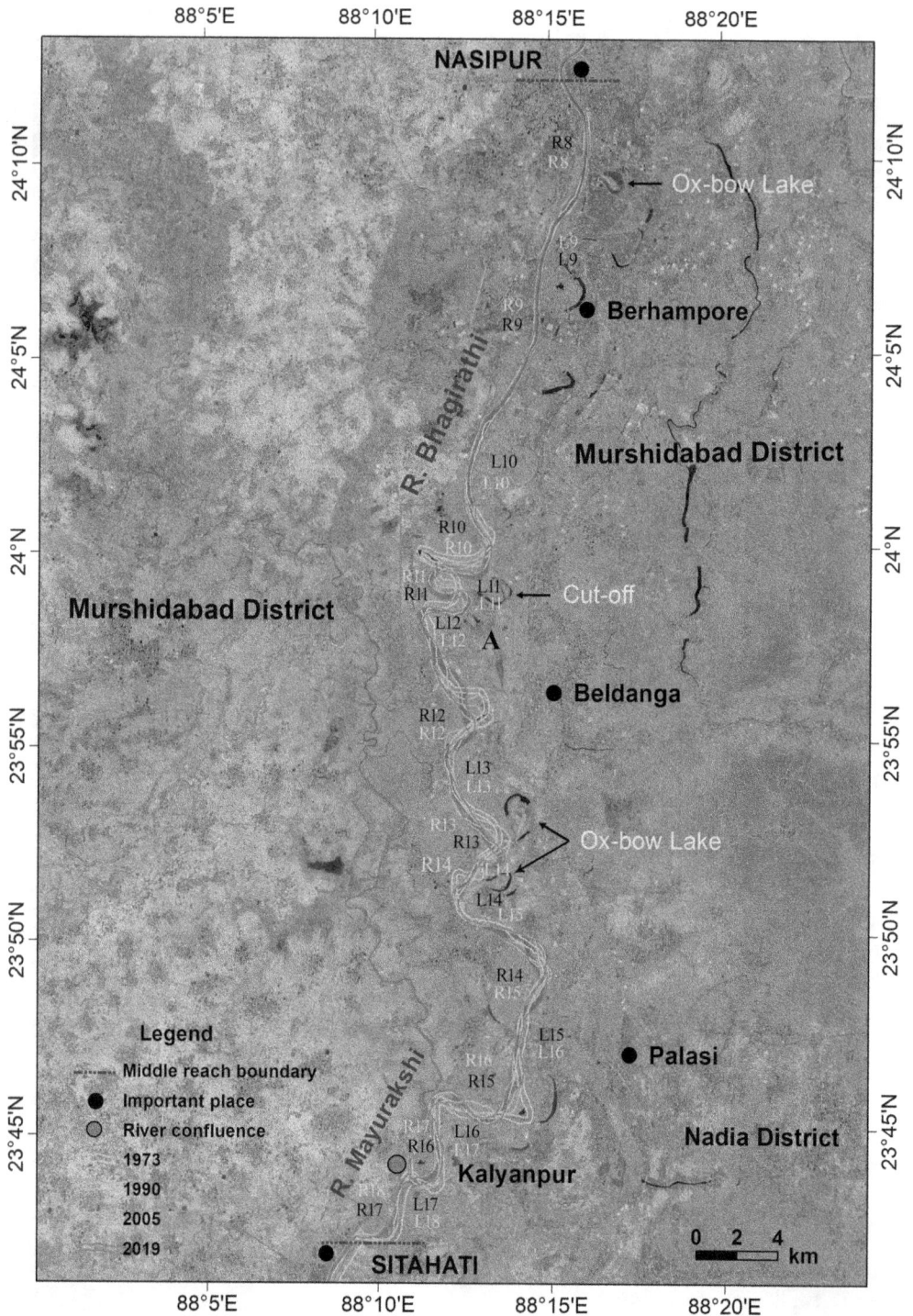

FIGURE 2.14 Channel oscillation in the middle reach during 1973–2019. (*Note*: L9–L17 and R8–18 indicate the number of the left bank and right bank meander loops, respectively; Loops shown in grey represent 1973 and loops in black for 1990, 2005 and 2019.) (*Source*: MSS-path 149, row 44, dated 22 February 1973 and path 149, row 43, dated 17 January 1973; TM-path 139, row 43 dated 6 February 1990, 15 February 2005 and path 138, row 44 dated 30 January 1990, 24 February 2005; OLI-path 138, row 44 dated 15 February 2019 and path 139, row 43 dated 6 February 2019 and Sentinel 2A T45QXG dated 5 January 2019.)

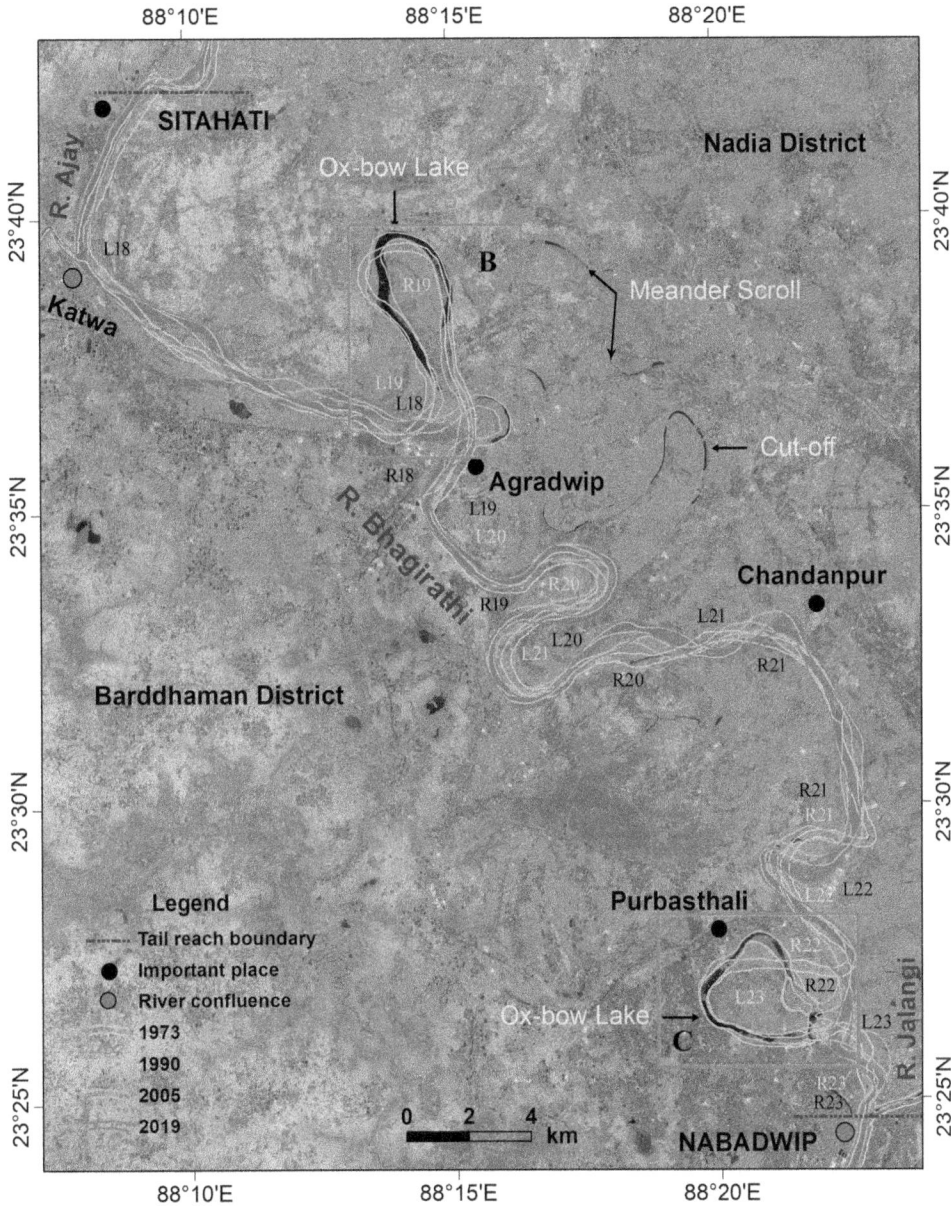

FIGURE 2.15 Channel oscillation in the tail reaches during 1973–2019. (*Note*: L18–L23 and R18–R23 indicate the number of the left bank and right bank meander loops, respectively. Loops shown in grey represent 1973 and loops in black for 1990, 2005 and 2019. However, upper black L18 represents 2005 and 2019 and lower black L18 represents 1990 only; upper black R21 for 2005 and 2019 and lower black R21 for 1990 only.) (*Source*: MSS-path 149, row 44, dated 22 February 1973; TM-path 138, row 44 dated 30 January 1990, 24 February 2005; OLI-path 138, row 44 dated 15 February 2019 an Sentinel 2A T45QXG and T45QXF dated 05 January 2019.)

This is due to the combined effects of the FBP in the lean months and the contribution of the Ajay–Mayurakshi River during the monsoon season. During 1973 and 2019, channel instability and meander progression were quite notable. To trace out the evolution of the ox-bow lakes, three locations near Beldanga, Agradwip and Nabadwip were treated through windows A, B and C, respectively (Figure 2.16).

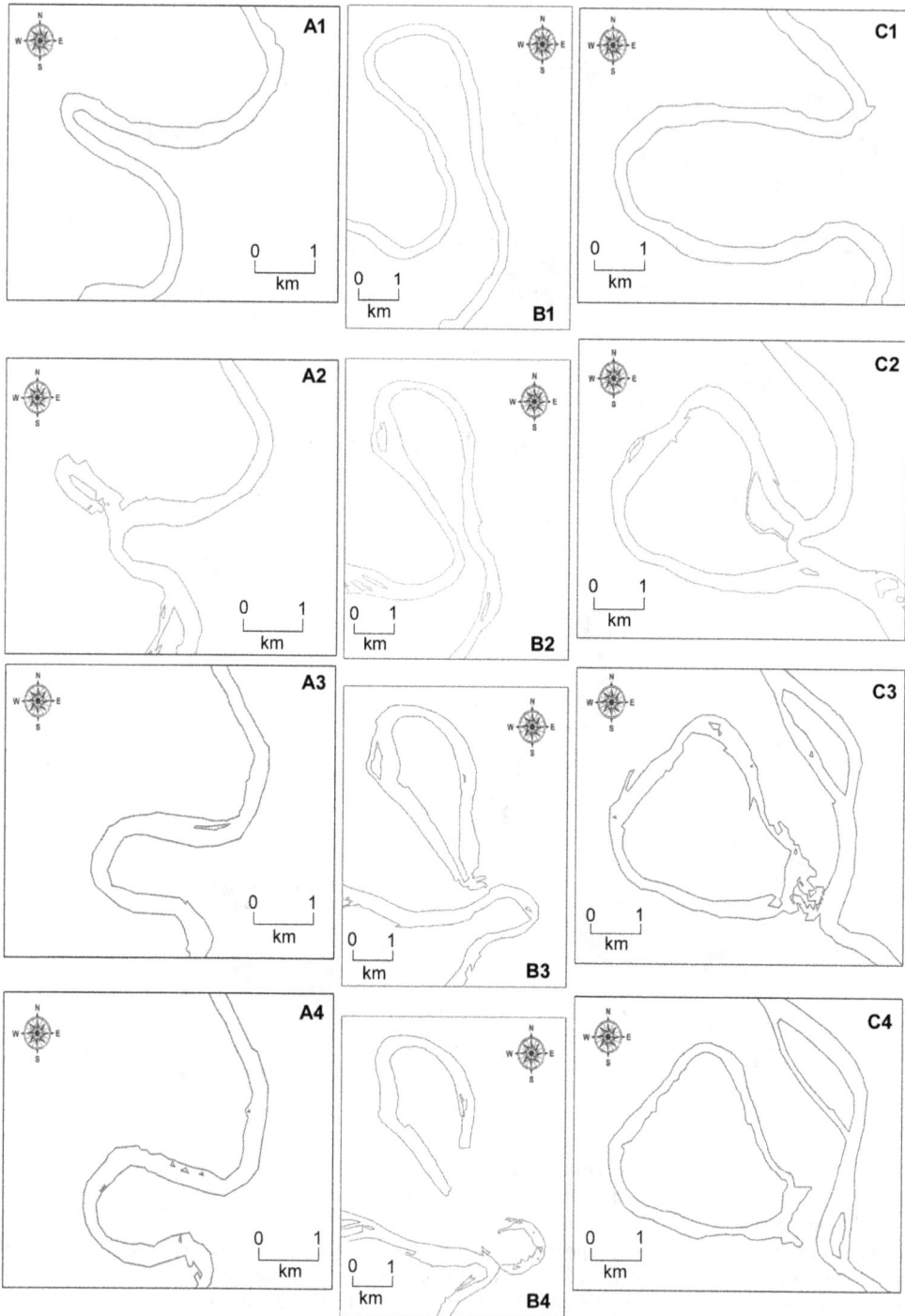

FIGURE 2.16 Window analysis of channel cut-off for the years 1973 (A1, B1, C1), 1990 (A2, B2, C3), 2005 (A3, B3, C3) and 2019 (A4, B4, C4). (*Note*: windows 'A1–A4' are prepared from MSS-path 149, row 43, dated 17 January 1973; TM-path 139, row 43 dated 6 February 1990, and 15 February 2005; OLI-path 139, row 43 dated 6 February 2019 and windows 'B1–B4, and C1–C4' are based on MSS-path 149, row 44, dated 22 February 1973; TM-path 138, row 44 dated 30 January 1990, 24 February 2005; OLI-path 138, row 44 dated 15 February 2019.)

In the year 1989, there was a chute cut-off in the lower reach of the channel. This cut-off is popularly known as Mayapur–Sankhapur cut-off. This cut-off occurred all of sudden. Since the time of Rennell (1788), neck distance was maintained for it variably from 5.62 km in 1788 to 2.20 km in 1975 just before the operation of FBP. But after the operation of FBP, it got sudden flood impulse, and in the year 1989, due to huge monsoon influx, it surfaced channel cut-off. Hence, actually this cut-off was a chute cut-off triggered by the flood impulse of 1989 (Basu et al., 2005). This cut-off has reduced the channel length by 9 km making the channel more straightened (Islam and Guchhait, 2017b).

In the year 1994, there was another cut-off nearby Bishnupur *Char*-Chakundi area which decreased channel length by 11.85 km. In 2008 in the same area, there was another cut-off, called *Char*-Chakundi cut-off II, reducing channel length by 3.6 km. Recurrent cut-off in the same area proves the channel dynamics in that reach by hydro-fluvial control and subsurface geology as explained in the following section. It should be mentioned that there is space-time specificity in the channel cut-off. From spatial perspective, three out of four cut-offs are confined to the lower reach of Bhagirathi (Katwa–Nabadwip). From the temporal perspective, all the four cut-offs occurred in the post-Farakka period by the supply of discharge of FBP as well as flood impulse. Superimposition of channels (Figure 2.17) shows that in the decade from 2004 to 2014, channel was oscillating but not at the pace immediately after the operation of FBP (Islam and Guchhait, 2017b).

Thus, FBP has initiated new hydrologic regime for which instability of Bhagirathi has emerged with a new dimension that attempts to adjust flood impulse in the form of cut-off. Besides the role of FBP on channel instability and meandering in general, role of Ajay-Mayurakshi River system and lithofacies needs explanation to account for the pattern of intense oscillation and meandering of the tail reach of the Bhagirathi River (Islam, 2016). The Ajay-Mayurakshi system supplies huge monsoon discharge (Ajay 1800 and Mayurakshi 1500 m³ s⁻¹) and coarse grained sediment (Ajay 1.39×10^6 t and Mayurakshi 2.52×10^6 t) annually which increase turbulence of the flow and bank

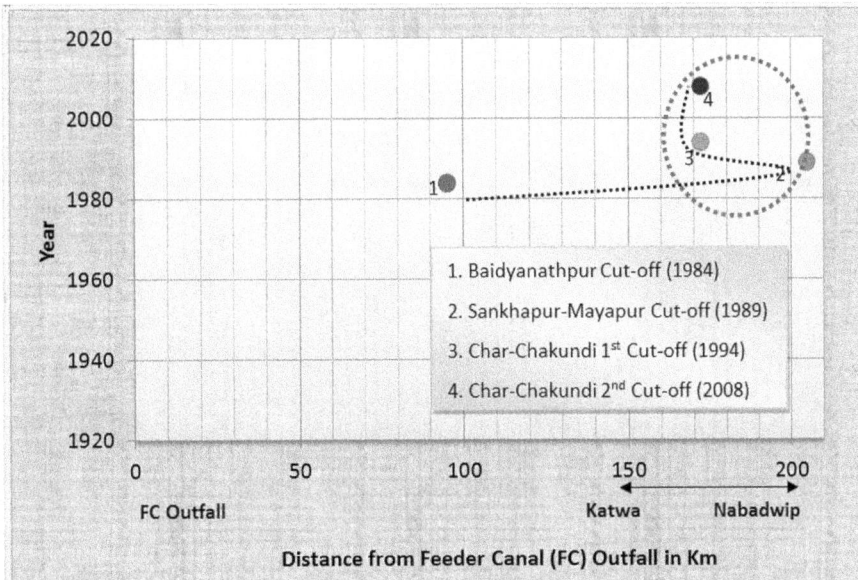

FIGURE 2.17   Space-time specificity in channel cut-off. (*Source*: Islam (2016).)

erosion (Guchhait et al., 2016). Moreover, due to the presence of the erosion resisting clay plug on the right bank restrict channel movement to the right bank of the floodplain while and erosion permitting silt and sand facies favour the free wandering of the channel on the left side of Bhagirathi floodplain (Guchhait et al., 2016).

b. Channel sinuosity and channel length

Channel sinuosity index (SI) has been computed following the algorithm (ratio of channel length to wavelength) of Leopold and Langbein (1966) for the river Bhagirathi encompassing 46 meander loops (23 on each bank) having SI > 1.05 for the years 1973, 1990, 2005 and 2019. The head reach contains 15 loops including eight on the left bank and seven on the right bank for all the observation years; middle reach contains 10 loops on the left bank and 11 loops on the right bank for 1973 while nine loops on the left bank and 10 loops on the right bank for the others years (1990, 2005, 2019) (Figure 2.13–2.15). Moreover, the tail reach also has five loops on each bank in 1973 while six loops on each bank for the other years (1990, 2005, and 2019). The present-day sinuosity portrays that tail reach compared to the head and middle reaches has higher sinuosity (Islam and Guchhait, 2017b). Besides, frequent changes in the SI in the tail reach of the river from one measurement year to the next imply dynamism in the channel evolution. In our study, the majority of the loops in the tail reach have portrayed this tendency explicitly (Figure 2.18).

For example, at the tail reach, the mean SI was observed as 2.30 in 1973 which came down to 1.85 in 2005 and again decreased to 1.80 in 2019. Moreover, the coefficient of variation (CV) of the SI was quite high (~56%) in the year 1973 (Table 2.2).

However, the SI of the head reach varies from 1.77 to 1.79 which indicates a relative stability with time. However, there is a spatial variation in the loop-wise SI as reflected by the relatively higher value of CV in the head reach in all the measurement years i.e. CV around 50%. Similarly, the middle reach is comparatively stable than the tail reach as indicated by the SI moving around 1.45 in different years. Moreover, this

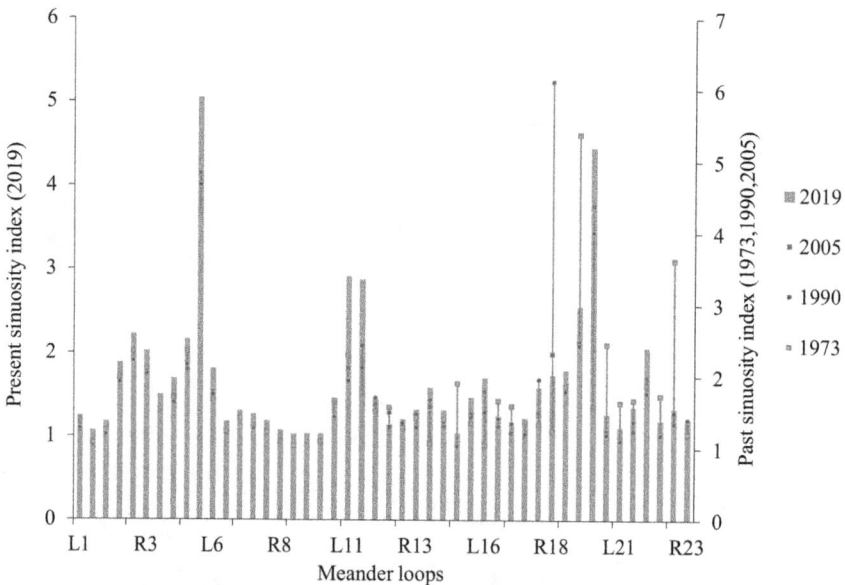

FIGURE 2.18 Loop-wise SI of the Bhagirathi River (prepared from MSS-path 149, row 43, dated 17 January 1973; TM-path 139, row 43 dated 6 February 1990, 15 February 2005; OLI-path 139, row 43 dated 6 February 2019 and MSS-path 149, row 44, dated 22 February 1973; TM-path 138, row 44 dated 30 January 1990, 24 February 2005; OLI-path 138, row 44 dated 15 February 2019).

TABLE 2.2
Dynamism in Channel Sinuosity during 1973–2019

	1973	1990	2005	2019
Head reach	1.79 (45.01)	1.77 (50.14)	1.78 (51.98)	1.79 (55.1)
Middle reach	1.42 (23.08)	1.4 (21.67)	1.41 (26.67)	1.43 (38.57)
Tail reach	2.30 (55.12)	2.15 (68.77)	1.85 (52.81)	1.80 (51.89)

Source: Computed by the author in 2019 (values within parentheses indicate CV of sinuosity index).

reach is relatively homogenous in loop characteristics as depicted by the lower CV of about 25% in the 1973, 1990 and 2005. However, a spatial heterogeneity is observed in the loop characteristics in the recent years (Table 2.2). The changes in the SI have a direct impact on the channel length. The rapidity in the evolution of a meander and its eventual cut-off reduces the channel length i.e. induce channel shortening. In this attempt, reach-wise channel length has been measured for the years 1973, 1990, 2005 and 2019 (Figure 2.19).

The results show that there is a clear spatiality in the channel shortening. The head reach is almost stable i.e. no change in channel length during 1973–2019. However, the middle reach shows a slight change, while the tail reach portrays huge reduction in the channel length in the post-Farakka period. Thus, the overall channel length has gradually decreased (1973: 234 km, 1990: 223 km, 2005: 216 km and 2019: 205 km). This is due to the effect of the increasing hydraulic impulse under new hydraulic regime of the FBP.

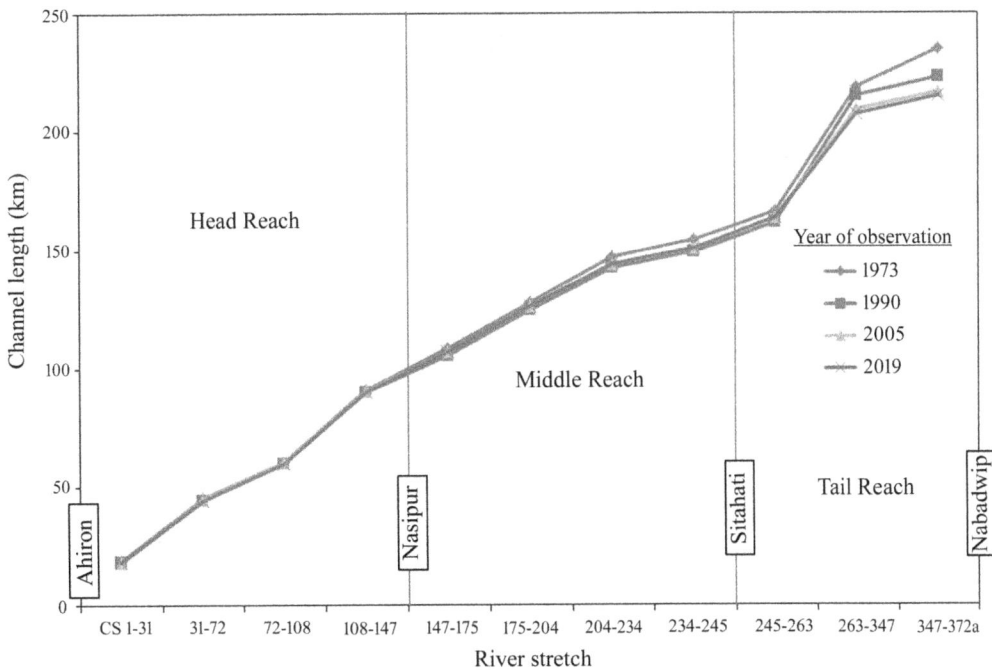

FIGURE 2.19 Channel shortening during 1973–2019. (*Note*: channel length is measured from Landsat images as mentioned in Figure 2.14 using measure tool of ArcGIS 10.2.)

c. Channel braiding and bar growth

Channel braiding implies an energy–load relationship. When a channel is not capable enough to carry its load, bed deposition starts which in the long run develops bars and islands, thereby forming a braiding or anastomosing channel. In the present case, braiding index (BI) has been computed using the methodology of Brice (1964) for the years 1990, 2005 and 2019. In 2019, all the three reaches portray an increasing BI from its previous period (1990). For example, tail reach has attained BI of about 0.4 in 2019 from about 0.3 in 2005 (Figure 2.20).

However, the head and middle reaches that have recorded a fall in the BI in the year 2005 have also portrayed an increase in the BI in 2019 (Figure 2.20). The increasing braiding tendency especially in the tail reach is due to the emergence of transport-limited condition (Islam and Guchhait, 2020). At present, there are nineteen mid-channel bars (head reach: 3, middle reach: 7 and tail reach: 9). Out of those bars, the evolution of the six bars has been traced during 1990–2019. The majority of the bars have shown an increasing area (Figure 2.20), while some have decreasing tendency. Moreover, there is a process transforming the mid-channel bars into the side attached bars operative in the Bhagirathi especially in the middle and tail reaches. The six bars selected from the lower reach and the lower-middle reach of the river are, namely, (1) Sarpakhia bar, (2) Manganpara bar, (3) Rajnagar bar, (4) Uday Chandrapur bar, (5) Rukunpur bar and (6) Kashthasali bar. Sarpakhia bar located at a distance of 71.12 km from the Feeder Canal outfall was germinated during 1994 covering an area of 0.036 km^2 which has increased the area of about 0.091 km^2 in the year 2006. Presently (2019), it has gained an area of about 0.261 km^2 (Figure 2.21a).

Similarly, Manganpara bar located at a distance of 160.33 km from the Feeder Canal outfall started its journey in the year 1994 with an area of about 0.3 km^2 which has at present (2019) gained 0.6 km^2 (Figure 2.21b). Besides, Rajnagar bar near Matiari located below the Ajay–Bhagirathi confluence at Katwa at a distance of 192 km from the Feeder Canal outfall (Figure 2.21c). This mid-channel bar first appeared in the year 1994 with a small area (0.354 km^2) which has gradually increased to 0.771 km^2

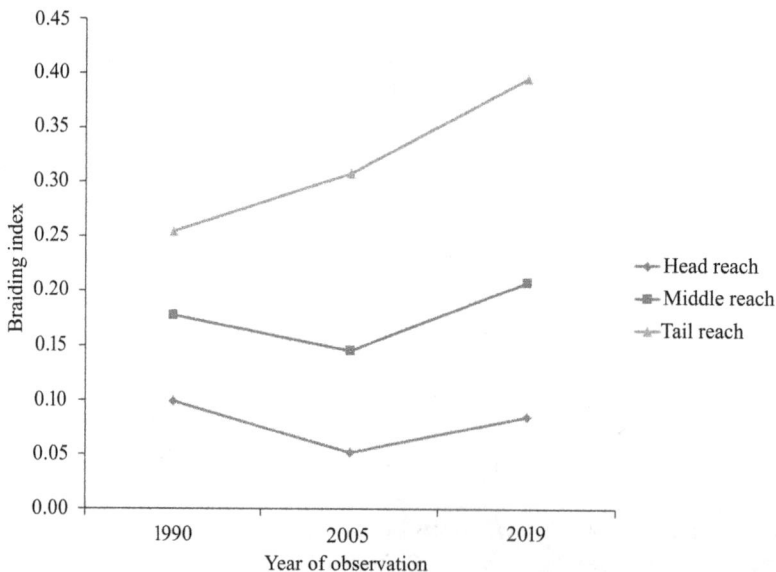

FIGURE 2.20 Nature of bar growth during 1990 and 2019 (computed from Landsat images as mentioned in Figure 2.14 in ArcGIS 10.2 platform).

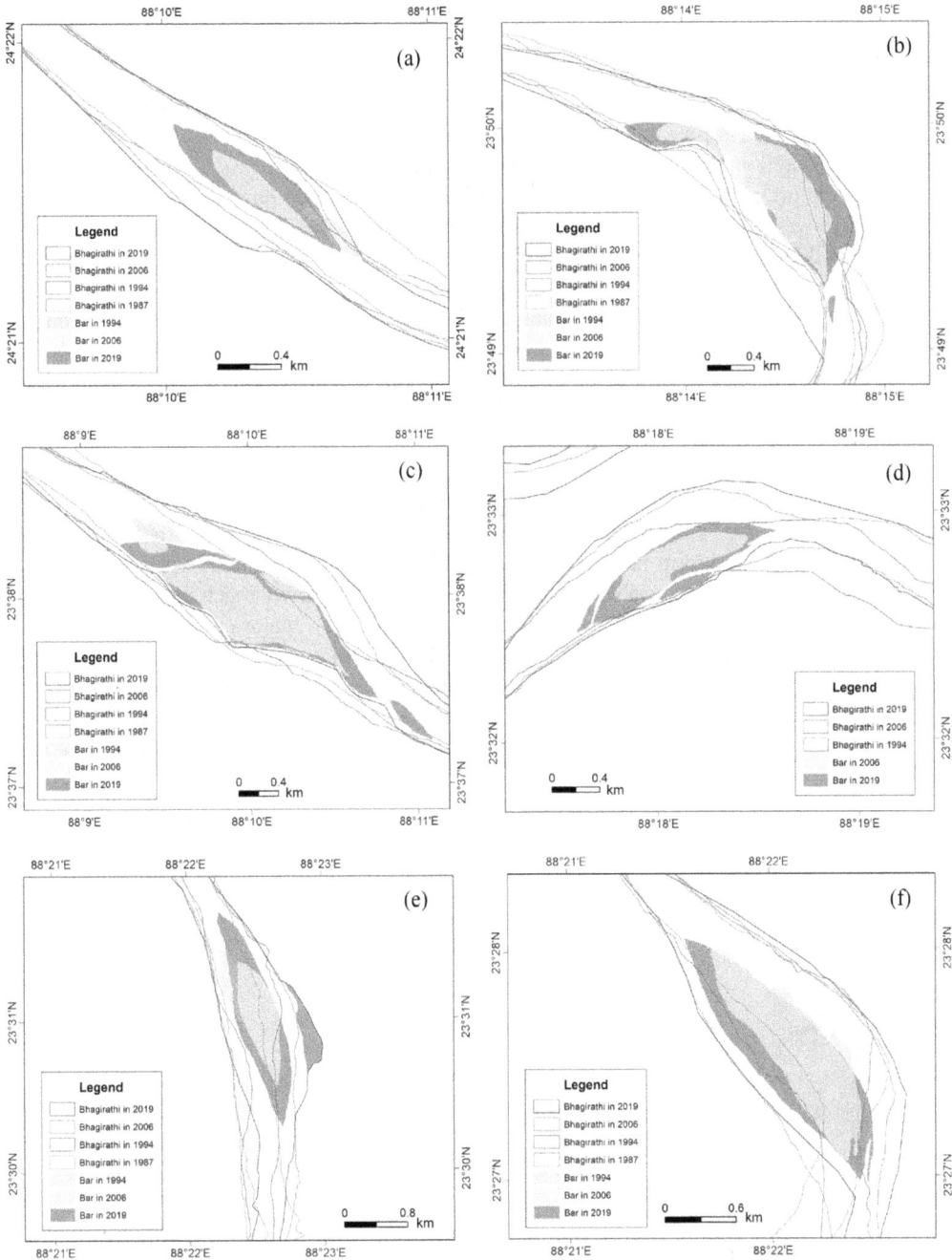

FIGURE 2.21 Windows showing the evolution of bar growth amid the Bhagirathi River, (a) Sarpakhia bar, (b) Manganpara bar, (c) Rajnagar bar, (d) Uday Chandrapur bar, (e) Rukunpur bar and (f) Kashthasali bar. (Based on Google Earth Images dated 19 November 2019, 24 October 2006, 31 December 1994 and 31 December 1987.)

in 2006 and now it has an area of about 1.057 km². Furthermore, Uday Chandrapur bar located at a distance of 223.96 km from the Feeder Canal outfall was initiated in the year 2006 with an area of about 0.261 km² which has at present (2019) gained 0.4432 km² (Figure 2.21d). Moreover, Rukunpur bar located at a distance of 233.83 km from the Feeder Canal outfall started its journey in the year 1994 with an area of about 0.271 km² which has at present (2019) gained 0.958 km² (Figure 2.21e). The Kashthasali bar is located near Nabadwip at a distance of 244.17 km from the Feeder Canal outfall (Figure 2.21f). This bar has also shown the progressive trend of area (1994: 0.385 km²; 2006: 0.765 km²; 2019: 1.092 km²).

IV. Erosion accretion sequence and reduced channel gradient

During pre-Farakka condition, volume of bank erosion was huge in monsoon months, but there was negligible erosion in the lean period due to the lack of flow of water. In the post-Farakka condition, the total volume of bank erosion has been substantially reduced, although erosion has become recurrent throughout the year (Islam and Guchhait, 2015, 2019). Immediately after commissioning of FBP, total volume of erosion was huge in the years 1976–1977 and 1977–1978 (Figure 2.22) as the river rejuvenated itself suddenly in the lean months due to extra input contributed by FBP. The total volumes of bank erosion of the river Bhagirathi for the years 1976–1977 and 1977–1978 were 2,720 × 10⁴ and 3,086 × 10⁴ M³, respectively (Parua, 1992).

After this phase (1976–1978), upper reach (Bhagirathi offtake to Nasipur, 13.5 km north of Berhampore) of the river Bhgairathi has been stabilised and adjusted to the flow. But the middle (Nasipur to Sitahati, 5.6 km north of Katwa) and lower reaches (Sitahati to Nabadwip) are not yet adjusted to morphogenic variables. It has been observed that lower reach of Bhagirathi is still experiencing severe bank erosion (Islam, 2011; Islam et al., 2012). Major erosion-prone belt in the lower reach of Bhagirathi is located in between Ajay–Bhagirathi confluence at Katwa and Jalangi–Bhagirathi confluence in Nabadwip. Recently, erosion at different reaches varied depending on the variability of flood discharge of monsoon months. Despite the general trend of variability, tail reach always holds the maximum annual erosion except for the year 2007. Share of bank erosion at the

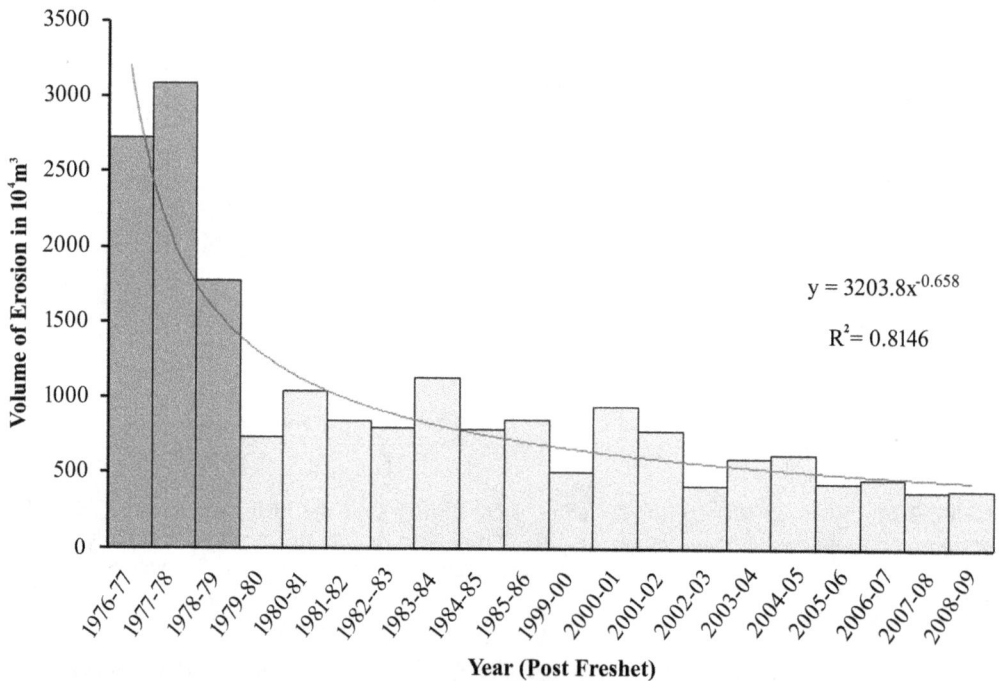

FIGURE 2.22 River bank erosion after commissioning of FBP in 1975 (computed from Parua (1992) and annual reports on river Bhagirathi of KoPT: 2003–2004 to 2008–2009).

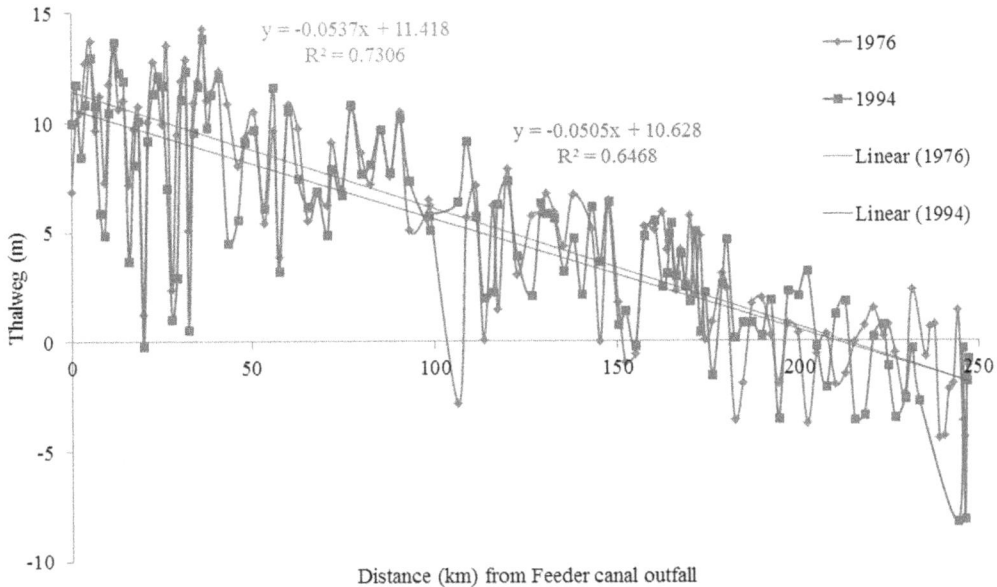

FIGURE 2.23 Changing channel gradient in the post-Farakka period (computed from Ghosh (2000)).

tail reach often exceeds the limit of 50% of total erosion. This clearly establishes severity of erosion in tail reach, covering almost 1/3 of the total length of the river. Nature of erosion in the tail reach, though fluctuating, is showing higher rate than the other two reaches (Islam and Guchhait, 2015). Furthermore, bed scouring and its deposition are notable in the post-Farakka episode. Due to hydraulic impulse, upper and middle reaches of the river also experienced some bed scouring in addition to bank erosion. The bank erosion was quite high compared to the bed scouring due to very gentle nature of the channel gradient (3°–5°). In the post-Farakka period, the overall channel gradient is less ($R^2 = 0.65$) than that of the pre-Farakka period ($R^2 = 0.74$). This is due to the erosion of the bank and bed materials at the upper stretches and its deposition on the river bed of the lower stretches (Figure 2.23).

This is due to inadequate discharge to carry its load. Bhattacharya (2000) claimed that a minimum of 40,000 cusecs of water will be needed to flush out the eroded materials. However, in post-Farakka period, Bhagirathi-Hooghly system never attained that figure.

2.3.2 Expansion of the Settlements and Urbanisation

2.3.2.1 Location and Growth

The deltaic part of West Bengal is densely populated for its hydrological and agro-ecological advantages. In the BRB, population is increasing by natural and migration-induced growth. In the decade of 2001–2011, there has been significant increase in population in the major urban centres of the basin. For example, Berhampore Municipality has registered a steady increase in population in different census years (1951: 55,613, 1961: 62,317, 1971: 92,889, 1981: 94,907, 1991: 115,144, 2001: 160,143, 2011: 195,223) (District Census Handbook, 2001, 2011). Similarly, Katwa Municipality had a population of 71,573 which has increased to 81,615 in 2011. Urban outgrowth and sprawl are notable i.e. a change in spatial urbanisation is striking. During the decade from 2005–2006 to 2015–2016, the settlement area has been increased in the buffer zone (the total buffer area 1,367.05 km^2) from 310.5 to 317.1 km^2 (Figure 2.24a and b). In other words, settlement area has recorded an increase from about 22.71% to 23.19%.

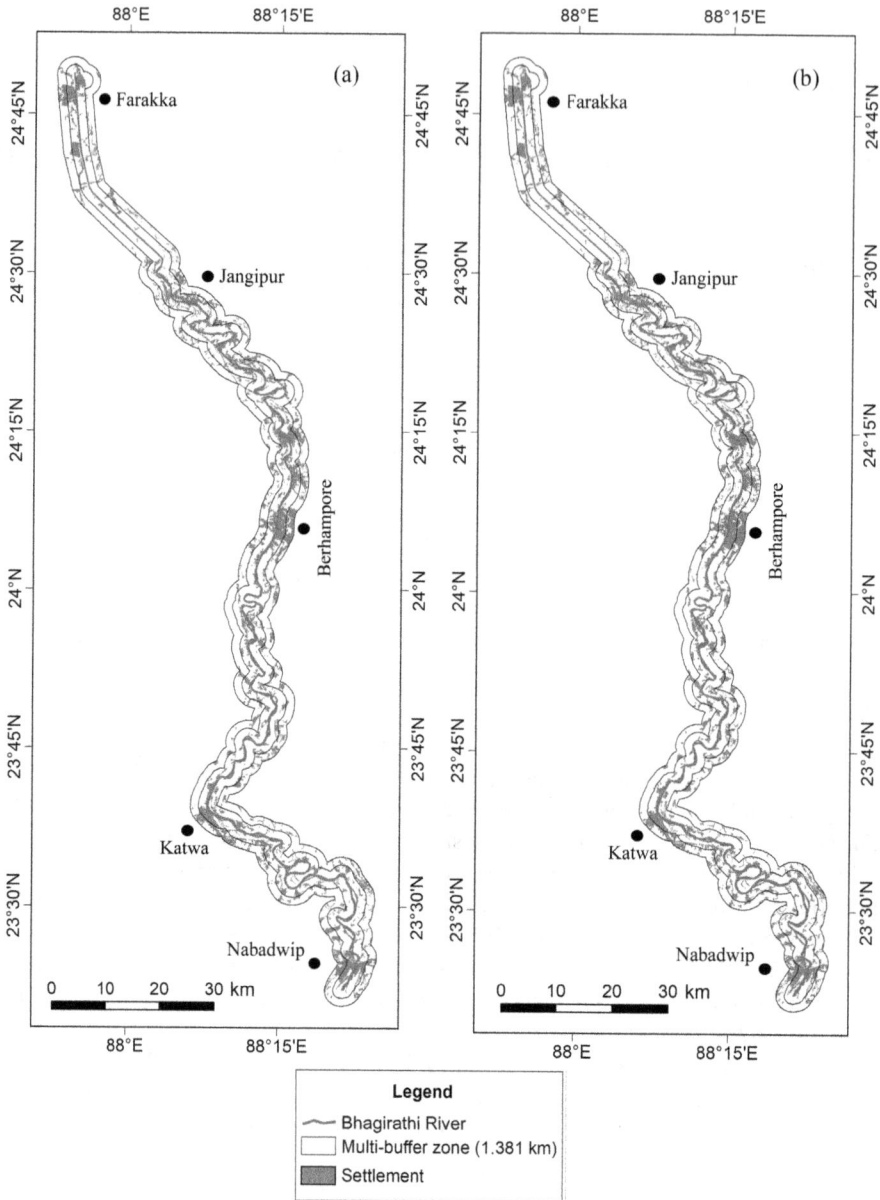

FIGURE 2.24   Changing urbanisation in the active belt of Bhagirathi River: (a) 2005–2006 and (b) 2015–2016. (Based on National Remote Sensing Centre (NRSC) Bhuvan WebGIS, 2005–2006 and 2015–2016.)

2.3.2.2   Urbanisation and Channel Changes

In general, run-off increases and lag time decreases due to increase in the impervious area of an urban setting. This induces a change in the sediment flux of the river which increases the physical and biological degradation of the river and the channel enlargement (Chin, 2006). Urbanisation is found to stabilise the channel movement. During 1973 and 2019, the channel oscillation was less along the major urban settlements compared to the other stretches. The left bank of the Bhagirathi adjacent to Berhampore city is more stable than its right bank dotted with some disperse rural settlements. The annual rate of bank line shifting on the left bank is about 0.39 m while 0.87 m for the right bank in this stretches (Figure 2.25a).

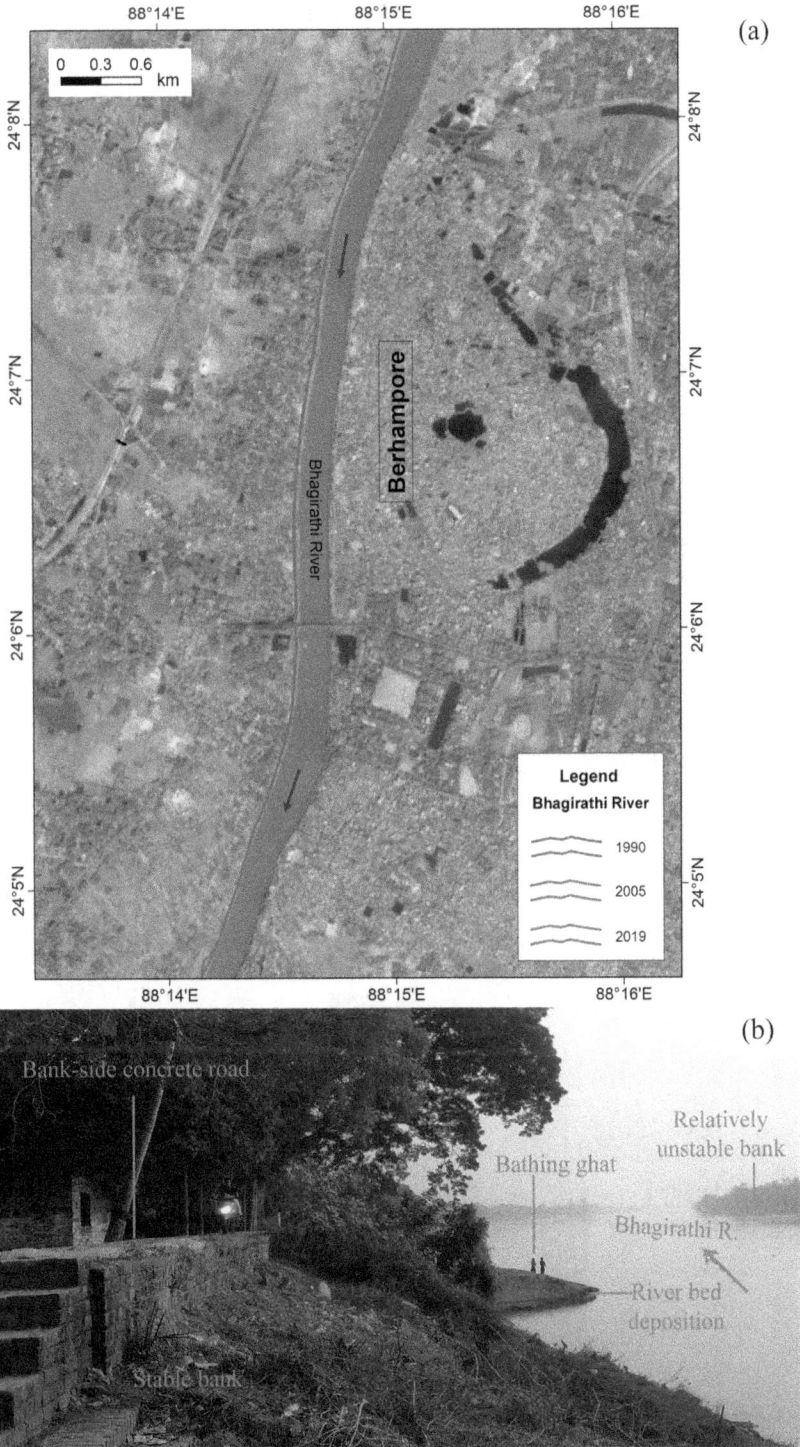

FIGURE 2.25 Urbanisation and channel stabilisation adjacent to Berhampore city (a) bank line shifting (prepared from MSS-path 149, row 43, dated 17 January 1973; TM-path 139, row 43 dated 6 February 1990, 15 February 2005; OLI-path 139, row 43 dated 6 February 2019 and Sentinel 2AT45QXG dated 05 January 2019; note green line for 1990 and violet for 2005) and (b) concrete urban bank at Berhampore. (Field Photograph, 23 November 2019.)

This is due to concretisation of the bank and forced stabilisation (Figure 2.25b). The concrete bank on the left bank is 5.94 km extended, while there is no such civil structure on the right. Moreover, similar stabilisation has been observed along Katwa Township on the right bank of the river (Figure 2.26). The rate of bank line shifting along Katwa is about 0.35 m, while on the left bank 1.97 m in the downstream of the Katwa reach, there is profuse channel oscillation and bar

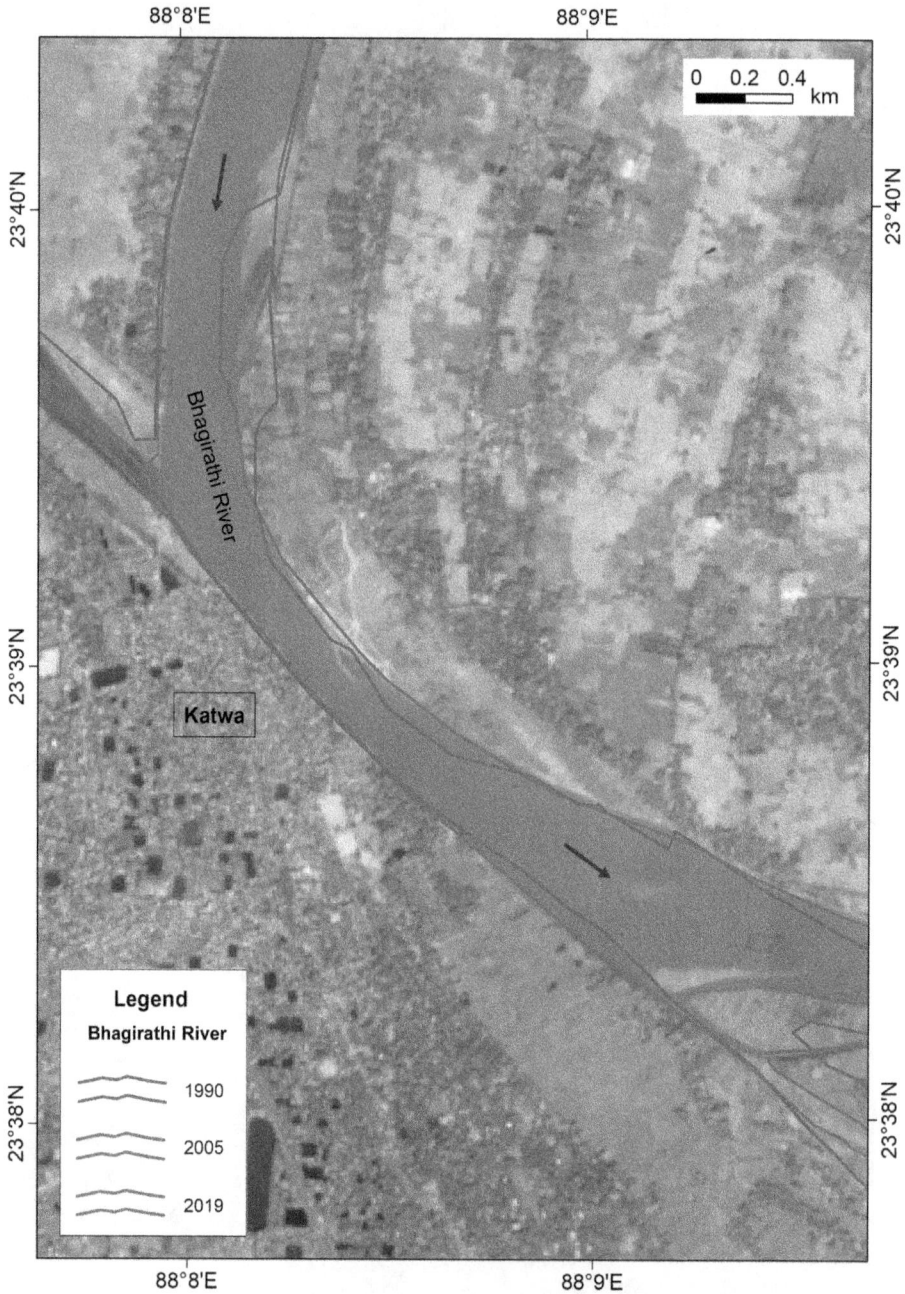

FIGURE 2.26 Impact of urbanisation on channel narrowing and stabilisation near Katwa town (prepared from MSS-path 149, row 44, dated 22 February 1973; TM-path 138, row 44 dated 30 January 1990, 24 February 2005; OLI-path 138, row 44 dated 15 February 2019 an Sentinel 2A T45QXG dated 5 January 2019; note green line for 1990 and violet for 2005).

growth. Furthermore, Nabadwip urban centre on the right bank has also impeded the rapid movement of the channel (Figure 2.27). Along Nabadwip, the annual rate of movement is 0.96 m on right and 1.36 m on left, while its upstream area has the rate of channel oscillation about 8.63 m on right and 15.94 m on left and downstream reach has the rate of 1.4 m on right and 4 m on left.

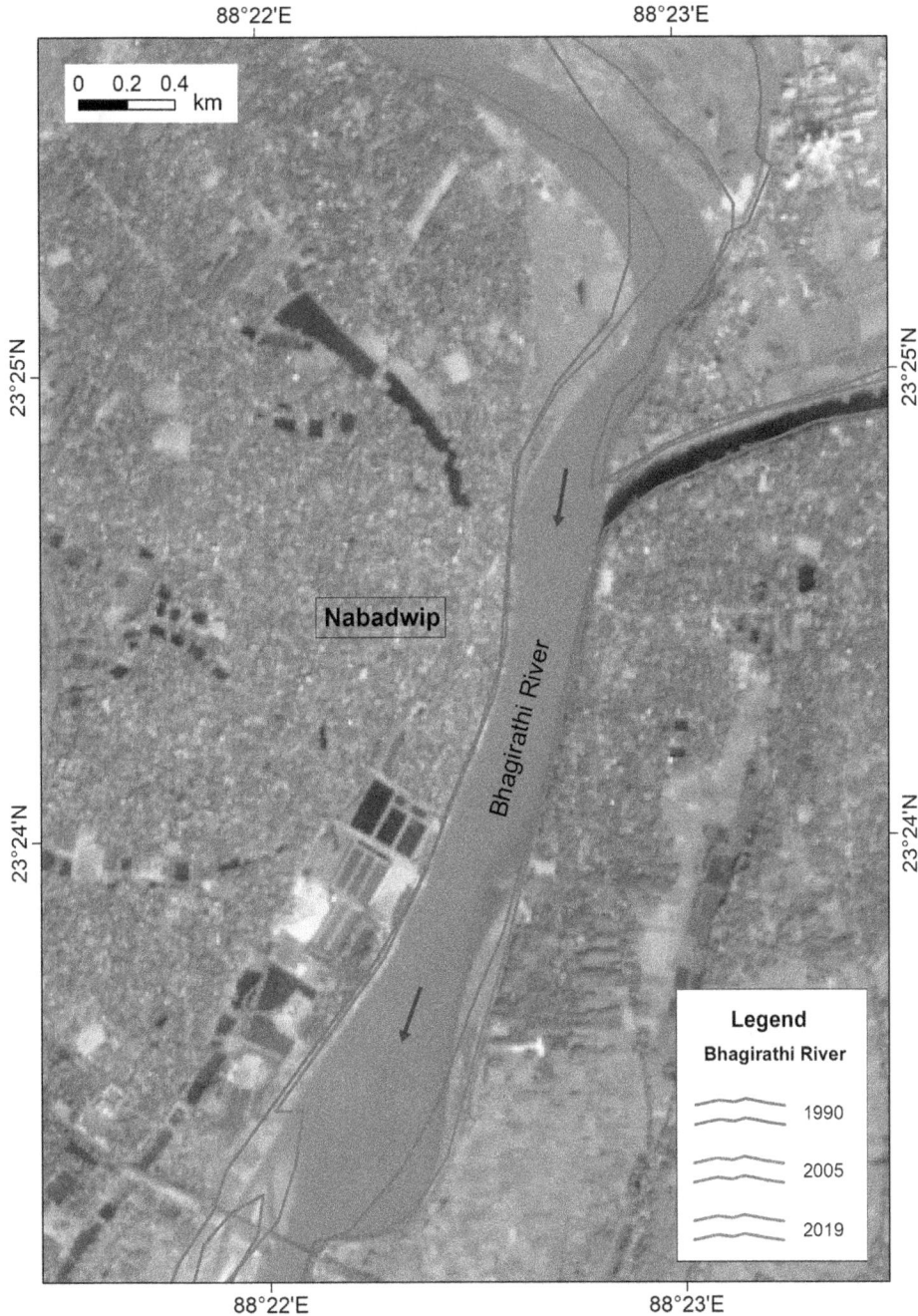

FIGURE 2.27 Impact of urbanisation on channel stabilisation near Nabadwip town (prepared from MSS-path 149, row 44, dated 22 February 1973; TM-path 138, row 44 dated 30 January 1990, 24 February 2005; OLI-path 138, row 44 dated 15 February 2019 and Sentinel 2AT45QXF dated 5 January 2019; note green line for 1990 and violet for 2005).

2.3.3   Brick Fields

2.3.3.1   Location and Development

Brick field industries and road stream crossings are the other aspects of the human interventions over the BRB. Along the banks of the river, the mushrooming of the brick fields is notable (Figure 2.28).

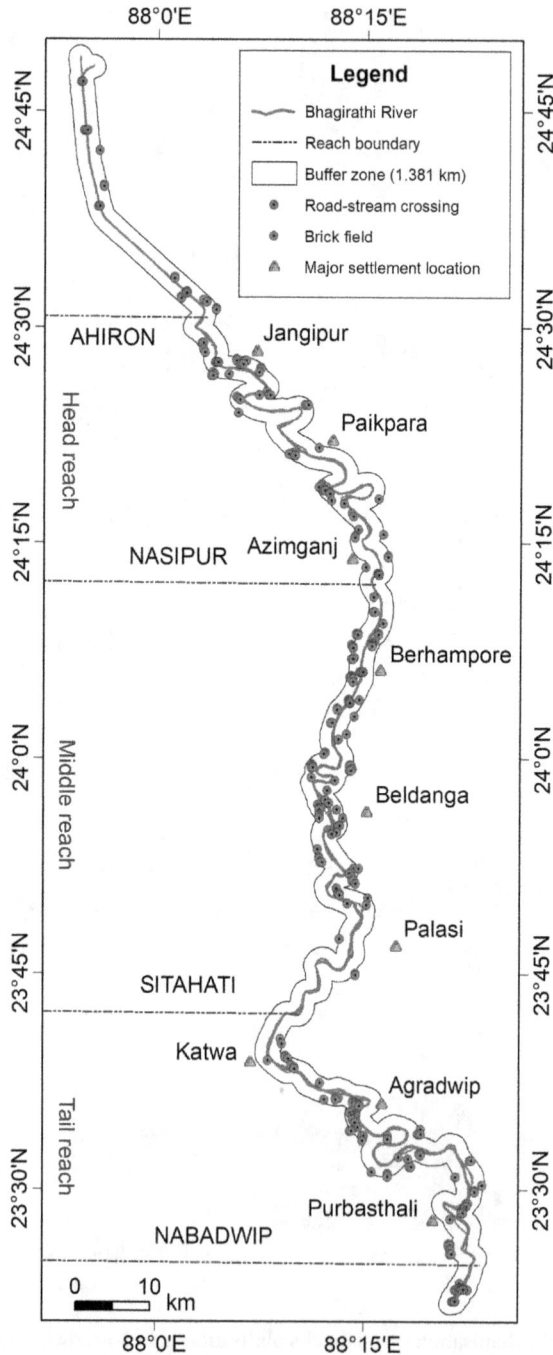

FIGURE 2.28   Location of brick field and road stream crossing (computed from Google Earth dated 31 December 1999 and 19 October 2019).

At present, there are 41 brick fields in the head reach (left bank: 16 and right bank: 25), 61 in the middle reach (left bank: 32 and right bank: 29) and 52 in the tail reach (left bank: 18 and right bank: 34).

2.3.3.2 Brick Field, River Bank and Sediment Flux

Brick field generally induces river bank erosion (Das, 2014). The higher the number of brick field along a river, the greater is the area of bank failure because of cutting of earth materials from the river bank (Figure 2.29). The ideal materials for the brick kiln industries are the silt or clay silt materials. The brick kilners extract the silty materials from upper horizons of the soils and the undermining of sandy layer from the bottom by the fluvial actions expedite the bank failure mechanisms along the Bhagirathi River. Moreover, they often create shallow depressions along the river bank to trap the silt during flood stage and extracting them afterwards for the industries (Figure 2.30). This way a brick kiln may increase the channel width with the passage of the time.

According to an estimate done by the District Land and Land Reforms Office (D.L. & L.R.O.), Nadia, Government of West Bengal in 2009, a kiln moves earth materials of about 133,110 ft^3 in a year. This much material is extracted from within that river basin. If this figure is multiplied by the number of the brick kilns located adjacent to river Bhagirathi as of 2019, it totals to about

FIGURE 2.29 Bank failure due to cutting of soil from the upper horizons near Nabadwip. (*Source*: Field Photograph, 2011.)

FIGURE 2.30 Shallow depression in the brick kiln area for entrapping sediment during flood along the left bank of Bhagirathi River, near Beldanga. (Field Photograph, 24 November 2019.)

20,498,940 ft³ materials movements, which is an increase from 6,522,390 ft³ in 2000. This earth material movement by the anthropogenic processes is a real concern while comparing this figure with annual sediment budget ($0.5–1.0 \times 106$ m³ per year) by natural river transport of the Bhagirathi River. This outlines the influence of the human on sediment transfer process of the Bhagirathi River belt.

Moreover, the morphological change along a bank dotted with kilns is also remarkable. To address the impact of brick kiln industry on the morphological changes along the Bhagirathi River, few cross sections were drawn. Generally, it is observed that from the bank of the river elevation increases landward or sometimes decreases after the natural levee is encountered (Figure 2.31a).

The natural landscaping is disrupted by excavation of shallow depression to entrap the sediment during flood. The nature of the cross sections portrays that from the bank of the Bhagirathi, there is a sharp fall in elevation and then abrupt rise where sediment cutting is not present (Figure 2.31b).

2.3.4 ROAD-STREAM CROSSING

2.3.4.1 Location and Growth

Over the Bhagirathi, there are four road crossings (head reach: 2; middle reach: 2). They are (1) Bhagirathi Bridge at Jangipur, (2) Nasipur Railway Bridge, (3) Ramendra Sundar Tribedi Bridge at Berhampore, and (4) a new bridge near Balarampur. Besides, the human activities such as bamboo fencing across the river and ship movements along the river also control the hydro-geomorphological aspects.

2.3.4.2 Road-Stream Crossing and Channel Changes

Road crossing across a river regulates flow characteristics of a channel, thereby inducing morphological changes (Gregory and Brookes, 1983). There are generally two types of structure, namely, (1) suspension bridge and (2) cantilever bridge. The former generally constricts the channel, while the latter directly modifies flow characteristics over Bhagirathi River. Majority of the crossings are cantilever having piers in between the two ends of the bridge. The piers play a vital role to modify flow pattern by obstructing the flow, thereby scouring in one side and deposition in between two piers. Morphological changes are observed below the Berhampore Bridge (Figure 2.32a).

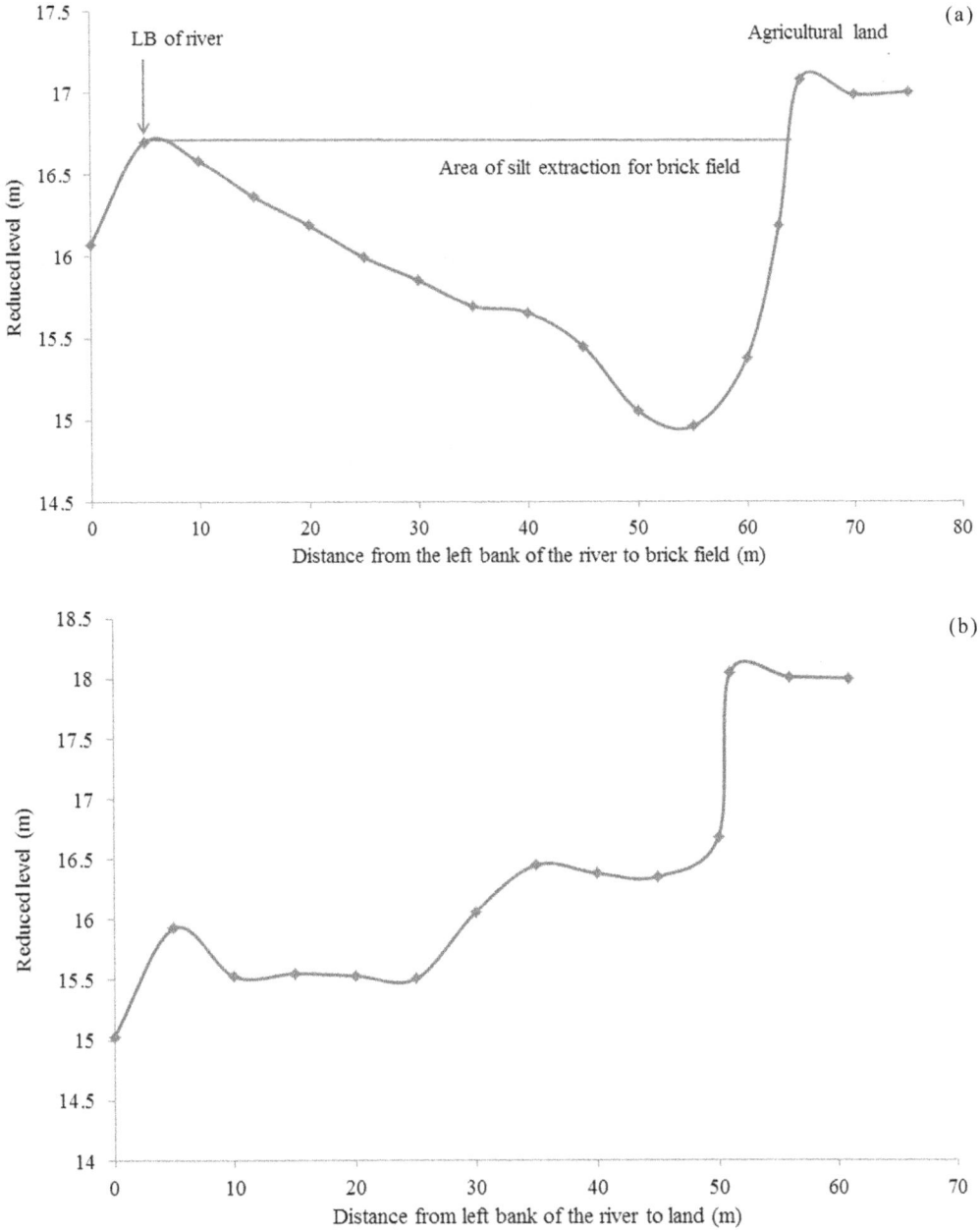

FIGURE 2.31 Alteration of topographic expression due to brick kiln industries. (a) Formation of artificial valley due to silt excavation and (b) artificial spur-like feature.

During 1976 and 1994, reversal thalweg was striking. Thalweg was oriented towards left bank during 1976, while right bank orientation of thalweg was found in 1994 (Figure 2.32b). The alternate pattern of bed scouring and bed deposition during 1976 was suspected to be a consequence of the bridge piers. However, the hydraulic impulse from the FBP has increased bed scouring at a much faster rate and increased the cross-sectional area, which has masked the effects of the road-stream crossing on channel morphology. Besides, the bridge near Hazarduari was completed in the year 2010 with the recent technology to carry the increasing loads (Figure 2.33).

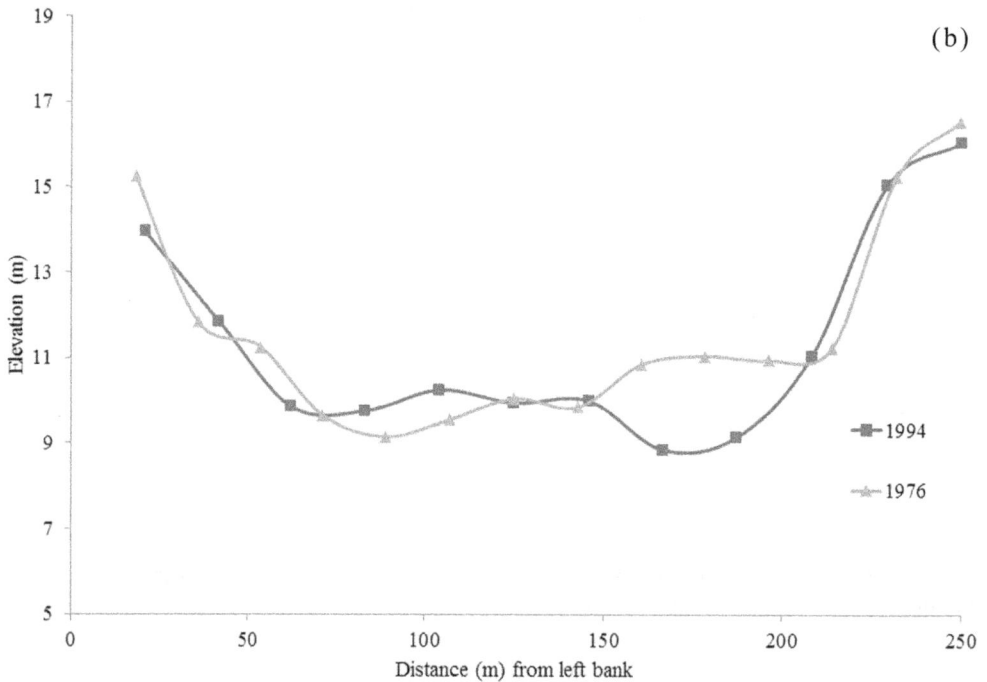

FIGURE 2.32   Bed scouring due to presence of road piers of Berhampore bridge: (a) pier location (Field Photograph, 23 November 2019) and (b) changing cross-sectional morphology. (*Source*: Based on Ghosh (2000).)

FIGURE 2.33 Nasipur railway bridge near Hazarduari, Murshidabad. (Field Photograph, 23 November 2019.)

In this bridge, two piers are directly within the channel which has impact on the local water level of the upstream of the piers. The typical geometry of these piers is that they have larger pier width at the bottom which has narrowed up. These piers have reduced the effective water area and, hence, an increase in the upstream water level called afflux. Besides, the river training works on both the banks also obstruct the flow during the high stage in monsoon period. Actually this boulder-strewn surface is prepared to restrict the channel oscillation and to save the bridge from the future probable collapse. The artificially created boulder bed offers flow resistance and hence modification of the hydraulic behaviour of the river Bhagirathi. Another bridge at Balarampur near Berhampore (Figure 2.34) is under construction with three piers amid the channel besides another pier for additional services.

FIGURE 2.34 New Bhagirathi bridge near Balaram under construction. (Field Photograph, 24 November 2019.)

FIGURE 2.35   Bamboo-made guide bank inducing sedimentation near Nabadwip. (Field Photograph, 2011.)

2.3.5 GUIDE BANK AND SEDIMENTATION

Guide bank is constructed to flow a river through a narrow channel without causing damage to the structure. Over the Bhagirathi River near *Prachin* (Old) Mayapur, Nabadwip, a bamboo-made guide bank was constructed across the river during 2011. The main objective of this structure is to divert the flow of the Bhagirathi towards a narrow passage on the right bank so that Inland Waterway Authority of India (IWAI) vessels can pass through that passage. Actually the width of the river in this stretch is about 1 km and the flow during the lean months is very less (around 25,000 cusecs). Naturally the dispersed flow cannot support the vessel movement which needs a minimum water depth of about 2–3 m. Therefore, man-induced sedimentation is triggered in one part of the channel and thus encouraging the flow to a narrow passage so that depth of the concerned area could be increased. Thus, the oblique alignment of the fencing constructed over the Bhagirathi River traps the sediment and deposits as a mid-channel bar (Figure 2.35).

2.3.6 SHIP MOVEMENTS AND BANK FAILURE

River Bhagirathi is a national waterway no. 1 since pre-independence period, and it was used for freight movements and services to the passengers. With time, navigability of Ganga River has reduced resulting in stopping of passenger's services through waterway from the Bay of

Bengal to Uttar Pradesh. However, this largest waterway is still used for freight movements. IWAI is maintaining a narrow passage of waterway (3 m width) for ship movement throughout the year. On Bhagirathi, regular movement of ship destabilises the river bank through the waves generated by them (Figure 2.36a).

Nanson et al. (1994) also noted that there is a significant positive correlation between the maximum wave height in a train induced by the ship movement and rate of bank failure of the Gordon River, Tasmania. Similarly, Kurdistani et al. (2019) also established bank instability with wave characteristics from a multivariate perspective taking the consideration of wave height, wave period and wave obliquity. It is pertinent to mention that the typical soil profile along the banks of river Bhagirathi is dominated by coarse sand material which is vulnerable to waves induced by the ship movements. When large ships move, an increase of 30–50 cm in the wave height accelerates the undermining action of the river leading to the collapse of the upper portion of the bank (Figure 2.36b).

FIGURE 2.36 Bank failure due to ship movement near Beldanga: (a) movement of vessel and (b) fall of cohesive bank. (Field Photograph, 26 November 2019.)

2.4 CONCLUSION

Structural interventions especially the FBP on the Bhagirathi system have altered the channel morphodynamics profusely. The radical shift in the fluvial regime of the river especially in the lean months (January–May) has been observed which has corresponding effects on the planform character and meander geometry. The major findings are increasing channel width in the post-Farakka period by about 21% and increasing cross-sectional area of the channel by 23%. The cubic capacity of the channel has increased in the upstream area as much as 51% while decreased at the tail reach (CS 347–CS 372a) by 57%. Moreover, channel oscillation especially in the lower reaches is notable, and the entire meander cut-off has been observed in the post-Farakka period mainly in the tail reach. Lower stretch is relatively more unstable as portrayed by the higher CV of the SI over the period 1973–2019. Besides, the higher BI in this stretch also marks the channel instability. Furthermore, bank erosion is recurrent along the banks of the river Bhagirathi especially in the middle and tail reaches. However, due to the unavailability of adequate discharge, bed load is not carried to the Bay of Bengal, and thus, bed deposition is found especially in the lower part of the river. This induces lessening the channel gradient and diminishing channel efficiency.

The above dimensions of the research point out that FBP in relation to channel dynamics is well explored especially by the attempts of Kolkata Port Trust. However, other small-scale interventions have also impacted on channel dynamics especially due to changes in the land use and land cover, construction of brick fields, road stream crossing and use of river as a national waterway. In this present attempt, these small-scale interventions have been outlined. However, these dimensions need to be addressed within a broader framework over longer time scale. There is a need to address the channel metamorphosis in the context of the dynamic equilibrium taking the consideration of the FBP and other small-scale channel modifications.

REFERENCES

Allison, M. A., Khan, S. R., Goodbred Jr, S. L., & Kuehl, S. A. (2003). Stratigraphic evolution of the late Holocene Ganges-Brahmaputra lower delta plain. *Sedimentary Geology, 155*, 317–342.

Bagchi, K. G. (1944). *The Ganges Delta*. Calcutta: Calcutta University.

Bandyopadhyay, S. (1996). Location of Adi Ganga paleo channel, S. 24 Parganas, West Bengal. *Geographical Review of India, 58*(2), 93–109.

Basu, S. R., Ghosh, A., & De, S. K. (2005). Meandering and cut-off of the river Bhagirathi. In S. C. Kalwar, ed., *Geomorphology and Environmental Sustainability* (pp. 20–37). New Delhi: Concept Publishing Company.

Bhattacharya, K. (2000). *Bangladesher Nadnadi o Parikalpana*. Kolkata: Vidyoday Library Private Ltd.

Brice, J. C. 1964. Channel patterns and terraces of the Loup Rivers in Nebraska. Geological Survey Professional Paper 422-D.

Chakraborty, S. (1970). Some consideration on the evolution of physiography of West Bengal. In A. B. Chatterjee, & S. Sengupta, eds., *Geology of the Southwestern Bengal*. Calcutta: Firma, KLM.

Chatterjee, S. N. (1989). *River System of West Bengal*. Kolkata: River Research Institute, Govt. of West Bengal.

Chin, A. (2006). Urban transformation of river landscapes in a global context. *Geomorphology, 79*, 460–487.

Colebooke, R. H. (1801). On the courses of Ganges through Bengal. *Asiatic Researchers, 7*, 1–31.

Das, B. C. (2014). Impact of in-bed and on-bank soil cutting by brick fields on Moribund deltaic rivers: a study of Nadia River in West Bengal. *The NEHU Journal, XII*(2), 101–111.

District Census Handbook. (2001). *Murshidabad District*, Directorate of Census operation, West Bengal, Part XII A & B Series 20.

District Census Handbook. (2011). *Murshidabad District*, Directorate of Census operation, West Bengal, Part XII B Series 20.

Hirst, F. C. (1915). *Reports on Nadia Rivers. Reprinted in River of Bengal* (Vol. III 2002). Calcutta: Gazetteers Dept.

Ghosh, A. (2000). *A Systematic Study of the Effect off Arakka Barrage on the Morphological Changes of the River Bhagirathi*. PhD Thesis. Kolkata: University of Calcutta.

Goodbred, Jr, S. L., & Kuehl, S. A. (1998). Flood plain processes in the Bengal Basin and the storage of Ganga-Brahmaputra River sediment: an accretion study using 137Cs and 210Pb geochronology. *Sedimentary Geology*, 121, 239–258.

Goodbred Jr, S. L., & Kuehl, S. A. (2000a). Enormous Ganges-Brahmaputra sediment load during strengthened early Holocene monsoon. *Geology*, 28(12), 1083–1086.

Goodbred Jr, S. L., & Kuehl, S. A. (2000b). The significance of large sediment supply, active tectonism, eustasy on the margin sequence development. Late Quaternary stratigraphy and evolution of the Ganges-Brahmaputra delta. *Sedimentary Geology*, 133(3–4), 227–248.

Gregory, K. J., & Brookes, A. (1983). Hydrogeomorphology downstream from bridges. *Applied Geography*, 3(2), 145–159.

Guchhait, S. K., Islam, A., Ghosh, S., Das, B. C., & Maji, N. K. (2016). Role of hydrological regime and floodplain sediments in channel instability of the Bhagirathi River, Ganga-Brahmaputra Delta, India. *Physical Geography*, 37(6), 476–510.

Islam, A. (2011). Variability of stream discharge and bank erosion – a case study on the river Bhagirathi. *Journal of River Research Institute River Behaviour and Control, 31*(1), 55–66.

Islam, A. (2016). River Bank erosion and its impact on economy and society a study along the left bank of River Bhagirathi in Nadia District West Bengal, an unpublished PhD thesis, Burdwan: The University of Burdwan.

Islam, A., & Guchhait, S. K. (2015). Is Severity of River Bank Erosion Proportional to Social Vulnerability? A Perspective from West Bengal, India. *LIFE AND LIVING THROUGH NEWER SPECTRUM OF GEOGRAPHY*, 35.

Islam, A., & Guchhait, S. K. (2017a). Analysing the influence of Farakka Barrage Project on channel dynamics and meander geometry of Bhagirathi river of West Bengal, India. *Arabian Journal of Geosciences*, 10(11), 245.

Islam, A., & Guchhait, S. K. (2017b). Search for social justice for the victims of erosion hazard along the banks of river Bhagirathi by hydraulic control: a case study of West Bengal, India. *Environment, development and sustainability*, 19(2), 433–459.

Islam, A., & Guchhait, S. K. (2018). Analysis of social and psychological terrain of bank erosion victims: a study along the Bhagirathi river, West Bengal, India. *Chinese geographical science*, 28(6), 1009–1026.

Islam, A., & Guchhait, S. K. (2019). Social engineering as shock absorbing mechanism against bank erosion: a study along Bhagirathi river, West Bengal, India. *International Journal of River Basin Management*, 1–14.

Islam, A., & Guchhait, S. K. (2020). Characterizing cross-sectional morphology and channel inefficiency of lower Bhagirathi River, India, in post-Farakka barrage condition. *Natural Hazards*, https://doi.org/10.1007/s11069-020-04156-9.

Islam, A., Laskar, N., & Ghosh, P. (2012). An areal variation of fluvial hazard perceptions of various social groups-A perspective from Rural West Bengal, India. *Indian Streams Research Journal*, 2(9), 1–9.

Kurdistani, S. M., Tomasicchio, G. R., D'Alessandro, F., & Hassanabadi, L. (2019). River bank protection from ship-induced waves and river flow. *Water Science and Engineering, 12*, 129–135.

Leopold, L. B., Langbein, W. B. (1966). River meanders. *Sci Am* 214(6):60–70.

Majumder, S. C. (1942). *Rivers of Bengal Delta*. Kolkata: Calcutta University.

Mirza, M. M. Q. (Ed.). (2006). *The Ganges Water Diversion: Environmental Effects and Implications* (Vol. 49). Dordrecht: Springer Science & Business Media.

Mukherjee, R. K. (1938). *The Changing Face of Bengal*. Calcutta: Calcutta University.

Nanson, G. C., Von Krusenstierna, A., Bryant, E. A., & Renilson, M. R. (1994). Experimental measurements of river-bank erosion caused by boat-generated waves on the Gordon river, Tasmania. *Regulated Rivers: Research & Management, 9*(1), 1–14.

Ness, R. (2004). *Belt Width Delineation Procedures*. Ontario: Toronto and Region Conservation Authority, Parish Geomorphic Ltd.

Oldham. (1870) President's address. *Proceedings, Asiatic Society of Bengal for February*, 1870, Calcutta.

Parua, P. (1992). *Stability of the Banks of Bhagirathi-Hooghly River System*. PhD Thesis. Kolkata: Jadavpur University.

Parua, P. K. (2009). Farakka Barrage and its alleged impact on floods and bank erosion problems of Malda and Murshidabad Districts of West Bengal. In P. K. Parua, ed., *Some Aspects about Farakka Barrage Project* (Vol. II, pp. 30–44). Berhampore: Shilpanagari Publishers.

Ray, A. (1999). Locational problems of the sixteenth century Bengal Coast. *Pratna Samiksha, Journal of the Directorate of Archaeology and Museum, 6* (8), 121–134.

Reaks, H. J. (1919). *Report on the Physical and Hydraulic Characteristics of the Rivers of the Delta, Appendix-II of the Report of the Hooghly River and its Head Waters.* Calcutta: The Bengal Secretariat Book Depot.

Rennell, J. (1788). *Memoir of a map of Hindoostan; or the Mogul's Empire.* London, printed by M. Brown.

Rudra, K. (2010). *Banglar Nadikatha.* Kolkata: Sahitya Samsad.

Rudra, K. (2011). *The Encroaching Ganga and Social Conflict: The Case of West Bengal, India.* Kolkata: Unpublished paper.

Sen, H. (2010). The drying up of river Ganga: an issue of common concern to both India and Bangladesh. *Current Science, 99*(6), 725–727.

Sherwill, W. S. (1858). *Report on Rivers of Bengal.* Calcutta: Savielle Printing and Publishing Co Ltd.

Singh, L. P., Parkash, B., & Singhvi, A. K. (1998). Evolution of the Lower Gangetic plain landforms and soils in West Bengal, India. *Catena, 33*(2), 75–104.

Umitsu, M. (1987). Late Quaternary sedimentary environment and landform evolution in the Bengal lowland. *Geographical Review of Japan, 2*(60), 164–178.

Willcocks, W. (1930). Lectures on the ancient system of irrigation in Bengal.

3 Anthropogeomorphology of the Lower Deltaic West Bengal with Special Reference to the Hugli River System

*Nabendu Sekhar Kar, Sayantan Das
and Sunando Bandyopadhyay*

CONTENTS

3.1 INTRODUCTION

During the past few centuries, humans have emerged as an effective geomorphological agent in local and global contexts. Their contribution in reshaping the earth surface is now equal or sometimes even greater than the other geomorphic factors. Although the energy released by human society is negligible compared to the endogenic processes, it is not incomparable with the exogenic processes operating on the earth in terms of reworking of surface materials (Szabó et al., 2010). Nearly one-third of the earth's continental surface ($149 \times 10^6 km^2$) is affected by direct or indirect anthropogeomorphic activity (Rózsa, 2007). This includes arable land and plantations ($15 \times 10^6 km^2$), grazing land ($35 \times 10^6 km^2$), forests ($38 \times 10^6 km^2$), built-up areas ($2 \times 10^6 km^2$), and other types of land uses (Loh and Wackernagel, 2004 in Rózsa, 2007). The human activities identified as responsible for modifying the natural landscape include agriculture, mining, industry, transportation, urban construction, sports and tourism, river and coastal management, and warfare, among others.

In the last few centuries, unprecedented population growth resulted in higher energy release, resulting in greater human-induced changes of the earth surface. Such tendencies of increasing human impact on the landscape are likely to affect the other natural processes (Szabó et al., 2010). Therefore, to prevent negative alterations in the anthropogenically modified environment, knowledge of surface changes made by the humans and its consequences have become absolute necessity for the geomorphologists. Hence, a new branch of geomorphological study has evolved, which is variously called as anthropic geomorphology, anthropogenic geomorphology and anthropogeomorphology. Anthropogenic geomorphology studies landform associations made by the human activity, investigates surface changes induced by these forms, predicts the corollaries of upsetting the natural equilibrium and makes recommendations for preventing damages. All these make anthropogenic geomorphology a discipline of applied character (Szabó et al., 2010). However, mere areal extension of human activities does not indicate increasing anthropogeomorphic impact as this primarily depends on influence of economic activities, population pressure and technological advancement that differ from one place to another.

The role of humans as a geomorphic agent was first explored by Marsh (1864) in his empirical work *Man and Nature; Or, Physical Geography as Modified by Human Action*, followed by Sherlock (1922) in his book *Man as a Geological Agent*. After World War II, several research papers dealt with this theme. Among them, the works done by Golomb and Eder (1964), and Brown (1970) are worth mentioning. Although human impact on geomorphic processes is recognised for a long time, a number of developments since 1969 have led to an increasing realisation of its importance (Goudie, 2018). These include (1) intellectual and policy-related changes, (2) technological developments that alter geomorphological processes, (3) demographic trends and (4) proliferation of techniques for the study of landform and process changes. A major work on anthropic geomorphology was published in 1983 by Nir through his book entitled *Man, A Geomorphic Agent*. In recent times, important contributions on this topic can be found in the writings of Hooke (2000), Lóczy and Pirkhoffer (2009), Szabó et al. (2010), Rózsa and Novák (2011) and Goudie (2018), which enlighten different anthropogeomorphological issues and define the scope and content of the subject.

An ideal way to measure the human impact on the natural landscape is by calculating the amount of earth moved by various anthropogeomorphological processes. However, calculation of such value is difficult. An estimation of global anthropic erosional rate was done by Nir (1983) for the year 1970. He assessed that the total amount of earth moved by humans was $173 \times 10^9 t \ yr^{-1}$ (Table 3.1), which is some 173 times greater than the sediment load carried by the Ganga–Brahmaputra River System (Milliman and Meade, 1983; Milliman and Syvitski, 1992). He concluded that on the basis of the anthropic erosion rate, agriculture can be singled out as the most significant landscape-modifying human activity. Nir (1983), till date, proposed the most useful model for the quantification

TABLE 3.1
Estimation of Global Anthropic Erosional Rates

Human Activity	Rate of Erosion ($10^9 t \ year^{-1}$)
Forest clearing	1
Grazing	50
Tilling the land	106
Mining	15
Roads, railways and urban construction	1
Total	173

Source: Nir (1983) and Rózsa (2007).
See Section 3.1.

of potential anthropogeomorphological impact, known as the index of potential anthropic geomorphology. His model was based on two parameters—the degree of development (DD; reflecting the rate of human impact) and the degree of perception (DP; concerning the perception of damage from the anthropogeomorphological processes). He proposed this index as a yardstick to specify how potential anthropogeomorphological processes can be harmful for a given region or country. The parameters used by him—the DD and the DP—are interrelated. Intensification of DD increases the chances of landscape modification by human intervention (Figure 3.5a). Interfering with the natural processes may bring about damages to the landscape and environment, which are irreversible and detrimental to the society. So, there should be restrictions and scientific management of such actions that lead to unfavourable results. This conscience is defined by Nir (1983) as DP (Figure 3.5b). DD is represented as the percentage of urban population (UP), while DP is expressed by the percentage of illiteracy (DI), since the level of illiteracy indicates lack of public DP. Nir (1983) determined the potential anthropogeomorphological processes by averaging these two parameters and multiplying the result with the combined score of two constants—K_c and K_r—representing climatic and relief conditions, respectively. The values of these constants range from 0.4 to 0.8 and from 0.2 to 0.8, respectively (Table 3.2). The result is then multiplied by 1/100, which converts into the I_{PAG} values ranging between 0 and 1. The pooled values of K_c and K_r represent the geomorphic sensitivity of a landscape to human impact, known as anthropic geomorphological sensitivity (Rózsa and Novák, 2011).

$$I_{PAG} = \frac{UP+DI}{2} \times \frac{1}{100} \times (K_c + K_r)$$

According to Nir (1983), when I_{PAG} value is less than 0.30, human geomorphological activities represent limited hazard. In contrast, when it lies between 0.30 and 0.49, the hazard is perceptible and some erosion control is necessary. If I_{PAG} value is greater than 0.50, the hazard has caused substantial damages and controlling measures are required.

Later, the human environmental impact was quantified by the equation forwarded by Erlich and Erlich (1990), considering it as the product of population pressure, per capita affluence and technological dependency. But the results obtained from this model can be highly generalised and may not reflect the dynamicity of the processes involved. Hooke (1994, 2000) applied another method to measure the human impact on landscape. He divided the anthropogeomorphic activities into intentional and unintentional processes (Table 3.3). Intentional processes are those which directly modify the landscape, the erosion rates of which are measurable in terms of quantification, e.g., mining, building, railway and road construction. Unintentional processes are indirect ones such

---

TABLE 3.2

Values of Climatic Condition (K_r) and Relief Condition (K_c) as Considered by Nir (1983)

K_c		K_r	
Climate	**Constant**	**Relief**	**Constant**
Equatorial	0.6	Plains	0.2
Monsoon–savanna	0.8	Hills	0.4
Arid and semi-arid	0.6	Plateaus	0.5
Temperate	0.4	Medium–high mountains	0.6
Cold	0.6	High (Alpine) mountains	0.8
Arctic	0.4		

See Section 3.1.

TABLE 3.3

Estimated Rates of Anthropic and Natural Geomorphological Activities

Geomorphic Agent	Earth Moved (10^9t yr^{-1})
Man: intentional based on GNP	30
Man: intentional based on energy consumption	35
Man: unintentional	99
Total anthropic	134
Rivers	53
Glaciers	4
Slope processes	1
Wave action	1
Wind	1
Mountain building	44
Deep ocean sedimentation	7
Total natural	111

Source: Hooke (1994, 2000), Haff (2003) and Rózsa (2007).
See Section 3.1.

as agriculture and grazing, which cannot be measured and distinguished easily from natural erosional processes. Hooke (2000) applied two parameters to determine the amount of annually moved earth for the entire world. He assumed that intentional anthropogeomorphic activity in a given region might have linear association with the Gross National Product (GNP) or with the energy consumption levels. He also estimated that about 130×10^9t yr^{-1} of earth can be moved by different human activities, which is even greater than the total earth moved by the natural agents working on the earth surface (Table 3.3). Therefore, at present, anthropic processes are anticipated to be the most dominant surface-modifying agents.

Direct human impact on topography can be marked and mapped by using the anthropogeomorphologic transformation index (R_{AG})—which was formulated by Lóczy and Pirkhoffer (2009) while working on the Hungarian landscape. It is defined as the ratio of the earth movement caused by human activities (Va) and the earth movement caused by natural processes (Vn).

$$R_{AG} = \frac{Va}{Vn}$$

R_{AG} values are assigned to determine and delineate spatial extent of different types of land uses which are variedly affected by human activities. Lóczy and Pirkhoffer (2009) identified six qualitative groups of such variation (Table 3.4).

Goudie (2018) stated that the human factor in geomorphological changes is now ubiquitous and firmly established. He identified 12 major global signatures of anthropic intervention that cause degeneration of the earth surface systems (Table 3.5).

Downs and Piégay (2019) examined the role of human impact in modification of river-catchment processes and resulting response of the river channel morphology across the world. Their study considered planform changes of 25 major rivers of various catchment sizes of the industrial nations during 1880–2005. They found that the changes due to human activity greatly accelerated between 1955 and 1990; after which, the trend is continuing at an increasing rate. They identified that dam construction, land-use alterations, bank protection and instream aggregate mining are the most common human-induced drivers of landscape change. Their effects predominantly involved narrowing of channels, terrace development, reduced bed sediment storage, lower activity rates and simplified channel geometries.

TABLE 3.4
Qualitative Groups of Human Impact

Sl. No.	Qualitative Groups of Human Impact	Selection Criteria/Techniques	Threshold Values
1	No human impact	Natural vegetation inhibits erosional processes	$R_{AG} = 0$, when $Va = 0$–1.5 tonnes ha^{-1} yr^{-1}
2	Accelerated areal erosion/accumulation due to agricultural activity	Volume determined by field measurement	$R_{AG} = 10$, when $Va = 5$–15 tonnes ha^{-1} yr^{-1}
3	Accelerated linear erosion/accumulation due to agricultural activity (gullies)	Volume of material estimated by width, depth and length of erosional gullies and hollow roads	$R_{AG} = 308$, when $Va = 400$ tonnes ha^{-1} yr^{-1}
4	Agricultural landscaping (terracing, drainage, etc.)	Volume of material estimated by measurements of terraced slopes	$R_{AG} = 25$, when $Va = 33$ tonnes ha^{-1} yr^{-1}
5	Quarries	Volume of material estimated by area of quarries (maps) and field measurements (height and shape of walls)	$R_{AG} = 10{,}385$, when $Va = 13{,}500$ tonnes ha^{-1} yr^{-1}
6	Human-controlled geo-environment: Built-up areas, permanent long-term surface management, *viz.* roads, railways	Natural geomorphological processes are completely hindered or compensated due continuous interventions by the stakeholders	$R_{AG} = \infty$ as Vn is absent

Source: After Lóczy and Pirkhoffer (2009).
See Section 3.1.

TABLE 3.5
The Great Anthropogeomorphological Transformations of Recent Decades

Transformation Type	Particulars
The great earth movement	Refers to the earth moved by man, estimated greater than the total earth moved by exogenetic processes (Table 3.1).
The great dust up	Refers to the great increase in atmospheric dust loading due to human alteration of landscape.
Great coastal change	Refers to the change in the behaviour of natural coastal processes by human actions resulting in coastal erosion, siltation of estuaries, etc.
Coral cover—the great degradation	Refers to the global degradation of coral covers globally due to human interventions such as pollution, tourism and sedimentation.
Oyster reefs—the great gastronomic disaster	Refers to the indiscriminate destruction of oyster reefs for harnessing food resource.
The great swamp disaster	Refers to the obliteration and reclamation coastal swamps and mangroves.
Great mass movements	Refers to accelerating mass movements and associated hazards due to modification in land use in the high areas.
Great sedimentation	Refers to the increase in sedimentation in water bodies due to greater earth movement by humans.
Annual floods	Refers to the occurrences of recurring floods due to human modification of hydrology.
Great shakes	Refers to the earthquakes generated by nuclear explosion or by water pressure in dams.
Great melting acceleration	Refers to the melting of glacial and polar ice in an accelerated fashion due to climate change and global warming.
Great proglacial lakes outburst	Refers to the increasing number of glacial lake outburst event due to climate change.

Source: After Goudie and Viles (2016) and Goudie (2018).
See Section 3.1.

On a regional scale, Owen (2017) identified human-induced climate change and construction activities as the major earth surface process of the Himalaya that also cause geohazards. Hudson et al. (2019) detected decline in sediment load and overall channel bed degradation in the Lower Mississippi, United States, occurring in response to the large-scale 20th-century engineering projects. The anthropic intervention in the fluvial processes of the US rivers was also documented by Wohl (2018). Zerboni and Nicoll (2019) stated that in North Africa, the natural processes have changed and became more human-dominated.

Due to increasing human intervention in landscape processes, it is now thought that geomorphologists must consider human factor as a separate earth system process, with the aim of proper management of negative outcomes of the changes (Goudie, 2018; Wohl, 2018).

Szabó (2010) presented a simplified version of Haigh's (1978) classification of anthropogeomorphic processes: (1) direct anthropogenic processes, which include constructive (e.g., urban constructions), excavational (e.g., mining and quarrying) and hydrological (e.g., river management techniques) practices; and (2) indirect anthropogenic processes, which cause acceleration of erosion and sedimentation (e.g., tillage), subsidence (e.g., mining subsidence), slope failure (e.g., quasi-natural landslides) and triggering of earthquakes (e.g., by river dams). Based on the above, Szabó (2010) identified three types of landforms associated with anthropic activities:

i. E-type or *Excavation* landforms produced by excavational processes. These are negative landforms mostly leading to material deficit. Excavation in anthropogeomorphology is the counterpart of natural erosion.

ii. P-type or *Planated* landforms formed by planation processes. This is a unique type of anthropogeomorphological process which cannot be compared with other types of natural geomorphic processes. Through planation, humans can destroy landforms created by them or by nature. This would include flattening of a coastal dune or a settlement unit or smoothening of surface by artificial accumulation like valley filling with debris. In this process, the slope of the surface gets reduced.

iii. A-type or *Accumulation* landforms produced by accumulation. It produces positive landforms, and it is the equivalent of natural aggradation, e.g., flood embankments constructed along the creeks or rivers.

Szabó and Dávid (2006) stated that the direct anthropogenic interventions can be segmented into primary and secondary processes. For example, terracing on a slope for agricultural purpose may be termed as primary, whereas accumulation of mining spoils, which is the by-product of the primary activity of mining excavation, can be regarded as secondary. The indirect anthropogeomorphic processes can also be subdivided into two categories—qualitative and quantitative. Qualitative indirect anthropogenic processes are those which are inherently quasi-natural, like anthropic alteration of the natural geomorphic processes acting on the earth's surface; such as occurrence of a landslide on a modified slope with large numbers of settlements. These impacts are describable in nature and the rate of change cannot be measured. For quantitative indirect anthropogenic processes, the human activities or the resultant landforms do not induce new processes but only change the extent and rate of already operating processes together with their consequences. For example, generation of additional runoff due to deforestation, leading to floods. Here the impact can be expressed by quantifiable measures.

Although many papers have addressed human modifications of landscape and geomorphic processes in India, noteworthy contribution on anthropogenic geomorphology *per se* is difficult to locate and there has been no attempt on this for the present area of study. In West Bengal, Lama (1994) worked on anthropogeomorphological modifications of Darjeeling town and their impacts.

Complete with examples and case studies, the present work deals with systematic analysis of anthropogenic geomorphology of the Lower Deltaic West Bengal (LDWB) region of eastern India comprising the active delta (~97% of LDWB) of the river Bhagirathi–Hugli (or Hooghly). In this chapter, the SoI (Survey of India) terminologies are used, i.e., *Hugli* for the tidal part of the river and *Hooghly* for the district of Indian state of West Bengal located in LDWB.

3.2 LOWER DELTAIC WEST BENGAL AND THE HUGLI RIVER SYSTEM

Defined on the basis of upstream penetration of high tide, the LDWB comprises $20,954\,km^2$ of area in 116 community development (CD) blocks of nine districts in the Indian state of West Bengal (Figure 3.1).

In 2011, the area carried a population of ~45 million at 1,306 persons km^{-2}. The region is shared by two river deltas, the Ganga–Brahmaputra (96.7%) and the Subarnarekha (3.3%). Its geomorphology is mostly determined by stages of delta building related to Holocene sea-level fluctuations, channel shifts and anthropogenic interventions.

The upstream (northern) limit of the LDWB is demarcated on the basis of tidal limits along its trunk river, the Hugli, and its tributaries besides the inlets and estuaries that open directly into the Bay of Bengal (Figures 3.2–3.4). The western boundary of the region is delineated by the edges of plateau-fringe palaeodeltas created by the western tributaries of the Hugli. Its eastern boundary is demarcated by the international boundary between India and Bangladesh.

The upper part of the LDWB is located within the non-tidal upper Ganga Delta, characterised by the presence of meandering rivers, palaeochannels and levees, constituting a floodplain. Here the landscape is mostly developed by non-tidal fluvial processes, even though the tidal surge travels

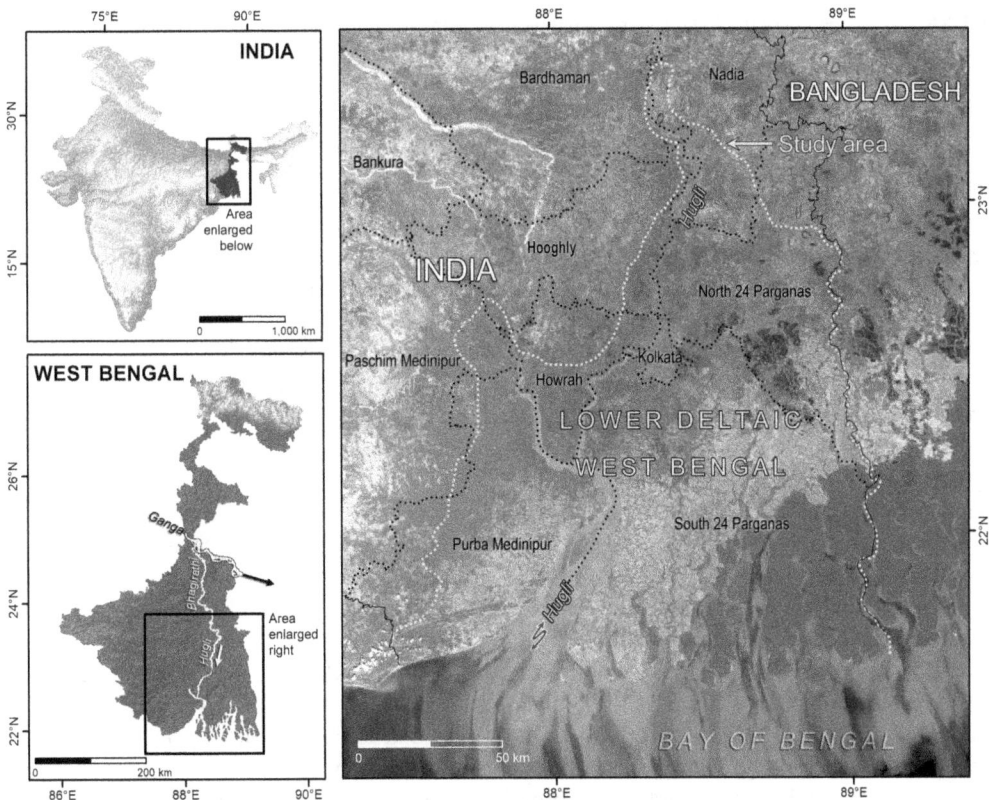

FIGURE 3.1 Location of the LDWB in the southwestern Ganga-Brahmaputra Delta. Its boundaries are demarcated by the tidal penetration along the distributaries of the Ganga and Political boundary of West Bengal (shown by yellow dotted line (white in print version)). Black dotted lines represent district boundaries. (Index maps on the left: SRTMGTOPO30DEM (1 km) of February 2000; FCC on the right: IRS– WiFS data of 3 March 1998.)

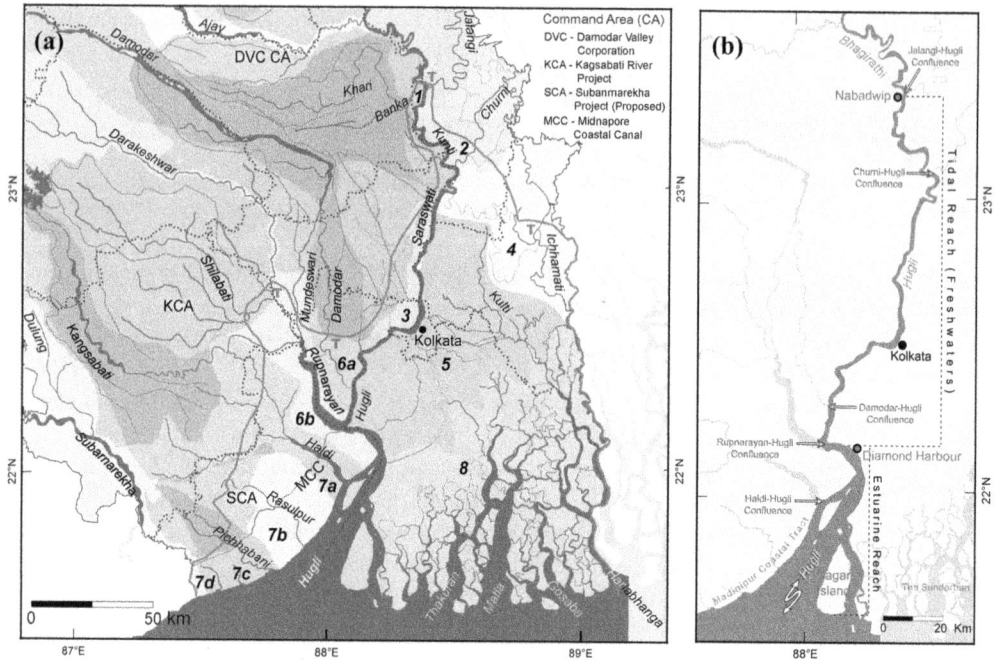

FIGURE 3.2　Location of the LDWB in the southwestern Ganga-Brahmaputra Delta. Its boundaries are demarcated by the tidal penetration along the distributaries of the Ganga and Political boundary of West Bengal (shown by yellow/white dotted lines in online/print version). Black dotted lines represent district boundaries. (Index maps on the left: SRTMGTOPO30DEM (1 km) of February 2000; FCC on the right: IRS– WiFS data of 3 March 1998).

through major channels such as the Rupnarayan, Damodar, Hugli, and Ichhamati (Figure 3.2). The lower part of the LDWB is characterised by tidally active lower Ganga Delta comprising mangrove-covered and reclaimed islands of the Indian Sundarban in the east of the Hugli, and, in the west, the Medinipur coastal plains (Bandyopadhyay et al., 2014). A distinct feature of the Medinipur strand plains are the chenier sand ridges of the eastern Subarnarekha Delta. The altitude of the LDWB varies from ~20 m in its northern levee tops to the sea level at the south. Based on physiographic characteristics, the region can be divided into five zones: (1) interior floodplains, (2) reclaimed Sundarban, (3) non-reclaimed Sundarban, (4) Medinipur coastal plains and (5) plateau-fringe palaeodeltas. Positions of the physiographic zones and drainage basins of the LDWB are shown in Figure 3.4.

The 511 km Bhagirathi–Hugli is the most important river of the deltaic West Bengal. From its off-take, the 241 km long upper non-tidal reach up to Nabadwip (Bhagirathi–Jalangi confluence) in Nadia District is known as the Bhagirathi, while the tidal reach of 270 km downstream of Nabadwip is known as the Hugli and drains the lower delta. As the headwaters of the Bhagirathi–Hugli are in decayed state since the beginning of the 20th century, the lower course of the river is maintained by the tides and its west bank tributaries apart from the 1975 Farakka Barrage Project (FBP), which diverts certain amount of water from the Ganga (Parua, 2010). The Hugli River is characteristically macrotidal with spring tide range of up to 5 m at Diamond Harbour (SoI, 2011). The tides move into the delta twice daily, maintaining the estuaries, creek networks and intervening mangrove islands in the Sundarban. The Hugli shows prominent time–velocity asymmetry in tidal movement, with the flood tide duration of 3 h within the 12.4 h tidal cycle (Sanyal and Chatterjee, 1995). This causes landward movement of sediments, most of which eventually get settled inside the channel (Ghotankar, 1972; Bandyopadhyay, 2000). Similar flood dominance is also seen in the estuaries of the Sundarban (Chatterjee et al., 2013) and the tidal inlets of the Medinipur coast.

In LDWB, the major west-bank tributaries of the Hugli are the Damodar, Rupnarayan and Haldi, while the Churni is the only major tributary from the east. Among the distributaries that have originated in this region, important are the Jamuna, Saraswati, Adi Ganga and Kulti–Bidyadhari. However, the first three channels have completely degenerated at present (Bandyopadhyay et al., 2015).

Based on the drainage characteristics and their anthropic modifications, eight drainage basins can be identified in the LDWB (Figure 3.2). Four of these are to the west of the Hugli: (1) Khari–Hugli, (3) Saraswati–Hugli, (6a–b) Damodar–Rupnarayan–Hugli, and (7a–d) Kangsabati–Haldi–Subarnarekha–Rasulpur–Pichhabani Coastal; and four to the east: (2) Churni–Hugli, (4) Ichhamati–Hugli, (5) Kulti–Hugli, and (8) Hugli–Sundarban coastal. The anthropogenic signatures are different for these basins. For example, in case of the dam-controlled Damodar–Rupnarayan–Hugli, the anthropic modifications are markedly different from the Hugli–Sundarban Coastal Basin (Figure 3.2), where the reclamation of deltaic islands and associated geomorphological aspects are distinct features.

A number of towns and cities of West Bengal including the megacity of Kolkata (erstwhile Calcutta) and two major industrial regions, the Hugli Industrial Region and the Haldia Industrial Region, are developed on the banks of the Hugli, which are served by two major riverine ports of eastern India, Kolkata and Haldia (Figure 3.3).

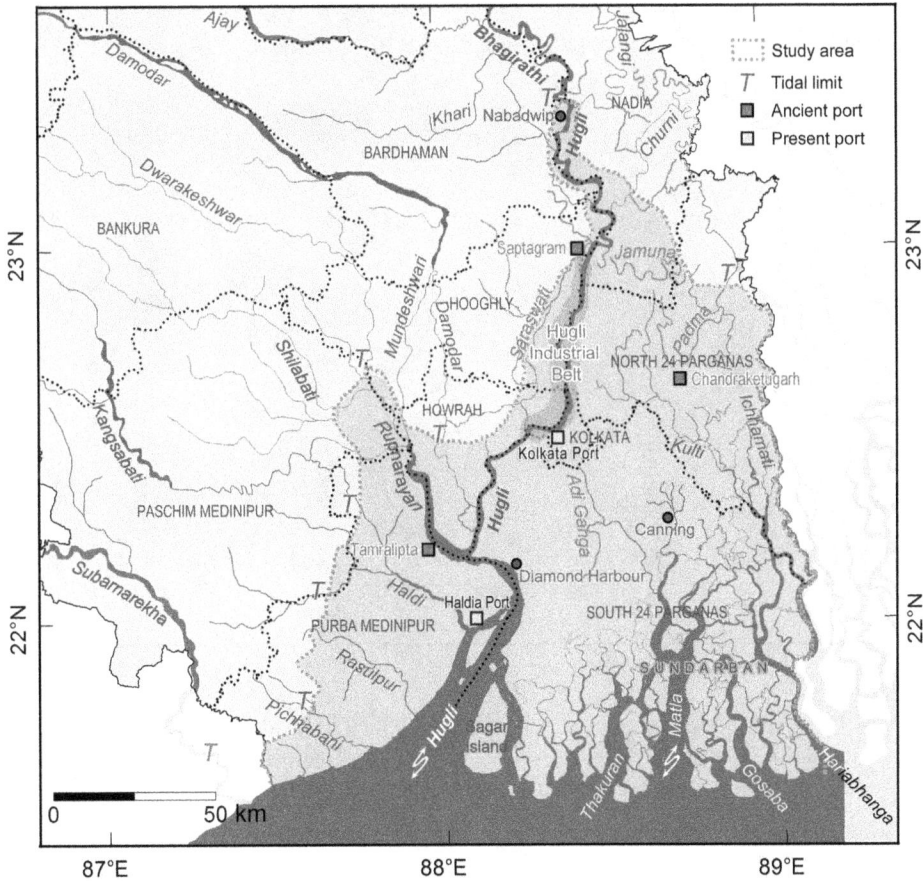

FIGURE 3.3 Important places, principal channels (blue (black in print version)) and palaeochannels (green (grey in in print version)) of the LDWB. Red (black in print version) and yellow (grey in print version) squares represent locations of the ancient and present ports, respectively. *T* (in purple (grey in print version)) represents tidal limit along the rivers. See Section 3.2 for details. (Hydrographic information based on NATMO, 1980, 1988. Modified after Bandyopadhyay et al. (2014).)

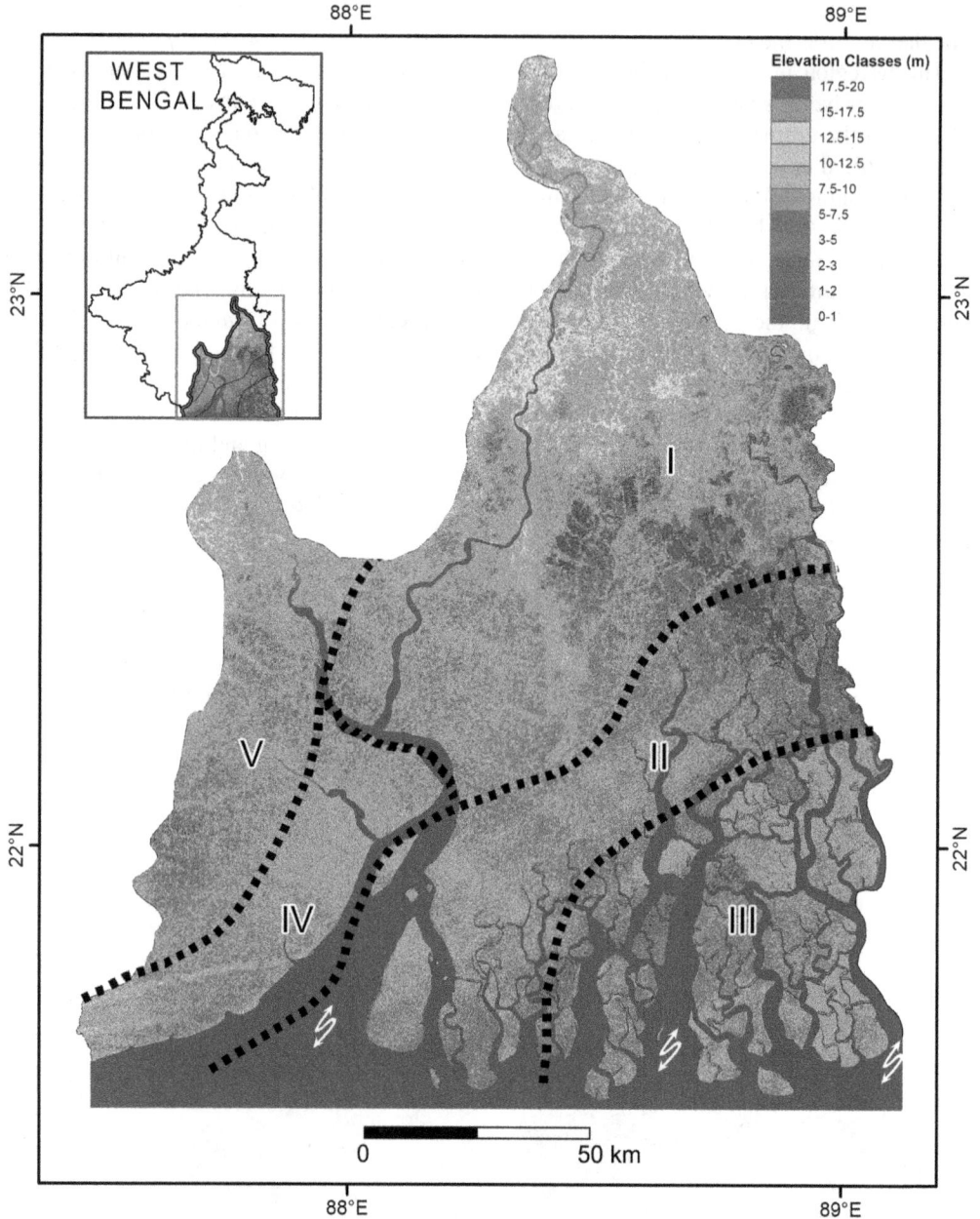

FIGURE 3.4 Physiographic divisions of the LDWB. Zone I: Interior floodplains; II: reclaimed Sundarban; III: non-reclaimed Sundarban; IV: Medinipur coastal plains; V: plateau-fringe palaeodeltas. See Section 3.2 for details. (1 arc-second (30 m) SRTM DEM, February 2000. Modified after Bandyopadhyay et al. (2014).)

The antiquity of anthropic modifications of natural geomorphic system in the LDWB dates back to the ancient and medieval periods (Hunter, 1875; Willcocks, 1930). The presence of ancient towns and ports in this basin such as Tamralipta, Chandraketugarh, Nabadwip and Saptagram still bear the traces of historical signatures of human alterations of natural landscape (Figure 3.3). The economic and urban development in the LDWB, especially along the Hugli, expedited during the British Raj

(Mukerjee, 1938). With the establishment of the first jute mill at Rishra in 1855, the Hugli industrial belt started functioning about 150 years ago (RM, 2018). This resulted in large-scale migration of people from different parts of eastern India. Besides the British, several colonies by other European countries were also established along the Hugli at Serampore (Danish), Chandernagore (French), Chinsurah (Dutch), and Bandel (Portuguese). Presently, the Hugli River and its surrounding areas are the most densely populated parts of India (CoI, 2011). In fact, the region is having one of the highest population densities in the entire country. In 2011, the population density of Kolkata District was 24,306 persons km^{-2} (CoI, 2011). The rural population density of this region is also very high, owing to the presence of fertile alluvial soil (Akter, 2015). The LDWB is traditionally well known for its agriculture as recurring floods used to bring fresh silts almost every year.

For the present study, the index of potential anthropic geomorphology (I_{PAG}), as proposed by Nir (1983) and explained in Section 3.1, is applied to quantify the anthropogeomorphological impact on the region. The I_{PAG} values of the LDWB are calculated on the basis of the 2011 block-level population and literacy statistics obtained from the Census of India (CoI, 2011). The combined value of K_c and K_r is taken as 1.0, as K_c for monsoon–savanna type region is 0.8 and K_r for plains is 0.2 (Table 3.2).

The anthropogenic transformation index values derived from measurements by Lóczy and Pirkhoffer (2009) (Table 3.4) are difficult to apply in the LDWB. It is difficult to categorise a number of land-use features like quarries or agricultural terraces from land-use mapping, as the concerned area is considerably large. Besides, many features such as swamps, natural levees, palaeochannels and coastal dune systems are not entirely human controlled. Therefore, for the present study, four groups of human-modified areas are identified from optical satellite images as (1) areas with little or no human impact, (2) moderate impact, (3) high impact and (4) extremely high impact. A mosaic of four Landsat-8 OLI images of 26 October 2018 (path-138, rows-44 and 45) and 18 November 2018 (path-139, rows-44 and 45) is used for extracting land-use classes in of the LDWB by maximum likelihood classification, which represent the four qualitative groups of human impact (Figure 3.7). Areal statistics of each category are calculated for each sub-basin of LDWB (Figure 3.2, Table 3.6).

Anthropogenic impacts on landscape vary according to the character of the human activities. In this study, the anthropogenic processes and landforms in the LDWB are identified and classified by adopting taxonomies as mentioned by Szabó and David (2006) and Szabó (2010) (Table 3.7).

TABLE 3.6
Sub-Basin-Wise Human Impact Levels in the LDWB (see Figure 3.2)

Sub-Basins (Arranged Approximately Southward)	Area (km²)	Human Impact Levels (% Area)			
		Little	Moderate	High	Extremely High
Khari–Hugli	137	0	13	77	10
Churni–Hugli	399	0	12	81	7
Saraswati–Hugli	867	0	17	44	39
Ichhamati–Hugli	1,538	0	9	88	3
Kulti–Hugli	3,746	0	22	46	32
Damodar–Rupnarayan–Hugli	2,188	0	16	81	4
Kangsabati–Haldi–Subarnarekha– Rasulpur–Pichhabani coastal	3,194	0	10	89	1
Hugli–Sundarban coastal	8,885	37	5	58	0

See Section 3.3.2 for details.

TABLE 3.7

Classification of Anthropogenic Processes and Landforms in Lower Deltaic West Bengal

Type of Intervention	Landform/Process Type	Direct		Indirect		Location/Example
		Primary	Secondary	Generally Measured in Forms (Primarily, Morphographic, Qualitative)	Generally Measured in Trends (Primarily Morphometric, Quantitative)	
Montanogenic (mining)	Excavation	Sand mining from river beds, brick kilns	—	River bank collapse	Change in river planform	Western distributaries of the Hugli River, e.g., Damodar, Kangsabati
	Planation	—	—			
	Accumulation	Induced tidal sedimentation	Brick kilns	Sediment trapping in sediment ponds	Obstructing sediment inputs to the coastal system	Lower Hugli Basin, Hugli–Ichhamati Basin
Industrogenic (industrial)	Excavation	—	—			
	Planation	Industrial parks/new industries	—	Raised planation	Obstructing natural surface runoff	Along national/state highways
	Accumulation	Dumping of raw materials	Coal dumps	Ash ponds	Discharge of sediments in rivers	Kolaghat Thermal Power Plant
Urbanogenic (settlements)	Excavation	Port, canal				Kolkata Port
	Planation	*Satellite towns/new towns	#Unscientific and non-planned parts of cities	*Raised planation #Waterlogging	*Obstructing natural surface runoff and groundwater recharge	*Bidhannagar, Rajarhat #Amherst Street and Behala in Kolkata
	Accumulation	Solid-waste disposal sites	—	Waste hills	Obstructing natural surface runoff and pollutant infiltration	Dhapa and Dum Dum landfills

(Continued)

TABLE 3.7 (Continued)
Classification of Anthropogenic Processes and Landforms in Lower Deltaic West Bengal

Type of Intervention	Landform/ Process Type	Direct		Indirect		Location/Example
		Primary	Secondary	Generally Measured in Forms (Primarily, Morphographic, Qualitative)	Generally Measured in Trends (Primarily Morphometric, Quantitative)	
Traffic (transport)	Excavation	*Road cut	#Culverts	*Roadside excavation channels (*Nayanjuli*)	*#Obstructing natural surface runoff; disastrous in case of floods	*National highways #Lower Damodar Basin
	Planation	Airfield	—	Elevated planation	Obstructing natural surface runoff	Netaji Subhash Chandra Bose International Airport
	Accumulation	*Elevated embankments for roads and railways	*Bridges, tunnels, fly-overs, subways #Multiple layers of concrete roads—built over the older ones	*Raised platforms (embankment) above flood levels #Waterlogging	*Obstructing natural slopes and surface runoff; disastrous in case of floods; change in river planform	*National Highways, Eastern and South Eastern Railways # Avenues in North Kolkata
Water management (hydrologic)	Excavation	*Drainage canals ^Artificial dredging ~Artificial water body	#Underground drainage networks	*Excavated channels for drainage ^Change in river planform ~Water body and ponds	*#~Modification of natural surface and sub-surface runoff	*Eastern drainage canal #Underground drainage network, Kolkata ^Hugli River near Haldia port ~Subhas Sarobar, Kolkata
	Planation	Water body fill up	—	—	Modification of natural surface runoff and waterlogging	Ward 94 (Tollygunge), Kolkata
	Accumulation	Marginal embankments	—	Reclaimed islands	Degeneration of tidal channel; rise of effective sea level; coastal floods	Reclaimed parts of the Indian Sundarban

(Continued)

TABLE 3.7 (Continued)
Classification of Anthropogenic Processes and Landforms in Lower Deltaic West Bengal

Type of Intervention	Landform/Process Type	Direct		Indirect		Location/Example
		Primary	Secondary	Generally Measured in Forms (Primarily, Morphographic, Qualitative)	Generally Measured in Trends (Primarily Morphometric, Quantitative)	
Agrogenic (agricultural)	Excavation	Irrigation canals	—	Excavated channels for irrigation	Modification of natural surface runoff	Midnapore canal
	Planation	Agricultural tillage	—	—	Modification of natural slope	Agricultural fields across the basin
	Accumulation	*River embankments and sluices ^Farmland dividers (*Aal*) ~Aquaculture ponds	—	*Flood control ^Waterlogging for paddy ~Destruction of ecosystem	*Obstructing natural river channel adjustment; disastrous when breached	*Along all major rivers ^Paddy fields ~Prawn farms (*Bheri*) in the Nayachar Island
Tourism and sports (recreational)	Excavation	Water bodies	—	Water body		Rabindra Sarobar, Kolkata
	Planation	Parks, Stadium	—	—	Modification of natural slopes and surface runoff	Eco Park, Rajarhat, Vivekananda Yuba Bharati Krirangan (*Salt Lake Stadium*)
	Accumulation	Strand concrete embankments and beautification	—	—	Obstructing natural river channel adjustment	Chandernagore strand, Haldia riverside

See Section 3.3.2 for details.

3.3 ANTHROPOGENIC SIGNATURES ON FORMS AND PROCESSES

3.3.1 LDWB: I_{PAG} ANALYSIS

In this study, the index of potential anthropogeomorphology (I_{PAG}) is derived and analysed for the LDWB, which is spread across 116 CD blocks in nine districts of West Bengal (Figure 3.5). The maximum anthropogenic impact on the natural landscape can be found in the central part of the region, which consists of densely populated Kolkata megacity and the Hugli industrial belt. In this part, both banks of the Hugli River are similarly affected by human activities, with all the CD blocks exhibiting I_{PAG} values of more than 0.40. This region is characterised by continuous built-up area and non-agricultural activities. The maximum I_{PAG} values can be found in Kolkata (block Id in Figure 3.6: 31, I_{PAG}: 0.60), and Uluberia municipal area in the Howrah District (block Id: 43, I_{PAG}: 0.65). There are a few blocks in the other parts of the basin, which also have significantly high

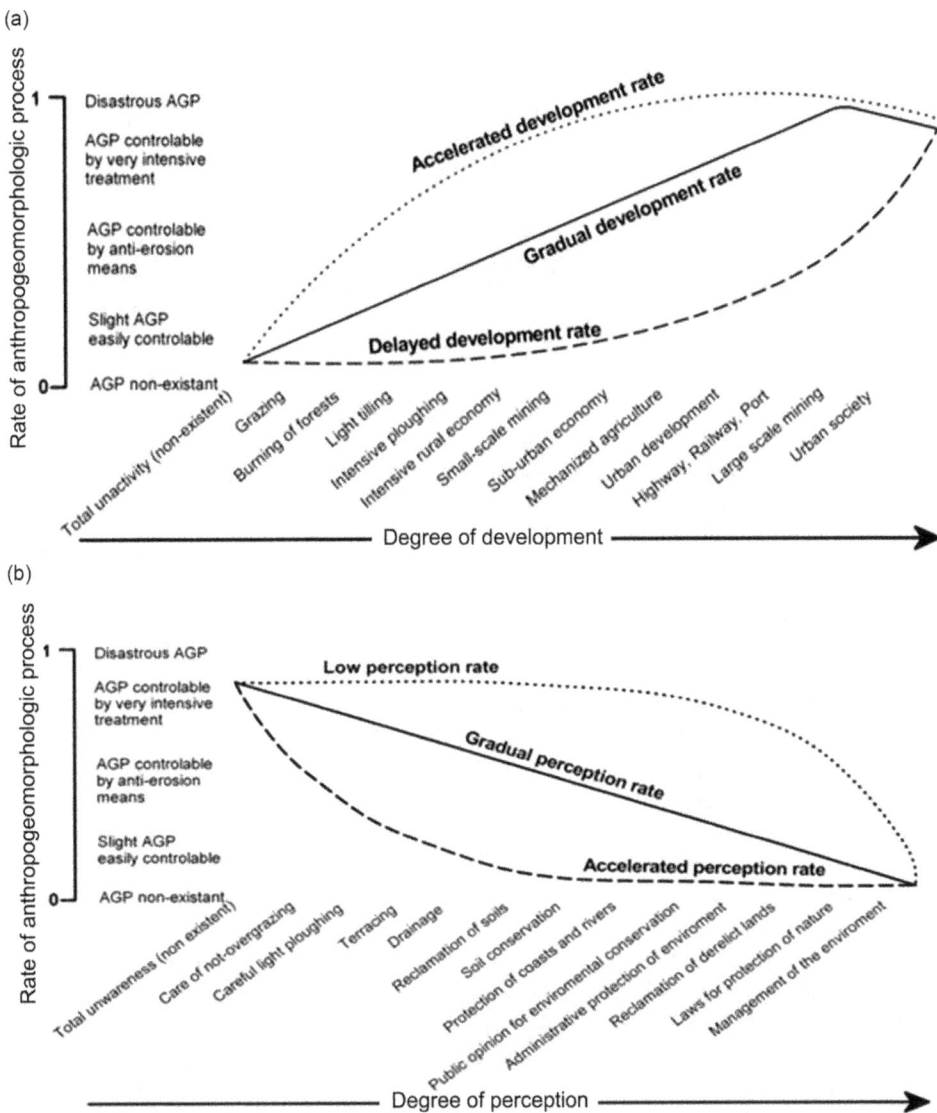

FIGURE 3.5   (a) Association between DD and anthropogeomorphic processes; (b) association between DP and anthropogeomorphic processes. See Section 3.1 for details. (After Nir (1983).)

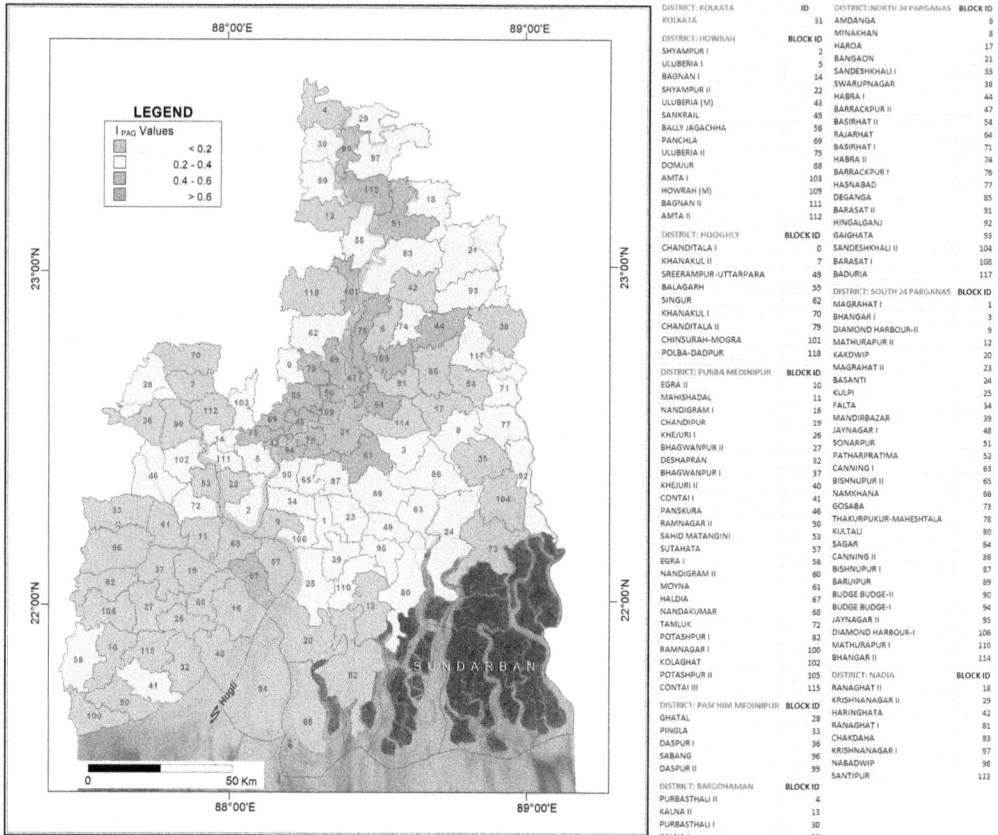

LEGEND

I_PAG Values
- < 0.2
- 0.2 - 0.4
- 0.4 - 0.6
- > 0.6

DISTRICT: KOLKATA	ID
KOLKATA	31

DISTRICT: HOWRAH	BLOCK ID
SHYAMPUR I	2
ULUBERIA I	5
BAGNAN I	14
SHYAMPUR II	22
ULUBERIA (M)	43
SANKRAIL	45
BALLY JAGACHHA	56
PANCHLA	69
ULUBERIA II	75
DOMJUR	88
AMTA I	103
HOWRAH (M)	109
BAGNAN II	111
AMTA II	112

DISTRICT: HOOGHLY	BLOCK ID
CHANDITALA I	0
KHANAKUL I	7
SREERAMPUR-UTTARPARA	49
BALAGARH	55
SINGUR	62
KHANAKUL I	70
CHANDITALA II	79
CHINSURAH-MOGRA	101
POLBA-DADPUR	118

DISTRICT: PURBA MEDINIPUR	BLOCK ID
EGRA II	10
MAHISHADAL	11
NANDIGRAM I	16
CHANDIPUR	19
KHEJURI I	26
BHAGWANPUR II	27
DESHAPRAN	32
BHAGWANPUR I	37
KHEJURI II	40
CONTAI I	41
PANSKURA	46
RAMNAGAR II	50
SAHID MATANGINI	53
SUTAHATA	57
EGRA I	58
NANDIGRAM II	60
MOYNA	61
HALDIA	67
NANDAKUMAR	68
TAMLUK	72
POTASHPUR I	82
RAMNAGAR I	100
KOLAGHAT	102
POTASHPUR II	105
CONTAI III	115

DISTRICT: PASCHIM MEDINIPUR	BLOCK ID
GHATAL	28
PINGLA	33
DASPUR I	36
SABANG	96
DASPUR II	99

DISTRICT: BARDDHAMAN	BLOCK ID
PURBASTHALI II	4
KALNA II	13
PURBASTHALI I	30
KALNA I	59

DISTRICT: NORTH 24 PARGANAS	BLOCK ID
AMDANGA	6
MINAKHAN	8
HAROA	17
BANGAON	21
SANDESHKHALI I	35
SWARUPNAGAR	38
HABRA I	44
BARRACKPUR II	47
BASIRHAT II	54
RAJARHAT	64
BASIRHAT I	71
HABRA II	74
BARRACKPUR I	76
HASNABAD	77
DEGANGA	85
BARASAT II	91
HINGALGANJ	92
GAIGHATA	93
SANDESHKHALI II	104
BARASAT I	108
BADURIA	117

DISTRICT: SOUTH 24 PARGANAS	BLOCK ID
MAGRAHAT I	1
BHANGAR I	3
DIAMOND HARBOUR-II	9
MATHURAPUR-II	12
KAKDWIP	20
MAGRAHAT II	23
BASANTI	24
KULPI	25
FALTA	34
MANDIRBAZAR	39
JAYNAGAR I	48
SONARPUR	51
PATHARPRATIMA	52
CANNING I	63
BISHNUPUR II	65
NAMKHANA	66
GOSABA	73
THAKURPUKUR-MAHESHTALA	78
KULTALI	80
SAGAR	84
CANNING II	86
BISHNUPUR I	87
BARUIPUR	89
BUDGE BUDGE-II	90
BUDGE BUDGE-I	94
JAYNAGAR II	95
DIAMOND HARBOUR-I	106
MATHURAPUR I	110
BHANGAR II	114

DISTRICT: NADIA	BLOCK ID
RANAGHAT II	18
KRISHNANAGAR II	29
HARINGHATA	42
RANAGHAT I	81
CHAKDAHA	83
KRISHNANAGAR I	97
NABADWIP	98
SANTIPUR	113

FIGURE 3.6   The index of potential anthropic geomorphology (I_{PAG}) of different CD blocks located in the LDWB. The population and literacy statistics pertaining to this were obtained from CoI (2011). Since any kind of human establishment is prohibited in the Sundarban National Park, the region has been kept away from the calculation of the I_{PAG} values. See Section 3.3.1 for details. (Background images: Landsat-8 OLI (Path-138, Row-45), 18 February 2015; Landsat-8 OLI (Path-139, Row-45), 27 February 2015.)

I_{PAG} owing to the presence of densely populated urban centres, *viz.* Habra-I in North 24 Parganas District (block Id: 44, I_{PAG}: 0.44); Ranaghat-I (block Id: 81, I_{PAG}: 0.46), Santipur (block Id: 113, I_{PAG}: 0.46) and Nabadwip (block Id: 98, I_{PAG}: 0.50) in Nadia District; and Haldia (block Id: 67, I_{PAG}: 0.45) in Purba Medinipur District. For example, the high I_{PAG} value for Haldia can be attributed to the presence of a functional port (Haldia Dock Complex under Kolkata Port Trust), an oil refinery and several petrochemical manufacturing units.

All the other CD blocks in the study area are characterised by rural landscape and dominance of farming activities. Most blocks in the districts of North 24 Parganas and South 24 Parganas exhibit I_{PAG} values between 0.20 and 0.40. These areas are the fluvially dominated northern parts of the LDWB and the tidally active reclaimed southern sections, respectively. On the other hand, 76% of the blocks located in the Purba Medinipur District (including five in the Medinipur coastal plains) and 70% of the blocks surrounding the Sundarban in the North 24 Parganas and South 24 Parganas Districts have I_{PAG} values below 0.20. The lowest I_{PAG} of 0.10 can be found in four blocks of the Purba Medinipur District—Bhagwanpur-II (block Id: 27), Deshapran (block Id: 41), Ramnagar-II (block Id: 50) and Contai-III (block Id: 115).

Taken together, 47 (40.5%) blocks of the LDWB have I_{PAG} levels below 0.20 (19 in Purba Medinipur, 8 each in North 24 Parganas and South 24 Parganas, 4 in Paschim Medinipur, 3 in Hooghly, 2 each in Howrah and Bardhaman, and 1 in Nadia District), 46 (39.7%) blocks have I_{PAG} levels between 0.20 and 0.40 (18 in South 24 Parganas, 8 in North 24 Parganas, 5 each in Purba Medinipur and Howrah, 4 in Nadia, 3 in Hooghly, 2 in Bardhaman and 1 in Paschim Medinipur District); the rest (23 blocks, 19.8%) possess I_{PAG} values of above 0.40 (seven in Howrah, five in North 24 Parganas, three each in South 24 Parganas, Hooghly and Nadia, one in Purba Medinipur and the Kolkata City).

3.3.2   LDWB: Extent of Anthropic Impact

The different qualitative groups of human impact obtained from the land-use land-cover patterns (Figure 3.7) indicate maximum concentration of anthropic activities in Kolkata and its suburban areas. The 'extremely high-impact group' is also spread across the adjacent districts of Howrah (urban centres: Howrah City, Uluberia), Hooghly (Bally, Serampore, Chandernagore, Chinsurah, Bandel), North 24 Parganas (Dum Dum, Barasat, Panihati, Barrackpore, Naihati), South 24 Parganas (Pujali, Sonarpur, Baruipur) and Nadia (Kalyani). Similar concentration can also be seen at a number of smaller urban centres of the LDWB, e.g., Tamluk, Haldia and Contai in Purba Medinipur District; Ghatal in Paschim Medinipur District; Diamond Harbour and Jaynagar in South 24 Parganas District; Habra and Basirhat in North 24 Parganas District; Ranaghat, Santipur and Nabadwip in Nadia District and Kalna in Bardhaman District. This impact group, mainly consisting of built-up areas, has completely modified the land-based hydrological regimes such as runoff and throughflow, and accounts for 6% of the land area of the LDWB.

Apart from the urban centres, the human-controlled geo-environment of the LDWB is dominated by farming and aquaculture activities that cover 66% of its area (high impact: manly agriculture). Modification of the geomorphology of this region is generally restricted to alteration of water distribution system through construction of canals, transportation corridors and embankments apart from formation of farmlands and aquaculture ponds. Within the LDWB's agricultural areas, somewhat lower levels of human impacts can chiefly be delineated in the chenier dunes of the Purba Medinipur District and along the natural levees and palaeochannels of the Adi Ganga (South 24 Parganas District), Ichhamati, Jamuna (North 24 Parganas) and Saraswati (Hooghly and Howrah). Isolated patches of farm forests and protected forests are also included in this category. It comprises 13% of the studied region (moderate impact: manly agriculture and plantation). Finally, little or no human impact is seen in the non-reclaimed, mangrove-covered tidal islands of the Sundarban (15% of LDWB). This connotes that about 85% of the LDWB comes under direct influence of human activities and is subjected to extensive anthropogeomorphic modifications.

The sub-basin-wise area coverage statistics of human impact zones are summarised in Table 3.6. The Kolkata–Hugli–Kulti and the Saraswati–Hugli basins are most affected by anthropic modifications as they contain most of the large urban centres of this area. The land-use data of 116 blocks covering the LDWB is also generated, showing the nature of human modification of landscape in each block (Figure 3.7).

A spatial comparison between human impact on the LDWB with its I_{PAG} reveals close correspondence between the two (Figures 3.6 and 3.7). As I_{PAG} levels depend on UP, the built-up units match closely with high I_{PAG} values (>0.40). The general assumption of the I_{PAG} index—i.e., the negative modification of landscape will be more in areas with high level of illiteracy—is seen in case of Uluberia municipal area which has the highest I_{PAG} score of the LDWB (0.65), where built-up area is much less compared to the megacity Kolkata (I_{PAG} = 0.60). Here the literacy factor defined the potentiality (probability) of human modification of natural landscape as shown by high I_{PAG}, whereas the classified image data is portraying the actual situation. In general, the impact areas primarily consisting of agricultural lands exhibit moderate and low I_{PAG} levels (below 0.40).

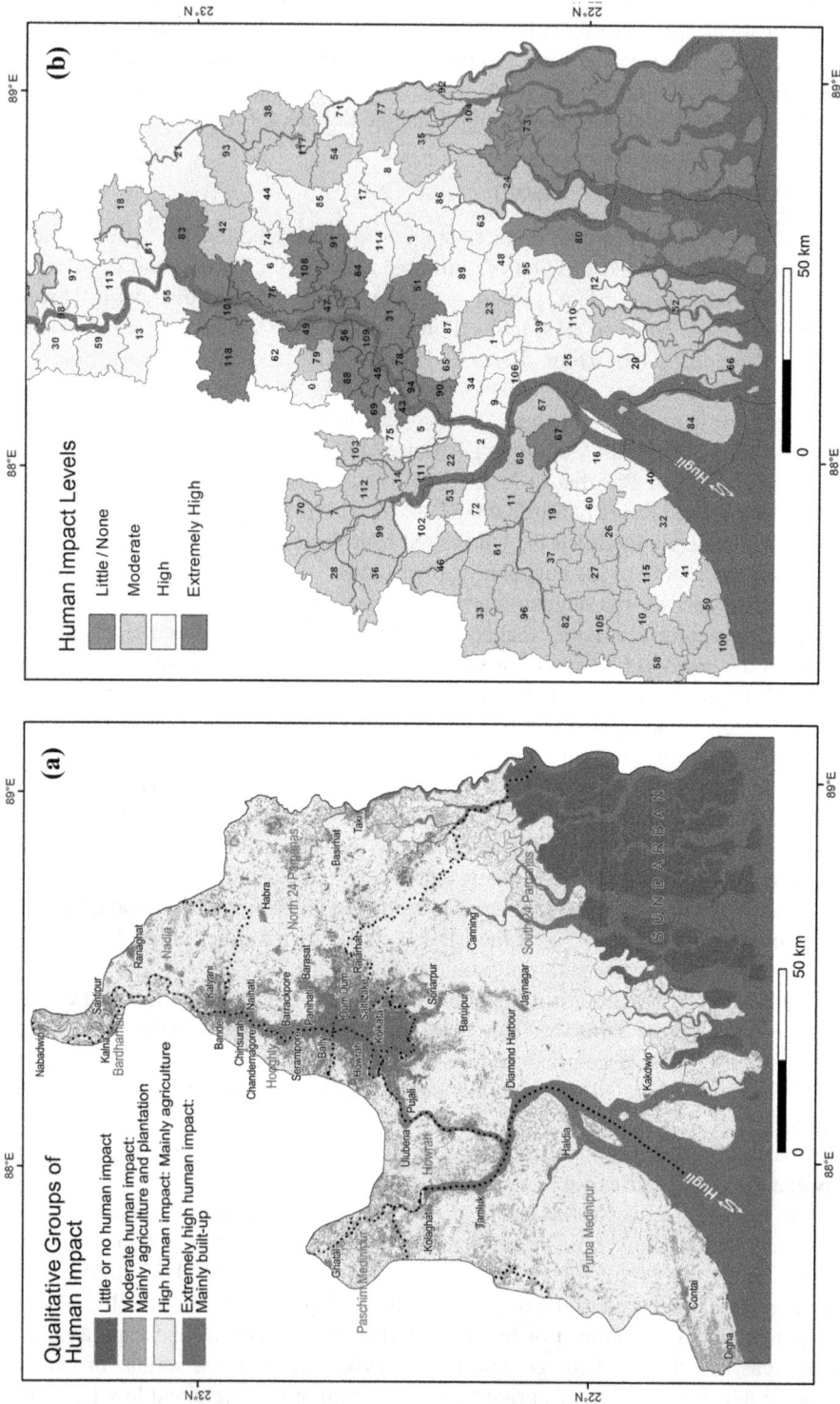

FIGURE 3.7 (a) The magnitude of human impact in different parts of the tidal LDWB attained by maximum likelihood image classification. The represented groups were obtained by amending the original index proposed by Lóczy and Pirkhoffer (2009). (Landsat-8 OLI image (Path-138, Rows-44 and 45), 26 October 2018; Landsat-8 OLI image (Path-139, Rows-44 and 45), 18 November 2018). (b) CD block-wise (see Figure 3.6 for block names) representation of the magnitude of human impact located in the LDWB. See Section 3.3.2 for details.

3.3.3 LDWB: NATURE OF ANTHROPIC IMPACT

The geomorphic impacts of the human activities in the LDWB can be classified into three primary groups: excavation, planation and accumulation (Table 3.7). More specifically, seven types of human interventions can be identified following the anthropic landforms and processes taxonomy suggested by Szabó et al. (2010): (1) montanogenic (mining), mostly affecting the river beds and banks; (2) industrogenic (industrial), characterised by the construction of highly mechanised entities that hinder surface and channel runoff; (3) urbanogenic (settlements), characterised by the formation of continuous impervious surface that may cause impoundment following heavy rains and hamper groundwater recharge; (4) traffic (transport), associated with raised surfaces like railway embankments; (5) water management (hydrologic), involving modification of natural drainage systems; (6) agrogenic (agricultural), causing the formation of excavated channels and agricultural bunds, modification of land use for aquaculture; and (7) tourism and sports (recreational), associated with modification of natural surfaces and water bodies. Each of these interventions is represented by different sets of direct and indirect anthropogenic processes, which are shown with examples in Table 3.7. The industrogenic, urbanogenic, traffic, tourism- and sports-related interventions are closely connected with the built-up geo-environment, whereas montanogenic and agrogenic interventions are primarily associated with agriculture-based geo-environment. In LDWB, both groups of the human-controlled geo-environments are linked with the water management-related interventions. The above types of anthropogeomorphic impacts on the LDWB are illustrated through case studies in the following section.

3.3.4 CASE STUDIES

3.3.4.1 Montanogenic Intervention: Brick–Kilns on the Lower Sections of the Hugli and Ichhamati Courses

As of 2019, a number of brick manufacturing units are situated on the Hugli and Ichhamati of the lower part of the LDWB, most of which dig narrow channels from the rivers into excavated areas adjacent to the river bank during the monsoons (June–September), where the incoming flow from the river deposits sediments. The channels are closed in the post-monsoon season when the 0.3- to 1-m thick accumulated sediments are quarried for brick making (Figure 3.8). As the kilns are located very close to the channels, they disturb the river bank morphology, sometimes leading to river bank erosion.

3.3.4.2 Industrogenic Intervention: Haldia Industrial Region

A number of large-scale industries are seen in the LDWB. Located near the confluence of the Haldi and Hugli Rivers (Figure 3.9), the 85 km² Haldia Industrial Region is fast emerging as a major industrial hub in the eastern India. Establishment of the hub is related to the development of the Haldia Dock Complex, which was founded in the 1960s to counter the navigational problems faced by the Kolkata Port. This region primarily depends on the port and consists of a large-scale petrochemical industrial unit as well as allied light industries since the late 1970s. There are more than 200,000 residents in this industrial township (CoI, 2011); more than one-third are directly or indirectly engaged with the activities related to the Haldia Dock Complex and industrial region. The last two decades saw doubling of the township population.

Physiographically, a part of the Medinipur Coastal Plain, Haldia and its surroundings were completely modified, resulting into the formation of excavated (port terminal waterway), planated (industrial units) and accumulated (riverfront embankments) surfaces.

3.3.4.3 Urbanogenic Intervention: Development of Settlements and Landfilling by Material Disposal

3.3.4.3.1 Salt Lake (Bidhannagar) and Rajarhat–New Town

Centred on Kolkata megacity, rapid urban growth is recorded in the LDWB since India's independence in 1947, which expedited in the last two decades resulting in emergence of a number

FIGURE 3.8 The tidally active Ichhamati course, surrounded by brick–kilns (orange). Oblique aerial photograph dated 12 September 2011. See Section 3.3.4.1 (Photo: Sunando Bandyopadhyay.)

of urban centres and satellite towns. Between 1980 and 2010, population and built-up area of the Kolkata region increased by of 64% (Cox, 2012) and 170% (Bhatta and Giri, 2012), respectively. During 2017–2018, the residential sales grew 280% in Kolkata, which is highest in the country (Moneycontrol, 2018).

The Salt Lake Township (*aka* Bidhannagar), as the name suggests, was built out of swampy marshlands and developed as a satellite city located immediately east of Kolkata. Some 10.2 km² of the low-lying swamps were filled up during the 1960s (Figure 3.10) with dredged materials from the Ghusuri Sand Bank of the Hugli (Ghosh and Sen, 1987). The original character of the landscape is now totally transformed into a brick and mortar setting hampering natural drainage. As a result, part of the township, especially the Sector V, called the IT capital of Eastern India, gets flooded during storms and in monsoon season (TT, 2015; NDTV, 2018). Though ways are adapted to avoid waterlogging, they involve huge cost and labour and are only partly successful (TT, 2017).

The 40 km² Rajarhat–New Town is another rapidly growing planned township in the North 24 Parganas District, located north-east of Kolkata, adjacent to Bidhannagar. Previously, the region was comprised of cultivable lands and water bodies. Between 1998 and 2005, these lands were acquired (Sengupta, 2008) and reformed in a premeditated manner, resulting in large-scale land-use changes. By comparing the land-use patterns of this region (Figure 3.11), it is found that the proportions of farmlands and water bodies have decreased significantly between 1990 (28.7% and 9.3%, respectively) and 2015 (17.9% and 5.3%, respectively). On the other hand, the proportion of built-up area increased rapidly during these years (5.1% in 1990 to 22.4% in 2015). Though Rajarhat–New Town is a recently developed planned city, several parts of this region undergo severe monsoon waterlogging lasting for months (ABP, 2015, 2018). The main concern here is illegal filling up of water bodies for real-estate development. It is reported that even the agency in charge of planning and development

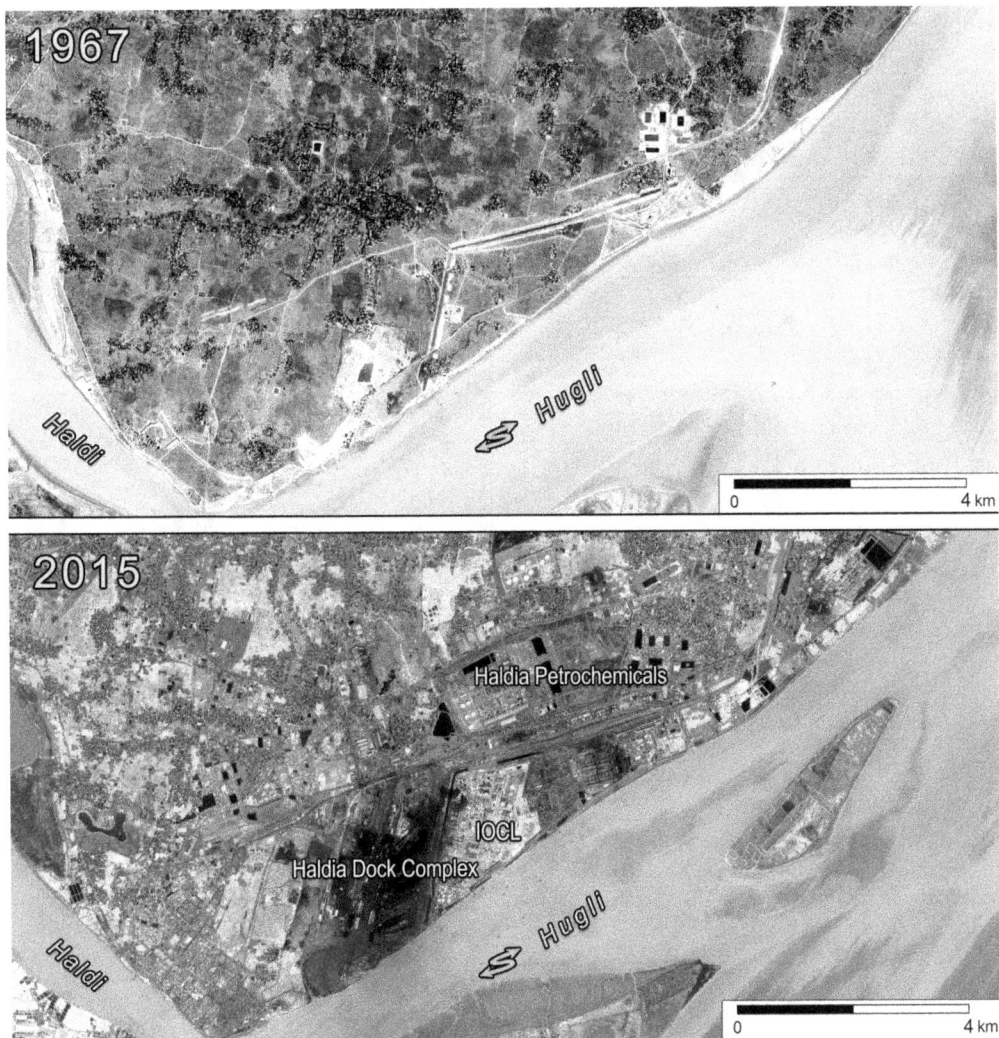

FIGURE 3.9 The Haldia Dock Complex and surroundings. The change is shown between 1967 and 2015. IOCL stands for Indian Oil Corporation Limited. See Section 3.3.4.2. (Corona KH4A Image (DS1038-2102DA-187) of 21 January 1967 and Resourcesat-2 L4fmx FCC (path-108, row-067, subscene-A, 09 March 2015.)

of Rajarhat Township—Housing Infrastructure Development Corporation (HIDCO)—unlawfully filled some 33 water bodies above $405\,m^2$ (Karmakar, 2014).

3.3.4.3.2 Dhapa and Dum Dum landfills

Dhapa, located on the eastern fringe of Kolkata, is the area where the solid wastes of the entire city are dumped (Figure 3.12). The 23-ha Dhapa landfill was initially set up in 1941 and the rate of disposal peaked up in the 1980s (Dutta, 2017; Roy, 2018). About 80% of the 5,000 metric tonnes of daily solid waste produced by the Kolkata residents used to be dumped into the Dhapa landfill which is now exhausted (Dutta, 2017). The garbage dump at Dhapa measures 35 m in relative height and $2 \times 10^6 m^3$ in volume (WBPCB, 2014). A similar type of solid-waste disposal landfills can be found in the northern suburbs of Dum Dum, where more than 500 metric tonnes of solid wastes are dumped daily in a 9 ha land since the late 1990s (Ghosal, 2016). Both of these sites obstruct natural surface runoff and facilitate pollutant infiltration into the ground water.

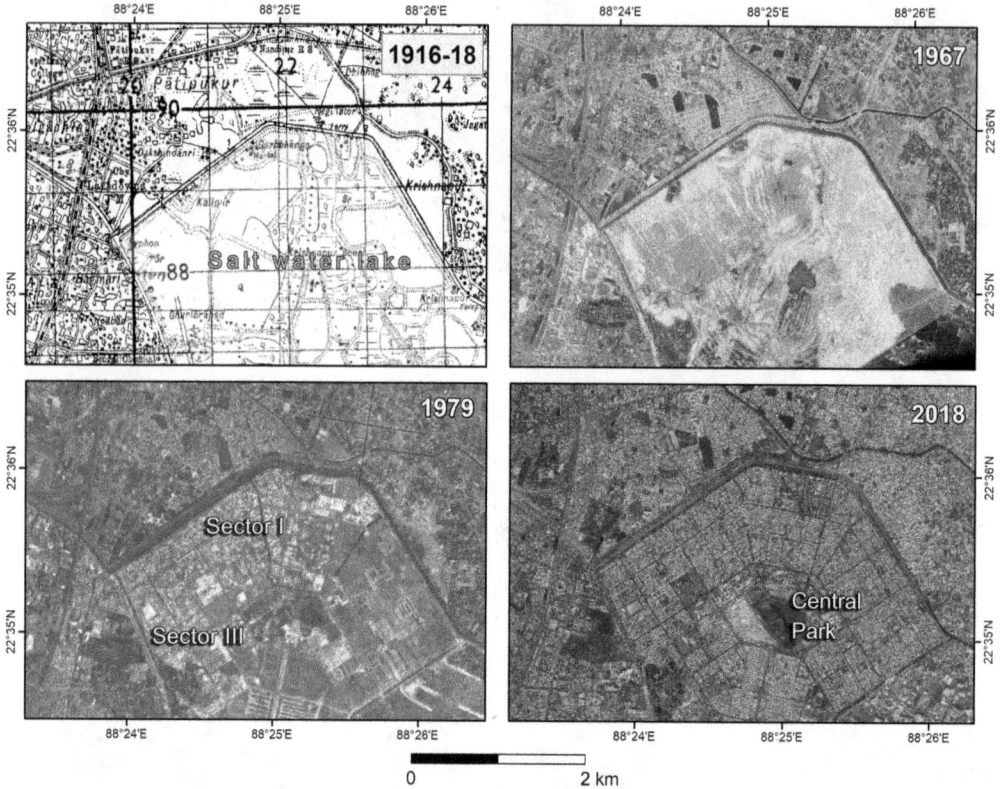

FIGURE 3.10 The transformation of the Salt Lake region in the last 100 years. 1916–1918: Presence of expansive swampy region south of the canal system; 1967: full-fledged land-filling of the swamps; 1979: development of sectors 1 and 3 (western part of the township), 2018: complete dominance of the built-up surfaces, barring the Central Park area at the centre. See Section 3.3.4.3.1. (1916–1918: Survey of India topographical map # 79B/6; 1967: Corona KH4A DS1038-2102DA 182B image, 21 January 1967; 1979: Corona KH9-15 DZB1215-500298L00-5001 image, 19 May 1979; 2018: Google Earth image, 8 May 2018.)

3.3.4.4 Transport (Traffic) Intervention: Railway Embankments

Opened in 1854 (ER, 2016), 361 km of railway tracks run parallel to the Hugli River in the LDWB, with 118 stations. The railway network serving Kolkata is a classic example of anthropogenic accumulation processes that involve the constructions of embankments, bridges, viaducts, fly-overs and subways. These accumulated landforms, by modifying natural slopes, greatly hindered natural drainage and made waterlogging commonplace along these tracks. The tracks especially lying between Tikiapara and Ramrajatala (South Eastern Railway, Howrah Division), between Bally and Bandel (Eastern Railway, Howrah Division) and between Sealdah and Dum Dum (Eastern Railway, Sealdah Division) are positioned up to 5 m higher than the surroundings. Many of the subways beneath these tracks remain waterlogged during the entire rainy season (Figure 3.13).

3.3.4.5 Hydrologic Interventions: Intervening the Riverine and Tidal Systems

3.3.4.5.1 Sediment Supply and Coastal Erosion

As mentioned earlier, the LDWB receives sediment-laden freshwater mainly through the Bhagirathi–Hugli River System. The rain-fed western tributaries of the Bhagirathi–Hugli originating from Chhotanagpur plateau used to carry a large amount of sediments during the monsoons; but their impoundment by the large river valley projects and several minor irrigational schemes (www.wbiwd.gov.in) markedly reduced their sediment contribution. For example, monsoonal sediment discharge

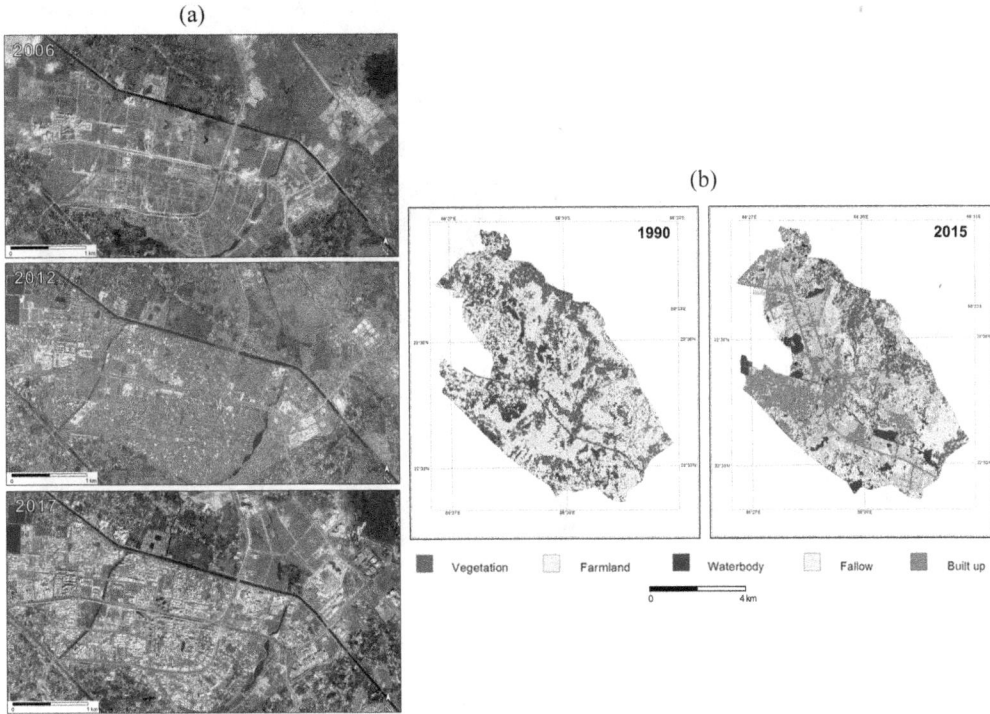

FIGURE 3.11 The recent changes in the Rajarhat–New Town region represented by Google images (a) and land-use land-cover classified maps (b). See Section 3.3.4.3.1. (Google Earth image of 24 October 2006, 3 May 2012 and 6 February 2017 (left panel); Landsat-4 TM image (Path-138, Row-44), 14 January 1990, Landsat-8 OLI image (Path-138, Row-44), 27 February 2015 (right panel).)

through the Damodar reduced by 40% between its pre-dam and post-dam years (Bhattacharyya, 1998). The major river projects on the Bhagirathi–Hugli tributaries include the Damodar Valley Corporation (DVC), Mayurakshi Reservoir Project and Kangsabati Reservoir Project. The low input from these headwaters may have contributed to the erosion of the LDWB coastline (Bandyopadhyay and Bandyopadhyay, 1996).

3.3.4.5.2 Navigational Dredging
The Kolkata and Haldia ports are two of the most important riverine ports of India. The river pilotage distance to Kolkata from open waters of the Bay of Bengal through the Hugli is 148 km, the longest in the world (Mitra, 2018). Since its inception in 1870, navigation to Kolkata through the Hugli is difficult because of the presence of several shoals and islands (Sen, 2019) (Figure 3.14). Although the navigational conditions of the Hugli improved somewhat after commissioning of the 1975 FBP, the channel is now maintained by regular dredging by Dredging Corporation of India (DCI) as the Hugli transmits an annual runoff of approximately 493 km^3 and carries about 616×10^6 t of suspended solids to the estuary mouth (Qasim et al., 1988 in Prasad et al., 2014). To maintain the ports at Kolkata and Haldia, average annual maintenance dredging is estimated at 25 MCUM (Dubey, 2011). During 2005–2015, DCI dredged some 228 MCUM of sediments (http://dredge-india.nic.in/projects1.html retrieved on 06 May 2019) and dumped them in selected in-channel and onshore/offshore disposal sites (Prasad et al., 2014). Such huge human alterations of the river bathymetry have major impacts on fluvio-tidal processes, estuarine hydrology and hydrographic changes such as erosion and accretion of the estuarine bars and islands.

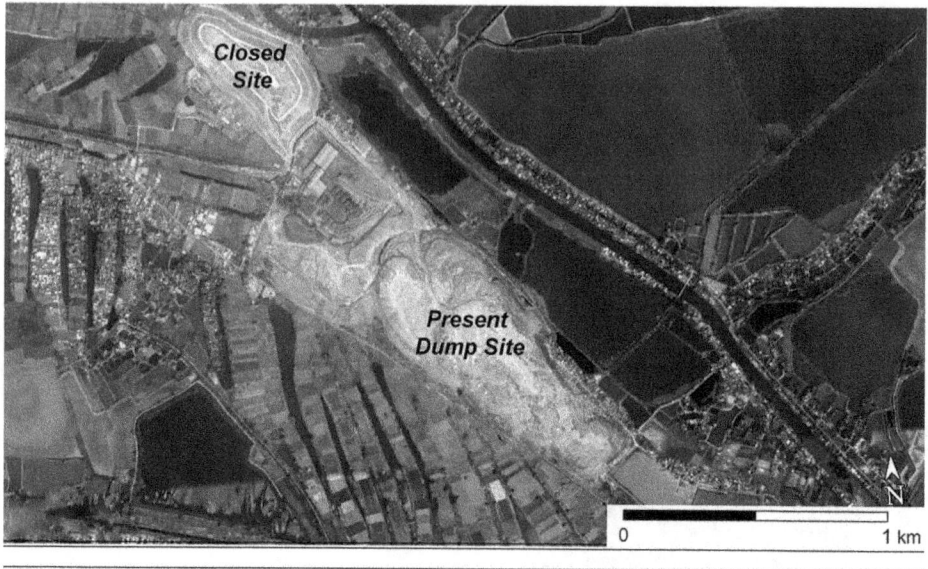

FIGURE 3.12   Google Earth image of 18 June 2019 showing the dump sites of the Dhapa landfills. The photograph below shows the enormous dump of urban waste in the background. See Section 3.3.4.3.2. (Photo: Sunando Bandyopadhyay.)

3.3.4.5.3   Shrinkage of Urban Wetlands

Nine percent (1,850 km²) of the LDWB comes under Kolkata urban agglomeration and its peri-urban areas. Huge transformations occurred in the Kolkata Metropolitan Area in the past few decades as agricultural lands and water bodies were replaced by urban land use. Between 1990 and 2015, a decrease of 10% of wetlands is recorded (Sahana et al., 2018) despite enforced restrictions (DoE, GoWB, 2012).

The 12.5-km² East Calcutta Wetlands, a Ramsar site, is famous for natural sewage water treatment and its ecological services to the city (Ghosh, 1999; WWF, 2011; Danda and Hazra, 2016).

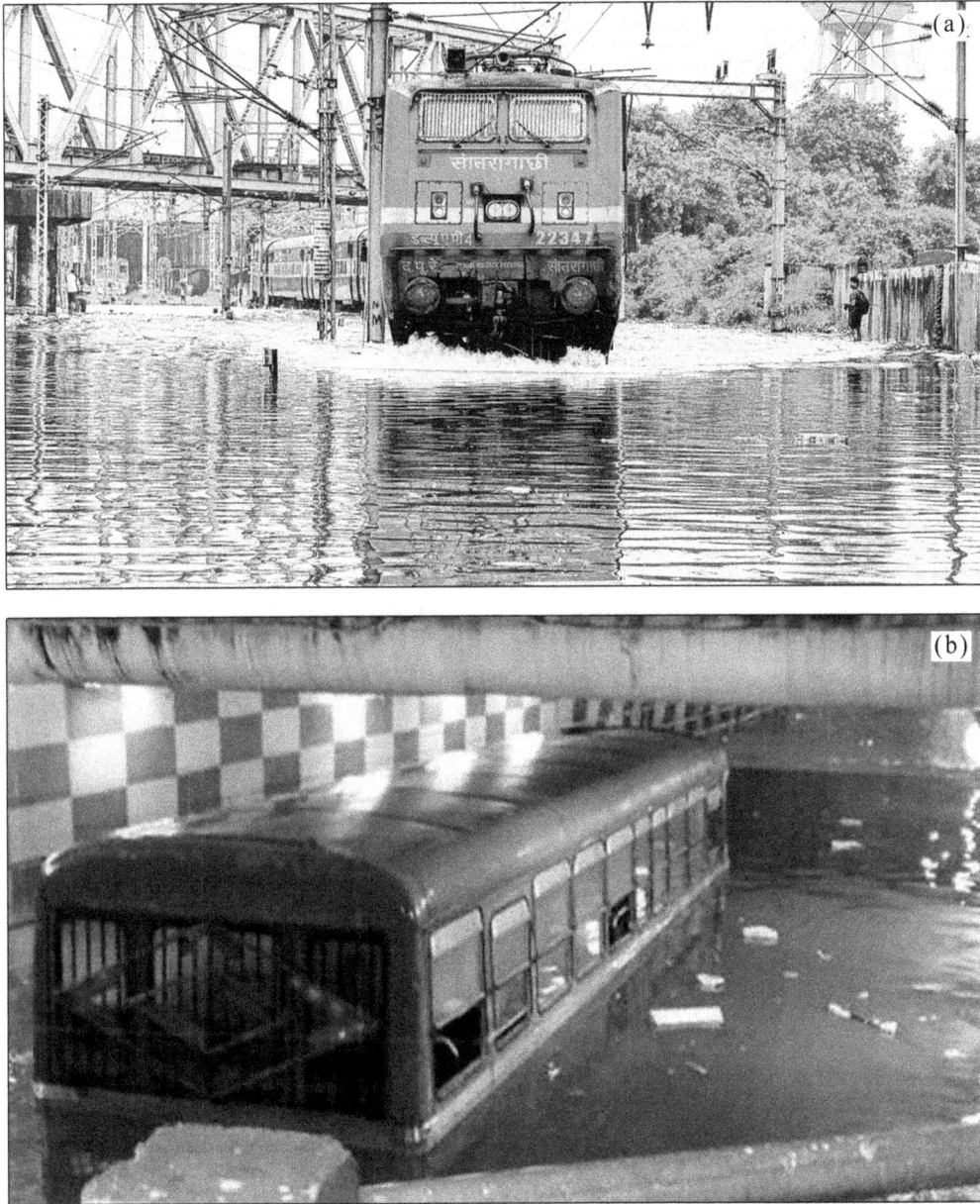

FIGURE 3.13 (a) Waterlogged railway track after heavy shower in Tikiapara, near Howrah railway station. (b) A bus stuck in Patipukur railway underpass due to waterlogging. See Section 3.3.4.4 for details. (TT (2016) and India Blooms (2017).)

The wetland is now increasingly threatened by real-estate developers and about 25,000 encroachers (Ray Chaudhuri, 2012; SANDRP, 2018). Its area is reduced by 23.4% between 2002 and 2017 (Das, 2018) (Figure 3.15). The conversion mostly affected the water bodies smaller than 3 ha. The standard practice is to convert a water body first into an agricultural land and then into a settlement area to avoid legal complications (Dey, 2015). Such disappearances of wetlands led to alteration of drainage, waterlogging (ABP, 2015, 2018) and absence of groundwater recharge (Sahu and Sikdar, 2011; Bhattacharya et al. 2012). In the other parts of the LDWB, similar conversions are also taking place around most urban areas (Barman, 2014; TNN, 2016; The Indian Express, 2016a).

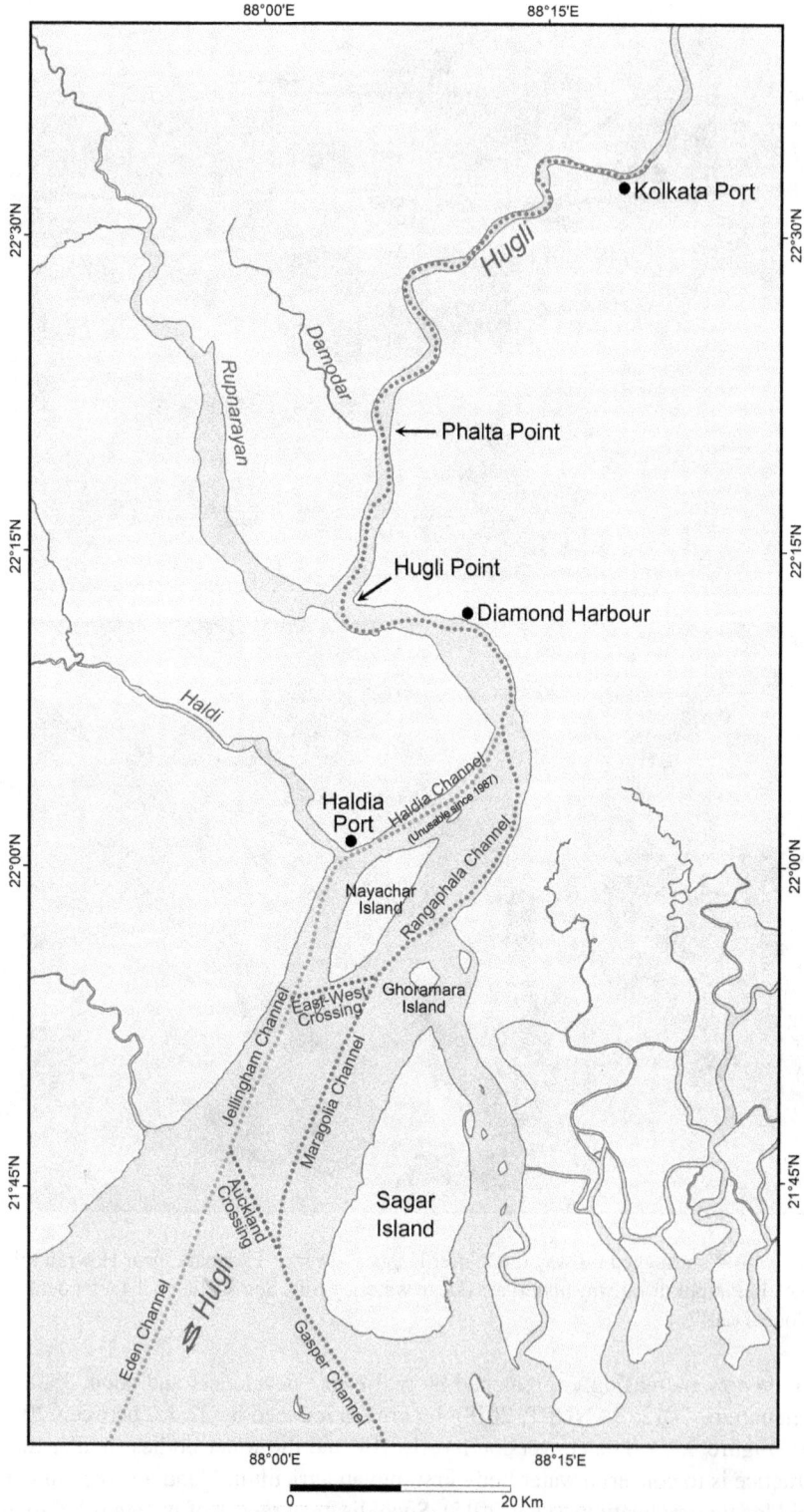

FIGURE 3.14 Pilotage to Kolkata port. The present pilotage is the longest in the world due to the presence of several shoals and islands. It is maintained by regular dredging, artificially altering the hydrography. See Section 3.3.4.5.2.

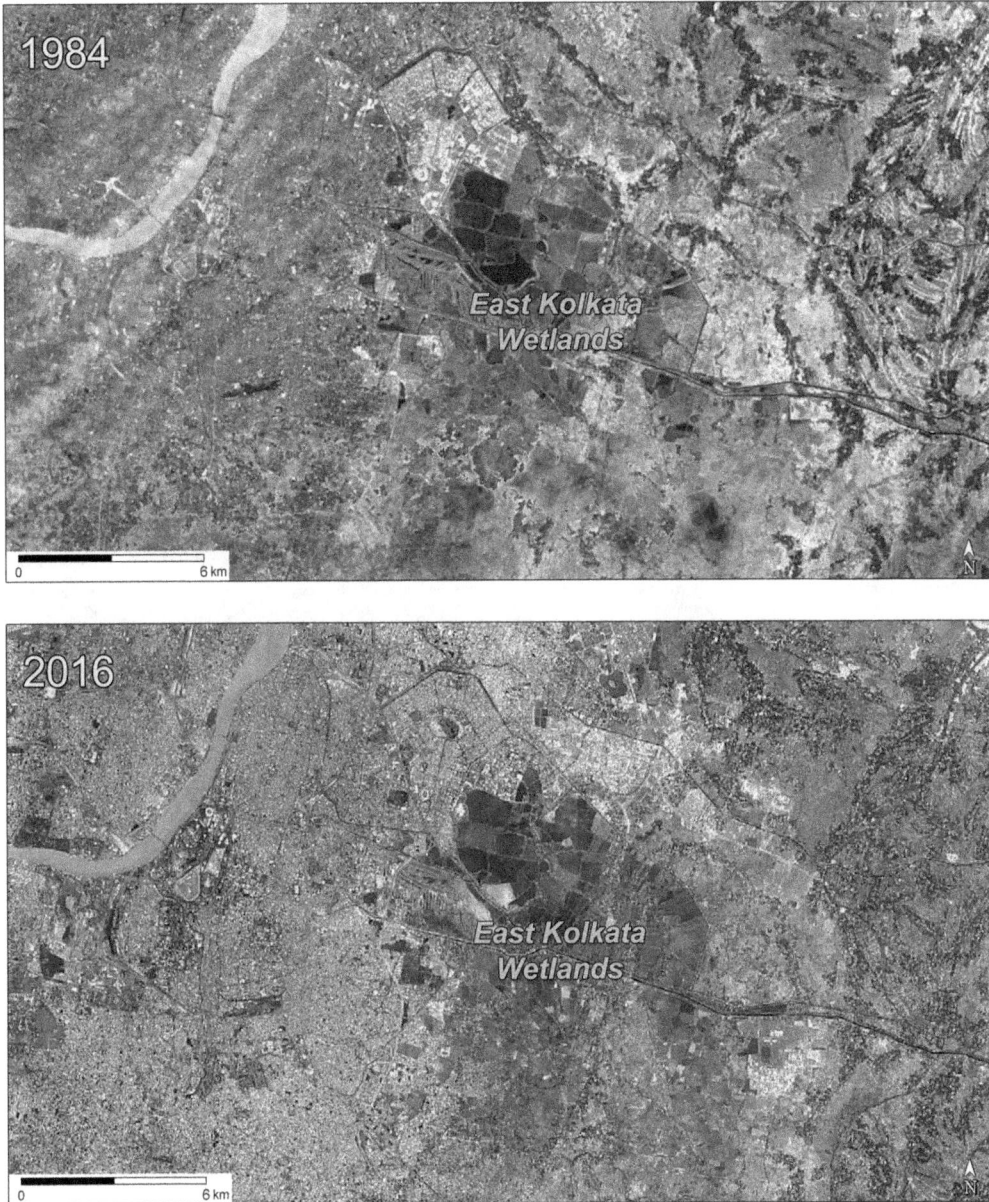

FIGURE 3.15 Encroachment of East Kolkata Wetlands between 1984 and 2016. See Section 3.3.4.5.3. (Google Earth image, 31 December 1984 and 31 December 2016.)

3.3.4.5.4 Waterlogging and Flood

Flood embankments were placed along most rivers in the LDWB by the early 20th century (Inglis, 1909). While this promoted agriculture and settlements by preventing annual inundations, the embankments induced in-channel siltation that elevated riverbeds above the level of the surrounding floodplains. In this way, waterlogging due to lack of drainage at the times of flood became a major problem in many low-lying areas of the LDWB. Three most important areas facing this kind of hazard are: (1) the Lower Damodar area comprising Khanakul-I, Khanakul-II and Dhaniakhali blocks of Hooghly District, and Amta and Udaynarayanpur of Howrah District; (2) the Ghatal region of Paschim Medinipur District; and (3) the Keleghai–Kapaleswari region comprising Sabang

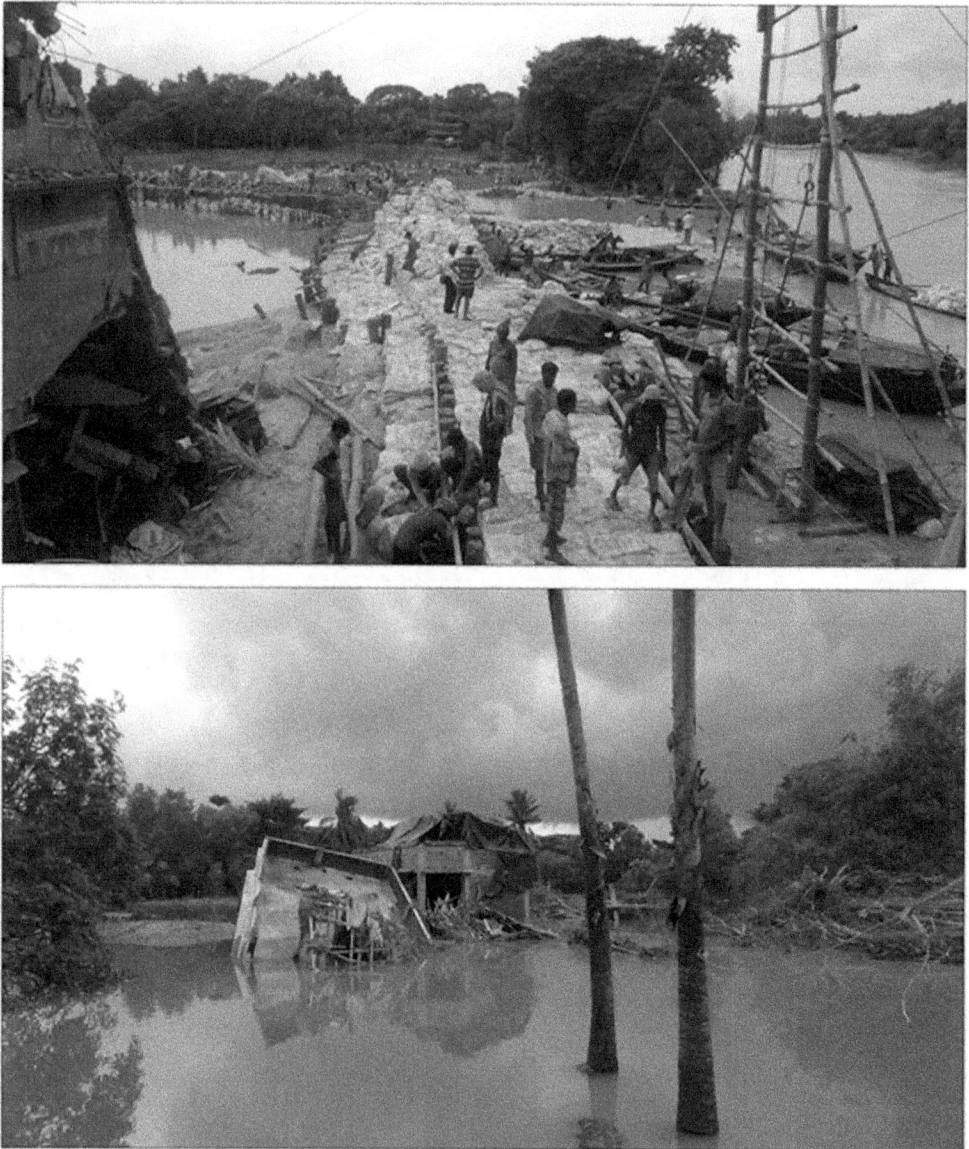

FIGURE 3.16   Embankment breaching at right bank of Silabati River and resulting flood in Pratappur, near Ghatal, due to heavy downpour followed by sudden release of water from Kangsabati Barrage on 25 July 2017. See Section 3.3.4.5.4. (Photo: Ina Dhar Roy Dasgupta.)

amd Pingla blocks of Purba Medinipur District (Figure 3.16). The duration of inundation in these areas is usually 1–3 months because of topography and slow outflow through the tidally active Hugli. The term 'man-made flood' is quite popular while deliberating floods of these regions (The Indian Express, 2016b; NDTV, 2017; Millenniumpost, 2017).

3.3.4.5.5   Artificial Waterways

The LDWB is traversed by a number of canal networks. The earlier canals were excavated by the colonial rulers followed by constructions under post-independence large-scale river projects. The canals can be divided into two categories: the irrigation canals and the drainage canals.

The irrigation canals are mostly found in the western part of LDWB and initiate from several dams and barrages built on the western tributaries of the Hugli. Following the natural slopes, they run parallel to the tributaries and ultimately meet the Lower Hugli. The oldest of them is the 137 km Midnapore–Orissa Coastal Canal, excavated in 1866 (Inglis, 1909; DoIW-GoWB, 2019a) that runs through the entire coastal LDWB west of the Hugli River. The two other important irrigation projects of the LDWB are the DVC and the Kangsabati Reservoir Project. The Subarnarekha Barrage Project is now in the proposal stage (Bandyopadhyay et al., 2014).

A large 2,494 km distributary network of irrigation canals, emanating from the Durgapur Barrage of the DVC, cover large areas of Bardhaman, Howrah and Hooghly Districts of the LDWB (DoIW-GoWB, 2019b). The network takes in and intervenes with other rivers like the Darakeshwar and distributaries of the Damodar such as the Mundeswari, *Kana* Damodar, *Kana* Nadi and Ghiya. A large section of the 3,218 km canals of the Kangsabati Reservoir Project also falls in the LDWB (DoIW-GoWB, 2019c).

The drainage canals of the LDWB were constructed to relieve flooding and removing urban wastewater. Some of the drainage canals of the Damodar–Rupnarayan–Hugli Basin of the western LDWB include the Bakshi, Deoantala, Gesopati, Dakatia, Madaria and Rajapur. In the east of the Hugli, a large number of canals were excavated primarily for Kolkata's urban wastewater and storm discharge. Most of these flow eastwards following the slopes from the ~6.5 m levees of the Hugli to the backswamps of the East Kolkata Wetlands (Bandyopadhyay et al, 2014). The oldest of these is the Tolly's Nullah, which was originally excavated as a transportation corridor in 1777. Others include the Circular–Beliaghata–New Cut–Kestopur Canal, Bhangar Canal, Tollygunge–Panchannagram (TP) Canal, Manikhali Canal, Begore Canal, Keorapukur Canal and Churial Canal (Paul, 2009) (Figure 3.17). Majority of these take the water to the Kulti Gang, 30 km east of the Hugli. Sluices, installed at the outfall of the canals, prevent backflow during high water and makes discharge of the city's waters impossible at certain times of the day.

3.3.4.5.6   *Reclamation of Sundarban Islands*

Reclamation of the southeastern part of LDWB, the Indian Sundarban, started in the first half of the 18th century under British colonial governance, to convert this mangrove swamp into revenue-earning farmlands (Richards and Flint, 1990). The reclamation process of the Sundarban involved embanking the major tidal channels and coastline, besides blocking the small creeks. These were done to prevent inundation of the low-lying areas by highest high tide or storm surge (Ascoli, 1921). Reclamation continued up to 1971, and out of 102 islands of the Indian Sundarban, 54 have been reclaimed, comprising an area of 5,364 km^2 (Mukherjee, 1983). The length of marginal embankments of the Sundarban amounts to 3,122 km (DoIW-GoWB, 2019d).

Typically 3 m high with 45° slopes on both sides, the marginal embankments prevented tidal spill, and the region remained lower than the high tide levels due to absence of vertical accretion from tidewater. The embanking also resulted in degeneration of small tidal creeks (Paul, 2002). Due to reclamation and subsequent reduction in tidal prism, channel siltation is observed, resulting in net rise of mean high water levels (Pethick and Orford, 2013; Bhattacharyya et al., 2013) (Figure 3.18).

An on-going reclamation event can be seen in Nayachar, a newly developed island in the Hugli estuary, off the Haldi outfall. It started accreting since 1948 and became the second largest island (48 km^2) on the estuary by 2008. Though legally protected from alteration, the island is reclaimed in phases, starting from 1980 with buffalo grazing, followed by development of a guide wall at its northern tip by Kolkata Port Trust and brackish water aquaculture project when the island was handed over to the State Fisheries Department by the Government of West Bengal in 1987. After proposals for constructing a chemical hub and a tourism centre did not materialise (Bandyopadhyay, 2008; Chakma and Bandyopadhyay, 2012), a huge number of people from the Purba Medinipur District occupied the island from 2015 and completely transformed the mangrove swamps of Nayachar into brackish water aquaculture ponds by placing mud dykes (Figure 3.19).

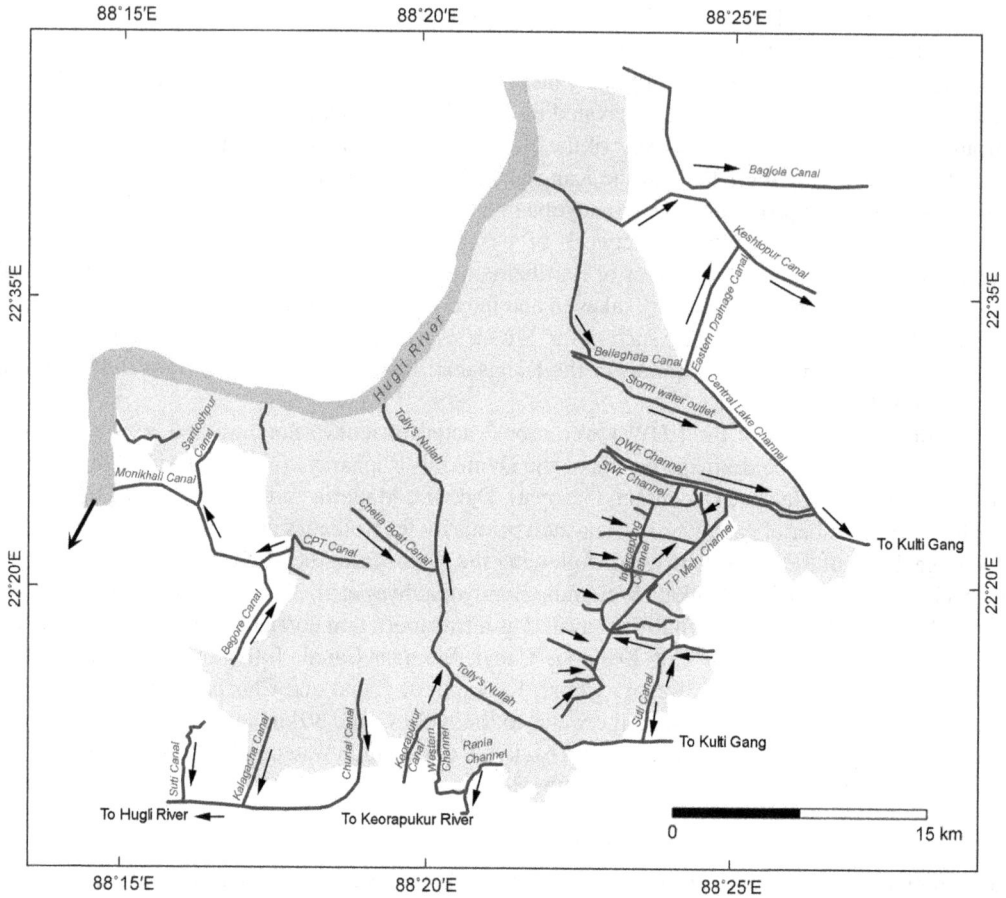

FIGURE 3.17 Drainage canal network of Kolkata. See Section 3.3.4.5.5. (Modified after KEIIP, SD31/2017-18: http://www.keiip.in/bl3/PDF/IEE-SD31.pdf.)

3.3.4.6   Agrogenic: Modifications Related to Agricultural and Aquaculture

Since time immemorial, the mainstay of the people of LDWB is agriculture, mostly paddy farming, covering 57% of its area. Preparation of the land for agriculture involved construction of low 0.3–0.5 m sized bunds to separate the plots. In floodplains, the arrangement of the plots often follows slopes of the accretion topography subtly reflecting the past positions of palaeochannels and point bars (Figure 3.20).

Except in the saline tracts of the Sundarban, cropping intensity in most parts of the LDWB is very high, harvesting three to four crops a year (BAES-DoSPI-GoWB, 2016). Although there are some major irrigation canal networks present, majority of the dry season irrigation comes from groundwater sources throughout LDWB (Jana, 2012) resulting in ground water depletion (Mukherji, 2007; Marwah, 2018) and, in the coastal tracts, salinisation (Chand et al., 2012; ADB, 2019).

A recent trend is the conversion of cultivated lands into aquaculture ponds in coastal districts of the LDWB. This is most noticeable in the blocks of Moyna, Contai, Deshapran and Egra of Purba Medinipur and the coastal islands located at the margin of the Sundarban mangroves (De Roy, 2013; Acharyya et al., 2015; Dutta et al., 2016; Ojha and Chakrabarty 2016; Duari, 2017; Giri, 2018). In Moyna, almost 90% of the cultivated lands are converted to freshwater aquaculture ponds between 2014 and 2017 (AITC, 2017; Ghosh, 2018) (Figure 3.21). Several reports affirm that such conversion increased salinity in the groundwater, affecting the nearby crop fields negatively (TT, 2012; Fiinovation, 2014).

FIGURE 3.18 A reclaimed part of Sundarban in the Basanti Island (north) and a non-reclaimed area in the Herobhanga Island, dominated by mangrove species (south). (Image: Google Earth, 16 February 2016. The photograph below, taken at Jatirampur Ghat, Gosaba, shows that the flood tide level is considerably higher than the embanked reclaimed region. See Section 3.3.4.5.6. (Photo: Sunando Bandyopadhyay.)

3.3.4.7 Recreational Interventions: Purba Medinipur Coast

The 60 km Medinipur coast stretches between the estuaries of the Subarnarekha and Rasulpur. Among a number of historically important settlements of this region, a development scheme involving 0.55 km^2 area was initiated by the Government of West Bengal at Digha, in Ramnagar-I Block, to promote tourism. This was extended to 4.45 km^2 in 1966 (DDA, 1997). In 1975, the 6.01 km^2 urban area of Digha township was created. At present, the township has emerged as one of the largest coastal resorts of Eastern India.

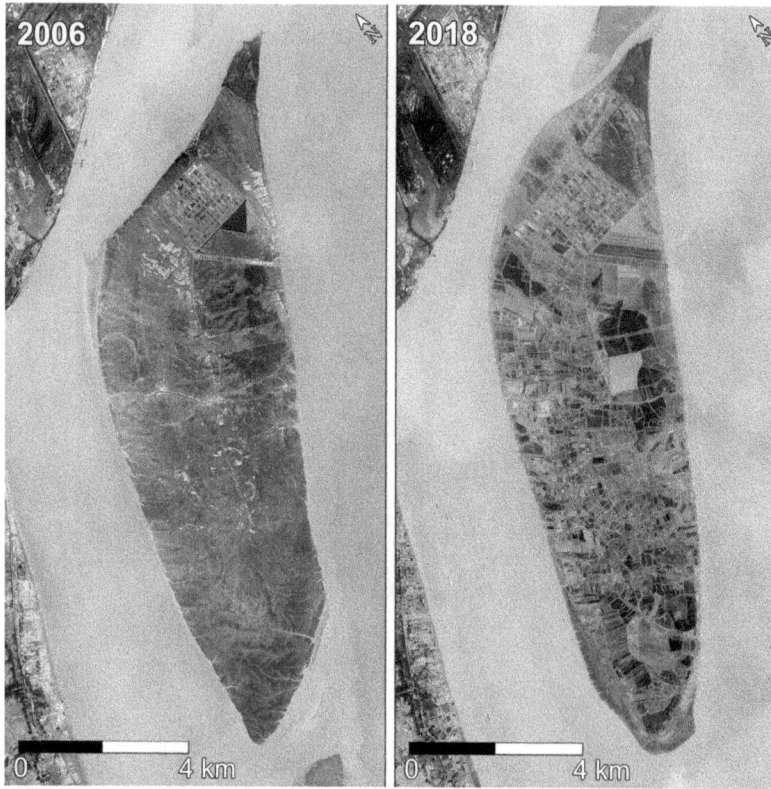

FIGURE 3.19   The changing landscape of the Nayachar Island between 2006 and 2018. See Section 3.3.4.5.6. (Images: Google Earth, 12 December 2006 and 25 November 2018.)

FIGURE 3.20   Arrangement of agriculture plots following the contours of accretion topography at adjoining the Hugli River. See Section 3.3.4.6. (Google Earth image, 09 March 2019.)

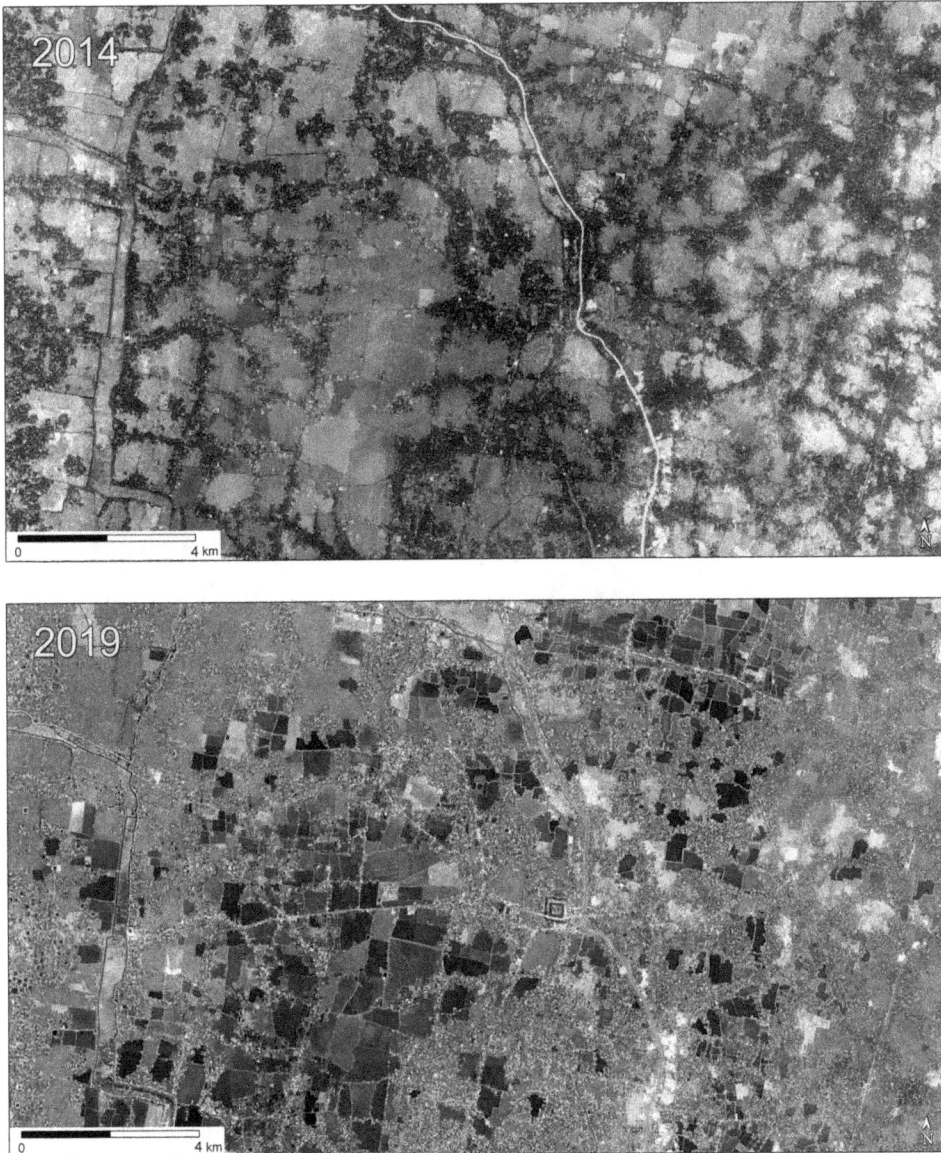

FIGURE 3.21 Almost 90% of the farmlands in the Moyna area, Purba Medinipur, were transformed into aquacultural ponds between 2014 and 2019. See Section 3.3.4.6. (Google Earth images, 04 February 2014 and 09 March 2019.)

Geomorphically, Digha is located on the two outermost ridges of the chenier dune system of the eastern Subarnarekha Delta, the height of which ranges between 5 and 15 m. Human modifications of the area mainly involve extensive flattening and landscaping of the dunes and filling up of some of the interdune wetlands (Figure 3.22). Soon after its establishment as a coastal resort, Digha ran into the problem of coastal erosion. To protect the town from inevitable destruction, a 3.7-km coastal embankment, paved by rectangular laterite blocks, was built between 1972 and 1982. This was extended in several stages to bring its present length to 6.87 km. Because beach lowering and coastal retreat work as complimentary processes at Digha (Niyogi, 1970), the dyke, by arresting coastal retreat, leads to extensive beach steepening and shrinkage of intertidal area in the eastern sector of the Digha town (Bandyopadhyay et al., 2009).

FIGURE 3.22 The upper photograph, taken in 1996, shows planation of dunes for extension of the Digha township. The photograph below, taken in 2005, shows destruction of a Digha township house located close to the retreating shoreline. See Section 3.3.4.7. (Photo: Sunando Bandyopadhyay.)

The other coastal resorts of the LDWB that extensively modified pre-existing landforms include Bakkhali (Namkhana Block), Mandarmani (Ramnagar-II), Gangasagar (Sagar) and Shankarpur (Ramnagar-I).

3.4 CONCLUSION

Sufficient evidences are already there to recognise the anthropogenic impacts as the responsible factor for significant geomorphic changes in LDWB and to remind the term 'Anthropocene'—coined by Crutzen and Stoermer (2000). It depicts a condition when natural and human forces become intertwined, so that the fate of one determines the fate of the other. As discussed in this study, such changes are occurring at a very rapid pace in LDWB at par with the region's growing population pressure and economically driven changes, neglecting the environmental concerns.

In view of Government of India's future policies aiming at economic growth of Eastern India led by Kolkata conurbation (NDTV, 2016; ET, 2016a), an excessive modification of landscape is expected to occur in LDWB. A report shows that by 2035, this part of India centred on Kolkata can be a $3 trillion economy (25% of total GDP of India) if development plans are properly implemented (ET, 2016b). Since per capita GDP annual growth rate of West Bengal doubled from 4.8% in 2015–2016 to 10.36% in 2017–2018 (BT, 2019), it is now contributing the lion's share of GDP (~40%) of the eastern region of India (ET, 2016b). The agronomy of this region is also growing rapidly with assistance of projects like BGREI (Bringing Green Revolution to Eastern India) operating since 2010–2011 (https://rkvy.nic.in/).

Following the thresholds suggested by Nir (1983), it is easier to control the alteration of physical landscape where the rate of anthropogeomorphological process is quite low (Figure 3.5a and b). In contrast, very intensive treatment is advised for the areas having a significantly higher rate of potential anthropogeomorphology. In LDWB, the blocks having I_{PAG} values above 0.40 would require some urgent measures to control the unrestrained human dominance, while in the blocks possessing the I_{PAG} levels below 0.20, the anthropic activities would be easier to control, or may not require any control at all. Introduction of anti-erosion measures can be suggested for the blocks with I_{PAG} levels between 0.20 and 0.4, which are mostly associated with the areas characterised by frequent river planform change and intensive agricultural systems.

It is observed that human modifications of LDWB in many cases lead to negative quasi-natural outcomes. Lack of environmental awareness, modifications of land-use overpowering the regulations, dominance of economic and political drive, engineering-led land development projects and little involvement of environmentalists and earth scientists in decision-making, all have worsened the situation. Considering the future population growth, economic development is of utmost necessity of this developing part of world, but it should be sustainable and endeavour to conserve the natural landscape.

REFERENCES

Anandabazar Patrika (ABP). 2015. City ward waterlogged for two months (in Bengali). https://www.anandabazar.com/calcutta/rajarhat-gopalpur-20-no-ward-waterlogged-from-two-months-1.198023 (Accessed September 23, 2019)

Anandabazar Patrika (ABP). 2018. Protest in Rajarhat against household waterlogging (in Bengali). https://www.anandabazar.com/calcutta/waterlogged-rajarhat-due-to-heavy-rain-protest-continues-1.838651 (Accessed September 23, 2019)

Acharyya, N., Bhattacharya, M., Das, P., Patra, B.C. and J. Bandyopadhyay. 2015. Thestatus of locally endangered fish faunal: A case study of Contai Sub-division, Purba Medinipur District; West Bengal. *American Research Thoughts* 1(5): 1337–1348.

Asian Development Bank (ADB). 2019. West Bengal drinking water sector improvement project, sovereign (public) project report, 49107-006. https://www.adb.org/sites/default/files/linked-documents/49107-006-sd-01.pdf (Accessed on May 7, 2019).

All India Trinamool Congress (AITC). 2017. Bengal's largest fisheries hub to come up at Moyna in Purba Medinipur District. http://aitcofficial.org/aitc/bengals-largest-fisheries-hub-to-come-up-at-moyna-in-purba-medinipur-district/ (Accessed on May 7, 2019).

Akter, N. 2015. Agricultural productivity and productivity regions in West Bengal. *The North Eastern Hill University Journal* 13(2): 49–61.

Ascoli, F.D. 1921. *A Revenue History of Sundarban from 1870 to 1920* (Reprinted in 2002: West Bengal District Gazetteers). Kolkata: Govt. of West Bengal.

Bureau of Applied Economics and Statistics, Department of Statistics and Programme Implementation, Government of West Bengal (BAES-DoSPI-GoWB). 2016. District wise estimates of yield rate and production of nineteen major crops of West Bengal during 2014–15, Annual Report.

Bandyopadhyay, S. 2000. Coastal changes in the perspective of long term evolution of an estuary: Hugli, West Bengal, India. In *Quaternary Sea Level Variation, Shoreline Displacement and Coastal Environments*, eds. V. Rajamanickam and M.J. Tooley, 103–115. New Delhi: New Academic Publishers.

Bandyopadhyay, S. 2008. Nayachar: Factories in place of mangroves. In *Nandigram and Beyond*, ed. G. Ray, 199–224. Kolkata: Gangchil.

Bandyopadhyay, S. and M.K., Bandyopadhyay. 1996. Retrogradation of the western Ganga-Brahmaputra delta, India and Bangladesh: Possible reasons. In *Proceedings of 6th Conference of Indian Institute of Geomorphologists, National Geographer*, ed. R.C. Tiwari, 31(1–2), 105–128. Allahbad: The Allahbad Geographical Society, University of Allahabad.

Bandyopadhyay, S., Kar, N.S., Das, S. and J. Sen. 2014. River systems and water resources of West Bengal: A review. In *Rejuvenation of Surface Water Resources of India: Potential, Problems and Prospects*, ed. R. Vaidyanadhan, Special Publication 3: 63–84. Bangalore: Geological Society of India.

Bandyopadhyay, S., Das, S. and Kar, N.S. 2015. Discussion: 'Changing river courses in the western part of the Ganga–Brahmaputra delta' by Kalyan Rudra (2014), Geomorphology, 227, 87–100. *Geomorphology* 250: 442–453.

Bandyopadhyay, S., Mukherjee, D. and D., Pahari. 2009. Coastal erosion and its management at Digha, Medinipur, West Bengal. In *Geomorphology in India*, ed. H.S. Sharma and V.S. Kale, 124–132. Allahabad: Prayag Pustak Bhavan.

Barman, M. 2014. Problems and prospects of inland fishery development in the district of South 24 Parganas, West Bengal. Unpublished Ph.D. thesis, Department of Commerce, The University of Burdwan.

Bhatta, B. and B. Giri. 2012. Urban growth of Kolkata from 1980 to 2040: A Remote Sensing perspective. *Proceedings of the Seminar on Geographical Appraisal of The City of Joy's Environmental Wellbeing*, Sarsuna College, Kolkata.

Bhattacharya, S., Ganguli, S., Bose, S. and A. Mukhopadhyay. 2012. Biodiversity, traditional practices and sustainability issues of East Kolkata Wetlands: A significance Ramsar site of West Bengal, (India), *Research & Reviews in BioSciences* 6(11): 340–347.

Bhattacharyya, K. 1998. Applied geomorphological study in a controlled tropical river _ the case of the Damodar between Panchet Reservoir and Falta. Unpublished Ph.D. thesis, Department of Geography, University of Burdwan, 78–83.

Bhattacharyya, S., Pethick, J. and K. Sensarma. 2013. Managerial response to sea level rise in the tidal estuaries of the Indian Sundarban: A geomorphological approach. In *Water Policy, Special Edition: The Ganges Basin Water Policy*, eds. W. Xun and D. Whittington, 15, 51–74. IWA Publishing. doi:10.2166/wp.2013.205.

Brown, E.H. 1970. Man shapes the earth. *The Geographical Journal* 136(1): 74–85.

Business Today (BT). 2019. West Bengal's untold story: An investment destination growing in stature, and significance. https://www.businesstoday.in/impact-feature/corporate/west-bengal-untold-story-an-investment-destination-growing-in-stature-and-significance/story/316813.html (Accessed on May7, 2019).

Chakma, N. and S. Bandyopadhyay, 2012. Swimming against the tide: Survival in the transient islands of the Hugli estuary, West Bengal. In *West Bengal: Geo-spatial Issues*, ed. N.C. Jana, 1–19. Bardhaman: The University of Burdwan.

Chand, B.K., Trivedi, R. K., Dubey, S. K. and M.M., Beg. 2012. *Aquaculture in Changing Climate of Sundarban*, Survey report on Climate Change Vulnerabilities, Aquaculture Practices & Coping Measures in Sagar and Basanti Blocks of Indian Sundarban, National Initiative on Climate Resilient Agriculture, West Bengal University of Animal & Fishery Sciences, Kolkata, India.

Chatterjee, M., Shankar, D., Sen, G.K., Sanyal, P., Sundar, D., Michael, G.S., Chatterjee, A., Amol, P., Mukherjee, D., Suprit, K., Mukherjee, A., Vijith, V., Chatterjee, S., Basu, A., Das, M., Chakraborti, S., Kalla, A., Misra, S.K., Mukhopadhyay, S., Mandal, G. and K. Sarkar. 2013. Tidal variations in the Sundarban estuarine system, India. *Journal of Earth System Science* 122(4): 899–933.

Census of India (CoI). 2011. Ministry of home affairs, Government of India. http://www.censusindia.gov.in/DigitalLibrary/Archive_home.aspx (Accessed on June 11, 2015).

Cox, W. 2012. The evolving urban form: Kolkata: 50 mile city, New Geography, http://www.newgeography.com/content/002620-the-evolving-urban-form-kolkata-50-mile-city (Accessed on May7, 2019).

Crutzen, P.J. and E.F. Stoermer. 2000. The anthropocene. *IGBP Global Change Newsletter* 41: 17–18.

Danda, A. and S. Hazra. 2016. East Calcutta Wetlands, Kolkata's unique heritage. http://calcuttapublic.org/East%20Kolkata%20Wetlands%20to%20Mayor_3.pdf (Accessed on May 7, 2019).

Das, C. 2018. An analytical study of temporal changes in landuse and landcover in the East Kolkata Westlands Since 2002, based on multidated satellite images. Unpublished Masters Dissertation, Department of Remote Sensing and GIS, Kumaon University.

Digha Development Authority (DDA). 1997. Land use & development control plan 1995–2011, Urban Development Department, Government of West Bengal, 1–73.

De Roy, S. 2013. Impact of fish farming on land relations: Evidence from a village study in West Bengal. *Indian Journal of Agricultural Economics* 68(2): 222–238.

Dey, D. 2015. Eco-system dependent livelihood and Urban land-use practices – The challenges for Kolkata and East Kolkata Wetlands. Unpublished Ph.D. thesis, Department of Economics, University of Calcutta.

Department of Environment, Govt. of West Bengal (DoE, GoWB). 2012. West Bengal wetlands and water bodies conservation policy.

Department of Irrigation & Waterways, Govt. of West Bengal (DoIW-GoWB). 2019a. Midnapore canal. https://wbiwd.gov.in/index.php/applications/midnapore_canal (Accessed on May7, 2019).

Department of Irrigation & Waterways, Govt. of West Bengal (DoIW-GoWB). 2019b. Barrage and irrigation system of DVC. https://wbiwd.gov.in/index.php/applications/dvc (Accessed on May7, 2019).

Department of Irrigation & Waterways, Govt. of West Bengal (DoIW-GoWB). 2019c. Kangsabati reservoir project. https://wbiwd.gov.in/index.php/applications/kangsabati (Accessed on May7, 2019).

Department of Irrigation & Waterways, Govt. of West Bengal (DoIW-GoWB). 2019d. AILA. https://wbiwd.gov.in/index.php/applications/aila (Accessed on May7, 2019).

Downs, P.W. and H. Piégay. 2019. Catchment-scale cumulative impact of human activities on river channels in the late anthropocene: Implications, limitations, prospect. *Geomorphology* 338: 88–104.

Duari, B. 2017. Changing agricultural resource and agricultural land use pattern through time by time: A critical case study of egra-I and egra-II block in Purba Medinipur, West Bengal. *International Journal of Scientific Research and Reviews* 6(4): 88–98.

Dubey, R.P. 2011. Modelling the sediment transport in a highly hydrodynamic estuary a case study of Hooghly. Unpublished Ph.D. thesis, School of Civil Engineering, KIIT University.

Dutta, D., Das, C.S. and A. Kundu. 2016. A geo-spatial study on spatio-temporal growth of brackish water aquaculture along the coastal areas of West Bengal (India). *Modelling Earth System & Environment* 2(61): 1–10.

Dutta, S. 2017. Kolkata's landfill crisis: The city's dependency on the sole landfill of Dhapa may soon result in an unresolvable crisis. https://swachhindia.ndtv.com/kolkatas-landfill-crisis-the-citys-dependency-on-the-sole-landfill-of-dhapa-may-soon-result-in-an-unresolvable-crisis-11618/ (Accessed on May 12, 2019).

Eastern Railway (ER). 2016. Historical perspective – The first journey. https://er.indianrailways.gov.in/view_section.jsp (Accessed on May 12, 2019).

Erlich, P.R. and A.H. Erlich. 1990. *The Population Explosion*. New York: Simon and Schuster.

The Economic Times (ET). 2016a. Government focused on sustainable growth of eastern India: Jayant Sinha. https://economictimes.indiatimes.com/news/economy/policy/government-focused-on-sustainable-growth-of-eastern-india-jayantsinha/articleshow/50514333.cms?from=mdr (Accessed on May 7, 2019).

The Economic Times (ET). 2016b. Eastern states led by West Bengal can be a $3 trillion economy by 2035: Report. https://economictimes.indiatimes.com/news/economy/indicators/eastern-states-led-by-west-bengal-can-be-a-3-trillion-economy-by-2035-report/articleshow/50706111.cms?from=mdr (Accessed on May 7, 2019).

Fiinovation. 2014. Shrimp Farming gets a boom in West Bengal as Farmland's become saline. http://www.fiinovation.co.in/news/shrimp-farming-gets-boom-west-bengal-farmlands-become-saline/ (Accessed on May 7, 2019).

Ghosal, S. 2016. Garbage mountain on fire, thousands despondent. http://timesofindia.indiatimes.com/articleshow/52593876.cms?utm_source=contentofinterest&utm_medium=text&utm_campaign=cppst (Accessed on May 7, 2019).

Ghosh, D. 1999. Wastewater utilisation in East Calcutta Wetlands. *UWEP Occasional Paper*, 3–19. The Netherlands: WASTE.

Ghosh, S. 2018. Fish farming in extended fields at Moyna, Dist. Purba Medinipur, West Bengal. *Science and Culture* 84(3–4): 126–128.

Ghosh, D. and S. Sen. 1987. Ecological history of Calcutta's wetland conversion. *Environmental Conservation* 14(3): 219–226.

Ghotankar, S.T. 1972. Tidal propagation in the Hooghly. In *Bhagirathi–Hugli Basin, Proceedings of Interdisciplinary Symposium*, ed. K. Bagchi, 127–132. Calcutta: University of Calcutta.

Giri, S. 2018. Decreasing agricultural land for increasing inland fisheries and its impact on the socio-economic development of Deshapran Block in Purba Medinipur coastal area, West Bengal, India. *International Journal of Humanities and Social Science Invention* 7(5): 1–7.

Golomb, B. and H.M. Eder. 1964. Landforms made by man. *Landscape* 14(1): 4–7.

Goudie, A. 2018. The human impact in geomorphology – 50 years of change. *Geomorphology*. doi: 10.1016/j. geomorph.2018.12.002.

Goudie, A.S. and H.A. Viles. 2016. *Geomorphology in the Anthropocene*. Cambridge: Cambridge University Press.

Haff, P.K. 2003. Neogeomorphology, prediction, and the anthropic landscapes. In *Prediction in Geomorphology, American Geophysical Union Geophysical Monograph Series*, eds. P.R. Wilcock and R.M. Iverson, 135, 15–26. Washington: American Geophysical Union.

Haigh, M.J. 1978. Evolution of slopes on artificial landforms. University of Chicago Geography Research Papers, 1–307.

Hooke, R.L. 1994. On the efficacy of humans as geomorphic agents. *GSA Today* 4(9): 217, 224–225.

Hooke, R.L. 2000. On the history of humans as geomorphic agents. *Geology* 28(9): 843–846.

Hudson, P.F., van der Hout, E. and M. Verdaasdonk. 2019. (Re)Development of fluvial islands along the lower Mississippi River over five decades, 1965–2015. *Geomorphology* 331: 78–91.

Hunter, W.W. 1875. *A statistical account of Bengal* (Reprinted in 2002: 24-Parganas, Statistical account of Bengal, 1(1): 9–15). London: Turbner & Co.

India Blooms. 2017. Over 40 passengers rescued unhurt from sinking bus on waterlogged Kolkata street. https://www.indiablooms.com/news-details/N/34016/over-40-passengers-rescued-unhurt-from-sinking-bus-on-waterlogged-kolkata-street.html (Accessed on June 18, 2019).

Inglis, W.A. 1909. *The Canals and Flood Banks of Bengal* (Reprinted in 2002: Rivers of Bengal, 5(1):62–83, West Bengal District Gazetteers). Kolkata: Govt. of West Bengal.

Jana, S.K. 2012. Irrigation development in West Bengal. *Land Bank Journal* 51(2): 41–55.

Karmakar, J. 2014. Spatial extension of Kolkata Metropolitan Area and its multidimensional outcomes: A case study of Rajarhat. In *History and Beyond: Trends and Trajectories*, eds. J. Mukherjee and C. Palit. New Delhi: Kunal Books.

Lama, S. 1994. Anthropo-geomorphology of Darjiling town. Urban Geomorphology of Darjiling Town. Unpublished Ph.D. Thesis, University of North Bengal, 170–199.

Lóczy, D. and E. Pirkhoffer. 2009. Mapping direct human impact on the topography of Hungary. *Zeitschrift für Geomorphologie* 53(3): 145–155.

Loh, J. and M. Wackernagel. 2004. *Living Planet Report 2004*. Gland, Switzerland: World Wildlife Fund International.

Marsh, G.P. 1864. *Man and Nature; Or, Physical Geography as Modified by Human Action* (Reprinted in 2003). Washington, DC: Washington University Press.

Marwah, M. 2018. *Mapping institutions for assessing groundwater scenario in West Bengal, India*. Working paper 411, The Institute for Social and Economic Change, Bangalore.

Millenniumpost. 2017. Havoc by man-made floods. http://www.millenniumpost.in/opinion/havoc-by-man-made-floods-256474http://www.millenniumpost.in/opinion/havoc-by-man-made-floods-256474 (Accessed on May 7, 2019).

Milliman, J.D. and R.H. Meade. 1983. Worldwide delivery of river sediment to the oceans. *Journal of Geology* 91: 1–21.

Milliman, J.D. and J.P.M. Syvitski. 1992. Geomorphic/tectonic control of sediment discharge to the ocean: The importance of small mountainous rivers. *Journal of Geology* 100(5): 525–544.

Mitra, I. 2018. Spatialization of calculability, financialization of space: A study of the Kolkata port. In *Logistical Asia: The Labour of Making a World Region*, eds. B. Nielson, N. Rossiter and R. Samaddar, 47–68. Singapore: Palgrave Macmillan.

Moneycontrol. 2018. Kolkata, Hyderabad sizzle as residential sales grow 25% in H12018: JLL report. https://www.moneycontrol.com/news/business/real-estate/kolkata-hyderabad-sizzle-as-residential-sales-grow-25-in-h12018-jll-report-2801991.html (Accessed on May 7, 2019).

Mukerjee, R.K. 1938. *The Changing Face of Bengal: A Study of Riverine Economy*, 166–168. Calcutta: Calcutta University.

Mukherjee, K.N. 1983. History of settlement in the Sundarban of West Bengal. *Indian Journal of Landscape System* 6(2): 1–19.

Mukherji, A. 2007. Implications of alternative institutional arrangements in groundwater sharing – Evidence from West Bengal. *Economic and Political Weekly* 42: 2543–2551.

National Atlas and Thematic Mapping Organisation (NATMO). 1980. *National Atlas of India* 1: Plate No. 30 (Patna) on 1:1M.

National Atlas and Thematic Mapping Organisation (NATMO). 1988. *National Atlas of India* 1: Plate No. 33 (Calcutta) on 1:1M, 2nd edition.

NDTV. 2016. Eastern India focal point of my development plan: PM Narendra Modi. https://www.ndtv.com/india-news/eastern-india-focal-point-of-my-development-plan-pm-narendra-modi-1414723 (Accessed on May 7, 2019).

NDTV. 2017. Flood situation in Bengal is man-made: Mamata Banerjee. https://www.ndtv.com/india-news/flood-situation-in-bengal-is-man-made-west-bengal-chief-minister-mamata-banerjee-1730149 (Accessed on May 7, 2019).

NDTV. 2018. After heavy rain in Kolkata, people face tough time going to work. https://www.ndtv.com/kolkata-news/after-heavy-rain-in-kolkata-people-face-tough-time-going-to-work-1890169 (Accessed on May 12, 2019).

Nir, D. 1983. *Man, a Geomorphological agent: An Introduction to Anthropic Geomorphology*. Dordrecht, Boston, London: Reidel.

Niyogi, D. 1970. Geological background of beach erosionat Digha, West Bengal. *Bulletin of the Geological Mining and Metallurgical Society of India* 43: 1–36.

Ojha, A. and A. Chakrabarty. 2016. Village level land use land cover change dynamics and brackish water aquaculture development: A case study in Desopran Block, West Bengal, India using multi-temporal satellite data and GIS techniques. *International Journal of Remote Sensing and Geosciences* 5(1): 13–18.

Owen, L.A. 2017. Earth surface processes and landscape evolution in the Himalaya: A framework for sustainable development and geohazard mitigation. *Geological Society Special Publications* 462(1): 169–188. London.

Parua, P.P. 2010. *The Ganga: Water use in the Indian Subcontinent*. Dordrecht: Springer.

Paul, A. K. 2002. *Coastal Geomorphology and Environment: Sundarban Coastal Plain, Kanthi Coastal Plain, Subarnarekha Delta Plain*. Kolkata: ACB Publication.

Paul, A. 2009. Assessment of waterlogging in CMC and its social impact. Unpublished Ph.D. thesis. Department of Geography, University of Calcutta.

Pethick, J. and J.D. Orford. 2013. Rapid rise in effective sea-level in southwest Bangladesh: Its causes and contemporary rates. *Global and Planetary Change* 111: 237–245.

Prasad, B., Swetha, M. and B. Subbareddy. 2014. Dredging maintenance plan for the Kolkata Port, India. *Current Science* 107: 1–12.

Qasim, S.Z., Sengupta, R. and T.W. Kureishy. 1988. Pollution of the seas around India. *Proceedings of Indian Academy of Sciences (Animal Sciences)* 97(2): 117–131, cited in Prasad et al., 2014.

Ray Chaudhuri, S., Mukherjee, I., Ghosh, D. and A.R. Thakur. 2012. East Kolkata Wetland: A multifunctional niche of international importance. *OnLine Journal of Biological Sciences* 12(2): 80–88.

Richards, J.F. and E.P. Flint. 1990. Long-term transformations in the Sundarban wetlands forests of Bengal. *Agriculture and Human Values* 7(2): 17–33.

Rishra Municipality (RM). 2018. Rishra municipality history. http://rishramunicipality.org/history/ (Accessed on April 4, 2019).

Roy, S. 2018. Calcutta landfill burning, city getting pollutant-rich smoke. https://www.telegraphindia.com/states/west-bengal/calcutta-landfill-burning-city-getting-pollutant-rich-smoke/cid/1678642 (Accessed on May 8, 2019).

Rózsa, P. 2007. Attempts at qualitative and quantitative assessment of human impact on the landscape. *Geografia Fisica e Dinamica Quaternaria* 30: 233–238.

Rózsa, P. and T. Novák. 2011. Mapping anthropic geomorphological sensitivity on a global scale. *Zeitschrift für Geomorphologie* 55(1): 109–117.

Sahana, M., Hong, H. and H. Sajjad. 2018. Analyzing urban spatial patterns and trend of urban growth using urban sprawl matrix: A study on Kolkata urban agglomeration, India. *Science of the Total Environment* 628–629: 1557–1566.

Sahu, P. and P.K. Sikdar. 2011. Threat of land subsidence in and around Kolkata City and East Kolkata Wetlands, West Bengal, India. *Journal of Earth System Science* 120(3): 435–446.

South Asia Network on Dams, Rivers and People (SANDRP). 2018. East India Wetlands review 2017: West Bengal bent on destroying world's largest natural sewage treatment plant. https://sandrp.in/2018/02/01/east-india-wetlands-review-2017-west-bengal-bent-on-destroying-worlds-largest-natural-sewage-treatment-plant/ (Accessed on May 8, 2019).

Sanyal, T. and A.K. Chatterjee. 1995. The Hugli Estuary: A profile. In *Port of Calcutta: 125 Years Commemorative Volume*, ed. S.C. Chakraborty, 45–54. Calcutta: Calcutta Port Trust.

Sen, R. 2019. *Birth of a Colonial City: Calcutta*. London: Taylor & Francis.

Sengupta, S. 2008. A history of the brutal Rajarhat land acquisition, Bengal's new IT hub. http://sanhati.com/excerpted/945/ (Accessed on April 6, 2019).

Sherlock, R.L. 1922. *Man as a Geological Agent: An Account of his Action on Inanimate Nature*. London: H.F. and G. Withersby.

Survey of India (SoI). 2011. *Hugli River Tide Tables*. Dehradun: Survey of India.

Szabó, J. 2010. Anthropogenic geomorphology: Subject and system. In *Anthropogenic Geomorphology: A Guide to Man-made Landforms*, eds. J. Szabó, L. Dávid and D. Lóczy, 3–10.

Szabó, J. and L. Dávid. 2006. Antropogén Geomorfológia (in Hungarian). University notes. Debrecen: Kossuth University Publishing House.

Szabó, J., Dávid, L. and D. Lóczy. 2010. *Anthropogenic Geomorphology: A guide to man-made landforms*. Dordrecht, Netherlands: Springer. doi:10.1007/978-90-481-3058-0.

The Indian Express. 2016a. Government won't tolerate illegal construction on wetlands: Mamata Banerjee. https://indianexpress.com/article/cities/kolkata/government-wont-tolerate-illegal-construction-on-wetlands-mamata-banerjee-2950384/ (Accessed on May 7, 2019).

The Indian Express. 2016b. Man-made flood: West Bengal mulling legal action against DVC, warns Mamata Banerjee. https://indianexpress.com/article/india/india-news-india/man-made-flood-west-bengal-mulling-legal-action-against-dvc-warns-mamata-banerjee-2992988/ (Accessed on May 7, 2019).

Times News Network (TNN). 2016. Mamata cracks whip on illegal wetland fill-up, The Times of India, Kolkata City. https://timesofindia.indiatimes.com/city/kolkata/Mamata-cracks-whip-on-illegal-wetland-fill-up/articleshow/53515768.cms (Accessed on May 7, 2019).

The Telegraph (TT). 2012. Salt water from fish ponds affects farming. https://www.telegraphindia.com/states/west-bengal/salt-water-from-fish-ponds-affects-farming/cid/426617 (Accessed on May 7, 2019).

The Telegraph (TT). 2015. Water world–Parts of the township stay inundated for hours after a shower and there is no solution in sight. https://www.telegraphindia.com/states/west-bengal/waterworld/cid/1469733 (Accessed on May 8, 2019).

The Telegraph (TT). 2016. Train pain on water tracks. https://www.telegraphindia.com/states/jharkhand/train-pain-on-water-tracks/cid/1333263 (Accessed on June 18, 2019).

The Telegraph (TT). 2017. End to Monsoon Misery–It has been a very wet start to the rainy season but the lanes and by lanes of SaltLake have not got inundated so far. https://www.telegraphindia.com/states/west-bengal/end-to-monsoon-misery/cid/1474451 (Accessed on June 18, 2019).

West Bengal Pollution Control Board (WBPCB). 2014. Dhapa dumpsite environmental and social assessment report: Final report, COWI A/S. http://documents1.worldbank.org/curated/en/383421497427377147/pdf/SFG3430-V1-EA-P091031-Box402914B-PUBLIC-Disclosed-6-14-2017.pdf (Accessed on June 18, 2019).

Willcocks, W. 1930. The "overflow irrigation" of Bengal. In *Lectures on the Ancient System of Irrigation in Bengal and its Application to Modern Problems*, 1–28. Calcutta: Calcutta University.

Wohl, E. 2018. Rivers in the anthropocene: The U.S. perspective. *Geomorphology*. doi: 10.1016/j.geomorph.2018.12.001.

World Wildlife Fund (WWF). 2011. Impact of Urbanisation on Biodiversity–Case studies from India: Report, WWF, India. https://wwfin.awsassets.panda.org/downloads/impact_of_urbanisation_on_biodiversity.pdf (Accessed on June 18, 2019).

Zerboni, A. and K. Nicoll. 2019. Enhanced zoogeomorphological processes in North Africa in the human-impacted landscapes of the anthropocene. *Geomorphology* 331: 22–35.

4 Floodplain Alteration of the Bagmari–Bansloi– Pagla River System

Debabrata Mondal

CONTENTS

4.1 INTRODUCTION

In the conviction of environmental determinism, geoscientists are mostly involved to highlight the human interference on natural environment such as land-use changes, modern agricultural practices and engineering structures along or across the large river, which brought changes into the riverine floodplain (Mondal and Pal, 2017; Pal and Talukdar, 2018; Talukdar and Pal, 2018; Saha, and Pal, 2019; Roy and Sahu, 2017, 2018; Islam and Guchhait 2017; Ghosh, and Guchhait, 2014, 2016). The construction of hydro-engineering projects at any part of the river course may lead to disruption of the normal flow which increases or decreases the stream transport capacity and thereby aggradation or degradation occurs at the lower reaches of the river.

There are numerous freshwater wetlands of Bengal Basin that are associated with the floodplains of the Ganga and its tributaries–distributaries. Consequently, all the floodplain wetlands experience episodic flooding, which exhibit considerable variability in their function and hydro-geomorphic configuration. Being situated at the tropical monsoon climate, the whole floodplain embodies an ephemeral and erratic appearance; consequently, these water bodies are sometimes called as seasonally flooded wetland. The Ramsar Convention on Wetland separately recognised these overlooked water bodies as a part of the 'temporary pool'. Due to their unpredictable erratic occurrence, these wetlands are often neglected in mapping and mostly mistreated in India. Here most of the people appeal the conspicuous permanent water body as wetland, and still the whole floodplain is itself an important wetland landscape having ample hydrological significances. A permanent wetland does not support straight infiltration as much as transient or seasonal wetlands maintain it. In fact, floodplains are the zone of complex interaction among stream, wetland and aquifer system. The seasonally flooded wetlands are situated at the riparian habitat under which a number of hyporheic zones are located as lotic surface water components. The spatio-temporal characteristics of these saturated zones are driven by floodplain geomorphology, river regime, and underground local and regional groundwater systems (Woessner, 2017). Spatio-temporal hydrological fluctuations of seasonally flooded freshwater wetlands, mainly due to engineering constructions, are largely modifying the underground hyporheic zones.

4.1.1 HUMAN-INDUCED FLOODPLAIN ALTERATION

Most of the world rivers and their floodplains are highly affected by human-induced flow regulation. After the 1960s, a good number of geomorphologists paid their attention to document the effects of any flow regulation activities. Such kinds of modification like any barrage construction have direct and indirect implications for the conversion of river morphology as well as flow regime of floodplain. Any barrage and its associated engineering assemblies interrupt the natural hydrological regime of river and its adjacent freshwater ecosystem. Such types of channel instability may change hydrological fluctuation in floodplain to a great extent by changing its inundation magnitude, frequency, duration, etc. (Mondal and Pal, 2018). At the same time, some other activities, such as channel obstruction, artificial river embankments, road or elevated railway path, disturb the natural connectivity between lotic and lentic water bodies (Ward and Stanford, 1995).

4.1.2 REMOTE SENSING AND FLOODPLAIN MAPPING

A high-temporal resolution and micro-level mapping of floodplain dynamics of the aforementioned poorly studied region are indispensable for monitoring which enable a thorough understanding of the segment-wise comprehensive change mechanism of floodplain waterlogging. In India, understanding such floodplain waterlogging has been possible with free archive of Landsat imageries with high-temporal resolution and multi-band facility. The spectral signature of the freshwater is easily distinguishable from other floodplain feature that helps us to identify wetland using index-oriented image classification techniques. As the floodplains are highly dynamic in nature, mapping with single image cannot adequately represent the definite extent of shallow freshwater wetland (Borro et al., 2014; Thomas et al., 2011; Wang et al., 2014). Using Modified Normalized Difference Water Index (MNDWI) derived from pre- and post-monsoon Landsat TM, ETM+ and OLI imageries, multitemporal analyses were performed. The current chapter is intended to explore the spatio-temporal inundation characteristics of the lower reach of Bagmari–Bansloi–Pagla basin area after accomplishment of Farakka Barrage and allied schemes over a 28-year period.

4.2 THE RIVER AND ITS BASIN

The lower reaches of Bagmari–Bansloi–Pagla basin, which is the sub-basin of Bhagirathi Rivers, are located between 24°19′ N to 24° 47′ N latitude and 87° 14′ E to 88° 04′ E longitude (~2,522 km²), covering northern part of Birbhum and Murshidabad districts (Figure 4.1). This study area is part of the Bengal plain situated at the rifted margin of Indian plate controlled by a Rajmahal and Malda–Kaliaganj faults. According to Morgan and McIntire (1959), tilting of the faulted tract and unequal subsidence are possible causes of episodic course change of the Ganga and Bhagirathi Rivers. During the last few centuries, rivers have experienced such massive change of course where floodplain wetlands are the remaining shot left. The study area comprises mainly Jurassic–Cretaceous metamorphic rock, Pleistocene older alluvium and recent alluvium.

Physiographically, the area is a depositional floor associated with fluvial deposition of Ganga–Bhagirathi and its right bank tributaries. The ground slope is smoothly inclined towards the east and south-east direction. The Ganga–Padma River enters this part of Bengal basin through the narrow gap of Rajmahal–Barind tract and follows the regional slope. The entire area is drained by the Bagmari, Pagla and Bansloi (first three right bank tributaries) originating from the Rajmahal hills, runs eastward and was debouched into the feeder canal and Bhagirathi River.

Numerous seasonally flooded and permanent freshwater wetland system predominating here are characterized by the presence of interconnecting channels, meander cut-offs, effluents and

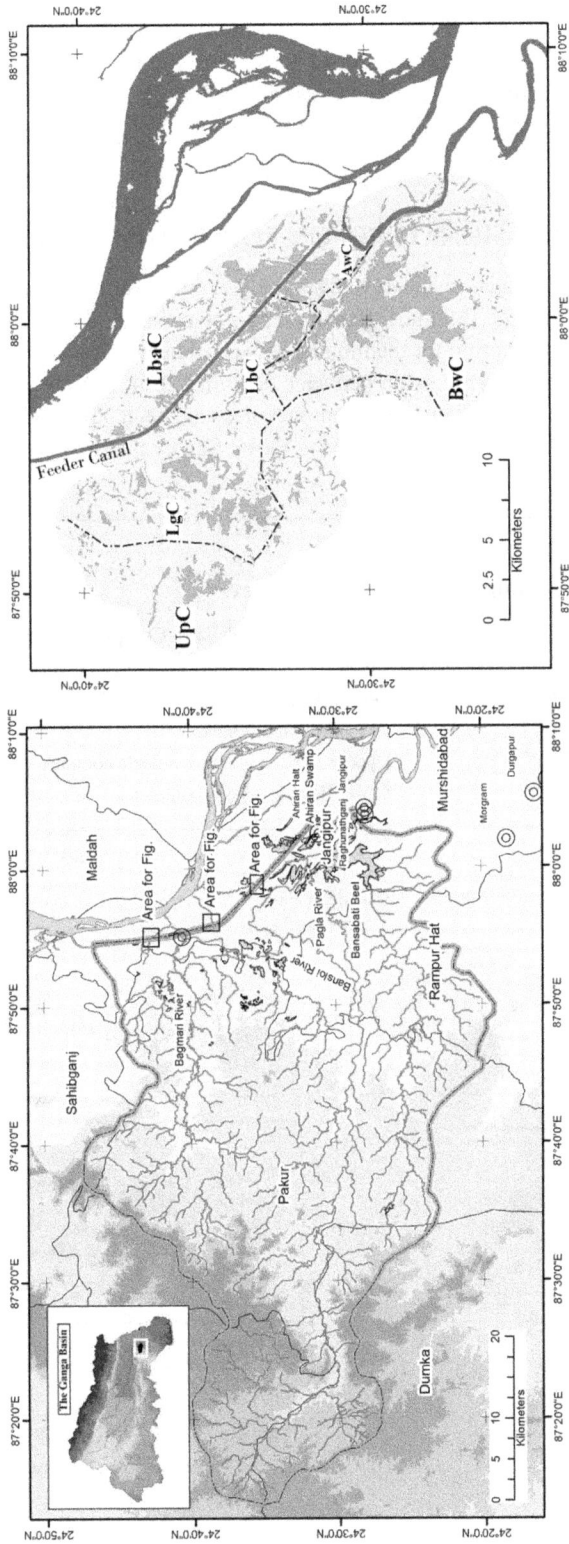

FIGURE 4.1 Location map of the study area.

associated overflow low-laying areas. Under the influence of the Indian monsoon, Bhagirathi and its right bank tributaries reach the annual peak discharge in monsoon season (June–September). Among all the wetland system, Ahiranbeel (comprising an area of ~400 ha) is renowned as an important bird sanctuary for colonial nesting of water birds during 2005–2006 by Ministry of Environment and Forests (MoEF) of Government of India.

For micro-level comprehensive analysis, I have divided the entire study area into six different wetland complexes.

4.3 ANTHROPOGENIC SIGNATURES ON FORMS AND PROCESSES

The produced mapping and related results reveal spatially explicit wetland inundation frequencies in six wetland complexes across the Bagmari–Bansloi–Pagla river basin (Table 4.1). Required satellite images (Landsat) have been acquired for the months between January and April as pre-monsoon and November and December as post-monsoon between 1987 and 2014. MNDWI (Xu, 2006) has been used to map the waterlogged area. Frequency approach has also been adopted for sorting the consistency in waterlogging area (Figure 4.2) (Borro et al., 2014; Sun et al., 2014; Pal and Talukdar, 2018).

TABLE 4.1
Six Deliniated Wetland Complexes

BwC	Bansabati Wetland Complex
LbC	Lower Bansloi complex
LbaC	Left bank of canal
AwC	Ahiran wetland complex
UpC	Upper reaches of Bagmari–Pagla complex
LgC	Lower Bagmari–Pagla complex

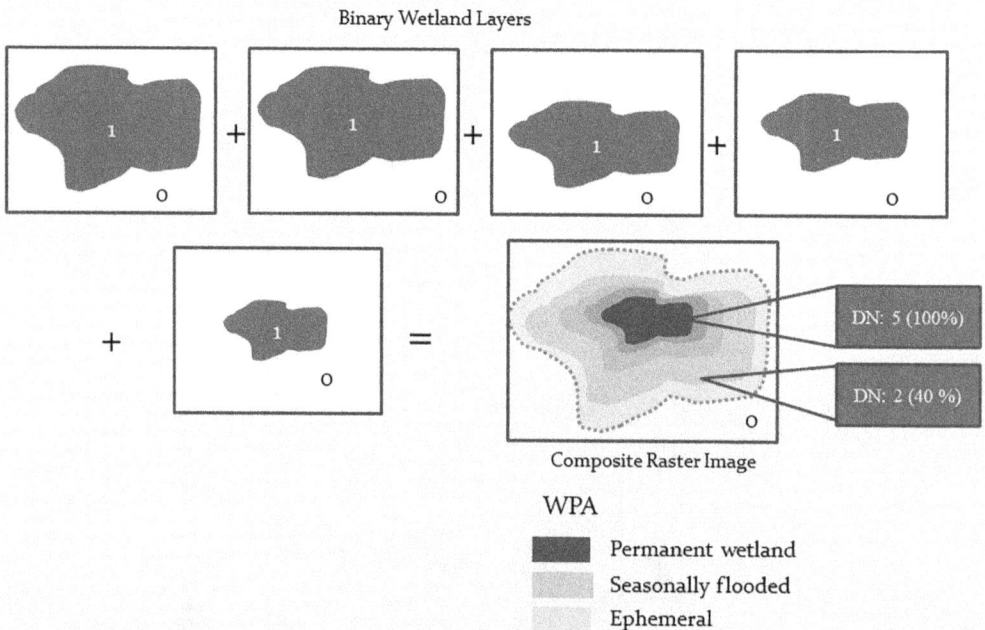

FIGURE 4.2 Schematic presentation of raster-based frequency approach.

Accordingly, the whole timeline from 1987 to 2014 has been split into two periods and seasons. Both the WPA (water presence area) maps (Figures 4.3 and 4.4) of the first time frame (period-i) represent the total waterlogging area with varying degree of water persistence that is about 4,132 ha in pre-monsoon and 6,475 ha in post-monsoon season, respectively. On the other hand, the total waterlogging areas in the latter 14 years aggregate with respect to the pre- and post-monsoon seasons are 2,751 and 5,729 ha, respectively. The phenomenon of widespread waterlogging in post-monsoon season proves that the area is highly controlled by the monsoon climate. The higher dynamics in waterlogging area between two seasons in the second study period signify the higher inter-annual variability of floodplain wetland landscape.

One thing I have observed is that there is a harsh decline in water persistence when change study was performed between the post-monsoons of period-i and period-ii. During the post-monsoon of period-i, about 49% of the total waterlogging area comprised areas under high degree of water

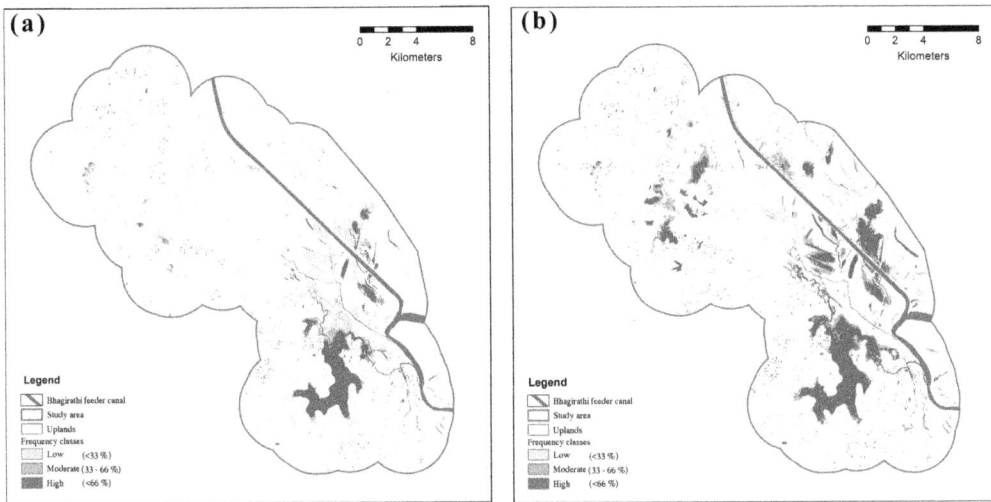

FIGURE 4.3 Spatial variation of persistent waterlogging area of period-i; pre-monsoon (a) and post-monsoon (b).

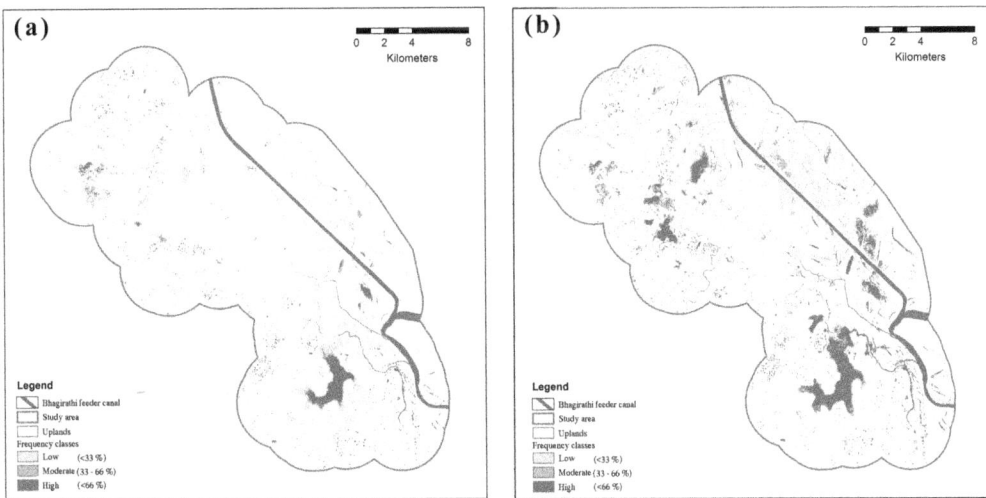

FIGURE 4.4 Spatial variation of persistent waterlogging area of period-ii: (a) pre-monsoon and (b) post-monsoon.

persistence (PWA: 66%). During period-ii, the high-frequency zones decreased dramatically (37% of the total waterlogging) along with increase in low water persistence zones (PWA: 33%) and that is about 45% to the total water inundation area.

As the main purpose of the current work is to identify the sensitive patches in terms of water presence within the floodplain, the whole study area has been divided into six wetland complexes in regard to comprehensively symbolise the spatio-temporal changes. From Figures 4.5 and 4.6, it can be observed that the maximum coverage of water surface area is in BwC and least in LbC. Among the six wetland systems, the LbaC, UpC, LbC and LgC are the highly variable in their areal extent, while BwC and AwC are relatively consistent. Shallowness due to silting process, agricultural encroachment, diminishing water supply from concerned river and somewhat relatively high relief are merely responsible for such phenomena. The widespread net deterioration in different wetland complexes provide some interesting pattern in three PWA classes. Once LbaC was positioned across river courses and after feeder canal construction, this adjacent floodplain was near to obsolete but abandoned low -lying region forming a small seasonally controlled wetland complex. The BwC and AwC are the topographic bowls at lower reach of Pagla and Bansloi Rivers. It receives water from both courses the Pagla–Bansloi and Bhagirathi Rivers during inundation period though

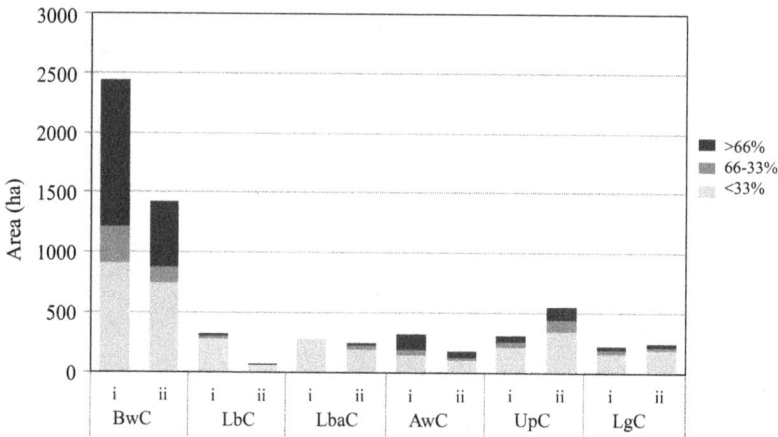

FIGURE 4.5 Wetland complex-wise area under various inundation zones for pre-monsoon (period-i and period-ii).

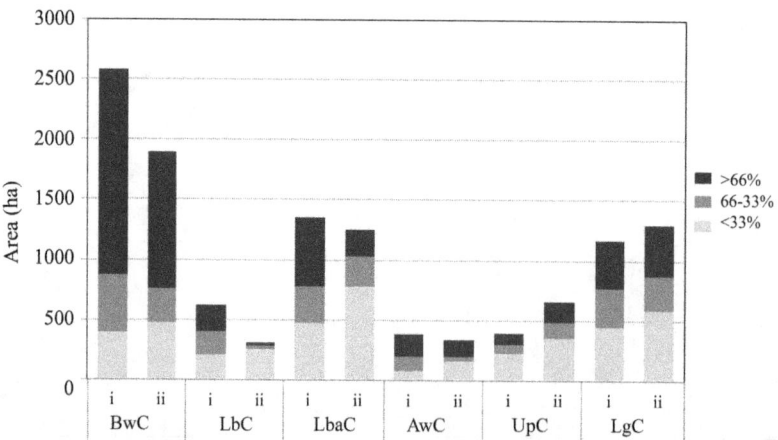

FIGURE 4.6 Wetland complex-wise area under various inundation zones for post-monsoon (period-i and period-ii).

the contribution of Pagla and Bansloi are quite insignificant. During the lean period, WPA with high-frequency class reveals its nature of permanency. The wetland system with relatively least hydrological disturbance in terms of water inundation frequency is observed in the AwC. It is one of the significant healthy marshy ecosystems considered as a winter habitat of some migratory birds.

4.3.1 SPATIO-TEMPORAL CHANGE IN PERSISTENT WATERLOGGING AREA

In order to identify the inter-annual variability of PWA, substation of the period-i from period-ii map has been done. Similarly, inter-seasonal changes of persistence waterlogging zones were detected by subtracting the PWA maps of the pre-monsoon maps from post-monsoon of both periods. The PWA change maps (Figures 4.7 and 4.8) and their attribute databases reveal the spatial variability of change magnitude (Table 4.2). During the last five decades, the Bagmari–Bansloi–Pagla basin has been affected by multi-spatial and -temporal anthropogenic interventions which have substantially modified its normal hydro-geomorphological settings. The anthropogenic control consists mostly of direct interventions, such as Bhagirathi feeder canal construction and road cum rail line, but also of indirect effects on floodplain hydro-dynamics, such as chemical fertiliser-based winter paddy cultivation and excessive groundwater withdrawal. About 400 years ago, the Bhagirathi River started deteriorating and was about to become mere a spill channel. Consequently, the Bhagirathi–Hugli River system gradually deteriorating and heavy silting in the river made the bottom of river shallower. At same time, there is gradual siltation at bifurcation point of the Bhagirathi River which reduced the water discharge which threatened the navigability of the Kolkata port (Mondal and Pal, 2017; Banerjee, 1999). Therefore, a multi-purpose Farakka Barrage Project was constructed to give the best technical solution of the problem. The project was conceived in the year 1858 but fully completed during 1975, where a 38 km long feeder canal was dug to divert 40,000 cusecs of water. Excavated feeder canal and its associated engineering structures hindered the natural flow of the rivers (Mondal and Pal, 2017). Accordingly, the right and left side of the canal converted into waterlogging area due to blockage of channel, and process water leaching also started from somewhat raised canal. This concurred with a significant second period inter-seasonal decreasing trend in wetland waterlogging area in contrast to a few stable zones in the period-i.

The altered flow regime disrupts the dynamic equilibrium of the three river system by accumulating huge sediment (Mondal and Pal, 2017) that ultimately converts the high wetland inundation frequency class to low-frequency zones. The inter-seasonal change statistics of period-I proved

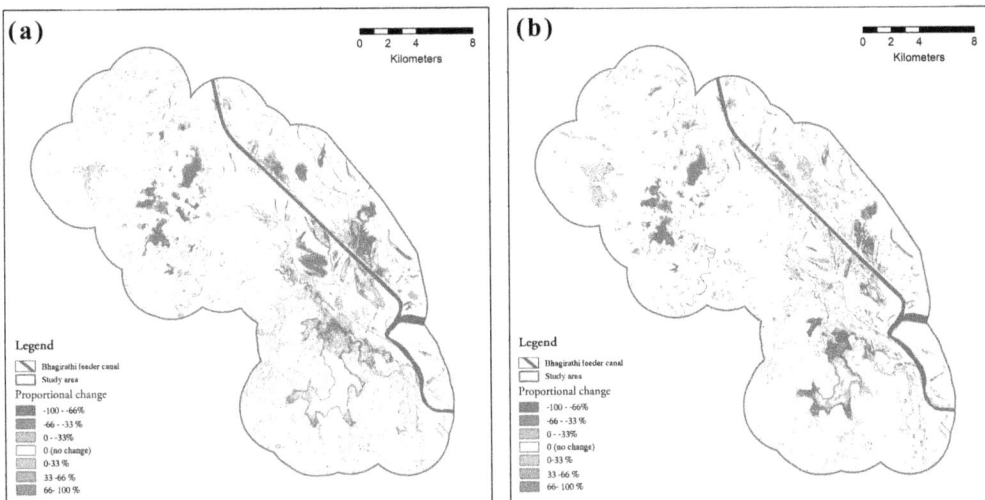

FIGURE 4.7   Inter-seasonal change map: (a) period-i and (b) period-ii.

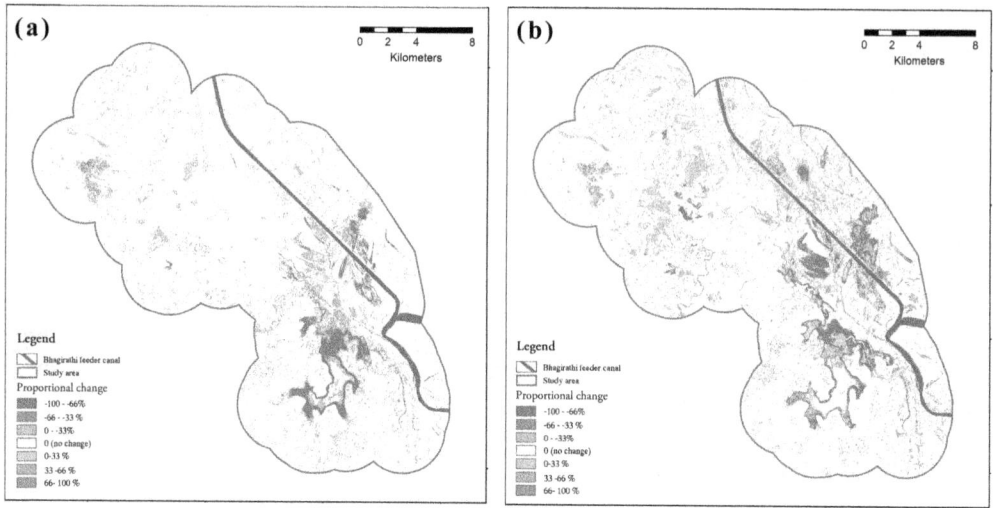

FIGURE 4.8   Inter-periodic change map: (a) pre-monsoon and (b) post-monsoon.

TABLE 4.2
Changes in Water Presence Frequency of Different Periods

Changes in Water Presence Frequency of Period-1 (Inter-Seasonal)						
Frequency Change Classes	BwC	LbC	LbaC	AwC	UpC	LgC
<−0.66	3.0	20.3	24.5	2.6	3.6	27.8
−0.33 to −0.66	16.8	33.2	25.1	17.6	8.7	29.9
−0.0 to −0.33	24.9	36.2	39.5	29.3	40.5	38.8
0	38.5	5.2	5.9	29.6	17.3	2.2
0 to 0.33	16.1	5.1	5.0	20.4	28.6	1.3
0.33 to 0.66	0.8	0.1	0.0	0.5	1.4	0.0
>0.66	0.0	0.0	0.0	0.0	0.0	0.0
Changes in Water Presence Frequency of Period-2 (Inter-Seasonal)						
<−0.66	15.6	7.4	10.5	11.6	3.0	26.7
−0.33 to −0.66	19.9	7.9	21.3	17.9	7.3	23.7
0.0 to −0.33	35.2	68.0	61.3	51.4	41.2	46.2
0	18.0	7.2	3.9	12.7	22.0	2.5
0 to 0.33	11.3	9.5	3.0	6.4	25.7	0.9
0.33 to 0.66	0.1	0.0	0.0	0.0	0.9	0.0
>0.66	0.0	0.0	0.0	0.0	0.0	0.0
Changes in Water Presence Frequency of Pre-Monsoon (Inter-Period)						
<−0.66	21.2	2.0	7.9	7.8	0.2	0.0
−0.33 to −0.66	14.2	7.2	21.2	22.3	2.1	1.6
0.0 to −0.33	38.4	74.3	46.7	48.5	18.5	31.3
0	16.4	6.1	7.4	12.5	20.7	28.2
0 to 0.33	7.8	10.3	14.5	8.5	42.1	35.6
0.33 to 0.66	1.6	0.0	1.6	0.4	11.8	2.7
>0.66	0.5	0.0	0.7	0.0	4.6	0.6

(Continued)

TABLE 4.2 (*Continued*)
Changes in Water Presence Frequency of Different Periods

Changes in Water Presence Frequency of Post-Monsoon (Inter-Period)					
<−0.66	11.9	19.0	7.7	3.3	1.0	2.0
−0.33 to −0.66	19.1	29.1	18.6	18.6	2.2	2.7
0 to −0.33	23.3	30.3	36.2	24.0	15.0	15.7
0	30.6	4.6	13.7	28.4	23.5	39.0
0 to 0.33	11.1	16.8	22.0	23.2	39.7	36.7
0.33 to 0.66	2.8	0.2	1.6	2.0	13.3	3.5
>0.66	1.1	0.0	0.4	0.4	5.3	0.3

the above facts. There are serious ecological implications for the Ahiran and Bansabati complex because of their ecological background. The drying tendency is acute in rest two complexes of lower Bonsai and left bank of the feeder canal, which sometimes makes it difficult to find their existence during the dry season.

4.3.2 LOSS OF FLOODPLAIN CONNECTIVITY

As with persistent waterlogging area, the current results document the floodplain alteration and channel forms and associated changes in longitudinal, lateral and vertical connectivity. In this low-lying area, long man-made levees were created on both sides of the Bhagirathi feeder canal

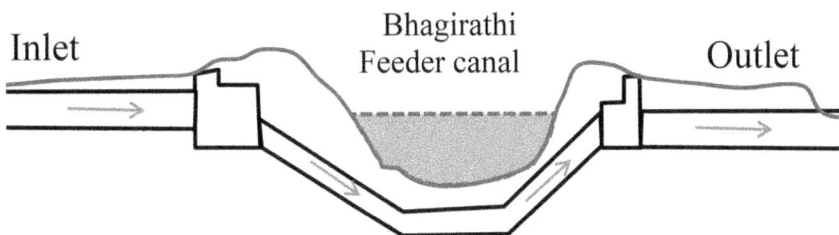

FIGURE 4.9 Arial and sectional view of the Bagmari syphon.

that traverse in north–south direction. The height of the canal was somewhat improved from the normal height to continue the water supply in Bhagirathi River, due to which the height of levees on both sides of canal, apart from the surroundings, has to be raised a lot. Such levees creates divisive barrier in the floodplain that hinders direct water exchange between left and right banks of the canal. Although one syphon (Bagmari syphon, Figure 4.9), inlets and regulators were constructed in several places to overcome this worse condition but that was not adequate. As a result, feeder canal and associated levees severely reduced lateral connectivity water, sediment, organic component and aquatic species, leading to loss of wetland habitat and water bird abundance over the study area.

4.4 CONCLUSION

The systematic mapping and change monitoring of waterlogging area of the present study showed that the feeder canal and its associated engineering works play significant role in alteration of floodplain. The cumulative impacts increased sedimentation in wetland bottom agricultural growth, and decline in groundwater table is also responsible for floodplain changes. In specific terms, the lower reaches of Bagmari–Bansloi–Pagla River are multiple-stress fragile floodplain ecosystem, and the study shows that its degree of inundation frequency changes from permanent to ephemeral category. On the other hand, during the past few decades, freshwater permanent wetlands in both abundance and area have experienced widespread changes over the study area. These dramatic floodplain modifications in terms of inundation regime transformation, their abundance and associated hydro-geomorphological instability support an uprising uncertainty and vulnerability of future water needs in the upcoming decades. Therefore, there is an urgent need to manage wetland ecosystem in a holistic and alternative approach for the endangered bird species. To this end, current mapping and related record can effectively support a decision regarding various inundation stability zones in floodplain on which the plan will be based on. Alternatively, it also provides a fundamental and essential monitoring root for continuous future study.

REFERENCES

Banerjee, M. (1999). *A Report on the Impact of Farakka Barrage on the Human Fabric*. South Asian Network on Dams, Rivers and People, New Delhi.
Borro, M., Morandeira, N., Salvia, M., Minotti, P., Perna, P., & Kandus, P. (2014). Mapping shallow lakes in a large South American floodplain: a frequency approach on multitemporal Landsat TM/ETM data. *Journal of Hydrology*, *512*, 39–52.
Ghosh, S., & Guchhait, S. K. (2014). Hydrogeomorphic variability due to dam constructions and emerging problems: a case study of Damodar River, West Bengal, India. *Environment, Development and Sustainability*, *16*(3), 769–796.
Ghosh, S., & Guchhait, S. K. (2016). Dam-induced changes in flood hydrology and flood frequency of tropical river: a study in Damodar River of West Bengal, India. *Arabian Journal of Geosciences*, *9*(2), 90.
Islam, A., & Guchhait, S. K. (2017). Analysing the influence of Farakka Barrage Project on channel dynamics and meander geometry of Bhagirathi river of West Bengal, India. *Arabian Journal of Geosciences*, *10*(11), 245.
Mondal, D., & Pal, S. (2017). Evolution of wetlands in lower reaches of Bagmari-Bansloi-Pagla rivers: a study using multidated images and maps. *Current Science (00113891)*, *112*(11), 2263–2272.
Mondal, D., & Pal, S. (2018). Monitoring dual-season hydrological dynamics of seasonally flooded wetlands in the lower reach of Mayurakshi River, Eastern India. *Geocarto International*, *33*(3), 225–239.
Pal, S., & Talukdar, S. (2018). Drivers of vulnerability to wetlands in Punarbhaba river basin of India-Bangladesh. *Ecological Indicators*, *93*, 612–626.
Roy, S., & Sahu, A. S. (2017). Potential interaction between transport and stream networks over the lowland rivers in Eastern India. *Journal of Environmental Management*, *197*, 316–330.
Roy, S., & Sahu, A. S. (2018). Road-stream crossing an in-stream intervention to alter channel morphology of headwater streams: case study. *International Journal of River Basin Management*, *16*(1), 1–19.

Saha, T.K., & Pal, S. (2019). Emerging conflict between agriculture extension and physical existence of wetland in post-dam period in Atreyee River basin of Indo-Bangladesh. *Environment, Development and Sustainability*, 21(3), 1485–1505. https://doi.org/10.1007/s10668-018-0099-x

Talukdar, S., & Pal, S. (2018). Wetland habitat vulnerability of lower Punarbhaba river basin of the uplifted Barind region of Indo-Bangladesh. *Geocarto International*, 1–30.

Thomas, R. F., Kingsford, R. T., Lu, Y., & Hunter, S. J. (2011). Landsat mapping of annual inundation (1979–2006) of the Macquarie Marshes in semi-arid Australia. *International Journal of Remote Sensing*, 32, 4545–4569.

Wang, J., Sheng, Y., & Tong, T. S. D. (2014). Monitoring decadal lake dynamics across the Yangtze Basin downstream of Three Gorges Dam. *Remote Sensing of Environment*, *152*, 251–269.

Ward, J. V., & Stanford, J. A. (1995). Ecological connectivity in alluvial river ecosystems and its disruption by flow regulation. *Regulated Rivers*, *11*, 105–119.

Woessner, W. W. (2017). Hyporheic zones. In F. R. Hauer & G. A. Lamberti (Eds.), *Methods in Stream Ecology* (Vol. 1, pp. 129–157). Elsevier, London.

Xu, H. (2006). Modification of normalised difference water index (NDWI) to enhance open water features in remotely sensed imagery. *International Journal of Remote Sensing*, *27*, 3025–3033.

5 Role of Human Interventions in the Evolution of Forms and Processes in the Mayurakshi River Basin

Aznarul Islam, Suman Deb Barman,
Mainul Islam and Susmita Ghosh

CONTENTS

5.1 INTRODUCTION

In the era of Anthropocene when every sphere of this terrestrial system is being transformed into an inorganic and unsustainable habitat for the sustenance of any living creature, it is imperative to evaluate the status of all the geophysical and anthropogenic stuffs that are incessantly engaged in sculpturing the physical landscape of this 'Blue Planet' (Davis and Turpin, 2015). The Mayurakshi, Ajay and the Damodar, having a peninsular rainfall regime, have been inducing flood since long, and hence, this system is often called the *Sorrow of Bengal* (Ghosh and Mistri, 2012). The geomorphic evolutions of these basins including the flood geomorphology are the results of human interventions at a small or large scale as discussed in Chapter 6 for Damodar and Chapter 7 for Ajay. The majority of the previous works concentrated on the floods of the Mayurakshi River Basin (MRB) that are mainly due to the breaching of embankment (Office of the District Magistrate, 2016). From the perspective of the anthropogeomorphology, it must be stressed that the MRB offers some exquisite examples of human interventions in the form of dams and barrages, embankments, sand mining and stone crushing, on-bed and off-bed agricultural practices and deforestation. Ghosh and Pal (2015) have shown that in the post-dam condition, flood peaks have been reduced while the low to medium magnitude flood has increased due to construction of the Massanjore and Tilpara barrage. Similarly, Pal (2017) has shown that due to construction of Tilpara barrage, backwater reach of the Kushkarani River, a tributary of Mayurakshi River, portrays a lower bed slope and higher rate of sedimentation of the tributary. Pradhan et al. (2019) showed that damming of the river Brahmani has portrayed an increasing lean season discharge and a decreasing peak of the suspended sediment concentration in the downstream segment. Besides, the braiding and the channel cut-off have increased in the post-dam period. Furthermore, Pal (2016) showed how the embankment along the river Mayurakshi has increased the problem of sediment deposition in the channel. Similarly, sand mining and stone crushing are other interventions noted in the MRB. Pal and Mandal (2019) observed that due to dust deposition from the stone crushing, channel aggradations occur besides the increasing sediment load and total dissolved solids (TDS) concentration and damaging of the water quality. Moreover, Chatterjee (1995) showed that the forest cover constituted by the tropical deciduous (*sal, segun, palash*) is in the degraded condition. Debanshi and Pal (2020) noted that the forest cover has decreased at an alarming rate (42%) during 1972–2016 which increased the soil and gully erosion in the MRB. In brief, it can thus be mentioned that the MRB bears the tell-tale marks of the human interventions which have triggered see-saw changes in the geomorphic landscape. Having a close peruse at the previous works, it becomes apparent that some areas still are virtually unexplored while others need treatment at depth. Thus, the major gaps in the present discourse are valley configuration by embankment, changes in channel morphology due to sand mining and changes in valley profile due to on-bed and off-bed agricultural practices. Therefore, the prime objective of the present chapter is to address these unexplored issues coupled with the other issues already encountered by the previous works related to an accelerated rate of human interventions in the evolutionary record of the MRB to prepare this work for baseline assessment from the perspective of the anthropogeomorphology.

5.2 THE RIVER AND ITS BASIN

Mayurakshi River having originated from Trikut Hill near Deoghar, Jharkhand, has debouched with the river Bhagirathi near Kalyanpur traversing a course of 250 km (Islam and Barman, 2020). The basin extends 23°37′43″ N to 24°37′36″ N latitude and 86°50′16″ E to 88°15′52″ E longitude covering an area of about 9,596 km². This basin is configured by the three main rivers, namely, Mayurakshi, Dwarka and Kuea. We have demarked 13 sub-basins having a minimum order of four (eight sub-basins of Mayurakshi—M1 for Bhurbhuri River, M2 for Dhobbi River, M3 for Matihara River, M4 for Tepra River, M5 for Bhamri River, M6 for Siddheswari River,

M7 for Kushkarani River, M8 for Pusharo *Nala*; three sub-basins of Dwarka—D1 for Dwarka River, D2 for Chila–Gharmora *Nala*, D3 for Brahmani River and two sub-basins of Kuea—K1 for Kopai River and K2 for Bakreshwar River) based on the 'V' rule of Patton and Baker (1976) (Figure 5.1). The river Mayurakshi has a gradual fall in relief towards downstream from its source to mouth. The MRB has thus been segmented into three reaches based on the relief characteristics and regime of the river. The head reach extends from the Trikut Hill to Silajuri, middle reach from

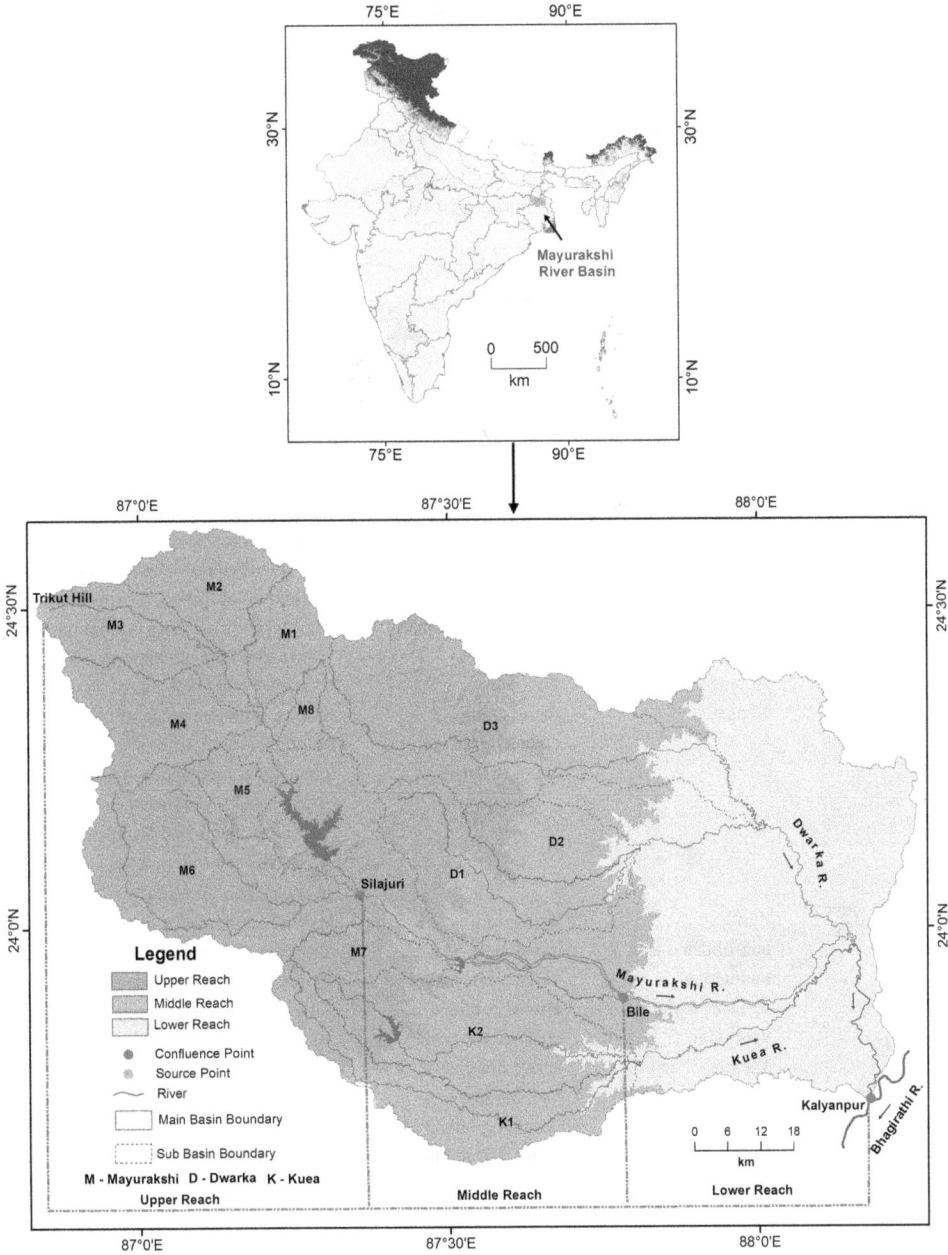

FIGURE 5.1 Location of the MRB and its components (prepared from Survey of India (SoI) 1: 50,000 toposheets and SRTM DEM 2000).

Silajuri to Bile and the tail from Bile to Kalyanpur where the river has its confluence with the Bhagirathi. The head reach dominated by higher relief accounts for about 101.7 km length of the river with 4,447.9 km² basin area, while the tail reach characterised by erosional plain traverses a distance of 97.5 km having an areal coverage of about 2,716.1 km². The middle reach accounting for about 56.2 km linear distance and 2,430.4 km² area exhibits characteristics of a plateau fringe.

Each reach has its own distinctive characteristics from the perspective of geology and physiographic expressions (Table 5.1).

Furthermore, the MRB is much diversified in its geological aspects, relief, drainage, soil and climatic characteristics briefly discussed in the following sections.

TABLE 5.1
Distinctive Features in Different Reaches of the Mayurakshi River

Reaches of Mayurakshi River	Landscape Settings Typology	Valley and Channel Characteristics	Distinct Hydro-geomorphological features
Head reach (Trikut Hill to Silajuri)	Uplands and dissected part of Chotanagpur Plateau, especially the parts of Deccan shields.	Moderately spaced valley form with confined channel and absolute absence of active floodplain; maintaining a channel gradient of about 7.6 m km⁻¹	Partial maintenance of ideal flow condition; at the forefront of Massanjore dam, channel congestion, formation of alternating channel bar with braided channel patterns and huge submerged landmass noticed; dominance of pediplain, pediment, residual hill, ridge, moderately dissected hill and valley, inselberg, pediment-corestone-tor composite, Rajmahal traps and gullied land
Middle reach (Silajuri to Bile)	Traversing over two hinterlands, such as remnants of Chotanagpur Plateau to moderately sloping *Rarh* Bengal	Partly narrow to widespread valley form, with partially confined to unconfined channel and active floodplain; maintaining a channel gradient of about 0.7 m km⁻¹	Prolific human modifications in the form of numerous barrages located at main stream and its tributaries, mining, intensive agricultural practice. As a result, most of the channels run drier throughout the year; channel–floodplain interactions interrupted in the pre- and post-monsoon periods; imperceptible merging of channel and floodplain with negligible valley slope; channel looking agricultural field and grazing belts; prominence of pediplain, pediment, pediment-pediplain complex, valley fill, lateritic upland, older floodplain, younger floodplain, lateral bar, channel bar
Tail reach (Bile to Kalyanpur)	Slightly consolidated and undulating surfaces at the west with alluvial Ganga plain in the east.	Narrow to moderately spaced valley floor with almost confined channel and mainly active floodplain; maintaining a channel gradient of about 0.05 m km⁻¹	All kinds of human activities ranging from drinking of river water to a preferable site of waste disposal found; due to paving and embankment in this reach, the channels most likely to appear as a very narrow and straight artificial canal. Sand mining leading to channel widening. Criss-crossed by spill-over channel; bed deposition from mound of sand extracted leading to very less height difference between floodplain and river bed; frequent inundation and formation of swampy land; palaeo levee, point bar, and meander scar

Source: Prepared by the authors.

5.2.1 GEOLOGY

The formation of the upper part of the river basin dates back to the deposition of Dharwanian sediments followed by Hercynian orogeny from Cambrian to Silurian period. Most of the upper parts of the basin are composed of Granitic Gneiss. Some other depositions such as Hornblende, Schist and Amphibolite; Charnockite and Augen Gneiss; Pegmatite and Migmatite; and Sandstone and Shale are found in small patches (Figure 5.2). The middle part of the river basin is mostly characterised by the deposition of lateritic soil and hard clays impregnated with caliche nodules. The lower catchment is mostly characterised by recent alluvial deposition of alternate layers of sand, silt and clay attributed by alluviation of river diversion, flooding and consequent shaping by twin action of Kuea (right bank tributary of Mayurakshi) and Dwarka (left bank tributary of Mayurakshi River).

5.2.2 BASIN MORPHOMETRY AND RELIEF

The basin area of the tributary rivers varies greatly from about as low as $73\,km^2$ (M8) to as high as $1,300\,km^2$ (D3) with the co-efficient of variation (CV) of about 76% (Table 5.2). The average basin size is about $450\,km^2$. Moreover, the basin perimeter varies from about 55 m (M8) to 270 m (D3). A similar observation is recorded for the basin length. Furthermore, the elongation ratio of the basins ranges from about 0.45 (K1) to 0.95 (M2) indicating a higher diversity of the basin shape (Table 5.2).

From the evolutionary perspective, most of the tributary basins are located in the mature or old stages as indicated by the lower value of the hypsometric integral (<0.3). Moreover, the relative relief is high (~330m) as it is maturely dissected by the fluvial process. However, relative relief mostly varies from one basin to another. For example, basin M7 portrays relative relief as low as 98 m, while the basin M6 has recorded relative relief as high as 604 m (Figure 5.3).

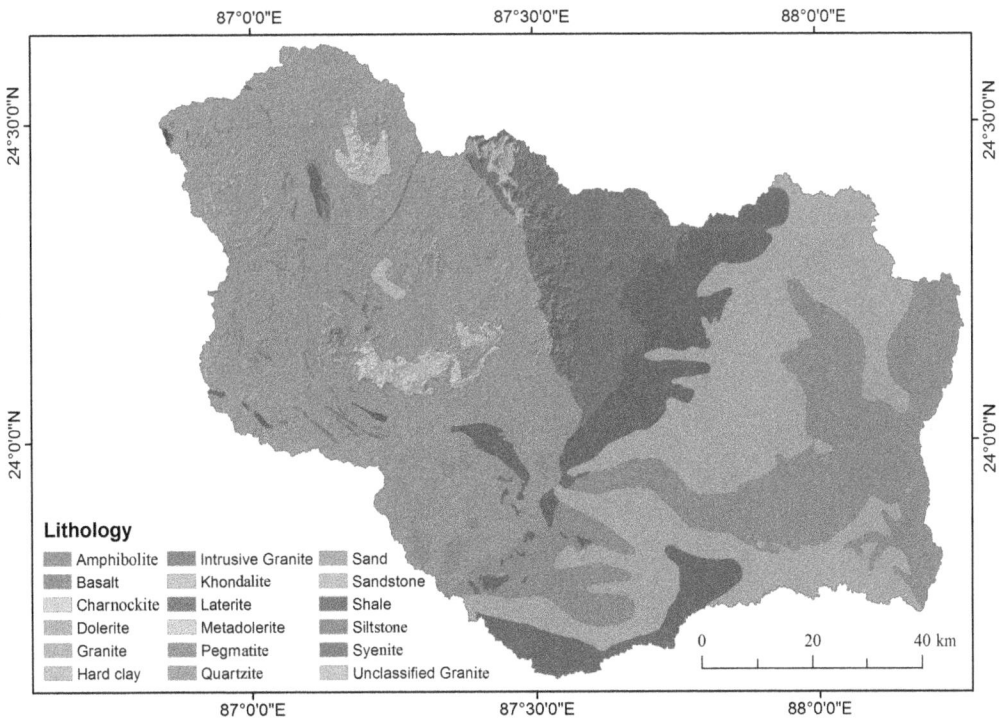

FIGURE 5.2 Geology of the MRB. (Prepared from District Resource Map of Birbhum, Murshidabad and Barddhaman of West Bengal and Devghar, Dumka, Sahibganj and Pakaur of Jharkhand, GSI.)

TABLE 5.2

Planform Characteristics

Basin Name	Basin Area (km²)	Basin Perimeter (km)	Basin Length (km)	Elongation Ratio
D1	557.07	180.63	42.48	0.63
D2	272.66	91.99	27.86	0.67
D3	1,262.94	268.75	72.77	0.55
K1	554.11	226.47	60.27	0.44
K2	726.16	184.55	54.49	0.56
M1	202.68	78.28	24.31	0.66
M2	321.61	105.54	21.82	0.93
M3	316.32	99.95	31.98	0.63
M4	381.70	123.91	34.33	0.64
M5	121.87	63.27	16.33	0.76
M6	874.45	179.15	49.42	0.68
M7	177.92	85.96	26.70	0.56
M8	72.68	55.11	16.56	0.58
Mean	449.40	134.12	36.87	0.64
SD	340.99	67.05	17.72	0.12
CV	75.88	49.99	48.07	18.35
Kurtosis	1.37	−0.51	−0.40	2.67
Skewness	1.25	0.74	0.75	1.05

Source: Prepared by the authors.

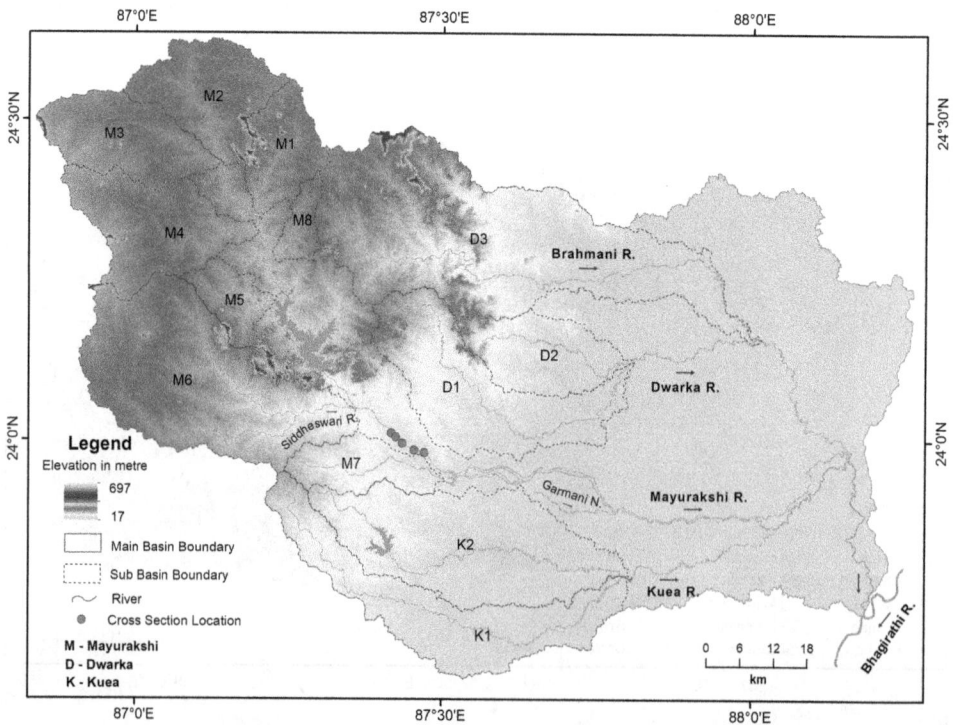

FIGURE 5.3   Relief of the MRB. (Prepared from SRTM DEM 30 m, 2000 and SoI 1: 50,000 toposheets.)

TABLE 5.3
Relief Characteristics

Basin Name	Hypsometric Integral	Relative Relief (m)	Dissection Index	Ruggedness Number	Average Slope (°)
D1	0.22	329.00	0.93	0.31	1.51
D2	0.17	301.00	0.93	0.28	1.54
D3	0.42	487.00	0.97	0.47	2.32
K1	0.33	137.00	0.86	0.12	0.75
K2	0.33	122.00	0.84	0.10	0.84
M1	0.23	354.00	0.73	0.26	3.13
M2	0.19	336.00	0.70	0.23	1.99
M3	0.20	563.00	0.79	0.45	1.93
M4	0.33	340.00	0.74	0.25	1.94
M5	0.17	297.00	0.72	0.21	2.66
M6	0.30	604.00	0.89	0.54	2.09
M7	0.42	98.00	0.62	0.06	1.18
M8	0.25	228.00	0.66	0.15	2.18
Mean	0.27	322.77	0.80	0.26	1.85
SD	0.09	158.30	0.11	0.15	0.68
CV	32.19	49.05	14.24	55.65	36.95
Kurtosis	−0.99	−0.47	−1.33	−0.50	−0.18
Skewness	0.46	0.34	0.05	0.57	0.02

Source: Prepared by the authors.

Moreover, the dissection index is quite high for the basin as a whole (0.80). Similarly, the sub-basins also have high values of the dissection index (Table 5.3). Furthermore, the average ruggedness number (0.26) portrays the undulations produced by the differential erosion of the basin. Regarding the mean slope of the basin, it is observed that the slope ranges from 1° to 4°.

Furthermore, the MRB bears the marks complex physiographic evolution and valley development. A study carried out by Chakrabarti (1985) indicates that the Mayurakshi and its tributary rivers have clear breaks in their long profiles i.e. characterised by knick points and terraces. Chakrabarti (1985) argues that under similar lithological formations especially in the upper stretches of the Mayurakshi River, three breaks at an elevations of 213.36, 182.88 and 152.40 m are not possible except changes in the base level of erosion. Moreover, the tributaries also have breaks at the similar altitude (Dwarka: 167.64, 91.44 m; Brahmani: 167.64, 137.16, 91.44 m; Siddheswari: 167.64, 91.44 m; Tepra: 213.36, 167.69 m; and Dhobbi: 213.36 m), which indicates the accordance of the erosion surfaces. Besides, the upper part of the river has been almost graded to their profile. However, there is a regional uplift in the middle of the profile of the Mayurakshi River (Figure 5.4). In brief, the MRB is profusely controlled by the regional tectonics containing at least four erosional surfaces.

5.2.3 Drainage Characteristics

The name Mayurakshi denotes transparent water. It has been derived from two words—*Mayur* meaning peacock and *Ankhi* meaning eye. In the post-monsoon period, the crystal clear water resembles peacock eye (Figure 5.5).

The MRB has an average drainage density of about 1.52 km/km². However, there is a spatial variation across the different basins. For example, basin K1 has a minimum density (0.60), while the basin M3 has the maximum density (1.93). A similar observation is recorded for the stream

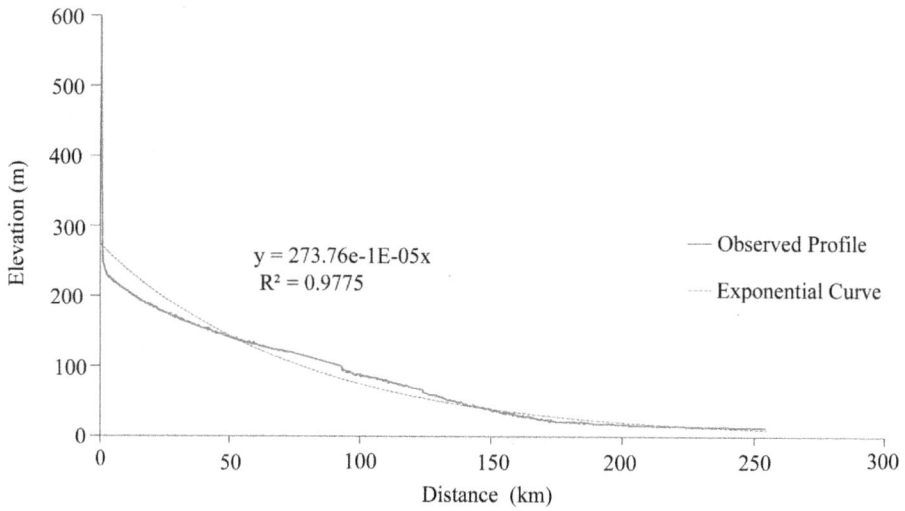

$$y = 273.76e\text{-}1E\text{-}05x$$
$$R^2 = 0.9775$$

FIGURE 5.4 Breaks in the long profile of the Mayurakshi River. (Based on SRTM DEM, 30 m, 2000.)

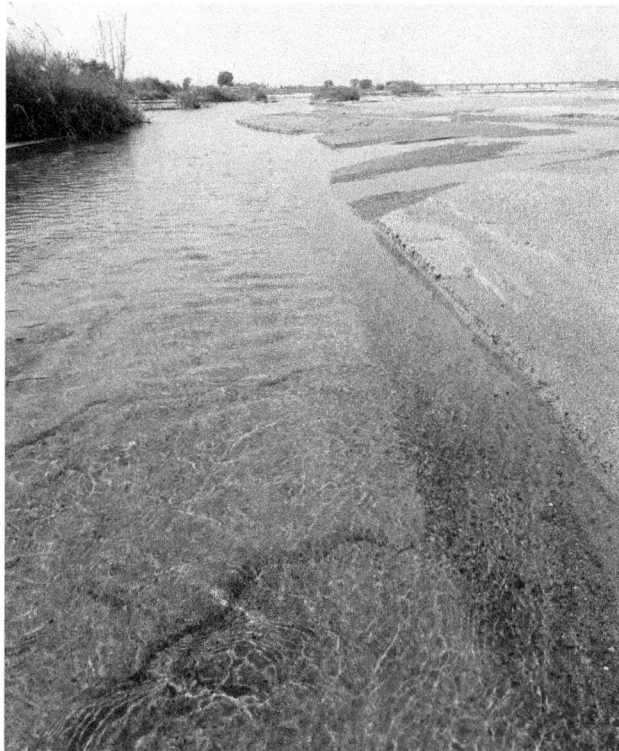

FIGURE 5.5 Crystal clear water of Mayurakshi River at Amjora, Jharkhand during post-monsoon period. (Field Photograph, 16 November 2019.)

frequency. Moreover, the average drainage texture of the MRB is about 2.85 with a minimum of about 0.25 recorded by the M1 and a maximum of about 4.93 recorded by the basin M3. Besides, the bifurcation ratio of this basin is also high (4.32) with a minimum of 2.95 (M5) and a maximum of about 6.76 (M7) (Table 5.4).

TABLE 5.4
Drainage Characteristics

Basin Name	Drainage Density (km/km^2)	Stream Frequency	Drainage Texture	Average Bifurcation Ratio
D1	1.64	2.03	3.33	5.85
D2	1.17	1.05	1.22	4.12
D3	1.45	1.49	2.16	3.50
K1	0.60	0.43	0.25	3.87
K2	0.82	0.65	0.53	4.50
M1	1.68	1.96	3.30	4.26
M2	1.67	1.99	3.31	3.49
M3	1.91	2.58	4.92	3.81
M4	1.89	2.49	4.71	5.32
M5	1.67	1.80	3.01	2.95
M6	1.93	2.21	4.27	4.41
M7	1.64	1.69	2.77	6.76
M8	1.67	1.95	3.26	3.32
Mean	1.52	1.72	2.85	4.32
SD	0.41	0.66	1.47	1.08
CV	27.23	38.28	51.50	25.06
Kurtosis	1.00	−0.03	−0.48	0.82
Skewness	−1.34	−0.83	−0.50	1.10

Source: Prepared by the authors.

5.2.4 SOIL AND CLIMATE CHARACTERISTICS

Textural variation of soils in the entire MRB is remarkably observed. The upper part of the river basin is a basaltic trap where dominant soil type is fragile coarse lateritic soil having sandy and sandy loam textural character (Figure 5.6). These soils are poorly aggregated and water-holding capacity is very low. In the middle part of the river basin is the *Rarh* area characterised by transported lateritic alluvium also known as residual soil. In the lower part, most of the areas are very prone to siltation due to occurrence of frequent floods (Chakrabarti, 1985). The texture of the soil observed is clay, clay loam, and loam which have high water retention capacity.

The upper stretch of the river is a part of plateau fringe influenced by typical heavy downpour in the monsoon months especially during September and October. This typical rainfall regime coupled with a low-pressure system developed over the Bay of Bengal induces floods in the basin.

5.3 ANTHROPOGENIC SIGNATURES ON FORMS AND PROCESSES

Fluvial geomorphological research in India has widely been concentrated on the river reaction to climate change and tectonic control in Quaternary Period (Jain et al., 2012). Present research of fluvial systems has exerted more emphasis on spatial variability and incorporated new conceptual considerations that include hierarchy, frequency, magnitude, threshold, equilibrium, non-linearity, connectivity, sensitivity, complexity and multidisciplinary (Jain et al., 2012; Gregory and Lewin, 2014). But in recent times, some works have been conducted on the area of contemporary river behaviour as a response to human interventions. For example, Sinha et al. (2012) and Jain et al. (2016) stated that anthropogenic influences cause flux or slope variability in the channel which alter the geomorphology and ecological sustainability of the river system. Similar tunes were also echoed

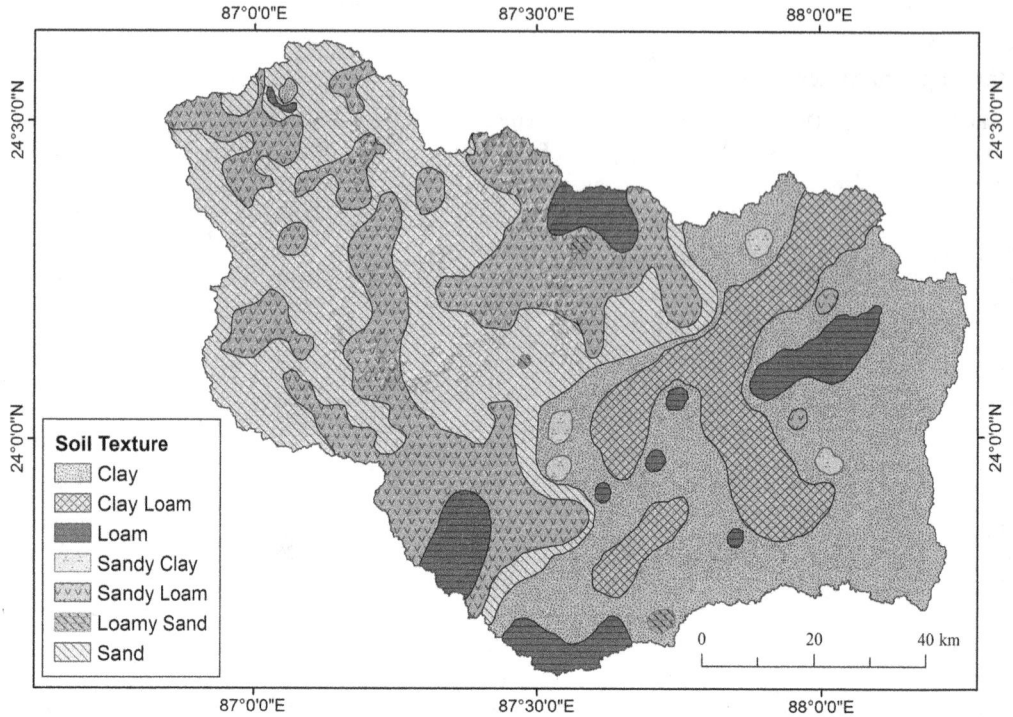

FIGURE 5.6 Soils of the MRB. (Based on Chakrabarti (1985).)

in the works of Ghosh and Guchhait (2014). They argued that anthropogenic hindrances occurred through a series of constructions of dams, reservoirs and embankments. In the present context, the major anthropogenic signatures include dams and barrages, embankments, check dams, sand mining and stone quarrying, road stream crossing and change in the land use and land cover (LULC) as discussed in the following sections.

5.3.1 DAMS, BARRAGES AND IRRIGATION CANAL

5.3.1.1 Location and Characteristics

As humans got civilised, the growing demands for natural resources especially for water had increased manifolds. As a result, the present-day manifestations of dams, barrages, and reservoirs are constructed to meet the thriving needs of the masses. The formation of dam has inflated secure water distribution by 28% which is expected to grow to 34% by 2025 (Rao et al., 2013). Thus, frequent construction of dams and reservoirs has grieved the followers of deep ecology. However, its beneficial effects in the directions of human developments are also stressed by the regional planners and development economists. Dam installation with time has exponentially increased especially in areas of hydrological aberration including extremely dry and high precipitation areas. At present, approximately 70% of the world's rivers are chocked by large reservoirs (Vörösmarty, 2003). International Commission on Irrigation and Drainage (ICID) in its report revealed that massive dams have enabled people to harnessing large water resource potential to meet the demands of rapidly growing societies around the globe. It encompasses food, fodder, fish production (aquaculture), drinking water, clothing fibres, sanitation, energy, industry, wildlife and other lucrative outputs coming from the dams (Role of Dams for Irrigation, Drainage and Flood Control, 1999–2000). Another report of International Water Management Institute (IWMI) indicated that the best irrigation efficiency has been achieved through extending irrigation areas by establishing more dams and storages. A survey conducted by IWMI in its 87 member countries showed that 95% of total irrigated area had contributed about 40% of the world's food production. In the context of the MRB, the tell-tale marks of the human interventions in the form of the dams and barrages since independence are prominent. The major objectives for the construction of the dams are to control the historically documented floods in the basins. All the major rivers including the Brahmani, Dwarka, Mayurakshi, Bakreshwar Kopai and Kuea are controlled by the massive hydraulic interventions i.e. Brahmani barrage over Brahmani River, Deucha barrage on the Dwarka River, Massanjore dam and Tilpara barrage on the Mayurakshi River, Bakreshwar weir on Bakreshwar River, Kopai barrage on Kopai River and Labhpur mini barrage on Kuea River (Figure 5.7). Actually, the dams and barrages were conceived as early as 1928 through the notion of the Mayurakshi Reservoir Project, and after independence, they were started in 1951 and implemented in 1955. This project has been completed in all respects in 1985. Massanjore dam (24°6′24″ N and 87°18′28.13″ E) located in Dumka district of Jharkhand is the first major intervention in the MRB. It is located at the distance of about 94 km from the source of the Mayurakshi River covering a catchment area of about 1,860 km² and a flood design level of 122.56 m. It was commissioned in 1955 and had cost of Indian National Rupees (INR) 16.10 crore (Mohammad, 2006). Subsequently, another massive project, Tilpara barrage (23°56′46″ N and 87°31′31″ E), was built on Mayurakshi River (Figure 5.8a) in 1951 in Birbhum district of West Bengal at a far downstream of the river (124 km from the source of the Mayurakshi River). Compared to the other dams and barrages, it covers the maximum catchment area of the MRB accounting for about 3,208 km² with a flood design of 63.44 m. It had cost of INR 1.11 crore. Furthermore, Brahmani barrage was constructed with an eye to controlling flood and irrigating vast agriculture land of the MRB. It is situated in Birbhum district (24°15′ 4″N and 22″ N 87°44′20″ E) at a distance of about 76 km from the source of the Brahmani River. It encompasses a catchment area of about 689 km² with flood design level of 48.82 m. Similarly, Dwarka, another tributary of

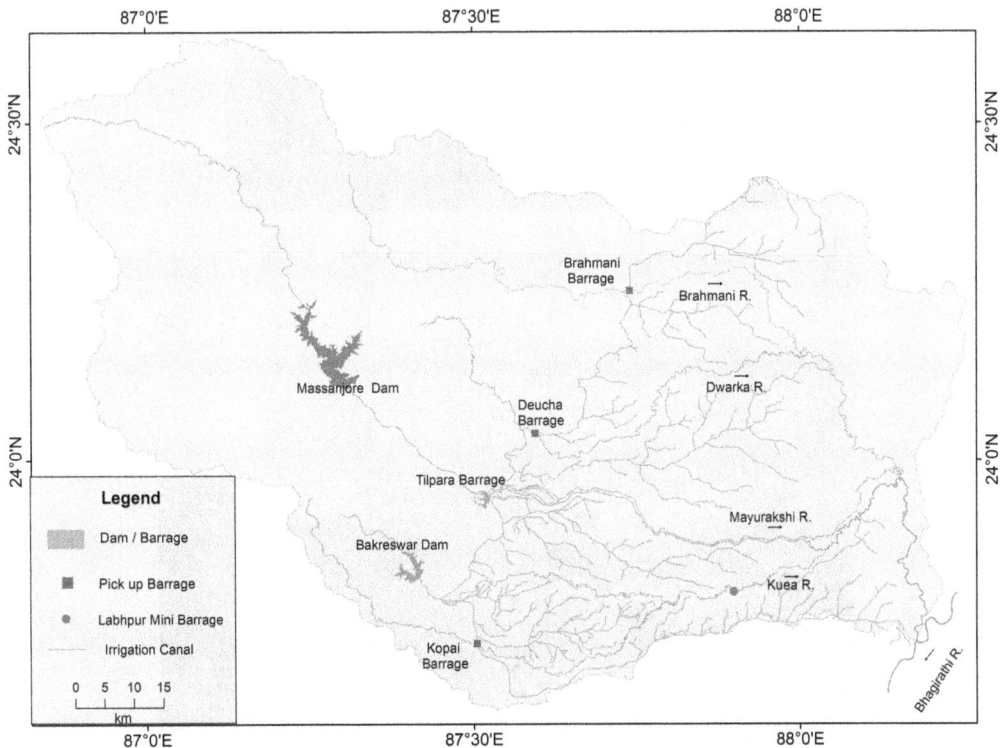

FIGURE 5.7 Location of dams, barrages and irrigation canal. (Prepared from NRSC Bhuvan. https://bhuvan-app1.nrsc.gov.in/thematic/thematic/index.php) and SoI 1: 50,000 toposheets.

Mayurakshi River, is controlled by a civil structure popularly known as Deucha barrage (24°2′24″ N and 87°35′41.78″ E) which is located in Birbhum district, West Bengal, at a distance of 43 km from the source of the Dwarka River covering an area of 303 km² with a flood discharge level of 54 m. Besides, in the *Rarh* tract of West Bengal, another two barrages have been constructed mainly for irrigation in the lower stretch of Mayurakshi which is a fertile tract of the basin. Bakreshwar weir is located at the distance from 23 km from the source of the Bakreshwar River in Birbhum district of West Bengal (23°49′34″ N and 87°25′1″ E) covering a catchment area of 1,403 km² with a flood design of 60.15 m. Moreover, Kopai barrage is located in the *Rarh* region (23°43′42.05″ N and 87°30′24.03″ E) over the Kopai River at a distance of 53 km from its source point. This barrage traverses a catchment area of about 212 km² with a flood design of 55.16 m. Thus, the majority of the dams and barrages are located in the intermediate stretches of the river except for the Massanjore which is located at the upstream segment of the Mayurakshi River. It is interesting to note that all the civil structures have been constructed over the tributaries along an axis (Figure 5.7).

The dams and barrages over the MRB have a great role to play to provide irrigation. Thus, irrigation canals are constructed from the reservoir of the dams and barrages in the lower part of the basin (Figure 5.8b).

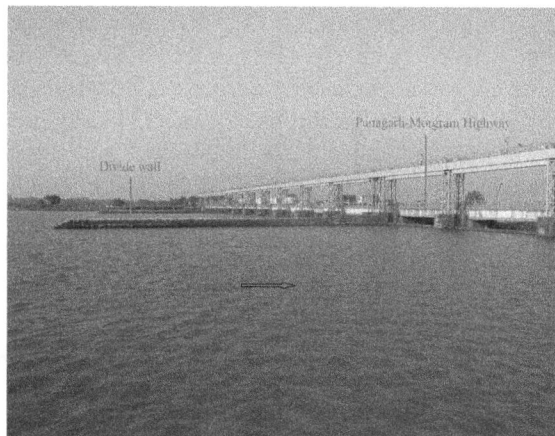

FIGURE 5.8A   Tilpara barrage during post-monsoon period. (Field Photograph, 29 December 2019.)

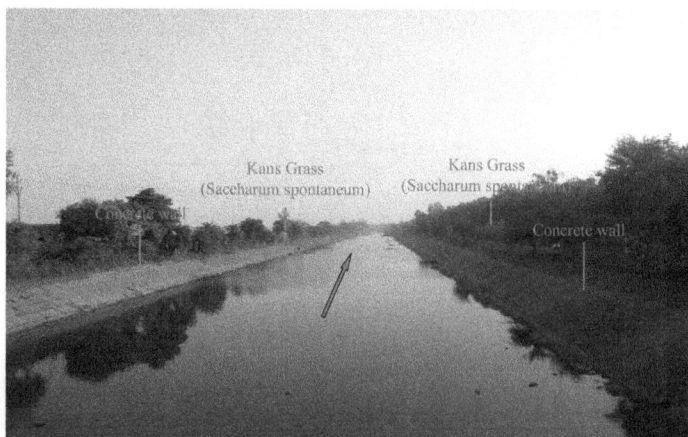

FIGURE 5.8B   Mayurakshi Bakreshwar main canal near Tilpara. (Field Photograph, 29 December 2019.)

The lower part of the MRB is more densely populated than the upper segment of the river. For example, Birbhum and Murshidabad districts of West Bengal located in the lower portion of the MRB have a population density of 771 and 1,334 persons km^{-2}, respectively. However, districts such as Dumka, Jamtara, Pakaur and Deoghar in the state of Jharkhand is the upper part of the MRB that portrays a low population density (351 persons km^{-2} for Dumka, 602 for Deoghar, 437 for Jamtara and 497 for Pakaur as per 2011 census). This is due to the natural endowment of fertile crescent, availability of river water and plain terrain. In brief, population density is greatly controlled by agricultural performances which depend on physiography, soil and water. However, as the river is monsoon-fed, availing water in the lean months creates hardship before the farmers. Thus, to address this recurrent challenge over the years, the Government has installed irrigation canals from the dam and barrages of the MRB (Ghosh, 2012). In this river basin, there are seven major irrigation canals serving the need of *rabi* and *kharif* crops. They are (1) Mayurakshi–Dwarka Main Canal (length: 16.62 km, command area: 226,720 ha for *kharif* and 20,250 ha for *rabi*), (2) Dwarka–Brahmani Main Canal having a length of 38.72 km, (3) Brahmani North Main Canal (length: 53.35 km), (4) Mayurakshi–Bakreshwar Main Canal (length: 22.53 km, 49,797 ha for *kharif* and 160,931 ha for *rabi*), (5) Bakreshwar–Kopai Main Canal (length: 11.27 km, command area: 87,449 ha), (6) Kopai South Main Canal (length: 76.47 km, command area: 69,626 ha) and (7) Bakreshwar Branch Canal (length: 18.11 km, command area: 4,938 ha). In brief, the irrigation canal has benefitted three districts of West Bengal, namely, Birbhum (2,209 km^2), Murshidabad (806 km^2) and Barddhaman (897 km^2).

5.3.1.2 Dams and Hydro-Geomorphological Changes

Controlling hydrological systems through the construction of dams, reservoirs and other engineering impoundments are not the notions of very recent origin; rather, it is an evolving part of human–nature interactions nurtured through different stages of human civilisation. Dams offer disruption to the flow of the river, and hence, longitudinal disconnection occurs due to damming.

i. Change in river regime

Dams immediately alter the discharge hydrograph of the channel which in the long run induces channel metamorphosis. The annual hydrograph portrays that there is a clear shift of the peak of the hydrograph towards right (Figure 5.9) and the lowering of the flood peak in the post-dam condition (1954–2013). This indicates that high-frequency low-magnitude flood events become more common compared to that of the pre-dam (1945–1953) period. In the pre-dam stage, the monsoon regime exclusively controlled the nature of flood which is now dam controlled. The rightward shift is also indicative of the capturing monsoon water while it rains and releases afterwards due to inability to store water beyond a certain limit which is diminishing day by day by siltation in the upstream stretch of the dams.

Furthermore, the flood frequency has substantially increased after the construction of the Massanjore dam in 1956 and Tilpara barrage in 1976 (Figure 5.10). However, the sharp rise is visible after the 1990s because the cubic capacity of the reservoir has diminished significantly.

ii. Channel morphology

This hydrological alteration induced huge change in channel morphology both in the upstream and downstream stretch of the river. Khatun and Pal (2017) observed that in the downstream of the Tilpara barrage, channel capacity decreased by about 25% due to significant decreases in the channel depth (14–71 cm). However, the overall width of the channel has increased by 3%–30% with an increase in the cross-sectional (CS) area of the 11% CS, while the decrease of the CS area for the 17% CS studied. This change in the CS morphology is essentially due to the variable discharge released from the barrage. For example, in the closing period, the velocity is as feeble as 0.1 m s^{-1} while during the opening stage, it gains the velocity as high as 0.7 m s^{-1}. The bed slope of the channel has

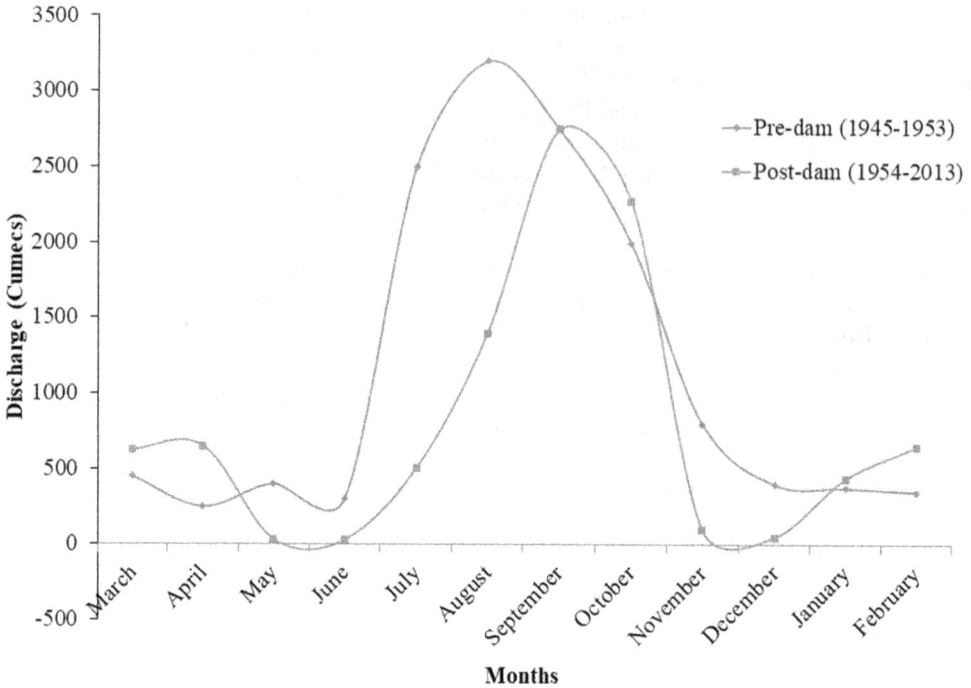

FIGURE 5.9   Annual discharge hydrograph in the pre-dam and post-dam situations.

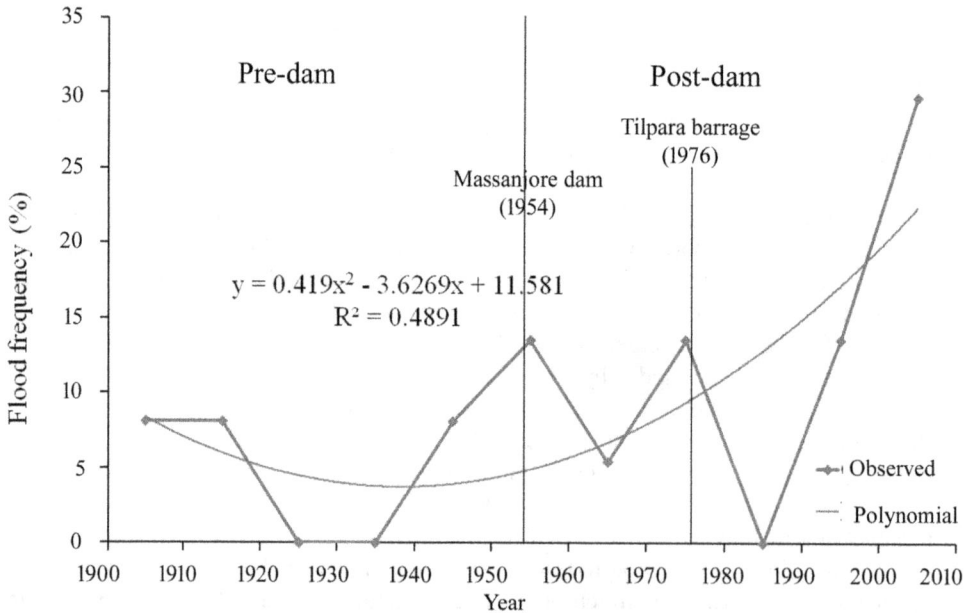

FIGURE 5.10   Changing pattern of flood frequency in the pre-dam and post-dam situations.

also declined from 3.15% to 1%. Formation of several obstacles across the streams like dams and barrages has also caused compulsory disconnectivity in the system. Numerous small dams constructed on the extra-peninsula and peninsular streams in India have re-budgeted the water and sediment supply capacity to a greater extent. However, massive dams were involved in more conspicuous disconnectivity on the sediment budget, as it is evident from the rivers of the Peninsula characterised by pronounced decline in sediment supply during the last few decades (Jain et al., 2016). Thus, the bank erosion was severe immediately after the construction of the barrage which backed the huge channel migration resulting in an increase of the sinuosity by 13%. In general, it may be mentioned that overall flood deposition rate has increased by 21 cm since 1971 and channel form has been more stabilised compared to the random and uncontrolled river regime. Moreover, the upstream stretches also have undergone radical changes in the channel morphology. For example, Khatun and Pal (2017) have observed that the width–depth ratio has increased in the upstream segment of the reservoir (Figure 5.11). And the rate of the increase is inversely proportional to the distance from the reservoir location due to the ponding action.

iii. Channel planform

Due to the construction of civil structures, sediment starvation of river happens which may reinforce bank erosion with changing channel sinuosity. Therefore, eroded floodplains are found at multiple reaches of the downstream. The sinuosity indices of three different reaches of Mayurakshi River during 1970 and 2019 are computed using the methodology Muller's sinuosity index (1968). Channel index (ratio of channel length and air length) indicates that the tail reach is the most sinuous reach of the river (1970: 2.31; 2019: 2.29) compared to the head (1970: 1.10; 2019: 1.01) and the middle reach (1970: 1.251; 2019: 1.249) of the river because of the free oscillation of the river over the alluvial tract than that of the hard granitic terrain. In general, it is seen that topographic sinuosity index (TSI) is higher than hydraulic sinuosity index (HSI) due to topographic features of floodplain area such as river bank configuration, valley alignments and neo-tectonic activity. However, it is observed that hydraulic control is increasing with time across the three stretches of the river, while the topographic control is falling. For example, during 1970, the HSI of upper reach was about 6% which has increased to about 29% in 2019. Similar observation is found for the middle (1970: 9%; 2019: 24%) and the tail reach (1970: 0.76%; 2019: 1.75%). Moreover, the TSI has decreased from about 95% to 71% from 1970 to 2019 in the upper reach. Similar results are observed for the middle (1970: 91%; 2019: 76%) and tail reach (1970: 99.24%; 2019: 98.25%).

However, the standard sinuosity index (SSI) has increased for the all stretches during the period of observation. The SSI has increased from 1.005 in 1970 to 1.026 in 2019 for the upper reach. Similar increase is also noted for the middle (1970: 1.019; 2019: 1.051) and tail (1970: 1.004; 2019: 1.009) reach (Table 5.5). In brief, the SSI signifies that the lower

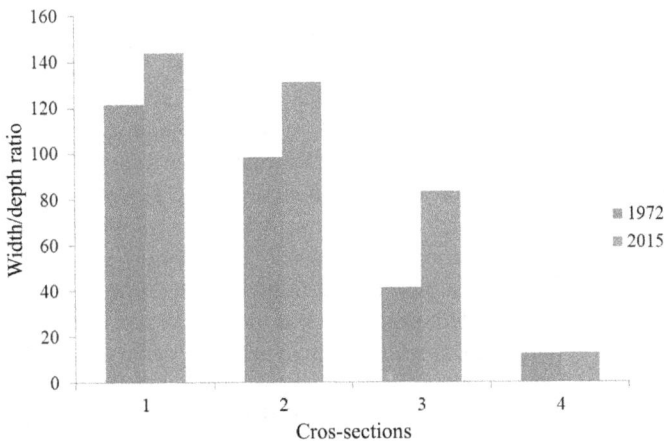

FIGURE 5.11 Changing width–depth ratio.

TABLE 5.5

Sinuosity Index

Reach	Year	Channel Length (km)	Valley Length (km)	Air Distance (km)	Chanel Index	Valley Index	HSI (%)	TSI (%)	SSI
Upper	1970	77.67	77.27	70.4	1.1033	1.0976	5.52	94.48	1.005
	2019	77.3	75.33	70.4	1.098	1.07	28.57	71.43	1.026
Middle	1970	63.34	62.15	50.6	1.2518	1.2283	9.33	90.67	1.019
	2019	63.23	60.19	50.6	1.2496	1.1895	24.08	75.92	1.051
Lower	1970	93.24	92.73	40.3	2.31	2.3	0.76	99.24	1.004
	2019	92.4	91.49	40.3	2.2928	2.2702	1.75	98.25	1.0099

Computed by the authors (2019).

reach is less sinuous than the other two reaches as the river is bounded by the embankments. It is noteworthy that a sharp rise in the HSI and a fall in the TSI are not normal under stable geology and climate within this short period. Actually, anthropogenic control in the form of massive dams and barrages over the MRB may be responsible for this change in the sinuosity. It may be argued that dams and barrages introduced a high-frequency low-magnitude flood event instead of high-magnitude low-frequency event of the pre-dam period. This situation has aggravated due to siltation in the reservoir and dwindling capacity of the reservoir day by day. As a result, fluvial hydraulics is now a control factor over the channel planform round the year instead of being seasonal.

Another important aspect of channel planform changes is braid channel. The braiding index is calculated using the formula of Brice (1964). The braiding index for both the head and tail reach is zero during the period of observation because of the absence of any bar. However, an extended longitudinal bar (length >10.15 m) is developed just below the Tilpara barrage in the middle reach. The braiding index is 0.3214 in 1970 and 0.3258 in 2019 for the middle reach. The length of the bar has increased by 0.12 km in the last 49 years. The longitudinal bar may be developed because of the development of the Tilpara barrage. When the capacity of the barrage exceeds, a huge amount of high-velocity water is released which bifurcates the river channel. This is due to the exceeding capacity of the single channel to carry an increasing load. Therefore, flowing more than 10 km, the bifurcating channels meet each other and the longitudinal bar is formed.

iv. Scouring, sedimentation and bar formation

Pal and Let (2011) had revealed that increasing anthropogenic stress on river morphology is the most solitary factor leading to this change. Dam construction is one of the most economical means by which humans can alter ecosystems (Rosenberg, 1997). Pal and Let (2011) argued that channel is more sensitive than its valley for it is exceedingly affected by immediate reductions of discharge while valley responds at much later date. The dam creates a longitudinal disconnection. The upstream area of the dam is used for construction of reservoir. Due to this artificial structure, water is stored at a certain height leading to formation of backwater. The channel was very narrow having forested area before the construction of Bakreshwar dam. When Bakreshwar dam was constructed in 1999, the forested area was cleared to form reservoir (Figure 5.12). The channel width and CS area of Bakreshwar River and its tributary immediately above the dam has inflated profusely. Whenever water jumps from a certain height, it creates some plunge pool-like feature created at the waterfall site. These depressions are not only created due to hydraulic impact but also sediment-free water tries to entrain the sediment of the locality (Figure 5.13).

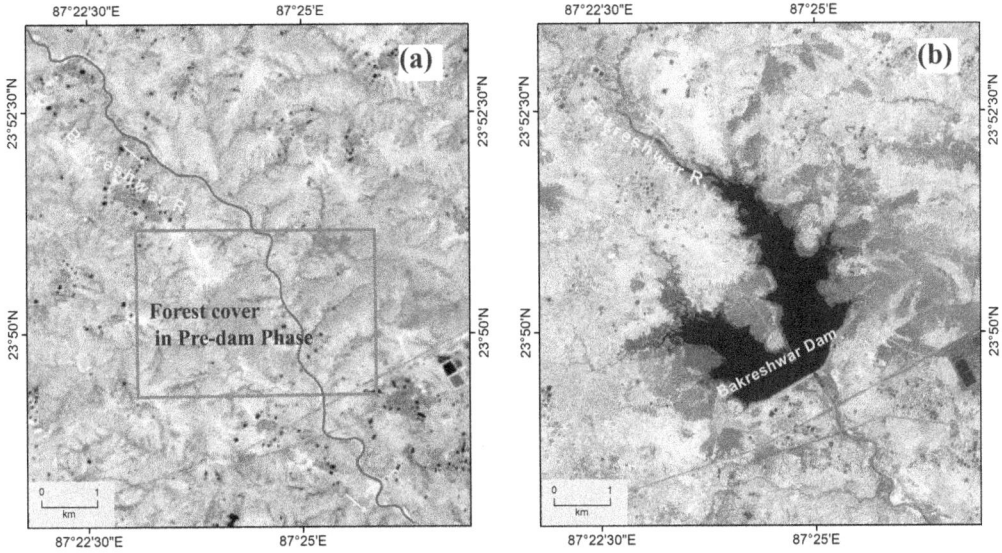

FIGURE 5.12 Ponding action due to the construction of the Bakreshwar barrage. (a) Pre-dam condition (1995; prepared from Landsat 5; path and row: 139, 43; dated 3 January 1995) and (b) post-dam condition (2019). (Prepared from Sentinel 2B; tiles no. T45QWG; dated 31 March 2019.)

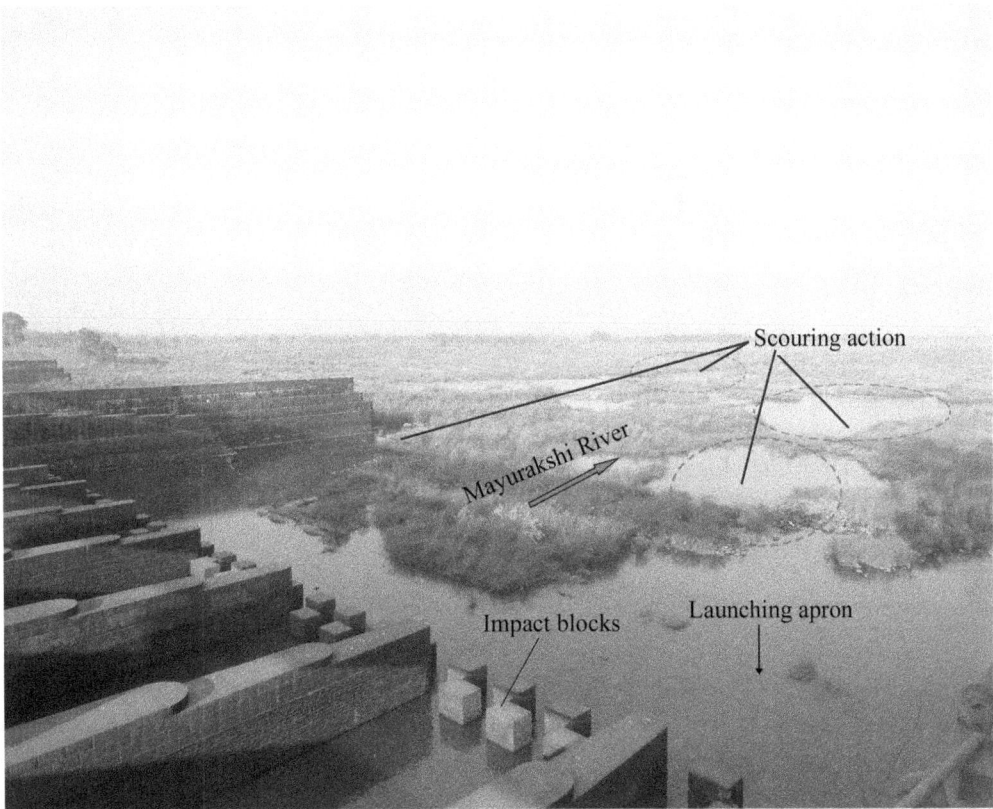

FIGURE 5.13 The effects of Tilpara barrage on the landscape diversity in the downstream. (Field Photograph, 29 December 2019.)

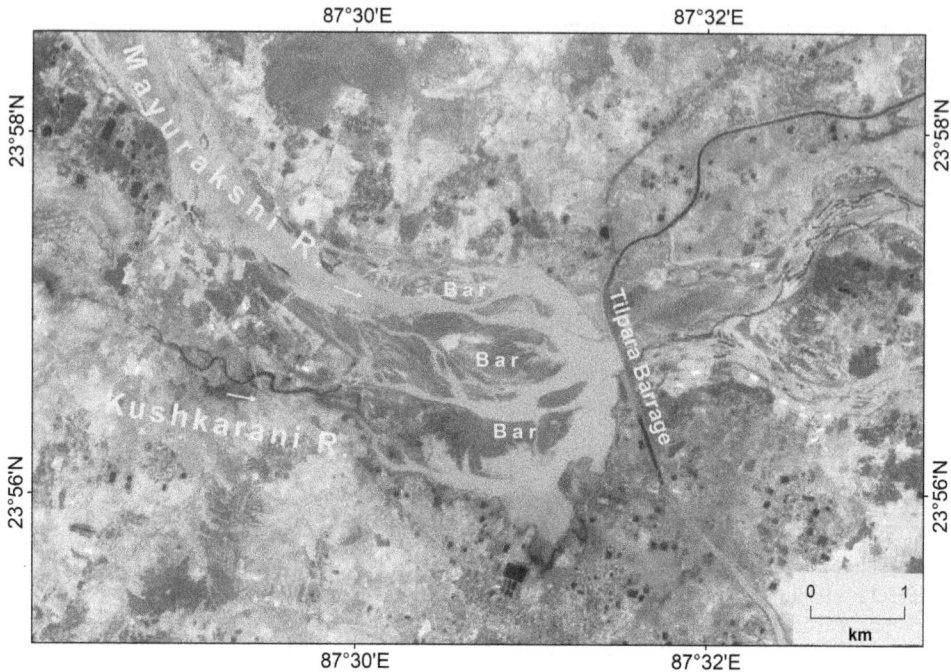

FIGURE 5.14 Impact of Tilpara barrage on the formation of backwater and channel bars. (Based on Sentinel 2B; tiles no. T45QWG; dated 31 March 2019.)

Sedimentation related to ponding action is a common phenomenon upstream of the dam. In case of Tilpara barrage due to ponding action of the river, sediment settles down and deposits on the bed of reservoir consequently forming bar in the upstream area (Figure 5.14). The observation in the recent year (2019) indicates that on the one hand bed deposition occurs on the upstream and on the other bed scouring is found immediately below the dam site (Figure 5.13).

5.3.2 CHECK DAMS

5.3.2.1 Location and Characteristics

Check dams are constructed to impede the flow of a river with different objectives including inducing infiltration, providing irrigation to the agriculture and checking erosion (Boix-Fayos et al., 2007; Renganayaki and Elango, 2013; Boix-Fayos et al., 2008). Check dams are often sought as an alternative to massive engineering projects such as dams and barrages. The developed nations like the United States have already started decommissioning of the dams for maintaining the environmental flow and sustainability (Warrick et al., 2015; Gosnell and Kelly, 2010; Muehlbauer et al., 2009). So, the small-scale check dams have gained widespread popularity among the regional planners. Besides, the low costs also make this project attractive among the developing nations. In India, check dams are widely used in the different river basins. The MRB also portrays an increasing number of check dams (341 for 2000 and 1,622 for 2019). There is a clear spatiality in the distribution of the check dams. The majority of the dams are found to concentrate on the upper (N = 1,496) and middle stretches (N = 126) of the basin (Figure 5.15).

The check dams of the basin are probably principally constructed with an eye to checking the gully erosion of the basin (Pal, 2017). Besides, some check dams are found to support the irrigation system. Check dams have a typical length of 18–25 m, width of 0.75–1.2 m and height of 1.2–1.6 m

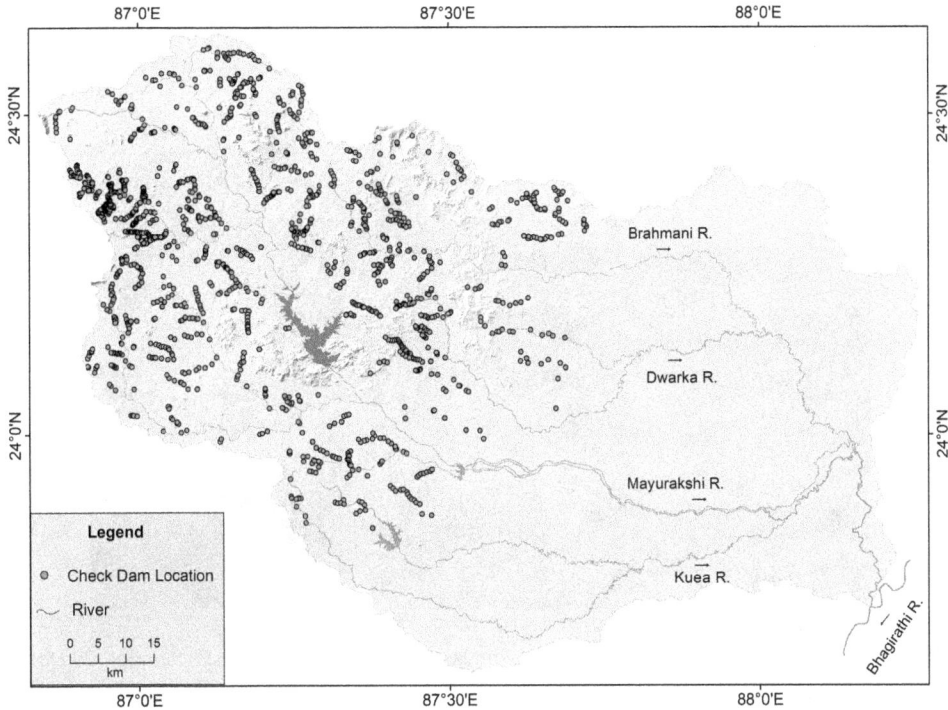

FIGURE 5.15 Location of the check dams. (Prepared from Google Earth Image, 28 March 2019, and 19 November 2019.)

as observed in the study area during the field investigation during 2019–2020. The typical geometry of these check dams varies widely in the study area. For example, the field investigation indicates that some check dams do not contain any pillar or lock gates, while some have several pillars (7) with a width ranging from 0.25 to 0.32 m with a height of pillar 1.68–1.7 m and the gap in between the pillar 2.2–2.9 m.

5.3.2.2 The Major Impact of Check Dams

Check dams like massive dams also trigger a longitudinal disconnection of river leading to sedimentation and scouring. However, check dams do not alter river morphology at massive scale. For example, a check dam located on Kushkarani River (23°58′53.47″ N and 87°24′45.36″ E) shows an upstream sedimentation of about 0.25 m and a scouring depth of about 0.4 m. Check dams also increase channel width and CS area (Figure 5.16).

To explore these dimensions, some CS measurements have been undertaken in the upstream and downstream area. The CSs are taken in such a way that constructed (concrete) and natural channel profile can be distinguished. For the first check dam (23°58′53.47″ N and 87°24′45.36″ E), upstream CS has been taken in the concrete area which shows almost a geometric profile shaping like rhombus and measuring CS area of 72.6 m^2 (Figure 5.17a). In the downstream segment, the CS is undertaken 120 m away from the construction site. This CS morphology is natural which portrays an irregular profile having thalweg on the left bank and an accretion on the river bed amid the channel and CS area measures 38.64 m^2 (Figure 5.17b). This portrays how check dam controls upstream and further downstream areas. For the second check dam (23°59′59″ N and 87°22′49″ E), one CS is taken from each of the upstream and downstream areas. Here, upstream CS is at the naturally configured area located at a distance of 300 m from the dam. This CS also portrays irregularity throughout the course and traverses an area 35.325 m^2 (Figure 5.17c). The downstream CS is taken immediately below the check dam that is at a construction site which resembles a geometric shape (rectangular) (Figure 5.17d).

FIGURE 5.16 Check dams and ponding action at Nimdaspur, West Bengal. (Field Photograph, 29 December 2019.)

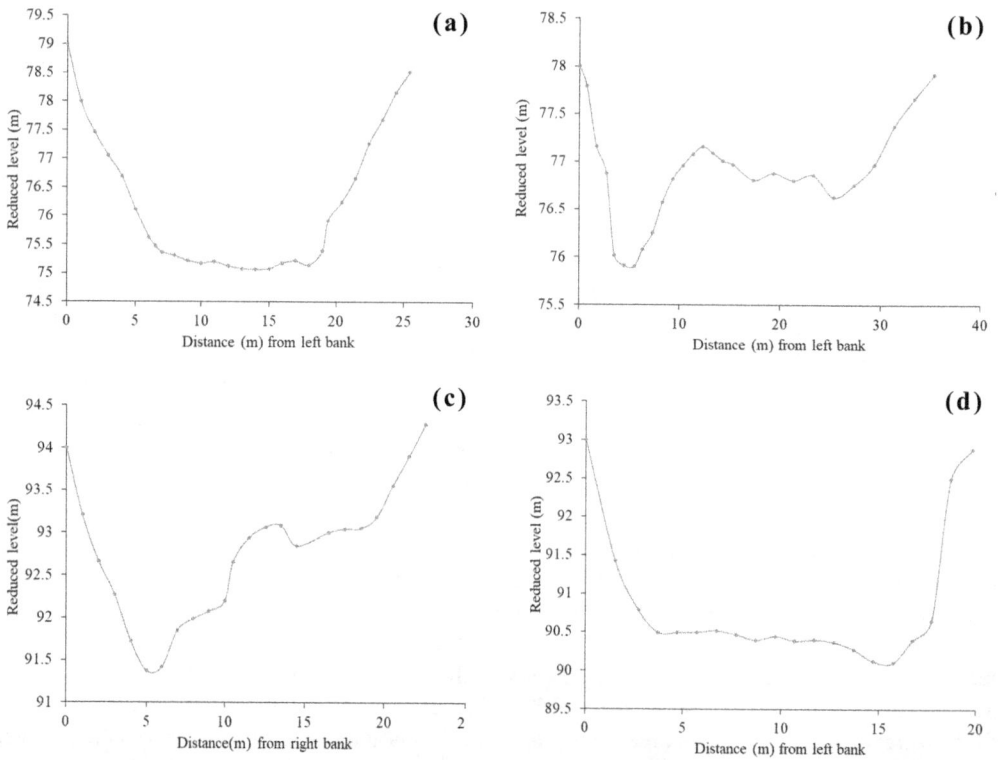

FIGURE 5.17 Changing channel morphology due to construction check dam. (a) Upper part of check dam 1, (b) lower part of check dam 1, (c) upper part of check dam 2 and (d) lower part of check dam.

5.3.3 EMBANKMENTS

5.3.3.1 Location and Characteristics

Since the pre-independence period, the MRB bears the marks of embankment mainly along the Mayurakshi River. The people constructed the earthen embankment on both sides somewhere and on a single side elsewhere (Figure 5.18). The main objective to construct the embankment was to save the crops from the effect of the devastating floods. Thus, embanking the river was quite wide at the upper stretch of the river compared to the lower counterpart. The wider distance between the left-bank and right-bank embankments is due to lesser land value in terms of agriculture or settlements compared to the lower floodplains of the basin. Moreover, the embankment is not continuous along the banks. Thus, the embankment along the left bank measures 68.6 km and along the right bank about 44.21 km for the Mayurakshi River.

Furthermore, the other rivers are also structurally controlled. For example, the Brahmani River has 46 km of binding; the river Dwarka has 41.02 km and *Kana–Mayurakshi* has about 33.2 km. Besides, recently the Government of West Bengal has started embanking the river especially in its lower stretch through a massive engineering project popularly called the Kandi Master Plan. Under this flagship project, concrete embankment is being constructed along the *Kana–Mayurakshi*, Mayurakshi, Hijuli, Kuea, etc. The major objective for constructing the embankment is to save the Kandi–Berhampore area from the flood and to increase the navigability of the rivers Bile, Jhum Jhum Khali *Khal*, Banki *Khal* and Jibati *Khal*. Thus, these embankments were quite high to contain the flood water within the valley of the rivers. The dimension of the embankment is briefed up in Table 5.6.

For constructing the embankments, 277.38 ha land have been acquired from the MRB. For the Mayurakshi River, 55.68 ha land is acquired from the left bank while 67.05 from the right bank. Similarly, 11.43 ha from the left bank and 36.26 ha from the right bank of the Bele–Dwarka; 12.81 ha from the left bank and 14.22 ha from the right bank of the Kuea–Babla River; 2.99 ha from the left bank and 21.33 ha from the right bank of the Banki River; 5.13 ha from the left bank and 3.04 ha from the right bank of the Jhum Jhum Khali *Khal* and 2.88 ha from the left bank and 3.13 from the right bank of the Swarup Khali *Khal* area have been acquired.

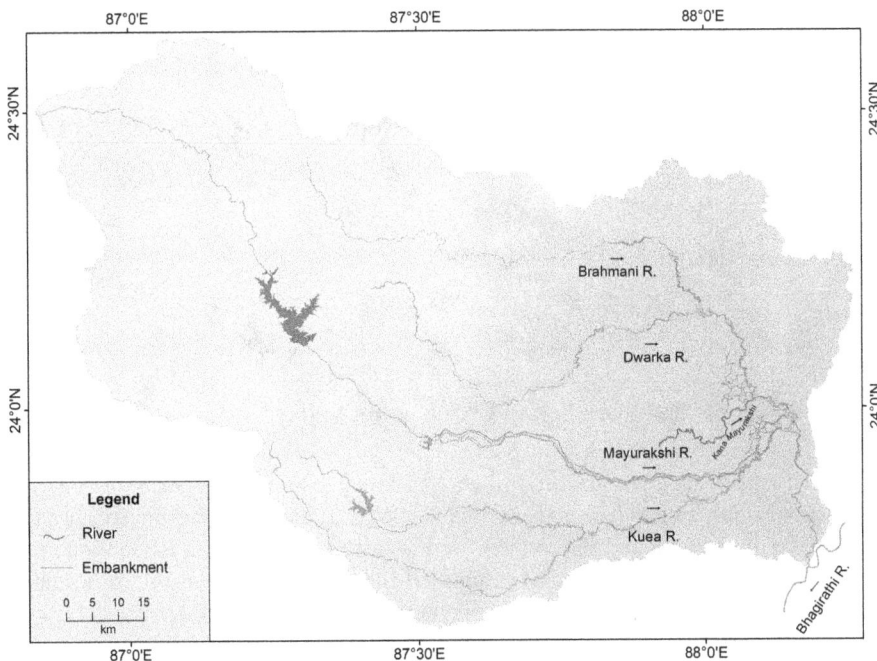

FIGURE 5.18 Location of embankment. (Adapted from Kandi Master Plan, 2012 and SoI 1: 50,000 toposheets.)

TABLE 5.6
Embankments in the Middle and Lower Stretches of the Mayurakshi River

Flood Return Period (years)	Name of the Embankment	Length of the Embankment (km)	Name of the River	Bank
25	Titidanga–Juran Kandi Embankment	26	Brahmani	Right bank
	Titidanga–Juran Kandi Embankment	20	Brahmani	Left bank
	Titidanga–Juran Kandi Embankment	5.96	Dwarka	
	Sehalai–Piprikuri–Thakuranichak Circuit Embankment	9.4	Bele and Kuea	Right bank and left bank
	Mayurakshi right bank (From Saithia rail line up to Birbhum border)	23.85	Mayurakshi	Right bank
	Mayurakshi right bank Embankment (Birbhum border to Kuea River outfall)	20.36	Mayurakshi	Right bank
50	Sherpur–Joragachi Embankment (Talgram to Ahirimondal)	24.1	Kuea	Right bank
	Sherpur–Joragachi Embankment (Ahirimondal to Jorgachi)	8.9	Babla	Right bank
100	Tilpara barrage to Gangedda Embankment (Tilpara barrage to Harishchandrapur)	59.46	Mayurakshi	Left bank
	Tilpara barrage to Gangedda Embankment (Harishchandrapur to Mondalpur)	9.14	Mayurakshi	Left bank
	Kandi–Ganggeda Embankment (Kandi (Laharpara) to Ranagram)	17.5	*Kana–Mayurakshi*	Right bank
	Kandi–Ganggeda Embankment (Ranagram to Bagchara)	9.14	Dwarka	Right bank
	Kandi–Ganggeda Embankment (Bagchara to Andulia)	7.29	Bele	Left bank
	Kandi–Ganggeda Embankment (Andulia to Chuator)	8	Chuator	Left bank
	Kandi–Indradangapara Embankment (Indradangapara to Ranagram)	25.92	Dwarka	Right bank
	Kandi–Indradangapara Embankment (Ranagram to Kandi)	12.5	*Kana–Mayurakshi*	Left bank
	Laharpara–Kandi Embankment (Laharpara to Kandi)	3.2	*Kana–Mayurakshi*	Right bank

Based on Kandi Master Plan (2012).

5.3.3.2 Embankments and Evolution of Channel and Valley

It is very obvious that breadth of a river channel partially explains the energy of the stream, its state of velocity, erosional capability, retaining as well as transporting capacity of water and sediments, its consistencies, etc. (Islam and Guchhait, 2017a). Dam and reservoir constructions have constrained in the way to expand the channel in downstream reaches but embankment construction along the active floodplain zones has amplified the restriction processes of normal channel widening procedures (Pal and Let, 2011). To prevent and minimise the disastrous effects of flood and increase short-term economic benefits, local people have built embankment as a strategy to restrict flood water within the natural bank of the valleys. However, this embankment induces several hydro-geomorphic problems as discussed in the following sections.

i. Channel and valley constriction

The embankment has a great role to play in the long-term evolution of the channel and the valley of the MRB. In general, the channel width and the valley width increase downwards. However, the Mayurakshi portrays a reverse trend—lower channel and valley width downstream (valley width 183 m and channel width 51 m) computed to its immediate (17.9 km) upstream (valley width 795 m and channel width 350 m). This narrowing valley configuration is principally controlled by the embankment on either sides of the river (Figure 5.19). Furthermore, breaching the earthen embankment is a common phenomenon especially during the monsoon months. This collapse of bank has immediate effects on the

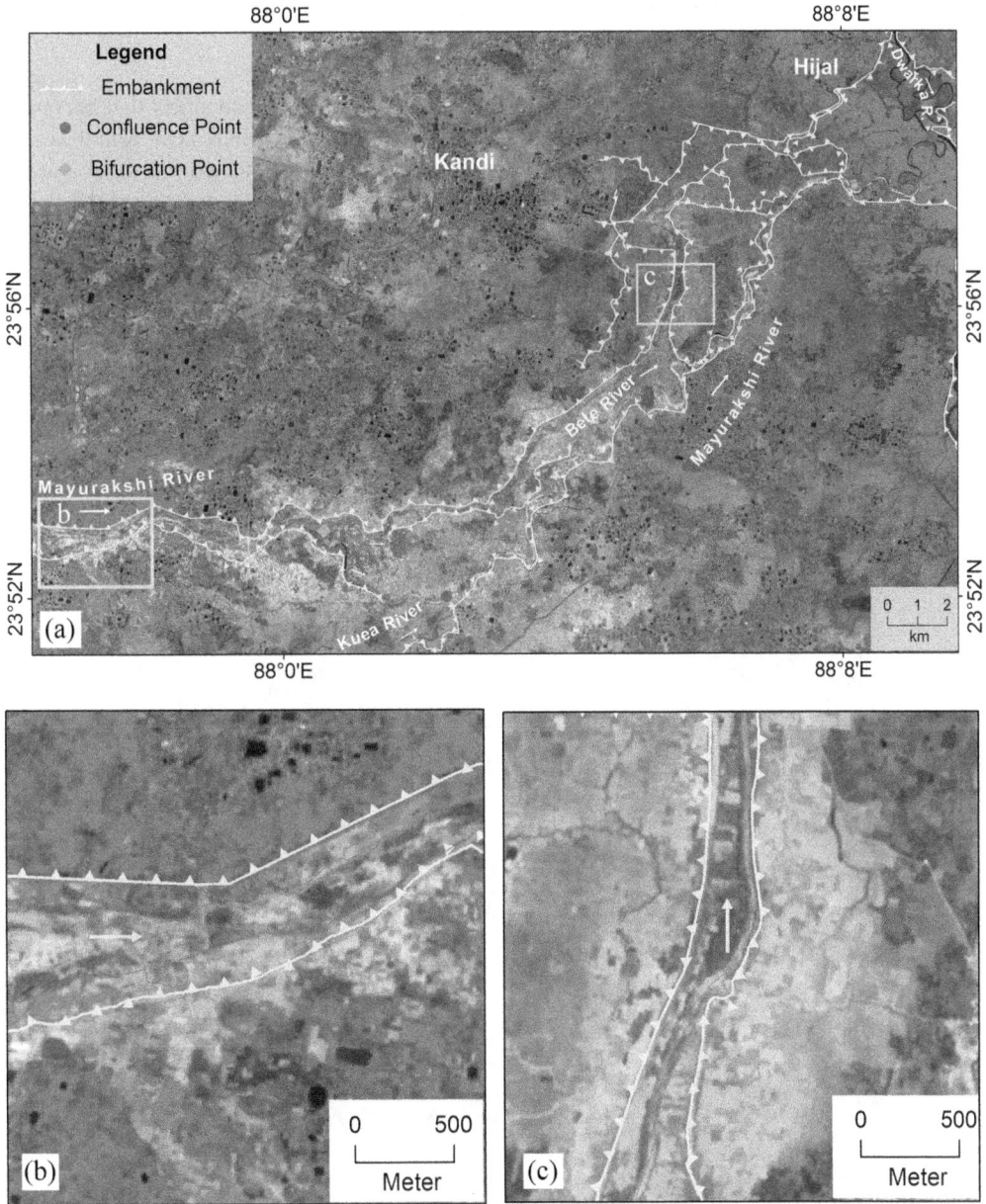

FIGURE 5.19 Channel width and embankment. (Prepared from Sentinel 2B; tiles no: T45QWG, T45QXG; Dated: 31 March 2019.)

increasing width–depth ratio due to bed deposition which is found to decrease the cubic capacity of the river and, hence, decrease the channel efficiency.

ii. Breaching of embankment and sedimentation

The extended embankments and other riverine hydraulic structures such as spurs and artificial levees have been formed along the river, that in turn could immensely intensify the flood magnitude many times and transport huge boulder and gravel-like materials instead of carrying fertile silt and other organic deposits along its bed (Figure 5.20). Hence, once which is considered as a gift of the nature is now considered as human fears. Saha (1935) expressed concern about the obnoxious effect of embankments on the hydrological and sedimentological character of the stream while illustrating the flood problems of Bengal (Table 5.7). While approaching the flood hazard of lower Damodar plain, Willcocks (1930) blamed the river-edge embankments as the 'Satanic chains'.

iii. Fragmentation of spill channels

In the active floodplains of the MRB, there are numerous spill channels that become active during the monsoon season. In the lower stretch of the MRB near Kandi area, the spill channel in between the two major river systems is connected via sluice gates constructed under the Kandi Master Plan. There are 55 sluice gates to control the flow

FIGURE 5.20 Breaching of embankment and sediment flux to the river Dwarka near Surkhali. (Field Photograph, 18 July 2018.)

TABLE 5.7
Average Depth of Sediment Deposition at Different Reaches of Mayurakshi River

Sl. No.	Location	Distance downstream of Tilpara Barrage (km)	Avg. Bed Level as on 2009 (M.G.T.S.)	Avg. Bed Level as on 1981 (M.G.T.S.)	Avg. Depth of Sedimentation (m)
1	Talbona	40.71	27.58	25.85	1.73
2	Peturi	44.21	25.46	23.57	1.89
3	Narayanpur	47.26	23.88	22.52	1.36
4	Hatisala	51.83	23.08	21.78	1.30
5	Panchachupi	56.40	20.04	18.70	1.34
6	Raha	64.02	18.69	17.03	1.66
7	Upstream of Bele Bridge	65.55	17.92	16.79	1.13

Source: Kandi Master Plan (2012). (Note: Avg. for average, and M.G.T.S for metre with respect to the Great Trigonometrical Survey datum)

in the spill channel via certain vents with a typical diameter ranging from 0.9 to 1.95 m (Table 5.8).

However, in the middle reach of the river, there are no such sluice gates to maintain the flow through the spill channels, and thus, they are now abandoned and dried (Figure 5.21). So, the fragmentation of the spill channels is triggered by the construction of the earthen embankment without vents.

iv. Breaching of embankment and sandsplay

Breaching of the earthen embankment is common during the rainy season and flood. As Mayurakshi River carries average annual sediment load of about 2.52 million tons (Rudra, 2010), the majority of this sediment is coarse (Guchhait et al., 2016) which are either deposited on bed or outflows to the Bhagirathi River system. However, the massive flood like the 2000 episode broke the embankment apart near Janardanpur and Tulia. Consequently, huge sand dispersed on the agricultural bed along the river (Figure 5.22). This sediment dispersal mechanism not only modifies the micro-structural fluvial elements but also badly impacts the agricultural practice for some years to come (Molla, 2011).

5.3.4 SAND MINING

5.3.4.1 Location and Characteristics

Mining is an integral function of contemporary societies and arrives in a wide range of geomorphic settings. It is one of the most important areas of advanced geomorphological research because the peculiar kinds of excavated and accrued landscapes generated through this process are vulnerable to severe geomorphic hazards. Mining landscapes encompassed greater than half million hectares of land in the globe in 1990 (Young et al., 1992) and are solely responsible for greater production of sediments than the construction of paved road, house building and agriculture. Mining-induced expulsion of vegetation, surface deposits, blasting and rising slopes put landscapes to a greater height to failures, decaying, floods, submergence and other acute geomorphic hazards which have been the prime concern in geomorphological studies. The efficient application of geomorphological knowledge is an important tool in combating or reducing the negative consequences of mining. Studies and research about mining landscapes provide insights to geomorphic awareness of landscape evolution (Haigh, 1978) and greater knowledge about hazardous landscapes. Like rapid pace of industrialisation, urbanisation also has increased the need for mining substances, and therefore, the importance of mining activities has increased substantially. Mining is not a very recent activity germinated in one or two decades ago, rather it's a prehistoric activity as evidenced by Plini's book

TABLE 5.8

Dimensions of the Sluice Gates in the Kandi Area

Sl. No.	Location	Vent (No.)	Diameter in m	River and Circuit Embankment
1	Badan Nalla Sluice at Rameshwarpur	2	0.9	Kandi–Indradanga Embankment
2	Barpuri Sluice at Sunderpur	4	1.2	Kandi–Indradanga Embankment
3	Bhugal Sluice at Ruhigram	2	0.9	Kandi–Indradanga Embankment
4	Bhermani Sluice	2	0.9	Kandi–Gandedda Circuit Embankment
5	Haribagan Sluice	1	0.9	Kandi–Gandedda Circuit Embankment
6	Pirtala Sluice	1	0.9	Kandi–Gandedda Circuit Embankment
7	Purandarpur Sluice	1	0.9	Kandi–Gandedda Circuit Embankment
8	Chorki Sluice	2	0.9	Kandi–Gandedda Circuit Embankment
9	Goliermukh Sluice	4	1.2	Kandi–Gandedda Circuit Embankment
10	Chand Khali Sluice	1	0.9	Kandi–Indradanga Embankment
11	Mondalpur Sluice	1	1.2	Kandi–Indradanga Embankment
12	Katkatagher Sluice	1	1.2	Kandi–Indradanga Embankment
13	Rajarghar Sluice	1	0.9	Kandi–Indradanga Embankment
14	Amtala Sluice (arch Sluice)	3	1.95	Kandi–Indradanga Embankment
15	Nabab–Bahadur Sluice	1	0.9	Kandi–Gandedda Circuit Embankment
16	Sarpmara Sluice	1	0.9	Kandi–Gandedda Circuit Embankment
17	Bagmari Sluice	4	1.2	Kandi–Gandedda Circuit Embankment
18	Lakshmantapur (New) Sluice	1	0.9	Bele River Left Embankment
19	Lakshmantapur (Old) Sluice	2	0.9	Bele River Left Embankment
20	Andulia Sluice	3	0.9	Bele River Left Embankment
21	Bagdanga Sluice	5	1.2	Chautoor River
29	Kasipur Sluice	1	0.9	Kuea River
30	Munuti Sluice	1	0.9	Mayurakshi River
31	Dewarbhastar Sluice	1	0.9	Mayurakshi Right Bank
32	Muhula Sluice	1	0.9	Mayurakshi Right Bank
33	Chaitpur Sluice	1	0.9	Mayurakshi Right Bank
34	Hatishala Sluice	1	0.9	Mayurakshi Right Bank
35	Sundupur Sluice	1	0.9	Mayurakshi Right Bank
36	Harishchandrapur Sluice	2	0.9	Mayurakshi Right Bank
37	Ruha Sluice	1	0.9	Mayurakshi Right Bank
38	Jakhri Sluice (Near Chopa Kolabagan)	1	0.9	Mayurakshi Right Bank
39	Dawlai Sluice	1	0.9	Mayurakshi Right Bank
40	Amlai Sluice	2	0.9	Kuea River (Right Embankment)
41	Srikrishnapur Sluice	2	0.9	Kuea River (Right Embankment)
42	Thakuranichak Sluice	1	0.9	Kuea River (Right Embankment)
43	Ranipur Sluice	1	0.9	Kuea River (Right Embankment)
44	Sunia Sluice	1	0.9	Kuea River (Right Embankment)
45	Angarpur Sluice	2	0.9	Kuea River (Right Embankment)
46	Sikarnala Sluice	2	0.9	Kuea River (Right Embankment)
47	Talgram Sluice	2	0.9	Kuea River (Right Embankment)
48	Gaysabad Sluice	2	0.9	Kuea River (Right Embankment)
49	Domanipara Sluice	1	0.9	*Kana–Mayurakshi River*

(Continued)

TABLE 5.8 (*Continued*)
Dimensions of the Sluice Gates in the Kandi Area

Sl. No.	Location	Vent (No.)	Diameter in m	River and Circuit Embankment
22	Naternala Sluice	1	0.9	Sehalali–Thakuranchak Circuit Embankment
23	Peprikari Sluice	1	1.2	Sehalali–Thakuranchak Circuit Embankment
24	Bagnerghar Sluice	1	0.9	Sehalali–Thakuranchak Circuit Embankment
25	Mondalpur Sluice	1	0.9	Rajarampur–Mandalpur Circuit Embankment
26	Rajarampur Sluice	1	1.2	Rajarampur–Mandalpur Circuit Embankment
27	Gorebill Sluice	1	0.9	Rajarampur–Mandalpur Circuit Embankment
28	Balichona Sluice	1	0.9	Kuea River
50	Katkata Sluice	3	0.9	*Kana–Mayurakshi* River
51	Choator	1	0.9	Bele River
52	Ahirimondal Sluice	2	0.9	Kuea–Babla Right Embankment
53	Joyrampur Sluice	2	0.9	Kuea–Babla Right Embankment
54	Sapmara Sluice	1	0.9	Bele Left Bank
55	Dhamalipara Sluice	1	0.9	*Kana–Mayurakshi* Left

Based on Kandi Master Plan (2012).

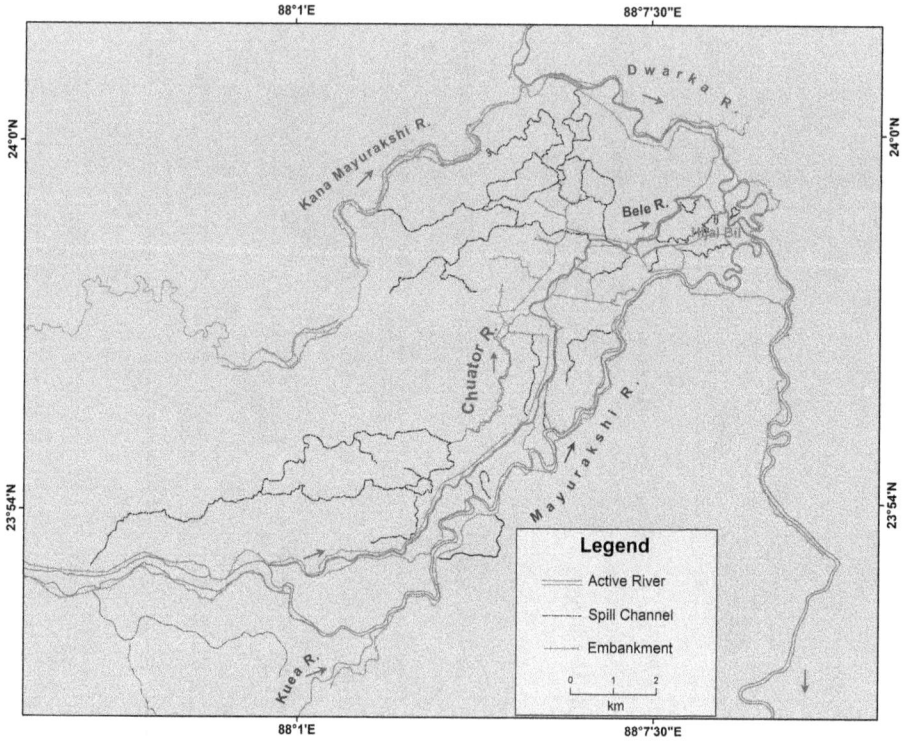

FIGURE 5.21   Effect of embankment on the fragmentation of the spill channel. (Prepared from SoI 1: 50,000 toposheets and Kandi Master Plan, 2012.)

FIGURE 5.22   Sandsplay due to breaching of the embankment along the Mayurakshi River near Janardanpur, West Bengal. (Google Earth Image, 7 October 2006.)

of *Natural Historia*. The extent and magnitude of mining activities grew substantially at the period of industrial revolution. Johnson and Lewis (1995) and Young et al. (1992) revealed that the utilisation of mineral resources has increased 10-fold from 1750 to 1900 and by 13-fold from 1900 to 1990. Mining is one of the most advanced activities in the present world that leads to material displacement by humans currently being greater than the materials translocated by all the natural processes (Hooke, 1994, 2000). Due to the rapid pace of urbanisation, the demands for mining materials have increased manifold, and to maintain this ever-increasing needs, large volumes, sand, gravel and pebbles are extracted along the river channels throughout the world especially the areas nearer to an urban centre. Places with no access or very little access to bedrock and glacial deposit, fluvial or riverine deposit is the best source of aggregate used in construction activities.

In the context of MRB, sand mining and stone crushing are dominant activities. The increasing demand for the *pakka* (concrete) building backed by the phobia to escape from flood-induced damages of building and *Pradhan Mantri Abas Yojona* (a scheme sponsored by the Central Government to provide *pakka* building to the poor). Naturally, demand for construction material especially locally available sand from river bed of the Mayurakshi River has exponentially increased. This increasing demand has induced illegal sand mining from the river bed. The extensive field observations indicate that from the source to the mouth of the Mayurakshi River, some tributaries like Siddheswari, Bhurbhuri, Dhobbi, Tepra,Kuea, Brahmani and Dwarka have explicitly shown the trend of discrete point sand mining and continuous linear stretch mining. Along the Mayurakshi River, there are 94 discrete point locations in the upper stretches of the basin. Similarly, 19 discrete locations in the middle stretches have been observed during 2019–2020 (Figure 5.23).

Furthermore, the linear stretch of about 122 km (53 km at middle stretch and 69 km at lower stretch) extending from Kenduli to Hijal has intensive mining activity. This clearly points out that

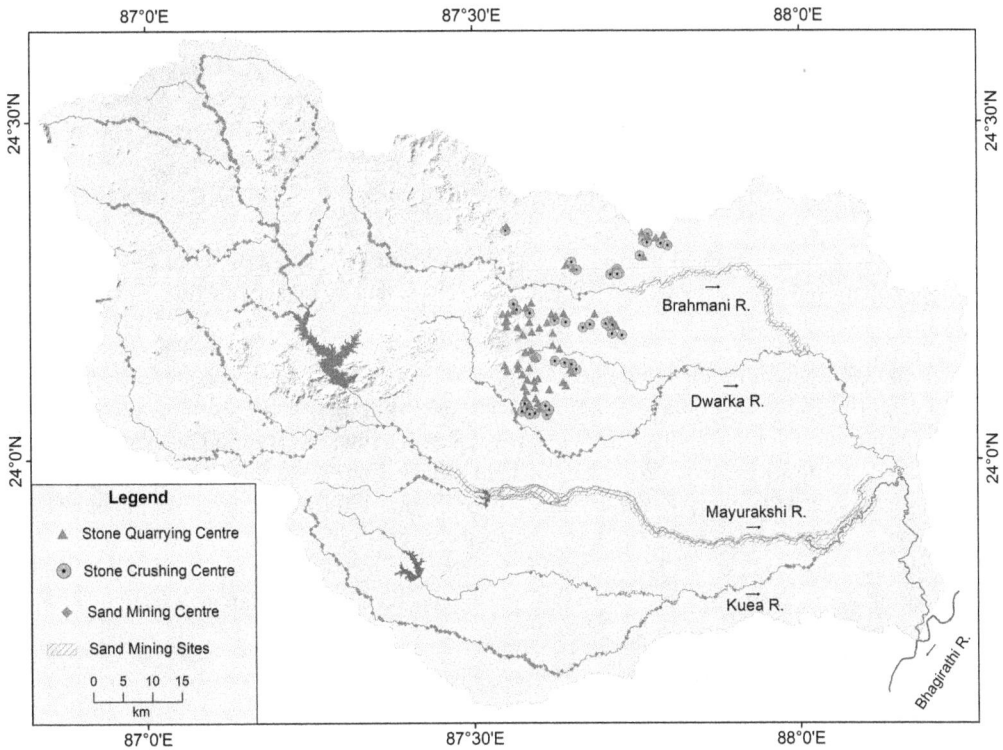

FIGURE 5.23   Location of sand mining and the stone crushing. (Prepared from Google Earth Images, 9 January, 10 January, 9 March, 28 March, and 30 April, 2019 and field survey 2019.)

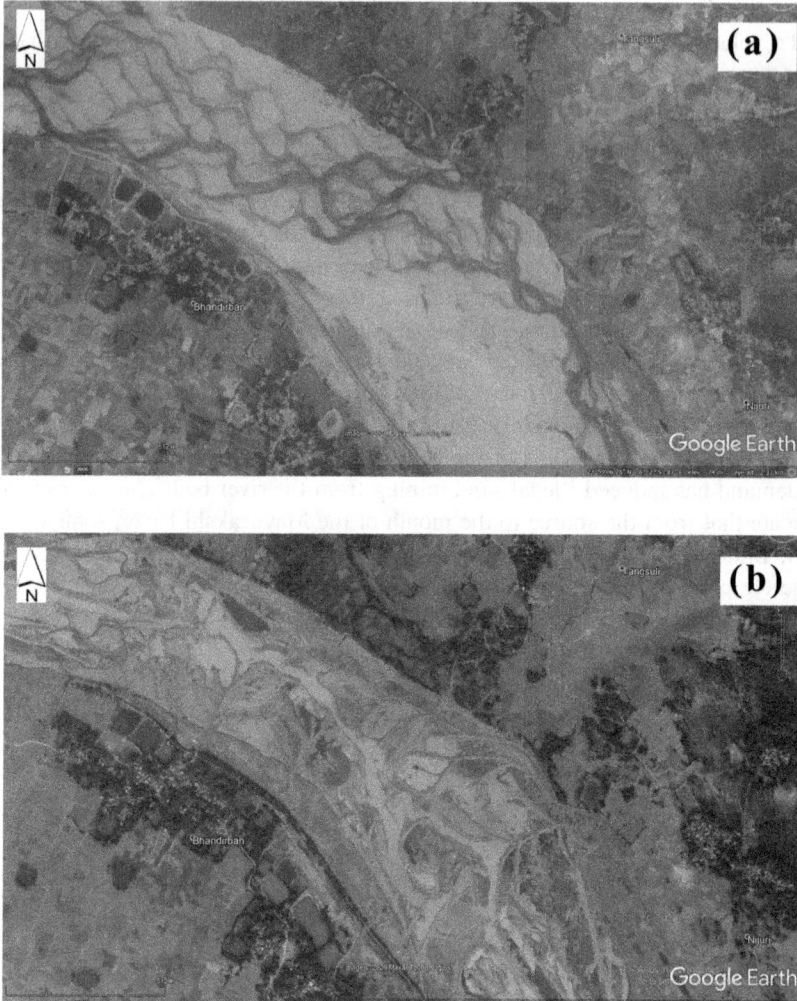

FIGURE 5.24 Magnitude of sand mining on physical condition of Mayurakshi River bed near Tilpara barrage. (a) Before sand mining (January, 2011) and (b) after sand mining (January, 2019).

the magnitude of sand mining is more in the lower part compared to the upper segment. Moreover, the tributaries like Siddheswari have 70 discrete locations while Brahmani has shown linear stretch about 87 km extending from Nalhati to Hazipur registering intensive mining activity. Regarding intensification of mining process with time, the field observation shows that during the last decade sand mining has reached to a peak. For example, the stretches of the Mayurakshi River near Tilpara barrage have been exposed due to illegal sand mining (Figure 5.24).

Moreover, the middle reach of Mayurakshi River near Langulia portrays the severity of channel bed alteration due to sand mining. The river is no longer a natural flow rather it has been a truck passage to carry a load of sand (Figure 5.25).

5.3.4.2 Sand Mining and Channel Morphology

Worldwide mining of coarse-grained alluvium from and along rivers causes geomorphic, hydrologic and biotic changes. Highly mechanised technology, such as excavators, hydraulic dredges and bulldozers enormously accelerate the rate of expulsion landscapes produced through mining leading to a number of hydro-geomorphic changes like upstream and downstream incision, lateral

FIGURE 5.25 Sand mining from the river belly near Langulia, West Bengal. (Field Photograph, 6 November 2019.)

channel erosion, bed armouring modification and disconnection of the floodplain consequently to channel incision, and water table lowering (Kondolf, 1994; Bravard et al., 1997).

i. Channel bed incision and pit formation

Newport and Moyer (1974) estimated that 10%–20% of sediments extracted in the United States in 1974 were taken from streams and its adjacent active floodplain areas. Besides altering the river form, the legacy of extracting sediments such as sand and gravel from rivers and its adjoining active floodplains includes a preponderance of pits, dug and water bodies of various size. Degradation, which can extend to channel incision of several metres, is a common response to mining the channel bed (Kondolf, 1994; Davis et al., 2000; Marston et al., 2003; Uribelarrea et al., 2003). In the MRB, such pits are common having diameter ranging from a metre up to few metres (Figure 5.26a). There are three types of in-stream sediment mining (Kondolf, 1994). They are dry-pit mining, wet-pit mining below the water level and bar skimming consisting of removing all the material in a gravel bar. In the MRB, dry-pit mining is more common in the upper and middle stretches of the river, while the wet-pit mining is commonly found in the lower reaches of the MRB. However, floodplain pit mining is not found in the MRB. Therefore, the in-stream mining activity has a specific impact on river system. Wet- and dry-pit mining directly alters the channel geometry producing a local sediment deficit. Besides, exposure of hard clay on the bed of Mayurakshi is also found during field investigations (Figure 5.26b). The scratch marks produced on the hard clay is due to the effects of the bulldozer while sand mining. Besides, the geomorphic implications of dredging on the river bed include not only the expulsion of sediments but also the destruction and manipulation of landforms, such as pools and riffles, ripple marks, channel thalweg and wider floodplain with active valley replenishment (Lagasse et al., 1980; Hay, 1985; Lagasse, 1986).

ii. Bank failure and channel widening

Planform and position changes include increased bank erosion, lateral migration and channel shifting when the river avulses into the floodplain. Mined reaches can widen from

FIGURE 5.26A Effect of sand mining on bed incision of Dwarka River near Indrani. (Field Photograph, 7 November 2018.)

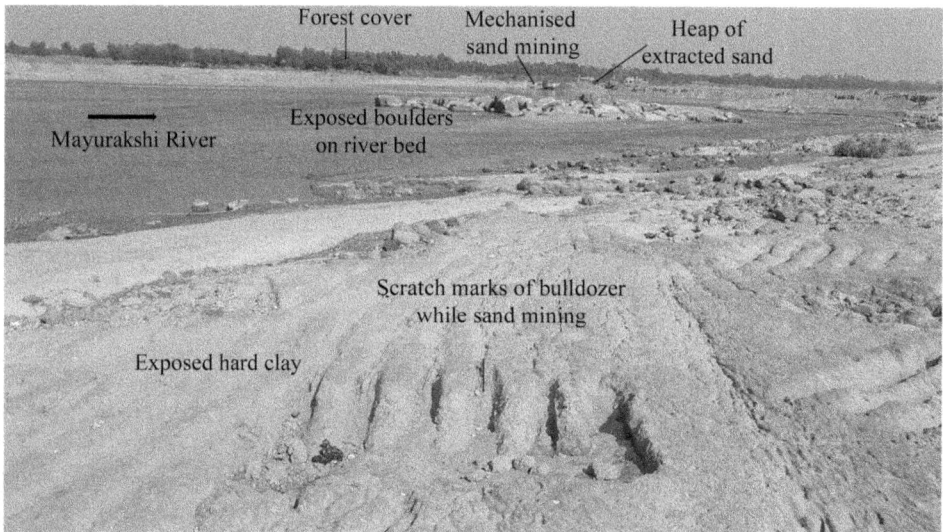

FIGURE 5.26B Exposure of granitic surface and hard clay due to sand mining near Langulia, West Bengal. (Field Photograph 16 November 2019.)

material extraction along the channel perimeter or by capturing the anthropogenic form of a pit through migration or avulsion (Mossa and Marks, 2011). In the MRB, bank failure due to sand mining is a common phenomenon because the mining of the sand from the river bed destabilises bank. Five CSs have been drawn in the areas of sand mining in the middle stretch of the MRB (Figure 5.3) which show channel widening. It occurs by the collapse mechanism when the upper part of layer becomes supportless due to undermining of the bottom layer (Figure 5.27). It is more common in case of the steep, bare surface slopes which are more prone to erosion and mass wasting processes.

FIGURE 5.27 Bank failure due to sand mining near Margram (mining of sand from the river bank which induces collapse of the bank for the vacuum created after sand mining). (Field Photograph, 12 October 2018.)

Due to bank failure, channel width increases with time. An extensive field survey carried out in the middle reach of the river Mayurakshi shows that the width–depth ratio for the selected CSs ranges from about 150 to 950 depending on the intensity of the mining activity (Table 5.9). The shallowing of the bed is due to sediment flux from the river bank failure.

iii. Ridge and furrow configuration of the valley

Channel and floodplain mining tend to manipulate and derange the channel bars as well as the CS form of the channel. The large-scale mechanisation processes of mining can drastically alter the shapes and forms of the river valleys and its adjoining floodplains, creating multiple pits and ponds within the channel. To depict this perspective, few CSs have been undertaken at both the mining and non-mining areas. It is to note that majority of mining area is located in the middle and lower stretches. That's why some CSs that were taken at the upper stream area represent naturally configured profile not influenced by mining action while CS taken at the middle reaches portrays channel configuration due to extensive sand mining (Figure 5.28). The natural configuration at the head reach portrays

TABLE 5.9

Cross-Sectional Morphology of the Middle Reach of the Mayurakshi River

Cross Section	Valley Width (m)	Channel Width (m)	Dmax (m)	Average Depth (m)	Width–Depth Ratio
1	520	50	2.38	0.66	143.9189189
2	290	30	0.4	0.23	238.6363636
3	393	18	0.85	0.35	952.1126761
4	300	80	0.56	0.338888889	658.5365854
5	280	90	0.32	0.17	450

Based on field survey (2019).

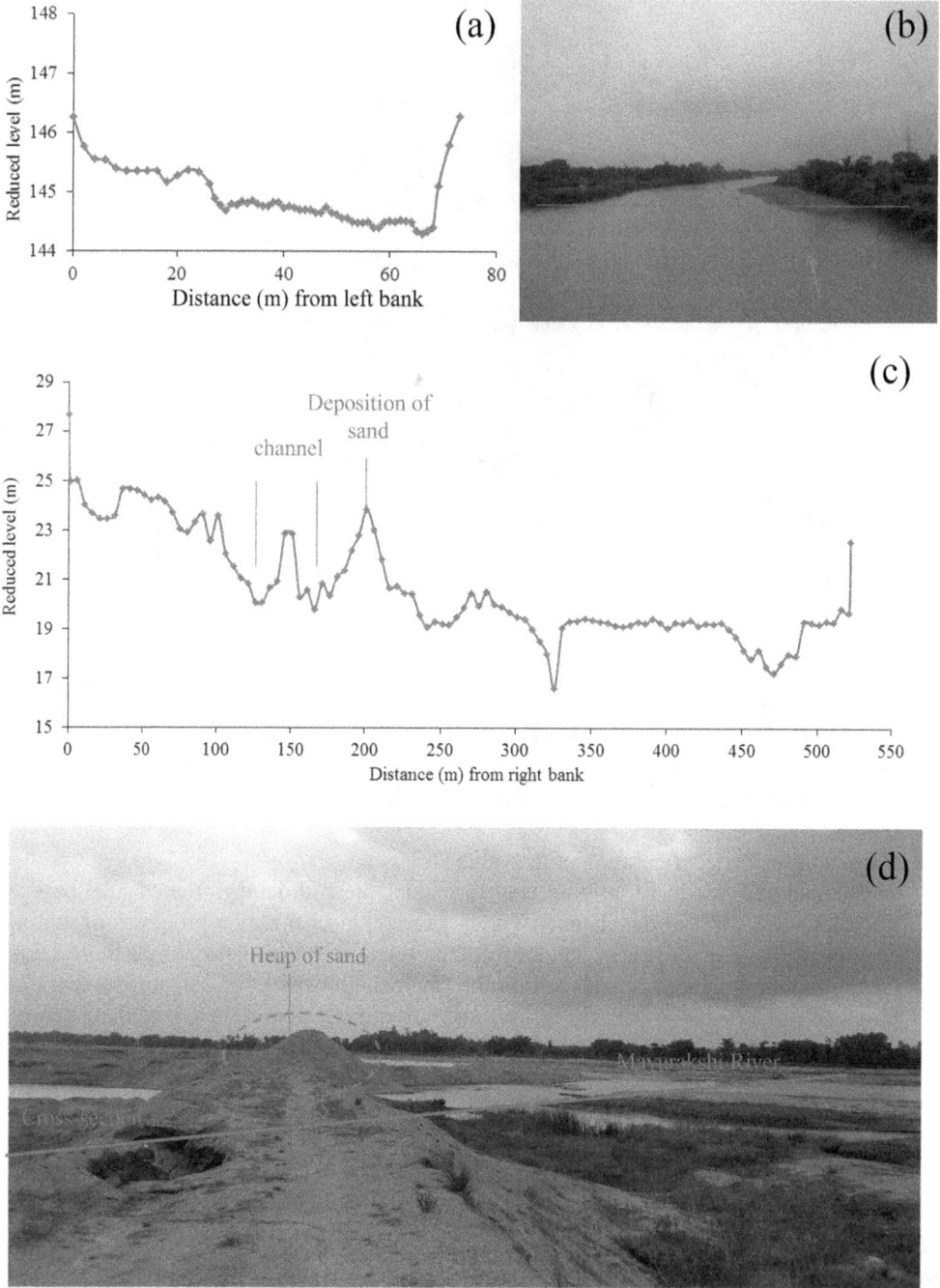

FIGURE 5.28   Sand mining and CS morphology, (a) and (b) for non-sand-mining area in the head reach, Gajamba, Jharkhand (28 September 2019), (c) and (d) for the sand mining area in the middle reach, Bhandirban, West Bengal. (Field Survey, 6 July 2019.)

a narrow channel (width 70 m) with a small sand bar at mid of the river. However, CS of the mining area denotes a ridge and furrow configuration of the CS. This typical formation is due to scouring, and hence, deep incised channel and the ridges mark the heap of sand mined from the river bed.

5.3.5 STONE QUARRYING AND STONE CRUSHING

5.3.5.1 Location and Characteristics

Stone quarrying and stone crushing are observed in the middle reach of MRB because this particular reach contains basaltic formation of Rajmahal trap (Chatterjee, 2010). In the MRB, there are 56 stone quarry centres and 42 stone crushing centres. The dominant areas for stone extraction include the hilly areas such as Makrapahari, Khalsa Pahari, Amchua Pahar, Salbona Pahar, Gamar Pahari, Gosain Pahari, Panchami and Litia Pahar. In temporal sequence, stone quarry activity has also increased. The river bed has been exposed, and numerous pond-like depressions in and around the river bed are increasingly observed (Figure 5.29). From 2007–2018, the pond area has increased from 12,817 to 184,554 m².

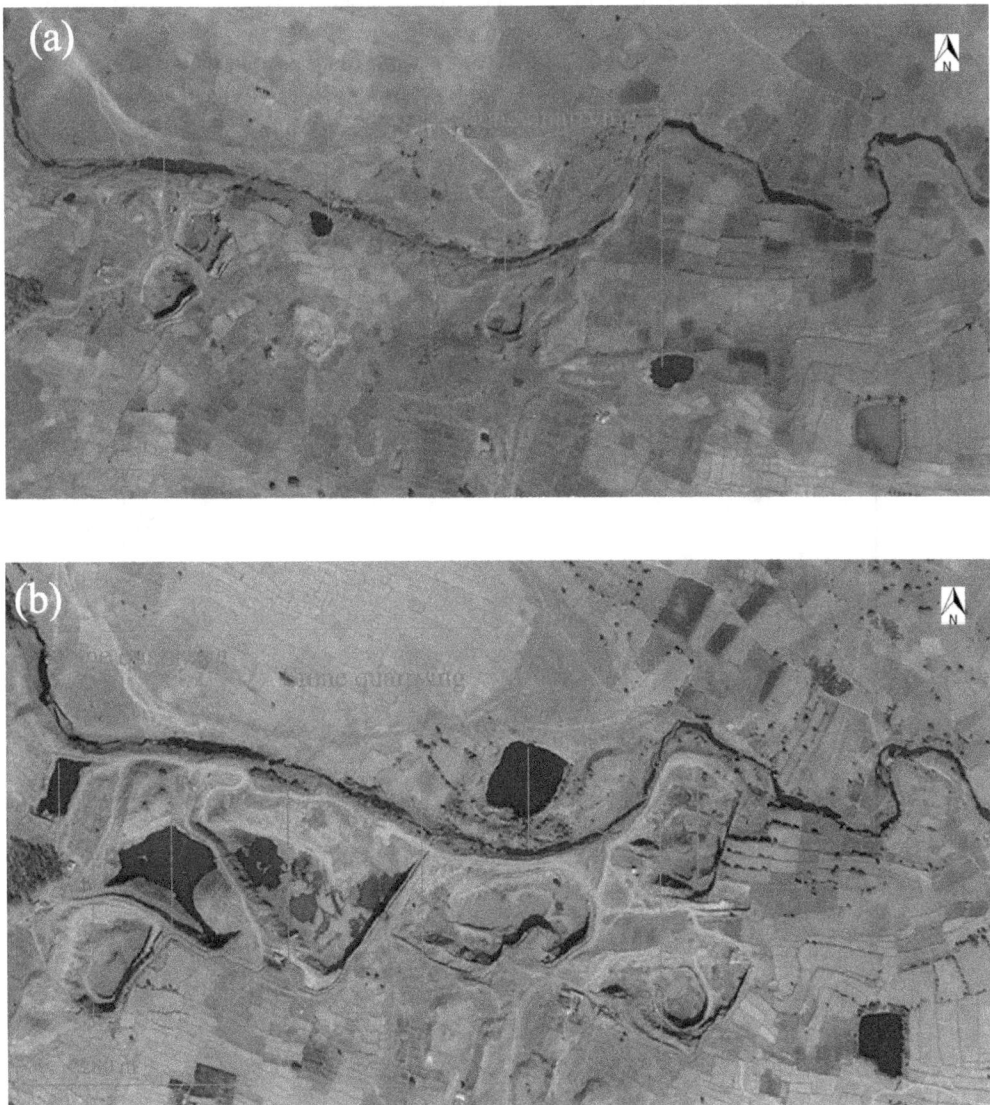

FIGURE 5.29 Extent of stone quarrying near Thakurpura, West Bengal. (Prepared from the Google Earth Images, (a) 27 February 2007 and (b) 23 October 2018.)

5.3.5.2　Stone Quarrying, Crushing and Channel Morphology

i. Stone quarrying and fingertip streams

The stone quarrying affects streams of the catchment area or the upstream area. The first-order streams are altered to a great extent due to the concentration of this activity in this stretch (Figure 5.30). Out of the total stone quarrying centres over the MRB, 34 centres are concentrated at an elevation of 130–65 m, 10 centres below 65 m and 12 centres above 190 m, respectively. Due to excavation of materials, fingertip streams are either bifurcated or obstructed to flow along its path. This may lead to the loss of the fingertip stream which may have corresponding effects on the basin morphometry and channel behaviours of the downstream reaches.

ii. Bed deposition and channel morphology

Stone quarrying and stone crushing are the dominant anthropogenic inputs that control both the long profile and cross profile of the river. The stone crushing centre produces huge quantity of stone chips (Figure 5.31a). A fraction of the stone chips flows from the crushing centre to the river by the overland flow (Figure 5.31b).

The deposition of stone chips in the long run modifies bed morphology of a river. Two cross profiles CS 1 and CS 2 and one long profile along the AB have been drawn adjacent to stone crushing and stone quarrying area at Haripur located in the middle stretch of the MRB (Figure 5.32).

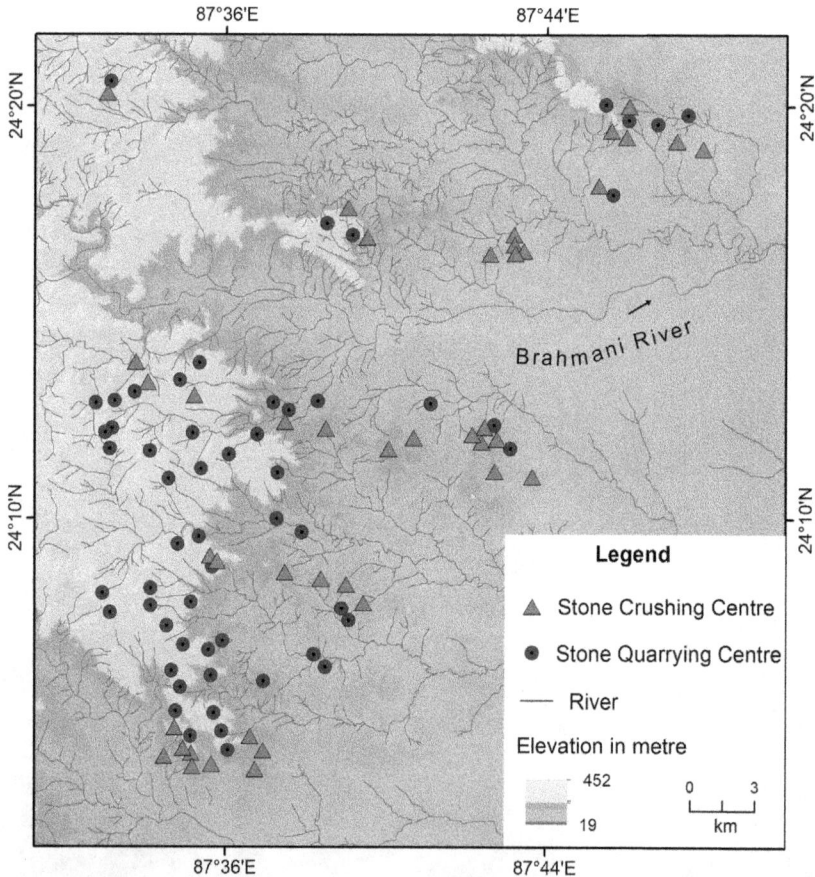

FIGURE 5.30　Stone crushing centre near the first-order streams. (Prepared from Google Earth Images, 23 October 2018, 24 October 2018, and 28 December 2018 and SoI 1: 50,000 toposheets.)

FIGURE 5.31 Impact of stone crushing on sediment flow. (a) Stone crushing centre and (b) Sediment flux by overland flow at Chirudih, Jharkhand. (Field Photograph, 7 July 2019.)

FIGURE 5.32 Location of CSs and long profile near stone crushing centre at Haripur. (Prepared from Google Earth Images, 23 October 2018)

The CS 1 shows that the CS morphology has experienced a radical change with time (Figure 5.33). The channel width (~19 m) is maintained with time. However, channel depth has decreased with time as portrayed by the lesser average depth of the actual profile (1.20 m) compared to that of the excavated profile (1.41 m). As a result, the CS area computed with reference to actual bed profile (26.7 m²) is much lesser than that of the excavated profile (27.9 m²) due to bed deposition. However, the nature of deposition follows the natural bed configuration and sources the materials. For the present profile, deposition inclined towards right bank (20–800 cm) because on the reverse bank there is no deposition.

Stream avulsion or channel shifting is a natural phenomenon arising out of the stream energy and sediment load behaviour. Generally, a channel finds a new path when its present course is maintained at a higher bed level compared to the surroundings. This natural phenomenon is also guided by the anthropogenic activities in some areas (Islam and Guchhait, 2017b). In the MRB, this type of shifting is commonly found. The CS 2 near stone crushing centre at Haripur shows that bed deposition occurs at such a pace that a channel has been abandoned and shifted to another location (Figure 5.34a). The CS morphology shows that bed elevation is at 68.89 m for the present course and 69.38 m for the abandoned course (Figure 5.34b). The anthropogenic interventions have choked up the present course to use as a roadway for carrying material. Therefore, the present course is diverted through the abandoned stone quarry field.

Generally, long profiles descend gradually or in step-like manner. Smoothly carved out concave long profile is found after a long time of erosion over a drainage basin. However, long profile may contain some perturbations due to tectonic or lithologic reasons. Furthermore, some anthropogenic controls may also lead to gain in the bed elevation. In the MRB, the

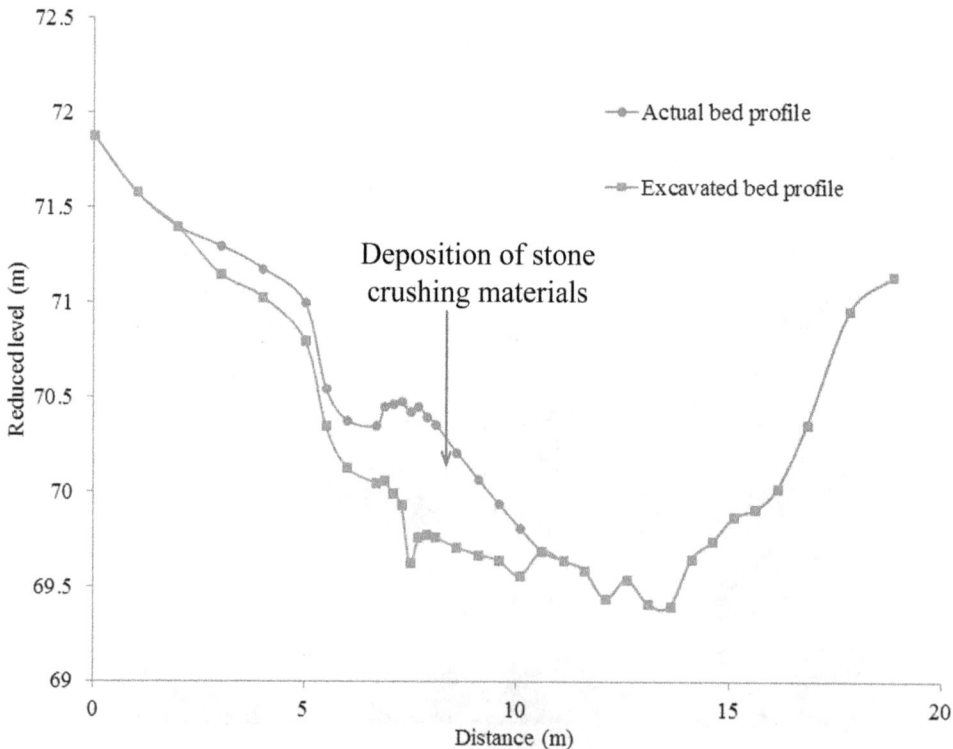

FIGURE 5.33   Impact of stone crushing on shallow channel bed on CS 1. (Based on field data, 2019)

FIGURE 5.34A Mechanism of channel abandonment by stone crushing. (Field Photograph, 17 November 2019.)

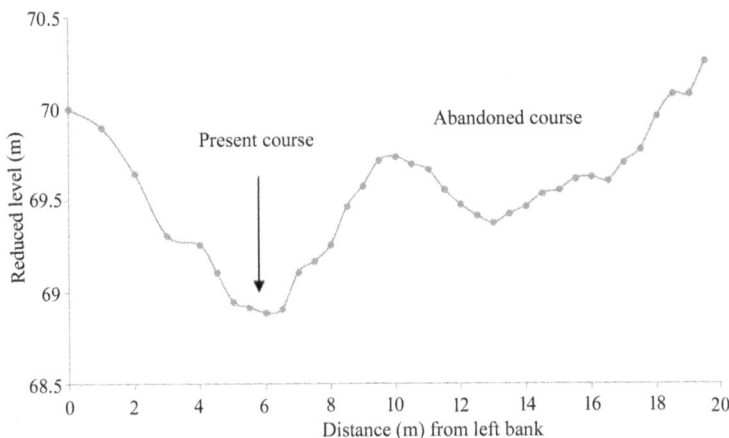

FIGURE 5.34B Impact of stone crushing channel abandonment on CS 2. (Based on field data, 17 November 2019.)

upper part of the Mayurakshi River near a stone crushing centre has portrayed a gain in the bed elevation of about 0.7 m within a very short distance of about 27 m (Figure 5.35). The field investigation establishes that a gradual rise in the long profile along the line AB is due to the deposition of boulder and stone chips on the river bed.

5.3.6 ROAD STREAM CROSSING

5.3.6.1 Location and Characteristics

Roads are often treated as the life line of a society and imply the level of societal development. The higher the road density, the higher the development of that region. In the MRB, there are three broad categories of road—National highway, State highway, PMGSY and municipality maintained road. There is a clear variation in road density in different stretches. The upper reach has road density 0.55 km/km², middle reach has 0.26 km/km² while tail reach has 0.23 km/km². This is due

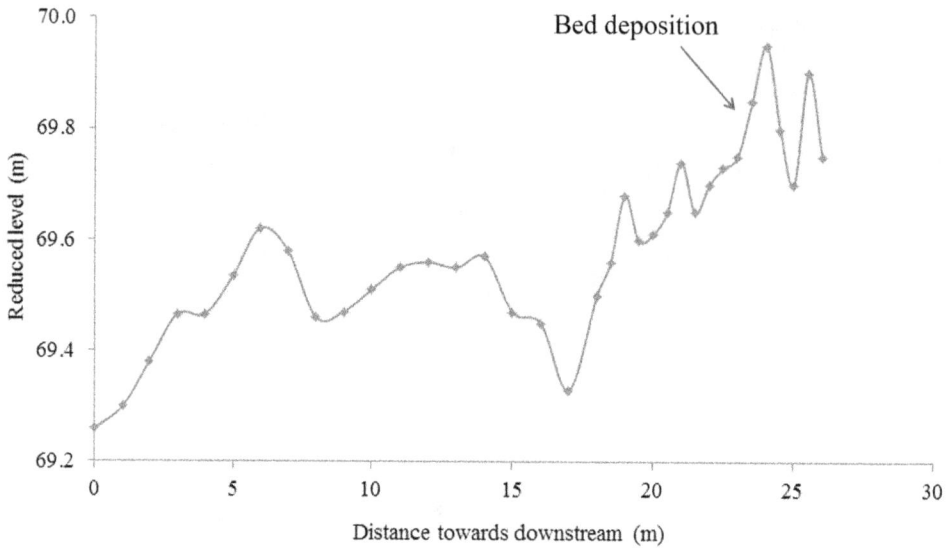

FIGURE 5.35　Stone crushing and long profile. (Based on field data, 2019.)

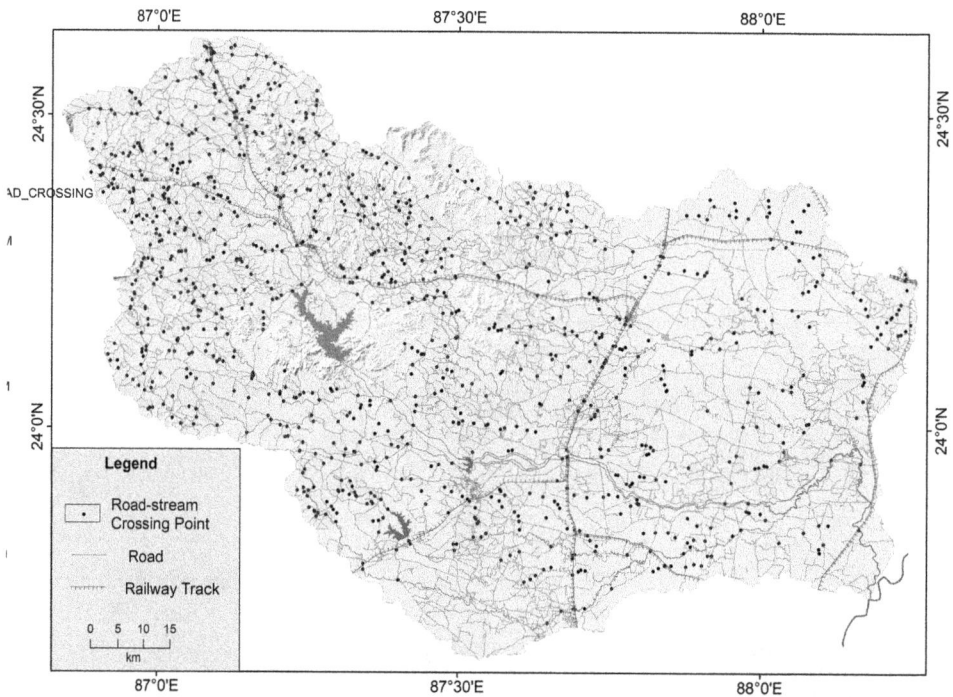

FIGURE 5.36　Pattern of the road stream crossing. (Prepared from open street map, 2019 and SoI 1: 50,000 toposheets.)

to the presence of palaeochannels, *beels* and other water bodies in the lower part of the MRB. As the MRB criss-crossed by numerous channels of different orders, a good number of road crossings are found in the MRB (Figure 5.36). In the upper reach, there are 626 road crossings, 205 in the middle stretches and 191 in the tail reaches. Road stream crossings have some typical geometry of the bridges e.g. box, pipe and cylinder is more in these stretches.

5.3.6.2 Road Stream Crossing and Channel Morphology

i. CS morphology

The nature of road stream crossing affects the morphological behaviour of a channel. The geometry of box, pier, and cylinder affects morphology in different ways. Road crossing having box pattern of pier is surveyed at Dhanyagram in the middle stretches of Mayurakshi River which shows that bed scouring and bed deposition occur in an alternate sequence (Figure 5.37a). Besides, pier-concentrated flow scour bed while in between the flow concentration i.e. in between piers bed deposition is observed. The cross profile portrays that scour depth is maximum 61.81 m (msl) at a distance of 10 m from the left bank where the minimum scour depth observed is 62.35 m (msl) at a distance of 11 m from the right bank. Similarly, in between the bed scouring, bed deposition is found at elevation of about 62.53 m (msl). The natural cross profile is transformed into step-like cross profile (Figure 5.37b).

FIGURE 5.37A Road stream crossing and channel bed morphology. (Field Photograph, 6 July 2019.)

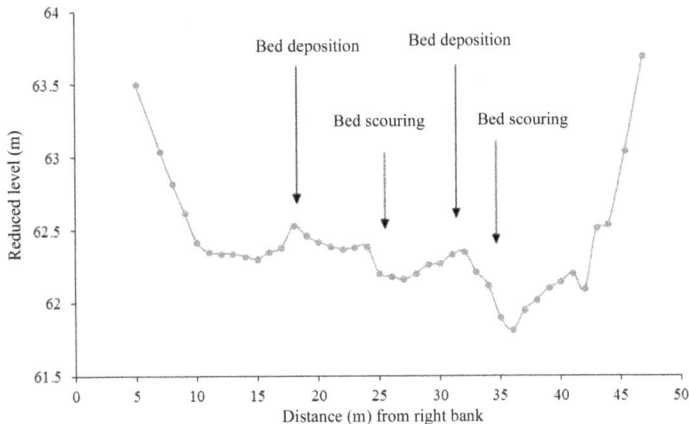

FIGURE 5.37B Impact of road crossing on channel morphology. (Based on field data, 2019.)

ii. Constricted waterway

In the uncontrolled catchment below the barrages of the Mayurakshi Reservoir Project, there are about 36 major road bridges/culverts and 15 rail bridges on different roads and railways interfering with natural drainage of the basin, out of which 12 bridges and culverts are found inadequate. It is the very basic principle of civil engineering that based on cost–benefit analysis, bridges are constructed where river width is at minimum. In the MRB, bridges have constricted the waterway. The rate of constriction is minimum (2.64%) at Badshahi road on River Kuea while maximum (58.19%) at Gramsalika–Kandi–Khagraghat road (Table 5.10). Therefore, river bridges tend to fall apart due to this inadequate waterway in the catchment.

Besides, there are several earthen and bamboo-made road stream crossing to maintain the village communication. On the River Dwarka near Surkhali, such a stream crossing has reduced the water surface and constricted the channel (Figure 5.38). During monsoon, these are often damaged and affect the sediment load of the river.

5.3.7 Brick Fields

5.3.7.1 Location and Characteristics

Brick fields are also boomed due to the same reason as that of the demand for sand. It is a basic building material prepared from good quality silt. In the MRB, mainly two types of brick fields have been observed during the field investigation i.e. a. small-scale brick field encompassing an area of about $1500\,m^2$ and b. large-scale brick field covering an area of about $45,000\,m^2$. The small-scale units are mainly concentrated in the upper reach while the large scale units are mainly found preferably along a river in the middle and lower reaches of the MRB because of the availability of silt in this particular area (Figure 5.39). The growth of brick field industry is alarming in the last decade. During 2006, the number was 79 (upper: 18, middle: 44 and lower: 17), and at present (2019), it is 440 (upper: 234, middle: 129 and lower: 77).

5.3.7.2 Brick Field and Channel Morphology

Brick field industry impacts the river channel morphology especially on the CS morphology. The main raw material of brick field is silt or clay which attracts brick kilners to mushroom on the banks of a river. According to an estimate of the District Land & Land Reforms Office (DL&LRO) of Nadia District in 2009, a single brick field can move $133,110\,ft^3$ materials per year. However, the earth material transferred is negligible for the small scale units compared to the large-scale brick fields. Therefore, concerning the large-scale units, in the Mayurakshi Basin, $8,119,710\,ft^3\ yr^{-1}$ earth materials could be moved in the year 2006, about $5856840\,ft^3\ yr^{-1}$ in the middle reach and about $2,262,870\,ft^3\ yr^{-1}$ in the lower reach. Besides, all the small-scale units concentrated in the upper reach also contribute some material movements. However, they could not be quantified due to unavailability of a standard measurement regarding sediment movement by a small-scale unit. In brief, these figures have alarmingly increased in recent years. The total earth material movement accomplished by anthropogenic process is increased to about $27420660\,ft^3\ yr^{-1}$ in the total basin area, about $17171190\,ft^3\ yr^{-1}$ in the middle reach and about $10249470\,ft^3\ yr^{-1}$ in the tail reach. Previously, the kilners collected silt or clay directly from the river bank which was responsible for the river bank erosion and increase in river width (Das, 2014), though recently extraction of silt from river bed has been banned. Hence, kilners collect the earth materials from somewhere else in the Mayurakshi basin. In brief, the earth movement is increased 294% in the middle reach and 453% in the tail reach only in last 13 years (2006–2019). Though majority of the kilners collect materials from the floodplains of the MRB, few brick fields still collect sediment from the river bed which greatly controls the CS morphology of the channel. The CS is drawn near the brick kiln which shows a shallow depression near left bank due to the extraction of silt from river bank by anthropogenic processes (Figure 5.40).

TABLE 5.10

Effect of Road Stream Crossing on Channel Constriction

Sl. No.	Road/Rail	Bridge Location	Catchment Area up to Bridge Point (sq. km)	Existing Waterway (m)	Required Waterway (m)	Constricted Waterway (m)	Channel Constriction (%)
1	Bolpur–Purandarpur–Dhullia Road	River Kopai	114	106.39	146.79	40.4	27.52
2	Bolepur–Purandarpur–Dhullia Road	River Bakreshwar	43.2	61.59	114.86	53.27	46.38
3	Rajgram–Nalhati–Saithia Bolepur Road and Sahebganj Loop Rail Line	River Brahmani	466.6	198.34	213.37	15.03	7.04
4	Rajgram–Nalhati–Saithia Bolepur Road and Sahebganj Loop Rail Line	River Mayurakshi	104	333.7	376.46	42.76	11.36
5	Suri–Ahmedpur–Kirnahar Road and Ahmedpur–Katwa Rail Line	River Kuea	464.99	128.34	185.17	56.83	30.69
6	Badshahi Road	River Brahmani	1,618.3	252.5	280.33	27.83	9.93
7	Badshahi Road	River Kuea	659.54	183.79	188.77	4.98	2.64
8	Gramsalika–Kandi–Khagraghat Road	River Dwarka	2,448.25	121.08	289.63	168.55	58.19
9	Kandi–Bharatpur–Salar Road	River Kuea	1,030.26	335.96	418.39	82.43	19.70
10	Khagraghat–Ramnagar Road/ Azimganj–Katwa Rail Line	River Uttarasan	–	111.73	140	28.27	20.19
11	Khagraghat–Ramnagar Road/ Azimganj–Katwa Rail Line	River Babla	–	154.3	221.26	67	30.28
12	Khagraghat–Ramnagar Road/ Azimganj–Katwa Rail Line	River Uttarasan	–	91.77	140	48.23	34.45

Source: Kandi Master Plan (2012).

FIGURE 5.38 Village road on the river of Dwarka connecting Surkhali and Bhatkhanda villages in Khargram, Murshidabad. (Field Photograph, 18 July 2018.)

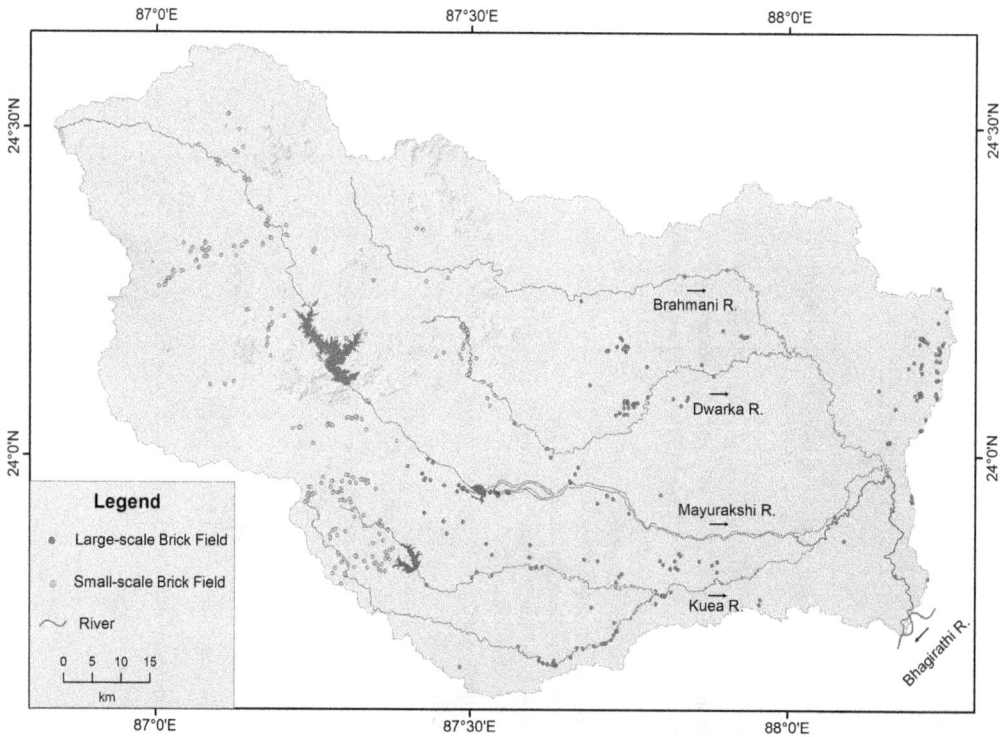

FIGURE 5.39 Location of the brick fields. (Prepared from the Google Earth Images, 8 January 2019, 10 January 2019, 7 February 2019, 9 March 2019, 28 March 2019.)

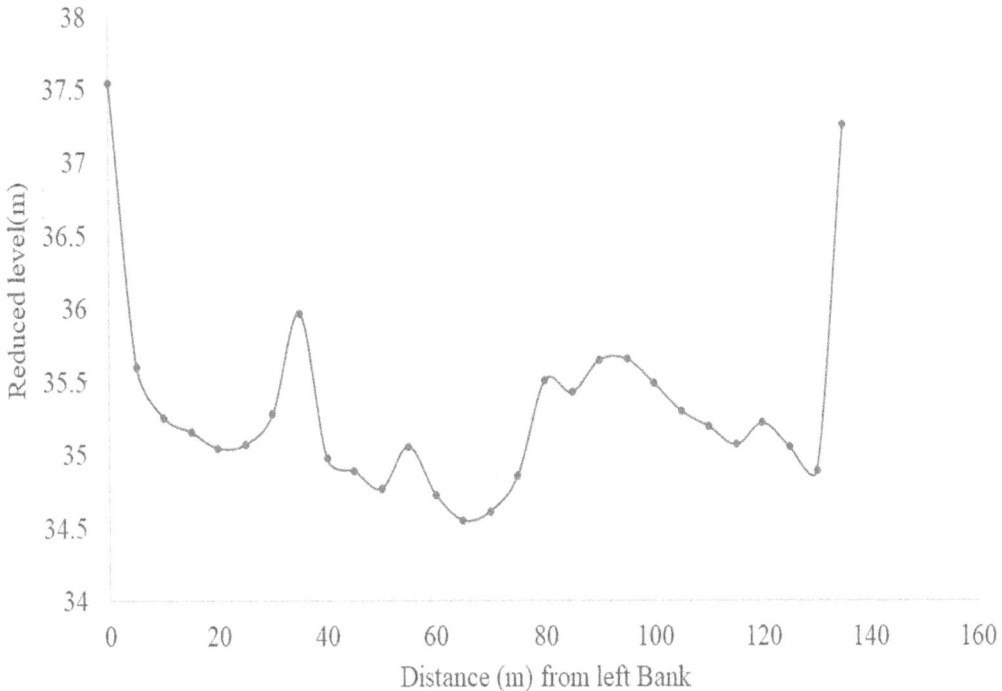

FIGURE 5.40 Impact of brick field on channel morphology. (Prepared from the field data, 2019.)

5.3.8 Land-Use, Land-Cover Changes

5.3.8.1 Dynamics of LULC

Land cover is the most important shroud that covers the entire face of the earth which protects the immediate unconsolidated surfaces from any detrimental influences of human entities. In a watershed context, land use, land cover is the most significant basis of functioning and maintaining the general equilibrium of the natural system. Like a steep slope of a mountain, LULC is also considered as a very sensitive ingredients in the way of maintaining aquatic ecosystem of a river basin. The LULC changes are the most determinant factors in the food and water supply as well as socio-economic advancement of a region. At the very prior stage of human development there was no such human domination pertaining to land. However, with the passage of time, corresponding changes in social structure as well as rapidly growing anthropocentric ideality brought the humongous changes in the model and form of LULC change. In the world of accelerating rise of human population and urbanisation, sustainable LULC changes are the only ways to feed these countless mouth and give habitation for their survival. Land-use land-cover change, environmental sustainability and human development are the three most significant parameters of global importance that can never be fit together.

In the context of MRB, profuse alteration of land-use land-cover feature has been observed during the last decade (2005–2006 to 2015–2016) (Figure 5.41a and b). At micro level, the changes in LULC feature over the MRB are noticed. The major LULC types which have registered an increase are barren land, mining area while the forest cover has declined sharply (Figure 5.41). Moreover, the settlement has also slightly increased in the MRB LULC area i.e. 0.141% for urban built-up area and 0.221% area for rural built-up area (Table 5.11).

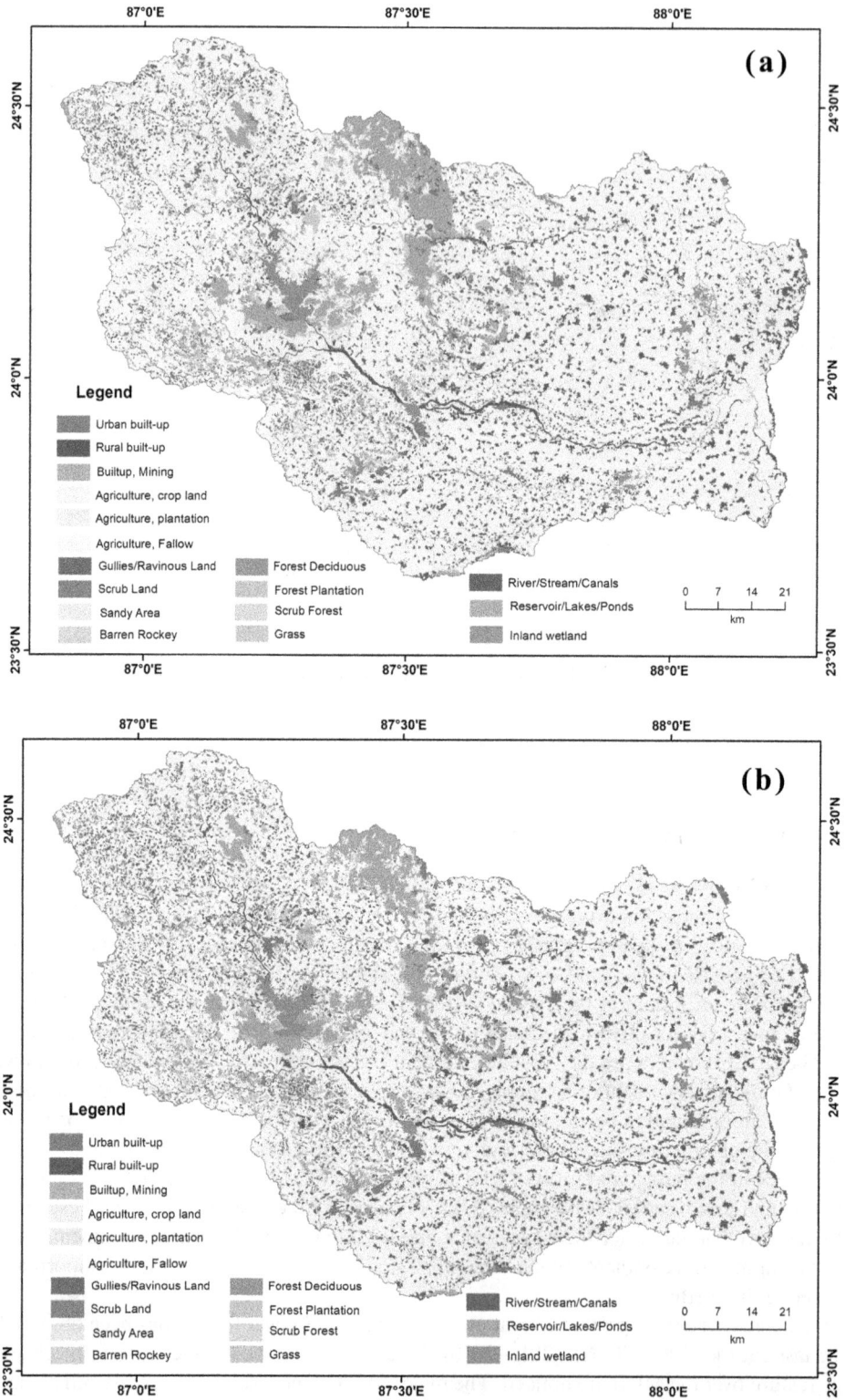

FIGURE 5.41 (a) LULC of the MRB in 2005–2006 and (b) LULC of the MRB in 2015–2016. (Based on NRSC Bhuvan, https://bhuvan-app1.nrsc.gov.in/thematic/thematic/index.php)

TABLE 5.11

Changes in LULC Types during 2005–2006 and 2015–2016

LULC Types	2005–2006 (Area in km²)	% of Area	2015–2016 (Area in km²)	% of Area
Plantation	14.193	0.148	21.650	0.226
Forest plantation	17.334	0.181	29.560	0.308
Deciduous	607.053	6.326	520.320	5.422
Mining	6.579	0.069	40.260	0.420
Reservoir/lakes/pond	115.922	1.208	110.154	1.148
Inland water body	50.985	0.531	27.540	0.287
River	154.391	1.609	143.540	1.496
Rural	782.338	8.153	803.540	8.374
Urban	49.019	0.511	62.520	0.652
Cropland	6,853.247	71.418	6,606.665	68.848
Fallow land	427.509	4.455	567.250	5.911
Sandy	1.444	0.015	2.580	0.027
Barren land	19.039	0.198	26.340	0.274
Gullied land	1.696	0.018	3.271	0.034
Scrub land	295.102	3.075	401.360	4.183
Scrub forest	161.036	1.678	189.200	1.972
Barren rocky	39.113	0.408	40.250	0.419
Total	9,596.000	100.000	9,596.000	100.000

Computed by the authors.

5.3.8.2 LULC and Channel Hydro-Geomorphology

The implication of LULC change on the hydrological regime of a basin is of utmost concern to the water resource managers. The impact of LULC change on channel hydrology was investigated by Garg et al. (2017). The result showed that outbalancing channel hydrology with variation in base flow parameters occurred at the basin outlet. Variation of evapotranspiration was also found at different LULC scenarios. The LULC changes are considered as the most driving forces in almost all hydrological processes including evaporation, groundwater recharge, overland flow and base flow, and a minute change in one of these ingredients can vigorously alter the whole process response system of the hydrological cycle. Upstream and downstream linkages found in the works of Harvey (1991), Kondolf et al. (2002) and Walling (1983) suggest that any minor changes in upstream LULC can lead to bring changes in the downstream channel morphology, water and sediment transport of a river basin. Recent explorations by Harvey (1991) and Hooke (2003) have exhibited that sediment generation, transport and distribution to a downstream river channel are not only a function of the overall physiographic unit it belongs to but also the mechanisms of spatial arrangement, synthesis and the intimate connectivity of various physiographic provinces.

Among various LULC changes, it is found that deforestation alone is the most responsible factor for changes in all hydrological processes. In developing countries like India, deforestation along with an accelerated pace of urbanisation are the major driving forces responsible for drastic changes in hydrological water balance in a basin unit. The hydrological character of a river basin is a mechanism of the interrelationship between catchment geomorphology (basin area, shape and size of the basin, physiography, channel pattern and slope, stream density, frequency and water storage capacity of the channel), hydrology and pattern of land-use land-cover change especially the distribution of vegetation in the basin (Loukas et al., 1996). The organisation, association and alterations of vegetation in the valley side and stream bottom are strongly interrelated to fluvial geo-hydrological processes and forms, which are strongly affected and maintained by dynamics of water discharge.

Deforestation or expulsion of grassy vegetation such as herb and shrub by cultivation, grazing, lumbering and burning reduces rainfall interception and allows splash erosion as well as encourage soil creep, sheet, rill and gully erosion. Explorative works on a micro-scale, bare surface catchment in New Hampshire revealed that summer discharge raised 40% while approximately 15 times more dissolved inorganic element from the catchment is now washed away (Likens et al., 1970). Hydrologic experiences of 219 streams in Russia also exhibit that little small, vegetated and forested watersheds have considerably very low overland flow and run-off but greater percolation and seepage which could maintain the conventional base flow during dry weather (Bochkov, 1970).

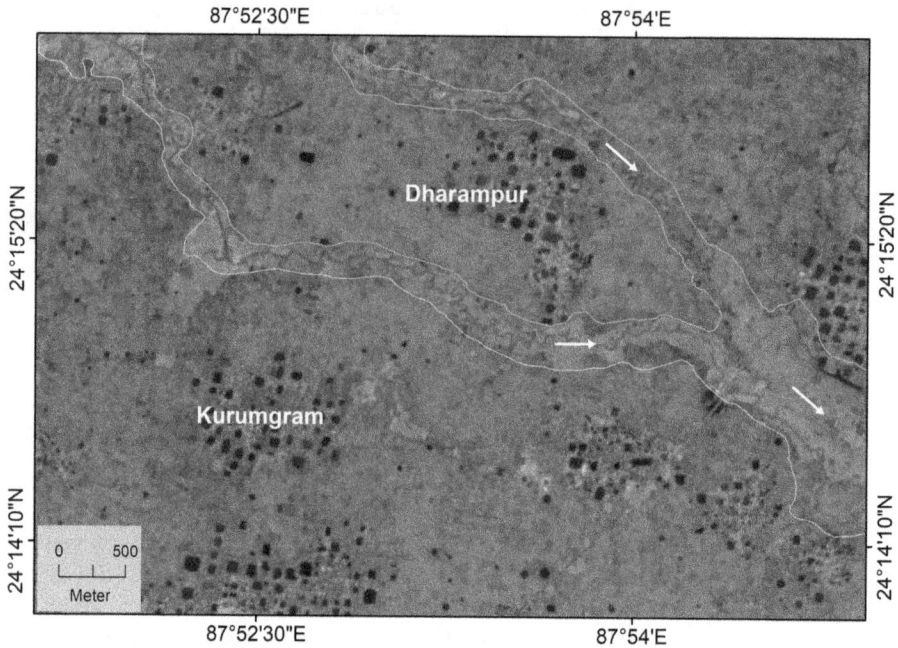

FIGURE 5.42 Channel occupied by agricultural practice near Dharampur. (Prepared from Sentinel 2B, 31 March 2019: tiles no: T45QWG.)

FIGURE 5.43A Practice of on-bed agricultural in the MRB on river bed near Amjua in the middle stretch. (Field Photograph, 29 December 2019.)

Channel geomorphology and sedimentology are mainly directed by rapidly growing socio-economic and demographic changes through intensification of farming practices in the low-lying fertile valleys and continuous expansion of agricultural lands towards the bed of the river. Thus, on-bed and off-bed changes in the LULC are observed. A common trend in the cut-off channel or spill-over channel is that the channel beds are occupied by agricultural practice. For example, a stretch of the Brahmani channel portrays that about 80% of the river bed is occupied by seasonal agriculture (Figure 5.42). A similar pattern is also found in the middle reach of Mayurakshi River in Amjua having an intensive practice of growing crops such as potato, mustard seed and vegetables. This characterises an ideal anthropic channel (Figure 5.43a).

FIGURE 5.43B Agricultural field within river valley at Ranagram. (Field Photograph, 2018.)

FIGURE 5.43C Agricultural practice on the bed of the Dwarka River near Ranagram. (Field Photograph, 2018.)

Furthermore, a stepped valley configuration is greatly controlled by the on-bed agricultural practice. Along the bank of the river, farmers create agricultural fields especially in the post-monsoon period and reap the harvest with minimum investment. Though from a farmers' point of view it may be profitable, channel morphology, sedimentology and ecological behaviour to a large extent are controlled. Along the banks of the Mayurakshi near Ranagram, farmers have steepened the bank slope that may induce bank erosion under favourable condition (Figure 5.43b). Similarly, along the bank of the Dwarka River near Ranagram, the agricultural fields are tilled and exposed (Figure 5.43c). When the rain comes, splash erosion and gully erosion may be operative inducing soil loss of that area and hence increasing the sediment budget of the river.

5.4   CONCLUSION

The evolution of the MRB is complex process involving sculpturing by long-term geological and climatological and anthropogenic factors. In the present study, it has been observed that naturally developed landscape over the periods has been altered by the human activities in the forms of dam, barrage, irrigation canal, check dam, embankment, sand mining, stone quarrying, brick field, changes in LULC and road stream crossing. The major findings of the study indicate the following.

1. There are six dams and barrages, namely, Massanjore, Tilpara, Deucha, Bakreshwar, Brahmani and Kopai which have had a great control over the fluvial hydraulic behaviour. In the pre-dam period, flood frequency was low and magnitude was high. In the post-dam condition, high-frequency low-magnitude floods are observed.
2. The massive dams and barrages are controlling channel morphology induced by the change of the fluvial regime. Channel sinuosity and braiding have increased in the post-dam period especially in the dam-dominated stretches. Channel width–depth ratio and CS area have also increased both in the downstream and backwater as portrayed by the Kushkarani River. Dams have also induced sedimentation in the upstream reservoir area and scouring in the downstream section. The upstream area portrays the growth of the mid-channel and point bars.
3. Check dams, though not massive control like the dams, influence CS morphology to a certain aspect. Both in the upstream and downstream of the check dams, channel has been concretised and hence geometric shapes of the CSs are found. However, areas farther up and down of the check dams show natural channel configuration.
4. The embankment is a long-standing anthropogenic imprint which has impeded the free evolution of the channel and river valley and thus the channel and valley tappers off in the downstream direction. Furthermore, the breaching of earthen embankment also has increased the rate of sedimentation.
5. Sand mining at an accelerating rate has been shaping the channel since the last decade. Natural channel configuration has been modified to a great extent as evident from higher channel width and CS area. Similarly, stone quarrying also has created depression on the river bed exposing the hard bed. Furthermore, channel diversion in the areas of stone quarrying and crushing centres is also notable.
6. Road stream crossing has created bed deposition and scouring in alternate sequence. Besides, collapse and subsidence of some road crossing induce sediment flux to the river.
7. Practice of on-bed agriculture has completely modified channel bed configuration. Virtually, channel bed is converted to agricultural fields in the post-monsoon period.

These dimensions of the human interventions indicate the present state of the art of the MRB in the Anthropocene. Kandi Master Plan has taken structural measures for flood abatement. However, alternative planning measures are absolutely necessary to restore the channel and valley characters. The present study may be helpful to frame alternative measures. Still, some in-depth study are required for the past, present and future direction of the basin evolution.

A comprehensive study to relate sand and stone mining to the river water quality and associated ecological stress is required. A pioneering effort is also needed to create a baseline study on environmental flow in relation to dams and barrages and other mode of human alternation. Changes in LULC on sediment flux of the river in different periods of the channel evolution may be investigated. Fluvial hydraulics in relation to mining activities may be undertaken. Morphological changes due to varying nature road stream crossing may be another domain of research.

REFERENCES

Bochkov, A. P. (1970). Forest influence on river flows. *Nature and Resources, 6*(1), 10–11.

Boix-Fayos, C., Barberá, G. G., López-Bermúdez, F., & Castillo, V. M. (2007). Effects of check dams, reforestation and land-use changes on river channel morphology: case study of the Rogativa catchment (Murcia, Spain). *Geomorphology, 91*(1–2), 103–123.

Boix-Fayos, C., de Vente, J., Martínez-Mena, M., Barberá, G. G., & Castillo, V. (2008). The impact of land use change and check-dams on catchment sediment yield. *Hydrological Processes: An International Journal, 22*(25), 4922–4935.

Bravard, J. P., Amoros, C., Pautou, G., Bornette, G., Bournaud, M., Creuzé des Châtelliers, M., & Tachet, H. (1997). River incision in south-east France: morphological phenomena and ecological effects. *Regulated Rivers: Research & Management: An International Journal Devoted to River Research and Management, 13*(1), 75–90.

Brice, J. C. (1964). *Channel Patterns and Terraces of the Loup Rivers in Nebraska.* US Government Printing Office, Washington, DC, USA.

Chakrabarti, B. (1985). A geomorphological analysis of the Mayurakshi River Basin. (Unpublished PhD thesis, The University of Burdwan).

Chatterjee, N. (1995). Social forestry in environmentally degraded regions of India: case-study of the Mayurakshi Basin. *Environmental conservation, 22*(1), 20–30.

Chatterjee, N. (2010). The basalt stone quarries of eastern India. *International Journal of Environmental Studies, 67*(3), 439–457. doi: 10.1080/00207233.2010.491253.

Das, B. C. (2014). Impact of in-bed and on-bank soil cutting by brick fields on Moribund deltaic rivers: A study of Nadia river in West Bengal. The NEHU Journal, *XII*(2), 101–111.

Davis, H., & Turpin, E. (2015). *Art in the Anthropocene: Encounters Among Aesthetics, Politics, Environments and Epistemologies.* Open Humanities Press, London.

Davis, J., Bird, J., Finlayson, B., & Scott, R. (2000). The management of gravel extraction in alluvial rivers: a case study from the Avon River, southeastern Australia. *Physical Geography, 21*(2), 133–154.

Debanshi, S., & Pal, S. (2020). Assessing gully erosion susceptibility in Mayurakshi river basin of eastern India. *Environment, Development and Sustainability, 22*, 883–914, https://doi.org/10.1007/s10668-018-0224-x.

Garg, V., Aggarwal, S. P., Gupta, P. K., Nikam, B. R., Thakur, P. K., Srivastav, S. K., & Kumar, A. S. (2017). Assessment of land use land cover change impact on hydrological regime of a basin. *Environmental Earth Sciences, 76*(18), 635.

Ghosh, A. (2012). Impact of Mayurakshi irrigation canal system on the socio economic aspects of its command area, an unpublished PhD thesis, Santiniketan: Visva-Bharati.

Ghosh, K. G., & Pal, S. (2015). Impact of dam and barrage on flood trend of lower catchment of Mayurakshi river basin, eastern India. *European Water, 50*, 3–23.

Ghosh, S., & Guchhait, S. K. (2014). Hydrogeomorphic variability due to dam constructions and emerging problems: a case study of Damodar River, West Bengal, India. *Environment, Development and Sustainability, 16*(3), 769–796.

Ghosh, S., & Mistri, B. (2012). Hydrogeomorphic significance of sinuosity index in relation to river instability: a case study of Damodar River, West Bengal, India. *International Journal of Advances in Earth Sciences, 1*(2), 49–57.

Gosnell, H., & Kelly, E. C. (2010). Peace on the river? Social-ecological restoration and large dam removal in the Klamath basin, USA. *Water Alternatives, 3*(2), 362.

Gregory, K. J., & Lewin, J. (2014). *The Basics of Geomorphology: Key Concepts.* Sage, London.

Guchhait, S. K., Islam, A., Ghosh, S., Das, B. C., & Maji, N. K. (2016). Role of hydrological regime and floodplain sediments in channel instability of the Bhagirathi River, Ganga-Brahmaputra Delta, India. *Physical Geography*, 37(6), 476–510.

Haigh, M. J. (1978). Micro-erosion processes and slope evolution on surface-mine dumps at Henryetta, Oklahoma. *Oklahoma Geology Notes, (Oklahoma Geological Survey)*, 38(3), 87–96.

Harvey, A. M. (1991). The influence of sediment supply on the channel morphology of upland streams: Howgill Fells, Northwest England. *Earth Surface Processes and Landforms*, 16(7), 675–684.

Hay, A. (1985). Scientific method in geography. In R. J. Johnston (ed.), *The Future of Geography* (pp. 129–142). Methuen, London.

Hooke, J. (2003). Coarse sediment connectivity in river channel systems: a conceptual framework and methodology. *Geomorphology*, 56(1–2), 79–94.

Hooke, R.L., 1994. On the efficacy of humans as geomorphic agents. *GSA Today*, 4(9), 224–225.

Hooke, R. L. (2000). On the history of humans as geomorphic agents. *Geology*, 28(9), 843–846.

Islam, A., & Guchhait, S. K. (2017a). Search for social justice for the victims of erosion hazard along the banks of river Bhagirathi by hydraulic control: a case study of West Bengal, India. *Environment, Development and Sustainability*, 19(2), 433–459.

Islam, A., & Guchhait, S. K. (2017b). Analysing the influence of Farakka Barrage Project on channel dynamics and meander geometry of Bhagirathi river of West Bengal, India. *Arabian Journal of Geosciences*, 10(11), 245.

Islam, A. & Barman, S. D. (2020). Drainage basin morphometry and evaluating its role on flood-inducing capacity of tributary basins of Mayurakshi River, India. *SN Applied Sciences*. https://link.springer.com/article/10.1007/s42452-020-2839-4

Jain, V., Tandon, S. K., & Sinha, R. (2012). Application of modern geomorphic concepts for understanding the spatio-temporal complexity of the large Ganga river dispersal system. *Current Science(Bangalore)*, 103(11), 1300–1319.

Johnson, D. L., & Lewis, L. A. (1995). *Land Degradation: Creation and Destruction*. Blackwell, Oxford, UK.

Khatun, S., & Pal, S. (2017). Categorization of morphometric surface through morphometric diversity analysis in Kushkarani River Basin of Eastern India. *Asian Journal of Physical and Chemical Sciences*, 2, 1–19.

Kondolf, G. M. (1994). Geomorphic and environmental effects of instream gravel mining. *Landscape and Urban Planning*, 28(2–3), 225–243.

Kondolf, G. M., Piégay, H., & Landon, N. (2002). Channel response to increased and decreased bedload supply from land use change: contrasts between two catchments. *Geomorphology*, 45(1–2), 35–51.

Lagasse, P. F. (1986). River response to dredging. *Journal of Waterway, Port, Coastal, and Ocean Engineering*, 112(1), 1–14.

Lagasse, P. F., Winkley, B. R., & Simons, D. B. (1980). Impact of gravel mining on river system stability. *Journal of Waterways and Harbors Division*, 106(ASCE 15643).

Loukas, A., Quick, M. C., & Russell, S. O. (1996). A physically based stochastic-deterministic procedure for the estimation of flood frequency. *Water Resources Management*, 10(6), 415–437.

Likens, G. E., Bormann, F. H., & Johnson, N. M. (1970). Nitrification: importance to nutrient losses from a cutover forested ecosystem. *Science*, 163(3872), 1205–1206.

Mohammad, S. (2006). Irrigation Projects in Birbhum District, Published in *PaschimBanga* (in Bengali), Birbhum Special Issue, Govt. of West Bengal, pp. 168–169.

Marston, R. A., Bravard, J. P., & Green, T. (2003). Impacts of reforestation and gravel mining on the Malnant River, Haute-Savoie, French Alps. *Geomorphology*, 55(1–4), 65–74.

Muehlbauer, J. D., LeRoy, C. J., Lovett, J. M., Flaccus, K. K., Vlieg, J. K., & Marks, J. C. (2009). Short-term responses of decomposers to flow restoration in Fossil Creek, Arizona, USA. *Hydrobiologia*, 618(1), 35–45.

Molla, H. R. (2011). Embankment and changing micro-topography of lower Ajoy basin in Eastern India. *Ethiopian Journal of Environmental Studies and Management*, 4(4).

Mossa, J., & Marks, S. R. (2011). Pit avulsions and planform change on a mined river floodplain: Tangipahoa River, Louisiana. *Physical Geography*, 32(6), 512–532.

Newport, B. D., & Moyer, J. E. (1974). *State-of-the-Art: Sand and Gravel Industry*, Technical Series Report 660. US Environmental Protection Agency, Corvallis.

Office of the District Magistrate. (2016). *District Disaster Management Plan*. Govt. of West Bengal, Murshidabad District.

Pal, D. K. (2017). Soils of the Indo-Gangetic alluvial plains: historical perspective, soil-geomorphology and pedology in response climate change and neotectonics. In *A Treatise of Indian and Tropical Soils* (pp. 71–90). Springer, Cham.

Pal, S. (2010). Changing inundation character in Mayurakshi River basin: a spatio temporal review. *Practising Geographer*, *14*(1), 58–71.

Pal, S. (2016). Impact of Massanjore dam on hydro-geomorphological modification of Mayurakshi river, Eastern India. *Environment, Development and Sustainability*, *18*(3), 921–944.

Pal, S. (2017). Impact of Tilpara barrage on backwater reach of Kushkarni River: a tributary of Mayurakshi River. *Environment, Development and Sustainability*, *19*(5), 2115–2142.

Pal, S., & Let, S. (2011). Channel morphological trend and relationship assessment of Dwarka River, Eastern India. *Global Journal of Applied Environmental Science*, *1*(3), 221–232.

Pal, S., & Let, S. (2013). Channel leaning or channel fattening and Quasi Misfit stream generation. *International Journal of Advanced Research in Management and Social Sciences*, *2*(1), 12–29.

Pal, S., Let, S., & Das, P. (2012). Assessment of channel width disparity of the major rivers within Mayurakshi River Basin. *International Journal of Geology, Earth and Environmental Sciences*, *2*(3), 1–10.

Pal, S., & Mandal, I. (2019). Impacts of stone mining and crushing on environmental health in Dwarka river basin. *Geocarto International*, 1–29.

Patton, P. C., & Baker, V. R. (1976). Morphometry and floods in small drainage basins subject to diverse hydrogeomorphic controls. Water *Resources Research*, *12*(5), 941–952.

Pradhan, C., Chembolu, V., & Dutta, S. (2019). Impact of river interventions on alluvial channel morphology. *ISH Journal of Hydraulic Engineering*, *25*(1), 87–93.

Rao, K. N., Subraelu, P., Nagakumar, K. C. V., Demudu, G., Malini, B. H., & Rajawat, A. S. (2013). Geomorphological implications of the basement structure in the Krishna-Godavari deltas, India. *ZeitschriftfürGeomorphologie*, *57*(1), 25–44.

Renganayaki, S. P., & Elango, L. (2013). A review on managed aquifer recharge by check dams: a case study near Chennai, India. *International Research Journal of Engineering and Technology*, *2*(4), 416–423.

Role of Dams for Irrigation, Drainage and Flood Control, ICID Position paper, (1999–2000).

Rosenberg, D. M., Berkes, F., Bodaly, R. A., Hecky, R. E., Kelly, C. A., & Rudd, J. W. (1997). Large-scale impacts of hydroelectric development. *Environmental Reviews*, *5*(1), 27–54.

Rudra, K. (2010). Dynamics of the Ganga in West Bengal, India (1764–2007): implications for science–policy interaction. *Quaternary International*, *227*(2), 161–169.

Saha, M. N. (1935). *Collected works of MN Saha*. University of Calcutta, Kolkata.

Sinha, R., Jain, V., Tandon, S. K., & Chakraborty, T. (2012). Large river systems of India. *Proceedings of the Indian National Science Academy*, *78*(3), 277–293.

Uribelarrea, D., Pérez-González, A., & Benito, G. (2003). Channel changes in the Jarama and Tagus rivers (central Spain) over the past 500 years. *Quaternary Science Reviews*, *22*(20), 2209–2221.

Vörösmarty, C. J., Meybeck, M., Fekete, B., Sharma, K., Green, P., & Syvitski, J. P. (2003). Anthropogenic sediment retention: major global impact from registered river impoundments. *Global and Planetary Change*, *39*(1–2), 169–190.

Walling, D. E. (1983). The sediment delivery problem. *Journal of Hydrology*, *65*(1–3), 209–237.

Warrick, J. A., Bountry, J. A., East, A. E., Magirl, C. S., Randle, T. J., Gelfenbaum, G., … Duda, J. J. (2015). Large-scale dam removal on the Elwha River, Washington, USA: source-to-sink sediment budget and synthesis. *Geomorphology*, *246*, 729–750.

Willcocks, W. (1930). *Lectures on the Ancient System of Irrigation in Bengal*. University of Calcutta, Kolkata.

Young, R. P., Maxwell, S. C., Urbancic, T. I., & Feignier, B. (1992). Mining-induced microseismicity: monitoring and applications of imaging and source mechanism techniques. *Pure and Applied Geophysics*, *139*(3–4), 697–719.

6 Anthropogeomorphological Signatures over the Ajay River Basin

Suvendu Roy

CONTENTS

6.1 INTRODUCTION

The initial imprint of human civilisation over the lower Gangetic Basin was almost five to four thousand years ago (3 ka–2ka BP),which has been recognised mainly within the lower Ajay River Basin (ARB) in the name of *Pandurajardhibi*, which is presently known as *Panduk* village of Purba Barddhaman district (Ghosh and Chakrabarti,1968; Datta,2005; Rajaguru et al., 2011; Roy, 2012a; Roy and Sahu, 2016a). Geoarchaeological studies have explored that the region has rich cultural settlements since the Chalcolithic period (Rajaguru et al., 2011), which were mainly agricultural communities with some pockets of hunting and gathering society (Figure 6.1). Since then, the basin carries numbers of signature of human activities in the forms of deforestation, embankment, dams, settlement, food production etc., which might interact with the geomorphology of the basin.

The ARB with more than three million people is still a less regulated watershed than the neighbour basins such as Damodar and Mayurakshi in terms of the number of dams constructed, the degree of water extraction for irrigation and industry, the location of urban centres beside the trunk rivers, coal and sand mining, landscape alternation by industries, etc. While the basic topography of the basin indeed emerges from the natural driving forces, anthropogenic activities are also playing a significant role to modify the nomenclature of the basin rapidly since the mid-20th century. The present chapter explores such emerging anthropogenic activities and their effects quantitatively and qualitatively with an extensive literature review, field experience and advanced remote sensing from anthropogeomorphological approach. The study might be a helpful document for geoscientists as well as for natural resource planners during the period of Anthropocene.

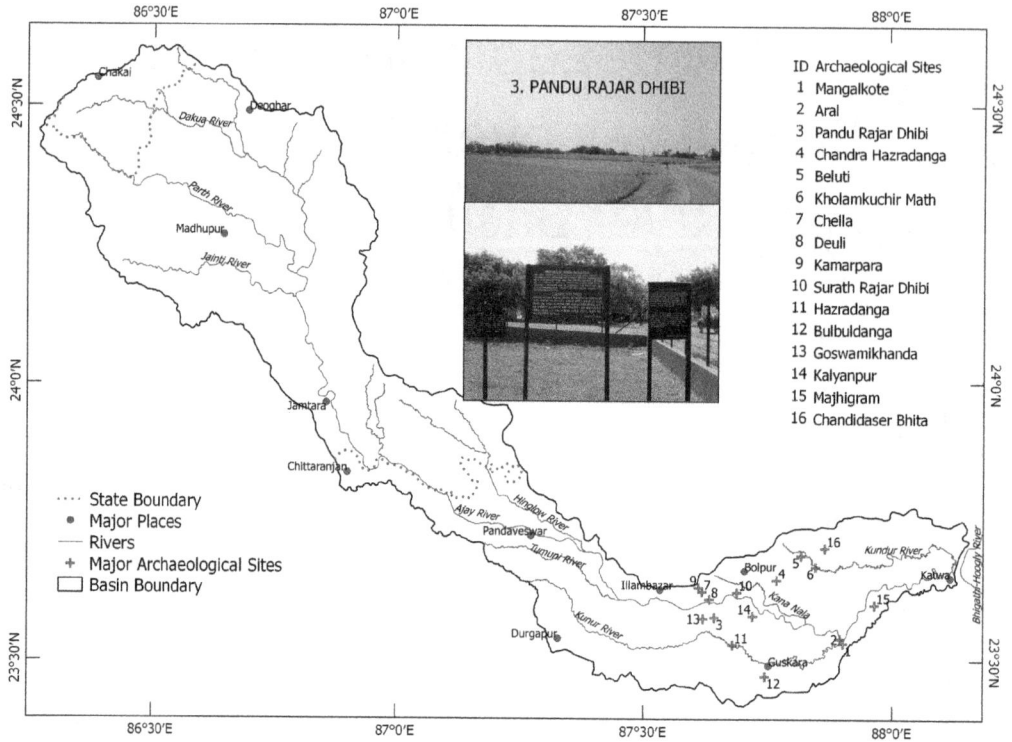

FIGURE 6.1 Major archaeological sites of the Chalcolithic period over the lower ARB as per Datta (2005) and Roy (2012a).

6.2 THE RIVER AND ITS BASIN

Ajay River, an important tributary of Bhagirathi-Hooghly River, is extending its long and narrow catchment area from Chhotanagpur Plateau to Gangetic Alluvial Plain. The basin area spreads between latitude 23°25′N to 24°35′ N and longitude 86°15′E to 88°15′ E and is occupying an area of about 6,050 km², of which ~53% area comes under the State of Jharkhand, ~6% in the Bihar and ~41% in the West Bengal. Based on the state boundary, the basin could be divided into two major parts: (1) 'Upper Ajay Basin' (UAB) including the Jharkhand and Bihar and (2) 'Lower Ajay Basin' (LAB) within the West Bengal (Figure 6.2).

The trunk river of the basin is flowing almost 299 km in the east-south-easterly direction from the eastern fringe of Chhotanagpur Plateau to the confluence point with Bhagirathi-Hooghly River at Katwa, West Bengal, located at about 216 km upstream from Kolkata (Niyogi, 1984). Seven major tributaries joins the trunk river at its different length between source to confluence, which are Dakua (48 km), Parth (80 km), Jayanti (87 km), Hinglow (180 km), Tumuni (192 km), Kunur (252 km) and Kundur (293 km) rivers. The average annual discharge capacity of the river is about 2,036 million m³ (Niyogi, 1984).

The basin area comes under three major geological units and comprises a variety of rocks (Figure 6.3). The consolidated formation of high-grade metamorphic Archaean Gneiss has covered the upper part of the basin area, which is almost 61% of entire basin. The region is also characterised with high density of lineaments in different directions and with a number of faults (Niyogi,1984). About 8% of the middle basin area is covered by semi-consolidated formation of Gondwana basin. The region mainly comprises the coal seams of both districts in addition to some sedimentary formation of sandstone, grit, conglomerate, shale and ironstone shale. Rest of the basin area (~31%) is mainly covered by the Quaternary sediments of marine–estuarine–fluviatile origin, within which a

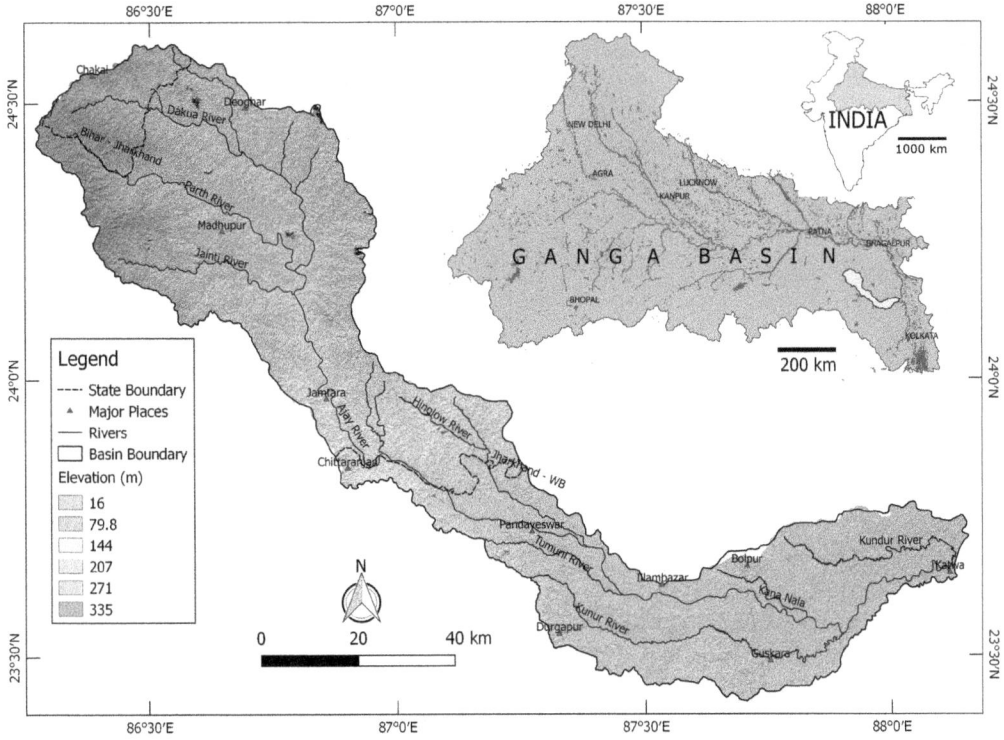

FIGURE 6.2 Location of the ARB in the lower Ganga Basin of India.

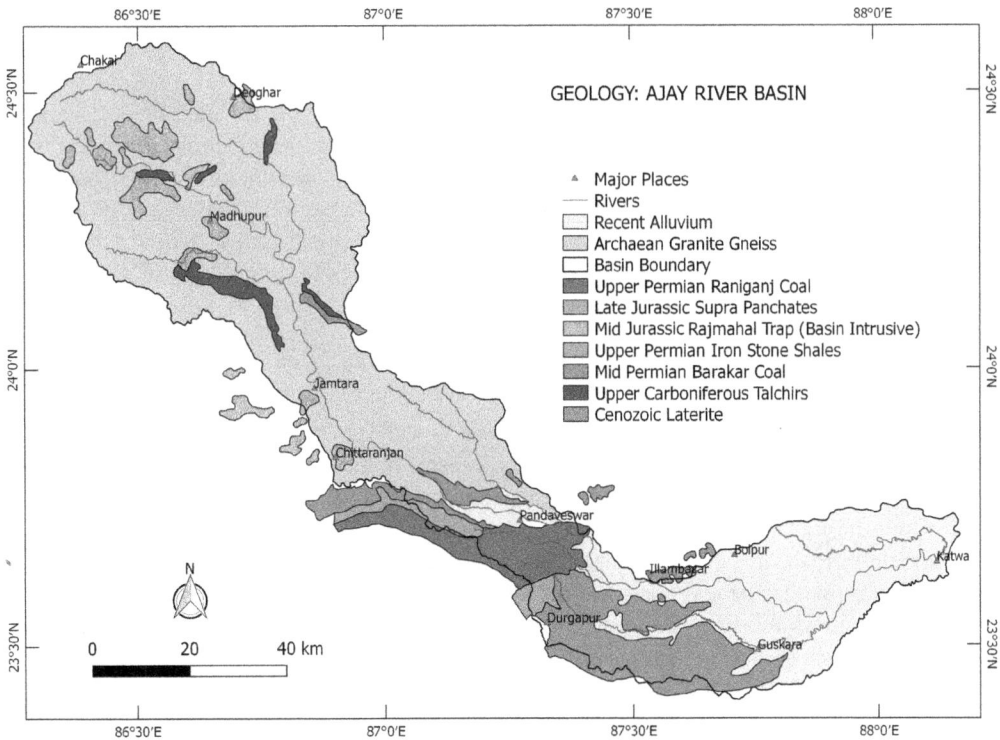

FIGURE 6.3 Geology of the ARB. (Modified after Niyogi (1984).)

significant amount is covered by Cenozoic Laterite in the middle of the region. The thickness of the alluvium increases towards east and southeast.

The maximum elevation (335 m) and the minimum elevation (16 m) of the basin have been observed at the extreme western upland of the catchment and at the confluence zone, respectively. Niyogi (1984) has classified the basin into three major physiographic parts as follows: (1) dissected erosional plain with monadnocks in the upper part of the basin; (2) erosional plain with broad swells and depression mainly over the Gondwana basin; and (3) riverine aggradation plain with extensively developed alluvial fans, which merge with the Bhagirathi Delta proper. As per the classification of Bagchi and Mukherjee (1979), these three regions may also be renamed as 'plateau proper' (>120 m), 'plateau fringe' (36–120 m) and 'marginal plain' (>36 m), respectively.

The primary climate of the basin is monsoon type, and 85% of rainfall occurs mainly from mid-June to October. The mean annual rainfall amount is 1,380 mm with a mean annual temperature of 25.8°C (IMD, 2014). However, over the basin area, a significant variation in rainfall amount has been observed annually as well as monthly.

The type of soils and their texture over the basin are clearly associated with the lithological characteristics and pedogenic processes (Niyogi, 1985). Red–yellow and red soils with sandy-loam to loamy texture have been observed on the Archaean Gneiss, and the areal coverage of these soil types is about 40% and 25%, respectively. The downstream region or lower basin area are mainly covered by younger alluvial, older alluvial and lateritic soils, which are about 6%, 19% and 10% of the total basin area, respectively (Niyogi, 1985).

The land-use/land-cover scenario of the basin reveals the intensive nature of human interference with 70%–80% of agricultural land and up to 10% of agricultural waste at block level even in the upper most areas. The basin area is covered by dry peninsula type of *Sal* forest, i.e. only 9.9% on average, which is temporally lost at very high rate. The residential area has covered almost 5% of basin with six to seven major urban centres.

From the administrative aspect, the basin comes under three states of India, which has been chosen as the primary basis to classify the ARB into two parts, namely, UAB and LAB. As per the Census of India (2011), the UAB comprises four districts of Jharkhand and a small part of Bihar and LAB is composed of two districts of West Bengal. However, the entire basin comes under highly populated part of India, whereas the districts (Jamui: 568 person km^{-2}; Giridih: 493 person km^{-2}; Deoghar: 602 person km^{-2}; Jamtara: 437 person km^{-2}) in the UAB are characterised with low population density in comparison with the districts (Barddhaman: 1,099 person km^{-2} and Birbhum: 771 person km^{-2}) of LAB. The difference in topography and soil characteristics between UAB and LAB might be the major causes behind this disparity. Nevertheless, a significant increase in the population density of Jharkhand state from 2001 (338 person km^{-2}) to 2011 (414 person km^{-2}) is becoming crucial problem for the geomorphological alternation in the upper part of the ARB.

6.2.1 Problems of Flooding and Sand-Splay

Flood in the West Bengal (WB) is now an annual feature and its effectiveness has increased year by year. At present, about 42.30% of the total geographical area of the state is susceptible to flood (IWD-GoWB, 2016). According to Rudra (2008), throughout the West Bengal, the downstream of 26 river basins is frequently affected by flood in the rainy season, and particularly floods come here from the end of September to the mid of October. The two fundamental causes of West Bengal flood are as follows: (1) the narrow channel width and very gentle longitudinal slope of channel bed of all major rivers and (2) enormous volume of water from huge precipitation (Table 6.1) and rapid flow of water from upper catchment due to plateau landscape of impermeable rocks. Other important causes of flood in the study area are breaching of the embankment (Mukhopadhyay, 2010), discontinuous embankment (Molla, 2011) and topographical depression in the right bank of the Ajay River (Roy, 2012b).

The devastative nature of flooding in the lower ARB is not a recent phenomenon. As per the archaeological investigation, because of ancient floods most of the archaeological sites of Chalcolithic

TABLE 6.1

Rainfall Data during Last Five Major Floods in West Bengal (Data in Millimetres)

Districts	23/09/1956–27/09/1956	08/09/1956–12/09/1956 and 30/09/1959–03/09/1959	31/08/1978–03/09/1978 and 27/09/1978–01/10/1978	26/09/1995–28/09/1995	18/09/1995–21/09/1995	Annual Rainfall (Average)
Barddhaman	NA	229/231	76.70/162	NA	369	1,271
Birbhum	321.33	121/206	102.40/636	347	800	1,234
Murshidabad	256.25	141/370	N.A.	276	470	1,338
Nadia	296	203/331	87.50/349	NA	352	1,401
Hugli	276.33	291/247	NA	NA	352	1,516
Medinipur	133	193/192	351/248	NA	90	1,428
N & S 24 Parganas	284.33	276/298	154/646	151	160	1,428
Haora	331	NA/254	NA	NA	122	1,676
Purulia	NA	142/168	345/208	NA	169	1,307
Bankura	NA	201/202	281/471	NA	50	1,271

Source: Irrigation Department of West Bengal.
N.A., Not Available.

period are now buried under 2–3 m of alluvial sediments (Roy, 2012a). The major recorded flood years are 1956, 1959, 1970, 1971, 1973, 1978, 1984, 1995, 1999, 2000, 2005 and 2007. Among these, most devastating flood years are 1978, 1995, 1999 and 2000. Frequent floods are generally observed below Illambazar particularly after the confluence point of the Hinglo river, which used to suffer in almost all flood years (Mukhopadhyay, 2010). A study shows (Mukhopadhyay, 2010) that during the floods, out of 619 *mouzas* (small administrative unit) of 12 C.D. Block of lower ARB, 493 *mouzas* were affected by flood entirely or partially, which is about 79.65% of the total area. The data related to flood affected areas over the lower ARB shows the progressive trend since 1956 (Table 6.2).

Due to the breaching of embankment and flood, the most prominent effect is sand-splay as post-flood hazard. The number of breaching points is rapidly increasing towards downstream. It has been also noticed that the breaching points are more developed in the right bank of river, and its number is also increased with time: in 1978, it was 12; in 1999, it was 22; in 2000, it was 25; and in 2005, it was 8. The process of sand-splay in lower ARB has covered a significant amount of agricultural land associated with the loss of cultivated area of different mouzas (Table 6.3).

6.3 ANTHROPOGENIC SIGNATURES ON FORMS AND PROCESSES

6.3.1 DAMS

A global-level database on dams by the International Commission on Large Dams (ICOLD) has listed that more than 45,000 dams are above 15 m high and a total of 88,000 dams present globally and affecting over half of the world's large rivers (GRanD, 2011). While all aspects of a river system are influenced by dam directly or indirectly, the well-recognised influence has been observed on the flow regime of all catchments (Marren et al., 2014). For example, Fitzhugh and Vogel (2011) have estimated that across the United States, there are more than 25% reduction in the average annual flood of 55% of large rivers and an average reduction in peak discharge of 67% rivers of the entire nation.

The River Ajay was a dam-less drainage basin till the end of the 1970s. However, with increasing demand for water for agriculture across the highly populated region of India, number of dams, weirs and check dams have been constructed and plan to construct at the different tributaries and trunk

TABLE 6.2

Flood-Affected Area of Lower Ajay River Basin from 1956 to 2007

| Year | Affected Area in sq km | % of Total Area | Affected Number of Mouzas | | | Extent of Sand-Splay in Hectares | Maximum Extension of Sand-Splay from River Embankment (Distance in km) |
			Entirely Affected	Partially Affected	Total		
1956	680	24.14	153	32	185	231.45	0.38
1959	584.34	20.74	120	27	147	269.63	0.38
1970	812.24	28.83	186	36	222	693.48	0.47
1971	642.71	22.81	130	31	161	762.11	0.78
1973	639.02	22.68	124	36	160	1,193.2	1.12
1978	1,680	59.64	307	67	374	3,421.32	2.42
1984	305.72	10.85	78	20	98	865.55	0.68
1995	1,380.82	48.99	227	49	276	1,245.67	1.40
1999	1,408	49.98	237	60	297	2,567.23	2.12
2000	1,488	52.82	263	106	369	3,788.25	2.57
2006	764.23	27.12	152	46	198	2,143.56	1.35
2007	972.79	34.53	214	79	293	2,421.57	1.76

Sources: Department of Irrigation and Waterways, Govt. of West Bengal, 2010; Mukhopadhyay (2010).

TABLE 6.3

Sand-Splay Cover and Loss of Cultivated Land due to Post-Flood Hazard

Name of the Mouza	Area Covered by Sand-Splay (% of Total area)	% of Land Loss to the Total Cultivated Area	Name of the Mouza	Area Covered by Sand-Splay (% of Total Area)	% of Land Loss to the Total Cultivated Area
Bhedia	33.58	18.4	Itanda	42.43	39.89
Brahmandihi	21.66	32	Nabagram	16.43	30.7
Malocha	38.08	25.67	Natunhat	18.6	23.99
Maliara	44.03	21.52	Bira	17.3	24.82
Basudha	32.07	23.59	Narenga	20.32	29.84
Gitgram	58.92	30.8	Srikrishnapur	38.2	51.92
Natungram	36.74	44.66	Husainpur	48.25	26.66
Rasulpur	33.82	36.58	Vepura	44.46	40
Haripur	38.12	43	Pandura	21.39	17.08

Sources: Burdwan and Birbhum Zilla Parishad Office and Handbooks, 2001; Mukhopadhyay, 2010.

river of Ajay Basin. The major purposes of all the dams and weirs are irrigation and flood control. In comparison to the Damodar Basin, the number and size of dams in ARB are less and small. At present, ARB has an active major dam at Sikatia (Deoghar, Jharkhand) with a flood discharge capacity of 10,000 cumecs and one medium range weir at Daruwa River (WRIS-India Web-GIS, 2018) (Table 6.4). In particular, all the minor check dams have been constructed at the edge of the upper catchment, and the area administratively comes under the state of Bihar, while similar potentiality is present in the Jharkhand and West Bengal states also (Figure 6.4). Such result indicates a government-level disparity in planning for the micro-level water resource management.

Hence, at present, the basin has five major projects (two under construction) of water management for the purpose of irrigation and flood control (Table 6.4). Among them, four projects are situated

TABLE 6.4

Description of Major and Minor Dams and Weirs over the Ajay River Basin

Sl. No.	Name of Dams/ Weirs	Across the River	State/ DistrictLat– Long	Purpose	Year of Complete	Length (m)	Height (m)/Detail of Gate	Volume (TCM)/ Spillway Capacity (cumec)	Area under Irrigation Service ('000 hectares)
1	Ajay Barrage (Sikatia) Project	Ajay (Upper)	Jharkhand/ Deoghar	Irrigation	2004	275	170 (major)	10,000cumec (Flood discharge capacity)	41.34
2	Hinglow Dam	Hinglow (Lower)	WB/Birbhum	Irrigation	1976	1,158	12/(16×8.3) of 15 gates	3,000/1,631	12.65
3	Punasi Dam	Dakua(Upper)	Jharkhand/ Deoghar	Irrigation	Under construction	NA	N.A. (major)	NA	24 (proposed)
4	Burhai Dam	Parth(Upper)	Jharkhand/ Deoghar	Irrigation	Under construction	NA	NA	NA	NA
5	Daruwa Weir	DarhwaNadi (upper)	Jharkhand/ Deoghar	Irrigation	NA	60.96	NA (medium)	601 cumec (flood discharge capacity)	1.63

Source: India WRIS WebGIS, 2018.

at the upper part of ARB, at Jharkhand in particular, and single in the lower ARB. As a result, a significant amount of cultivated land of the upper basin comes under the irrigation facilities, which might be increased agricultural production. However, the lower basin area is less facilitated by its own water rather than the water from Damodar and Mayurakshi basins through the canal system of Durgapur Barrage and Mayurakshi Dam, respectively (Figure 6.4).

6.3.2 MINING

Mining involves the extraction of fossil fuel, metallic ores (gold, copper, zinc, lead, etc.) and aggregates (sand and gravel) (Charlton, 2008). The direct effect of mining activity is input of huge sediments from large particles to suspended materials to the river system, which disturbed the entire system. The scientific work on the impact of mining (gold) on fluvial geomorphology has been initiated by Gilbert (1917) in the Sacramento River, California. The study of Gilbert (1917) shows that large-scale gold mining has had a long-lasting effect in the basin and induced to input about one billion m³ of sediments during the period of 1853–1884. In-stream huge sedimentation, as the immediate effect, has been shown in the tributaries, and thereafter, the major floods have shifted those sediments to the trunk rivers and reduced the carrying capacity of water, thus extending the areal coverage of flood inundation. Graf (1979) has also worked on the effect of gold and silver mining on the landscape alternation of Central City District of Colorado. Hettler et al. (1997) have also shown that the intensive copper–gold porphyry mining increases the suspended sediments load about 5–10 times of natural background over the Ok Tedi/Fly River system in Papua New Guinea, in addition to the changing chemical composition of the river water.

The study basin also experienced the effects of three active surface coal mines, which directly and indirectly affect the fluvial system as well as landscape of the ARB (Figure 6.5; Table 6.5). The major two working coal mines, e.g., Chora Colliery and Sonepur Bazaria - Khottadih Colliery in the LAB are operated by Eastern Coalfield Limited (ECL), a subsidiary of Coal India Limited

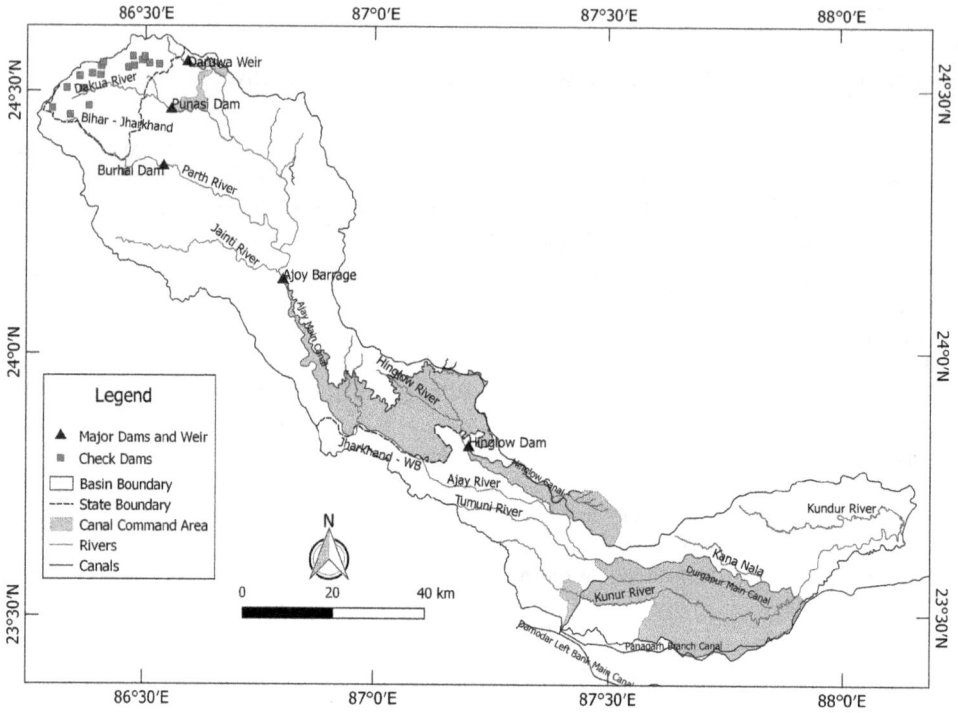

FIGURE 6.4 Allocation of major dams, weirs and check dams as significant anthropogenic signatures over the ARB, including the networks of canals and their command area. (Google Earth, India WRIS WebGIS, 2018.)

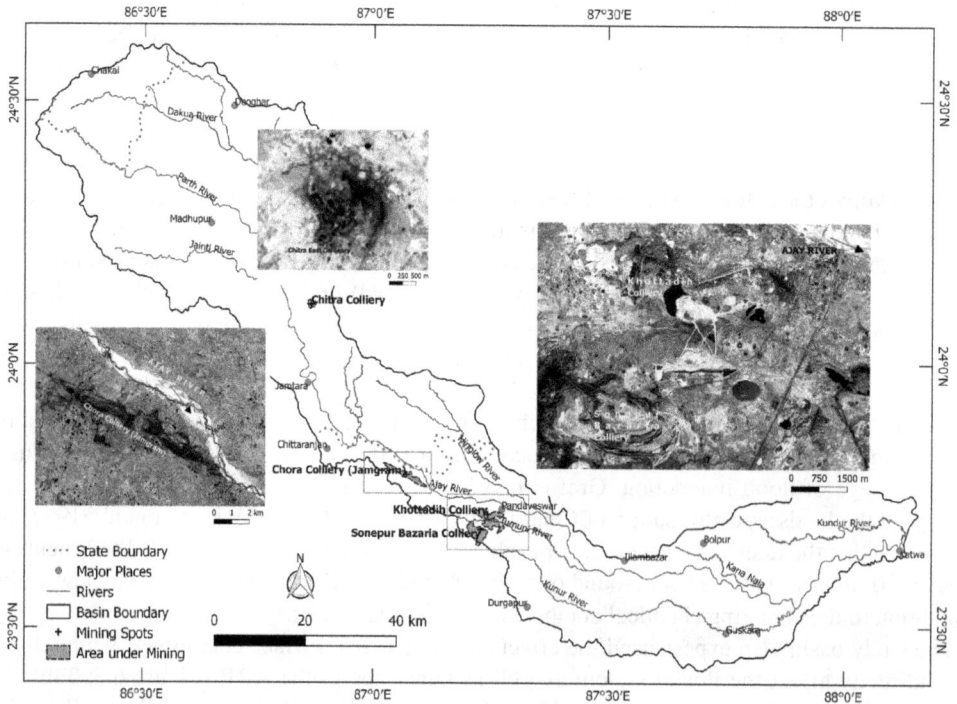

FIGURE 6.5 Major open cast mining spots over the ARB.

TABLE 6.5

Description of Major Surface Coal Mining Areas over the Ajay River Basin

Sl. No.	Name	Location	Started Year; Recent Target of Output	Type of Mining	Area (km²)	Distance from Ajay River
1	Chitra East Coal Mine	Deoghar, Jharkhand	1975; 2.5 MTY	Opencast	1.55	4.25 km
2	Chora Colliery, Jamgram Coal Co. Pvt. Ltd.	Jamgram, Barabani, Barddhaman, West Bengal	1994		~09	2.5 km to 50 m
3	Sonepur Bazaria Colliery and Khottadih Colliery	Pandaveswar, Barddhaman	1990		~18	3 to 4 km

Source: Compiled from different maps and literatures by Author (2018).

(CIL) and the area is renowned as Raniganj Coalfield of Eastern India. In the process of opencast mining, these coalfields are mainly excavated and remove the overburden rocks and soil layers and dump them beside the mine. Annual Report of ECL (2011–2012) stated that in every one million tonne of coal production through opencast mining, the mining authority should remove on an average 2.54 million cubic metre overburden. The dumping process is one of the major anthropogenic activities to alter topography and local landscape with positive (dumped area) and negative landforms (excavated area). Manna and Maiti (2014) have observed that the depth of quarries goes up to 95 m in some places with an average area of 2.4 km², whereas the height of the dumped ridges are more than 60 m at some peaks with an average area of 1.5 km² (Figure 6.6a).

In the case of Sonepur Bazari opencast mining, Manna and Maiti (2014) have estimated that the mining lease boundary is about 12.06 km² and already 5.03 km² is defaced by 2010. At present, two large dumped ridges have been observed with an average volumetric measure of 12 and 57 million m³, and the height of these two ridges are 130 and 103 m, respectively (Manna and Maiti, 2014). The distance between the trunk river and mining spots ranges from 2 to 4.5 km, while the distance with tributaries is very less, which is an essential matter of concern as a major source point of sediments to the river system (Figure 6.6b and c).

6.3.3 EMBANKMENTS

Embankment (also called artificial levees, flood banksor dykes) means the barrier of the river channel to protect the overflow of the excess river water and increase the water discharge capacity. It may be made by concrete or mud. It is highly used to measure floods. Flood is caused due to inadequate capacity within the bank of rivers to contain the high flows brought down from the upper catchment due to concentrated heavy rainfall. To protect the areas from flooding, provision of marginal embankment along river banks and littoral areas was an age-old practice perhaps since the dawn of civilisation (Mukhopadhyay and Dasgupta, 2010). As per Goudie and Viles (2016), the first embankment was constructed by Indus Valley (Harappan) Civilisation in Pakistan and North India about 4,600 years ago. The Nile River had also been embanked by Egyptian Civilisation about 3,000 years ago. However, the earliest recorded embankment in West Bengal was constructed during the Sultani Period (1213–1519) (Molla, 2011).

The total length of flood-protected embankment in West Bengal is about 10,400 km, and particularly in Burdwan district, it is ~600 km in length and for Birbhum district its length is ~300 km (IWD-GoWB, 2019). Of these embankments in both districts, maximum part is situated along the Ajay River side. An eminent engineer Majumdar (1941) has warned about the long-term effects of embankment and stated that "construction of embankment as flood controlling measures would be like mortgaging the future generation". However, the practice is still continuing with repairing

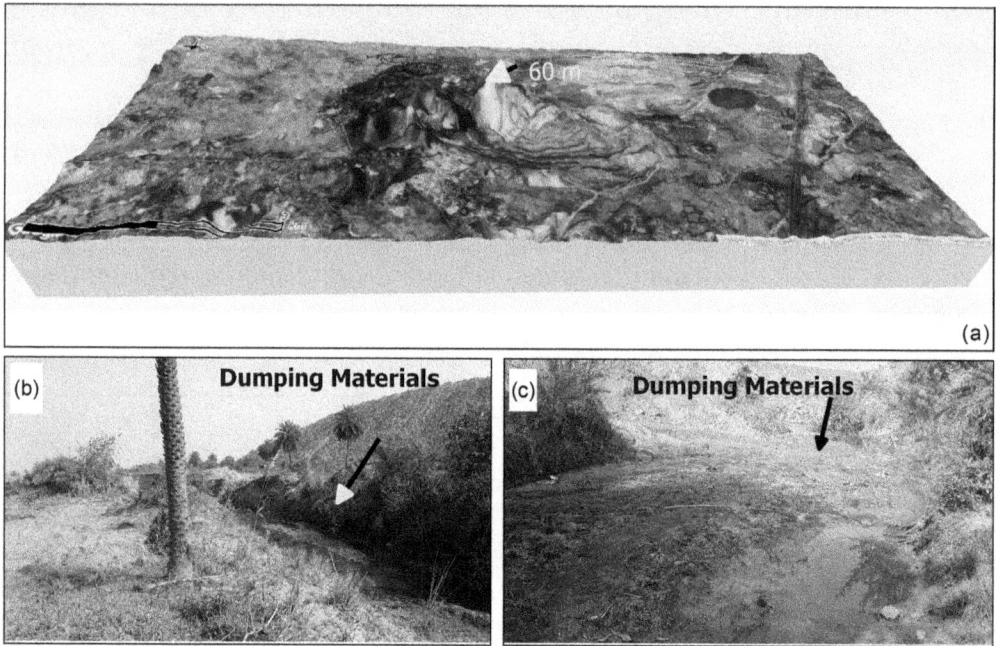

FIGURE 6.6 (a) 3D view of Sonepur Bazari opencast mining showing the surface configuration of local landscape with more than 60 m of dumped ridges; (b and c) sites of coupling between loose dumped materials with the tributaries of Tumuni River. (Picture Courtesy: Sourav Mukherjee, Department of Geography, The University of Burdwan.)

and reconstruction at the flood-affected part of embankment and at the discontinuous parts, respectively.

Flooding behaviour of the entire ARB shows that flood rarely occurs in its upper catchment, whereas the lower section is frequently affected by havoc floods. Therefore, embankment is mainly constructed along the lower section of Ajay River, mainly below the Illambazar section (Figure 6.7). Previous documents show the period of pre-embankment during the 18th and 19th centuries, the channel of Ajay River was navigable with significant depth and containing large volume of discharge. The flood intensity was normal, and there was extended floodplain with rich ecological importance (Mukhopadhyay, 2010). However, the unusual floods and associated severe losses of life and property, crops and houses during the beginning of the 20th century (1913, 1914) have induced English District Officers to take immediate actions against the flood and help the riverine people. A partial attempt had been made to construct embankment along the Ajay River in association with the local Zamindars. However, the embankment was not technically sound as well as scattered in nature, which was basically made by silt and clay. Therefore, these were frequently subjected to breach, causing devastating floods in the basin area. Although 'Bengal Embankment Act' was formed in 1873 to monitor the embankments along the Bengal Rivers to protect the cultivated lands and increase the revenue collection, no significant improvement has been observed up to the end of British Period (1947).

After independence, the Directorate of Irrigation and Waterways Department, Government of West Bengal has taken the control of embankment along the Ajay River, and the flood scenario has significantly changed over the lower ARB. The embankments have been remodelled and start to make continuous in nature. A recent statistic shows that the total length of these embankments is about 136 km, of which about 81 km present along the right bank and 55 km on the left side (Molla, 2011). The embankment on the right and left banks protected about 37,040 hectares and 29,785 hectares of cultivated land, respectively (Molla, 2011).

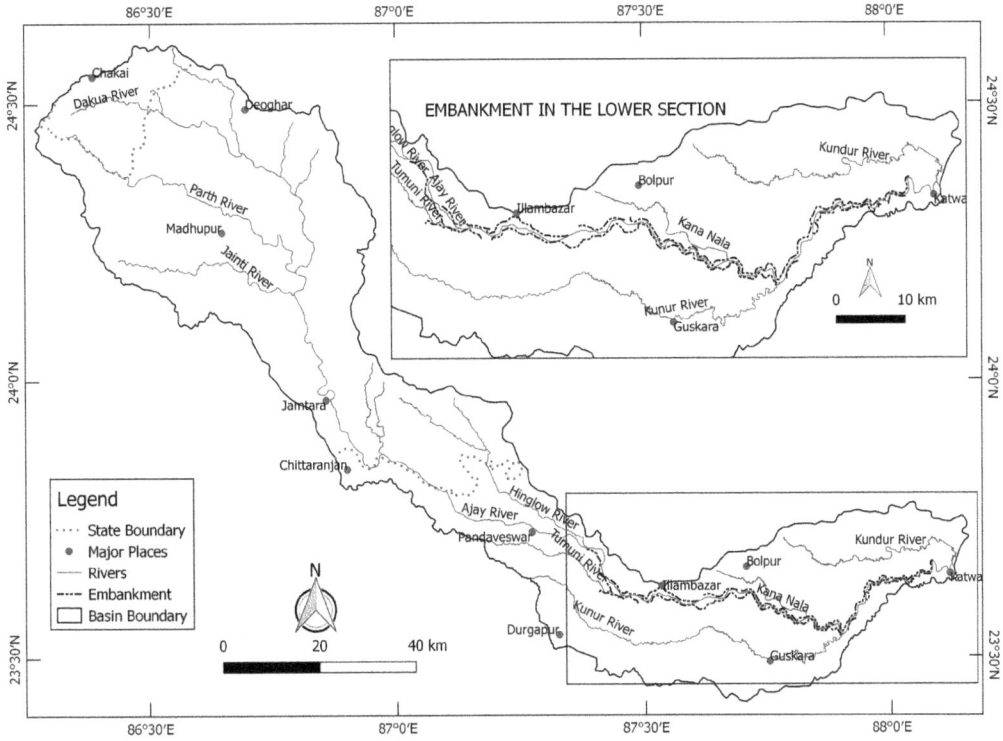

FIGURE 6.7 Allocation of embankment along the Ajay River, mainly in its lower section.

6.3.4 Transport Networks

The transport network is a key element of every cultural landscape and works as the artery system of society. The expansion of transport network is an indicator of societal development and advancement of technology. Jimenez et al. (2013) observed a significant increase in road length during the past decades over the world. However, the development of linear infrastructures in terms of roads and railways has increased the human interferences to the geomorphological processes over the earth landscape, specially the drainage pattern and runoff generation (Montgomery, 1994; Tarolli et al., 2018). These have modified the natural hill-slope profiles (Luce, 2002), increased the rate of soil erosion, increased the sediment yields (Wemple and Jones, 2003; Sidle and Ziegler, 2012), reduced the connectivity in fluvial system (Blanton and Marcus, 2014) and directly modified the channel morphology at the point of road-stream crossing (RSC) (Roy and Sahu, 2017, 2018).

Among the effects, when road networks behaving like a barrier for fluvial connectivity is crucial for Ajay River. Riverine connectivity refers to the flow, exchange and pathways of organism, sediments, waters, nutrients and energy throughout the watershed. Generally, three levels of discontinuity are generated by transport infrastructure. Road and railways along the river are creating lateral disconnection between main channel and its adjacent floodplain, whereas bridges and culverts make longitudinal disconnection between upstream and downstream (Blanton and Marcus, 2014). Roads in the floodplain could significantly alter the exchange of flood water between channel and floodplain as a dyke along the floodplain (Tarolli and Sofia, 2016). Nevertheless, the construction of bridges and culverts across the channel has influenced the flow directly and increased the velocity, which facilitated channel degradation and widening processes significantly (Roy and Sahu, 2018).

As per the Open Street Map available online, the total length of the transport networks is about 4,300 km and transport network density of the basin is about 0.71 km/km[1]. However, about 29% area of the basin has no transport network (Figure 6.8). The grid-based (6.25 km^2) mapping shows over the

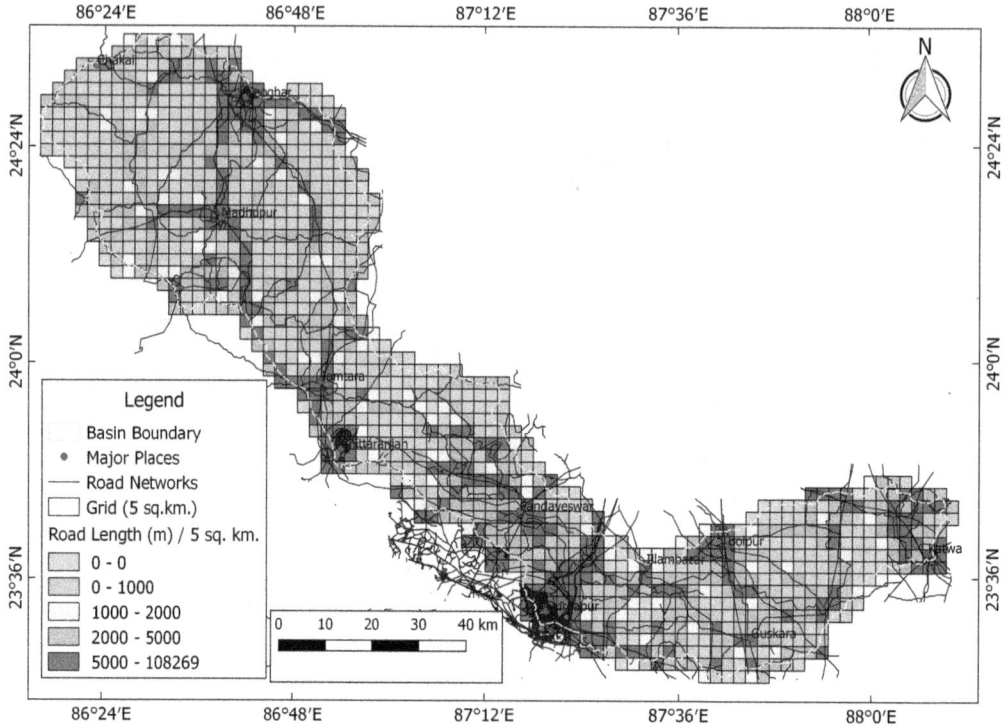

FIGURE 6.8 Transport network density map of the ARB. (*Source*: The vector file of the transport networks extract from OpenStreetMap Foundation (OSMP©) through WebGIS.)

basin about 7%, 8%, 35% and 21% area have been characterised by 0–1,000 m/6.25 km², 1,000.01–2,000 m/6.25 km², 2,000–5,000 m/6.25 km² and >5,000 m/6.25 km², respectively (Figure 6.8).

The density map of the transport network reveals that the maximum density zones have been concentrated in and around the major urban centres. As expected, the relief condition positively influenced the network density of the basin because the presence of undulating plateau in the upper catchment is major cause of lower network density over here, whereas the alluvial plain of the lower ARB is characterised by relatively high density (Figure 6.8). Geospatial technology helps to extract all interaction points between major rivers and roads over the ARB, which are basically denoted as RSC. The heat map based on the vector of RSC shows that the density of RSC is positively associated with the location of major urban centres. A major hotspot of RSC has been identified at the Durgapur surrounding (Figure 6.9a). The proximity or near-distance analysis also shows that the distance between road and stream networks is very close in the downstream area, especially in the Durgapur Municipal Corporation. However, the distance between these two networks is high over the upstream of ARB due to rugged topography (Figure 6.9b).

6.3.5 Deforestation

Deforestation, as a major land-use/ land-cover change indicator, plays crucial role in the process of anthropogenic alternation of river system. This alternation may not be observed from direct river channel structure changes; it may also come from alternation of the hydrological regime of river basins, such as soil and water loss through runoff, interception, infiltration and soil moisture under different forest cover (Gupta, 1980). Several studies have been carried out to evaluate the effects of basin-scale deforestation on hydrological alternation (Hewlett and Helvey, 1970; Nagasaka and Nakamura, 1999; Wilk et al., 2001; Zabaleta and Antiguedad, 2013) in addition to the alternation

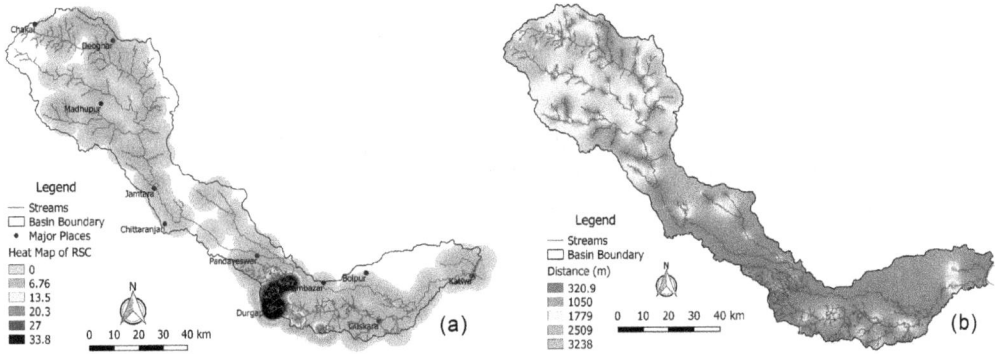

FIGURE 6.9 (a) Heat map showing the density distribution of RSC; (b) proximity map showing the variation of distance between road and stream networks over the ARB.

of chemical and biological characteristics (Mullen and Moring, 1988; Garman and Moring, 1991; Schnitzler, 1997).

As expected, deforestation is also an alarming threat for the ARB, mainly in the upper section of the basin area. The rugged and dissected topography of the upper ARB is characterised by vast barren/uncultivated scrubland, which has increased since 2005, as per the thematic service of Bhuvan-ISRO web GIS platform (http://bhuvan.nrsc.gov.in/gis/thematic/index.php) (Figure 6.10). However, a majority section of the basin is under agricultural land, which is expanding temporally with the conversion of forested land to the cultivated land (Roy and Sahu, 2016c).

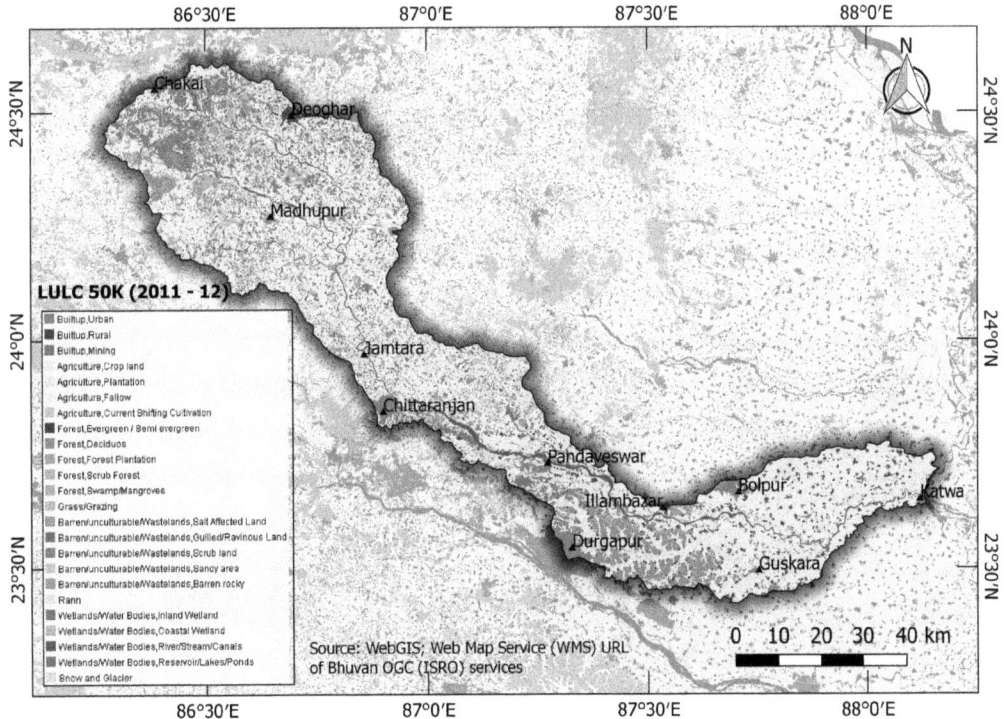

FIGURE 6.10 Land-use/land-cover map of the ARB and its surroundings (2011–2012). (Bhuvan—ISRO, WMS Layer, 2018.)

Micro-level observation through the high-resolution remote sensing image of Google Earth also helps to visualise the rapid rate of deforestation in the upper ARB and the relationship between deforestation and development of stream networks over the bare surface (Figure 6.11). Figure 6.11 shows that since 2006 a vegetated land has been transformed into clear-cut bare surface. The temporal change of surface configuration has also revealed the rapid enlargement of drainage system. The stream with shorter tributaries that has extended its tributaries' length, width and depth through rapid head-water erosion with a huge amount of soil loss.

6.3.6 Urbanisation

In recent times, almost every river system in the world has been influenced by human activities directly and/or indirectly. Morphological adjustment of river channels has effected by the development of urbanisation within the floodplain area. There are several researchers who investigated the impact of urbanisation on river systems from different aspects, such as changes in sediment yield and concentration and hydrological changes throughout the world (Chin, 2006). Over the last 50 years, it has been found that the research interest of geoscientists is increasing on the impact of urban development on river processes and forms (Chin, 2006). Chin (2006) stated after reviewing more than 100 published studies on the impact of urbanisation on river systems over the past five decades, where it is found that the majority of works (58 studies) are based on the quantification of morphological change within channels and watersheds. Out of these 58 investigations, 27 (47%) documents were collected from the United States and 13 (22%) from the British rivers, although some of the important studies happened in other countries such as Australia, Nigeria, Malaysia, Canada, Singapore, Zimbabwe, France and Israel.

Percentage cover of the built-up area of the ARB has notably changed in the past few decades. In the 1990s, about 29% of basin was covered by built-up area, which has been increased about 37% in the recent decades. The major urban centres are Durgapur, Bolpur, Chittaranjan (famous for the locomotors industry)and Katwa in the lower part of basin, and Deoghar and Madhupur in the upper basin area. The outlines of these urban centres are also expanded significantly over time. For example, the area of the Durgapur Urban Centre significantly has increased since 1984 (Figure 6.12). Such progression in urbanization has reducing the proximity between channel and built-up areas and developed number of point source of sediments, pollutants and water from impervious area.

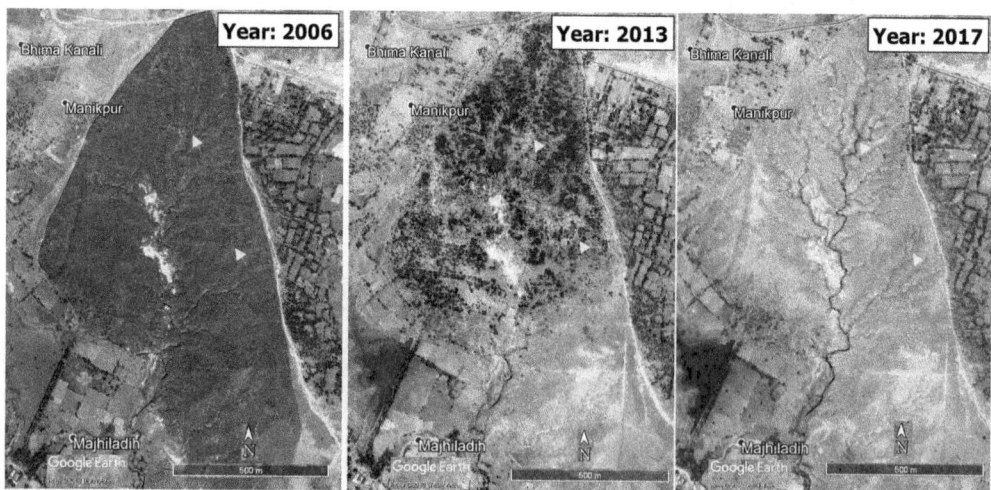

FIGURE 6.11 A sample site at the upper catchment of ARB has been clear-cut during last 10 years; therefore, the network of micro-watershed is also extended (marked arrows help to point out the changing scenario).

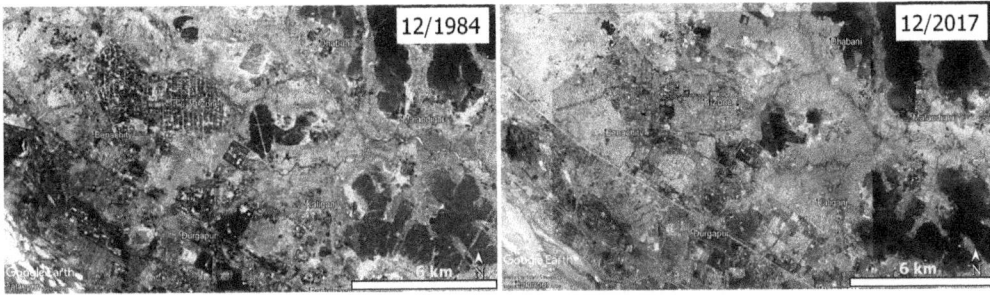

FIGURE 6.12 Temporal expansion of built-up area in and around the Durgapur Municipal Corporation. The edges of the surrounding forest patches are also degraded. (Google Earth, 2018.)

6.3.7 GEOCHEMICAL ALTERNATION OF AJAY RIVER WATER

Singh and Kumar (2017) have identified a number of pathways for heavy metal's entry in the system of ARB. Laboratory-based analysis shows the mean values of heavy metal pollution index (HPI) and pollution index (PI) are above the critical index and strong loadings, respectively, due to higher values of Cd, Pb and Fe. In addition, the assessment of human risk revealed that the high load of Cd, Pb and Fe in water body could harm the population. However, the physiochemical analysis of river water from different sample sites of Ajay River have concluded that the water is suitable for irrigation, riverine healthy ecology, and sustainable due to the limited influence of major anthropogenic activities directly to the river water (Kumar et al., 2015). According to Singh and Kumar (2017), natural and anthropogenic causes like weathering of minerals, lithological characteristics, industrial and agriculture processes and urbanisation have controlled the water chemistry in this basin.

6.3.8 MAJOR FLUVIO-GEOMORPHOLOGIC ISSUES FROM ABOVE SIGNATURES

a. The most direct impact of all anthropogenic activities over the basin has been observed on the morphology as well as the hydrological characteristics of the trunk river. The cross-section values of the lower Ajay River are showing typical channel characteristics from Pandaveswar to confluence point (Table 6.6). Fluvio-geomorphic rules say the width of any natural river increases towards the downstream for containing huge volume of water from its cumulative upper catchment area (Leopold and Maddock, 1953; Knighton, 1987). However, in case of the Ajay River, channel width decreases in the downstream direction (Table 6.6).

TABLE 6.6

Channel Dimensions of the Ajay River from Pandaveswar to Kogram

Cross-Section Site Name	Local Elevation from MSL (m)	Distance from Confluence (km)	Channel Width (w) (m)	Maximum Depth (d) (m)	Width–Depth Ratio (w/d)
Pandaveswar	65	130.28	1,026	11.89	86.29
JaydebKenduli	59	111.00	719	10.51	68.41
Bankati	58	106.27	518	9.39	55.17
Ramchandrapur	50	85.12	345	8.58	40.21
Bhedia	44	77.08	473	7.02	67.38
Paligram	36	59.42	273	11.48	23.78
Kalyanpur	29	50.31	316	10.68	29.59
Kogram	27	44.34	221	11.78	18.76

Source: GIS-based calculations from ASTER GDEM (30 m) and Sentinel 2A images (15 m), 2018.

The changes of channel depth towards downstream of the Ajay River are also not normal phenomena because of no clear relationship (r = 0.14) with channel width; whereas, there should be a direct negative relation between them (Leopold et al., 1964). Channel bed is incised at some places below the Pandaveswar because of insufficient channel area to contain peak discharge. Several breaching points are also developed during the floods because of low channel width and high unit of stream power. In an empirical work on the estimation of average unit stream power at a cross section (alluvial), Baker and Costa (1987) pointed that channel width is inversely related to the stream power at per unit area. As a result, in the Ajay River, downstream deceased width has increased the probability of bank erosion and embankment breaching.

b. An empirical study by Molla (2011) claimed that the establishment of embankment has been changing the micro-topography of the Ajay River and its topography due to rapid siltation and breaching. In particular, the study shows that the height of the present river bed is higher than the adjacent floodplains. The survey result of height difference between channel bed and floodplain at Bhedia, Mangalkote, Hussainpur and Palita are 0.68, 0.74, 0.91 and 0.52 m, respectively (Molla, 2011).

c. Due to the existence of embankments on both sides, river water cannot spill over during flood events, and consequently, suspended sediment gets silted on the river bed after sudden fall in flood water velocity (Kale, 2003). As a result, the river beds are temporally rising and fail to sustain previous-scale flood discharge and increase flood frequency and magnitude. The temporal rising of flood water level at different gauge stations also signifies the sedimentation problem (Figure 6.13). Due to a discontinuity of right bank embankment at some places e.g. after Kogram village, there is no embankment up to the Joykrishnapur village, and a huge volume of entrapped flood water suddenly gets released in that gap and quickly spreads over the confluence zone mainly in Mangalkote Block, Barddhaman, consequently generating havoc flood condition in the area.

d. The direct effect of the embankment on basin hydrology has been clearly perceived from the frequency analysis of the last 17 major floods since the 1900s. The percentage of flood frequency shows about 72% has occurred after 1956, the period of embankment construction (Table 6.7 and Figure 6.14).

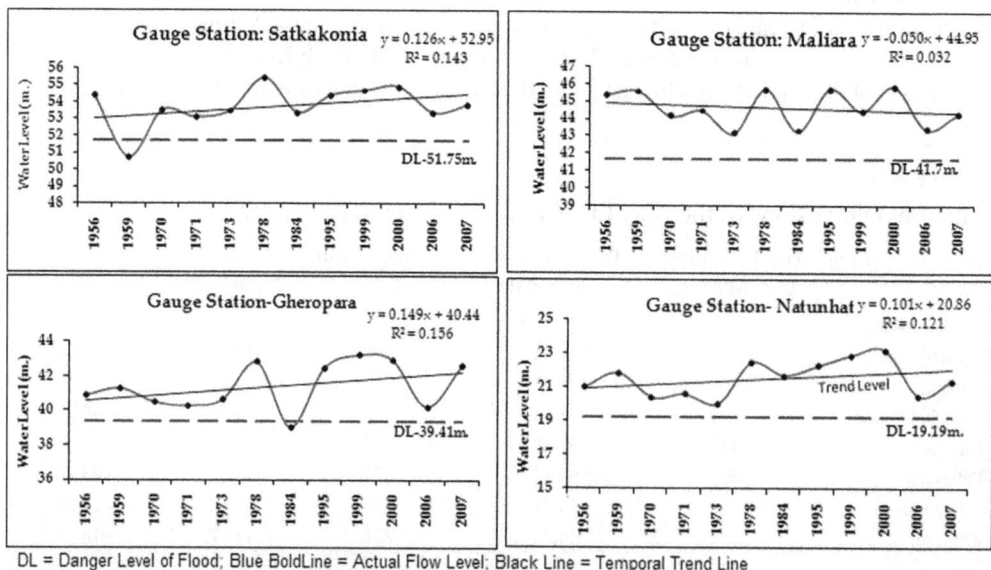

DL = Danger Level of Flood; Blue BoldLine = Actual Flow Level; Black Line = Temporal Trend Line

FIGURE 6.13 Temporal trend of water level of Ajay River at selected gauge stations. (Roy and Sahu, 2016b.)

TABLE 6.7

Trend of Flood Concentration before and after the Construction of Embankment

| Range of Time | Flood Frequency | | Cumulative Flood Frequency | |
	Frequency	% of Frequency	Frequency	% of Cumulative Frequency
1900–1910	0	0	0	0
1911–1920	2	12	2	2
1921–1930	1	6	3	4
1931–1940	1	6	4	5
1941–1950	1	6	5	6
1951–1960	2	12	7	8
1961–1970	1	6	8	10
1971–1980	3	18	11	13
1981–1990	1	6	12	14
1991–2000	3	18	15	18
2201–2007	2	12	17	20
Total	17	100	84	100

Source: Department of Irrigation and Waterways, Govt. of West Bengal, 2010 and Perception survey of local people.

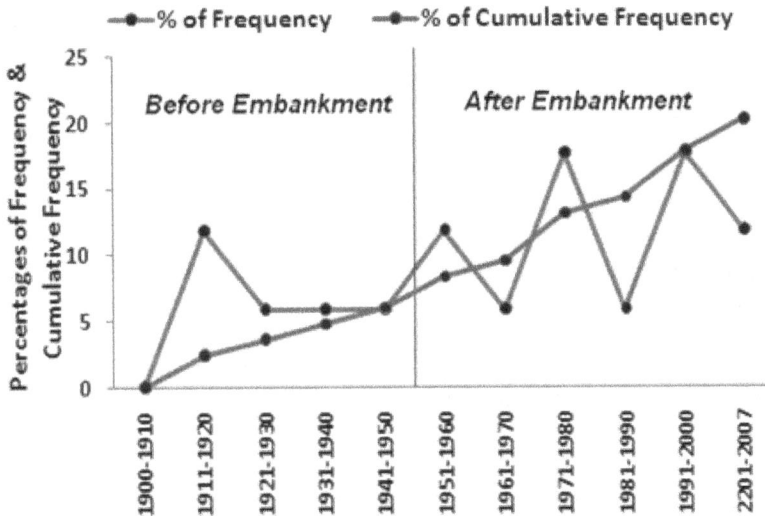

FIGURE 6.14   Showing the trend of flood frequency from 1956 to 2007.

e. Agricultural productivity has been also significantly decreased due to the vast sand-splay over the floodplain (Table 6.8). In addition, the chemical properties of riverine soil have also been changed after the major floods (Table 6.9).

f. The riverside mining activity has increased the input of sediment to the river system and accelerated the process of siltation of the study river. Such a process has been evidently observed at the confluence zone of Ajay and Tumuni Rivers because the Tumuni basin has been severely affected by two major collieries of opencast coal mining over an expanded area. The basin is experiencing direct input of dumping materials to the river water, which are carried to the trunk river (i.e. Ajay). As a result, micro-geomorphology of

TABLE 6.8
Decreasing Rate of Rice Production (kg ha⁻¹)

Name of the Mouza	Before Sand-Splay 1998	After Sand-Splay 2000 (Marginal Area)
Bhedia	3,750	1,500
Natungram	4,000	750
Maliara	3,000	800
Srikrishnapur	4,500	1,500
Narenga	3,600	1,312
Gomra	3,375	500

Source: Mukhopadhyay(2010).

TABLE 6.9
Nutrient Status of the Soil before and after Flood of 2000

Name of the Mouza	Ph		N_2 (kg ha⁻¹)		P_2O_5 (kg ha⁻¹)		K_2O (kg ha⁻¹)	
	Before	After	Before	After	Before	After	Before	After
Natungram	6.6	7.5	200	49.5	90	7.9	294	46
Gitgram	7	7.3	250	30.5	85	4.2	240	65
Maliara	7.1	7.1	280	39	21.2	4.2	316	59
Bhedia	6.9	6.9	330	26.4	45.5	3.6	305	72
Srikrishnapur	7	8.2	300	35	70	5.5	220	70

Source: Technical Report, vol. 7, Dept. of Soil Science, PalliShikshaBhavana, Visva-Bharati; Mukhopadhyay (2010).

the confluence zone of Ajay and Tumuni River has been changed over time (Figure 6.15). Specially, in this zone, a drastic change in the bar formation has been observed. The recent image (04 November 2017) shows a large stable island has been formed and which is also induced to increase the length of Tumuni River.

The detailed investigation of Figure 6.14 also helps to notice the other human activities like sand mining through the in-stream transport routes and development of settlement very close to the trunk river.

g. The effect of the transport network on the channel dynamics over the ARB has been successfully carried out through three case studies by Roy and Sahu (2016b, 2017, and 2018). Roy and Sahu (2016b) worked on the longitudinal disconnection of Ajay River at three selected bridge sites, viz. near Pandaveswar, Illambazar and Bhedia. The result of this study shows the multi-thread river turns into single-thread river since 1990 with changing in-stream geomorphology through the stabilisation of bar with permanent vegetation and channel thalweg wandering (Figure 6.16).

In Roy and Sahu (2017), the application of geoinformatics and field investigation over the Kunur River Basin (tributary of Ajay River) helps to find out a significant growth of the total road length and the number of RSCs over the last five decades (1970s–2010s). In particular, the growth of rural unpaved roads is extraordinary (1,922%), which are the major sources of fine to coarse sediments to the river system. As a result, the distance between channel and road networks is decreasing over time and the event like coupling between stream and road network becomes a serious problem of the basin.

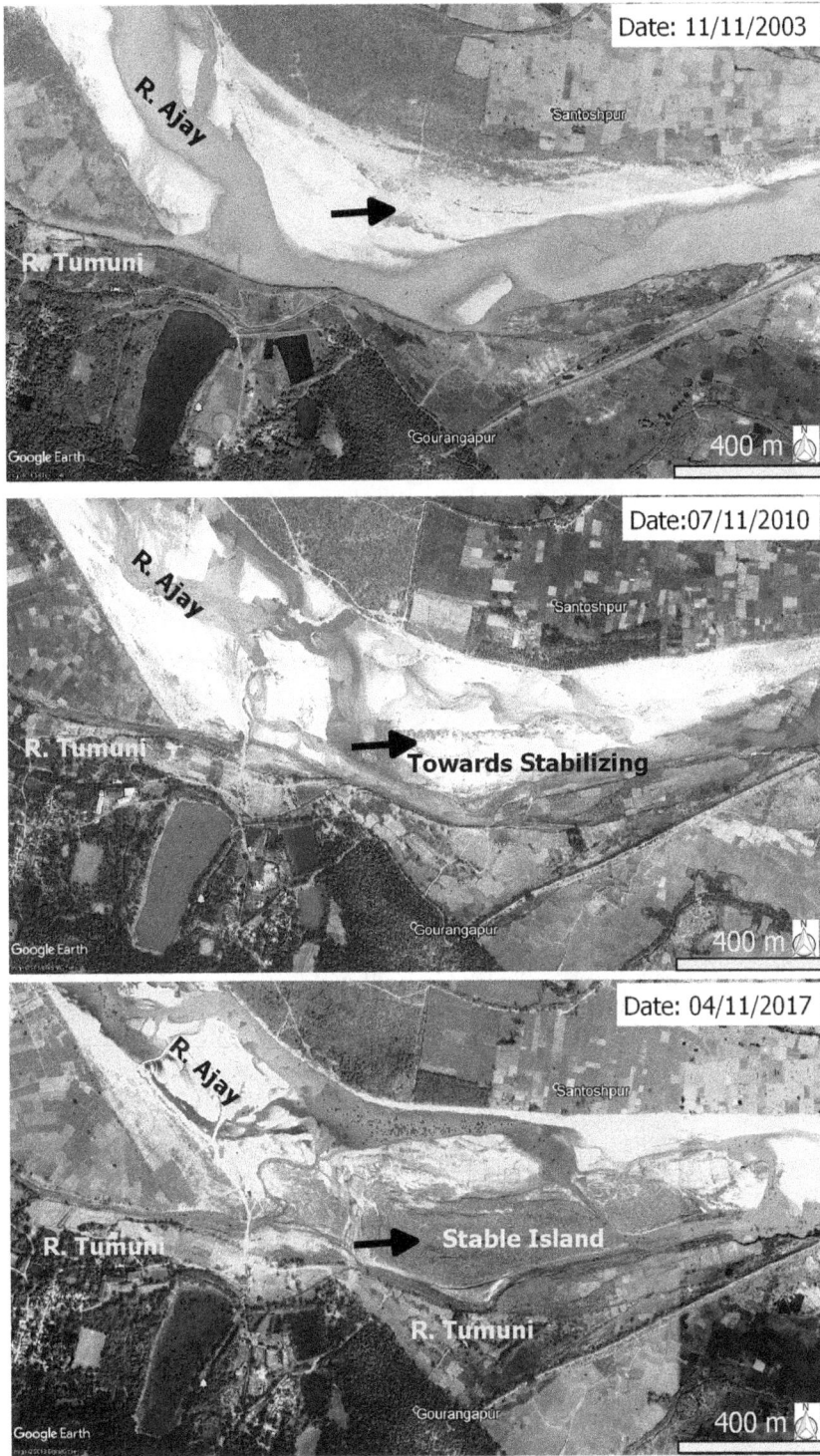

FIGURE 6.15 Micro-geomorphological changes at the confluence zone of Ajay and Tumuni Rivers. The black arrow indicates the growth of mid-channel bar and its temporal stabilising with the supply of sediments from the Tumuni basin. As a result, the channel length of Tumuni River has increased and the exact confluence point moved towards downstream. (Google Earth Image, 2018.)

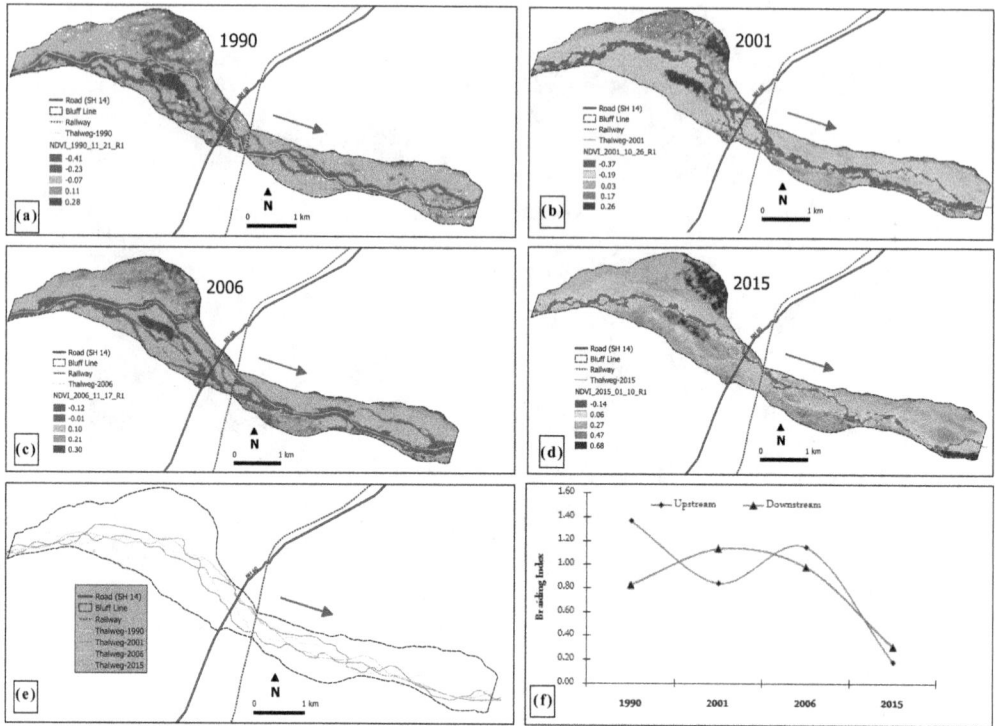

FIGURE 6.16 Temporal study at a bridge site of Ajay River near Pandaveswar showing the multi-thread river turning into single-thread river with decreasing barding index and thalweg wandering. (Roy and Sahu, 2016b.)

In Roy and Sahu (2018), a pure field survey-based work shows the effect of in-stream engineering through the installation of RSC on the alternation of channel morphology and hydraulics. The result shows the immediate upstream and downstream of the RSC significantly affected by alerted channel flow, where a significant increase in channel width (30%), depth (17%) and cross-section area (54%) has been estimated in comparison with the 50 m upstream from the crossing site. One dimensional steady flow modelling through Hydrologic Engineering Centre-River Ayalysis System (HEC-RAS) software on the longitudinal profiles of channel bed, water surface and energy gradients has shown the formation of knickpoint, deep scour and supercritical flow below the crossing sites.

h. A scholarly effort by Roy and Sahu (2016c) shows the influence of land-use land-cover changes on the deformation of channel morphology of headwater streams of Kunur River Basin, which is an important downstream tributary of the Ajay Basin. The application of applied statistical techniques on the field-based cross-section data has helped to estimate the changes of channel configuration from the forested area to agricultural area. Due to the transformation of land cover from forested to agriculture land, the channel widths (269%) and cross-section area (78%) have been increased significantly, whereas agricultural channels become shallower (40%) than would be predicted from forested streams.

6.4 CONCLUSION

- The basin has a wide range of anthropogenic interventions, which are altering the catchment landscape as well as the morphology of channels. The influences were more prominent during the last phase of the 20th century and at the beginning of the 21st century, especially by the dams, embankment, mining and land-use changes.

- The interaction between transport infrastructures and channel networks is an appalling issue to analyse for this basin as well as for the other plainland rivers because of its control on the connectivity between channel and floodplain and along the in-stream reaches.
- The process of urbanisation over the basin is a point of concern. Among the urban centres, Durgapur is the more sensitive to influence the fluvial system of the ARB through Kunur River Basin, due to its very close allocation with the stream lines, higher number of RSC and huge volume of wastewater.
- In-stream siltation is a major problem for the trunk river and also for the tributaries due to increasing sediment yield from the mining area and deforested areas. As a result, the channel bed height is rising temporally and inducing to increase flood height and its frequency.
- Geochemical alternation of the river water is still not a big issue for this basin. However, it should be in our concern to monitor the effect of increasing pesticides in agricultural land and development of different industries, especially wastewater from the paper and fertiliser industries, over the catchment area.
- The basin carries a good geoarchaeological research evidences to study the evolution of floodplain geomorphology since the Paleolithic period, specially the rate of sedimentation during the Quaternary period in correspondence with the past land-use land-cover change using sophisticated dating techniques, e.g. Optical-Stimulated Luminescence (OSL) dating.
- After the 1970s, the basin has experienced some major dams and irrigation projects. However, the effect of those in-stream engineering projects on the in-stream geomorphological and extra-stream landscape alternations has not been well documented with proper field-based research work. The lack of hydrological data of this basin is also an important cause of poor research on the hydrological regime alternation.
- The effect of mining on channel siltation could be perceived from the remote sensing data and field observation. However, the influences of mining on the changing sediment yield capacity of the mining-affected channels have not scientifically estimated, and no study has shown the channel morphology of those streams before and after mining practices.
- The embankment and transport lines are significantly disconnected the trunk river from its adjoining floodplain as well as the upstream and downstream reaches, which must have significant effect on the alternation of riverine geomorphology and ecology. Detail and intensive scientific work is urgently required to figure it out.
- The data available from the Department of Forestry, Government of West Bengal, shows no changes in the forest cover over the lower ARB, especially in the Barddhaman district. However, the field visit and direct observation have clarified that the native forest cover has been degrading temporally and transformed by the eucalyptus trees, which directly influenced the rate of soil erosion and channel sedimentation.
- Climate change is also an alarming issue for all river basins around the world. However, no valuable research work has been carried out regarding the control of climate change on hydrology and related geomorphology of this basin as well as other basins of the lower Gangetic region. Climate change is also triggering the effect of flood and erosion combinly with human interventions in hydrological system, especially through the construction of major and minor dams and irrigation projects and different engineering control (Goudie and Viles, 2016). Sterling et al. (2013) reported that changes in land use and land cover by altering evapotranspiration will have a direct effect on the future stream flow, and at global scale, it will definitely be increased. At the same time, Arnell and Gosling (2013) show that more than two-thirds results of climatic modelling on the global scale runoff assessment have projected a significant growth in annual average runoff.

REFERENCES

Arnell NW, Gosling SN (2013) The impacts of climate change on river flow regimes at the global scale. *Journal of Hydrology* 486: 351–364.

Bagchi K, Mukherjee KN (1979) *Diagnostic Survey of Rarh Bengal, Part –I, Morphology, Drainage and Flood: 1978*. Calcutta: Department of Geography, University of Calcutta.

Baker VR, Costa JE (1987) Flood power. In Mayer, L., Nash, D. (Eds.), *Catastrophic Flooding*. Boston: Allen and Unwin, pp. 1–21.

Blanton P, Marcus WA (2014) Roads, railroads, and floodplain fragmentation due to transportation infrastructure along rivers. *Annals of the Association of American Geographers* 104(3): 413–431. doi:10.1080/00045608.2014.892319.

Census of India (2011) Population Enumeration Data (Final Population). Access from https://censusindia.gov.in/2011census/population_enumeration.html on 03/07/2019.

Charlton R (2008) *Fundamentals of Fluvial Geomorphology*. London: Routledge, Taylor & Francis e-Library.

Chin A (2006) Urban transformation of river landscapes in a global context. *Geomorphology* 79: 460–487.

Datta A (2005) Subsistence strategies of the Chalcolithic people of West Bengal: an appraisal. *Indo-Pacific Prehistory Association Bulletin* 25(3): 41–47.

Fitzhugh TW, Vogel RM (2011) The impact of dams on flood flows in the United States. *River Research and Applications* 27(10): 1192–1215.

Garman GC, Moring RJ (1991) Initial effects of deforestation on physical characteristics of a boreal river. *Hydrobiologia* 209: 29–31.

Ghosh AK, Chakrabarti DK (1968) Prehistoric metal stage in West Bengal. *Bulletin of the Cultural Research Institute* 7: 114.

Gilbert GK (1917) *Hydraulic-Mining Debris in the Sierra Nevada*. United States Geological Survey Professional Paper. 105. Washington, DC: Government Printing Office.

Goudie AS, Viles HA (2016) *Geomorphology in the Anthropocene*. Cambridge, UK: Cambridge University Press.

Graf WL (1979) Mining and channel response. *Annals of the Association of American Geographers* 69(2): 262–275.

GRanD (2011) Global Reservoir and Dam (GRanD) database. Technical documentation – Version 1.1. Retrieved from http://www.gwsp.org/products/grand-database.html.

Gupta RK (1980) Consequences of deforestation and overgrazing on the hydrological regime of some experimental basins in India. *In the Proceedings of the influence of man on the hydrological regime with special reference to representative and Experimental basins*, IAHS-AISHPubl.no.130.

Hettler J, Irion G, Lehmann J (1997) Environmental impact of mining waste disposal on a tropical lowland river system: a case study on the Ok Tedi Mine, Papua New Guinea. *Mineralium Deposita* 32: 280–291.

Hewlett JD, Helvey JD (1970). Effects of forest clear-felling on the storm hydrograph. *Water Resources Research* 6: 768–782.

Indian Meteorological Department (IMD) (2014). *IMD District Wise Normals*. Barddhaman: Govt. of India.

IWB-GoWB: Irrigation and Waterway Directorate-Govt. of West Bengal (2016). Annual Flood Report for the Year 2016. Advance Planning, Project Evaluation, and Monitoring Cell, Jalsampad Bhavan, Salt Lake, Kolkata.

IWD-GoWB: Irrigation and Waterways Department- Govt. of West Bengal (2019). Annual Report 2018 – 19. Irrigation and Waterways Department, Jalsampad Bhavan, Salt Lake, Kolkata Govt. of West Bengal. Accessed from https://wbiwd.gov.in/uploads/admin_yearly_report/Annual%20Report%202018-19_I&WD.pdf.

Jimenez MD, Ruiz-Capillas P, Mola I, Pérez-Corona E, Casado MA, Balaguer L (2013) Soil development at the roadside: a case study of a novel ecosystem. *Land Degradation & Development* 24(6): 564–574.

Kale VS (2003) Geomorphic effects of monsoon floods on Indian rivers. *Natural Hazards* 28: 65–84.

Knighton AD (1987) River channel adjustment-The downstream dimension. In *River Channels: Environment and Process*, Oxford: Basil Blackwell, 95–128.

Kumar B, Singh UK, Padhy PK (2015) Character and classification of Ajay River water through index analysis and chemometrics. Abstract of IWA Symposium on Lake and Reservoir Management.P.8.

Leopold LB, Maddock TJ (1953) *Hydraulic Geometry of Stream Channels and Some Physiographic Implications*. U. S. Geological Survey Professional Paper (vol. 252). Washington, DC: United States Government Printing Office, p. 55.

Leopold LB, Wolman MG, Miller JP (1964) *Fluvial Processes and Geomorphology*. 1st Ed. San Francisco, CA: Freeman and Co.

Luce CH (2002) Hydrological processes and pathways affected by forest roads: what do we stillneed to learn? *Hydrological Processes* 16: 2901–2904. doi:10.1002/hyp.5061.

Majumdar SC (1941) *Rivers of Bengal Delta*. Department of Irrigation and Waterways, Govt. of West Bengal, Calcutta.

Manna A, Maiti R (2014) Opencast coal mining induced defaced topography of Raniganj coalfield in India -remote sensing and GIS based analysis. *Journal of the Indian Society of Remote Sensing* 42(4): 755–764.

Marren PM, Grove JR, Webb JA, Stewardson MJ (2014) The potential for dams to impact lowland meandering river floodplain geomorphology. *The Scientific World Journal* 2014(309673): 1–24. doi: 10.1155/2014/309673.

Molla HR (2011) Embankment and changing micro-topography of lower Ajoybasin in Eastern India. *Ethiopian Journal of Environmental Studies and Management* 4(4): 50–61.

Montgomery DR (1994) Road surface drainage, channel initiation, and slope instability. *Water Resources Research* 30: 1925–1932.

Mukhopadhyay S (2010) A geo-environmental assessment of flood dynamics in lower Ajoyriver including sand splay problem in Eastern India. *Ethiopian Journal of Environmental Studies and Management* 3(2): 96–110.

Mukhopadhyay SC, Dasgupta A (2010) *River Dynamics of West Bengal* (vol. II) Applied Aspect. Kolkata: Prayas Publishers and Book Sellers, p. 96.

Mullen DM, Moring JR (1988) Partial deforestation and short-term autochthonous energy input to a small new England stream. *Water Resources Bulletin: American Water Resources Association* 24(6): 1273–1279.

Nagasaka A, Nakamura F (1999) The influences of land-use changes on hydrology and riparian environment in a northern Japanese landscape. *Landscape Ecology* 14: 543–556.

Niyogi M (1984) Water resources of the Ajay basin – a geographical – hydrological study. Ph.D. Thesis, Department of Geography, University of Calcutta.

Niyogi M (1985) Ground water resource of the Ajay Basin. In: Chatterjee SP (ed), *Geographical Mosaic-Professor K.G. Bagechi Felicitation*. Calcutta: Manasi Press, pp. 165–182.

Rajaguru SN, Deotare BC, Gangopadhyay K, Sain MK, Panja S (2011) Potential geoarchaeological sites for luminescence dating in the Ganga Bhagirathi-Hugli delta, West Bengal, India. *Geochronometria* 38(3): 282–291.

Roy S (2012a) Locating archaeological sites in the Ajay River Basin, West Bengal: an approach employing the remote sensing and geographical information system. *Proceedings of 14th Annual International Conference and Exhibition on Geospatial Information Technology and Application*, Indian Geospatial Forum, pp. 1–10.

Roy S (2012b) Spatial variation of floods in the lower Ajay River Basin, West Bengal: a geo-hydrological analysis. *International Journal of Remote Sensing and GIS* 1(2): 132–143.

Roy S, Sahu AS (2016a) Palaeo-path investigation of the lower Ajay River (India) using archaeological evidence and applied remote sensing. *Geocarto International* 31(9): 966–984. doi:10.1080/10106049.2015.1094526.

Roy S, Sahu AS (2016b) Effect of longitudinal disconnection on in-stream bar dynamics: a study at selected road-stream crossings of Ajay River. In: Das BC, Ghosh S, Islam A, Ismail M (eds), *Neo-Thinking on Ganges-Brahmaputra Basin Geomorphology*. Springer International Publication, pp. 81–97.

Roy S, Sahu AS (2016c) Effect of land cover on channel form adjustment of headwater streams in a lateritic belt of West Bengal (India). *International Soil and Water Conservation Research* 4(4): 267–277. Elsevier Publication. doi:10.1016/j.iswcr.2016.09.002.

Roy S, Sahu AS (2017) Potential interaction between transport and stream networks over the lowland rivers in Eastern India. *Journal of Environmental Management* 197: 316–330. doi:10.1016/j.jenvman.2017.04.012.

Roy S, Sahu AS (2018) Road-stream crossing an instream intervention to alter channel morphology of headwater streams: case study. *International Journal of River Basin Management* 16(1): 1–19. doi:10.1080/15715124.2017.1365721.

Rudra K (2008) *BanglarNadikatha (in Bengali)*. Kolkata: SahityaSamsad.

Schnitzler A (1997) River dynamics as a forest process: interaction between fluvial systems and alluvial forests in large European river plains. *The Botanical Review* 63(1): 40–60.

Sidle RC, Ziegler AD (2012) The dilemma of mountain roads. *Nature Geoscience* 5, 437–438.

Singh UK, Kumar B (2017) Pathways of heavy metals contamination and associated human health risk in Ajay River basin, India. *Chemosphere* 174: 183–199. doi:10.1016/j.chemosphere.2017.01.103.

Sterling SM, Ducharne A, Polcher J (2013) The impact of global land-cover change on the terrestrial water cycle. *Nature Climate Change* 3: 385–390.

Tarolli P, Sofia G (2016) Human topographic signatures and derived geomorphic processes across landscapes. *Geomorphology* 255: 140–161.

Tarolli P, Sofia G, Wenfang CAO (2018) The geomorphology of the human age. In: *Encyclopedia of the Anthropocene*, pp. 35–43. doi:10.1016/B978-0-12-809665-9.10501-4.

Wemple BC, Jones JA (2003) Runoff production on forest roads in a steep, mountain catchment. *Water Resources Research* 39(8): 1220.

Wilk J, Andersson L, Plermkamon V (2001) Hydrological impacts of forest conversion to agriculture in a large river basin in northeast Thailand. *Hydrological Processes* 15: 2729–2748. doi:10.1002/hyp.229.

WRIS-India Web-GIS (2018) Major and minor irrigation project of India. Retrieved fromhttp://www.indiaw-ris.nrsc.gov.in/wrpinfo/index.php?title=Major_%26_Medium_Irrigation_Projects.

Zabaleta A, Antiguedad I (2013) Streamflow response of a small forested catchment on different timescales. *Hydrology and Earth System Sciences* 17: 211–223. doi:10.5194/hess-17-211-2013.

7 Responses of Fluvial Forms and Processes to Human Actions in the Damodar River Basin

Sandipan Ghosh and Rahaman Ashique Ilahi

CONTENTS

7.1 INTRODUCTION

The Anthropocene represents the time since human impacts have become one of the major external forcing on natural processes and humans have interacted with rivers from the time of ancient civilisations (CWC, 2000; Sinha et al., 2005). The Indian sub-continent, which hosts many large and perennial rivers with significant hydrological and geomorphic diversity, is also home to an ancient civilisation and is currently one of the most populated regions on the globe (Banerji, 1950; Chatterjee, 1967; Choudhury, 1995; Jain et al., 2016). Now one of the major research concerns is the development of hydrology–morphology–ecology relationship in the river system and the assessment of the anthropogenic disturbances on this or part of this relationship. Interest and focus on how humans change geomorphic systems accelerated in the late 20th century as concerns mounted about global environmental change and growing population pressures (Holdren and Ehrlich, 1974). Anthropogenic geomorphology focuses on many key aspects of geomorphological processes within the Anthropocene. Anthropogeomorphology deals with the human impact on earth's landforms where human actions transform, correct and modify natural processes by increasing or decreasing their rate of action and by causing the rupture of certain equilibrium which nature will try to reconstitute in different ways (Goudie and Viles, 2016).

The major river basins of India are considered as the important repository of the Anthropocene hydrologic and climatic changes. It has been found that the fluvial systems are the most sensitive elements of earth's surface, and any shift in climatic parameters, environmental conditions and human interferences instigates a rapid response from the fluvial systems (Reed, 2002; Alila and Mtiraoui, 2002; Sridhar, 2008; Kale et al., 2010; Mujere, 2011; Ewemoje and Ewemooje, 2011). When a river carries much more water than usual, it is said to be *in spate*, a term of uncertain origin but that may be derived from the Dutch verb *spuiten*, meaning 'to flood' (More, 1969; Benson, 1968; Allaby, 2006). In nature, a flood is an unusual high stage of a river due to runoff from rainfall and/or melting of snow in quantities too great to be confined in the normal water surface elevations of the river or stream, as the result of unusual meteorological combination (Raghunath, 2011). 'Flood' means that the flow in the river is in such an excess as to raise the level of the river at places so that it overflows the banks and rises to a level higher than the adjacent countryside, thus inundating the areas adjacent to the channel (Sarma, 2002; Roy and Mazumder, 2005). In this regard, when we consider the flood of alluvial tract of West Bengal, the flood is regarded as extreme hydro-meteorological annual phenomena of monsoon season (June to September) when the excessive surface runoff and overflow of alluvial rivers submerge the surrounding low-lying areas of Barddhaman, Hooghly and Howrah districts. Then, the surface drainage systems of lower Damodar Basin and flood regulation of DVC (Damodar Valley Corporation) dams failed to cope up with it. Quaternary to Recent floodplains of Damodar River has certain degree of 'flood risk' which involves quantification of the probability that flood hazard will be harmful and the tolerable degree of risk depends upon what is being risked, flood management being much more important than land utilisation for agriculture (Bell, 1999).

The flood history of Damodar River is very much popular to the deltaic part of West Bengal, because this river is notified as *Sorrow of Bengal* due to havoc destruction by annual floods, particularly in the lower floodplains which are heavily populated and connected through dense communication and transport networks (Satakopan, 1949; Saha, 2005). If we consider floods as extreme hydrological phenomenon and as quasi-dynamic equilibrium fluvial process, then it is not surprising to us that to attain the present form of lower Damodar floodplain, the flood aggradation and degradation are considered as the prime processes. But in reality, settlements beside a river, like Damodar River, can be a mixed blessing, for once in while the river may overflow its banks and exact a heavy toll of property losses, income losses and sometimes losses of life as well. In some cases, man has learned to live with such periodic inundations of the fertile floodplain (Bhattacharyya, 2011). The causes of floods can be natural, but human interference intensifies many episodes of flood. The decision to live in a floodplain, for a variety of perceived benefits, is one that is fraught with difficulties. The increase in flood damage is related to the increasing number of people living in floodplain regions (Nagle, 2003).

Floods, a popular theme of research as far as the monsoon-dominated rivers (viz. Narmada, Tapti, Kosi, Ganga, Kaveri, Godavari, Krishna, Brahmaputra, Luni, Pennar, Ajay, Damodar, etc.) of India are concerned, are investigated with the perspectives of flood hydrology, flood geomorphology, palaeohydrology and hydro-meteorology. Large floods of Indian river basins and their temporal variations have been studied by Grade and Kothyari (1990), Kale et al. (1993), Rajaguru et al. (1995), Kale (1998, 1999, 2003, and 2005), Sinha and Jain (1998), Goswami (1998), Rakhecha (2002), Nandargi and Dhar (2003), Dhar and Nandargi (2003), Jha and Smakhtin (2008), Mukhopadhyay (2010) and Jha and Bairagya (2012). In West Bengal, the lower segment of Damodar River is taken as main spatial unit of study by many researchers and scholars of different disciplines to analyse the flood ferocity and propensity. The flood history of Damodar during the period 1817–1917 can be traced from the E. L. Glass report submitted to the then Bengal Government as observed at Ranjganj (Bhattacharyya, 2011). The pre-dam and post-dam temporal characters of floods and results of dam construction are well analysed by Glass (1924), Goodall (1945), Krik (1950), Pramanik and Rao (1952), Bagchi (1977), Roy et al. (1995), Mishra (2001), Sengupta (2001) and Chandra (2003). The causes, factors, extent and trend of floods and phases of changing courses of Damodar River are studied, investigated and assessed minutely by Sen (1985 and 1991), Majumder et al. (2010), Bhattacharyya (2011) and Ghosh and Guchhait (2014 and 2016).

Remembering and analysing the past flood events of Lower Damodar River, the very common but unsolved problem of Damodar River Basin (DRB) is to manage the excessive inflow water of runoff from upper catchment through the existing engineering structures of DVC flood regulation system. It is a ground reality that incapable flood storage system, low canal consumption, declining carrying capacity of river, drying up of palaeochannels, mounting unscientific embankments, etc. are the chief causes of recent floods and these are not capable to cope up with huge volume of runoff. The essential part of this study is the precise estimation and forecast of runoff in relation to flood risk of Lower Damodar River. So following anthropogeomorphic perspective, we have considered the following four prime objectives of study as follows:

1. Understanding the hydrogeomorphic settings of DRB
2. Analysing the dam-induced changes in river morphology
3. Analysing the influence of human and monsoon on river hydrology
4. Estimating potential runoff and identifying dominant factors of flood risk to understand the magnitude of flood discharge

7.2   THE RIVER AND ITS BASIN

The Damodar River, or the *Deonad Nadi* as it is known in its upstream sector, is a sub-system of the Lower Ganges River system of India. The local meaning of the word Damodar is 'womb' or *Udar*, which is 'full of fire' (Bhattacharya, 2011). This implies that the Damodar flows through a coal-rich area. The river rises in the Chotanagpur Plateau approximately at 23°37′ N and 84°41′ E (Bhattacharya, 2011). The main tributaries are the Barakar, Tilaiya and Konar. Below the confluence of the Barakar and Damodar, there are a few insignificant tributaries such as the Nunia and Sali. Once the main distributaries were the Khari, Banka, Behula and Gangur, but now they look more like independent rivers. Near Palla, the river takes a sharp southerly bend. Below Jamalpur, the river bifurcates into the Kanki–Mundeswari and the Amta Channel–Damodar and joins the Hooghly River (also spelled Hugli) at Falta some 48.3 km south of Kolkata (Ghosh and Guchhait, 2014). The DRB is a sub-basin and part of the Ganges River spreading over an area of about 23,370.98 km² in the states of Jharkhand and West Bengal. The sample study area (Figure 7.1) includes the lower reach of Damodar River between Rhondia (Galsi I Block) and Paikpara (Jamalpur Block). Its latitudinal extension ranges from 23°00′ to 23°22′10″ N and longitudinal extension ranges from 87°28′23″ to 88°01′00″ E.

At Anderson Weir (near Rhondia Village), the estimated catchment area coverage of Damodar River is 19,920 km². The total stretch of Damodar between Rhondia and Paikpara is approximately 82 km. This reach crosses Sonamukhi and Patrasayer blocks (Bankura), Kanksa, Galsi I, Galsi II, Barddhman I, Khandoghosh, Raina, Memari I and Jamalpur blocks (Barddhman). The slope category for 5–30 m of relief of Lower Damodar Valley is flat lands of 0°–0°30′, having 7.74% of basin area, and slope category for upper 30–60 m of relief is gentle sloping plain of 0°30′–1°00′, having 3.57% of basin area (Betal, 1970). The river had caused 16 large magnitudes of major floods in its basin during 1823–1943 though after constructions of dams and reservoirs, the ferocity of flood had reduced but the number of floods had been increased (Majumder et al., 2010). The floods of 1823, 1840, 1913, 1935, 1941, 1958, 1959 and 1978 had peaks of more than 16,992 m³s⁻¹. A peak flow of about 18,678 m³s⁻¹ has been recorded three times: in August 1913, September 1935 and October 1941 (Bhattacharyya, 2011). To mitigate flood, the DVC came into existence on 7 July 1948 by an act of the Constituent Assembly (Chandra, 2003). In the first phase, only four dams, viz. Tilaiya (1953), Konar (1955), Maithon (1957) and Panchet (1959), were constructed by DVC. Then two more reservoir, Tenughat (1978) on Damodar under the control of Jharkhand and Durgapur Barrage (1955) under the control of West Bengal, are constructed, respectively (Chandra, 2003) (Table 7.1).

The study about floods always incorporates an interdisciplinary approach of earth science which includes the method and techniques of hydrology, hydro-climatology and fluvial geomorphology. We have collected secondary data and information mostly from Damodar Planning Atlas (1969),

(a)

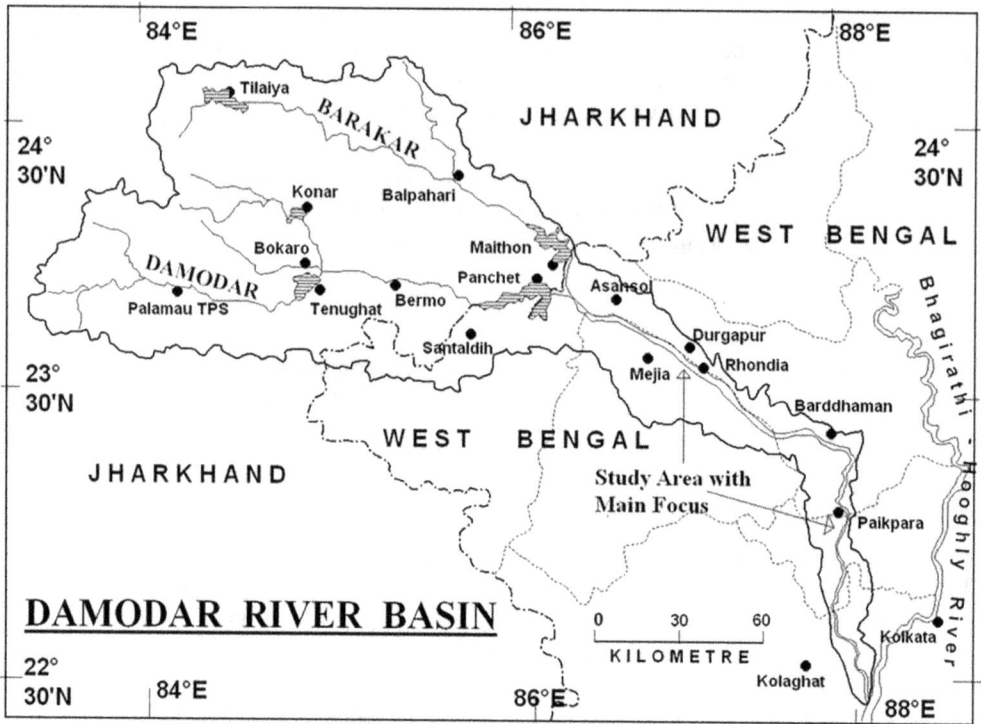

(b)

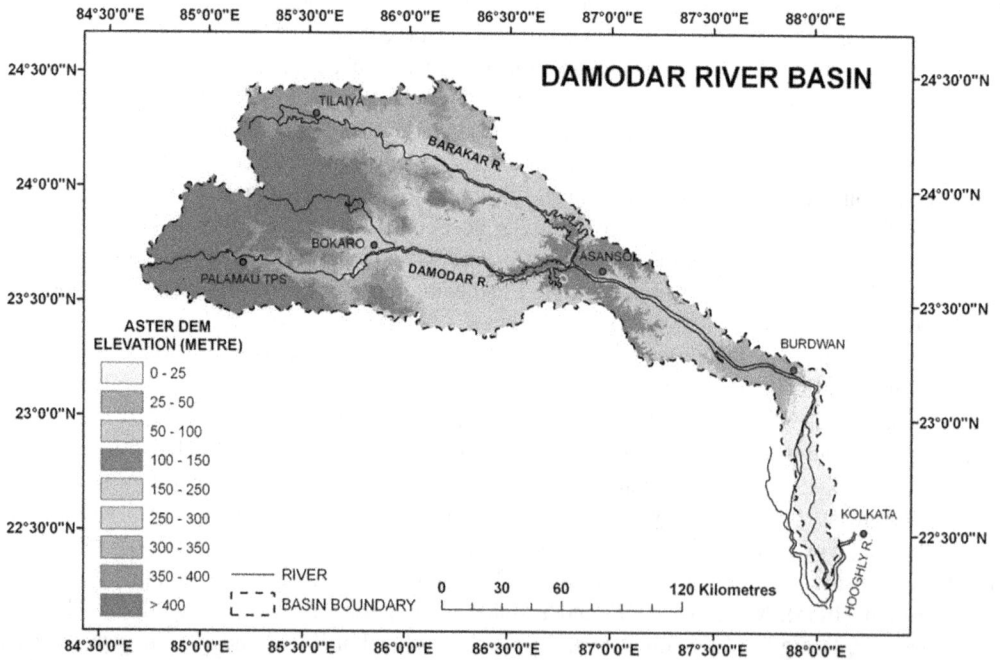

FIGURE 7.1   (a) Spatial extent of DRB and (b) ASTER elevation map of the basin.

TABLE 7.1
Areal Coverage of Damodar River Basin in Jharkhand and West Bengal

Sl. No.	District	Total Area (km²)	Area in the Basin (km²)	% Area of District in the Basin	% Share in the Basin
		Jharkhand Sub-Region			
1	Palamu	12,677	736.84	5.81	3.15
2	Ranchi	18,311	910.33	4.97	3.9
3	Hazaribagh	11,152	6,631.56	59.47	28.38
4	Giridih	6,908	5,376.81	77.83	23.01
5	Dhanbad	2,996.8	2,996.8	100	12.82
6	Santhal Pargana	14,129	571.05	4.04	2.44
Sub-total		-	17,223.39	-	73.7
		West Bengal Sub-Region			
1	Purulia	6,259	1,383.28	22.1	5.92
2	Bankura	6,881	1,564.67	22.74	6.69
3	Burdwan	7,028	2,113.61	30.07	9.04
4	Hooghly	3,145	359.87	11.44	1.54
5	Howrah	1,474	726.16	49.29	3.11
Sub-total		-	6,147.59	-	26.3
Grand total		-	23,370.98	-	100

Source: Bhattacharyya (2011).

book entitled *Lower Damodar River, India: Understanding the Human Role in Changing Fluvial Environment* (Bhattacharyya, 2011), websites of Irrigation and Waterways Department of West Bengal, West Bengal State Marketing Board, Indian Meteorological Department of India, Geological Survey of India (GSI), Google Earth, Shuttle Radar Topographic Mission (SRTM) data and different research articles relating to Damodar River. We have also employed the maps of National Atlas and Thematic Mapping Organisation (NATMO), Survey of India (SOI) toposheets (73 M/7, M/11, M/12, M/15, M/16, N/13 and 79 A/4) and GSI. After collecting secondary and primary data, we have employed statistical analysis, viz. mean, standard deviation, skewness, correlation, regression, probability distribution, t- and chi-square tests, etc. for hydrological interpretation of streamflow data in Microsoft Excel 2003 and SPSS 14.0. Geographic Information System (GIS) (i.e. MapInfo Professional 9.0) is used to represent spatial data into organised format or thematic map. Finally, gathering the results, we have made significance of adopted methods, interpretation of outcomes and conclusion to chalk out the major findings of this analysis. The employed quantitative expressions are summarised as follows:

To measure the annual runoff volume of Damodar River at Rhondia (West Bengal), Dhir et al. (1958) developed an empirical equation (Jha and Smakhtin, 2008):

$$\text{Annual Runoff Volume}\left(\text{in million m}^3\right) = 13,400 \text{ Annual Precipitation}\left(\text{in cm}\right) - 5.75\,10^5$$

Employing linear regression and t-test, we have established a significant rainfall–runoff equation to predict the trend of runoff on the basis of actual rainfall and runoff data (1934–1950) at Rhondia. The equation is expressed as follow:

$$\text{Runoff}\left(Y\right) = 0.819 \text{ Rainfall}\left(X\right) - 530.21$$

Converting annual peak discharge (m^3s^{-1}) of Damodar to $1\,m^3s^{-1}$ day in m^3 (i.e. maximum volume of water flow passing through the measuring station) and dividing it with drainage area (km^2) above Rhondia, we have obtained potential contributing runoff (mm) after satisfying the storage of dams. The expression is drawn as follows (Reddy, 2011):

$$\text{Potential Contributing Runoff} = \left(86,400 \text{ Peak Discharge}\right)10^4 \big/ \text{Drainage Area } 10^6$$

According to Gumbel's probability distribution, the exceedence probability (P) is given by (Gumbel, 1941; Reddy, 2011)

$$P = 1 - e^{-e-y}$$

where y is called reduced variates given by

$$y = \left[(X - X_{mean}) + 0.45\,\sigma / 0.7797\,\sigma\right]$$

where X_{mean} is the mean of sample, X is the observed data and σ is standard deviation of sample.

If X_T denotes the magnitude of the flood with return period of T years [$T = 1/(1 - e^{-e-y})$],

$$X_T = X_{mean} + K_T\,\sigma$$

where K_T is the frequency factor ($y_T - y_{mean}/\sigma_n$), y_T is the reduced variate, y_{mean} is the mean of reduced variate and σ_n is standard deviation of reduced variates (calculating these from table value of Gumbel's distribution).

Pearson Type III distribution is a skew distribution with limited range in the left direction, usually bell shaped (Griffs et al., 2007; Raghunath, 2011; Sathe et al., 2012). Logarithmic Pearson Type III distribution (LP3) has the advantage of providing a skew adjustment, and if the skew is zero, the log Pearson distribution is identical to the log normal distribution (Vogel and McMartin, 1991; Rao and Hamed, 2000; Stichellout et al., 2006; Raghunath, 2011; Millington et al., 2011; Zakaullah et al., 2012; Whitfield, 2012). The probability density function for type III (with origin at the mode) is

$$f(x) = F_o(1 - x/a)^c \exp(-cx/2)$$

where $c = 4/\beta - 1, a = (c/2)(\mu_3/\mu_2), \beta = \mu_3^2/\mu_2^3$

$$f_o = (n/a)\left[c^{c+1}/e^c\,\Gamma(c+1)\right]$$

where
  μ_2 = the variance
  μ_3 = third moment about the mean = $\sigma^6 g$
  E = the base of the Napierian logarithms
  Γ = the gamma function
  n = the number of years of record
  g = the skew coefficient
  σ = the standard deviation

The United State Water Resource Council (1967), Hann (1977), Oberg and Mades (1987), and CWC (2000) adopt this distribution to achieve standardisation of procedures.

$$\text{Mean} - \log x_{mean} = \Sigma\,\log x/n$$

$$\text{Standard Deviation} - \sigma_{\log x} = \left[\Sigma\left(\log x - \log x_{mean}\right)^2 \big/ (n-1)\right]^{1/2}$$

$$\text{Skew Coefficient} - g = n \Sigma \left(\log x - \log x_{mean}\right)^3 \Big/ (n-1)(n-2)\left(\sigma_{\log x}\right)^3$$

The values of x for various recurrence intervals are computed from

$$\log x = \log x_{mean} + K_T \sigma_{\log x}$$

where K_T is the frequency factor, tabulated in respect of skew coefficient and recurrence intervals (Hann, 1977).

Then, we have employed 'confidence limits' to the distribution of peak discharge. From the sample (n), we do not get the exact value for the discharge, but mean confidence limits give us a close interval within which the peak discharge mean falls with 99% and 95% probabilities (CWC, 2000; Mahmood, 2008; Raghunath, 2011).

$$X_{mean} - 3\,\sigma/\sqrt{n} \text{ to } X_{mean} + 3\,\sigma/\sqrt{n} \text{ for 99\% confidence limit}$$

$$X_{mean} - 2\,\sigma/\sqrt{n} \text{ to } X_{mean} + 2\,\sigma/\sqrt{n} \text{ for 95\% confidence limit}$$

Pearson's 'chi-square test' (χ^2) is the best known of several chi-squared tests—statistical procedures whose results are evaluated by reference to the chi-squared distribution. Its properties were first investigated by Karl Pearson in 1900 (Kidson and Richards, 2005; Griffs et al., 2007; Mahmood, 2008). A test of goodness of fit establishes whether or not an observed distribution differs from a theoretical distribution. It tests a null hypothesis stating that frequency distribution of certain events observed in a sample is consistent with a particular theoretical distribution (Mahmood, 2008). The expression of the tests statistic is

$$\chi^2 = \sum_{i=1}^{k} \frac{(O_i - E_i)^2}{E_i}$$

where O_i = observed value or frequency (actual peak discharges of the consecutive years)

E_i = expected value or frequency (estimated theoretical value of peak discharge of the consecutive years)

k = number of data points

i = 1, 2, 3, ... k

The reduction in degrees of freedom (df) is calculated as p = s = 1, where s is the number of parameters used in fitting the distribution. Degree of freedom will be (n–p), where n is the number of categories or cells in table.

To estimate the carrying capacity (m^3) of a particular reach and the threshold level of flood discharge ($m^3 s^{-1}$), we have calculated mean cross-sectional area (m^2) and mean maximum length of the reach (m) of three selected channel segments of Damodar River, viz. (1) Rhondia to Jujuti segment, (2) Jujuti to Chanchai segment and (3) Chanchai to Paikpara segment, using SRTM data and Global Mapper 13.0 software (3D path profile tool). Bankfull volume of a segment is expressed in $1\,m^3 s^{-1}$-day (m^3). If this total volume of water passed from a reach having a rate of $1\,m^3 s^{-1}$, it is regarded as $1\,m^3 s^{-1}$-day = $1 \times 24 \times 60 \times 60 = 86,400\,m^3$; so, it can be assumed that in a day, $86,400\,m^3$ of water is passed having a flow rate of $1\,m^3 s^{-1}$ in a particular segment of the river. Reddy (2011) and Ghosh and Guchhait (2014) have developed the following expressions:

$$\text{Carrying capacity} = \text{Mean cross} - \text{sectional area} \times \text{Mean maximum length of a reach}$$

$$\text{Bankfull flood discharge} \left(m^3\ s^{-1}\right) = \text{Carrying capacity of a reach} \times 1\,m^3\ s^{-1} / 86,400$$

7.3 ANTHROPOGENIC SIGNATURES ON FORMS AND PROCESSES

7.3.1 DAMODAR VALLEY PROJECT—A MAJOR ANTHROPOGENIC FACTOR

Conceived in 1945 on the lines of Tennessee Valley Authority (TVA), the DVC was formed in 1948 under the leaderships of W. L. Voorduin, Dr. Meghnad Saha, Dr. Nalinikanta Basu, Raja Udaychand of Burdwan, etc. (Bhattacharya, 1959). After the devastating flood of 1943, the British Bengal Government set up a 'Damodar Flood Enquiry Committee', which advised that nothing less than a complete catchment basin scheme could prevent further destructive flooding by the Damodar in West Bengal (Krik, 1950; Saha, 1979). It was decided that the scheme had to be implemented through eight large dams at eight sites, viz. Tilaiya, Maithon and Balpahari on the Barakar River; Bokaro on Bokaro River; and Konar on the Konar River and the Panchet, Aiyer and Bermo detention dams (Krik, 1950; Chandra, 2003). But in the first phase, only four dams, viz. Tilaiya (1953), Konar (1955), Maithon (1957) and Panchet (1959), had been constructed by DVC (Figures 7.2 and 7.3). Then two more reservoir, Tenughat (1978) on Damodar under control of Jharkhand and Durgapur Barrage (including main left bank and right bank irrigation canals) (1955) under control of West Bengal, are constructed (Chandra, 2003).

Project targets had following quantifiable characteristics (Saha, 1979):

1. The dams were to provide a controlled storage capacity of 65,000 million m³ and this represents a capability of moderating a 28,000 m³ s⁻¹ flood to one of 7,00 m³ s⁻¹

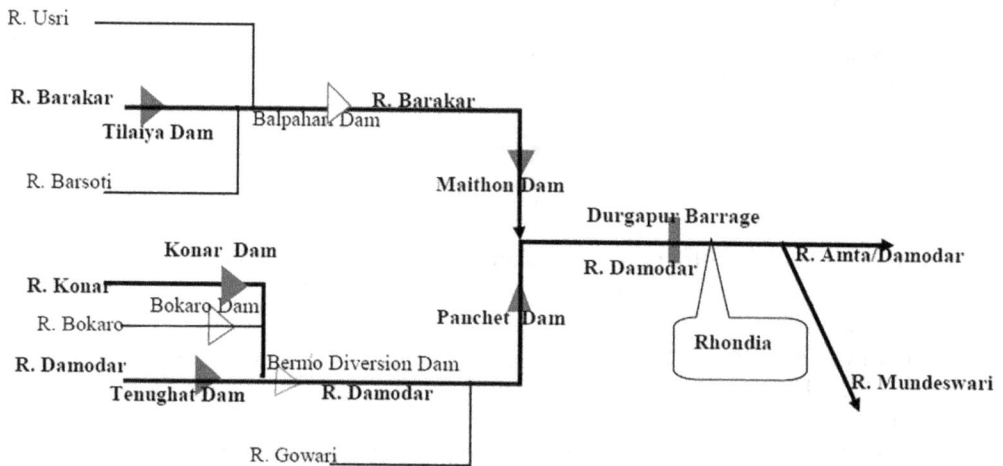

FIGURE 7.2   Present form of DVC-controlled structures on DRB (Ghosh and Guchhait, 2014).

FIGURE 7.3   (a) Panchet dam on Damodar River, (b) Maithon dam on Barakar River and (c) Anderson Weir on Damodar River at Rhondia (daily discharge measuring station in Lower Damodar Valley) (Ghosh and Guchhait, 2014).

2. The upstream dams were to ensure a minimum water release of $230\,m^3\,s^{-1}$ all the year to provide perennial irrigation to the irrigable area of 310,000 hectares

3. The valley's power generation-installed capacity of 137 MW in 1943 was to be interested to 350 MW by construction of a system

4. Finally, the DVC program envisaged a wide range of other planning activities such as silt control, soil conservation of upper valley, malaria control in the lower valley and navigation

7.3.2 SILTATION OF DVC RESERVOIRS

The proposed project of DVC was set forth to provide a controlled storage capacity of 6,500 million m^3, but only four dams and Tenughat reservoir provide a maximum storage capacity of 3,591 million m^3, only 55% of storage capacity originally envisaged (Saha, 1979). The last two terminal dams, Maithon and Panchet, are located close to break of topographical slope in the border of Jharkhand and West Bengal. So, the upstream tributaries of Damodar and Barakar bring heavy sediment-laden water to these reservoirs and the siltation of reservoirs is emerging as a major problem to affect the downstream flood control measures (Basu, 2011). The siltation rates of Maithon and Panchet reservoirs are 1,310.0 and $1,059.0\,m^3\,km^{-2}year^{-1}$, respectively. A recent study shows that the sedimentation of Panchet (on the basis of Landsat TM images of 1990 and 2005) occurs approximately at a rate of 0.041–$0.047\,cm\ year^{-1}$, whereas it was 0.033–$0.034\,cm$ per year in 1990 (Majumder et al., 2012). If we consider the overall capacity (including dead zone, live zone and flood zone), the losses of capacities of Maithon and Panchet reservoirs are 22.1% (up to 1994) and 14.1% (up to 1996), respectively (Table 7.2) (Rudra, 2002; Bhattacharyya, 2011).

The temporal variation of flood moderation by Panchet and Maithon dams shows that there is now a declining trend of flood controlling performance. It is a reality that DVC dams cannot attain their previous capacity to accommodate flood water due to siltation. In the post-dam period, when DVC dams suddenly had released excess water to save structure of dams (Table 7.3), the floods occurred in 1959, 1978, 1995, 1999, 2000, 2003, 2006 and 2007. Recently due to deep cyclonic rainfall (145 mm on 7 to 8.08.2011) over Jharkhand and Chhattisgarh from 7 August 2011, Tenughat, Panchet and Maithon reservoirs released 454, 510 and $481\,m^3\,s^{-1}$, respectively. On 8 August 2011 that had a cumulative effect on Durgapur Barrage when $1,275\,m^3\,s^{-1}$ water was released through main channel and canals (Ghosh and Mistri, 2013). As a result, on 13 August 2011, the numbers of total flood-affected blocks of West Bengal were reached up to 15 and 16 million people were directly affected from flood inundation (Ghosh and Mistri, 2013).

TABLE 7.2

Maithon and Panchet Reservoirs: Loss of Capacity (in thousands ac ft, 1 ac ft = 1,233.481 m^3) in Different Years

	Zones	1955 Original	1963	1965	1971	1979	1987	1994
Maithon	Flood zone (480–495 ft)	309.8	312.6	313.2	308.7	306.4	301.7	296.4
	Loss in %		-	-	0.3	1.1	2.6	4.3
	Overall (up to 495 ft)	969.5	918.3	909	881	844.1	795.2	755.2
	Loss in %		5.3	6.2	9.1	12.9	18.0	22.1
		1956 Original	**1962**	**1964**	**1966**	**1974**	**1982**	**1996**
Panchet	Flood zone (410–445 ft)	885.7	883.4	880.3	877.9	871.7	872.7	855.4
	Loss in %		0.3	0.6	0.9	1.6	1.5	3.4
	Overall (up to 445 ft)	1,281.7	1,217.5	1,203.4	1,195.9	1,164	1,131.4	1,101.1
	Loss in %		5.0	6.1	6.7	9.2	11.7	14.1

Source: Rudra (2002).

TABLE 7.3

Regulation of 1978 Flood with and without Dams

	With 4 DVC Dams (m³ s⁻¹)		Without Dams (m³ s⁻¹)	
	3 hourly peak inflow	3 hourly peak outflow	4 hourly peak inflow	4 hourly peak outflow
At Maithon and Panchet	21,917	4,615	26,958	26,968
At Durgapur	10,732	10,732	33,414	33,414

Source: Chandra (2003).

7.3.3 ASSESSMENT OF FLOOD MODERATION BY DVC DAMS

Huge-volume monsoonal runoff water (June–September) regularly created unaccountable pressure to the last two terminal dams which are at last compelled to release water to save the dam structure. Here we have found that using Dhir et al. (1958) empirical equation and mean annual rainfall data of Tilaiya, Tenughat, Maithon, Panchet and Durgapur (rain gauge stations on DRB), the mean annual runoff potentiality is approximately 1,008,344 million $m^3 year^{-1}$ (runoff depth of 50.62 cm), which signifies the gigantic volume of water pressure and high runoff coefficient (greater than 0.55) of upper catchment (Table 7.4). We have noticed that for 4 days (18–20 September 2000) rainfall of 359.17 mm over Damodar Basin, approximately 430,000 cusec (12,176 $m^3 s^{-1}$) of water inflowed into Panchet and Maithon reservoirs combinedly (17–26 September 2000) and two dams released water up to 205,000 cusec (5,805 $m^3 s^{-1}$) having flood moderation capability of 49% (Rudra, 2002, pp. 333 and 335). Examination of actual inflow and outflow data for two terminal dams at Maithon

TABLE 7.4

Moderation of Combined Flow by Maithon and Panchet Dams during Peak Inflow

Date	Inflow (m³ s⁻¹)	Outflow (m³ s⁻¹)	Flood Moderation (m³ s⁻¹)	% of Moderation
16/09/1958–17/09/1958	15,717.60	4,956	10,761	68.464651
1/10/1959–2/10/1959	17,646.36	8,156.16	9,487	53.761796
27/9/1960	9,855.36	2,605.44	7,249.92	73.563218
2/10/1961–3/10/1961	14,643.12	4,559.52	10,053.6	68.657499
2/10/1963–3/10/1963	12,772.32	3,426.72	9,345.6	73.170732
24/10/1963–25/10/1963	13,168.8	2,577.12	10,591.68	80.430108
29/7/1964–30/7/1964	10,563.36	2,208.96	8,354.4	79.088472
3/9/1970–4/9/1970	8,269.44	2,265.6	6,003.84	72.602740
16/7/1971–18/7/1971	12,007.68	5,125.92	6,881.76	57.311321
12/10/1973–13/10/1973	16,652.16	4,956	11,696.16	70.238095
26/9/1975–27/9/1975	9,742.08	3,143.52	6,598.56	67.732558
27/9/1978	21,919.68	4,616.16	17,303.52	78.940568
27/8/1980–28/8/1980	9,657.12	4,219.68	5,437.44	56.304985
29/9/1995	17,360.16	7,080	10,280.16	59.216966
1999	10,300	3,400	6,900	66.990291
19/8/2000–21/8/2000	10,800	5,700	5,300	49.074074
24/9/2006	14,400	6,900	7,500	52.083333
25/9/2007	11,100	7,500	3,600	32.432432

Data Source: Bhattacharyya (2011).

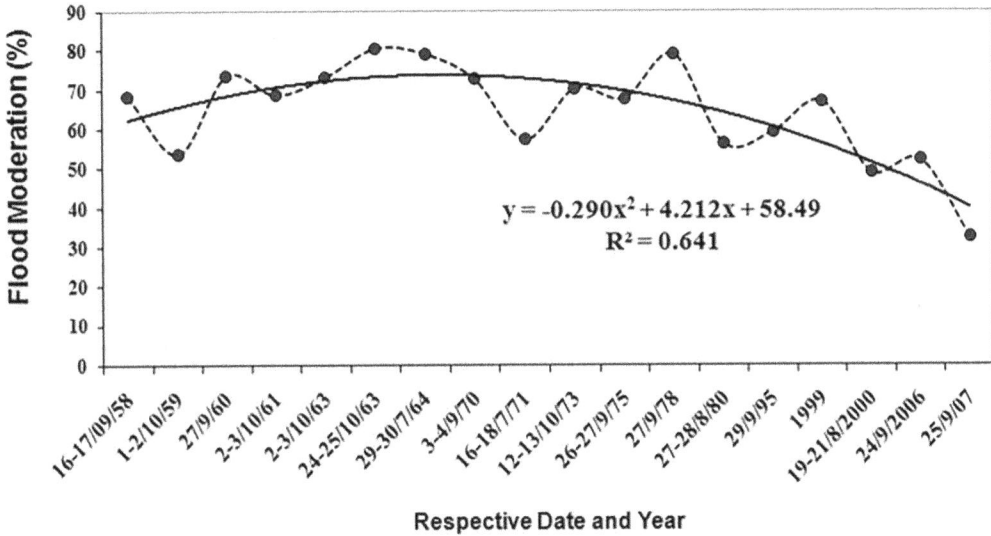

FIGURE 7.4 Temporal trend of flood moderation capacity of Panchet and Maithon dams.

and Panchet shows (Figure 7.4) that flood moderation has been achieved during the past years (Table 7.4). As siltation of river bed and reservoirs is the ongoing problem of flood regulation, if all eight dams had been constructed as per design, then DVC can get ample chance to moderate flood flow up to 7,079 $m^3 s^{-1}$ at Rhondia. The temporal variation of flood moderation graph shows (Figure 7.4) that there is a declining trend of flood moderating performance with time ($Y = -0.2908 x^2 + 4.2128 x + 58.496$) till 2007. In spite of flood moderation by the DVC dams, the floods occurred in 1959, 1978, 1995, 1999, 2000, 2006 and 2007, demonstrating that the lower valley is still vulnerable to sudden floods (Bhattacharyya, 2011).

The funnel-shaped basin, with a wide upper catchment and a narrow lower catchment (bottle neck location, elbow shape near Burdwan), will generally have phenomenal increase of peak discharge or streamflow in the lower catchment with considerable time lag as is evidenced in the Damodar Basin. In the pre-dam period, the peak discharges are observed in the month of August (on an average of 1,238 $m^3 s^{-1}$), but after construction of dams and flood regulation shifts, the monsoonal peak discharge from August to September (on an average of 1,247 $m^3 s^{-1}$) has low variation from the former. The dams temporarily store inflow runoff and streamflow till late August, but due to continuation of heavy rainfall and critical reservoir storage limit, the dams are compelled to release water in September when the soils of West Bengal gained full moisture and excess water adds with this streamflow. For that reason, now the flood probability or chances are more common between September and October.

It was estimated by Voorduin and DVC that if all the proposed dams were established, the system could moderate the peak discharge up to 7,075 $m^3 s^{-1}$ (611,280,000 $m^3 m^3 s^{-1}$-day) at Rhondia and any discharge above it would cause flood at downstream section. As the mean cross-sectional area of river is decreased downstream (from 12,290 to 7,077 m^2), the bankfull volume of reach also is declined at downstream section (from 346,568,177 to 133,259,910 m^3). Now we have estimated that threshold levels of peak discharge are 4,011, 2,366 and 1,542 $m^3 s^{-1}$, respectively, for the selected reaches of Lower Damodar River (Rhondia to Paikpara). So, now any bankfull discharge above 4,011 $m^3 s^{-1}$ at Rhondia is considered as threshold level of flood discharge for lower catchment. As the reach below Amta is not capable of carrying a discharge of 1,415 $m^3 s^{-1}$ (Chandra, 2003, p.8), it is estimated that a discharge of 2,000 $m^3 s^{-1}$ or more (having 5 years of return period) has 73% probability of occurrence. That's why in post-dam period, the riverine tracts of Barddhaman,

Hooghly and Howrah districts had experienced high magnitude of floods in 1958, 1961, 1976, 1978, 1995, 1999, 1987, 2000, 2006 and 2007. Now the situation is more critical because east and south-easterly flowing distributaries or palaeochannels of Lower Damodar, e.g. Khari, Banka, Behula, Gangur, and *Kana* Damodar, are completely detached from the main Damodar by embankments and agricultural activities. So, excessive volume of water does not get a chance to distribute through these channels.

7.3.4 Changing Channel Pattern

To understand the influence of dams on the river, two indexes have been used: (1) sinuosity index and (2) braiding index. Mueller's sinuosity index includes the hydraulic sinuosity (i.e. that freely developed by the channel uninfluenced by valley–wall alignment) and topographic sinuosity (i.e. that imparted by the geometry of the valley) (Mueller, 1968).

The Standard Sinuosity Index (SSI) varies from 1.04 (pre-dam) to 1.16 (post-dam) between Rhondia and Paikpara (Table 7.5 and Figure 7.5). The most important fact is that in pre-dam period (1922–1943), Hydraulic Sinuosity Index (HSI) of four reaches is constantly decreased from 48% to 13%, whereas Topographic Sinuosity Index (TSI) is increased from 51% to 87%. It signifies that in pre-dam phase (prevailing almost natural condition), sinuosity of Damodar River is mainly controlled by the topographic factors of floodplain, viz. fault-guided valley alignment (Singh et al., 1998), cohesive bank materials, less number of bars, coarse bed configuration or neo-tectonic activity, not by solely monsoonal flood discharge and other hydraulic factors. But in post-dam period (especially between 2001 and 2006), the sinuous pattern of Damodar River is achieved from downstream to upstream direction. The most significant fact is that HSI is radically increased in all four reaches (62%–93%) in comparison to TSI (7%–38%). Surprisingly between Idilpur and Chanchai, HSI has contributed 92% influence on sinuosity, and between Chanchai and Paikpara, it is 85%.

Geomorphologically, the straight stretches often occur in conjunction with or between bends or along braided reaches (Knighton, 1998). Even stretches with straight embanked banks have a sinuous thalweg pattern with asymmetrical shoals and point bars alternating along either bank, just like Lower Damodar River. As studies done by Sen (1985, 1991), Garde (2006), Ghosh (2011) and Bhattacharyya (2011), now the infrequent bankfull discharge, variable flow velocity, fluctuating monsoonal flow regime, high width–depth ratio, high hydraulic radius of channel, channel bed roughness, turbulent monsoonal river flow including eddies, coarse sand bed and vegetation growth on bars and banks, low sediment supply due to trap efficiency of upper catchment reservoirs, sand quarrying, elevated concrete embankments and flow diversion through canals are the

---

TABLE 7.5

Temporal Variation of Mueller's Sinuosity Indexes in Damodar River

Years	Bend I (Rhondia to Kashpur)			Bend II (Kashpur to Idilpur)			Bend III (Idilpur to Chanchai)			Bend IV (Chanchai to Paikpara)		
	SSI[a]	HSI[b] (%)	TSI[c] (%)	SSI	HSI (%)	TSI (%)	SSI	HSI (%)	TSI (%)	SSI	HSI (%)	TSI (%)
1922–1943	1.05	48	51	1.08	39	61	1.02	34	66	1.02	13	87
1969–1974	1.17	92	08	1.10	51	49	1.02	12	88	1.04	32	68
1990	1.10	59	41	1.09	60	40	1.02	72	28	1.02	13	87
2001	1.11	65	35	1.15	60	40	1.10	74	26	1.01	04	96
2006	1.18	71	29	1.16	62	38	1.13	93	07	1.15	85	15

[a] Standard Sinuosity Index, [b] Hydraulic Sinuosity Index and [c] Topographic Sinuosity Index.

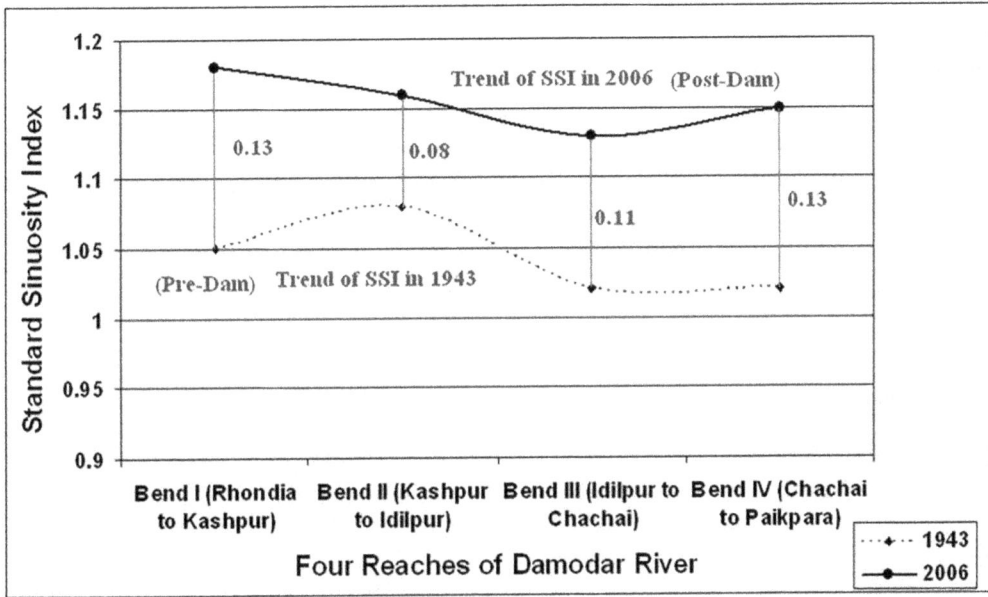

FIGURE 7.5 Significant deviation of SSI (increasing value of sinuosity) in the four segments of Damodar River (Rhondia to Paikpara) between pre-dam and post-dam periods.

dominant hydraulic factors to deviate channel sinuosity in the downstream section (Ghosh and Mistri, 2012).

Brice's Braiding Index (BI) is a measure of the sum of island or bar lengths in a reach, and hence of the increase in bank length that results from braiding (Brice, 1964). The degree of braiding as displayed in the satellite images is well established by braid–channel ratio (BR) which is devised by P.F. Friend and R. Sinha (1993). Landsat TM images of 1990 and 2006 are used to calculate BR and BI in the selected segments, viz. (1) Rhondia to Kashpur, (2) Kashpur to Idilpur and (3) Idilpur to Chanchai. BIs of 1990 and 2006 range from 5.31 to 2.87 and 4.40 to 3.30, respectively (Table 7.6). Similarly, BR of 1990 has the highest value of 2.75 on bend II and the lowest value of 2.04 on bend III. Again BR of 2006 has the highest value of 2.13 on bend I and the lowest value of 1.37 on bend III. As the values of BR do not reach to unity (1), it identifies that three reaches have glimpses of braiding pattern. The decreasing temporal variations of BI and BR are due to water level or stages of active channel area in 1990 and 2006. It appears that braiding of Damodar River is a type of hydraulic adjustment that may be made in a channel possessing a particular or coarse bank material in response to a debris load too large to be carried by a single channel (Garde, 2006; Bhattacharyya, 2011). Braiding and avulsion are interlinked in the alluvial floodplains. It is observed that Damodar River had shifted its course further south (>1 km) near Rhondia and Idilpur, forming mature islands and bars on left side (Figure 7.6).

7.3.5 FLUVIAL AGGRADATION AND DEGRADATION

The palaeoclimatic studies of India indicate that in last hundred years, the periods of aggradation of peninsular rivers are linked to periods of weaker south-west monsoon and reduced sediment supply (Singhvi and Kale, 2009). As the natural flow is regulated by DVC dams, the transportational energy of Damodar River is lost gradually in the immediate downstream of Durgapur Barrage and it also affects the sedimentation profile of bars (Bhattacharyya, 2011). Few intra-bedded brown clayey silt and sandy silt deposits of recent Hooghly Morpho-stratigraphical Unit (HMU) (Acharyya and

TABLE 7.6

Index of Braiding (1990 and 2006) in Selected Bends of Lower Damodar River

	Bend I (Rhondia to Kashpur)		Bend II (Kashpur to Idilpur)		Bend III (Idilpur to Chanchai)	
Year	BI[a]	BR[b]	BI	BR	BI	BR
1990	5.31	2.47	4.72	2.75	2.87	2.04
2006	4.4	2.13	3.91	2.08	3.03	1.37

[a] Braiding Index and [b] braid–channel ratio.

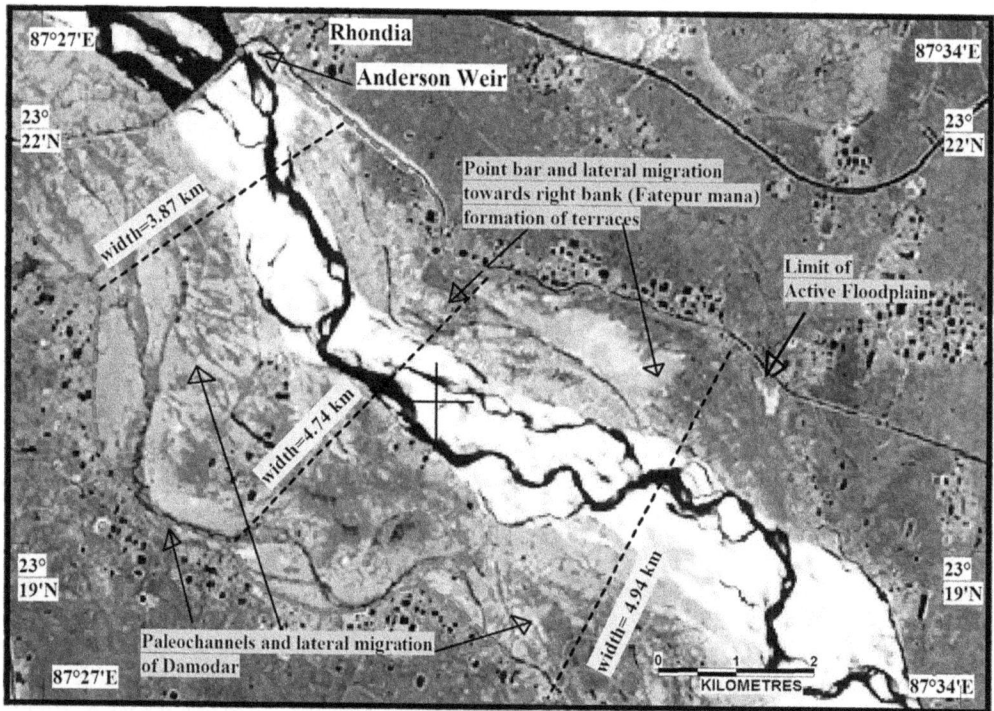

FIGURE 7.6 Glimpses of widely active floodplain (up to 5 km width), spill channels, avulsion, alluvial terraces (now used as arable land) and mature point bars showing the high degree of lateral oscillation and aggradation of Damodar River immediate downstream of Rhondia (FCC of Landsat ETM+, 2005).

Shah, 2007) signify infrequent extreme flood events (high-level discharge), but thick deposit of clay and mud carries the indication of slow deposition (post-dam waning flood deposits) in more sub-humid climate. Disappearance of iron nodules and high proportion of silt in top lithofacies of Fm (Miall, 1985) signifies the younger (post-dam period) fluvial deposits under stagnant water condition (suspended load) in the mature point bars of left bank. The facies of fine sand with clay partings (Sh) and small-size gravel base (Miall, 1985) carry the sign of lateral accretion and avulsion towards right bank of Damodar. It is evident that from Anderson Weir of Rhondia, the river had opened up much wider and it had been maintained its active channel towards far left side in 1922–1943. After dam construction, the river now shifts towards far right side (evident in Landsat images), creating alluvial terraces and point bars on left bank side (Figure 7.6). Near Deulpara, in 2006 and 1973, thalweg shifts far north (left side) compared to the position in 1922–1943. Near Silla, thalweg again shifts towards right bank (meandering channel way) in 2006. For that reason, Damodar River gives birth of many longitudinal bars (locally named as *char*) and islands (locally named as *mana*) in mostly middle-left side, viz. Kasba *Mana*, Fatepur *Mana* and Sadpur *Mana*.

Near Satyanandapur, in 1972, river was more constricted towards its right side, but in 1990 and 2006, channel shifts to north nearer of left-side embankment. After that, it tends towards right-side bank creating elongated Majher Char. The most vulnerable locations of bank erosion are Ghoradanga, Baikuthapur, Somsar, Beldanga, Dadpur (Bankura district) and Lakshmipur village (Barddhaman district) (Figure 7.7). The flow is restricted near Sadarghat due to construction of Krishak Setu (bridge). But after crossing the bridge, it opens up widely and erodes its right bank at Bangachha village. Then the river erodes its left bank near Hatsimul and Belna (Figure 7.8). Within its narrow active space of elevated embankments (from Barsul to Paikpara), the river alters its active

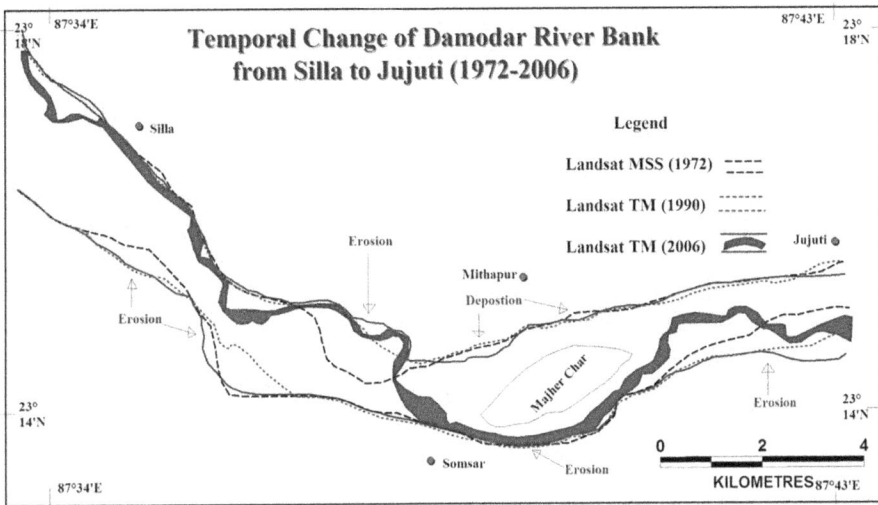

FIGURE 7.7 Temporal change of Damodar River banks from Silla and Jujuti (1972–2006).

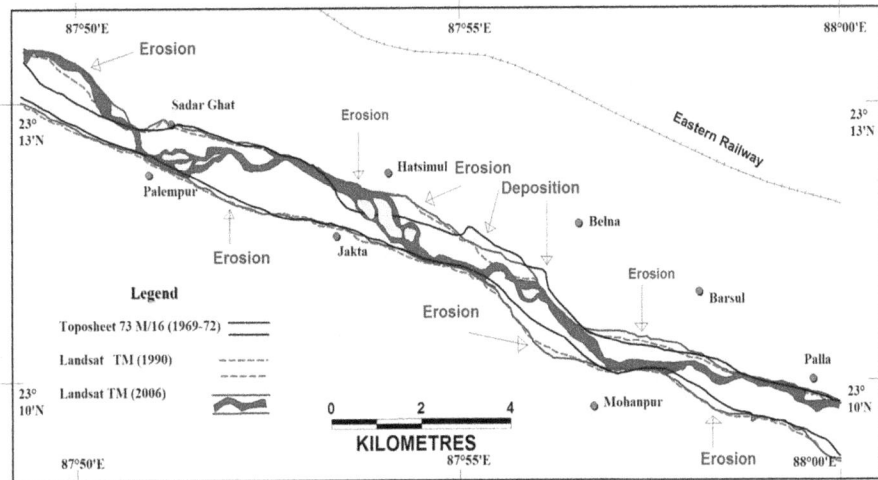

FIGURE 7.8 Temporal change of Damodar River banks from Sadarghat to Palla (1972–2006).

thalweg frequently in a meandering pattern and changes its aggradational and degradational processes in slip-off slope and cut-off slope, respectively. In 44 years (1972–2006), the land lost, due to bank erosion of Damodar, are 0.73 km^2 (Natu), 1.36 km^2 (Idilpur), 1.10 km^2 (Hatsimul), 0.12 km^2 (Jankull), 2.23 km^2 (Dadpur) and 1.93 km^2 (Mohanpur) respectively (Table 7.7). The present study reveals that the river between Rhondia and Barddhaman shows dominance of aggradational landforms, braiding, erosion of right bank and valley widening, but downstream of Barsul and Chanchai the bank erosion, high sinuosity and narrowness are more pronounced (Figure 7.9).

From Chanchai, Damodar turns right side towards permanently southward, and surprisingly, its valley becomes narrower. Near Chanchai, there are signs of deposition, but on opposite side, right bank becomes wider in 2006 than 1972 due to erosion (Figure 7.9). The vulnerable sites of erosion are Berugram, Jamalpur, Jot Shriram, Fatehpur, and Haragobindapur villages and the bifurcated course near Paikpara completely dried up in 2006 due to sedimentation.

Below Silla village, Damodar turns easterly to Belkash, having wide valley. Here river shifts several times developing distributary of Bodai Nadi (meet with Sali Nadi) and few stagnant water courses on right side (near Deulpara, Deora, Tashuli, Panchpara, etc.). Near Satyanandapur, in 1972, river was more constricted towards its right side, but in 1990 and 2006, channel shifts to north nearer to left-side embankment (Figure 7.10 and Table 7.7). After that, it tends towards right-side bank

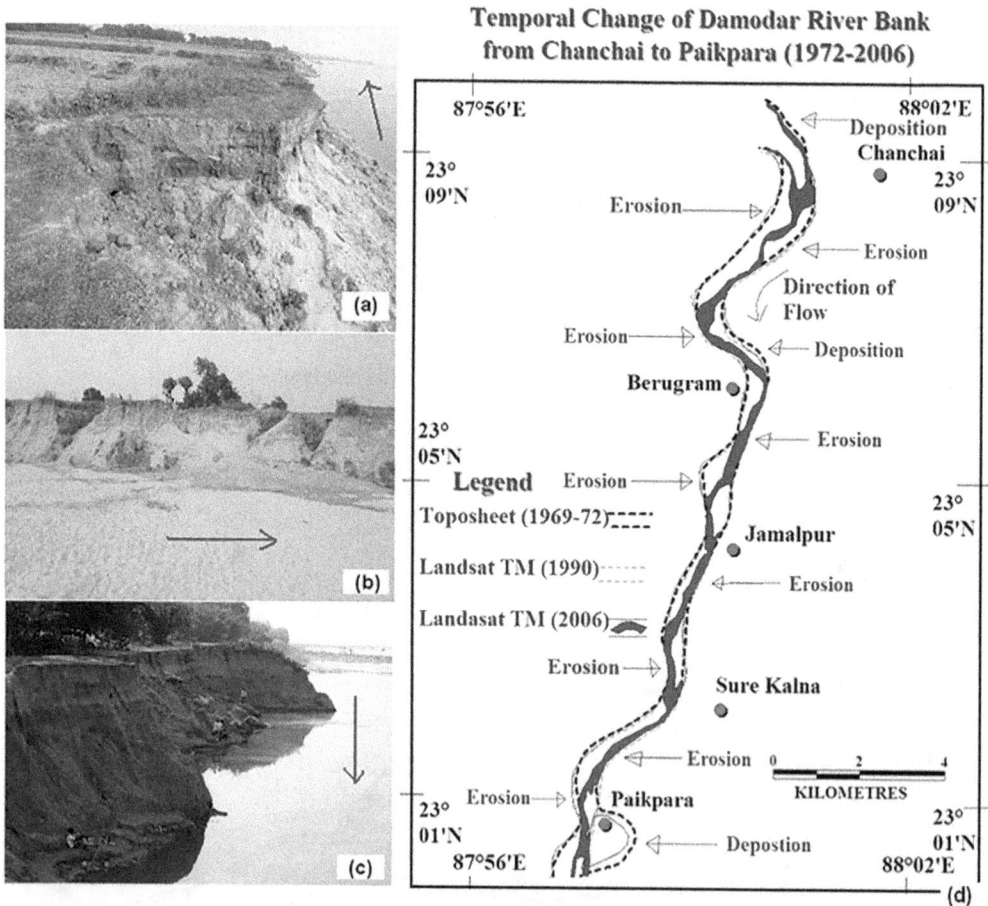

FIGURE 7.9 Glimpses of bank erosion at Idilpur (a), Belkash (2) and Amrun (c). (Note: arrow indicates direction of flow) and (d) temporal change of Damodar River banks from Chanchai to Paikpara (1972–2006).

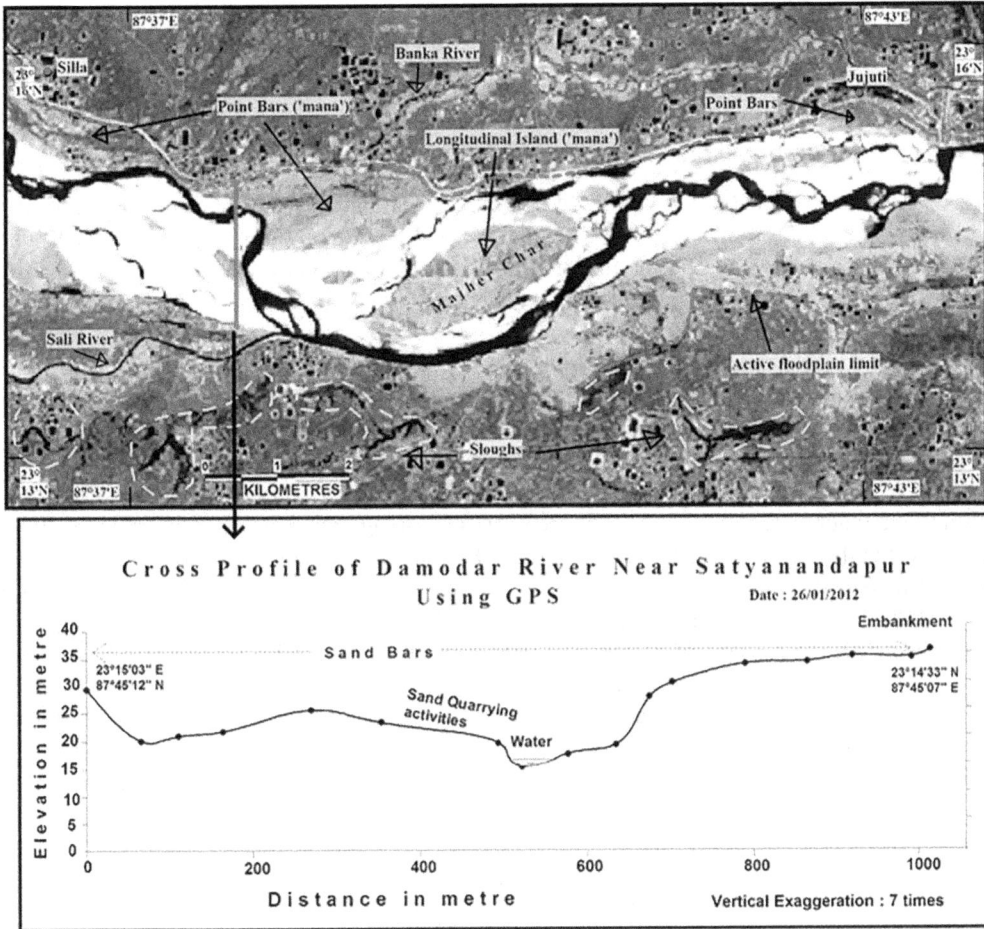

FIGURE 7.10   Development of wide floodplain, sloughs, point bars, islands and braiding showing the degree of lateral migration and aggradation of Damodar River (Landsat ETM+, 2000) and GPS cross-profile showing site of sand quarrying activities and left-bank erosion near Satyanandapur.

TABLE 7.7
Calculated Eroded Area of Banks and Rate of Bank Erosion from 1972 to 2006

	Estimated Eroded Area in km²			
Nearest Village	1972–1990	Average Rate of Erosion (km² yr⁻¹)	1990–2006	Average Rate of Erosion (km² yr⁻¹)
Natu (right bank)	0.5410	0.027	0.1943	0.012
Idilpur (left bank)	0.7106	0.035	0.6467	0.040
Hatsimul (left bank)	0.8104	0.041	0.2942	0.018
Jankull (right bank)	0.0895	0.005	0.0273	0.002
Dadpur (right bank)	1.7110	0.086	0.5758	0.036
Mohanpur (left bank)	1.4170	0.071	0.5142	0.032

creating elongated Majher Char. The most vulnerable locations of bank erosion are Ghoradanga, Baikuthapur, Somsar, Beldanga, Dadpur (Bankura district) and Lakshmipur village (Barddhaman district).

7.3.6 DAM-INDUCED CHANGES IN FLOOD HYDROLOGY

With the introduction of dam building, the hydrological regime and flood behaviour of a river is deliberately altered through control flow regulation. Flood hydrology is concerned with the quantitative relationship between rainfall and runoff (i.e. passage of water on the surface of the Earth) and, in particular, with the magnitude and time variations of flood. The flow of any stream is determined by two entirely different sets of factors, the one depending upon the climate with special reference to the precipitation and the other upon the physical characteristics of the drainage basin. Rivers do not, however, remain at a high stage throughout the monsoon season. It is only after a spell of heavy rains, which may last for a period of several days that large volume of runoff is generated in the catchments and the rivers experience floods annually (Kale, 2003). Monsoon disturbances are the important synoptic systems that cause floods in the peninsular rivers, during the south-west monsoon season (June to September). On average, six depressions form in the Bay of Bengal during the 3 months of July to September (Rao, 2001). These disturbances generally move west–north–westwards along the lower Gangetic plain to Chotanagpur Plateau after their formation at the head of the Bay of Bengal (Pramanik and Rao, 1952; Rao, 2001). The tracks of depressions are following the lower segment to upper catchment of DRB. It is well known that heavy rainfall occurs in the south-western sector of the monsoon depression due to strong convergence in that sector (Rao, 2011). It has been found that the high flash-flood magnitudes of 1950–1957 and 1958–1969 are associated with 118 and 217 numbers of cyclones (which includes cyclonic disturbances—wind speed 17 knots or more, cyclones—34 knots of more and severe cyclones—48 knots or more) of Bay of Bengal, respectively, but low flood magnitude of 1988–1995 is related with only 95 numbers of cyclones (Bhattacharyya, 2011).

7.3.6.1 Understanding Flood Propensity in Monsoon Period

Lower segment of the basin is affected by the floods due to upstream heavy rainfall and huge runoff volume generated in the upper catchment (i.e. western part of Maithon and Panchet dams) which consists of two drainage systems: (1) Damodar drainage system and (2) Barakar drainage system. It is estimated that Damodar catchment receives monsoon rainfall (June to October) of 855.57–1,043.55 mm and Barakar catchment receives monsoon rainfall of 840.54–1,079.81 mm annually (Goyal and Ojha, 2010). At present, significant inter-annual and intra-seasonal variabilities in the observed monsoon rainfall are displayed over this so-called sub-humid region, resulting in recurrent droughts and floods (Lal et al., 2001). It is projected that 10%–15% of rainfall will be increased in area-average monsoon rainfall over the Indian sub-continent, but the date of onset of summer monsoon over central India could become more variable in future (Lal et al., 2001). More intense rainfall spells are also projected in a warmer atmosphere, increasing the probability of extreme rainfall events and flash floods in the sun–mountainous or plateau scarp region (Singh et al., 1974; Lal et al., 2001).

The main hydro-climatic factors which influence runoff and flood of Damodar can be categorised as (Wisler and Brater, 1959): (1) type of rainfall; (2) rainfall intensity; (3) duration of rainfall on basin (intense short period or prolong duration); (4) direction of storm, depression and cyclone movement; and (5) antecedent rainfall and soil moisture. Heavy incessant and prolonged rainfall for long period is the basic cause of floods because enormous amount of water gets collected on the surface flowing as runoff. Higher magnitude of rainfall coupled with a larger catchment area, like Damodar Basin (23,370.98 km^2), leads to a greater volume of runoff. We emphasise the following influences of monsoonal rainfall and catchment characteristics on flood risk of lower Damodar Basin:

1. More or less same monsoonal rainfall pattern is observed both in the districts of upper catchment and lower catchment (Table 7.8). Within the Damodar Command Area, the upper and middle parts of the basin receive 1,209 mm rainfall annually and lower parts of the basin receive 1,329 mm (Chandra, 2003). Therefore, large fan-shaped contributing area of Damodar annually receives huge rainfall and due to low-infiltration capacity and high-runoff coefficient of rocky-deforested upper catchment, 70%–80% of rainfall converted into surface runoff. Due to high drainage density of upper catchment, huge volume of runoff appears as excessive streamflow, which finally contributes to the main Damodar River, below Panchet and Maithon Dams.

2. Rains of long duration and direction of monsoonal depressions (generally south to northwest from Gangetic West Bengal) severally aggravate huge volume of inflow water into the reservoirs of DVC. During late monsoon, the reservoirs including the rivers remain bankfull and groundwater of alluvial West Bengal is recharged to the fullest extent. A sudden cloudburst or strong depressional rainfall at this juncture frequently causes a disastrous flood (Rudra, 2002). The period from September to early October is now considered as the cruellest one for Lower Damodar River, because all the major floods of 1956, 1958, 1959, 1978, 1995 and 2000 appeared in this session (Table 7.9).

3. The eastern parts of Jharkhand and Gangetic West Bengal are recommended as one of major rainstorm zones and flood-prone areas of India (Kale, 2003, 2005). It has been found

TABLE.7.8

Sub-Divisional Monsoonal Rainfall Covering the Damodar River Basin

Year	Rainfall (mm) in Jharkhand Sub-Division	Rainfall (mm) in Gangetic West Bengal Sub-Division
2005	1,105.0	1,136.0
2006	1,092.0	1,126.0
2007	1,093.0	1,127.0
2008	1,092.5	1,291.7
2009	1,002.6	971.5
2010	1,084.5	1,140.6
2011	1,101.5	1,394.7
1951–2000 (mean normal rainfall)	1,091.9	1,167.9

Source: Ghosh and Guchhiat (2016).

TABLE.7.9

Rainfall Recorded during Monsoonal Major Floods in the Damodar Valley (DV) and Barddhaman District (BD)

Year	Month	Date	Total Storm Rainfall (mm)
1913 (DV)	August	5–11	314
1935 (DV)	August	10–15	288
1958 (DV)	September	14–16	166
1959 (DV)	September–Oct	30 September–7 October	231
1978 (DV)	September	26–29	184
1995 (BD)	September	26–28	369
2000 (BD)	September	18–21	800

Source: Ghosh and Guchhiat (2014).

Annual Normal Rainfall Pattern of Damodar Plain
Barddhaman District (1901-50)

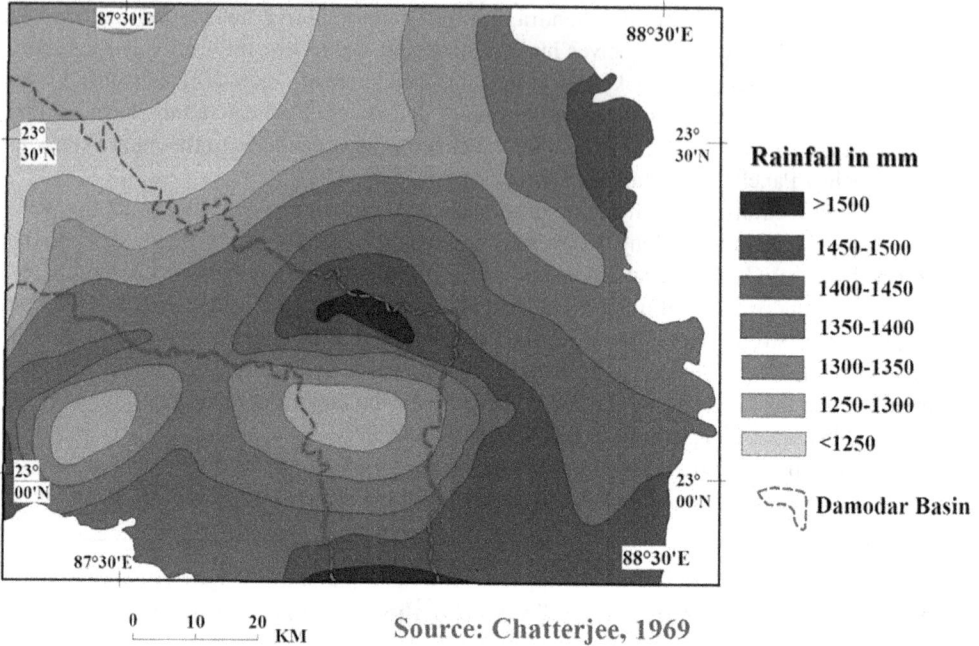

FIGURE 7.11   Spatial distribution of mean annual rainfall in lower segment of DRB (Ghosh and Guchhait, 2016).

that on an average of 25 years, the average monsoonal rainfall in 6 hours ranges between 14 and 16 cm in the parts of eastern Jharkhand and western West Bengal (Pramanik and Rao, 1952; Roy et al., 1995).

4. When Jharkhand division (73.70% share in the basin) experiences intense rainfall of wide coverage, at that time, Gangetic West Bengal (26.30% share in the Basin) is already oversaturated by continuing monsoonal depressions and lower catchment receives more than 1,400 mm rainfall (Figure 7.11). Then uncontrolled inflow runoff of upper catchment (released at last by Durgapur Barrage) integrates with further additional rainfall of Gangetic West Bengal and enormous volume of water passes through congested, silted and oversaturated Lower Damodar River which is further narrowing down next to downstream of Barddhaman town. So, the overflow of river and additional runoff of lower catchment turns the inundated situation into misery.

7.3.6.2   Estimating Surface Runoff

The above analysis (Table 7.10 and Figure 7.12) provides us only a generalised idea about potentiality of runoff to be generated in the periods of 1901–1950 and 2006–2010. We have now established a linear regressional equation dealing with 17 years of estimated actual rainfall and runoff data (1934–1950) at Rhondia and a statistical significant test is employed to make that equation reliable and applicable to Damodar Basin. Here we have assumed that the calculated runoff is generated within the basin area depending on the land-use–land-cover conditions, soil moisture, channel storage, rainfall intensity, evapotranspiration and geological character of surface. So this rainfall–runoff relation reflects the hydrogeomorphic characteristics of Damodar Basin and flood risk of downstream region in particular. Again establishing this significant regression equation, we can predict potential runoff depth of uncalculated years which will forecast the flood flow at Rhondia.

TABLE 7.10

Establishing Annual Rainfall–Runoff Relation for Damodar Basin at Rhondia (1934–1950)

Year	Mean Annual Rainfall (mm)	Mean Annual Runoff (mm)	Calculations
1934	1,088	274	$a = 530.21$
1935	1,113	320	$b = 0.819$
1936	1,512	543	Runoff $(Y) = 0.819$, rainfall $(X) = 530.21$
1937	1,343	437	Coefficient of determination $(R^2) = 0.63$
1938	1,103	352	
1939	1,490	617	Standard deviation of $Y = 158$ mm
1940	1,100	328	Standard error of estimate of $Y = 96$ mm
1941	1,433	582	
1942	1,475	763	Calculated t value of $b = 3.25$
1943	1,380	558	Tabulated t value (0.01% confidence) = 2.95
			Degree of freedom (df) = $(n - 2) = 15$
1944	1,178	492	
1945	1,223	478	Product moment correlation coefficient (r) = 0.794
1946	1,440	783	Calculated t value of r = 4.01
1947	1,165	551	Tabulated t value (0.01% confidence) = 2.95
1948	1,271	565	
1949	1,443	720	Null hypothesis is rejected and the equation is very much significant
1950	1,340	730	

Data Source: Ghosh and Guchhiat (2016).

Observing the above calculations and two graphs, we have been able to establish a meaningful rainfall–runoff relation of Damodar Basin at Rhondia. From 't-test' of b value and r value, we have obtained that both slopes of positive increasing trend line and correlation between rainfall and runoff are very much significant, rejecting the null hypothesis. Also the distribution is positively skewed (0.63). With fluctuation of rainfall, runoff also fluctuated, but from 1934 to 1950, runoff has an increasing positive annual trend ($Y = 0.8194\,X - 530.21$). On an average, estimated runoff is 548.89 mm, having standard deviation of 153.67 mm. From the trend of 1934–1950, we have found that average annual runoff–rainfall ratio is 0.41 (0.25–0.54) which means that 41% of annual rainfall is converted into runoff in this area depending on different geographical conditions. It is a real fact that the largest part of runoff is generated at the time of peak monsoon (June to September) alone (when excess rainfall becomes direct runoff to the streams after fulfilment of surface storage) (Figure 7.13).

7.3.6.3 Test of Homogeneity and Stationarity in Flood Frequency Analysis

In this test, two samples of size p and q with p<q are compared. The combined data set of size N = p+q is ranked in increasing order. The Mann–Whitney (M–W) test considers the quantities V and W in the following equation (Rao and Hamed, 2000):

$$V = R - p(p+1)/2$$

$$W = pq - V$$

R is the sum of the ranks of the elements of the first sample (size p) in the combined series (N). V represents the number of times an item in sample 1 follows an item in sample 2 in the ranking, and

(a)

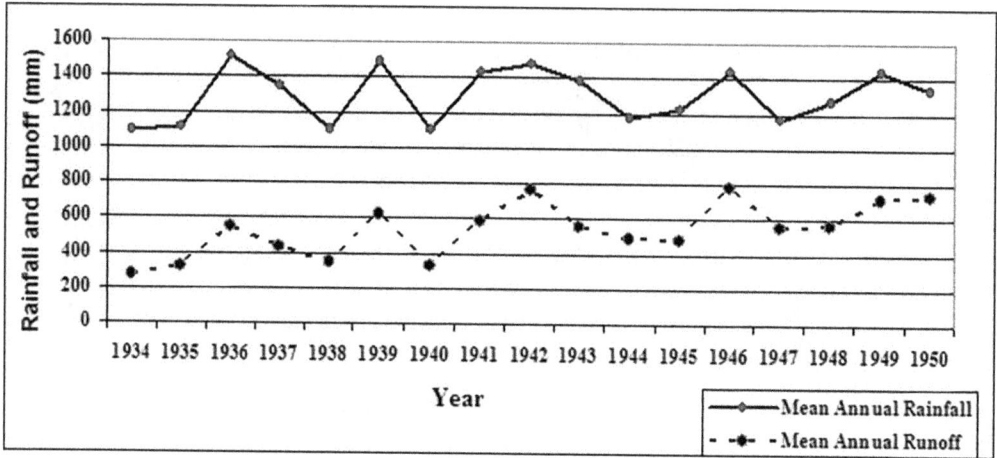

(b)

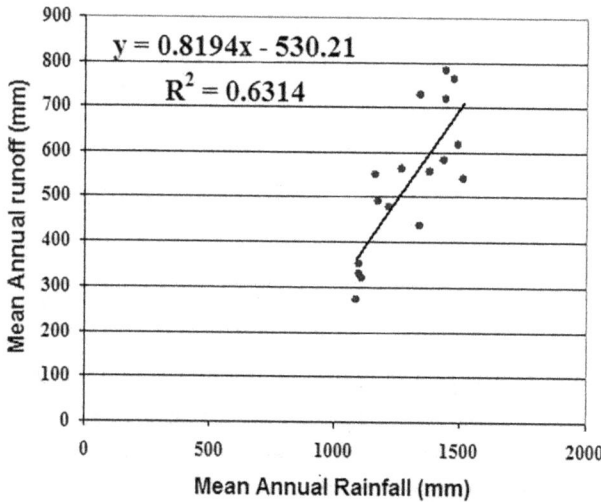

FIGURE 7.12 (a) Year-wise mean rainfall and runoff pattern at Rhondia from 1934 to 1950 and (b) increasing regressional trend line between rainfall and runoff of Damodar Basin.

similarly, W can be computed for sample 2 following sample 1 (Rao and Hamed, 2000). The M–W statistics U is defined by the smallest value of V and W, and U is approximately normally distributed with mean $U_{mean} = pq/2$ and variance var (U) (Rao and Hamed, 2000). The statistics u is used to test the hypothesis of homogeneity at significance level α by comparing it with the standard normal variate for that significance level.

$$u = (U - U_{mean})/[var(U)]^{1/2}$$

It this case, we have subdivided the annual peak discharge data of pre-dam (1934–1957) and post-dam (1958–2007) periods into two parts and then ranked them in increasing order to get the values of R, V and W in both periods (Table 7.11). As the dam construction radically influences the river discharge, so it is obvious that two series do not come from the same distribution with same mean. So, we have applied test of homogeneity and stationarity individually in the database of pre-dam and post-dam periods.

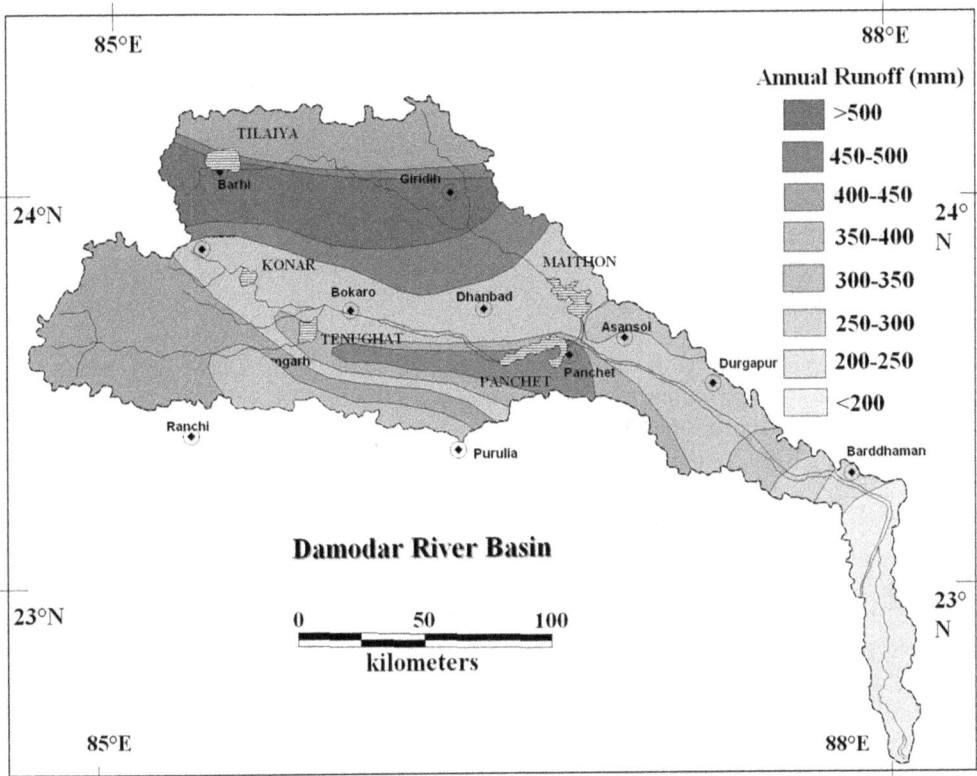

FIGURE 7.13 Average annual runoff pattern in the DRB.

TABLE 7.11

Test of Homogeneity and Stationarity for Annual Peak Discharge Data (Pre-Dam and Post-Dam Periods) of Damodar River

	U	U_{mean}	R	V	W	var (U)	u
Case 1. Pre-dam period (p = 13 and q = 11)	56	71.5	178	87	56	298	−0.89
Case 2. Post-dam period (p = 26 and q = 24)	286	312	637	286	338	265	−1.59

The pre-dam test value u = −0.89 is less than the critical value at 5% significance level $u_{0.025}$ = 1.96. The post-dam test value u = −1.59 is less than the critical value $u_{0.025}$ = 1.96. So in both cases, the data can be considered to be homogeneous and stationary at 5% level of significance. The series of pre-dam and post-dam should be treated separately to draw significant conclusion.

7.3.6.4 Pre-Dam and Post-Dam Changes in Annual Peak Discharge

The overall average monsoonal discharge (1934–2007) is 2,592 $m^3 s^{-1}$, and it shows numerous deviations throughout the 20th century. In the pre-dam period, the river had behaved more normally in peak monsoonal season, having a total monsoonal average discharge of 3,342 $m^3 s^{-1}$ (combining average discharge of June, July, August and September). But after dam constructions of DVC that falls to 2,191 $m^3 s^{-1}$ due to flood discharge regulation and storage in monsoon. To measure the degree of symmetry of discharge distribution, we have calculated skewness value. In both periods, the skewness is positive and the skewness value of post-dam period (1.21) is more or less same with the skewness of pre-dam period (1.24). E. L. Glass's Report on 1918 submitted to the Government of Bengal assessing the flood condition during the 61-year period from 1857 to 1917 at Raniganj (upstream of Rhondia) estimated the occurrence of extremely abnormal flood discharge in 1865 (12,742 m^{3}s^{-1}), 1866 (11,894 m^{3}s^{-1}), 1877 (14,158 m^{3}s^{-1}), 1913 (18,406 m^{3}s^{-1}) and 1916 (11,185 m^{3}s^{-1}) (Bhattacharyya, 2011). We have employed the 'confidence limits' to the distribution to get the exact value for the discharge mean confidence limits which give us a close interval within which the discharge mean falls with 99% and 95% probabilities (Table 7.12 and Figure 7.14).

It is estimated that any peak discharge above 7,075 m^{3}s^{-1} (250,000 cusec) at Rhondia has maximum chance of furious floods at downstream section (Sen, 1991). Pre-dam and post-dam mean annual peak discharges show a sharp contrast, because up to 1957 mean annual peak discharge is 8,378 m^{3}s^{-1}, but it comes down to 3,522 m^{3}s^{-1} (Figure 7.15). The severity of floods was more common in pre-dam phase, because confidence limit of 99% probability of occurrence ranges from 6,081 to 10,676 m^{3}s^{-1}. Due to flow regulation and storage, it comes down to the aforesaid limit of 2,574–4,470 m^{3}s^{-1}. In spite of that, Lower Damodar Valley is not safe from inundation because it is estimated from SRTM 90 m digital elevation data (2006) that now carrying capacity or threshold level of peak flow varies from only 4,011 to 1,413 m^{3}s^{-1} (from Rhondia to Paikapara). Regulated flow, loss of river energy, checking lateral expansion by embankments and excessive siltation of bed are the prime factors of declining carrying capacity of Lower Damodar River.

Between 1934 and 1957, we have noticed that at Rhondia extremely abnormal high floods above 7,075 m^{3}s^{-1} had occurred in 1935, 1938, 1939, 1940, 1941, 1942, 1943, 1946, 1947, 1949, 1950, 1951, 1953, 1954 and 1956 (Rudra, 2009; Bhattacharyya, 2011). Almost with 62.5% of frequency, lower Damodar Basin has experienced moderate to high flood phenomenon from 1934. The exceptional peak discharge of 18,112 m^{3}s^{-1} was measured on 12 July 1935 at Rhondia. So from the graph (Figure 7.15) we can easily understand how floods were more common to Lower Damodar River and why the river is called *Sorrow of Bengal*. In post-dam phase 4 years with long-time interval had witnessed extreme flood discharge of 8,792 m^{3}s^{-1} (02 October 1959), 10,919 m^{3}s^{-1} (27 September 1978), 7,305 m^{3}s^{-1} (24 September 2006) and 8,883 m^{3}s^{-1} (25 September 2007), respectively (Rudra, 2009; Bhattacharyya, 2011). Since peak discharge above 5,000 m^{3}s^{-1} had occurred in 1973, 1976, 1993 and 2000, Barddhaman, Hooghly and Howrah districts of West Bengal had experienced havoc floods in those years.

TABLE 7.12

Important Statistics of Peak Flow and Flood Discharge in Damodar River at Rhondia

Extreme Flood Flow (m^3 s^{-1}) at Rhondia	Date		Pre-Dam (1934–1957) Annual Peak Flow (m^3 s^{-1})	Post-Dam (1958–2007) Annual Peak Flow (m^3 s^{-1})
18,112	12 July 1935	Mean	8,378	3,522
12,002	06 September 1938	SD	3,752	2,235
17,942	10 October 1941	CV	0.45	0.63
10,811	10 August 1942	Skewness	1.24	1.21
11,012	11 September 1957	Confidence limit: 99%	6,081–10,676	2,574–4,470
10,919	27 September 1978	Confidence limit: 95%	6,846–9,910	2,890–4,160
		C_S	1.24	1.21
		C_f	6.52	1.16

Data Source: Ghosh and Guchhiat (2014).

Note: SD, standard deviation; CV, coefficient of variation; C_S, coefficient of skew and C_f, coefficient of flood.

(a)

(b)

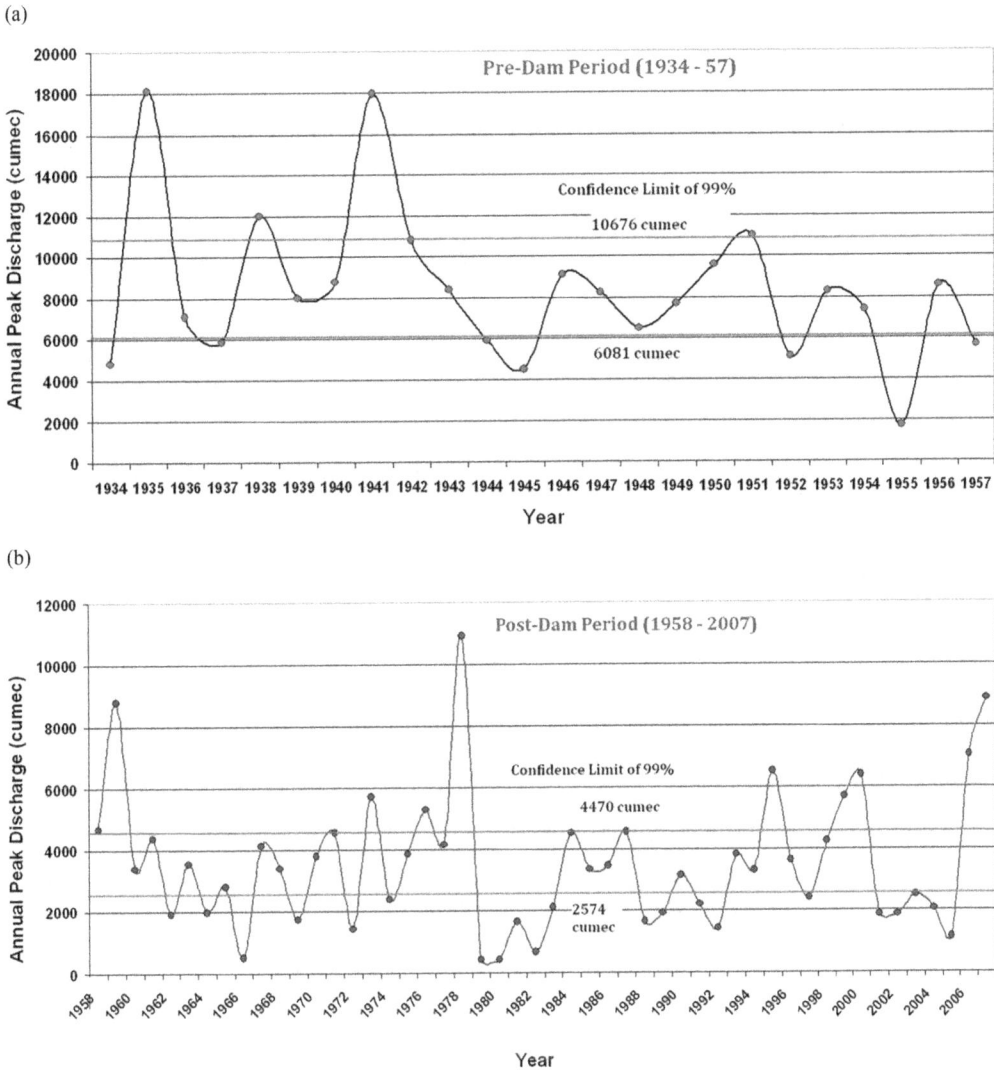

FIGURE 7.14 (a) Pre-dam (1934–1957) and (b) post-dam (1958–2007) variability and comparison of annual peak discharge at Rhondia and confidence limits of discharge to be occurred.

We have prepared the hydrographs (Figure 7.15) based on the year-wise and month-wise stream-flow data (1934–2007), and it is clear that Damodar River experiences a 'Tropical Single Maximum with a Long Low Water' (A_W) regime or hydrological region at Rhondia (Beckinsale, 1969). In the pre-dam period, the peak discharges are observed in the month of August (on an average 1,238 $m^3 s^{-1}$), but after construction of dams, the peak shifts from August to September (Figure 7.15) (on an average 1,247 $m^3 s^{-1}$). The important thing is that the main reason of that shift is the installation of flood storage system of DVC. The dams temporarily store inflow runoff and streamflow till late August, but due to continuation of heavy rainfall and critical reservoir storage limit, the dams are compelled to release water in September when the soils of West Bengal gain full moisture and excess water adds with this streamflow. That's why now the flood probability or chances are more common between September and October. By comparing average monthly discharge of pre-dam period with hydrographs of three flood years (1943, 1950 and 1953), we can understand that from

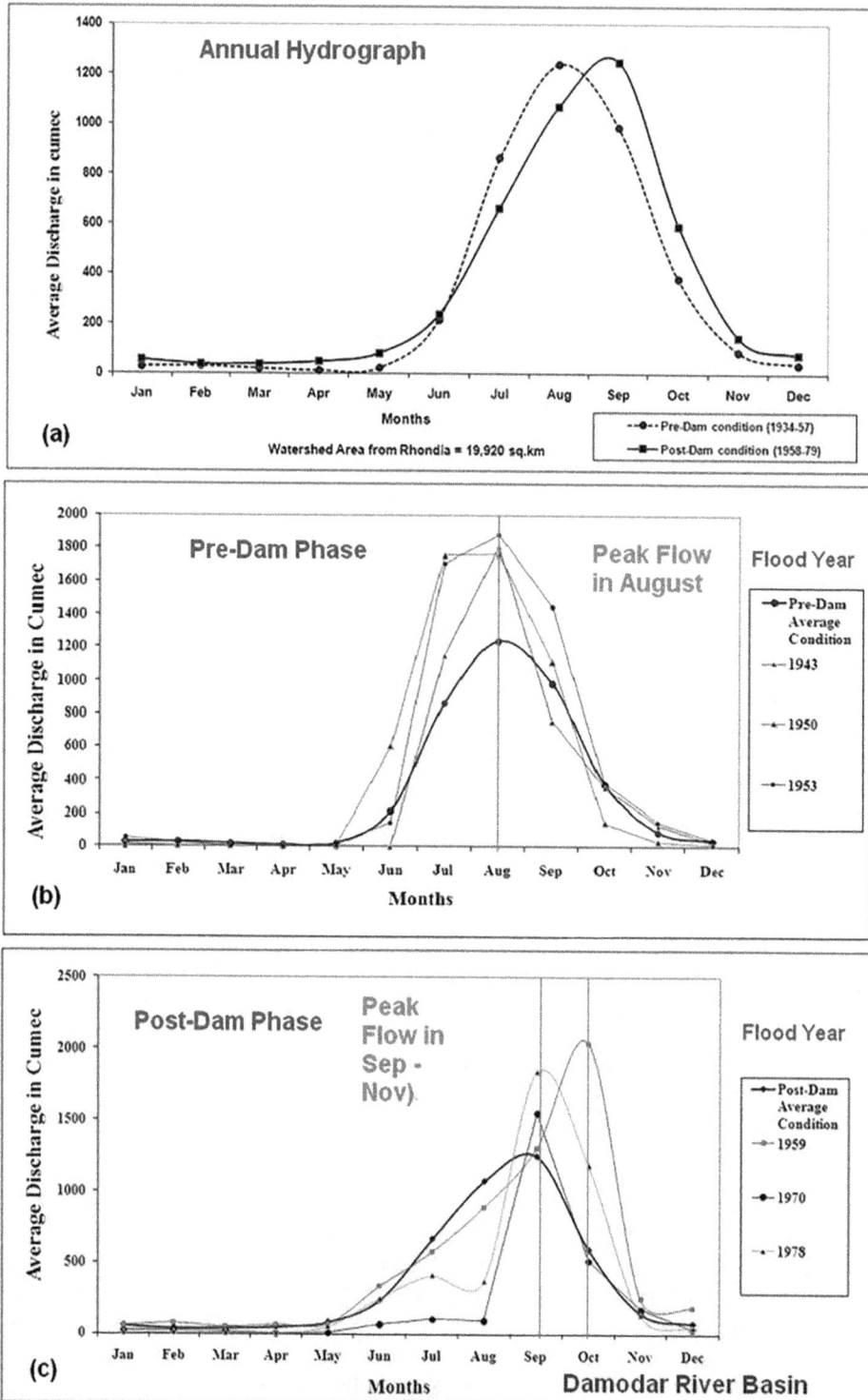

FIGURE 7.15 (a) Pre-dam and post-dam variation in annual flood hydrographs; (b) pre-dam condition of annual average discharges in flood years of 1943, 1950 and 1953 (one annual maxima in August) and (c) post-dam annual average discharges in flood years of 1959, 1970 and 1978 (two annual maxima in September and October).

July to August, Damodar River had high amount discharge at Rhondia. But after the installation of dams and Durgapur Barrage, the scenario is totally changed in two aspects:

1. The annual post-dam hydrograph and hydrographs of the flood years of 1959, 1970 and 1978 are characterised by high amount of peak discharge between September and October. The discharge is influenced by lag time and uncontrolled outflow from the dams and barrage
2. The deviations from average to peak values (>1,800 m^{3}s^{-1}) of three flood years are sharply high from the observation of pre-dam hydrographs. The late monsoonal peak values suggest that the regulation of streamflow is not controlled totally by those constructions

There is marked difference of annual peak flow between pre-dam (1933–1956) and post-dam periods (1958–2007). A peak flow of above 18,000 m^{3}s^{-1} has been recorded three times in August 1913, September 1935 and October 1941 (Saha, 1979; Bhattacharyya, 2011). In post-dam period, the highest combined inflow at Maithon and Panchet dams was recorded as 21,070 m^{3}s^{-1} on 27 September 1978. At present, extremely high-magnitude floods have disappeared, but the frequency of low-magnitude floods still remains high in lower Damodar Basin (Figure 7.16).

7.3.6.5 Flash Flood Magnitude Index

In old records, Damodar River has always been referred to as a river of sorrow. W. W. W. Hunter in 1876 writes that, during floods, the rainwater used to pour off the hills through hundreds of channels with such suddenness that water heaped up to form dangerous head waves known as *harkaban* (flash flood) (Lahiri-Dutt, 2006; Bhattacharyya, 2011). The absolute flood discharge of a river is not as important with regard to geomorphic change as the ratio between peak discharge and mean annual discharge. Floods likely to result in significant geomorphic change are those that produce discharges many times above that normally experienced by the river, i.e. those with high maximum peak discharge to mean annual discharge (Kochel, 1988). Beard (1975), Kochel (1988) and Kale (2003 and 2006) have applied a parameter called the Flash-Flood Magnitude Index (FFMI) which is employed here for annual peak discharge analysis of lower Damodar Basin to understand the change in pre-dam and post-dam periods at Rhondia (Table 7.13).

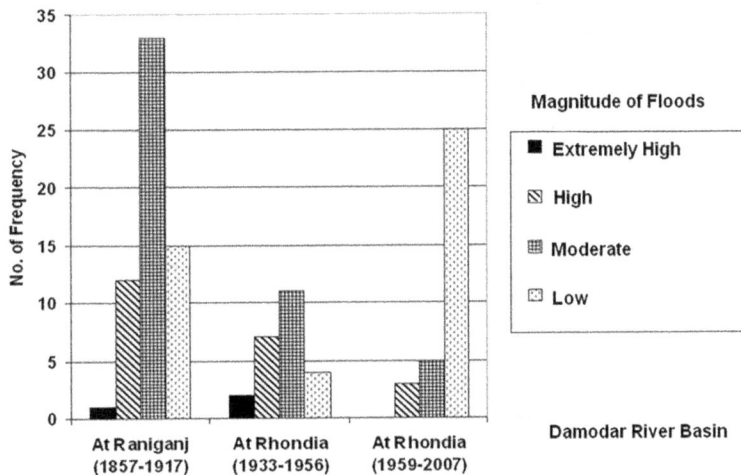

FIGURE 7.16 Three-phase variation of different magnitude of floods in Damodar River at Raniganj and Rhondia to understand the trend of occurrence (1857–2007)—extremely high-magnitude floods (above 12,744 m^{3}s^{-1}), high-magnitude floods (8,496–12,744 m^{3}s^{-1}), moderate-magnitude floods (5,664–8,496 m^{3}s^{-1}) and low-magnitude floods (below 5,664 m^{3}s^{-1}). (Bhattacharyya (2011).)

TABLE 7.13
Calculation of FFMI for Damodar River at Rhondia (1934–2007)

Year	X_m	Date of X_m	M	X	FFMI
1934–1941	4.258	12/07/1935	2.422	1.836	0.482
1942–1949	4.254	10/08/1942	2.577	1.677	0.402
1950–1957	4.042	11/09/1951	1.321	2.721	1.057
1958–1969	3.944	02/10/1959	1.164	2.780	1.104
1970–1977	3.724	13/10/1973	2.392	1.332	0.253
1978–1987	4.038	27/09/1978	2.190	1.849	0.488
1988–1995	2.549	29/09/1995	2.278	0.272	0.011
1996–2003	3.805	23/09/2000	2.370	1.435	0.294
2004–2007	3.949	25/09/2007	2.300	1.649	0.907

The FFMI is calculated from the standard deviation of the logarithms of annual maximum discharge as illustrated (Kochel, 1988).

$$FFMI = X^2 / (N-1)$$

where $X = X_m - M$
     X_m = annual maximum discharge
     M = mean annual discharge
     N = number of years of record
     X, X_m and M = logarithms.

It is a good measure of the variability of flood frequency measured as an index of flood flashiness. The index nearer to 1.0 or above indicates the high degree of flash flood with some geomorphic change within the basin (Kochel, 1988). Log deviation from mean discharge to extreme peak value is exponentially ($Y_c = 0.1525\ x^{2.04}$) related to FFMI. It means the episodes of high degree of deviation between X_m and X are associated with high FFMI and extreme floods in Damodar River, having coefficient of determination of 0.96 (Figure 7.17). FFMI gained maximum value between 1950 and 1957 (pre-dam phase) and consecutively between 1958 and 1969 (post-dam phase). Then 33 years (1970–2003) of record shows that FFMI was significantly lower than the pervious; that's why in these years no extreme floods (except 1978) had occurred. But the last one (2004–2007) is approaching towards FFMI of 1 which signifies forthcoming flood episodes following the years of 2006 (7,035 m³s⁻¹) and 2007 (8,853 m³s⁻¹). In general, streams with extremely high FFMI occur in semi-arid region, but in this monsoonal climate, we have observed high degree of flashiness. The fluctuation of FFMI (1934–2007) is associated with variability of southwest monsoon and strong tropical depressions in lower Gangetic plains. It is important to note that the floods of peak flow of 8,496 m³s⁻¹ or more had occurred 37 times between 1823 and 2007 (Bhattacharyya, 2011). Average FFMI of Damodar River is getting value of 0.56 which is superior to the world average of 0.278 and other Indian Rivers, viz. Narmada, Ganga, Godavari, Brahmaputra, Teesta and Tapi (Kale, 2003).

7.3.6.6   Flood Frequency Using Gumbel's and log Pearson Type III Distribution

In pre-dam period (1934–1957), the peak discharge of 18,112 m³s⁻¹ (1935) has attained a return period of 50 years with 2% probability of occurrence (Table 7.2 and Figure 7.3). But the peak discharges of 12,002 m³s⁻¹ (1938), 11,012 m³s⁻¹ (1951) and 10,811 m³s⁻¹ (1942) have attained only 6.7 years (15%), 4.9 years (20.4%) and 4.6 years (22%) return periods, respectively. This signifies the

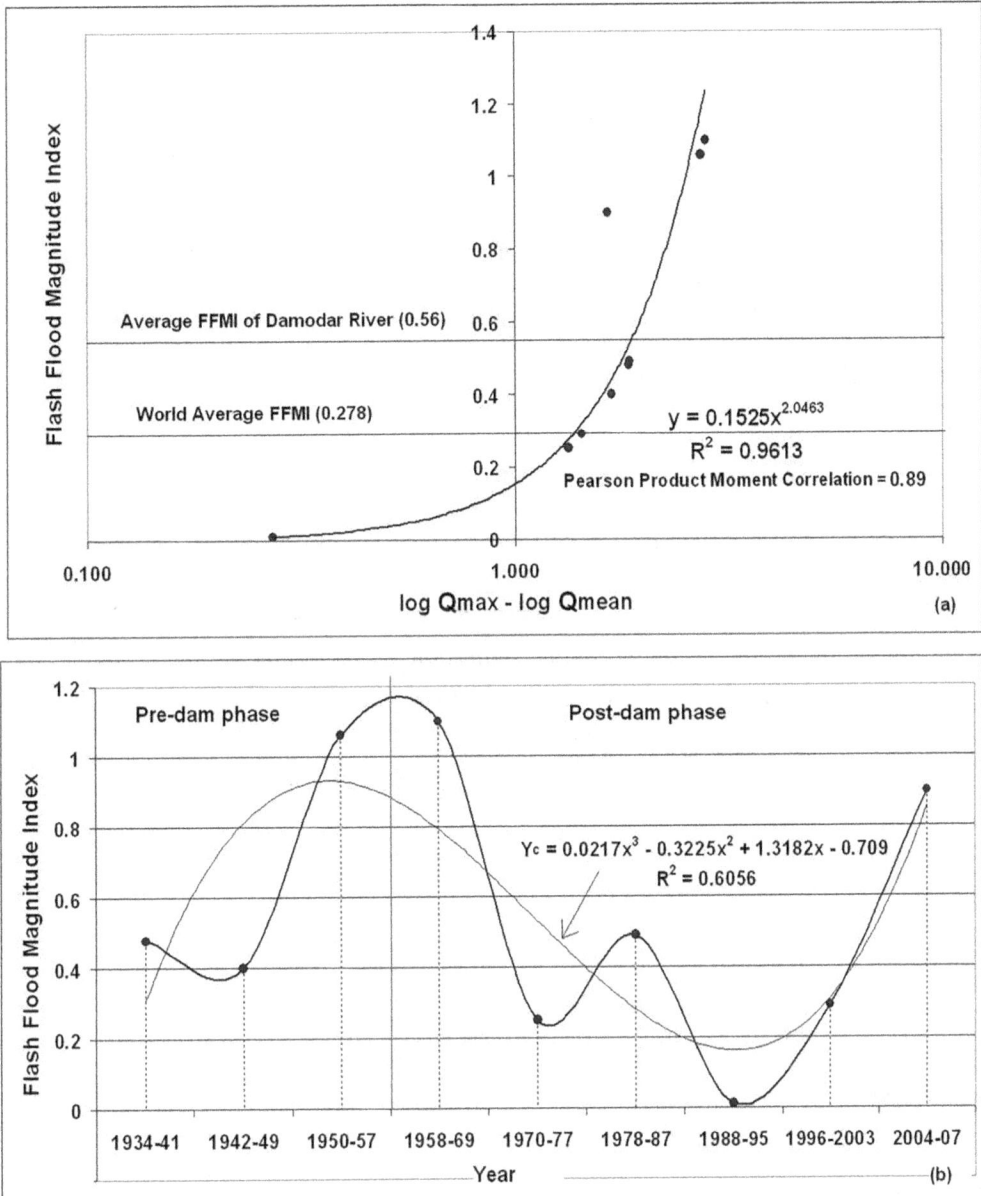

FIGURE 7.17 (a) Deviation from mean discharge to extreme peak value is exponentially related to FFMI, i.e. high degree of log deviation is ultimately transformed into high magnitude of flash floods in Lower Damodar River; and (b) temporal pattern of FFMI showing dominance of flash floods in three episodes, viz., 1950–1957, 1958–1969 and 2004–2007 and successive three calm periods up to 2003.

quick occurrence of high magnitude of flood in the lower valley between 1934 and 1955. It is estimated that any discharge above 7,075 m³ s⁻¹ (Standard Project Flood) at Rhondia creates flood flow of moderate–high magnitude. So on this basis, we have found the ferocity and high frequency of flood discharge in uncontrolled Damodar River up to 1957. Surprisingly, the return period of 7,075–9,561 m³ s⁻¹ discharge is only 1.7–3.2 years (59%–31% probability). In this unsymmetrical distribution, the general slope of the fitted curve is given by coefficient of variation ($C_V = \sigma / X_{mean}$)

which is 0.45 for pre-dam period (nearly 45° of steep slope). It indicates the short return period of high magnitude of annual maximum discharge. The departure from the straight line is given by the coefficient of skew

$$C_S = \sum (X - X_{mean})^3 / (n-1)\sigma^3$$

It is 1.24 (positively skewed distribution) and it indicates a more or less same range in the magnitude of floods. The coefficient of floods [$C_f = X_{mean}/(A^{0.8}/2.14)$] indicates the general magnitude of the floods in a particular river (A = 19,920 km² at Rhondia gauge station); hence, it fixes the height of the curve above the base. It is estimated as 6.52 for pre-dam period. From this distribution, we have found that the mean discharge of that period is 8,378 m³s⁻¹ (greater than 7,075 m³s⁻¹) with standard deviation of 2,235 m³s⁻¹ (Table 7.14 and Figure 7.18).

Implementation of Damodar River Valley Project has radically altered the peak discharge value (Table 7.15 and Figure 7.19) through controlled regulation of water from last two terminal big dams (Panchet on Damodar River and Maithon on Barakar River) and Durgapur Barrage. From

TABLE 7.14
Flood Frequency Analysis of Damodar River at Rhondia (1934–1957)

Rank	Year of Peak Flow	Peak Flow (m³ s⁻¹)	X–Mean	Gumbel's Distribution Return Period (years)	Exceedence Probability (%)	Log Pearson Type III Return Period (years)	Exceedence Probability (%)	Remarks	
1	1935	18,112	9,734	50	2	25.00	4	n = 24	
2	1941	17,942	9,564	47	2.13	12.50	8	Mean	8,378
3	1938	12,002	3,624	6.7	14.9	8.33	12	SD	3,752
4	1951	11,012	2,634	4.9	20.4	6.25	16	CV	0.44784
5	1942	10,811	2,433	4.6	21.7	5.00	20	C_S	1.24
6	1950	9,561	1,183	3.2	31.3	4.17	24	C_f	6.52
7	1946	9,133	755	2.9	34.5	3.57	28		
8	1940	8,773	395	2.6	38.5	3.13	32		
9	1956	8,579	201	2.5	40	2.78	36		
10	1943	8,384	6	2.3	43.5	2.50	40		
11	1953	8,287	−91	2.3	44.4	2.27	44		
12	1947	8,235	−143	2.2	45.5	2.08	48		
13	1939	7,989	−389	2.1	47.6	1.92	52		
14	1949	7,696	−682	2	50	1.79	56		
15	1954	7,407	−971	1.9	54.1	1.67	60		
16	1936	7,075	−1,303	1.7	58.8	1.56	64		
17	1948	6,500	−1,878	1.5	66.7	1.47	68		
18	1944	5,905	−2,473	1.4	71.4	1.39	72		
19	1937	5,876	−2,502	1.4	74.1	1.32	76		
20	1957	5,658	−2,720	1.3	76.9	1.25	80		
21	1952	5,120	−3,258	1.2	83.3	1.19	84		
22	1934	4,793	−3,585	1.2	87	1.14	88		
23	1945	4,514	−3,864	1.1	90.9	1.09	92		
24	1955	1,714	−6,664	1	100	1.04	96		

Source: Bhattacharyya (2011).
Note: SD, standard deviation; CV, coefficient of variation; C_S = coefficient of skew and C_f, coefficient of flood.

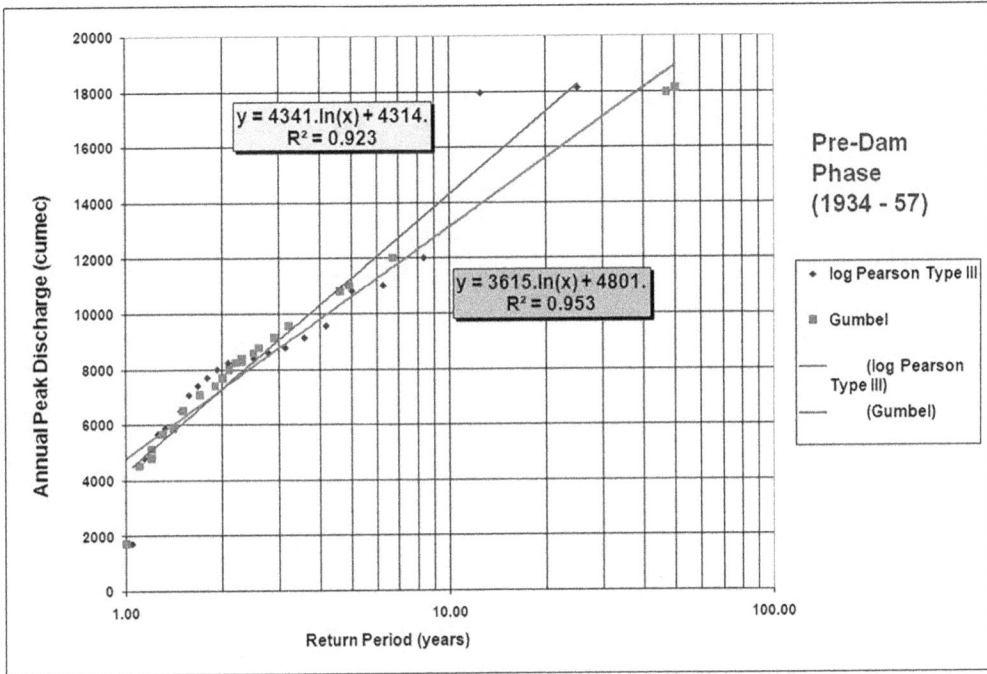

FIGURE 7.18 Fitting Gumbel's distribution to pre-dam trend of annual peak discharge and estimated return period with empirical logarithmic equations of flood prediction.

TABLE 7.15
Flood Frequency Analysis of Damodar River at Rhondia (1958–2007)

				Gumbel's Distribution		Log Pearson Type III			
Rank	Year of Peak Flow	Peak Flow $(m^3 s^{-1})$	X–Mean	Return Period (years)	Exceedence Probability (%)	Return Period (years)	Exceedence Probability (%)	Remarks	
1	1978	10,919	7,397	124	0.81	51.00	1.96	n	50
2	2007	8,883	5,361	39	2.56	25.50	3.92	Mean	3,522
3	1959	8,792	5,270	37	2.7	17.00	5.88	SD	2,235
4	2006	7,035	3,513	14	7.14	12.75	7.84	CV	0.63454
5	1995	6,522	3,000	11	9.09	10.20	9.80	C_S	1.21134
6	2000	6,387	2,865	10	10	8.50	11.76	C_r	1.163
7	1973	5,726	2,204	7	14.3	7.29	13.72		
8	1999	5,690	2,168	6.5	15.4	6.38	15.68		
9	1976	5,297	1,775	5.5	18.2	5.67	17.64		
10	1958	4,682	1,160	4	25	5.10	19.60		
11	1987	4,567	1,045	3.8	26.7	4.64	21.56		
12	1971	4,556	1,034	3.7	27	4.25	23.53		
13	1984	4,512	9,90	3.7	27.4	3.92	25.50		
14	1961	4,371	849	3.4	29.4	3.64	27.45		
15	1998	4,249	727	3.2	31.3	3.40	29.41		
16	1977	4,156	634	3.1	32.3	3.19	31.37		
17	1967	4,138	616	3.1	32.8	3.00	33.34		

(Continued)

TABLE 7.15 (*Continued*)

Flood Frequency Analysis of Damodar River at Rhondia (1958–2007)

Rank	Year of Peak Flow	Peak Flow (m³ s⁻¹)	X–Mean	Gumbel's Distribution		Log Pearson Type III		Remarks
				Return Period (years)	Exceedence Probability (%)	Return Period (years)	Exceedence Probability (%)	
18	1975	3,855	333	2.7	37	2.83	35.30	
19	1993	3,816	294	2.7	37.7	2.68	37.25	
20	1970	3,782	260	2.6	38.5	2.55	39.21	
21	1996	3,627	105	2.4	41.7	2.43	41.17	
22	1963	3,542	20	2.4	42.6	2.32	43.13	
23	1986	3,455	−67	2.3	44.4	2.22	45.10	
24	1968	3,391	−131	2.2	45.5	2.13	47.06	
25	1960	3,389	−133	2.2	45.9	2.04	49.02	
26	1985	3,317	−205	2.2	46.5	1.96	50.10	
27	1994	3,298	−224	2.1	47.6	1.89	52.94	
28	1990	3,146	−376	2	50	1.82	54.90	
29	1965	2,811	−711	1.8	57.1	1.76	56.86	
30	2003	2,496	−1,026	1.6	62.5	1.70	58.82	
31	1997	2,407	−1,115	1.5	65.4	1.65	60.78	
32	1974	2,392	−1,130	1.5	66.2	1.59	62.74	
33	1991	2,184	−1,338	1.4	70.4	1.55	64.70	
34	1983	2,098	−1,424	1.4	71.9	1.50	66.67	
35	2004	2,058	−1,464	1.4	73	1.46	68.63	
36	1964	1,977	−1,545	1.3	74.6	1.42	70.59	
37	1989	1,933	−1,589	1.3	75.2	1.38	72.55	
38	1962	1,926	−1,596	1.3	75.8	1.34	74.51	
39	2001	1,859	−1,663	1.3	76.3	1.31	76.47	
40	2002	1,859	−1,663	1.3	76.9	1.28	78.43	
41	1969	1,740	−1,782	1.3	78.7	1.24	80.39	
42	1981	1,635	−1,887	1.2	81.3	1.21	82.35	
43	1988	1,632	−1,890	1.2	82	1.19	84.31	
44	1992	1,443	−2,079	1.2	84	1.16	86.27	
45	1972	1,434	−2,088	1.2	84.7	1.13	88.23	
46	2005	1,140	−2,382	1.1	88.5	1.11	90.19	
47	1982	666	−2,856	1.1	94.3	1.09	92.15	
48	1966	502	−3,020	1	95.7	1.06	94.11	
49	1980	421	−3,101	1	96.6	1.04	96.08	
50	1979	413	−3,109	1	97.1	1.02	98.04	

Source: Bhattacharyya (2011).

Note: SD, standard deviation; CV, coefficient of variation; C_S, coefficient of skew and C_f, coefficient of flood.

the calculation, we have found a mean discharge of 3,522 m³s⁻¹ (less than 7,075 m³s⁻¹) with a standard deviation of 2,235 m³s⁻¹. For 50 years of record (1958–2007), C_V is 0.63 (much higher than previous). It indicates large range in the magnitude of floods than pre-dam period. C_S is more or less same, but C_f is reduced to 1.12 only. It indicates the lower magnitude of floods. Based on 50 years of data, peak flood discharge of 10,919 m³s⁻¹ (1978) has attained a return period of 124 years with only 0.81% probability of occurrence, but it was only 4–5 years (22% probability) in pre-dam period. It signifies the performance of DVC in flood regulation. The analysis tells that

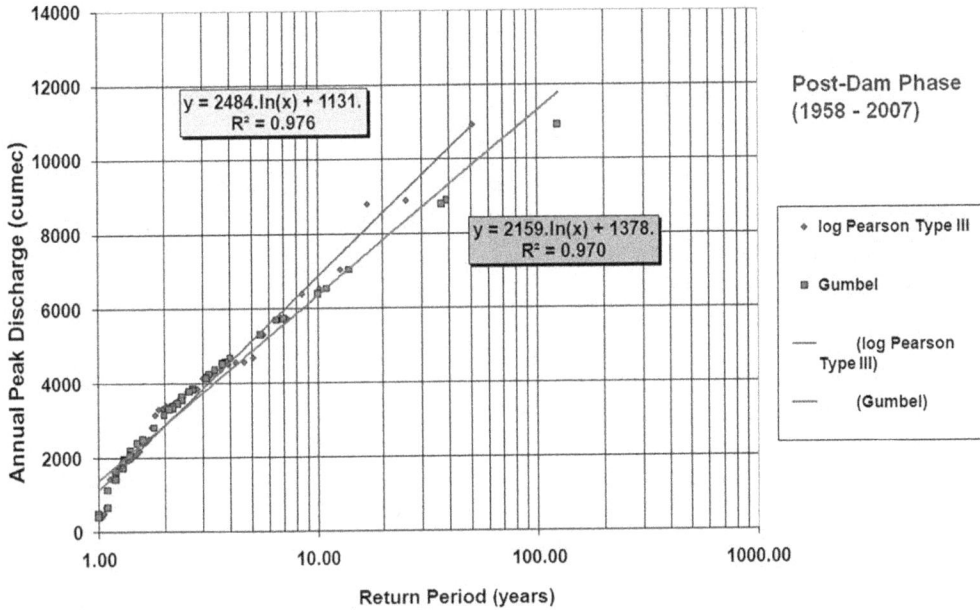

FIGURE 7.19 Fitting Gumbel's distribution to post-dam trend of annual peak discharge and estimated return period with empirical logarithmic equations of flood prediction.

flood discharge of 7,035–8,883 m³s⁻¹ has attained a return period of 14–39 years, having 7%–2.5% probability (Figure 7.19). The return period of 2–2.7 years (50%–37% probability) is associated with a peak discharge of 2,811–3,855 m³s⁻¹. At last, it is found that presently the annual peak discharge of Damodar follows the equations:

$$\text{Gumbel Distribution} - Q_T = 3522 + 2235\,K$$

$$\log \text{Pearson Type III Distribution} - Q_T = 3.451 + 0.3176\,K_T$$

We have successfully fit the extreme values to Gumbel's Probability Distribution and LP3, and it is required to test whether this distribution is significant or not to project future floods. Getting the theoretical or expected peak discharge of the consecutive years (1934–2007) from the methods, we have calculated χ^2 for pre-dam and post-dam period. The null hypothesis is that there is no difference between the observed and theoretical values, and Gumbel's method and LP3 fit the data significantly. The results of χ^2 testing suggest that the null hypothesis is accepted in both pre-dam and post-dam periods with up to 99.99% confidence level. Because in Gumbel's distribution, the calculated χ^2 (pre-dam: 11.2012 and post-dam: 2.8488) is much less than the theoretical χ^2 (0.05 significance level: 30.14 and 0.01 significance level: 36.19 in pre-dam period and 0.05 significance level: 61.66 and 0.01 significance level: 69.66 in post-dam period). Again in LP3, the calculated χ^2 (pre-dam: 16.9017 and post-dam: 7.9521) is much less than the theoretical χ^2. So, there is very limited difference between observed and theoretical peak discharge values in both methods. It can be said that the future prediction of peak flood discharge of different return period can be possible or may provide good results for flood forecasting in Lower Damodar River.

7.3.6.7 Projecting Maximum Flood Discharge

As the discharge of a river is very uncertain and random phenomenon in respect of the monsoonal rainfall, runoff and the carrying capacity of reservoirs, it is very difficult to predict the exact peak discharge of a certain return period for flood forecasting and management. We have compared the annual flood series results of Gumbel's distribution and LP3, but now we have focused on the predictable flood discharges of variable return periods, viz. 2, 5, 10, 25, 50, 100 and 200 years, using Gumbel's method, LP3, Chow's method and stochastic method. The resultant equations are as follows:

1. Gumbel's Distribution – $Q_T = 3522 + 2235\ K$
2. log Pearson Type III Distribution – $Q_T = 3.451 + 0.3176\ K_T$
3. V. T. Chow's method – $Q_T = a + b\ X_T$
   where Q_T = annual flood peak of T return period

$$X_T\ (\text{frequency factor}) = \log\big(\log T/\log T - 1\big)$$

a and b = parameters estimated by the method of moments from the observed data

$$\text{Pre-dam equation} - Q_T = 3942.21 - 7490.61\ X_T$$

$$\text{Post-dam equation} - Q_T = 3141.17 - 4345.08\ X_T$$

4. Stochastic method – One of the well-known stochastic equations based on annual flood data using Poisson probability law, and theory of sums of random number of random variables is

$$Q_T = Q_{min} + 2.3\big(Q_{mean} - Q_{min}\big)\log(n_f.n\,/\,T)$$

where n_f = number of recorded floods, counting only one for the same flood peak occurring in different years
Q_{min} = minimum peak discharge in a series
Q_{mean} = mean peak discharge

$$\text{Pre-dam equation} - Q_T = 1714 + 15327.4\ \log T$$

$$\text{Post-dam equation} - Q_T = 413 + 7150.7\ \log T$$

Except the results of stochastic method, the results of other three methods provide very close estimation in predicting flood discharge (Table 7.16). It is noticed that the predictable discharges of Gumbel's and Chow's methods are very close to each other. In pre-dam period, four estimated discharges of 2-year return period are 7,815 m³s⁻¹ (Gumbel), 8,110 m³s⁻¹ (LP3), 7,845 m³s⁻¹ (Chow) and 6,328 m³s⁻¹ (Stochastic) which are high above the critical limit (7,079 m³s⁻¹ at Rhondia, mentioned by DVC). The annual flood series of pre-dam period predicts that 100-year floods will be above 21,000 m³s⁻¹ and only 5-year flood will be greater than 10,000 m³s⁻¹. The massive floods of 1935 (18,112 m³s⁻¹), 1938 (17,942 m³s⁻¹), 1941 (12,002 m³s⁻¹), 1942 (10,811 m³s⁻¹) and 1951 (11,012 m³s⁻¹) signify the high frequency of abnormally large floods in Lower Damodar River.

7.3.6.8 Estimating Current Carrying Capacity of River

But now the river is controlled by large dams, so the situation is completely different than pre-dam period. The predictable discharge of 2-year return period will be above 2,500 m³s⁻¹ but less than 3,200 m³s⁻¹. If the annual flood series follows the trend of post-dam period, the 5-year flood will be above 5,300 m³s⁻¹ and 100-year flood will be above 11,000 m³s⁻¹. The only year of 1978 witnessed the peak discharge of 10,919 m³s⁻¹. But the lower Damodar Basin is not completely safe from annual floods. The prime reason is the declining carrying capacity of lower reach due to over siltation of

TABLE 7.16

Comparison of Projecting Flood Discharges with Variable Return Periods

Pre-Dam	Gumbel			Log Pearson Type III		V.T. Chow's Method		Stochastic Method
Return Period (years)	X_T	K_T	Q_T	K_T	Q_T	X_T	Q_T	Q_T
2	−0.521	−0.15	7,815	0.136	8,110	−0.521	7,845	6,328
5	−1.014	0.894	11,732	0.855	11,429	−1.014	11,538	12,427
10	−1.334	1.572	14,276	1.161	13,216	−1.334	13,935	17,041
25	−1.751	2.456	17,592	1.435	15,066	−1.751	17,058	23,140
50	−2.057	3.105	20,028	1.588	16,181	−2.057	19,350	27,754
100	−2.360	3.747	22,437	1.710	17,140	−2.360	21,620	32,368
200	−2.662	4.387	24,838	1.809	17,989	−2.662	23,882	36,982
2	−0.521	−0.157	3,170	0.136	3,119	−0.521	3,177	2,566
5	−1.014	0.821	5,357	0.855	5,273	−1.014	5,320	5,411
10	−1.334	1.455	6,776	1.161	6,592	−1.334	6,710	7,564
25	−1.751	2.283	8,625	1.435	8,054	−1.751	8,522	10,409
50	−2.057	2.891	9,982	1.588	9,016	−2.057	9,851	12,562
100	−2.360	3.491	11,326	1.710	9,863	−2.360	11,168	14,714
200	−2.662	4.091	12,665	1.809	10,593	−2.662	12,480	16,867

river bed. So it is necessary to estimate the return periods of present bankfull discharges of the different segments of Lower Damodar River. To estimate the bankfull discharge (m^3s^{-1}), we have calculated the bankfull volume of three selected channel segments of Damodar River, viz. (1) Rhondia to Jujuti segment, (2) Jujuti to Chanchai segment and (3) Chanchai to Paikpara segment.

Importantly, we have found three bankfull discharges of aforesaid segments which are 4,011, 2,366 and 1,542 m^3s^{-1}, respectively. Considering the DVC-mentioned critical discharge limit of 7,079 m^3s^{-1}, it is very essential to find out the probable return periods of the four bankfull discharges (flood discharges) using the post-dam annual flood series and aforesaid methods (Table 7.17). From the table, it has been found that the return period of 7,079 m^3s^{-1} discharge ranges from 8.55 to 14 years at Rhondia. Similarly, the other return periods of three discharges range from 1.18 to 3.18 years. So, it is clear that the threshold level of peak discharge (with short time span) is very small in respect of the post-dam annual flood series and carrying capacity of Damodar River. For that reason whenever the last two terminal dams (Panchet and Maithon) released excess water, the riparian tracts of Barddhaman, Hooghly and Howrah districts (covering lower Damodar Basin) had

TABLE 7.17

Predicted Return Periods of Probable Bankfull Discharges in Damodar River

Bankfull Discharge in $m^3 s^{-1}$	Estimated Return Period in Years			
	Gumbel	LP3	Chow	Stochastic
7,079	14.0	13.2	14.2	8.55
4,011	2.90	2.92	2.91	3.18
2,366	1.45	1.56	1.48	1.87
1,542	1.21	1.18	1.23	1.44

experienced monsoonal floods in 1958, 1961, 1976, 1978, 1995, 1999, 1987, 2000, 2006 and 2007, having very low magnitude of peak discharge in comparison to pre-dam period.

7.4 CONCLUSION

The floods, as experienced by present advanced human civilisation of India, were also observed frequently by the inhabitants of Harappa and Mohenjo-Daro Civilisations, as the flood of Indus River. The causes of floods can be natural; however, human interference intensifies many floods. The decision to live in a Damodar floodplain, for a variety of perceived benefits, is one that is fraught with difficulties. So to save from difficulties and risk, we should adopt flood frequency analysis for management purpose. Here we have successfully employed the statistical techniques to understand annual flood discharge character of Damodar River. It is identified that in pre-dam period, the flood peaks were high but the duration was small. The installation of dams has moderated the peaks but increased the duration of floods. In many cases, the dams of DVC authority control the flood situation minimally through water storages, but the upper catchment dams are repeatedly unable to control the heavy inflow of water due to excessive siltation. Due to high runoff coefficient of rocky terrain (0.8–0.9), the upper catchment of Damodar (lies in Jharkhand) yields excessive runoff in peak monsoon than does lower catchment. The upstream tributaries add to water discharge of Damodar River within the upper contributing drainage net, but lower distributing net of Damodar is narrow and congested by embankments, facing huge pressure of inflow water in rainy season. So it is inevitable that when huge volume of runoff water pass through narrow and shallow Lower Damodar River, the extreme flood occurred. The universal truth is that we can stop and predict flood discharge accurately but here flood frequency analysis of extreme values has provided a statistically significant tool to project potential flood flow of a given return period in post-dam phase.

A careful spatio-temporal study of Lower Damodar River and DVC flood regulation system has brought into focus so many new-fangled things of typical anthropogeomorphology of this study area which has a direct interaction or interplay with developmental processes of human civilisation. It is clearly observed that existing drainage system and DRB were forced to enter into new phase of equilibrium after the establishment of DVC (1948–1958). In the present phase, the whole basin and river bed is the product of complex hydro-geomorphic and rigorous anthropogenic processes, though the modification of floodplain was first set forth by British Rulers through installing embankments. In many cases, DVC controls the flood situation minimally through water storages, but the upper catchment dams repeatedly do not control the heavy inflow of water. So it is evitable that when huge volume of water would pass through narrow and shallow Lower Damodar River, flood will occur. If we observe the shape and geographical entities of Damodar Basin, we have found that the distributaries of Damodar River system are flooded whenever the water contributed by the upstream does not find easy passage to Hooghly River due to drainage congestion, burdens of roads, railways, canals and finally tidal behaviour of lower reach. In other words, the larger the ratio of contributing drainage net and distributing drainage net, the greater the chance of flooding, which has exactly happened to DRB. If we look closely at the flood phenomenon, we have noticed that man's affinity for riverine locations from very past is escalating the chance of flood vulnerability and this tradition is still evidenced in Lower Damodar River. Now the fallacy is completely wiped out—the dams too often generate flood and create many anthropogeomorphic problems to river and its adjoining floodplain. The idea that Nature can be conquered and it carries within it the seed of human destruction. The onus of adaptation lies on the human society and not on nature. So to fulfil the objectives of DVC and the dream of Dr. Meghnad Saha, the main focus should be placed on alternative flood management of Lower Damodar Valley, improving the carrying capacity of Damodar, Khari, Banka, Gangur, Behula, *Kana* Damodar, etc. channels. Nature, man and science can again make this prosperous valley into a smiling garden using flood water as resource.

REFERENCES

Acharyya, S.K. and B.A. Shah. 2007. Arsenic-contaminated groundwater from parts of Damodar fan-delta and west Bhagirathi River, West Bengal, India: influence of fluvial geomorphology and Quaternary morphostratigraphy. *Environmental Geology* 54: 489–501.

Alila, Y. and A. Mtiraoui. 2002. Implications of heterogeneous flood-frequency distributions on traditional stream-discharge prediction techniques. *Hydrological Processes* 16: 1065–1082.

Allaby, M. 2006. *Floods.* New Delhi: Viva Books Private Limited.

Bagchi, K. 1977. The Damodar valley development and its impact on the region. In: *Indian Urbanization and Planning: Vehicle of Modernization*, ed. Allen, G.N. and A.K. Dutt, 232–241, New Delhi: Tata McGraw Hill.

Banerji, S.K. 1950. Problems of river forecasting in India. *Proceedings of National Institute of Sciences of India* 16(6): 25–33.

Basu, S. 2011. *Soil Erosion and Environmental Problems.* New Delhi: APH Publishing Corporation.

Beard, L.R. 1975. Generalized evaluation of flash-flood potential. Technical Report – University of Texas Austin, Center for Research Water Resources CRWR-124: 1–27.

Beckinsale, R.P. 1969. River regimes. In: *Water, Man and Earth*, ed. Chorley, R.J., 455–470. London: Methuen and Co.

Bell, F.G. 1999. *Geological Hazards: Their Assessment, Avoidance and Mitigation.* London: E and Fn Spon.

Benson, M.A. 1968. Uniform flood-frequency estimating methods for Federal Agencies. *Water Resource Research* 4(2): 891–908.

Betal, H.R. 1970. Identification of slope categories in Damodar Valley, India. In: *Selected Papers Vol-1-Physical Geography*, ed. Chatterjee, S.P. and S.P. Dasgupta, 4–6. Calcutta: National Committee for Geography.

Bhattacharya, K. 1959. *Bangladesher Nad-Nadi o Parikalpana (Rivers of Bengal and Planning).* Calcutta: Bidyadoya Library Ltd.

Bhattacharyya, K. 2011. *The Lower Damodar River, India: Understanding the Human Role in Changing Fluvial Environment.* New York: Springer.

Brice, J.C. 1964. Channel patterns and terraces of the Loup River in Nebraska. Geological Survey Professional Paper 422-D, Washington DC, D2–D41.

Central Water Commission CWC. (2000). *Integrated Hydrological Data Book.* New Delhi: Central Water Commission of India.

Chandra, S. 2003. *India: flood management – Damodar River Basin.* Retrieved from www.apfm.info/pdf/case_studies/cs_india [accessed on 12 December 2013]

Chatterjee, S. P. 1967. *Damodar Valley Planning Atlas.* Calcutta: NATMO.

Choudhury, J. 1995. *Barddhaman Jelar Itihash o Smanskiti (History and Culture of Burdwan District).* Kolkata: Pustak Biponi.

Dhar, O.N. and S. Nandargi. 2003. Hydrometeorological aspects of floods. *Natural Hazards* 28: 1–33.

Dhir, R.D., P.R. Ahuja and K.G. Majumder. 1958. A study on the success of reservoir based on the actual and estimated runoff. *Central Board of Irrigation and Power* 68: 11–24.

Ewemoje, T.A. and M.A. Ewemooje. 2011. Best distribution and plotting positions of daily maximum flood estimation at Ona River in Ogun-Oshun River Basin, Nigeria. *Agricultural Engineering International Journal* 13(3): 1–13.

Friend, P.F. and R. Sinha. 1993. Braiding and meandering parameters. In: *Braided Rivers*, ed. Best, J.L. and Bristow, C.S., 105–111. Washington: Geological Society Special Publications No 75.

Ghosh, S. 2011. Hydrological changes and their impact on fluvial environment of the lower Damodar Basin over a period of fifty years of damming the mighty Damodar River in Eastern India. *Procedia Social and Behavioral Sciences* 19: 511–519.

Ghosh, S. and S.K. Guchhait. 2014. Hydrogeomorphic variability due to dam constructions and emerging problems: a case study of Damodar River, West Bengal, India. *Environment Development Sustainability* 16(3): 769–796.

Ghosh, S. and S.K. Guchhait. 2016. Dam-induced changes in flood hydrology and flood frequency of tropical river: a study of Damodar River of West Bengal, India. *Arabian Journal of Geosciences* 9: 90.

Ghosh, S. and B. Mistri. 2012. Investigating the causes of floods in Damodar River of India: a geographical perspective. *Indian Journal of Geomorphology* 17(1): 37–49.

Ghosh, S. and B. Mistri. 2013. Performance of DVC in flood moderation of lower Damodar River, India and emergent risk of flood. *Eastern Geographers* 19(1): 55–66.

Glass, E.L. 1924. Floods of the Damodar River and rainstorms producing them. *Minutes of the Proceedings* 217(1924): 333–346.

Goodall, M.R. 1945. River valley planning in India: the Damodar. *The Journal of Land and Public Utility Economics* 21(4): 371–375.

Goswami, D.C. 1998. Fluvial regime and flood hydrology of the Brahmaputra River, Assam. In: *Flood Studies in India*, ed. Kale. V.S., 53–76. Bangalore: Geological Society of India.

Goudie, A.S. and H.A. Viles. 2016. *Geomorphology in the Anthropocene*. Cambridge: Cambridge University Press.

Goyal, M.K. and C.S.P. Ojha. 2010. Analysis of mean monthly rainfall runoff data of Indian catchments using dimensionless variables by neural networks. *Journal of Environmental Protection* 1: 155–171.

Grade, R.J. 2006. *River Morphology*. New Delhi: New Age International Limited.

Grade, R.J. and U.C. Kothyari. 1990. Flood estimation in Indian catchments. *Journal of Hydrology* 113: 135–146.

Griffs, V.W., M. Asce and J.R. Stedinger. 2007. Log Pearson type 3 distribution and its application in flood frequency analysis: distribution characteristics. *Journal of Hydrologic Engineering* 12(5): 482–491.

Gumbel, E. J. 1941. The return period of flood flows. *The Annals of Mathematical Statistics* 12(2): 163–190.

Hann, C.T. 1977. *Statistical Methods in Hydrology*. Ames: Iowa State University Press.

Holdren, J.P. and P.R. Ehrlich. 1974. Human population and the global environment. *American Scientist* 62: 282–292.

Jain, V., R. Sinha, L.P. Singh and S.K. Tandon. 2016. River systems in India; the Anthropocene context. *Proceedings of the Indian National Science Academy* 82(3): 747–761.

Jha, R. and Smakthin, V. 2008. Review of methods of hydrological estimation at ungauged sites in India. International Water Management Institute. Retrieved from www.iwmi.cgiar.org/publications/Working_Papers/.../WOR130 on 26 December, 2011 at 9:45 am.

Jha, V.C. and H. Bairagya. 2012. Floodplain planning based on statistical analysis of Tilpara Barrage discharge: a case study on Mayurakshi River Basin. *Caminhos de Geografia* 13(43): 326–346.

Kale, V.S. 1998. Monsoon floods in India: a hydro-geomorphic perspective. In: *Flood Studies in India*, ed. Kale, V.S., 229–256. Bangalore: Geological Society of India.

Kale, V.S. 1999. Long period fluctuations in monsoon floods in the Deccan Peninsula, India. *Journal of Geological Society of India* 53: 5–15.

Kale, V.S. 2003. Geomorphic effects of monsoon floods on Indian rivers. *Natural Hazards* 28: 64–84.

Kale, V.S. 2005. Fluvial hydrology and geomorphology of monsoon-dominated Indian rivers. *Revista Brasileira de Geomorfologia Ano* 6(1): 63–73.

Kale, V. S. (2006). Floods in India: Their Frequency and Pattern. In: *Coping with Natural Hazards: Indian Context*, ed. Valdiya, K.S., 104–123. Hyderabad: Orient Longman.

Kale, V.S., S. Mishra, Y. Enzel, L. Ely, S.N. Rajaguru and V.R. Baker. 1993. Flood geomorphology of the Indian peninsular rivers. *Journal of Applied Hydrology* 6: 49–55.

Kidson, R. and K.S. Richards. 2005. Flood frequency analysis: assumptions and alternatives. *Progress in Physical Geography* 29(3): 392–410.

Kinghton, D. 1998. *Fluvial Forms and Processes*. London: Arnold.

Kochel, R.C. 1988. Geomorphic impact of large floods: review and new perspectives on magnitude and frequency. In: *Flood Geomorphology*, ed. Baker, V.R., R.C. Kochel and P.C. Patton, 169–187. New York: Wiley.

Krik, W. 1950. The Damodar Valley – "Valley Opima". *Geographical Review* 40(3): 415–443.

Lahiri-Dutt, K. 2006. *State and the community in water management case of the Damodar Valley Corporation, India*. Retrieved from http://www.wepa-db.net/pdf/0612sympo/paper/Kuntala_Lahiri-Dutt.pdf [accessed on 12 December, 2012].

Lal, M., T. Nozana and S. Emori. 2001. Future climate change: implications for Indian summer monsoon and its variability. *Current Science* 81(9): 1196–1207.

Mahmood, A. 2008. *Statistical Methods in Geographical Studies*. New Delhi: Rajesh Publication.

Majumder, A., S. Ghosh, A. Dasgupta and D. Seth. 2012. Analyzing reservoir sedimentation of Panhet Dam, India using remote sensing and GIS. *PANCHAKOTeSSAYA* 2(3): 82–95.

Majumder, M., P. Roy and A. Mazumder. 2010. An introduction and current trends of Damodar and Rupnarayan River network. In: *Impact of Climate Change on Natural Resource Management*, ed. Jana, B.K. and M. Majumder, 461–480. New York: Springer.

Miall, A.D. 1985. Architectural element analyses: a new method of analyses applied to fluvial deposits. *Earth Science Reviews* 22: 261–308.

Millington, N., S. Das and S.P. Simonovic. 2011. The comparison of GEV, log Pearson type III and Gumbel distribution in the upper Thames River Watershed under global climate models. Water Resource Research Report 007, The University of Western Ontario.

Mishra, D.K. 2001. Living with floods: people's perspective. *Economic and Political Weekly* 36(2): 2756–2761.

More, R.J. 1969. The basin hydrological cycle. In: *Water, Man and Earth*, ed. Chorley, R.J., 67–75. London: Methuen and Co.

Mueller, J.R. 1968. An introduction to the hydraulic and topographic sinuosity indexes. *Annals of the Association of American Geographers* 58(2): 371–385.

Mujere, N. 2011. Flood frequency analysis using the Gumbel distribution. *International Journal on Computer Science and Engineering* 3(7): 2774–2778.

Mukhopadhyay, S. 2010. A geo-environmental assessment of flood dynamics in lower Ajoy River inducing san splay problem in eastern India. *Ethiopian Journal of Environmental Studies and Management* 3(2): 96–110.

Mukhopadhyay, S.C. 2007. Contemporary issues in geography with particular emphasis on the flood hazards of West Bengal. In: *Contemporary Issues and Techniques in Geography*, ed. Basu, R. and S. Bhaduri, 77–110. Kolkata: Progressive Publishers.

Nagle, G. 2003. *Rivers and Water Management*. London: Hodder and Stoughton.

Nandargi, S. and O.N. Dhar. 2003. High frequency floods and their magnitudes in the Indian rivers. *Journal of Geological Society of India* 61: 90–96.

Oberg, K.A. and D.M. Mades. 1987. Estimating generalized skew of the log Pearson type III distribution for annual peak floods in Illinois. USGS Water Resources Investigations Report 86–4008: 1–42.

Pramanik, S.K. and K.N. Rao. 1952. Hydrometeorology of the Damodar Catchment. Retrieved from www.iahs.info/redbooks/a036/036060.pdf on 27 November, 2012 at 3:25 pm.

Raghunath, H.M. 2011. *Hydrology- Principles, Analysis and Design*. New Delhi: New Age Internationals Publishers.

Rajaguru, S.N., A. Gupta, V.S. Kale, S. Mishra, R.K. Ganjoo, L.L. Ely, Y. Enzel, Y. and V.R. Baker. 1995. Channel form and processes of the flood-dominated Narmada River, India. *Earth Surface Process and Landforms* 20: 407–421.

Rakhecha, P.R. 2002. Highest floods in India. *Proceedings of a symposium on the extreme of the extremes: extraordinary floods*. IAHS Publication. 771: 167–172.

Rao, A.R. and K.H. Hamed. 2000. *Flood Frequency Analysis*. Boca Raton: CRC Press.

Rao, G.N. 2001. Occurrence of heavy rainfall around the confluence line in monsoon disturbances and its importance in causing floods. *Journal of Earth System Science* 110(1): 87–94.

Rao, K.N. 1951. *Hydrometrological studies in India*. Retrieved from http://iahs.info/redbooks/a042/04223.pdf [accessed on 28 November, 2012].

Reddy, P.J.R. 2011. *A Textbook of Hydrology*. Bangalore: University Science Press.

Reed, D.W. 2002. Reinforcing flood-risk estimation. *Philosophical Transactions: Mathematical, Physical and Engineering Sciences* 360(1796): 1371–1387.

Roy, D., S. Mukherjee and B. Bose. 1995. Regulation of a multipurpose reservoir system: Damodar Valley, India. Retrieved from http//iahs.info/redbooks/a230/iahs_230_0095 on 13 October, 2012 at 12:10 pm.

Roy, P.K. and A. Mazumder. 2005. Hydrological impacts of climatic variability on water resources on the Damodar River Basin, India. In: *Regional Hydrological Impacts of Climatic Change: Impact, Assessment and Decision Making*, ed. Wagener, T., 147–156. London: IAHS Publication.

Rudra, K. 2002. Floods in West Bengal, 2000- causes and consequences. In: *Changing Environmental Scenario*, ed. Basu, S., 326–347. Kolkata: acb Publications.

Rudra, K. 2009. *Banglar Nadikatha (Tales of Rivers of Bengal)*. Kolkata: Sahitya Samsad.

Saha, M.K. 2005. River flood forecasting and preparedness- Lower Damodar Valley in West Bengal. In: *River Floods: A Socio-Technical Approach*, ed. Rahim, K.M.B., M. Mukhopadhyay and D. Sarkar, 244–249. Kolkata: acb Publications.

Saha, S.K. 1979. River basin planning in the Damodar Valley of India. *Geographical Review* 69(3): 273–287.

Sarma, S.S. 2002. Floods in West Bengal- nature, analysis and solution. In: *Changing Environmental Scenario*, ed. Basu, S., 315–325. Kolkata: acb Publications.

Satakopan, V. 1949. A report on the rainfall studies made in connection with the unified development of the Damodar River. *Memoirs of the Indian Meteorological Department* 27(6): 16–17.

Sathe, B.K., M.V. Khire and R.N. Sankhua. 2012. Flood frequency analysis of upper Krishna River Basin catchment area using log Pearson Type III distribution. *ISOR Journal of Engineering* 2(8): 68–77.

Sen, S.K. 1962. *Drainage Study of Lower Damodar Valley*. Calcutta: DVC Publication.

Sen, P.K. 1985. The genesis of floods in the Lower Damodar catchment. In: *The Concepts and Methods in Geography*, ed. Sen, P.K., 71–85. Burdwan: The University of Burdwan.

Sen, P.K. 1991. Flood hazards and river basin erosion in the lower Damodar Basin. In: *Indian Geomorphology*, ed. Sharma, H.S., 95–108. New Delhi: Concept Publishing Co.

Sengupta, S. 2001. *Rivers and floods*. Retrieved from www.breakthrough-india.org/archives/flood.pdf [accessed on 13 April 2012].

Singh, P., A.S. Ramanathan and V.G. Ghanekar. 1974. Flash floods in India. Retrieved from www.hydrologie.org/redbooks/a112/iahs_112_0114 [accessed on 4th July 2020]

Singh, L.P., B. Parkash and A.K. Singhvi. 1998. Evolution of the lower Gangetic Plain landforms and soils in West Bengal. *Catena* 33: 75–104.

Sinha, R. and V. Jain. 1998. Flood hazards of north Bihar River, Indo-Gangetic Plains. In: *Flood Studies in India*, ed. Kale, V.S., 27–52. Bangalore: Geological Society of India.

Singhvi, A. K. and V.S. Kale. 2009. *Paleoclimate Studies in India: Last Ice Age to the Present*, IGBP-WCRP-SCOPE-Report Series: 4. New Delhi: Indian National Science Academy.

Sinha, R., V. Jain, G. Prasad Babu and S. Ghosh. 2005. Geomorphic characterization and diversity of the fluvial systems of the Gangetic plains. *Geomorphology* 70: 207–225.

Sridhar, A. 2008. Fluvial palaeohydrological studies in western India: a syntheses. *Earth Science India* 1(1): 21–29.

Stichellout, E., A.G. Roy and F. Petit. 2006. Comparison of impacts of dams on the annual maximum flow characteristics in three regulated hydrologic regimes in Quebec (Canada). *Hydrological Processes* 20: 3485–3501.

The CGIAR Consortium for Spatial Information (CGIAR – CSI). 2006. *SRTM 90 m Digital Elevation Data*. http://srtm.csi.cgiar.org/ [accessed on 10 January, 2011].

Vogel, W.R. and D.E. McMartin. 1991. Probability plot goodness of fit and skewness estimation procedures for the Pearson type 3 distribution. *Water Resources Research* 27(2): 3149–3158.

Whitfield, P. H. 2012. Floods in Future Climates: A Review. *Journal of Flood Risk Management, Wiley*. Retrieved from http://onlinelibrary.wiley.com/doi/10.1111/j.1753-318X.2012.01150.x/pdf [accessed 22 August 2012].

Wisler, C.O. and E.F. Brater. 1959. *Hydrology*. New York: John Wiley and Sons.

Zakaullah, S.M.M., I. Ahmad and G. Nabi. 2012. Flood frequency analysis of homogenous regions of Jhelum River Basin. *International Journal of Water Resources and Environmental Engineering* 4(5): 144–149.

8 Anthropogenic Impact on Forms and Processes of the Kangsabati River Basin

Shambhu Nath Sing Mura and Ananta Gope

CONTENTS

8.1 INTRODUCTION

Both human being and river need each other. On the one hand, river as natural system provides the base for the civilisation of society; lifeblood of humanity, and on the other, river does not get its fullness without imagination, emotion and intuition of human mind. Great poet Tagore (2016, pp. 97–98) once expressed the inner meanings of river and human mind in a letter written on 22th *Poush* [December], 1329 [1922] in Bengali Calendar:

> I love river. Shall I say why? The land on which I live does not move…the river flows days and nights. It has its own echo. Its rhythm corresponds with rhythms of our movements. The flow of our conscious mind has similarity with the flow of the river-so that we have deep intimacy with river.

In the language of McCully (1998, p. 9), "All land is a part of a watershed or river basin and all is shaped by the water which flows over it and through it". Generally, a drainage basin is an area that congregates water from precipitation and delivers it to a larger stream, a lake or an ocean (Chorley et al., 1984). It is actually the source area of precipitation which provides water to the stream channels in various conduits (Leopold et al., 1964). Indeed, drainage basin is an open system, into which and from which energy and matter flow, and its boundaries are always well defined (Bloom, 2003). As a dynamic living entity, river is constantly modifying earth materials and forms in combination with multitude of factors. The morphology and dynamics of rivers are dependent on various environmental variables such as climatic elements, lithological characteristics, topographic slope aspects, vegetal cover and landuse–landcover pattern. "All these variables interact in a complicated feedback mechanism to determine the configuration of a river and the processes and rates of their operation in a specific fluvial system" (Morisawa, 1985, p. 181). In conjunction with these, human beings deliberately alter the quantity and quality of water in rivers and streams by direct channel manipulation, modification of basin characteristics, urbanisation and pollution (Goudie, 1981). The Kangsabati river basin is an important geomorphic unit for investigation because of its varied flow pattern, contrasting lithological and geomorphological characteristics. After emanating from the outer edge of Chotanagpur plateau and flowing over the *Rarh* plateau-fringe topography and extensive flat alluvial plain, the Kangsabati river ultimately falls in Hooghly river which is popularly known as Bhagirathi-Hooghly. In the Kangsabati basin tripartite geomorphic units, the granitic plateau, lateritic plateau-fringe and the alluvial plain have given rise to three-fold divisions of landforms with contrasting characteristics (Figure 8.1). Of late, by changing the surface charac- teristics i.e., by constructing dams, barrages, weirs, artificial embankments, cross bridges, mining sands and gravels from river beds, altering landcover and landuse, human beings have invited an unwanted change in the river regime, sediment supply, and rate of degradation, morphology and landform characteristics in different segments of the river basin. Therefore, the renowned statement, propounded by noted geomorphologists Gregory and Walling (1971, p. 291) "...the inclusion of man as a geomorphological agent affecting the drainage basin system" has become an established real- ity in today's world. The present study makes an endeavour for comprehensive explanation of the character of landforms that developed as a resultant effect of the interaction of the endogenetic and exogenetic processes modified by human beings in the different segments of the Kangsabati river basin. The study also emphasises upon the study of the geomorphological processes as well as the resultant landforms of Kangsabati basin being affected by the tectonic processes as well as shaped by anthropogenic mechanisms. In the present study, besides interpreting the geomorphic processes and forms of the Kangsabati river basin, due weightage has been given on the anthropogenic activi- ties in determining the geomorphologic aspects of the selected basin.

The major parts of the Kangsabati river basin are located in the districts of Puruliya, Bankura, Paschim Medinipur and Purba Medinipur in West Bengal and only a small portion of it in the

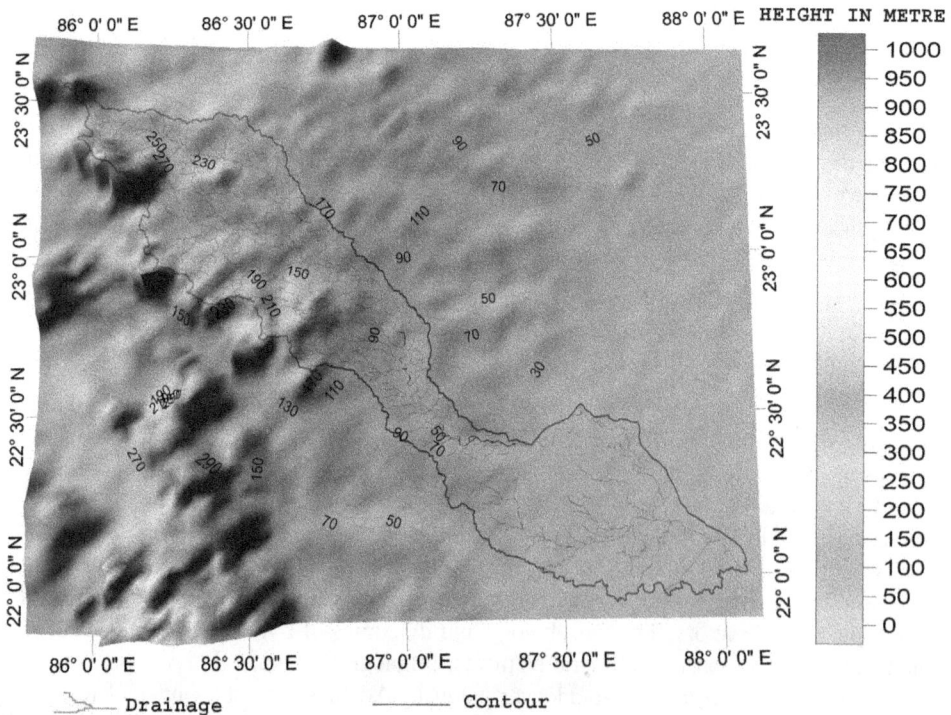

FIGURE 8.1 3D view of the study area and its environs. (Authors' compilation.)

district of Singhbhum in Jharkhand, India. This area extends from 21°55′36″ to 23°29′04″ N latitudes and 85°57′25″ to 88°05′53″ E longitudes. In an elongated shape, the basin covers an area of about 8,993 km². The perimeter of the basin is 852.04 km with linear lengths of 271.03 km (Figure 8.2). In the upper reaches, the river drains over the undulating plateau surfaces of Puruliya, in its middle parts rolling *Rarh* plain surfaces [plateau-fringe segment] of Bankura and Paschim Medinipur and in its lower reaches on alluvial plains in Purba Medinipur district. The river ultimately joins as Haldi river to the river Bhagirathi-Hooghly. The Kangsabati basin lies between the Damodar and Rupnarayan river basin in the north and east and the Subarnarekha river basin in the south and west.

The very title of the topic itself justifies the relationship between the physical elements of a specified basin area and the anthropogenic changes brought upon by the phenomena. In this case, the selected area of study is the fluvial forms and processes of Kangsabati river modified by human action. It is true

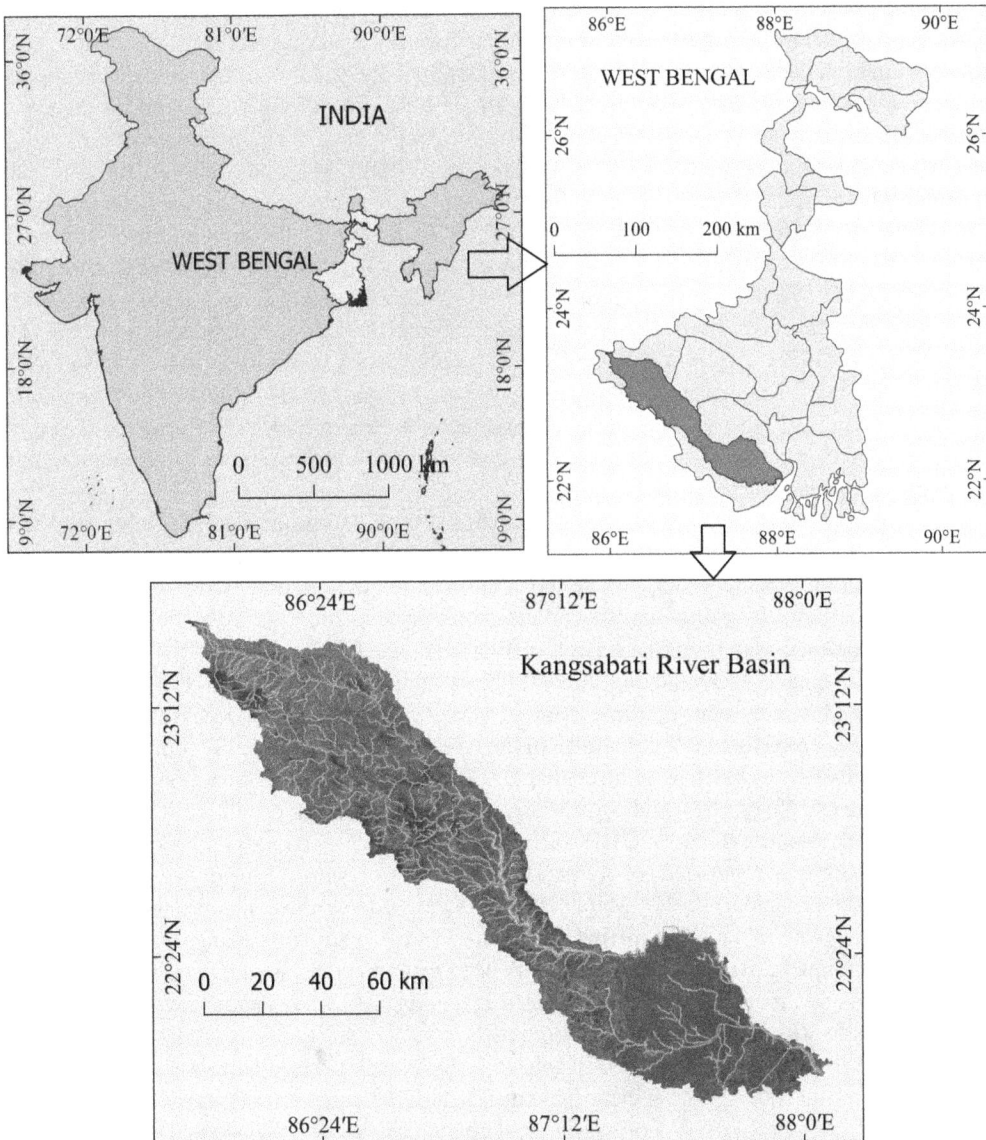

FIGURE 8.2 Location map of the study area.

that if any changes are imposed upon the physical elements such as land, water, soil and vegetation on a river basin through human interference, then automatically the impact will exert on the geomorphologic aspects of the basin. As such, to bring out the salient aspects of the anthropogenic imprints on landforms and processes shaping them, a detailed account of the previous literature concerning human–nature interaction in the study area and its impacts for carving river forms and processes deserves special attention.

The literatures, reports and records available in this context may be explained in a time frame ranging from 1794 to till date. The geographical aspects of the study area and its surroundings were mentioned in the writings of eminent scholars such as Colebrooke, Gastrell, Oldham, Forsyth, Hunter, Coupland, Culshaw and Mitra. Some of the literary reviews reflect the themes concerning to a particular issue; in some cases, they are related to description of physical elements such as topography, soil, rivers and vegetation of the study area; some others were directed to the analysis of human community, landuse–landcover, societal structure, social customs, economy, etc. In few cases, the two-way relationships between physical elements of the space and the community were established.

The geomorphological aspects of Bengal delta, existing irrigation systems, flood nature, problems of irrigation, decaying nature of the rivers of Bengal due to human interferences in the past and lack of proper initiative in the present, rhythmic behaviour of river flow and its relation to Bengal's economic history, prosperity or adversity of people have been analysed meticulously by eminent scholars such as Willcocks (1930), Mukherjee (1938), Majumdar (1942), Bagchi (1944) and Roy (1949). Empirical and descriptive information on the study area have been gathered from various literary descriptions based mainly on the physical state of the area and life style of the people which have been reflected in those writings. These include: *District Gazetteers*, 1908, 1911, 1968, 1985, 1995; *District Census Handbooks* 1961–2001, *District Human Development Report, Bankura* [2007], *Bangladesher Nad-Nadi O Parikalpana* [Rivers of Bangladesh and Planning] by Bhattacharya (1959), *Paschim Banger Nad-Nadi* [The Rivers of West Bengal] by Basu (2002), *Banglar Nad-Nadi* [The Rivers of Bengal] by Bandopadhyay (2007) and *Banglar Nadikatha* [The Story of Bengal Rivers] by Rudra (2008). Only the research-oriented writings of the last four writers need special attention because they are important concerning close connotation to the present issue, at least descriptively.

The geomorphological aspects of the Kangsabati basin and surrounding areas have been studied mainly on the basis of Toposheets by Niyogi, Roy and Munsi, Mukhopadhyay, Chakraborty and Ghosh (cited in Banerjee, 1983). Banerjee (1983) studied in detail the geomorphological characteristics of the basin by using Toposheets and intensive field surveys. "Morphometric Analysis of Kangshabati-Darkeswar Interfluves Area in West Bengal, India using ASTER DEM and GIS Techniques" was done by Gayen et al. (2013). Mittal et al. (2014) have tried to evaluate the hydrologic alteration caused by dam construction and climatic changes in the Kangsabati river basin. Their analysis indicated that flow variability in the river has been significantly reduced due to dam construction with high flows being reduced and low flows during pre-monsoon months considerably enhanced. Datta and Roy (2017) studied the characteristics of flow pattern in the Kangsabati catchment area. Using remote sensing and GIS techniques, morphometric analysis and prioritisation of sub-watersheds for four watersheds of Kumari and Kasai river basins of Puruliya district have been studied by Das and Bhandari (2017). Mahala (2018) made a comparative analysis of drainage basin morphometric analysis of mountain-plain [Kosi, Bihar] and plateau-plain [Kangsabati, WB] regions of tropical environment. Actually, no attempt has been made to correlate the dynamicity of change in fluvial forms and processes of a basin induced by endogenetic–exogenetic as well as anthropogenic mechanisms which the present work suggests.

The methods and techniques applied in this work have been derived from the objectives itself. The objectives of the study make it clear that river as a living entity carves its landforms depending on tectonic, geomorphic and geo-hydrological characteristics of the basin and, simultaneously, the landform characteristics of the basin are also influenced by the anthropogenic activities. In this study, the authors have attempted to evaluate the fluvial forms and processes of Kangsabati river modified by anthropogenic activities. Therefore, this study comprises the study of geological framework, relief characteristics, identification of geomorphic processes and forms, drainage network analysis, detailed morphometry of the basin area for the quantification of drainage networks, relief and slope and anthropogenic mechanisms such as construction of dams, barrages, weirs, embankments, extraction of sands and gravels from river channel, landcover and landuse changes in the entire basin and their possible impacts on the health of the river. To fulfil the above-mentioned objectives, US Army Map Service Topographic Map, 1955 [India and Pakistan, 1: 250,000], SRTM, LANDSAT and LISS-IV 2018 data have been used. By using QGIS 2.18, MapInfo 10.0, Ilwis 3.3 academic and Golden Surfer 13 various maps, graphs and charts have been prepared. This is followed by interpretations to reach its final conclusion.

8.2 THE RIVER AND ITS BASIN

The river Kangsabati, mentioned as *Kapila* by Kalidasa in his *Raghubansam* (Bandopadhyay, 2007) and now locally known as *Kansai* or *Kassai* and *Cossye*, is a right bank tributary of the Bhagirathi-Hooghly river system. It originates from Jabarban peak on the Ghoramara hill located east of Chotanagpur plateau [23°32′30″ N and 85°56′30″ E] and flowing through the districts of Puruliya, Bankura, Paschim Medinipur and Purba Medinipur ultimately falls into the Hooghly river. The entire basin has been divided into four distinct segments to represent the channel gradient (Figure 8.3).

The *Kansai* river flows in a south-east direction after originating from the Jabarban peak of the Ghoramara hill. It receives the combined waters of Girgiri *Nala* and Sahar *Jhor* confluenced at Begunkudar [23°12′ N and 86°04′ E]. The *Kansai Nala* becomes *Kansai* river from Girgiri *Nala*–Sahar *Jhor* confluence which flows eastwards in a relatively wider channel filled with sand and gravel, water being flown only in rainy season. A number of non-perennial first- and second-order streams descend from the northern face of Baghmundi hills and ultimately meet the course of

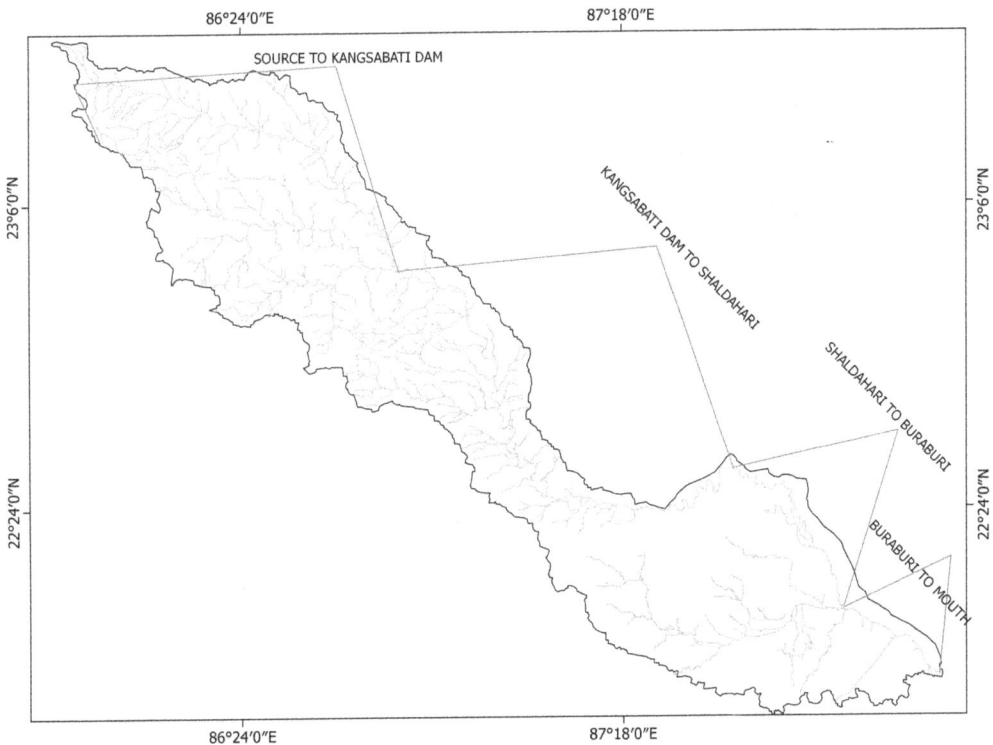

FIGURE 8.3 Longitudinal profile segments of Kangsabati river. (Cartosat-1, 2018.)

Kansai. The important right bank tributaries of *Kansai* from west to east are: Goura *Nala*, Chagha *Nala* and Bandhu *Nala* [the combined water of Chunmuti *Nala*, Burudih *Nala* and Sarambisi *Nala*]. The *Kansai* becomes a wild and perennial channel about 4 km from this confluence, receiving flood waters of Gobri *Jhor* flowing north-east. Up to this point, the left bank tributaries of *Kansai* are not so prominent and none of them exceed a length of 10 km. After crossing Puruliya town, located about 3 km in the north, Kansai flows in an easterly direction with a narrow channel scarred by gullies. It receives only the non-perennial Patoli *Nala* from left in this area and flows in south-easterly direction as a wide perennial channel meandering between Puncha and Manbazar Police Stations and ultimately merges in the Kangsabati Reservoir.

Kumari, the most important right bank tributary of Kangsabati river, originates from the eastern slope of the Baghmundi uplands covered by dense *sal* [shoria robusta] forest and receives two tributaries: the east flowing Kunwari *Nala* at an altitude of 266 m and south-east flowing Hanumanta *Nala* at an altitude of 230 m. Kumari then flows south-south-east direction almost in a straight channel and then bends towards east almost as a less steep sinuous channel. The Kumari also receives water from numerous right bank tributaries of which two most important are: the Nangasai *Nala* from the interfluve of Kumari–Subarnarekha basins, and Jam *Nala*, the combined water from the perennial streams of the Kumir *Nala*, Totko *Nala* and Jamuna *Nala* from Dalma range. Besides, the Kumari also receives the perennial Kulandari *Nala* flowing along the Puruliya–Bankura administrative boundary (Figure 8.4).

The river Kangsabati and its numerous tributaries in Puruliya do not offer any scope for navigation or large-scale canal irrigation due to topographic characteristics. Consequently, the district suffers less from flood hazards except occasional flash flood occurring in few monsoon months. Locally, this flood is known as *Hurpah Ban* (*District Gazetteer, Puruliya*, 1985). The river Kangsabati receives the popular name *Kansai* as it enters into the district of Bankura near the village Bhedua in Khatra Police Station (Basu, 2002). It flows in a south-easterly direction performing the administrative boundary of Khatra and Ranibandh Police Stations and thereafter flows in south-easterly direction through Raipur

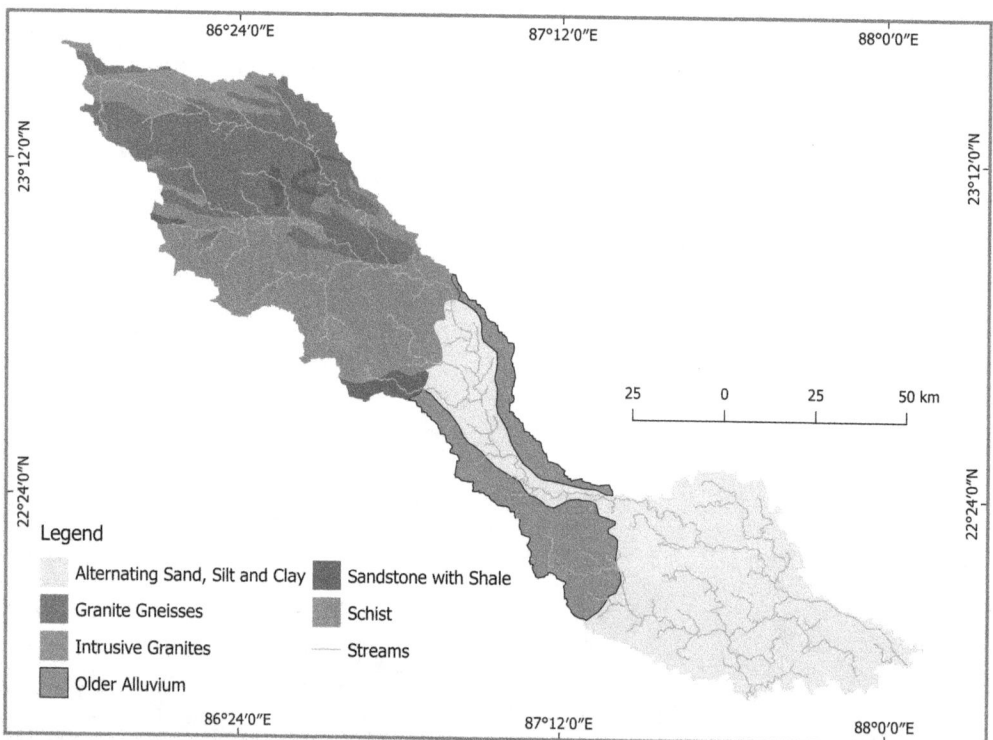

FIGURE 8.4   Drainage network of Kangsabati river basin. (Cartosat-1, 2018 and GSI, 1990.)

Police Station. In this part, the river forms several picturesque small waterfalls. The gradient of the river channel from the source to the Kangsabati dam is 1 in 407 (Figure 8.5). The Kangsabati river leaves Bankura district at the southernmost corner of Raipur Police Station.

The water course of Kangsabati had been used in the recent past to transport timber logs coppiced from forests near Raipur of Bankura and to be collected in Medinipur (*District Gazetteer: Puruliya*, 1985). This function has completely been ceased now because of the forests in the areas have disappeared due to large-scale deforestation. This action has a direct impact on the reduction of channel flow as eroded materials have filled up the channel reducing the navigability of the river.

The river Kangsabati enters into the north-west part of Paschim Medinipur from Bankura as an important tributary of the river Haldi. In Paschim Medinipur district, Kangsabati receives Bhairabbanki and Tarapheni as right bank tributary and in Purba Medinipur district it receives Kaliaghai from right.

The Bhairabbanki river originates from the forest clad highland of Ranibandh Police Station of Bankura district. Then it flows swiftly in a south-east direction through Raipur Police Station of Bankura and again the combined flow from Bhairabbanki and Tarapheni meets Kangsabati in Jhargram Police Station of Paschim Medinipur district. Both the rivers have barrages over them under the Kangsabati Reservoir Project. The river Kangsabati in this section flows in a tortuous course first to south and then to east near the Medinipur town which is situated on its left bank. The gradient of river channel from dam site to Shaldahari is 1 in 1,484 (Figure 8.6).

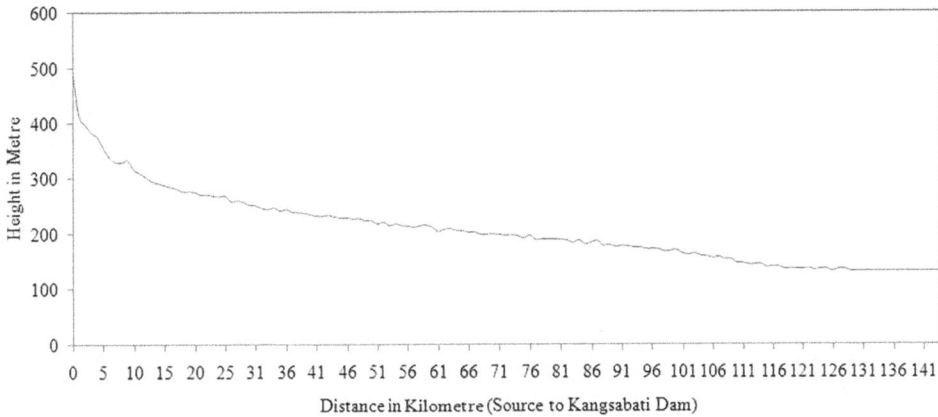

FIGURE 8.5   Longitudinal profile of the river channel from source to Kangsabati dam. (Google Earth, 2018.)

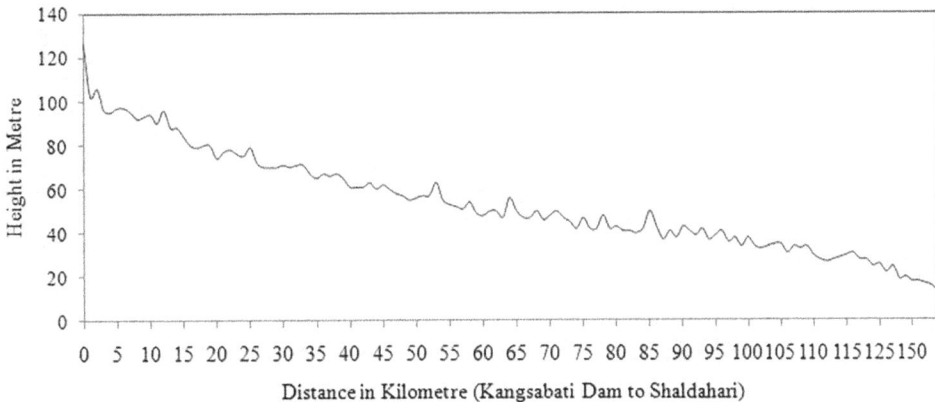

FIGURE 8.6   Longitudinal profile of the river channel from Kangsabati dam to Shaldahari. (Google Earth, 2018.)

The configuration of river channel has become narrow outside the Medinipur town. After flowing towards east in a tortuous course, it bifurcates into two as *Old Cossye* [the northern branch] and *New Cossye* [the southern branch] at Kapastikri of Paschim Medinipur district. At Daspur of Paschim Medinipur district, the *Old Cossye* again bifurcates into two: one known as Palaspaikhal flows further east and the other flows in south-easterly direction. Both the branches ultimately fall into the Rupnarayan river. *Old Cossye* is also connected with Shilabati river through a small channel known as Kankikhal. After flowing towards south-easterly direction, the *New Cossye* meets with Kaliaghai river at Dheubhanga of Purba Medinipur district and forms Haldi river. Kherai and Bakshikhal are the two principal tributaries of *New Cossye* river.

From Medinipur town to Panskura town, the Kangsabati river flows in a meandering course and becomes navigable for boats up to Panskura in rainy season (O'Malley, 1911). But this navigability has been restricted only to the parts experiencing high tides of sea water. Artificial embankment has been made of a considerable length on the right bank of Kangsabati to check inundation of water. But this has caused siltation of the river bed. In addition, siltation in the lower course has been increased periodically due to upsurge of tide. Thus, the depth of the channel has decreased through time. This fact has resulted in the overflow of high-tide water crossing the embankment, sometimes breaching the same. Sudden spill of water overflowing in this part caused devastating flood almost every year inviting considerable damage to crops and other property. In this part, channel gradient of the river from Shaldahari to Buraburi is 1 in 10,048 (Figure 8.7).

After originating in the west of Paschim Medinipur district and flowing in an easterly direction through the Police Stations of Narayangarh and Sabang, the Kaliaghai, another right bank tributary, joins Kangsabati near Jalpai. Kaliaghai receives water from the small streams of Kapaleswari and Chandia (*District Gazetteer, Medinipur*, 1995). Then the combined flow inundates a vast area such as Patashpur, Sabang and Bhagabanpur in every rainy season. The channel gradient from Buraburi to mouth of the river is 1 in 11,248 (Figure 8.8).

The arrangements of drainage lines are mainly controlled by geologic structure, lithological characteristics and local slope aspects of the drained area. The observed drainage patterns in different parts of the basin are represented in Table 8.1.

8.2.1 GEOLOGY AND STRUCTURAL CHARACTERISTICS

The study area and its environs include the eastern part of the Chotanagpur plateau (O'Malley, 1908), greater portion of which is constituted by rolling surface merged with rugged uplands on the west and plains in the east. The residual hills, locally known as *dungri*, bear the testimony of the ancient plateau which has now been changed into present landscape after planation. The basement

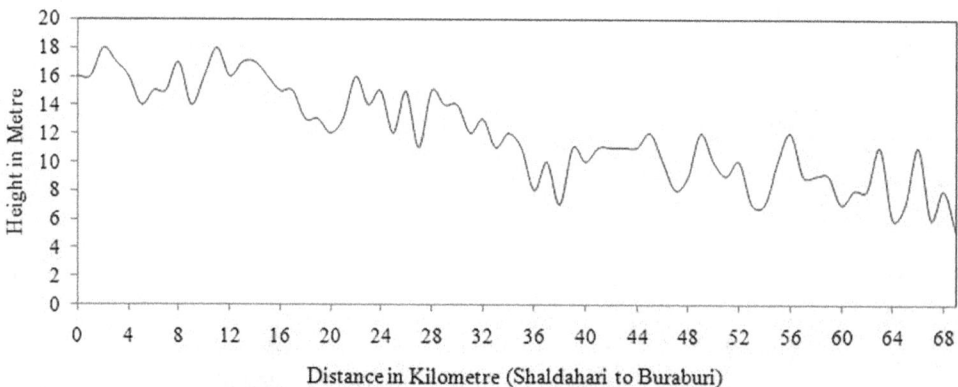

FIGURE 8.7 Longitudinal profile of the river channel from Shaldahari to Buraburi. (Google Earth, 2018.)

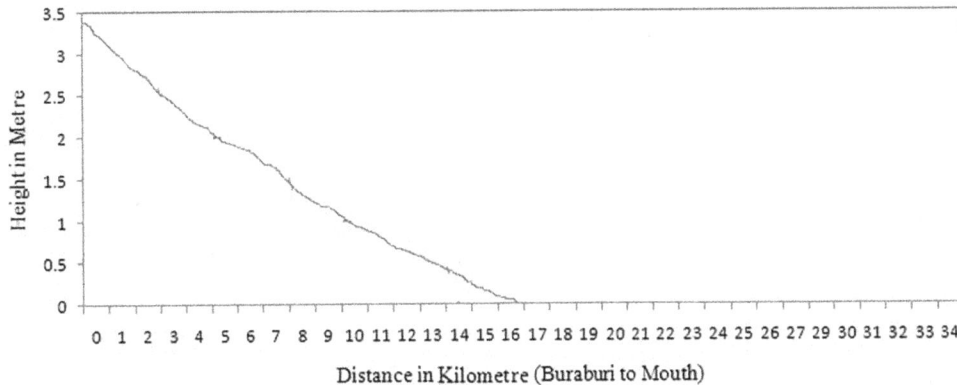

FIGURE 8.8 Longitudinal profile of the river channel from Buraburi to the mouth of the Kangsabati river. (Google Earth, 2018.)

TABLE 8.1
Principal Drainage Patterns of Kangsabati River Basin

Type	Segments	Control	Example
Dendritic	Plateau and plateau-fringe	Homogenous crystalline rocks	Eastern slopes of Baghmundi hill
Parallel	Plateau	Steeply sloping surface	Northern slopes of Baghmundi hill
Radial	Plateau	Residual hill	Barabhum–Balarampur area
Pinnate	Alluvial plain	Easily erodible materials particularly on horizontal homogenous rocks	Lower reaches of the basin
Annular	Plateau	Eroded hills in alternate hard and soft rocks	Upper reaches, at Baghmundi–Dalma segment
Centripetal	Plateau-fringe	Low land	Kangsabati dam area

rock of the study area and its surroundings are composed of Archaean granites and gneissose rocks (Mitra, 1953). Several outcrops of Eocene basalts are visible along the course of Kangsabati and its tributaries flowing through this area. Most of the surface area is composed of superficial Pleistocene alluvial deposits (*District Gazetteer, Bankura*, 1968). Metamorphic rocks are well exposed in several places which are cut up by many veins of granites. Haematites are also seen in and around the area under study. Similarly, mica-schist occurs around the south bank of Kumari [the principal right bank tributary of the Kangsabati]; phyllite and schists of mica and hornblend are found in the *Kansai*–Kumari interfluve (*District Gazetteer, Puruliya*, 1985).

Granites are highly micaceous with crystalline masses of whitish to reddish colour. The superficial deposits covering the study area and its surroundings are Pleistocene alluvial deposits, highly oxidized and leached, and form the part of the classical landscape of *Rarh* region. The newer alluvium of Sub-Recent to Recent age occurs in patches along the course of the Kangsabati and along the course of the right bank tributary of Kumari. A few insignificant patches of laterites, possibly of Pleistocene to Sub-Recent origin, are marked in some parts of the area [Field survey, 2017]. Laterites are formed by sub-aerial weathering effect of the prevailing monsoon climate having alternative spell of wet and dry seasons. The area also contains calcareous nodules, locally known as *ghuting*.

From Figure 8.9, it is observed that the stratigraphy of the rocks that compose the entire basin comprises three geological periods:

i. Azoic to Quaternary

   The rock groups included within this period are granite and gneisses stretching over an area of 1,774.978 km², of which intrusive granites cover 483.83 km², sandstone with shale 207.26 km² and sandstone with schist 1,963.132 km², respectively. This area has confined mainly the upper catchment part of the basin. The lithological characteristics of the area comprise faulted structures, joints, fractures, foliation and lateritic surface which yield low ground water prospects.

ii. Proterozoic to Quaternary

   The middle part of the basin is composed by older and younger alluvium which covers 1,057.80 km² of the basin area. Duricrust, mature deltaic plain, para-deltaic land surface have been developed which yield good ground water prospects.

iii. Upper Palaeozoic to Quaternary

   Older to younger alluvium with alternating sand, silt and clay covers 3,506 km² of the basin area. This part is mainly located in the lower reaches of the basin. This part is characterised by para-deltaic fan surface, deltaic plain with levee and flood basin zone. It has moderate to large ground water prospects.

8.2.2 Surface Elevation and Forms

The study area and its surroundings exhibit unique characteristics in its physical features, completely different from the other parts of the state of West Bengal. Geologically, the basement rock of the area in the upper reaches of the basin is formed of Precambrian and Archaean rocks and the surface rocks are intrusive in origin, being granite, gneissose and schistose in composition (Figure 8.9). These rock types, in combination with climatic weathering process, have given rise to some unique geomorphic surface features as well as laterites and lateritic red soils.

Physiographically, the area may be described as a zone of transition between the young alluvial plains of Ganges delta and the ancient Chotanagpur plateau. The upper part of the basin area located in the district of Puruliya is a part of Ranchi peneplain in terms of its structure and relief (*District Gazetteer, Puruliya*, 1985). The landscape of this lower peneplain is wavier and represented by isolated remnants above a gently sloping eroded platform having an altitude of 100 and 450 m from mean sea level. From south-west to east, the basin has been divided into three distinct topographic zones: plateau, plateau-fringe and alluvial plain. In the upstream part of the drainage basin, degradation is more active, resulting in patches of locally dissected badlands [*Khowai*] upon which residual hills, locally known as *dungri* [monadnocks], are very common. Most of the residual hills are more or less rounded in shape due to exfoliation weathering on igneous or metamorphosed rocks. Exfoliation weathering process on gneissic rocks [called *Bengal gneiss*] results dome-shaped hills, core stones and tors. These picturesque low hills are the outliers of the ancient Chotanagpur plateau. The topographic expression is characterised by a well-defined ridge and valley topography with multi-storeyed terraces of varying altitudes, gullied-surfaces, ravines, knolls and isolated residual hillocks with narrow floodplains that indicate a poly-cyclic nature of complex landscape of the basin (Figure 8.10).

The study area is traversed by Kangsabati river and its numerous tributaries. Kangsabati, the principal river, flows from north-west to south-easterly direction following the regional slope of the land. A large number of rivulets have intersected this gently undulating plain resulting into a number of parallel strips.

Figure 8.11 shows six elevation zones of the basin; less than 100 m elevation zone occupy 51.09% [4,594.53 km²] of the total area and are mostly covered by older and younger alluvium. The second zone with elevation ranges from 100 to 200 m occupying 19.36% [1,741.04 km²] of the basin area

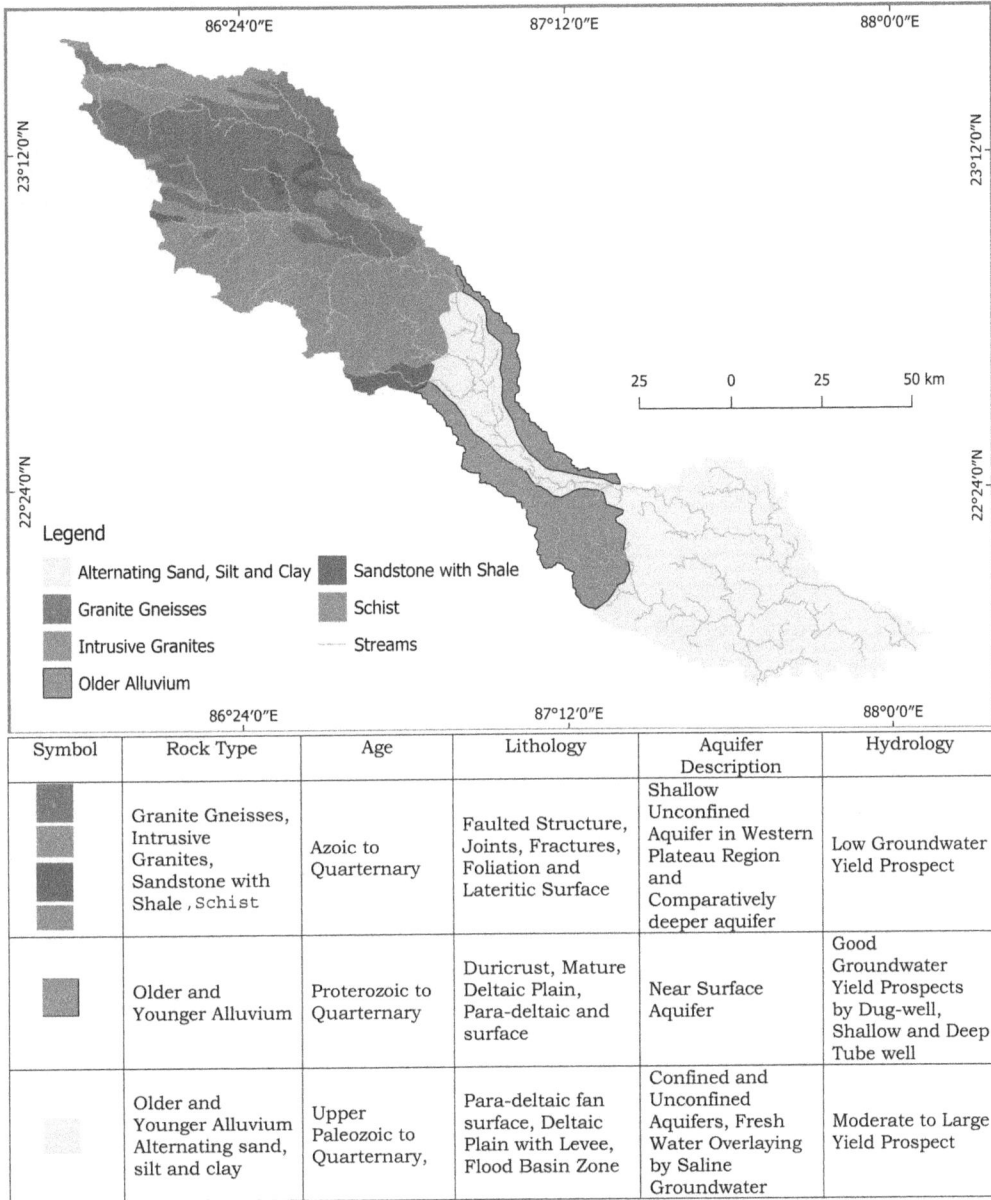

Symbol	Rock Type	Age	Lithology	Aquifer Description	Hydrology
	Granite Gneisses, Intrusive Granites, Sandstone with Shale , Schist	Azoic to Quarternary	Faulted Structure, Joints, Fractures, Foliation and Lateritic Surface	Shallow Unconfined Aquifer in Western Plateau Region and Comparatively deeper aquifer	Low Groundwater Yield Prospect
	Older and Younger Alluvium	Proterozoic to Quarternary	Duricrust, Mature Deltaic Plain, Para-deltaic and surface	Near Surface Aquifer	Good Groundwater Yield Prospects by Dug-well, Shallow and Deep Tube well
	Older and Younger Alluvium Alternating sand, silt and clay	Upper Paleozoic to Quarternary,	Para-deltaic fan surface, Deltaic Plain with Levee, Flood Basin Zone	Confined and Unconfined Aquifers, Fresh Water Overlaying by Saline Groundwater	Moderate to Large Yield Prospect

FIGURE 8.9 Lithological characteristics of Kangsabati river basin. (GSI, 1990.)

and are mainly covered by lateritic soil. The third elevation zone comprises 200–300 m elevations occupying 26.28% [2,363.36 km²] area of the basin and are covered by granite, gneiss and schists. Almost 1.86% [167. 27 km²] area falls between 300 and 400 m elevation zone, 0.81% [72. 84 km²] area falls within 400–500 m elevation zone and the rest 0.60% [53.96 km²] area falls above 500 m elevation zone and compose mainly of granite and gneiss rocks. Figure 8.12 shows slope direction of the basin, and Table 8.2 shows % distribution of slope direction.

The basin under study lies in tropical humid monsoon climatic zone with alternative dry and wet seasons. In the plateau-fringe section, the topography is wavier and the area testifies as prolonged processes of erosion being experienced by intensive chemical weathering processes with sheet wash, rill and gully erosion. The lateritic uplands of the basin especially some parts of the

FIGURE 8.10   Topographic expressions of the upstream part of Kangsabati river basin.

TABLE 8.2
Percentage Distribution of Slope Direction of the Kangsabati Basin

Slope Direction	N	NE	E	SE	S	SW	W	NW
% Distribution	13.21	9.74	24.99	10.14	18.19	6.84	11.64	5.26

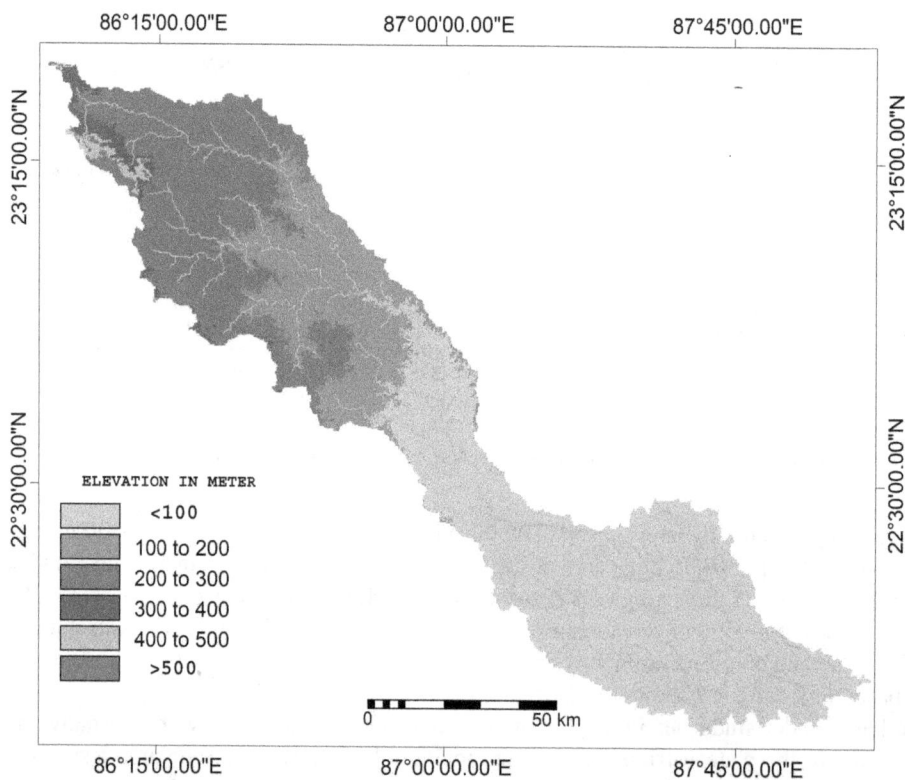

FIGURE 8.11   Surface elevation zones of Kangsabati river basin. (Cartosat-1, 2018.)

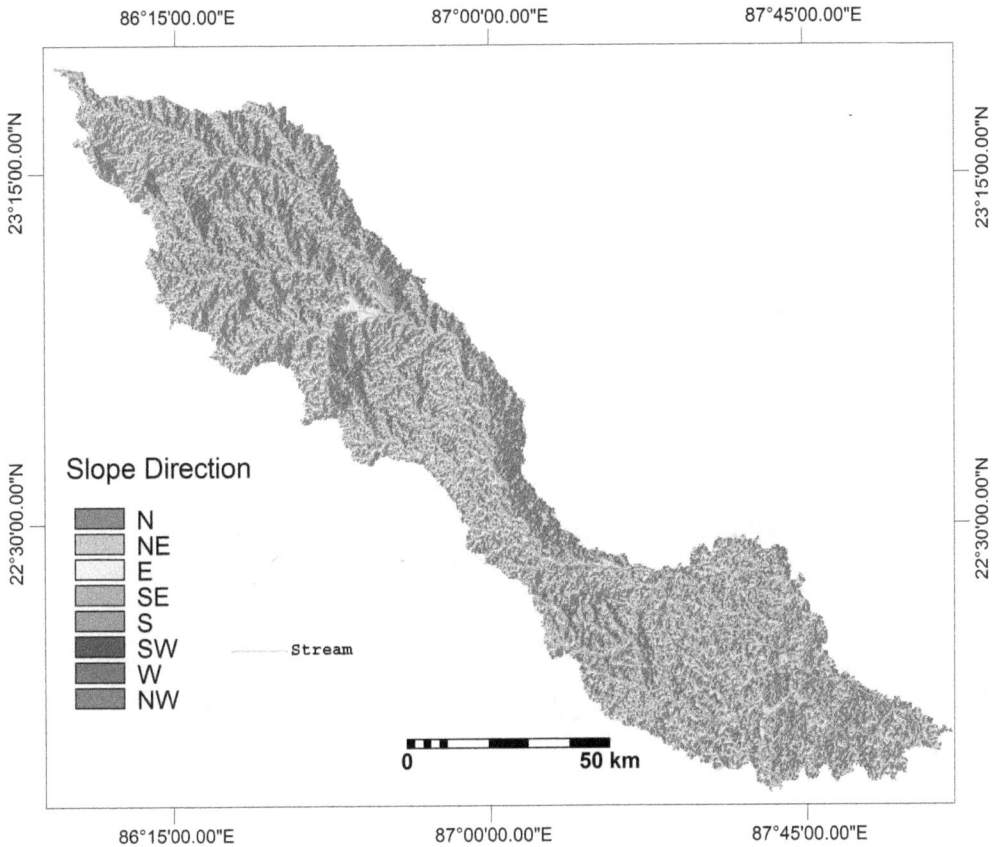

FIGURE 8.12 Slope direction zonation map of the Kangsabati river basin. (Cartosat-1, 2018.)

northern bank of Kangsabati river are highly affected by rill and gully erosion causing siltation and sedimentation in the valley floor (Figure 8.13). The magnitude of this erosion depends on the rate of infiltration and percolation of water into the soil, slope aspects, shears stress for erosion, vegetation cover and anthropogenic factors such as overgrazing and deforestation.

The south-eastern part of the basin comprises a part of the Gangetic delta. In this part, the delta building process and the process of floodplain formation both are active. The elevation of this part is lying below 20 m and the river gradient is very gentle i.e., 1 in 11,248. The wide meandering course of the Kangsabati and Kaliaghai [a right bank tributary] is associated with meander belt deposits, back swamp deposits, braided stream deposits and deltaic plain deposits which ultimately give birth to a large number of micro-level features such as channel bars, point bars, natural levees, floodplain, scour routes, meander scars, neck cut-off, chute cut-off, sand plug, clay plug and oxbow lakes. The form of these micro-level geomorphic riverine features has been modified by direct human interventions, such as erection of road networks, rail lines, artificial levees, temporary and permanent culverts, bridges and practicing agriculture on the floodplain (Figure 8.14).

8.2.3 MORPHOMETRIC ATTRIBUTES

A drainage basin is an organised section of the land surface, and its geometric features are functionally related with each other (Chorley and Kennedy, 1971). Drainage basin as the fundamental unit of geomorphic analysis has been emphasised by Horton (1932 and 1945), Leopold and Maddock (1953), Strahler (1957), Leopold, Wolman and Miller (1964), Shreve (1966), Chorley (1969), Gregory

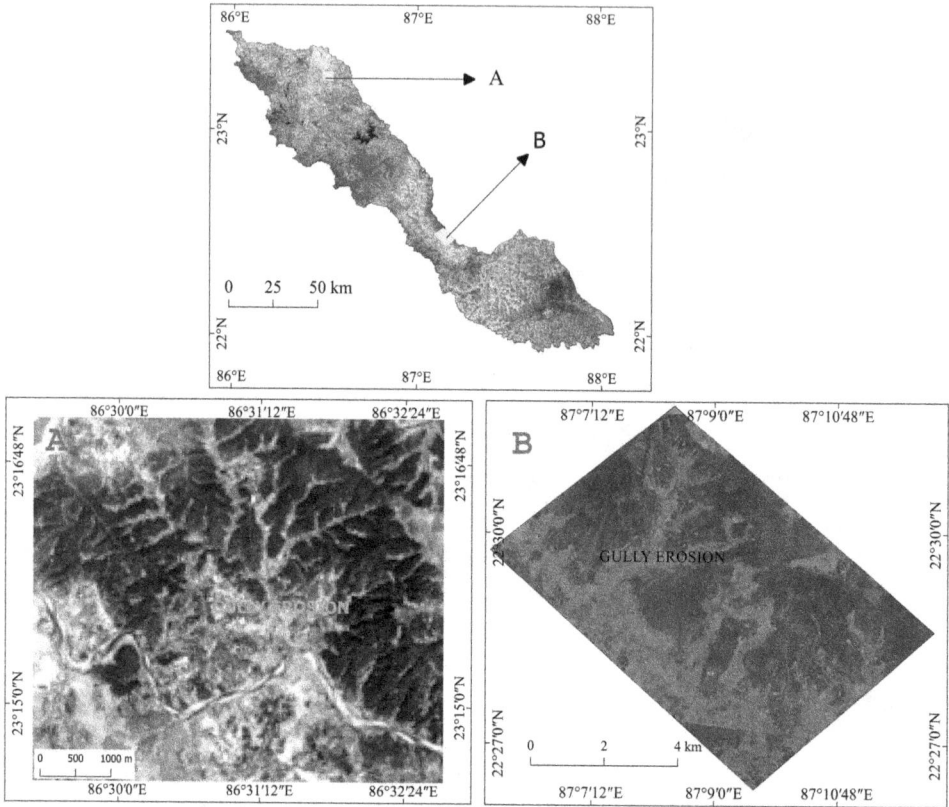

FIGURE 8.13   Rill and gully erosion in lateritic parts of the Kangsabati basin. (Resourcesat-1: LISS-III and Google Earth, 2019.)

and Walling (1973), Gregory (1976), Richards (1982), Chorley et al. (1984) and Morisawa (1985) that attracted a large number of geomorphologists throughout the world to comprehend the nature of landscapes, drainage networks and fluvial dynamics of a drainage basin. In the present case study, the areal and linear properties as well as shape factor of Kangsabati river basin as a whole are represented in Figure 8.15 and Table 8.3.

8.3   ANTHROPOGENIC SIGNATURES ON FORMS AND PROCESSES

In the recent past, human beings have emerged as an important agent of geomorphic change in the drainage basin along with endogenetic–exogenetic processes. Knighton (1998) categorised human-induced channel changing processes of river into two broad groups:

    i. Direct or channel phase changes

       This is caused by river regulation through the construction of water storage reservoirs and diversion of water through canals. Channel modifications are caused by bank stabilisa-tion, channel straightening and stream gravel extraction.

    ii. Indirect or land phase changes

       This is caused by landuse changes through removal of vegetation especially deforesta-tion, afforestation, changes in agricultural practices, building construction, urbanisation and mining activity. Land drainage systems are also maintained by agricultural drains and storm-water sewerage systems. Among the issues, the impacts of dam on river are more imperative than others.

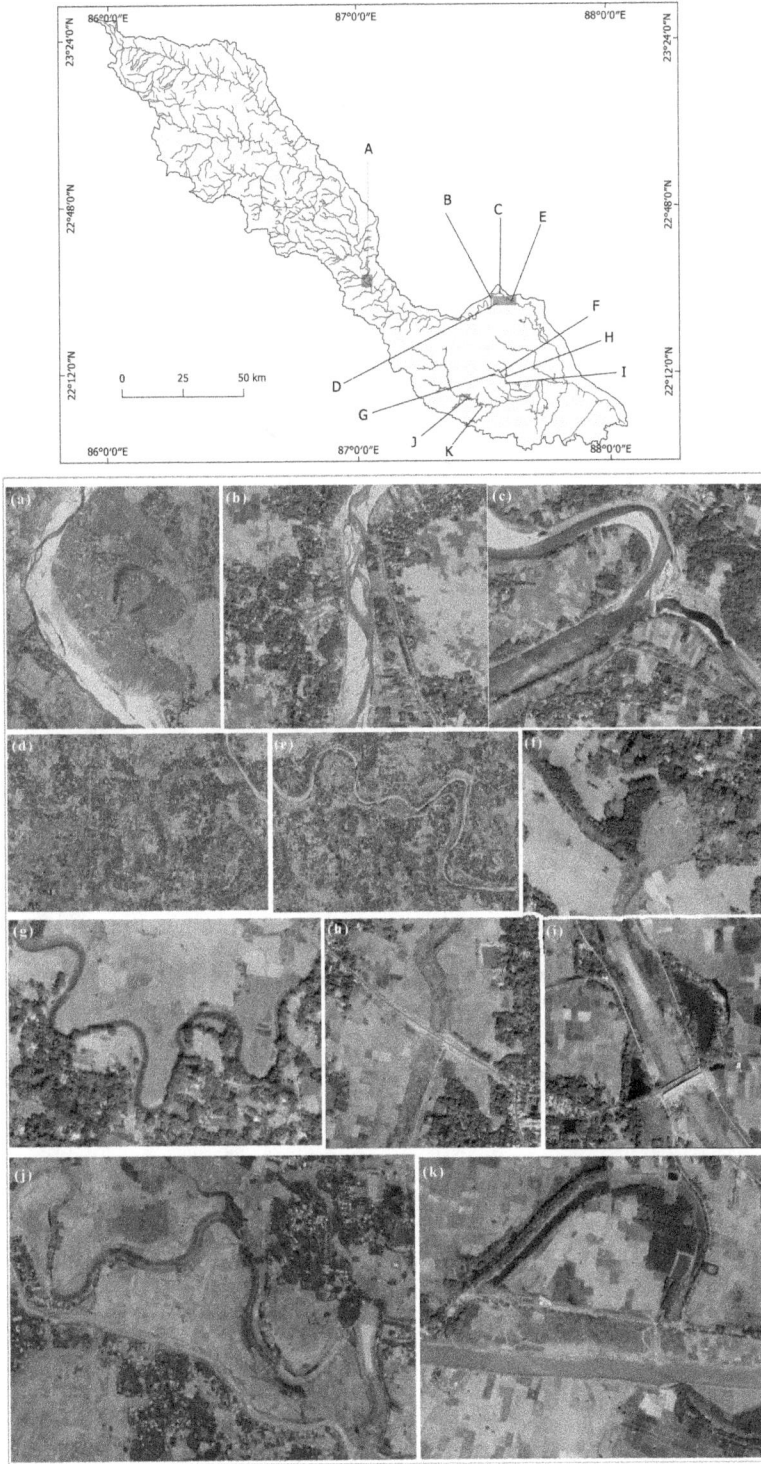

FIGURE 8.14 Micro-level modified floodplain features of Kangsabati basin. (Google Earth, 2019.) [(a) oxbow lake, (b) braided Channel pattern, (c) river channel condition after cross bund construction,(d) signature of palaeochannel, (e) embankment along the river bank, (f) signature of river cut-off and oxbow lake, (g) meandering river channel, (h) channel condition after bridge construction, (i) back swamp deposits, (j) dry cut-off river channel and (k) wet cut-off river channel.]

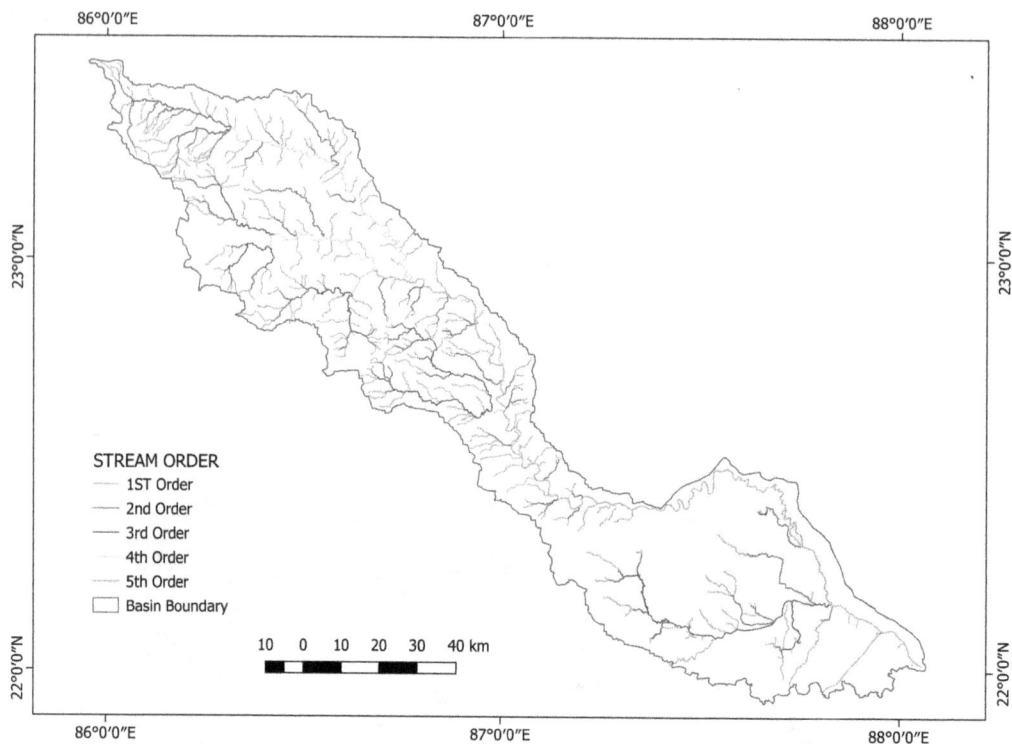

FIGURE 8.15   Stream order of Kangsabati river basin (After Strahler (1952)). (Cartosat-1, 2018.)

TABLE 8.3
Morphometric Properties of the Kangsabati River Basin

		Mathematical Expression	Source	Calculated Value		
Linear aspect	Stream order	-	Strahler (1952)	5th order		
	Stream number	$N_u = R_b^{s-u}$	Horton (1945)	402		
	Stream length	$L_u = L_1 R_L^{u-1}$	Horton (1945)	416.65 km		
	Law total stream lengths	$\sum L_u = L_1 R_b^{s-u} R_L^{u-1}$	Horton (1945)	2,692.8 km		
	Stream length ratio	$R_L = L_u/L_{u-1}$	Horton (1945)	Stream Order	Stream Length in km	R_L
				1	1,377	-
				2	540.2	0.39
				3	332.6	0.62
				4	154.7	0.47
				5	288.3	1.86
	Bifurcation ratio	$R_b = N_n/N_{n+1}$	Strahler (1957)	Stream Order	No. of Streams	R_b
				1	303	-
				2	75	4.04
				3	15	5.00
				4	8	1.88
				5	1	8.00
	Mean bifurcation ratio	-	-	4.73		

(Continued)

TABLE 8.3 (*Continued*)
Morphometric Properties of the Kangsabati River Basin

		Mathematical Expression	Source	Calculated Value
Areal aspects	Basin perimeter	P	-	852.04 km
	Basin area	$A_u = A_1 R_a^{u-1}$	Schumm (1956) (Cited in Morisawa, 1985)	8,993 km^2
	Constant of channel maintenance	$Lof = 1/D_d$	Horton (1945)	3.33
	Stream frequency	$F_s = N/A$	Horton (1945)	0.04 km^2
	Drainage density	$D_d = L/A$	Horton (1945)	0.30 km/km^2
	Texture ratio	$T = N_1/P$	Horton (1945)	0.36
Shape factor	Form factor	$F = A/L^2$	Horton (1932)	0.122
	Shape	$S = L^2/A$	US Corps of Engineers	8.168
	Circulatory ratio	$C = 4\pi A/P^2$	Miller (1953)	0.01
	Basin elongation	$E = 2\sqrt{(A/\pi)}/L$	Schumm (1956)	0.22
	Lemniscate ratio	$K = L^2/4A$	Chorley et al. (1957)	2.042

Indeed, a river is the channelised flow of water and a dam is a barrier which checks that flow. But from the geographical point of view, one can consider a river as something more than water flowing through its channel. Its ever-shifting beds and banks, the marshes, the meadows, dead channels, palaeo-channels, back waters of its floodplain and the ground water below can be seen as a part of river and the river as part of them. A river carries not only water through its channel towards down slope but also sediments, dissolved minerals and the nutrient rich in organic matter either dead or alive (McCully, 1998). It has now been observed that human interference is also becoming increasingly relevant in the river regulation by means of constructing dams, barrages, weirs, embankments and changing pattern of landuse and landcover in a river basin. The damming of the rivers in different parts of the world has brought a profound change in the course as well as watershed areas. Nothing alters a river, as like as a dam alters. A wild river is always dynamic. It changes its course through erosion and deposition. A dam is static. It tries to bring a river under control by regulating its seasonal pattern of floods and slow flows. Downstream morphology of river bed and banks, delta, estuary, coastline and biodiversity is altered as a result of change in the upstream from river valley to reservoir. A dam not only traps sediments and nutrients but also alters the river's temperature and chemistry. It upsets the geological processes of erosion and deposition through which the river sculpts the surrounding land (McCully, 1998).

Impacts of large dams are manifold, and most of them are deep rooted. Some of the impacts are direct and often known to common people, while some are indirect but often extreme. Although impacts on physical and economic environment are perceptible and logically quantifiable, the social impacts, though imperceptible, are more severe and deep rooted, which takes on the body and mind of the sufferers. The physical impacts sometimes affect negatively on the economy, and economic sufferings affect the ethno-cultural particularities of the sufferers. All the time, it implies that the impact is essentially social in character. Likewise, physical–ecological questions of the impacts of the dam ultimately negatively affect the human–ecological questions. Nearly all cases, a large dam itself is considered as an element of physical, ecological, economic and social concern. The WCD (2000, pp. 15, 16) observes that "large dams generally have a range of extensive impacts on rivers, watersheds and aquatic ecosystems and have led to irreversible loss of species". Throughout the world, large dams have altered 60% of the length of the world's large river systems and have caused a rapid loss of fresh water bio-diversity (WCD, 2000). Large dams in populated areas, particularly in the third world countries, have submerged an enormous area of rich forests and fertile agricultural lands in their catchment areas. As a result, a remarkable number of poor people have been evacuated without taking any proper resettlement and rehabilitation strategy for greater interest of the group living in the command areas. Large dams have flooded hundreds and thousands of kilometres of valuable lands, including irreplaceable habitats for endangered species and the farmlands of the rural poor. By observing the massive impacts of dam on human beings, once Church (1968, pp. 1) noted that

> Few of man's modifications of the landscape can initiate such profound physical, economic and social changes as dams … by harnessing water, we are taming a natural element, one of man's most fundamental assets, an essential of civilization, yet one in short supply.

The same view is noted by Skalak et al. (2013, p. 63) that "One of the greatest influences that humans have had on the fluvial landscape is the construction of dams". In terms of energy input–output, river basin is considered as an open system. In fluvial systems, the shape, size, morphological characteristics of valley, bed and channel of the river change due to the alteration of sediments and discharge by a dam. Grant et al. (2003, p. 210) noted that artificial dams along the course of river "alter two critical elements of the geomorphic system: the ability of the river to transport sediment and the amount of sediment available for transport". Whenever a dam is constructed on the course of any river, it has naturally brought perceptible changes in the ecology and environment of the area. Impact of any change does not only affect the physical state of an area, but the living organisms must have to experience those changes with their life and survival. Dam construction is the direct conviction in the ecological principles on the free flow of a river. In the language of Shiva (1991, p. 188), "Violence to the river is violence to the communities inhabiting a river basin". The essence of a river is to flow, while the essence of a dam is to remain still or static. Thus, a dam not only dissects the integrity of a river ecosystem but also changes the unique ecosystem into four distinct zones:

  i. A demand zone in the command area
 ii. The supply zone in the catchment area
iii. The riparian downstream zone
 iv. A transition zone lying between the catchment-command zone

After the construction of the dam, actually the river is taken away from the people who are living in the downstream parts. It means that their very source of drinking water, irrigation water and water for domestic needs is taken away. The ground water recharge is affected as the river dries up below the dam to a considerable extent. The biodiversity and fisheries are destroyed in the midstream, downstream and deltaic part of the river basin. The concentration of pollution in the downstream parts of the river increases due to stoppage of fresh water flow and the navigation of the downstream communities ceases. Sudden release of water from the dam increases severity of flash flood in the immediate downstream part of the dam as well as in the lower courses of the river. The geomorphologic behaviour of the river water i.e., the eroding power is altered due to trapping of the silts in the reservoir, and finally, the salinity ingress increases due to reduced flow of fresh water which ultimately affects the soil and ground water in the coastal zone (Thakkar, 2009; WCD, 2000; McCully, 1998). The Kangsabati dam was constructed just above the confluence of Kangsabati river and its tributary Kumari with two major objectives: irrigation and moderation of flood (Kangsabati CADA, 2007). Generation of electricity was not considered in its plan. The dam has been raised over both the rivers and joined subsequently to form a single reservoir. Canals have been constructed on both the banks. The left bank canal in its way meets with the Silabati river where a barrage has been constructed and then crosses it to irrigate lands between Silabati and Dwarakeswar rivers. In right bank, similar barrages have been constructed over Bhairabbanki and Tarapheni rivers to irrigate the lands in Jhargram and Medinipur South Sub-Divisions of Paschim Medinipur district. In this scheme, there were also provisions for supporting riparian rights for the existing irrigation system from Medinipur ancient canal system which was operated since 1872 to moderate peak flood in the lower valley of this river and to provide drinking water during the time of dry spells. Although the project was formally approved by the Government of India in the year 1961–1962 (Kangsabati CADA, 2007), the construction work was commenced in March, 1956. In the first phase, a dam was raised over the Kangsabati river. Irrigation started from 1966 on the *kharif* lands in the areas where canal network had been established since then. After that, the dam was completed over Kumari [a right bank tributary of the river Kangsabati] in 1973–1974 and both the dams were then connected to form a single reservoir named 'Kangsabati Reservoir' (Figure 8.16).

Anthropogenic impact study on river could not be done in proper way due to several reasons, of which inadequacy of data on almost all aspects was a reality. This is also true for the present study. However, when it is very much hard to avail the scientific information without any recorded data, complete assessment for anthropogenic impacts on river is difficult. Thus, assessment of the anthropogenic impacts on the forms and process of Kangsabati river may be presented in a descriptive way along with quantitative approach wherever possible. In the present case study, the impacts of dams on fluvial forms and process are more pronounced than others.

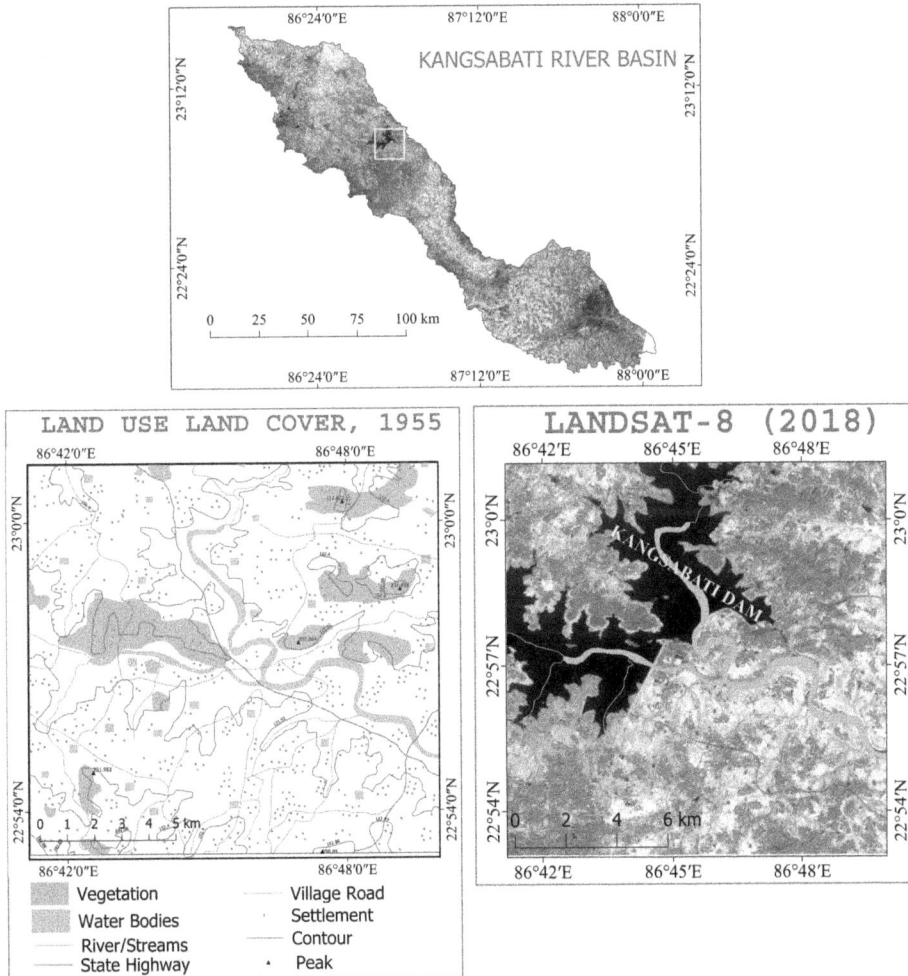

FIGURE 8.16 Parts of Kangsabati river and its surroundings in 1955 and 2018. (US Army Toposheet, 1955 and Google Earth, 2019.)

8.3.1 IMPACT OF THE DAM IN THE UPSTREAM AND PERIPHERY AREA

Establishment of the Kangsabati dam and its reservoir has submerged a sizeable extent of land situated just upside the dam wall. The dam has submerged 33,760 acres [13,660 ha] of land in total (Kangsabati CADA, 2007). It is the second-largest earthen dam in India next to Hirakund. Total length of the dam, excluding intermediate hillocks, is 10.4 km [6.50 miles]. Huge amount of stone slabs, boulders and earth/soils were used to construct the dam walls. Large amount of earth cutting was needed to supply those filling materials, for which, *dungri* or residual hillocks and relatively uplands were cut down to a considerable depth, converting the uplands into stair-like structures. Thus, earth cutting and levelling ultimately made the uplands [*tanr* and *dungri*] as the flat lands sometimes suitable to grow poor-quality, short-lived crops consumed by the local poor people living contiguous to the upstream section of the dam. This activity has brought a change into the character of the land in the surrounding areas of dam. Even after the completion of the dam, the people who lost their precious good quality agricultural lands, orchards and homesteads under the water of the reservoir, apply labour and effort to reclaim new lands on the periphery of the dam water for agricultural purposes. A considerable extent of relatively highlands were further converted into terraced-level lands suitable only to grow short-duration, low-water demanding, poor-quality crops like local pulses (Field Survey, 2018). The present status of landuse and landcover in and around the dam area is represented in Figure 8.17 and Table 8.4.

There are numerous igneous outcrops throughout the periphery of the dam, as the whole area is formed of old crystalline granitic and gneissic rocks. Local people dig up boulders by excavating these outcrops and break up the boulders into small stone chips to be sold out to the local agents. The rock boulders have a considerable depth or root which are dug up with mechanical efforts. Thus, after digging up and excavation, this area has turned into useless wasteland where neither agriculture nor forestry is possible. This activity causes accelerated soil erosion. In the peak season, when the reservoir is fully filled up with water, the wind blowing over the water surface forms waves of considerable force which hit the lands located at the margin of the reservoir water. Severe soil erosion is thus caused to those arable lands, the top soil and nutrients are washed down to the reservoir bed. This can be considered as an important impact of water storage.

Impedimentation over a river means stagnation caused in its flow. This stagnation directly affects the functioning of a river. The water stored in the reservoir checks the flow and dynamicity of the river, consequent upon the stagnation of the river and tributaries of the upstream. The whole area is made up of hard crystalline and impermeable hard rock which is characteristically against the percolation of water to the lower layers, and hence, recharge of the ground water becomes difficult. One of the most important effects of the reservoir is that the stored water slows down the inflow of the upstream rivers which is most helpful for unloading the gravels, silts and sediments carried down from the catchment areas. The upstream catchment part being deforested is susceptible to easy erosion and denudation. The eroded materials, not always able to reach the reservoir bed, are released in the beds and channels of the river where the flow of water is slow. In addition, the bed of the reservoir being filled up at a much faster rate than estimated and the possibility of percolation of water to the sub-surface aquifer is slowed down and the longevity of the dam is reduced. Thus, the surface hydrology of the upstream part experiences a deviation from its natural function. As the height of the water increases due to siltation, the water of the reservoir follows the chances of lateral movement, increasing the sub-surface water table of the areas contiguous to the reservoir. On the other hand, as the water for irrigation is released through the canals, the height of the water table consequently comes down, keeping a balance to the level of the reservoir water. As a result, the sub-surface water table, once alimented with the rise of the level of reservoir water, quickly seeps out with the decrease of the level of the reservoir water. Thus, rise and fall of the water level in the reservoir directly affect the sub-surface hydrology of the areas surrounding the upstream periphery of the reservoir.

LAND USE LAND COVER, 2018

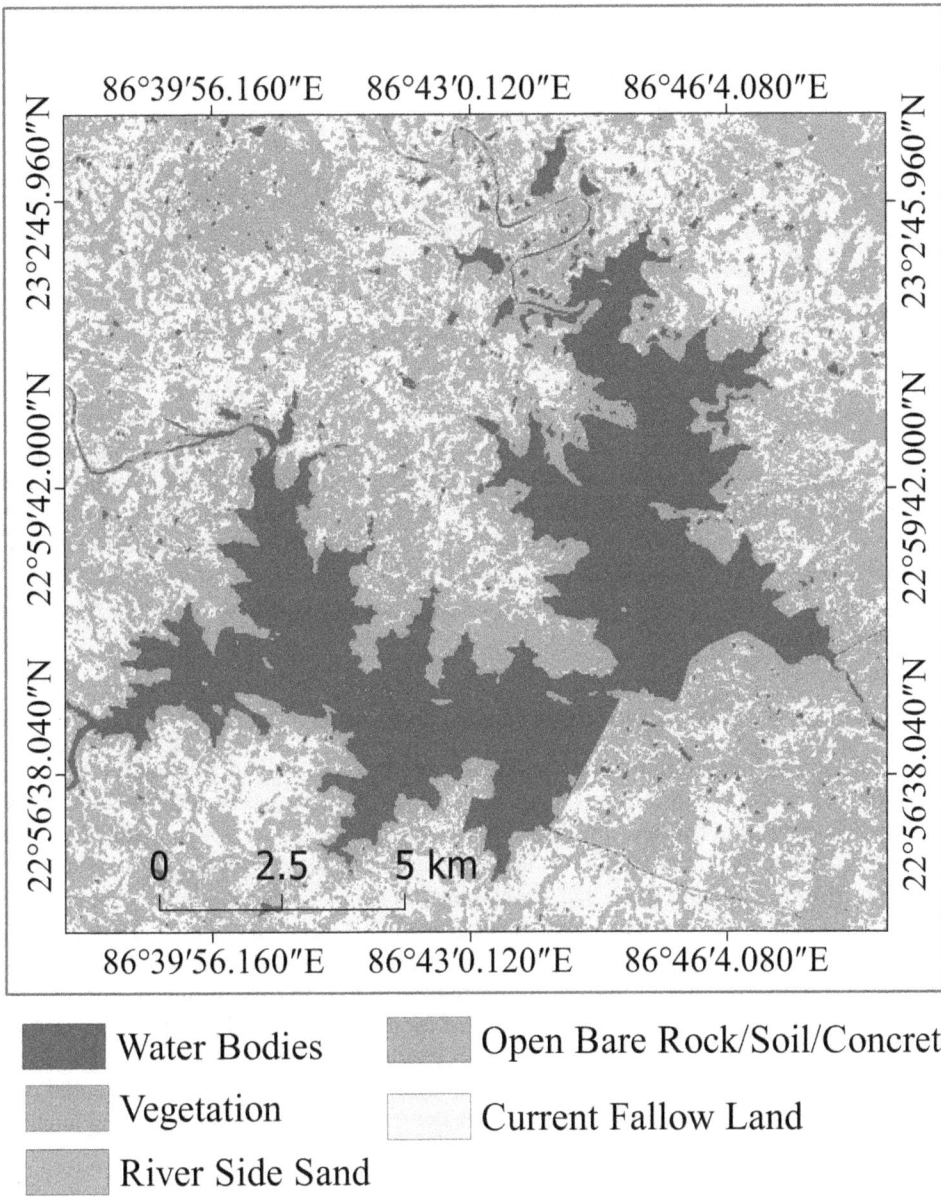

FIGURE 8.17 Landuse–landcover map of Kangsabati dam and its surroundings. (Landsat-8, 2018.)

8.3.2 CONCOMITANT PROBLEM OF RESERVOIR SEDIMENTATION IN THE UPSTREAM SECTION

During the planning of the project, an estimation of the silt deposit was attempted in 1953 by River Research Institute, West Bengal. The average silt concentration of suspended load during the period from June to December 1953 was estimated to be about 0.671 gm l⁻¹ (*Modernisation of Kangsabati Reservoir Project*, 1988).

Kangsabati Reservoir sedimentation studies were carried out by Irrigation and Waterways Department, Government of West Bengal in three consecutive periods of 1970–1971, 1976–1977

TABLE 8.4

Landuse–Landcover of the Kangsabati Dam and Its Surroundings, 2018

Sl. No.	Landuse–Landcover		Area (km²)	% of Area
1	Surface water bodies	Dam	45.16	16.52
		Others	13. 38	4.89
2	Vegetation		48.57	17.77
3	Open bare rocks/soils/concrete		64	23.41
4	Current fallow land		72.49	26.52
5	Degraded land		29.77	10.89
	Total		273.37	100

and 1993–1994. During the first period of survey in 1970–1971, it is noted that 3.99% of dead storage and 1.30% of live storage capacity of the reservoir has been lost. This rate of loss during the second survey period of 1976–1977 was 6.97% and 2.51% correspondingly. In the third phase of survey 1993–1994, it has been estimated that 26.18% of dead storage and 9.52% of live storage capacity of the reservoir has been lost. Although the rate of reservoir sedimentation is not very severe in present study, yet in course of field survey [2005–2018] in the upstream periphery of the dam, it has been observed that the reservoir level is rising upwards in monsoon months in sequential way (Figures 8.18 and 8.19).

8.3.3 IMPACT OF THE DAM IN THE DOWNSTREAM AND PERIPHERY AREA

The most outstanding feature in the immediate downstream section is drying up the Kangsabati river channel due to dam construction. Some parts of the old channel of Kansai-Kumari below the dam site have been converted into agricultural land. This conversion has changed natural land into culture one, conversion of a wet land into arable land, destroying the natural aquatic environment. But this alteration has put economic benefit for a few numbers of economically backward scheduled caste and scheduled tribe people. The emergency outlet, artificially constructed with hydraulic engineering to release the excess water or to avoid the danger of the breaching of the dam, is an important feature related to the physical environment of the immediate downstream area. This outlet has been extended for a considerable distance whose end point has been connected with the original but dried-up course of the Kangsabati river. Though this type of outlet construction is a rule for any river dam, the channel has been constructed on the brittle and decomposed granitic and schistose rocks where soil cover is very thin and anchorage

FIGURE 8.18   Traditional agriculture is being practised on the silted reservoir bed by dam-affected people.

FIGURE 8.19 Reservoir water gradually swallows the agricultural land in the upstream periphery area.

of vegetation roots is very slow and shallow. Thus, in every rainy season, when the dam is totally filled up with water, a huge volume of water is released through this outlet causing severe erosion in this part, even people living adjacent to the outlet faced trauma of flooding [Field Survey, 2009]. The incision of the channel bed during peak flows of sediment-free hungry water develops a coarse-grained rocky surface layer in the artificial outlet and adjacent river bed just downstream part of the dam (Figures 8.20 and 8.21).

Before the construction of the dam, a large part of the area was covered by dense *sal* [shoria robusta] forest. A sizeable part of that forest was destroyed for construction of the outlet channel, by the side of which old degraded vegetations are still evident. The member of tribal community mainly the *kharia* [Sabar], who had to be displaced, from Pareshnath *mouza* on which the dam wall has been constructed, has been rehabilitated in Barda mouza located below the dam (Figure 8.22). This *mouza* was partly covered by *sal* forests, which have been cleared for reclamation of agricultural land. Thus, in the immediate downstream part, a sizeable area of forest has been destroyed, obviously, which is a direct effect of the dam.

The impedimentation made over the river has disturbed the natural supply of water, as water below the dam has been artificially and totally stopped. Besides, as vegetation is one of the agent of holding soil moisture and water, their disappearance have affected both the surface water and the water just below the soil. The alteration of previous river bed into agricultural field is another cause of reduction in the humidity of the soil and stopping of the flow of water, which was the source of replenishment of sub-surface water. Thus, absence of forest and absence of river jointly affected the water resource of these downstream parts. As per field survey [2008–2018], it has been observed that all the dug wells, ponds and the natural courses of river to a considerable extent become dried up in the summer creating acute shortage of useable water for the whole areas, even the deep tube wells established by the government sometimes fail in the summer and the depth of the sub-surface water is now receding gradually as perceived by the local people.

FIGURE 8.20   Coarse-grained rocky surface layer in the river channel just below the dam. (Google Earth, 2018.)

FIGURE 8.21   Degraded parts of the river bank just below the artificial outlet of Kangsabati dam.

FIGURE 8.22   Dam-affected tribal people [Sabar tribe] in the downstream area.

8.3.4   Overall Anthropogenic Impact in the Downstream Part of the River

Gregory (1997, p. 3) mentioned that "Fluvial geomorphology is the branch of earth science that is particularly concerned with rivers and with their present behaviour, the effects that they have in contemporary scenery, and the ways in which they have developed in the past". In a similar vein, Schumm [1977] communicated that "In reality, the fluvial system is a physical system with a history" (Knighton, 1998, p. 261). Therefore, a perceptive of past history of rivers and present behaviour can provide a hint how rivers form, and geomorphic framework might change further in near future. In his provocative paper entitled *Downstream Hydrologic and Geomorphic Effects of Large Dams on American Rivers*, William L. Graf (2006, pp. 336, 337) mentioned that

> Writing half a century ago in *Man's Role in Changing the Face of the Earth*, Leopold [1956, p. 646] predicted that dams would someday become so numerous on American rivers that they would be the primary factor in controlling the characteristics of river channels.

This is true not only for American rivers but all over the world rivers where dams have made fragmentation and disruption in the natural course of the rivers. By constructing dams, human beings have transformed these natural life blood arteries into quasi-natural systems. The downstream geomorphologic impacts of dams are appropriately reflected in the writings of a host of researchers including geologists, engineers, geomorphologists and ecologists in the recent past. Reservoirs formed behind the dam wall drown existing river channels, trap sediments and alter regular flows of the river. Immediate downstream reaches of the river respond to altered flow regimes and reduced sediment supply in varied ways (Mittal et al., 2014; Ronco et al., 2009; Pizzuto, 2002; Williams and Wolman, 1984; Petts, 1979). The configuration of channel widths downstream from the dams may be narrowed, widened or remain constant, depending on the site over which river flows (Williams and Wolman, 1984). To make a comparison of channel width in pre-dam and post-dam situation of the river just below the Kangsabati dam, two maps [US Army Toposheet, 1955 and Google Earth, 2019] have been used. To measure the linear cross-channel length, four points [A–D] have been chosen in both the channels (Figure 8.23). The cross-channel lengths in four points show decreasing tendency of channel width in post-dam condition (Table 8.5).

Brandt (2000) demonstrated that the downstream geomorphologic effects of dam vary profoundly depending on location, environment, substrate, released water and sediment. Besides, the effects

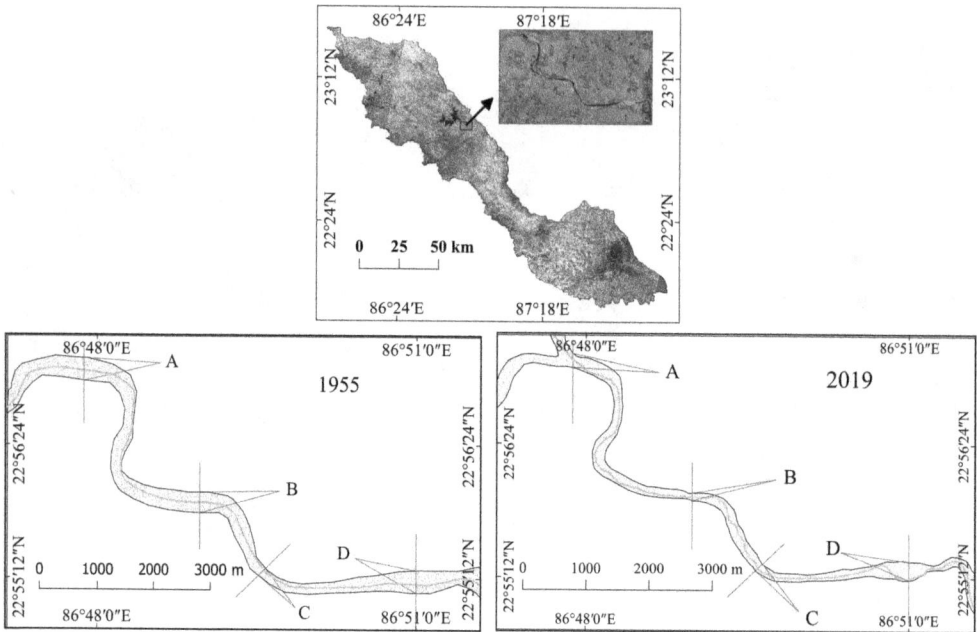

FIGURE 8.23　Changing configuration of channel width of Kangsabati river just below the dam. (US Army Toposheet, 1955 and Google Earth, 2019.)

TABLE 8.5

Channel Width of Kangsabati River just below the Dam (1955 and 2019)

Section Points	Channel Width in m [US Army Toposheet, 1955]	Channel Width in m [Google Earth, 2019]
A	380	240
B	340	120
C	170	160
D	380	300

may differ depending on the type of bed and bank material and the grain sizes of the transported material. Magilligan and Nislow (2005) analysed pre- and post-dam hydrologic changes after dam construction. They noted that by exerting timing, magnitude and frequency of low and high flows, dams alter previous hydrologic regime of river. By using historical maps of different periods and hydrological data, Li et al. (2007) noted that after a lapse of 50 years anthropogenic activities cause minor channel widening, riverbed incision and frequent bank failure that enhanced channel bar formation in the middle reaches of Yangtze river. Reduction of average annual flow, changing pattern of aggradations and degradation from the river mouth to upstream course of the lower Sakarya river are intensively studied by Isik et al. (2008). By using metrics: sediment supply and transport capacity, post-dam stream competency and potentiality of channel incision and relative reduction in flood magnitude, Schmidt and Wilcock (2008) noted that the channels change results from the imbalance between sediment supply and stream flow. Lin (2011) analysed positive and negative ecological impacts of dam in riverine ecosystems. An in-depth study of the morphological effects of damming was carried out by Ronco et al. (2009) in the lower Zambezi river. They demonstrated that an undisturbed river was affected by the construction of the Kariba and Cahora Bassa dam in minor way. But the construction of dams has produced erosion on delta area. After the construction of Three Gorges

Dam on Yangtze river, the water level in the river decreased appreciably in time and some parts of the riverbed had turned from being sediment sinks to sediment sources (Dai and Liu, 2012).

The channel morphology of a river basin is controlled by geologic and climatic variables as well as by human activities. The study area is situated within the regime of monsoon climate characterised by alternative dry and wet seasons. The maximum amount of rain occurs in the summer months. Annual rainfall in the catchment area of the river is 130 cm of which 80% amounts of rainfall occur in the months from middle of June to middle of September, July being the rainiest month of the year and December being the month of the lowest rainfall. The pattern of rainfall is erratic in nature and shows occasional quantitative variation from year to year. As a result of such rainfall pattern, the Kangsabati river has achieved transient characteristics in its flow. The river has very sluggish flow year round and a high peak flow on a certain monsoon months. In the Kangsabati catchment, hydrological study gives evidence that during the period of every 2 or 3 years, there is a chance of drought condition and subsequently there is a high flow year (Table 8.6 and Figures 8.24 and 8.25). These altered flow cause flash floods in the upper reaches of the river above the dam site, and in the lower reaches of the river, devastating floods occur at an interval of 2 or 3 years by the joint action of rain water and discharge water from the dam.

River flood either overbank or flash flood in Kangsabati basin in devastative way occurred in the years 1978, 1986, 1991, 1995, 2000 and 2013 (State Disaster Management Plan, 2014). To reduce the flood peaks to a more or less harmless magnitude, 200,000 acre feet space in the reservoir has been kept to detain flood water during monsoon months. This space is used to moderate a peak flood discharge of 10,627–6,376 cumecs. Figures 8.26–8.28 and Tables 8.7a–8.9 show inflows and outflows of water [through spillway] of Kangsabati dam during specific periods of 17–23 August 2007, 1 June to 30 September 2015 and 1 June to 30 September 2017, respectively. As the part of basin area of 2,500 km² lie below the dam, flood in the lower reaches of the basin occurs due to the

TABLE 8.6
Inflow to Kangsabati Reservoir in Pre-Dam and Post-Dam Conditions

Pre-Dam Inflow to Reservoir Site

Post-Dam Inflow to Reservoir

Pre-Dam Situation	Inflow to Kangsabati Reservoir Site [ha m^{-n}]	Post-Dam Situation	Inflow to Kangsabati Reservoir [ha m^{-1}]	Post-Dam Situation	Inflow to Kangsabati Reservoir [ha m^{-1}]	Post-Dam Situation	Inflow to Kangsabati Reservoir [ha m^{-1}]
1940–1941	160,936.08	1972–1973	278,148	1986–1987	190,970	2000–2001	200,776
1941–1942	303,802.88	1973–1974	78,376	1987–1988	122,935	2001–2002	125,326
1942–1943	357,743.04	1974–1975	166,393	1988–1989	176,401	2002–2003	125,956
1943–1944	218,077.62	1975–1976	82,206	1989–1990	228,670	2003–2004	346,797
1944–1945	212,757.11	1976–1977	154,159	1990–1991	107,514	2004–2005	90,767
1945–1946	118,617.78	1977–1978	314,013	1991–1992	138,403		
1946–1947	157,933.78	1978–1979	39,126	1992–1993	155,316		
1947–1948	156,778.5	1979–1980	122,592	1993–1994	234,441		
1948–1949	184,955.67	1980–1981	119,662	1994–1995	276,196		
1949–1950	155,518.62	1981–1982	49,648	1995–1996	159,422		
1950–1951	322,271.8	1982–1983	93,657	1996–1997	253,442		
1951–1952	124,812.33	1983–1984	224,023	1997–1998	103,754		
1952–1953	122,131.97	1984–1985	177,048	1998–1999	210,999		
1953–1954	251,002.45	1985–1986	160,253	1999–2000	71,748		

Source: Kangsabati CADA, 2007.

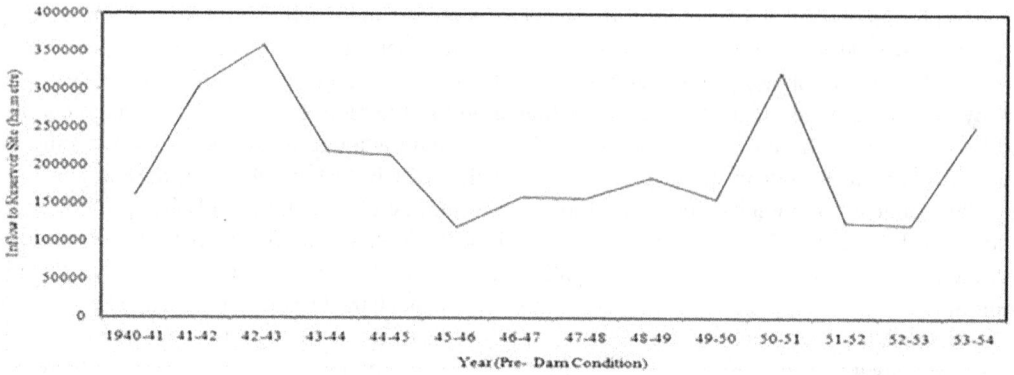

FIGURE 8.24 Inflow to reservoir site before dam construction [1940–1954]. (Kangsabati CADA, 2007.)

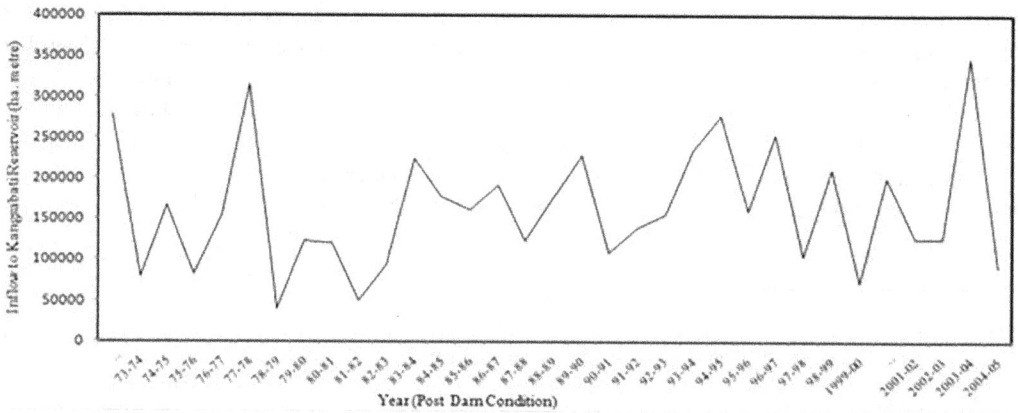

FIGURE 8.25 Inflow to Kangsabati Reservoir during 1973–2005. (Kangsabati CADA, 2007.)

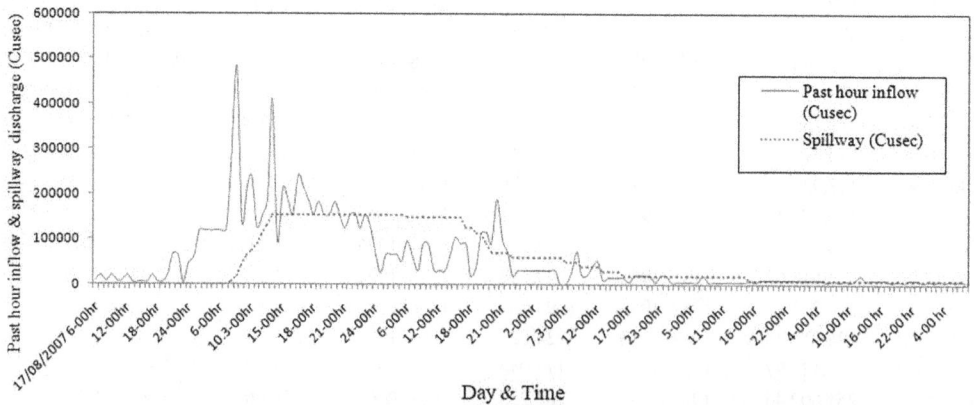

FIGURE 8.26 Inflows and outflows through spillway of Kangsabati Reservoir during 17–23 August 2007. (*Sechpatra,* Govt. of West Bengal, 2008.)

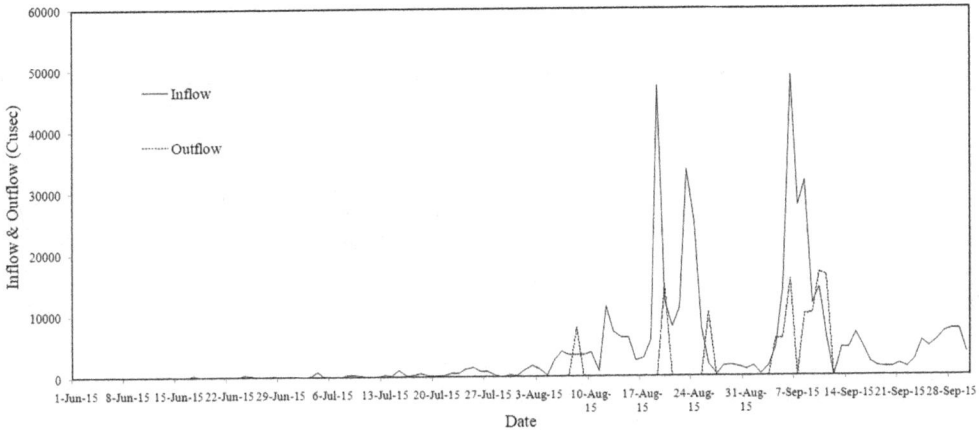

FIGURE 8.27 Inflows and outflows of Kangsabati dam from 1 June to 30 September 2015. (Annual Flood Report, Irrigation and Waterways Department, Govt. of West Bengal, 2016.)

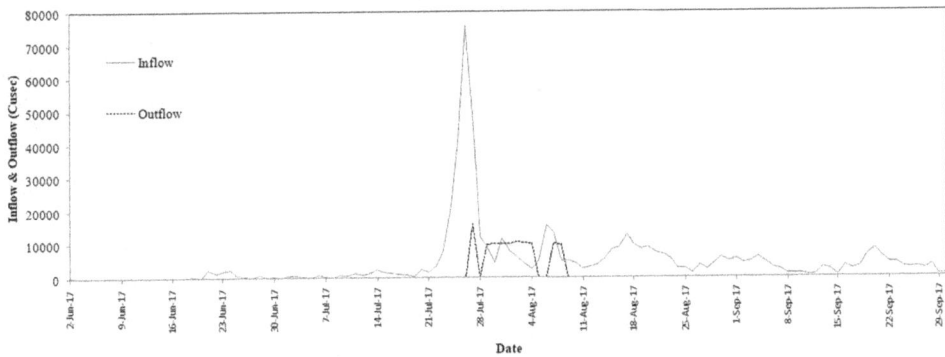

FIGURE 8.28 Inflows and outflows of Kangsabati dam from 2 June to 30 September 2017. (Annual Flood Report, Irrigation and Waterways Department, Govt. of West Bengal, 2017.)

effect of rainfall and minimal discharge of water [50,000 cusecs] from the dam. Besides, by making obstruction, the permanent and temporary constructed cross bridges and cross bunds aggravate the problem of flood hazard in the lower reaches of the river basin.

Ghatal CD block of Paschim Medinipur and Panskura CD block of Purba Medinipur district suffer from inundation due to dam discharge water matching with high tide in river Rupnarayan. The Kaliaghai river is mainly responsible for floods in Sabang CD block of Paschim Medinipur district. The lower portion of the river Haldi is affected by over bank spills and drainage problem during the monsoon as the entire stretch of 42 km of the river falls under the tidal influence of river Hooghly (Annual flood report, 2016, 2017).

Landuse–landcover change results in increased flood frequency and severity. Actually, effects of landuse–landcover change on floods are most pronounced at small scale and for frequent flood magnitudes. The landuse–landcover maps used in some parts of the study area were derived from the Resourcesat-1: LISS-III and Landsat-5[TM] imagery employing remote sensing and GIS techniques. The landuse–landcover change from 1987 to 2015 in some parts of the lower reaches in the study area is shown in Figure 8.29 and Table 8.10. Obviously, the proportion of built-up area has increased by 5.29%, whereas, agricultural land, fallow land and area of water bodies have remained more or less same between 1987 and 2015. The area under orchard fields decreased by 4.37% during this period. The natural flood mitigating *bils* [swamps] have been encroached for raising summer rice. The results signify an increase in river flood due to change in landuse–landcover pattern.

TABLE 8.7A

Inflows and Outflows through Spillway of Kangsabati Reservoir during 17 August 2007 to 19 August 2007

Date	Time	Past Hour Inflow (Cusec)	Outflow Through Spillway (Cusec)	Date	Time	Past Hour Inflow (Cusec)	Outflow Through Spillway (Cusec)	Date	Time	Past Hour Inflow (Cusec)	Outflow Through Spillway (Cusec)
17 August	6 hr	6,851	0	18 August	1 hr	61,015	0	19 August	1 hr	25,418	153,503
2007	7 hr	21,484	0	2007	2 hr	119,088	0	2007	2 hr	66,231	153,503
	8 hr	6,851	0		3 hr	118,563	0		3 hr	66,231	153,503
	9 hr	21,484	0		4 hr	118,033	0		4 hr	66,231	153,503
	10 hr	6,851	0		5 hr	118,033	0		5 hr	51,597	153,503
	11 hr	6,851	0		6 hr	118,033	0		6 hr	95,497	147,874
	12 hr	21,484	0		7 hr	118,066	0		7 hr	60,598	147,874
	13 hr	5,992	0		8 hr	264,394	6,976		8 hr	31,331	147,874
	14 hr	5,061	0		9 hr	484,212	20,022		9 hr	89,867	147,874
	15 hr	5,061	0		9.30 hr	138,093	41,185		10 hr	89,867	147,874
	16 hr	5,061	0		10 hr	217,790	63,713		11 hr	31,336	147,874
	17 hr	19,694	0		10.3 hr	240,318	75,832		12 hr	31,336	147,874
	18 hr	5,061	0		11 hr	126,218	91,382		13 hr	31,336	147,874
	19 hr	5,061	0		12 hr	150,444	111,176		14 hr	60,601	147,874
	20 hr	19,694	0		13 hr	187,305	131,268		15 hr	104,500	147,874
	21 hr	67,543	0		14 hr	409,926	153,503		16 hr	89,867	147,874
	22 hr	62,538	0		14.3 hr	95,497	153,503		17 hr	89,867	125,970
	23 hr	1,754	0		15 hr	212,565	153,503		18 hr	18,863	125,970
	24 hr	46,961	0		15.3 hr	183,298	153,503		18.3 hr	38,697	113,642
					16 hr	154,031	153,503		19 hr	114,166	113,642
					19.3 hr	114,166	89,988				
					20 hr	90,512	69,866				
					20.3 hr	187,452	69,866				
					21 hr	99,656	69,866				
					21.3 hr	70,390	69,866				
					22 hr	20,562	59,237				
					23 hr	30,495	59,237				
					24 hr	30,495	59,237				

Source: *Sechpatra* [Government of West Bengal, 2008].

TABLE 8.7B

Inflows and Outflows through Spillway of Kangsabati Reservoir during 20 August 2007 to 23 August 2007

Date	Time	Past Hour Inflow (Cusec)	Outflow through Spillway (Cusec)	Date	Time	Past Hour Inflow (Cusec)	Outflow through Spillway (Cusec)	Date	Time	Past Hour Inflow (Cusec)	Outflow through Spillway (Cusec)
20 August 2007	1 hr	30,495	59,237	21 August 2007	1 hr	6,161	20,299	22 August 2007	1 hr	9,561	9,073
	2 hr	30,495	59,237		2 hr	6,161	20,299		2 hr	9,561	9,073
	3 hr	30,495	59,237		3 hr	6,161	20,299		3 hr	9,561	9,073
	4 hr	30,495	59,237		4 hr	6,161	20,299		4 hr	9,561	9,073
	5 hr	30,495	59,237		5 hr	6,161	20,299		5 hr	9,561	9,073
	6 hr	30,495	58,524		6 hr	6,161	20,130		6 hr	5,072	9,063
	7 hr	976	58,524		7 hr	20,618	20,130		7 hr	9,556	9,063
	7.3 hr	488	50,824		8 hr	5,985	20,130		8 hr	5,077	9,063
	8 hr	25,659	50,824		9 hr	5,985	20,130		9 hr	9,556	9,063
	9 hr	73,372	50,824		10 hr	5,985	20,130		10 hr	5,077	9,063
	9.3 hr	22,053	39,778		11 hr	5,985	20,130		11 hr	9,556	9,063
	10 hr	20,136	39,778		12 hr	5,985	20,130		12 hr	19,710	9,063
	11 hr	40,273	39,778		13 hr	5,985	20,130		13 hr	9,556	9,063
	12 hr	51,256	39,778		14 hr	5,985	20,130		14 hr	9,556	9,063
	12.3 hr	11,007	29,872		15 hr	5,985	20,130		15 hr	9,556	9,063
	13 hr	15,183	29,872		15.3 hr	10,309	9,073		16 hr	9,556	9,063
	14 hr	15,734	29,872		16 hr	4,781	9,073		17 hr	9,556	9,063
	15 hr	15,734	29,872		17 hr	9,561	9,073		18 hr	5,077	9,063
	16 hr	15,734	20,299		18 hr	9,561	9,073		19 hr	9,556	9,063
	17 hr	6,161	20,299		19 hr	9,561	9,073		20 hr	5,077	9,063
	18 hr	20,794	20,299		20 hr	9,561	9,073		21 hr	9,556	9,063
	19 hr	20,794	20,299		21 hr	9,561	9,073		22 hr	9,556	9,063
	20 hr	20,794	20,299		22 hr	9,561	9,073		23 hr	9,556	9,063
	21 hr	20,794	20,299		23 hr	9,561	9,073		24 hr	5,077	9,063
	22 hr	6,161	20,299		24 hr	9,561	9,073	23 August 2007	1 hr	9,556	9,063
	23 hr	20,794	20,299						2 hr	5,077	9,063
	24 hr	20,794	20,299						3 hr	9,556	9,063
									4 hr	5,077	9,063
									5 hr	9,556	9,063
									6 hr	5,077	9,063
									7 hr	9,454	9,063
									8 hr	5,179	0

Source: *Sechpatra* [Government of West Bengal, 2008].

TABLE 8.8

Inflows and Outflows of Kangsabati Dam from 01 June 2015 to 30 September 2015

Date	Inflow (Cusecs)	Outflow (Cusecs)	Date	Inflow (Cusecs)	Outflow (Cusecs)	Date	Inflow (Cusecs)	Outflow (Cusecs)
1 June 2015	0	0	12 July 2015	0	0	22 August 2015	11,214	0
2 June 2015	0	0	13 July 2015	464	0	23 August 2015	33,642	0
3 June 2015	0	0	14 July 2015	232	0	24 August 2015	25,077	0
4 June 2015	0	0	15 July 2015	1159	0	25 August 2015	7,926	0
5 June 2015	0	0	16 July 2015	232	0	26 August 2015	1,986	10,492
6 June 2015	0	0	17 July 2015	464	0	27 August 2015	347	0
7 June 2015	0	0	18 July 2015	696	0	28 August 2015	1,740	0
8 June 2015	0	0	19 July 2015	229	0	29 August 2015	1,893	0
9 June 2015	0	0	20 July 2015	228	0	30 August 2015	1,555	0
10 June 2015	0	0	21 July 2015	229	0	31 August 2015	1,228	0
11 June2015	0	0	22 July 2015	686	0	1 September 2015	1,697	0
12 June 2015	0	0	23 July 2015	696	0	2 September 2015	451	0
13 June 2015	0	0	24 July 2015	1,391	0	3 September 2015	1,697	0
14 June 2015	0	0	25 July 2015	1,622	0	4 September 2015	4,748	6,152
15 June 2015	0	0	26 July 2015	915	0	5 September 2015	14,330	6,190
16 June 2015	0	0	27 July 2015	928	0	6 September 2015	49,025	15,879
17 June 2015	464	0	28 July 2015	461		7 September 2015	27,856	0
18 June 2015	0	0	29 July 2015	0	0	8 September 2015	31,905	10,089
19 June 2015	0	0	30 July 2015	457	0	9 September 2015	11,903	10,238
20 June 2015	0	0	31 July 2015	229	0	10 September 2015	14,544	16,921
21 June 2015	0	0	1 August 2015	1,144	0	11 September 2015	6,376	16,508
22 June 2015	0	0	2 August 2015	1,854	0	12 September 2015	0	0
23 June 2015	0	0	3 August 2015	1,392	0	13 September 2015	4,662	0
24 June 2015	457	0	4 August 2015	286	0	14 September 2015	4,665	0
25 June 2015	309	0	5 August 2015	2,733	0	15 September 2015	7,111	0
26 June 2015	0	0	6 August 2015	4,172	0	16 September 2015	4,678	0
27 June 2015	0	0	7 August 2015	3,660	0	17 September 2015	2,242	0
28 June 2015	232	0	8 August 2015	3,660	8059	18 September 2015	1,632	0

(Continued)

TABLE 8.8 (*Continued*)

Inflows and Outflows of Kangsabati Dam from 01 June 2015 to 30 September 2015

Date	Inflow (Cusecs)	Outflow (Cusecs)	Date	Inflow (Cusecs)	Outflow (Cusecs)	Date	Inflow (Cusecs)	Outflow (Cusecs)
29 June 2015	0	0	9 August 2015	3,616	0	19 September 2015	1,516	0
30 June 2015	0	0	10 August 2015	4,031	0	20 September 2015	1,423	0
1 July 2015	0	0	11 August 2015	994	0	21 September 2015	2,032	0
2 July 2015	0	0	12 August 2015	11,416	0	22 September 2015	1,421	0
3 July 2015	232	0	13 August 2015	7,289	0	23 September 2015	2,613	0
4 July 2015	928	0	14 August 2015	6,380	0	24 September 2015	5,764	0
5 July 2015	0	0	15 August 2015	6,380	0	25 September 2015	4,786	0
6 July 2015	0	0	16 August 2015	2,686	0	26 September 2015	5,914	0
7 July 2015	0	0	17 August 2015	3,022	0	27 September 2015	7,218	0
8 July 2015	457	0	18 August 2015	6,044	0	28 September 2015	7,540	0
9 July 2015	464	0	19 August 2015	47,396	0	29 September 2015	7,541	0
10 July 2015	229	0	20 August 2015	12,427	15036	30 September 2015	3,847	0
11 July 2015	0	0	21 August 2015	8,062	0			

Source: Annual Flood Report, Irrigation and Waterways Department, Government of West Bengal, 2016.

TABLE 8.9

Inflows and Outflows of Kangsabati Dam from 01 June 2017 to 30 September 2017

Date	Inflow (Cusecs)	Outflow (Cusecs)	Date	Inflow (Cusecs)	Outflow (Cusecs)	Date	Inflow (Cusecs)	Outflow (Cusecs)	Date	Inflow (Cusecs)	Outflow (Cusecs)
1 June 2017	175	0	11 July 2017	1,343	0	20 August 2017	9,192	0	29 September 2017	623	0
2 June 2017	351	0	12 July 2017	672	0	21 August 2017	7,498	0	30 September 2017	935	0
3 June 2017	0	0	13 July 2017	1,343	0	22 August 2017	6,905	0			
4 June 2017	0	0	14 July 2017	2,351	0	23 August 2017	5,698	0			
5 June 2017	0	0	15 July 2017	1,679	0	24 August 2017	2,647	0			
6 June 2017	0	0	16 July 2017	1,343	0	25 August 2017	2,636	0			
7 June 2017	0	0	17 July 2017	1,008	0	26 August 2017	1,220	0			
8 June 2017	0	0	18 July 2017	672	0	27 August 2017	3,658	0			
9 June 2017	0	0	19 July 2017	336	0	28 August 2017	2,439	0			
10 June 2017	0	0	20 July 2017	2,351	0	29 August 2017	4,268	0			
11 June 2017	0	0	21 July 2017	1,679	0	30 August 2017	6,207	0			
12 June 2017	0	0	22 July 2017	3,358	0	31 August 2017	4,878	0			
13 June 2017	0	0	23 July 2017	8,394	0	1 September 2017	5,488	0			
14 June 2017	0	0	24 July 2017	21,537	0	2 September 2017	4,268	0			
15 June 2017	0	0	25 July 2017	41,231	0	3 September 2017	4,878	0			
16 June 2017	0	0	26 July 2017	75,898	0	4 September 2017	6,097	0			
17 June 2017	0	0	27 July 2017	49,074	16019	5 September 2017	4,668	0			
18 June 2017	0	0	28 July 2017	12,406	0	6 September 2017	3,050	0			
19 June 2017	232	0	29 July 2017	9,296	9998	7 September 2017	2,522	0			
20 June 2017	0	0	30 July 2017	4,520	10180	8 September 2017	1,138	0			
21 June 2017	2,318	0	31 July 2017	11,819	10175	9 September 2017	1,147	0			
22 June 2017	1,391	0	1 August 2017	8,071	10045	10 September 2017	1,156	0			
23 June 2017	1,855	0	2 August 2017	6,222	10579	11 September 2017	570	0			

(Continued)

TABLE 8.9 (*Continued*)
Inflows and Outflows of Kangsabati Dam from 01 June 2017 to 30 September 2017

Date	Inflow (Cusecs)	Outflow (Cusecs)	Date	Inflow (Cusecs)	Outflow (Cusecs)	Date	Inflow (Cusecs)	Outflow (Cusecs)	Date	Inflow (Cusecs)	Outflow (Cusecs)
24 June 2017	2,318	0	3 August 2017	4,402	10366	12 September 2017	830	0			
25 June 2017	464	0	4 August 2017	2,720	10063	13 September 2017	2,817	0			
26 June 2017	232	0	5 August 2017	5,146	0	14 September 2017	2,489	0			
27 June 2017	0	0	6 August 2017	15,901	0	15 September 2017	618	0			
28 June 2017	695	0	7 August 2017	13,574	10260	16 September 2017	3,502	0			
29 June 2017	0	0	8 August 2017	5,068	9982	17 September 2017	2,501	0			
30 June 2017	232	0	9 August 2017	5,072	0	18 September 2017	3,086	0			
1 July 2017	0	0	10 August 2017	4,405	0	19 September 2017	6,739	0			
2 July 2017	464	0	11 August 2017	2,755	0	20 September 2017	8,565	0			
3 July 2017	457	0	12 August 2017	3,219	0	21 September 2017	6,163	0			
4 July 2017	232	0	13 August 2017	3,639	0	22 September 2017	4,331	0			
5 July 2017	0	0	14 August 2017	5,561	0	23 September 2017	4,321	0			
6 July 2017	696	0	15 August 2017	8,266	0	24 September 2017	3,049	0			
7 July 2017	232	0	16 August 2017	8,968	0	25 September 2017	2,541	0			
8 July 2017	0	0	17 August 2017	12,850	0	26 September 2017	2,845	0			
9 July 2017	696	0	18 August 2017	9,837	0	27 September 2017	2,320	0			
10 July 2017	464	0	19 August 2017	8,526	0	28 September 2017	3,403	0			

Source: Annual Flood Report, Irrigation and Waterways Department, Government of West Bengal, 2017.

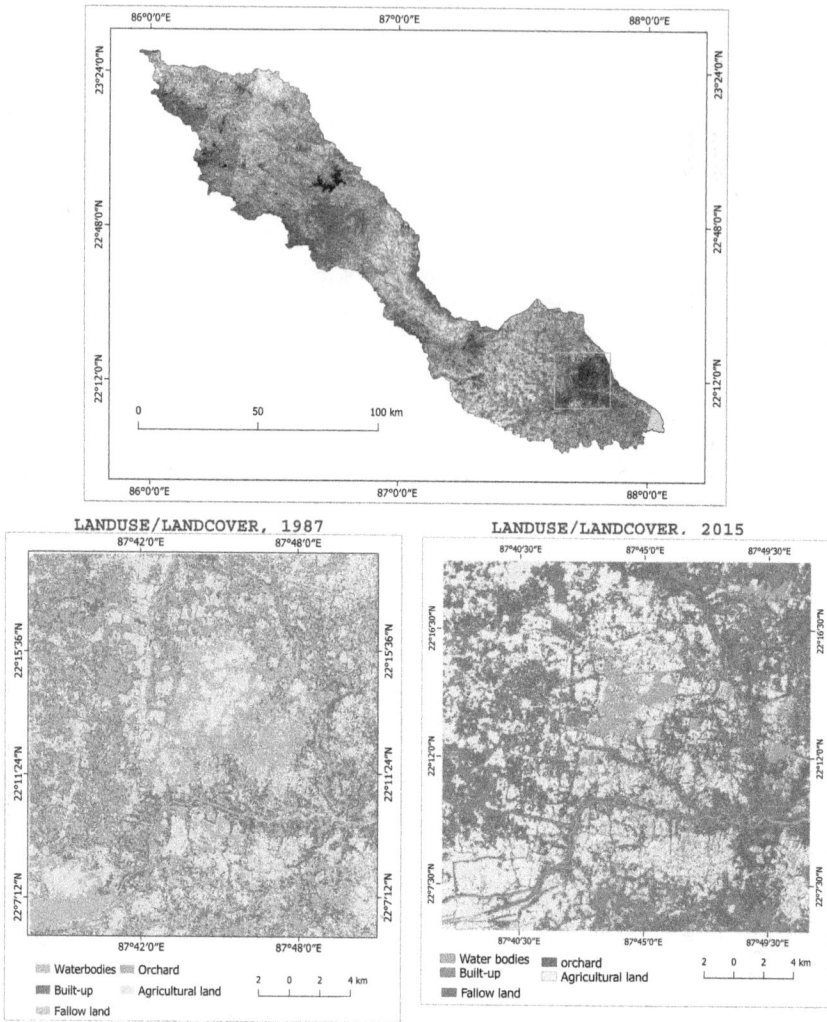

FIGURE 8.29 Comparative picture of landuse–landcover in selected parts of the basin. (Resourcesat-1: LISS-III, 1987 and Landsat-5[TM], 2015.)

TABLE 8.10

Comparative Picture of Landuse-Landcover in Selected Parts of the Basin

Landuse	Percentage of Land	
	1987	2015
Waterbodies	7.61	7.352
Orchard	21.61	17.24
Built-up area	23.98	29.27
Agricultural land	46.34	45.12
Fallow land	0.45	1.02

Source: Resourcesat-1: LISS-III and Landsat-5[TM].

Sediment is essential for the sustenance of rivers. Within a drainage basin, sediment is contributed to rivers by water erosion, landslides, soil creep, bank erosion and riverbed erosion (Collins and Dunne, 1990). Human being has been acting as an important agent of fluvial dynamics by extracting sands and gravels from the river bed. Environmental problems occur when the rate of extraction of sand, gravel and other materials exceeds the rate of deposition. Extraction of gravels from river bed causes bed degradation and can induce lateral bank erosion by increasing the height of banks, which are then more prone to undercutting and failure (Collins and Dunne, 1990). Sand and gravel mining disturbs the equilibrium of a river channel because it interrupts material load moving within a dynamic system and triggers an initial morphological response to recover the balance between supply and transport. The gravel and sand mining activities have several impacts on the river environment. It leads to changes in river channel form, increases the velocity of flow in river which destroys flow regime and eventually erodes the river banks. Channel widening causes shallowing of the streambed, producing braided or anastomosing flow or sub-surface inter-gravel flow in riffle areas. By removing sediment from the channel, instream gravel mining disrupts the pre-existing balance of load–volume relationship, typically inducing incision upstream and downstream of the extraction site. Excavation of pits in the active channel alters the equilibrium profile of the streambed, creating a locally steeper gradient upon entering the pit (Kondolf, 1997). Rinaldi et al. (2005) have classified the effects of sediment mining on channel morphology and environment into three broad categories: **Morphological effects** [upstream incision, downstream incision, lateral channel instability, bed armouring, reactivation of inactive channels, impacts on infrastructures and effects on coastal zone]; **Hydrological effects** [water table lowering, effects on frequency of inundation, change in tidal hydrodynamics within estuaries] and **Ecological and environmental effects** [loss of riparian and aquatic habitats, effects on fish populations, wildlife and invertebrates and aesthetic degradation of the fluvial landscape]. Sand aquifer helps in recharging the water table while sand mining causes sinking of water tables in the adjacent areas. Besides, sand mining activities also transform the riverbeds into large and deep pits which consequently lower the ground water table. Furthermore, aesthetic splendor in the course of river is deteriorated due to sand and gravel mining activities. Extraction of sands at low rates by bar skimming process typically results in a wider, shallower streambed, leading to increased water temperatures, modification of pool-riffle distribution and alteration of inter-gravel flow paths (Kondolf, 1997).

In case of Kangsabati river, gravel mining activities have been restricted in the upstream section, whereas sand mining activities are visible in all the three sections: upstream, midstream and downstream. Processes of instream and floodplain sand mining have been practised from the river. Instream sand mining activities are being practiced by means of pit excavation and bar skimming, while mining of sands is also visible in different pockets from overbank flood deposits (Figure 8.30).

By constructing dams, pick-up barrages, embankments, culverts and bridge crossings, the residents of the basin area have altered channel conditions of Kangsabati river. To compare the change of channel layout position in pre-dam and post-dam conditions in the downstream part of the dam, US Army toposheet, 1955, and satellite image Landsat-8, 2018, have been used. Both are georectified and superimposed using QGIS. Figure 8.31 shows channel layout from Medinipur town to Shaldahari. Due to altered flow, sediment supply and constructed embankment in the right side of the river, the position of channel layout has shifted towards north [leftward]. Figure 8.32 shows channel layout from Shaldahari to Buraburi. In this part, the channel position has shifted towards north-east. Flood-prone lower Kangsabati has been chained with a series of embankments to protect adjacent riparian tract from flood hazards. Embankment-induced enforced sedimentation at the river bed and contribution of sediments from unprotected sections makes the river channel shallower. Figure 8.33 shows some part of meandering channel pattern in the lower reaches of the river. The calculated value of sinuosity index [as per Brice, 1964] of pre-dam channel is 2.318, while in case of the post-dam channel, it is 2.440. Due to excessive sedimentation in some parts of the channel, the calculated value of sinuosity index has slightly increased in post-dam condition. Flood tendency is continuously increasing due to shallowing of the river bed.

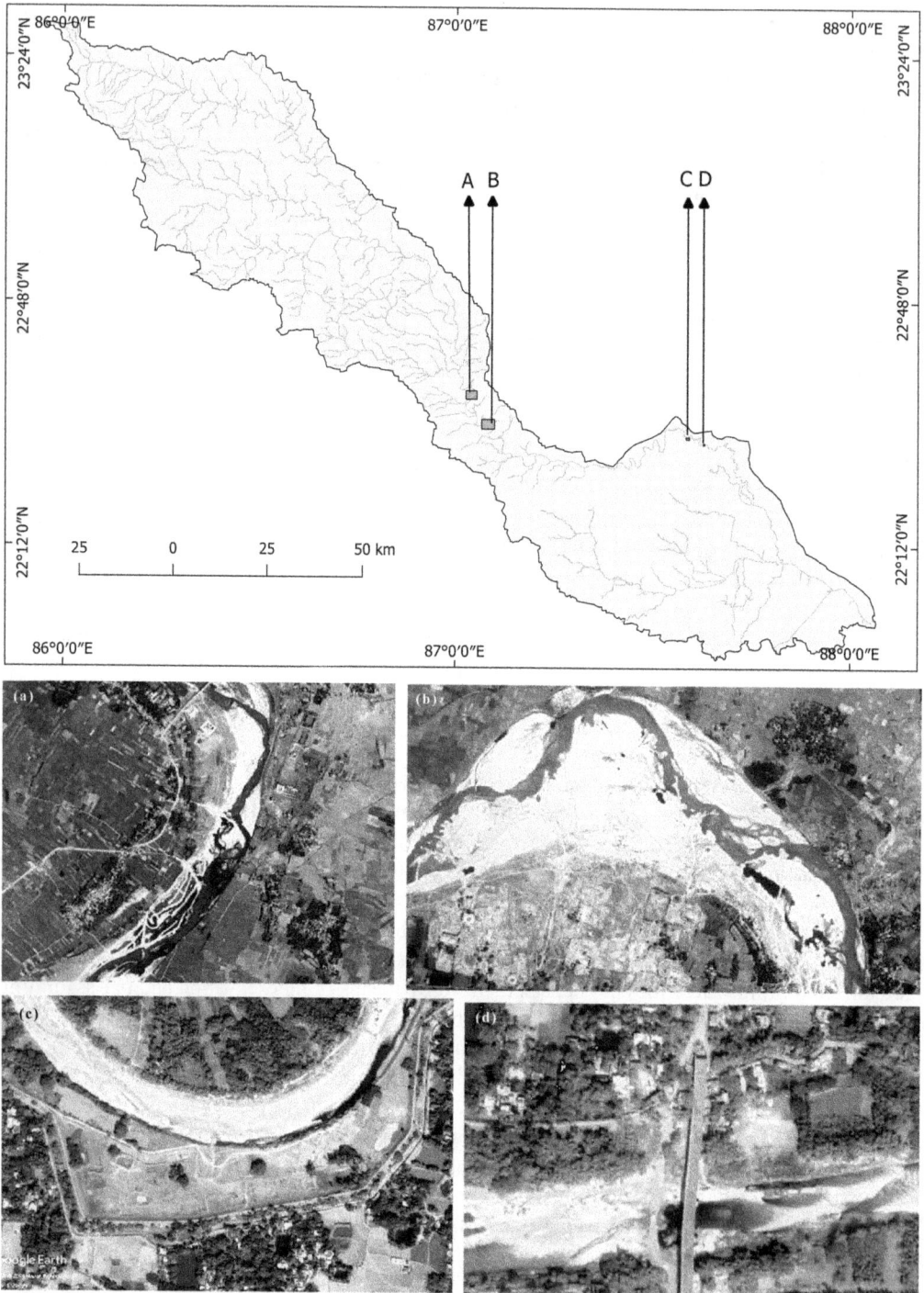

FIGURE 8.30 Selected sand mining places in Kangsabati river. (Google Earth, 2019.) [(a) Kanaipal *Bali Khadan*, (b) Salchaturi *Bali Khadan,* (c) Ban Basuri *Bali Khadan* and (d) Talberya Ghunghuri *Bali Khadan.*]

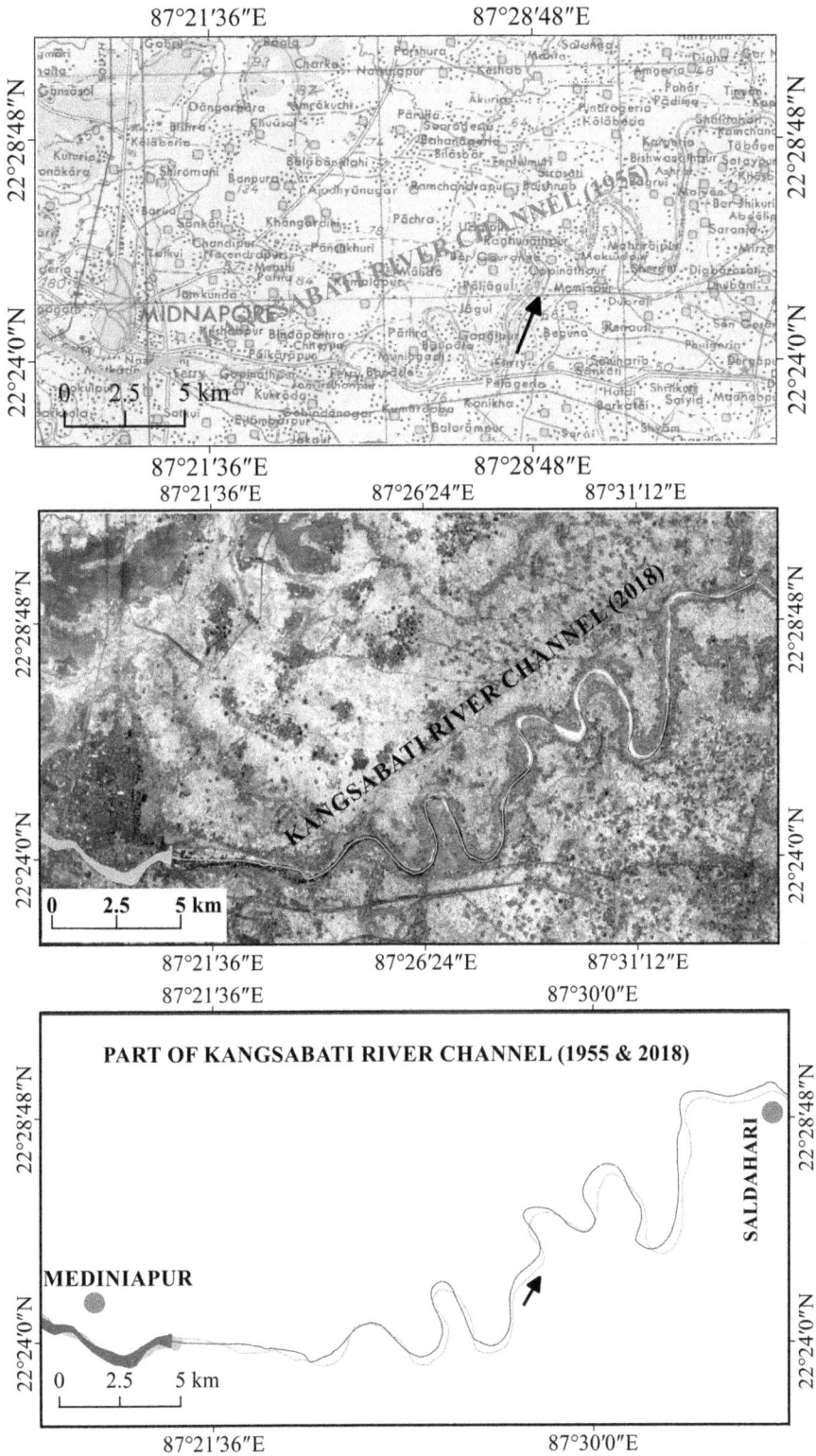

FIGURE 8.31 Shifting channel layout from Medinipur town to Shaldahari of the Kangsabati river. (US Army Toposheet, 1955 and Landsat-8, 2018.)

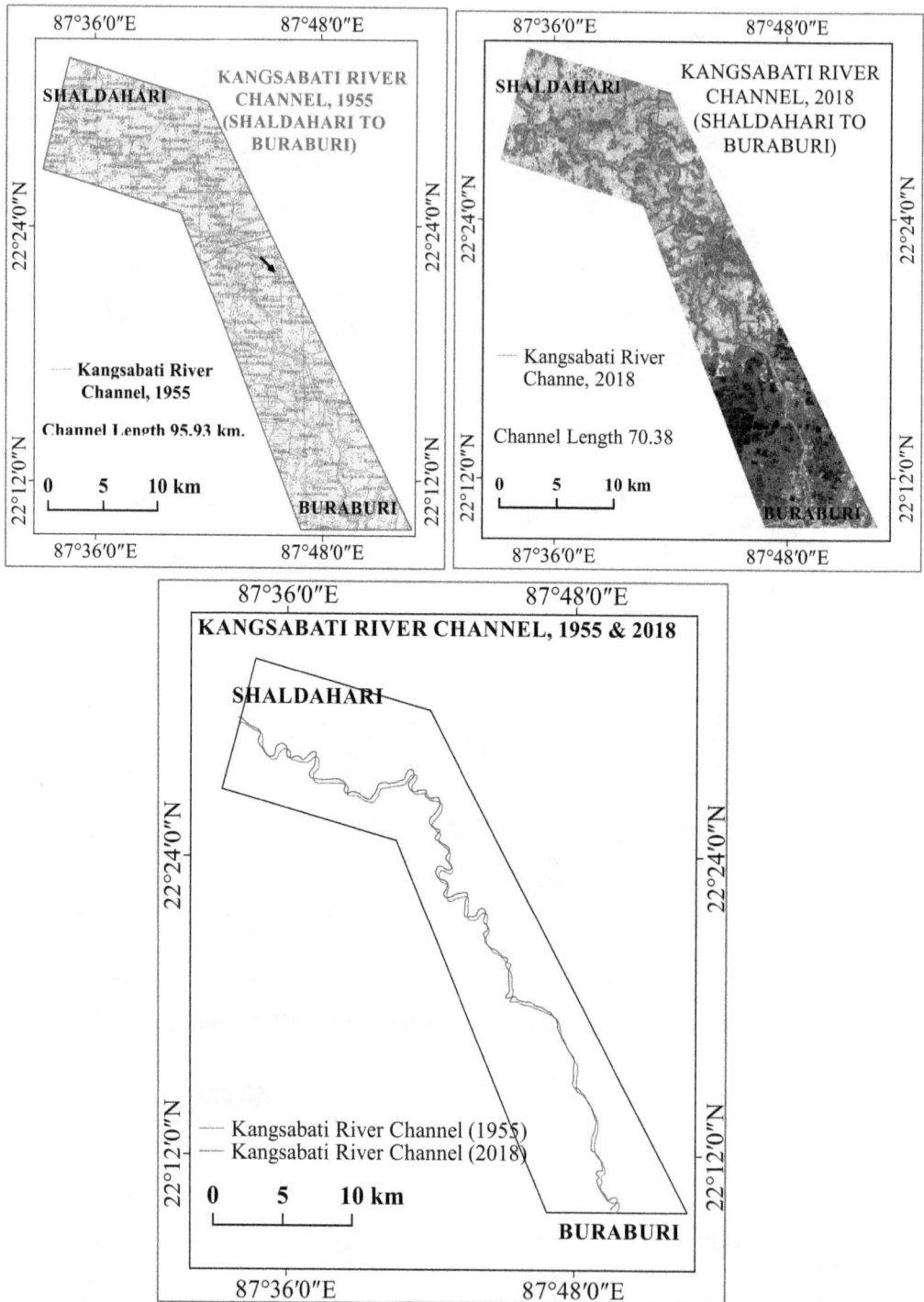

FIGURE 8.32 Shifting channel layout from Shaldahari to Buraburi of the Kangsabati river. (US Army Toposheet, 1955 and Landsat-8, 2018.)

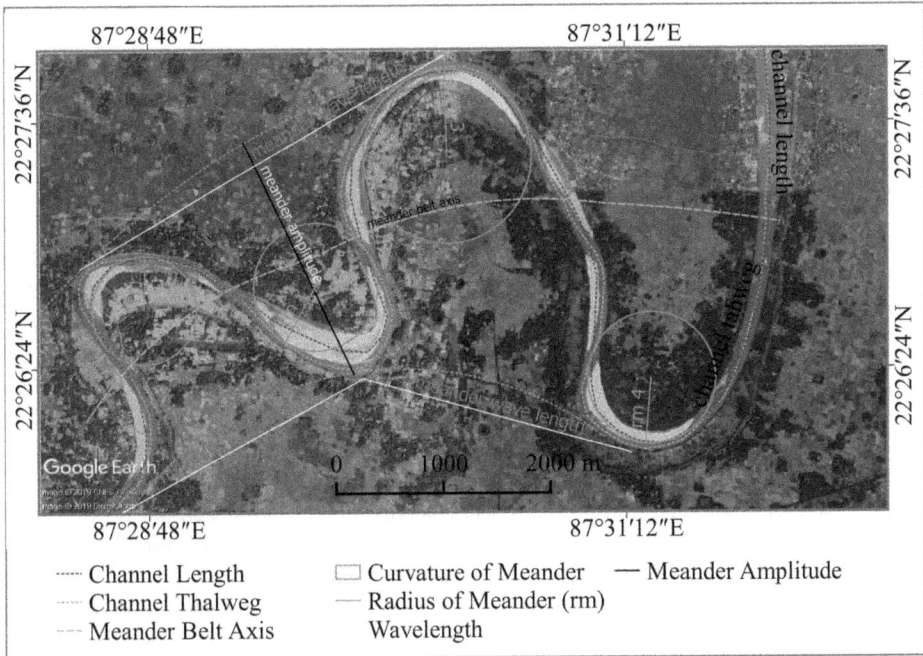

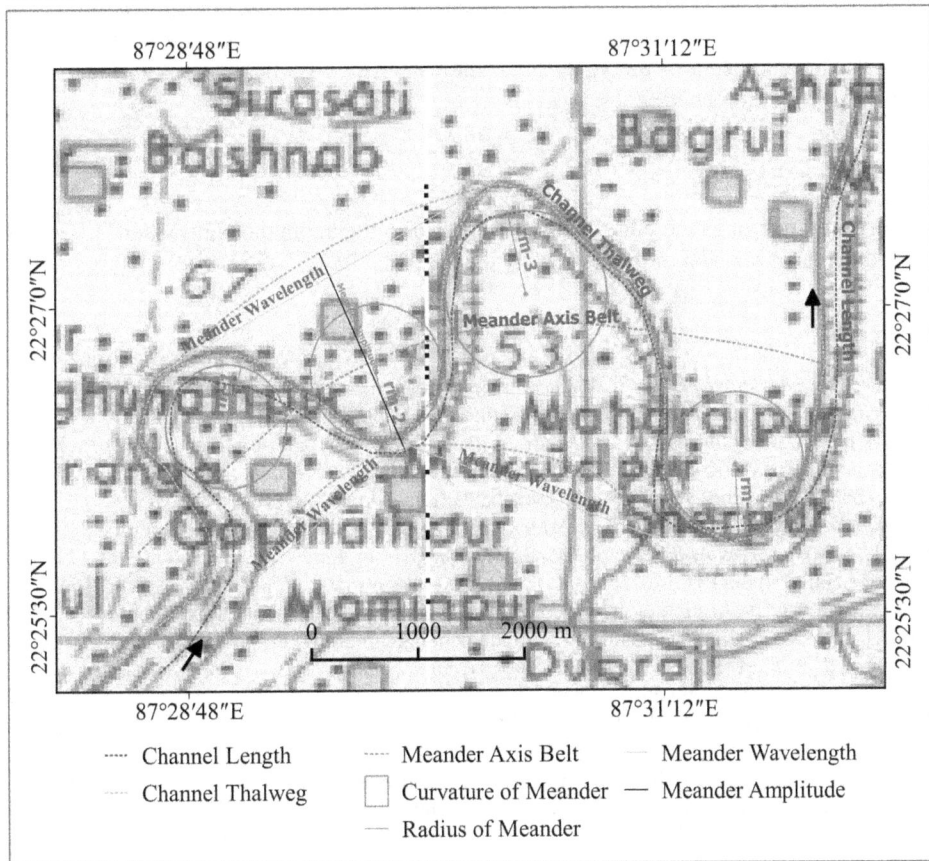

FIGURE 8.33 Changing sinuosity index of a specific part of Kangsabati river. (US Army Toposheet, 1955 and Google Earth, 2018.)

8.4   CONCLUSION

Change is a universal phenomenon. River changes over time due to either natural or human-induced mechanisms, but the nature of river channel change solely depends on input–output relations of the basin. Rivers always try to make an adjustment with the changing environmental conditions. Fluvial systems at all times have a tendency to maintain a dynamic equilibrium in its sediment load and movement, water discharge and channel geometry instead of being affected by tectonic, climatic and anthropogenic mechanisms. Construction of dams and pick-up barrages in the free flow of a river alters the sediment load and discharge of water. Likewise, altered landuse and landcover pattern and mining of sands and gravels in the basin area also transform bed load capacity of the river. As a result, geomorphologic behaviour of the river channel in the upstream and downstream parts of the basin changes. In the case of present study, the river forms have been modified mainly by the construction of dams and barrages across the upper reaches of the channel and embankments along the lower reaches of the channel below Medinipur anicut dam. Side by side, landuse and landcover change, sand and gravel mining activities and construction of cross bridges [permanent and temporary] have brought perceptible change in the channel form and flood frequency. Artificial channelisation process that involves channel modification through realignment, deepening, widening and straightening can give results of channel shortening which can increase channel slope and flow velocity of the river. This can give relief from acute flood to the inhabitants in the lower reaches of Kangsabati river basin. Although slightly increasing tendency of sinuosity index value and a minor change of the position of river channel within a time span of 63 years have been identified in the lower reaches in the study area, doubts may remain due to the use of the two different scale maps. To get an orderly picture of the river channel change by anthropogenic mechanisms in the study area, more intensive studies of pre-dam and post-dam situations of the river are needed.

ACKNOWLEDGEMENT

The authors are grateful to the editors of the book and to the anonymous reviewer for helpful and critical comments in revision of this essay.

REFERENCES

Bagchi, K. (1944). *The Ganges Delta*. Calcutta: University of Calcutta.

Bandopadhyay, D. (2007). *Banglar Nadnadi*. Kolkata: Dey's Publishing.

Banerjee, K. (1983). *Geomorphology of the Kangsabati Basin West Bengal* [Unpublished Ph.D. Thesis]. Burdwan: The University of Burdwan.

Basu, A.K. (2002). *PaschimbangerNadnadi*. Kolkata: Mitra and Ghosh Pvt. Ltd.

Bhattacharya, K. (1959). *BangladesherNad-Nadi O Parikalpana*. Kolkata: Bidyodoya Library.

Bloom, A.L. (2003). *Geomorphology: A Systematic Analysis of Late Cenozoic Landforms*. New Delhi: Prentice Hall of India Pvt. Ltd.

Brandt, S.A. (2000). Classification of geomorphological effects downstream of dams. *Catena*, 40, 375–401.

Chorley, R.J. (1969). The drainage basin as the fundamental geomorphic unit. In R. J. Chorley (Ed.), *Water, Earth and Man: A Synthesis of Hydrology, Geomorphology and Socio – Economic Geography* (pp. 77–99). London: Methuen and Co. Ltd.

Chorley, R.J. and Kennedy, B.A. (1971). *Physical Geography: A Systems Approach*. London: Prentice-Hall International Inc.

Chorley, R.J., Schumm, S.A. and Sugden, D.E. (1985). *Geomorphology*. London: Methuen and Co. Ltd.

Church, R.J.H. (1968). A geographical view. In N. Rubin and William M. Warren (Eds.), *Dams in Africa: An Inter-Disciplinary Study of Man-Made Lakes in Africa* (pp. 1–12). London: Frank Cass and Company Ltd.

Collins, B. and Dunne, T. (1990). *Fluvial Geomorphology and River-Gravel Minning: A Guide for Planners, Case Studies Included*. Sacramento: California Department of Conservation-Division of mines and Geology. https://archive.org/details/fluvialgeomorpho98coll.

Dai, Z. and Liu, J.T. (2013). Impacts of large dams on downstream fluvial sedimentation: An example of the Three Gorges Dam (TGD) on the Changing (Yangtze River). *Journal of Hydrology*, 480, 10–18.

Das, M. and Bhandari, G. (2017). Watershed prioritization of Kumari and Kasai river basins of Purulia district, West Bengal. *International Journal of Advances in Mechanical and Civil Engineering*, 4(6), 67–70.

Datta, D. and Roy, C. (2017). A holistic approach for determining the characteristic flow on Kangsabati catchment. *IJCER*, 7(3), 49–56.

Gayen, S., Bhunia, G.S. and Shit, P.K. (2013). Morphometric analysis of Kangshabati-Darkeswar interfluves area in West Bengal, India using ASTER DEM and GIS techniques. *Journal of Geology & Geosciences*, 2(4), 1–10. doi: 10.4172/2329-6755.1000133.

Goudie, A. (1981). *The Human Impact on the Natural Environment*. Oxford: Blackwell Publishers Ltd.

Government of West Bengal. (1961). *District Census Handbook: Puruliya*. Calcutta: The Superintendent, Government Printing.

Government of West Bengal. (1968). *West Bengal District Gazetteers: Bankura*. Kolkata: Bengal Secretariate Book Depot.

Government of West Bengal. (1985). *West Bengal District Gazetteers: Puruliya*. Kolkata: Bengal Secretariate Book Depot.

Government of West Bengal. (1988). *Modernisation of Kangsabati Reservoir Project*. Kolkata: Irrigation and Waterways Department.

Government of West Bengal. (1995). *West Bengal District Gazetteers: Medinipur*. Kolkata: Bengal Secretriate Book Depot.

Government of West Bengal. (2007). *Annual Report of Kangsabati Command Area Development Authority* [2004–2005]. Bankura: Administrator, Kangsabati C.A.D.A.

Government of West Bengal. (2007). *District Human Development Report: Bankura*. Kolkata: Development and Planning Department.

Government of West Bengal. (2008). *Sechpatra* [January]. Kolkata: Irrigation and Waterways Department.

Government of West Bengal. (2014). *Natural Disaster: Flood*. West Bengal Disaster Management & Civil Defence Department. http://wbdmd.gov.in/pages/flood2.aspx.

Government of West Bengal. (2016). *Annual Flood Report*. Irrigation and Waterways Department. https://www.wbiwd.gov.in/uploads/ANNUAL_FLOOD_REPORT_2016.pdf.

Government of West Bengal. (2017). *Annual Flood Report*. Irrigation and Waterways Department. https://wbiwd.gov.in/uploads/ANNUAL_FLOOD_REPORT_2017.pdf.

Graf, W.L. (2006). Downstream hydrologic and geomorphic effects of large dams on American rivers. *Geomorphology*, 79, 336–360.

Grant, G.E., Schmidt, J.C. and Lewis, S.H. (2003). A geological framework for interpreting downstream effects of dams on rivers. *Water Science and Application*, 7, 209–225. http://wpg.forestry.oregonstate.edu/sites/wpg/files/bibliopdfs/03GrantetalDeschutes.pdf.

Gregory, K.G. (1997). Introduction. In K.G. Gregory (Ed.), *An Introduction to the Fluvial Geomorphology of Britain* (pp. 1–18) *Geological Conservation Review Series, No.13*. London: Chapman and Hall.

Gregory, K.J. (1976). Changing Drainage Basins. *The Geographical Journal*, 142(2), 237–247.

Gregory, K.J. and Walling, D.E. (1971). Field measurement in the Drainage Basin. *Geography*, 56(4), 277–292.

Gregory, K.J. and Walling, D.E. (1973). *Drainage Basin Form and Process: A Geomorphologic Approach*. London: Edward Arnold Publishers Ltd.

Horton, R.E. (1932). Drainage-Basin characteristics. *Transactions, American Geophysical Union*, 350–361. https://hydrology.agu.org/wp-content/uploads/sites/19/2016/06/Horton-1932-DRAINAGE-BASIN-CHARACTERISTICS-Eos_Transactions_American_Geophysical_Union.pdf.

Horton, R.E. (1945). Erosional development of streams and their drainage basins: hydro physical approach to quantitative morphology. *Geological Society of America*, 56, 275–370.

Isik, S., Dogan, E., Kalin, L., Sasal, M. and Agiralioglu, N. (2008). Effects of anthropogenic activities on the Lower Sakarya River. *Catena*, 75, 172–181.

Knighton, D. (1998). *Fluvial Forms and Processes: A New Perspective*. London: Arnold.

Kondolf, G.M. (1997). *Hungry Water: Effects of Dams and Gravel Mining on River Channels*. California: University of California Berkeley, Department of Landscape Architecture and Environmental Planning. https://www.wou.edu/las/physci/taylor/g407/kondolf_97.pdf.

Leopold, L.B. and Maddock, T. (1953). *The Hydraulic Geometry of Stream Channels and Some Physiographic Implications*. Geological Survey Professional Paper 252. Washington, DC: United States Government Printing Office. https://pubs.usgs.gov/pp/0252/report.pdf.

Leopold, L.B., Wolman, M.G. and Miller, J.P. (1964). *Fluvial Processes in Geomorphology.* Sanfransisco, CA: W.H. Freeman & Co.

Li, L., Lu. X. and Chen, Z. (2007). River channel change during the last 50 years in the middle Yangtze River, the Jianli reach. *Geomorphology*, 85, 185–196.

Lin, Q. (2011). Influence of Dams on river ecosystem and its countermeasures. *Journal of Water Resource and Protection*, 3, 60–66.

Magilligan, F.J. and Nislow, K.H. (2005). Changes in hydrologic regime by dams. *Geomorphology*, 71, 61–78.

Mahala, A. (2018). Drainage Basin morphometric analysis of mountain-plain (Koshi, Bihar) and Plateau-plain (Kangsabati, WB) regions of tropical environment: a comparative analysis. *Journal of Indian Geophysical Union*, 22(4), 419–429.

Majumdar, S.C. (1942). *Rivers of the Bengal Delta.* Calcutta: University of Calcutta.

McCully, P. (1998). *Silenced Rivers: The Ecology and Politics of Large Dams.* Calcutta: Orient Longman Ltd.

Mitra, A. (1953). *An Account of Land Management in West Bengal (1870–1950).* Calcutta: West Bengal Government Press.

Mittal, N., Mishra, A., Singh, R., Bhave, A.G. and van der Valk, M. (2014). Flow regime alteration due to anthropogenic and climatic changes in the Kangsabati River, India. *Ecohydrology & Hydrobiology*, 14(1), 182–191.

Morisawa, M. (1985). *Rivers: Form and Process.* London: Longman.

Mukherjee, R.K. (1938). *The Changing Face of Bengal: A Study in Riverine Economy.* Calcutta: University of Calcutta.

Petts, G.E. (1979). Complex response of river channel morphology subsequent to reservoir construction. *Progress in Physical Geography*, 3, 329–362. http://citeseerx.ist.psu.edu/viewdoc/download?doi=10.1.1.1028.3324&rep=rep1&type=pdf.

Pizzuto, J. (2002). Effects of dam removal on river form and process. *Bioscience*, 52(8), 683–691.

O'Malley, L.S.S. (1908). *Bengal District Gazetteers: Bankura.* Calcutta: Bengal Secretariat Book Depot.

O'Malley, L.S.S. (1911). *Bengal District Gazetteers: Midnapore.* Calcutta: Bengal Secretariat Book Depot.

Richards, K. (1982). *RIVERS: Form and Process in Alluvial Channels.* London: Methuen &Co.

Rinaldi, M., Wyzga, B. and Surian, N. (2005). Sediment mining in alluvial channels: physical effects and management perspectives. *River Research Applications*, 21, 805–828, John Wiley & Sons.

Ronco, P., Fasolato, G., Nones, M. and Silvio, G.D. (2009). Morphological effects of damming on lower Zambezi River. *Geomorphology*, 115, 43–55.

Roy, N.R. (1949). *Bangalir Itihas: Aadi Parba.* Kolkata: Dey's publishing.

Rudra, K. (2008). *Banglar Nadikatha.* Kolkata: Sahitya Samsad.

Schmidt, J.C. and Wilcock, P.R. (2008). Metrics for assessing the downstream effects of dams. *Water Resources Research*, 44, 1–19.

Shiva, V. (1991). *Ecology and the Politics of Survival: Conflicts Over Natural Resources in India.* New Delhi: Sage Publications.

Shreve, R.L. (1966). Statistical law of stream number. *The Journal of Geology*, 74(1), 17–37.

Skalak, K.J., Benthem, A.J., Schenk, E.R., Hupp, C.R., Galloway, J.M., Nustad, R.A. and Wiche, G.J. (2013). Large dams and alluvial rivers in the Anthropocene: the impacts of the Garrison and Oahe dams on the Upper Missouri River. *Anthropocene*, 2, 51–64.

Strahler, A.N. (1957). Quantitative analysis of watershed geomorphology. *Transactions, American Geophysical Union*, 38(6), 913–921. http://www.uvm.edu/pdodds/files/papers/others/1957/strahler1957a.pdf.

Tagore, R. (2016). *Bhanu Singher Patraboli.* Calcutta: Visva-Bharati Granthan Bivag.

Thakkar, H. (2009). Displacement in the name of development. In Ramaswamy R. Iyer (Ed.), *Water and the Laws in India* (pp. 414–431). New Delhi: SAGE Publications India Pvt. Ltd.

Willcocks, W. (1930). *Ancient System of Irrigation in Bengal.* Calcutta: University of Calcutta.

Williams, G.P. and Wolman, M.G. (1984). *Downstream Effects of Dams on Alluvial Rivers.* Washington, DC: U.S. Government Printing Office.

World Commission on Dams. (2000). *Dams and Development: A New Framework for Decision – Making.* London: Earthscan Publications Ltd.

9 Anthropogenic Impact on Channel and Extra-Channel Geomorphology of the Dwarkeswar River Basin

Sadhan Malik and Subodh Chandra Pal

CONTENTS

9.1 INTRODUCTION

At present, anthropogenic activities are a significant geomorphic factor (Szabó, 2010) and the modification of the Earth system by this activity becomes now so large that it has been suggested to be renamed the geological period from Holocene to Anthropocene (Crutzen and Stoermer, 2000; Zalasiewicz et al., 2011; Zalasiewicz et al., 2008). Several kinds of anthropogenic activities such as the construction of dam, bridge, artificial levees and embankments, and urbanisations are very crucial for the development of the society, although they have a negative impact on the river system (Gregory, 2006; James and Marcus, 2006; Jeje and Ikeazota, 2002; Khan and Islam, 2015). Numerous studies have reported numerous consequences of anthropogenic impact on earth's surface processes; for example, the Mekong River Delta, world's third largest delta, has been suffering from rapid erosion and subsidence due to dam construction, large-scale commercial sand mining, and groundwater extraction (Anthony et al., 2015). In addition, Poyang Lake, the biggest freshwater lake in China, has also been experiencing an intense and continued recession of water from 2003 due to the linkage between Three Gorges Dam (Mei et al., 2015). Nilsson et al. (2005) have stated that global river systems have been increasingly altered by dam construction to meet the water requirements and other purposes. Therefore, it is clear that the natural system influenced by not only anthropogenic activity but also human society itself is suffering from the reaction of human impact (Szabó, 2010). So, the study of anthropogenic impact on the river is very much important.

Studies has been conducted the changes of channel morphology due to anthropogenic impact such as dam construction (Lake, 2000; Bunn and Arthington, 2002; Nilsson et al., 2005; Thorp et al., 2006; Lehner et al., 2011; Humphries et al., 2013; Grill et al., 2015), deforestation (Sweeney et al., 2004; Neill et al., 2006; Boix-Fayos et al., 2007; Latrubesse et al., 2009; Ilha et al., 2019), water extraction (Thoms and Sheldon 2000; Rolls et al., 2017), sand mining (Surian and Rinaldi, 2003; Bravard et al., 2013; Brunier et al., 2014; Lai et al., 2014), and construction of spar or dyke along the river as well as across the channel (Malik and Pal, 2019; Hei et al., 2009; Elawady et al., 2001; Ohmoto et al., 2009; Giglou et al., 2018; Sukhodolov et al., 2002). Among them, 40% of global river water volume is being affected by the construction of dam (Grill et al., 2015).

West Bengal is one of the states of India, which is the home of more than 90 million populations (Census of India, 2011). It is also a riverine state where 42.3% of the total area is susceptible to flood and the reason varies from region to region (Irrigation and Waterways Directorate Govt. of West Bengal, 2013, 2016). After independence, this river is suffering from numerous anthropogenic activities such as river water extraction, embankments, spars construction, sand mining and medium-size dam construction in the upper catchment area of the basin, with the extension of the transport network and urbanisation. Side by side, several studies have shown that human activities are the major problem to the river system (Bandyopadhyay et al., 2018; Ebisemiju, 1991; Ghosh et al., 2016; Rudra, 2014). Most of their studies are discrete in nature; therefore, there is lacking regarding a compressive study that can give a clear picture of anthropogenic activity on a river basin and its channel. Apart from this, the rivers of south Bengal have a prolonged history of British colonisation with a narrow business mind, which was rarely been considered as a catastrophic anthropogenic impact, particularly for a drainage basin. This had not only extorted the natural resources from this region but also changed the crucial geo-ecological environment intensely.

In this study, we have focused on Dwarkeswar River (area 4,356.72 km².), which is one of the major rivers in South Bengal. In addition, the lower part of Dwarkeswar River is very susceptible to flood due to its hydrology, location, topography and prolonged changes in the land use and land cover (Rudra, 2016). On the other hand, the upper part of the basin is experiencing drought-prone environment. Therefore, it is very important to understand the role of anthropogenic activities on the channel morphology of Dwarkeswar River. Previous studies of this river basin were based on water resource availability (Bagchi and Majumdar, 2011; Bandyopadhyay et al., 2014), hydro-chemical characteristics of groundwater (Nag and Lahiri, 2011), floods (Bandyopadhyay et al., 2014; Biswas et al., 2015; Mondal et al., 2017), drainage network character (Pal and Debnath, 2012; Nag and Lahiri, 2011; Pan, 2013), nature of sedimentation (Maity and Maiti, 2016) and sand mining (Sinha, 2016). However, few works on a particular aspect of anthropogenic activities have been carried out previously, but most of the work is discrete in nature, as well as they are not been able to provide a comprehensive aspect of anthropogenic activities on the channel morphology of Dwarkeswar River.

Therefore, our main objectives in this study are to find out the effects of colonial deforestation, dam construction, land use land cover change, spar construction and sand mining and thereby assess the role of anthropogenic activity on Dwarkeswar River in a comprehensive manner, which can help the planners for planning purpose.

9.2 THE RIVER AND ITS BASIN

9.2.1 GENERAL DESCRIPTION OF THE STUDY AREA

Dwarkeswar River, which is also known as Dhalkishore (Survey of India, 1978), is one of the major rivers in the western part of West Bengal. Figure 9.1 shows the location of Dwarkeswar River, which is enclosed between 23°32′00″ to 23°40′25″ N latitude and 86°31′08″ to 87°47′58″ E longitude and occupies an area of 4,356.6 km² with a perimeter of 426.86 km. The circularity ratio of this basin is 0.30, indicating that this river is elongated basin. The basin asymmetry index of this river basin is 55.3. Dwarkeswar River originates near Tilboni hill of Chhota Nagpur Plateau in Puruliya district and enters Bankura district near Chatna. It mainly flows towards the south-east, and after entering into Hooghly district, it turns south near Arambag town. Its main tributary is Gandheswari River; rising from Bankura district meets Dwarkeswar near Bankura town. After receiving contributions from other minor tributaries such as Arkasha, Berai, Shankari, Beko Nala, Dangra Nala, Kumari Nala, Futuari Nala and Dudhbhaiya Nala, Dwarkeswar finally joins with Shilabati at Bandar near Ghatal town of Paschim Medinipur district to form river Rupnarayan (Bhattacharya et al., 1985; O'Malley, 1908). The maximum width of the basin is 40.80 km, and the maximum length of the basin is 151.51 km.

The total length of the stream is about 228.65 km. Figure 9.1 also shows that the regional elevation ranges between 438 and 10 m (Survey of India or SOI topographical sheets). Figure 9.2 shows that the long profile of this river is concave in nature and the value of the hypsometric integral (HI) is 21.20% indicating that the region is in the mature to the senile stage. The shape of the drainage

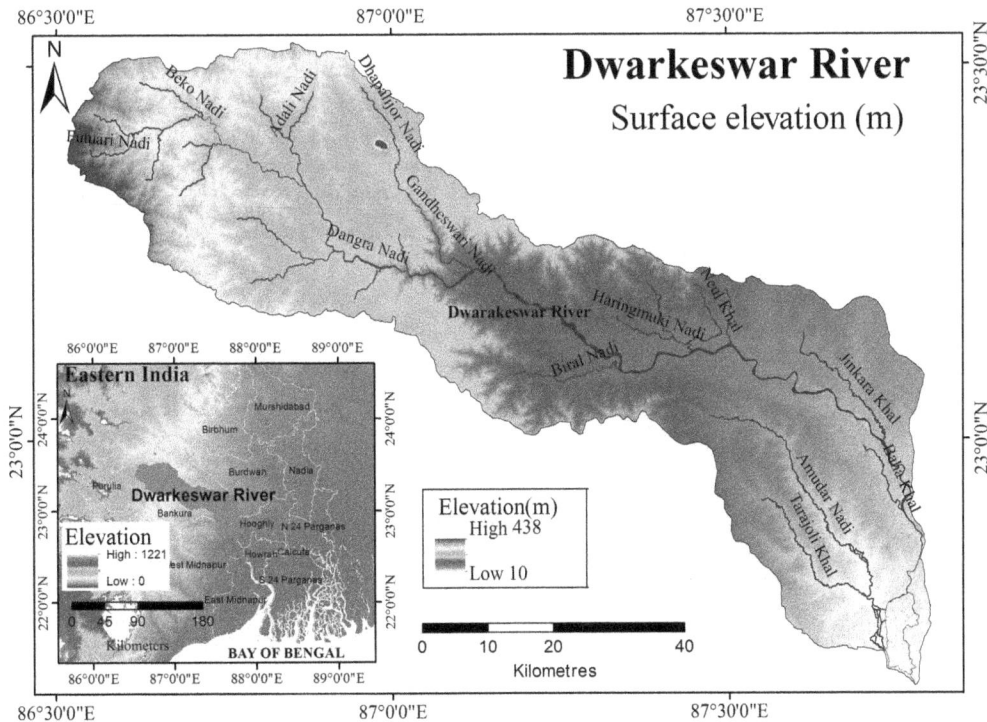

FIGURE 9.1 Location map of the study area.

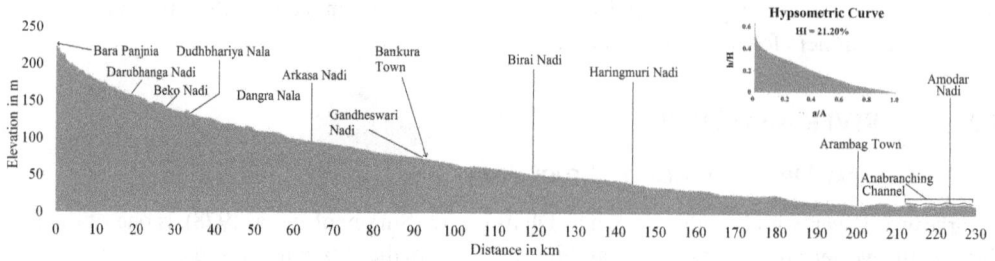

FIGURE 9.2   Long profile and hypsometric profile of Dwarkeswar River. (30 m SRTM DEM.)

basin is elongated in nature. The maximum width of the basin is 40.80 km, and the maximum length of the basin is about 159.84 km. The flow direction of this river is not in uniform. Figure 9.1 also demonstrates that the river has followed the direction of the north-west to south-east from its origin at Sindurpur near Tilabani to Bishnupur. After that, it has followed a curvature path with a radius of 31.7 km up the meeting point with Shilabati.

9.2.2   Drainage Basin Zonation

The Dwarkeswar River basin covers a diversified range of geological units. Figure 9.4a depicts that the upper portion of this basin belongs to the Chhota Nagpur Granite Gneiss Complex of Proterozoic age and the middle part belongs to the Laterite of Cainozoic age and to some extent Pleistocene Lower and Middle to Upper Sediment. The lower part of the river basin is associated with Holocene Sediment (Geological Survey of India, 1999). Such kind of diversified region of the western part of West Bengal is termed as *Rarh Bengal*, which extends up to the western and south-western border of West Bengal with Jharkhand (West Bengal Pollution Control Board (WBPCB), 2016). There are several types of classification available for this region such as classification of Hunter (1883 and 1877), O'Mally (1908), Bandyopadhyay et al. (2014) and report from West Bengal Pollution Control Board, (2016). Based on geological formation, relief aspect and channel characteristics, the Dwarkeswar River basin has been classified into three parts which have been shown in Figure 9.3, e.g., plateau proper (upper part), plateau fringe (middle part) and alluvial plain (lower part). Plateau proper is fundamentally the extension of Chhota Nagpur plateau, and it can be clearly found in the upper part of the basin with monotonous rolling topography and scattered residual hills of hard rock sometimes without vegetation. As shown in Figure 9.4a, geologically this area belongs to the Proterozoic Granitic Gneiss complex. Plateau fringe area or middle part of the basin is characterised by wide valleys and spurs dissected by transverse tributaries of Dwarkeswar River. Other dominant features of this region are lateritic tracts and its numerous rills and gullies formed by ephemeral first-, second- and third-order tributaries of this river (WBPCB, 2016).

9.2.3   Features of the Study Area Having a Potential Impact on Forms and Process

In this section, we have focused on those features which are having a potential impact on the fluvial forms and processes. Special focus has been given towards the anthropogenic activities such as urbanisation, extensive rail and road network expansion, small and minor dam construction, irrigation canal, embankment, extraction of river water for domestic and commercial use, construction of groyne for bank protection and agricultural practice on the river bed, which has been described next.

9.2.3.1   Physical Character of the Study Area

Figure 9.1 shows that the regional elevation of the Dwarkeswar River basin varies from 438 to 10 m (SOI topographical sheets, surveyed from 1972 to 1978). This basin is mainly belonging to the

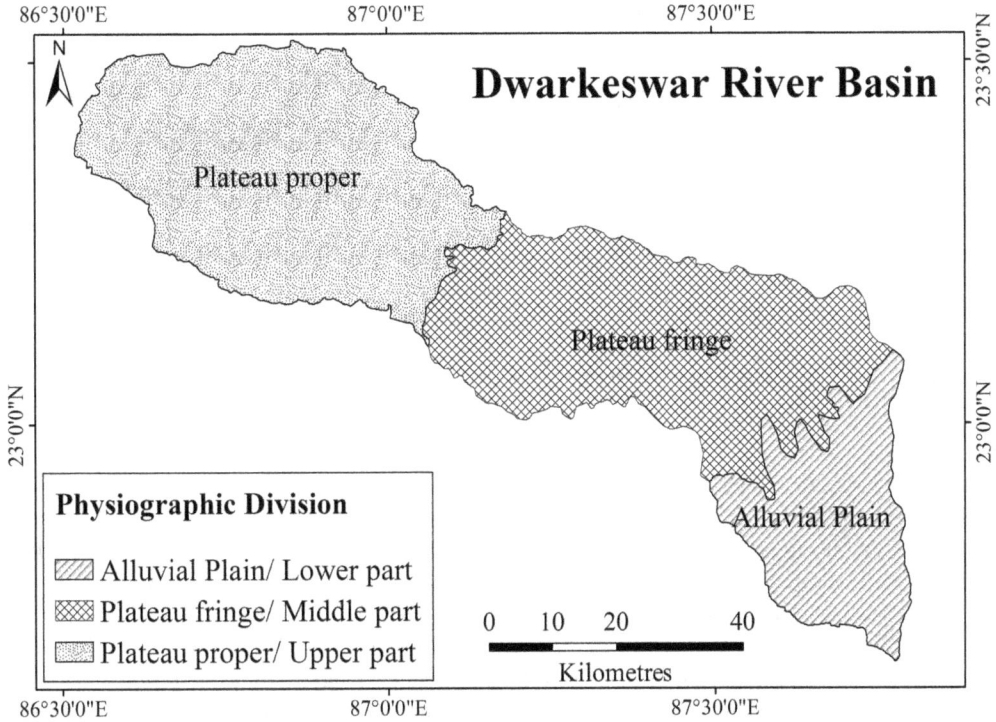

FIGURE 9.3 Physiographic division of the study area.

Rarh region literally meaning undulating terrain. The upper part of the basin mainly associated with the extended part of Chhota Nagpur plateau, and it can be clearly found in this area with rolling topography and scattered residual hills of hard rock sometimes without vegetation. Susuniya Pahar (438 m) and Tilaboni Hills (407 m) can be considered as a classical example of residual hill or monadnock, whereas the lower part of the basin is characterised by flood plain region. Agriculture is one of the most prominent activities in this alluvial part. Having a subtropical climate, a few portions of the middle part of the basin and entire upper part of the basin suffers from dry and drought-prone conditions during summer. Thereby, Dwarkeswar River and its tributaries dry up during the hot and cold season.

According to the National Bureau of Soil Survey and Land Use Planning (NBSS&LUP), red and yellow soils (Ochraquults, Rhodustults, Haplustults) can be dominantly found over the upper part of the basin. Lateritic soils (Plinthaquults, Plinthustults and Plinthudults) and red loamy soils (Haplustalfs, Paleustalfs and Rhodustalfs) can be found in the protected forest and its surrounding area located in the middle portion of the basin as shown in Figure 9.4b. The figure also shows that the lower part of the basin is associated with alluvial soils (younger and older) and its texture belongs to fine to fine-fine loamy (NBSS&LUP). The natural vegetation of this upper region is scattered dense mixed jungle mainly Sal and the middle to lower part of this basin mainly associated with the vast agricultural field. The average annual rainfall of this basin ranges below 1,300 mm in the westernmost part of the basin and increases towards the mouth and reached to above 1,500 mm year[-1] (Bandyopadhyay et al., 2014), whereas Dwarkeswar River became sluggish and the wide channel has become narrow with sands in the lower alluvial part of the basin.

9.2.3.2 Deforestation

Vegetation cover is one of the prominent aspects of the drainage basin. It is being generally considered that at least 30%–33% of vegetation cover is necessary for the natural flow of a stream

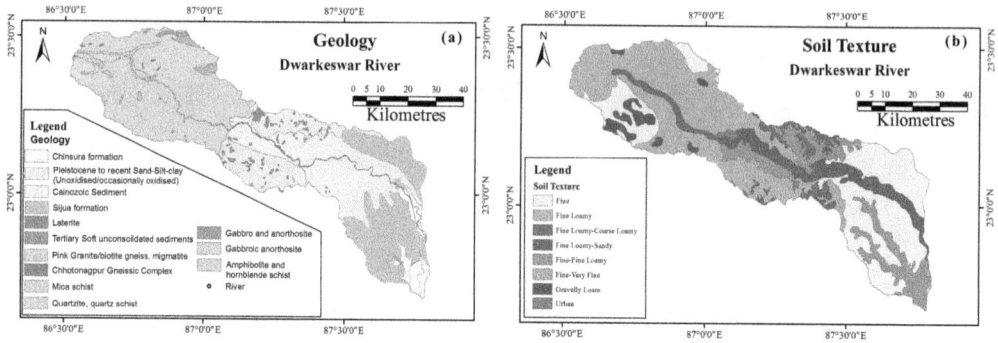

FIGURE 9.4   Surface geology (a) and soil (b) map of the study area.

(Fetriyuna et al., 2017). But, at present, this river basin represents only 11.56% vegetation cover. However, several historical and geographical accounts of this region indicate that the region had been covered with extensive thick Sal forest (Schlich, 1885; Duke, 1939; Dutta, 1960; Biswas, 1976; Panda, 1993; Siddique, 1996). According to the *Acbchanga Sutta* (5th century BC), the western districts moulded with a "pathless country" covered by impenetrable forests (Biswas, 1976 and Siddique, 1996). In *Bhavishya Purana* (15th–16th centuries), it was stated that 75% of the area was covered with forest (Biswas, 1976). Apart from these historical accounts, Schlich (1885) in his report on the Forest Administration in the Chhota Nagpur Division of Bengal stated that an extensive tract of mainly *Sal* forest was extended from Sone River through the Santal Parganas to the banks of the Ganga near Rajmahal on the east through Manbhum into Bankura and Medinipur in the south and through Singbhum and the tributary states of Chhattisgarh of Central India (Schlich, 1885). Census of India report of Birbhum (Census of India, 1961) and Siddique (1996) showed that Damodar and Dwarkeswar rivers were navigable with country boat at least up to the mid-18th century. But the British annexation into the western part of West Bengal started to expand the agriculture and thereby decreases the forest area started (Schlich, 1885; Duke, 1939; Sinha, 1962). Seton-Kerr (1877), Dickens (1869), Sinha (1962) and Siddique (1996) pointed out that vast forests in this area were rented out to private businesspersons. In addition, massive and irresponsible destruction of widespread forest area had been done for railways construction (Kharagpur–Gomoh Railway Branch line in 1902), highway bridges, warehouses and industrial establishments (Schlich, 1885; Duke, 1939; Dutta, 1960; Panda, 1993; Siddique, 1996; Census report of Bankura and Purulia 1961). Apart from this, there are ample of strong evidences to indicate that the whole range of policies of economic, social and political on resource use was anti-forest and anti-tribal (Schlich, 1885; Rasul, 1954; Duke, 1939; Sinha, 1962; Baske, 1976; Bhattacharya et al., 1985; Kulkarni, 1987; Siddique, 1996). Numerous letters from the second half of 19th century also depicted about increased revenue collection through cultivable lands which in turn demanded expansion of agricultural field into the forest areas (letters from D. Brandis, 1869; C.H. Dickens, 1868; A. Money, 1869; P. Dickens, 1869; Lieutenant Governor W.S. Seton-Kerr, 1867; H.H. Risley, 1893 and H.F.T. Maguire, 1897). W.W. Hunter in his writing *Annals of Rural Bengal* (1883) stated about the British merchants and servants about Bengal as "a great store of wealth to be looted with little effort". In this regard, Seton-Kerr in 1867 expressed deep concern and according to him "it was only a matter of time that the whole western districts should remain waste-producing nothing due to the destruction of forests and the local people would not accomplish this total change, that their livelihood would be endangered" (Siddique, 1996). The nature of deforestation over the period of time has been shown for 18th century in Figure 9.5a, 19th century in Figure 9.5b, 1945 in Figure 9.5c, 1972 in Figure 9.5d and 2019 in Figure 9.5e. Which is showing massive deforestation in plateau and plateau fringe area is a buried fact of the Dwarkeswar River basin. Vegetation cover decreased all over the basin, but

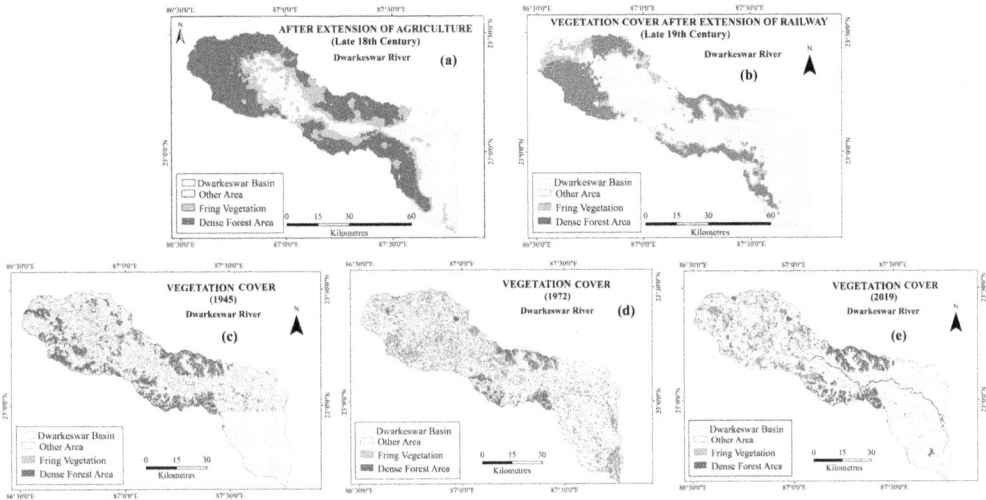

FIGURE 9.5 Nature of deforestation over the period of time: 18th-century (a) and 19th-century (b) vegetation cover map is a notional map modified after Siddique (1996), and the map of 1945 (c) and 1972 (d) are based on the topographical maps. The 2019 forest cover (e) was done from a satellite image.

TABLE 9.1
Land Use and Land Cover Change in the Different Physiographic Division of Dwarkeswar River

LULC and	Plateau Proper		Plateau Fringe		Alluvial Plain	
Year	1945	2018	1945	2018	1945	2018
Water body	3.05	5.43	2.41	2.00	4.60	2.44
Vegetation	25.91	19.11	29.20	13.00	9.53	2.56
Built-up area	4.67	14.45	4.76	19.18	9.05	20.35
Agriculture	66.37	61.01	63.63	65.83	76.82	74.65

Source: Calculated by the authors from topographical map and LULC map by ISRO.

drastic changes have been found in the middle part of the river basin, wherein Table 9.1 shows that decreases of vegetation for the year of 1945 was 29.20% and now it has decreased to 13% only.

9.2.3.3 Afforestation

Afforestation is another fact that can affect the channel flow properties of the basin. At present, several parts of the basin area have been considered as protected forest and afforested areas. Although, for revenue generations, the *Sal* trees from the protected forest are being cut down to date with 15–20 years of rotation, new trees are being planted as a replacement to deforested trees. The aboriginal forest has gone in this way, and the entire region has been afforested. Thereby, big tall *Sal* trees can be found only in few places (sacred places of tribal villages). When we have visually compared the trees of other surrounding places, we did not found any match with these trees, even if the trees of the protected forest were much smaller than the Sal trees in sacred places. Which is indicating work of afforestation after drastic deforestation? Apart from the protected forest, several areas of the basin experience afforestation of *Sonajhuri*. Figure 9.6 indicates that at present, 155.76 km² (3.58%) of the basin area is characterised by planting trees.

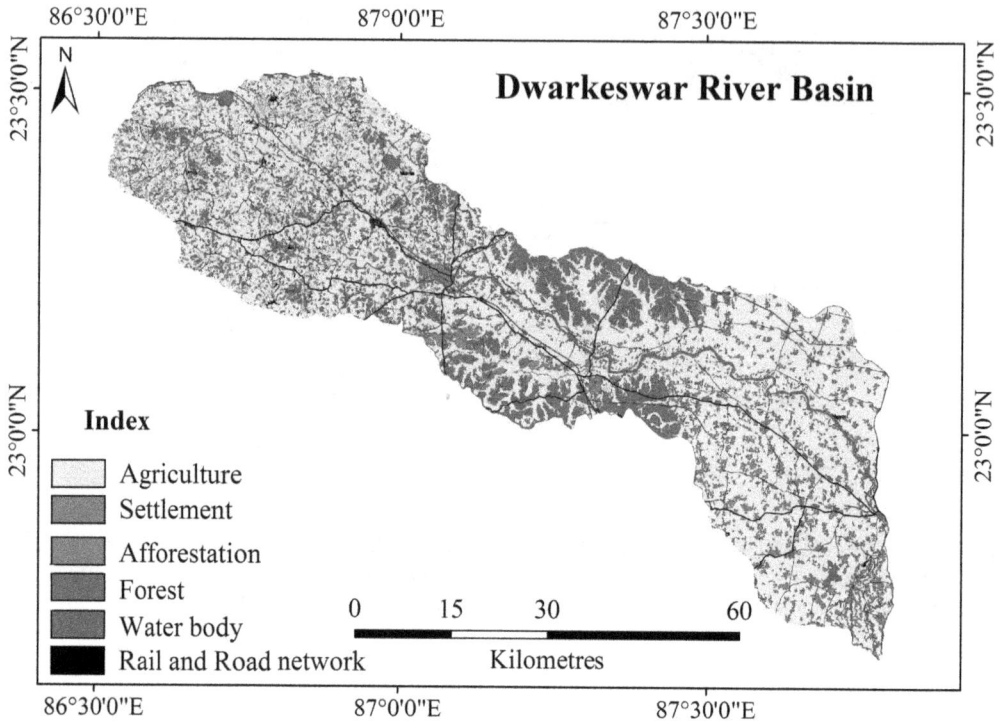

FIGURE 9.6 Location of afforestation (or indicate in the study area).

9.2.3.4 Land Use and Land Cover

Land use and land cover are some of the important variables for controlling the drainage basin. In the case of Dwarkeswar River, it has been found that from the past few decades, land use land cover has been changed drastically. It has been stated in Acbchanga Sutta that during 5th century BC, this region was covered by impenetrable forests (Biswas, 1976; Siddique, 1996).During 15th and 16th centuries, forest coverage has been reduced to 75% of this region, and there are few tribes left (Biswas, 1976.); therefore, till 15th and 16th centuries, other types of land use were very minimum, and this seems to be concentrated in a localised manner with a settlement. During the late 17th century, forest area decreased due to the expansion of agriculture (Schlich, 1885; Duke, 1939; Sinha, 1962). Construction of railways, highway bridges, warehouses and industrial establishments accelerated the rate of deforestation as well as the conversion of forest area to agricultural land in the late 19th century (Schlich, 1885; Duke, 1939; Dutta, 1960; Panda, 1993; Siddique, 1996). It has also been found during a conversation with local old people that during the 1970s, there was another episode of deforestation and extension of agriculture in the upper part of the drainage basin.

We have found that from 1945 to 2018, land use and land cover changed over the entire area of the basin, which has been shown in Figure 9.7a–c and Table 9.1. In this study, Table 9.1 and Figure 9.7c indicate that the water body covering area in upper part increased from 3% to 5.43%, and this was happened due to the construction of several small and minor dams as well as some water tank in the upper part of this basin. They also show that the vegetation cover decreased all over the basin, but drastic changes have been found in the middle part of the river basin, wherein in 1945, it was 29.20%, and now, it has decreased to 13% only. The built-up area has also increased all over the basin and middle part of the basin having the top position in it. Thereby, it is very much clear that drastic change in land use has occurred in this river basin.

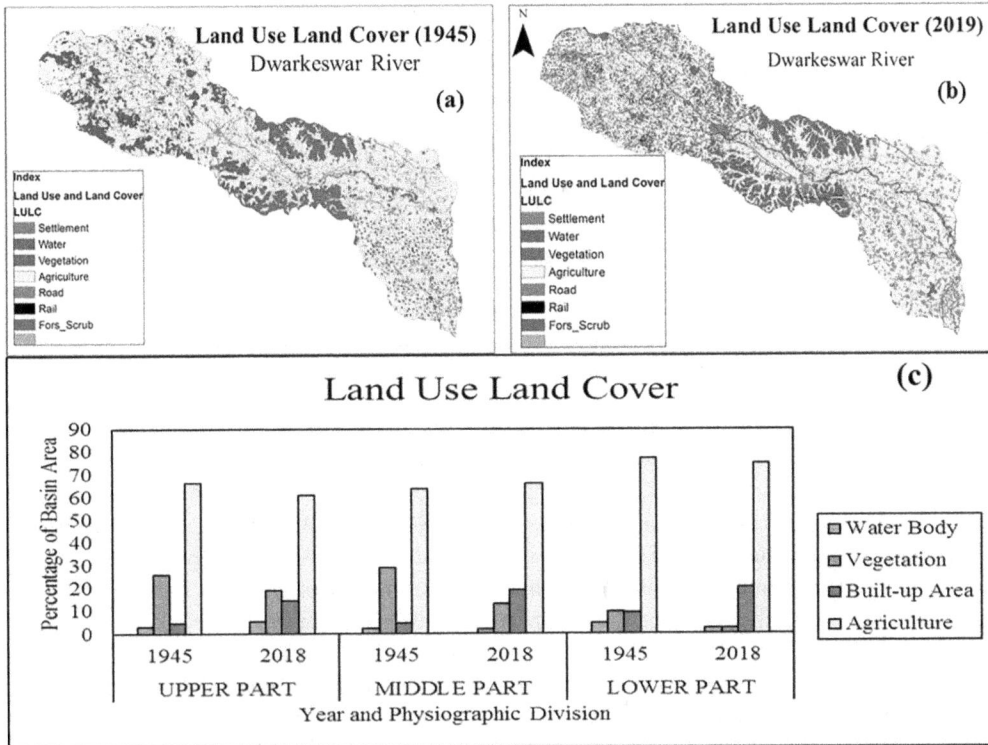

FIGURE 9.7 Land use and land cover of 1945 (a) and 2019 (b) of Dwarkeswar River and its change (c).

9.2.3.5 Transport Network and Bridge Construction

The transport network is an inevitable part of modern society. The transport network of this basin is well development of the transport network has been shown in Figure 9.8a–e. It has been found from Table 9.2 and Figure 9.8a–e that the transport network during 1779 was not much developed as of today. There were only a few roads, which have been demarcated by Rennet's map of Bengal

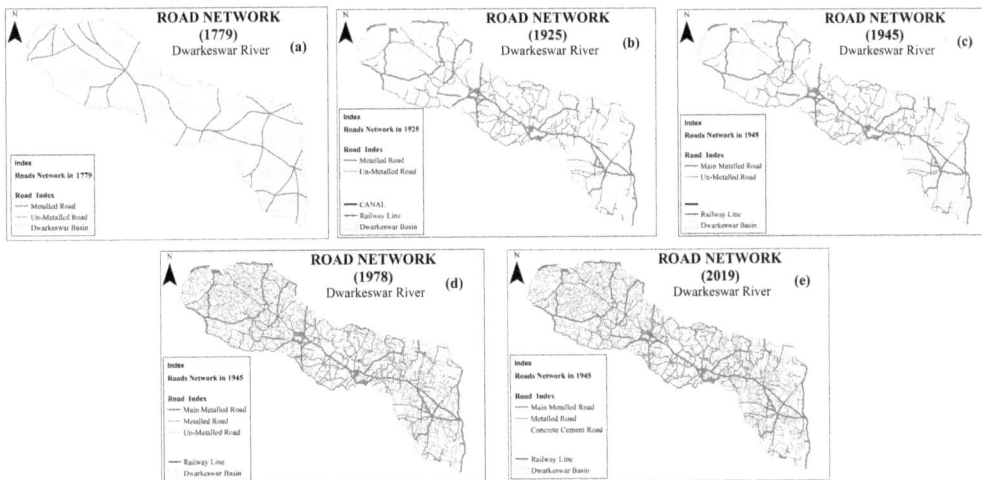

FIGURE 9.8 Spatial distribution of road network in times and the graph showing a quantitative representation of the length of the road in the years 1779 (a), 1925 (b), 1945 (c), 1978 (d) and 2019 (e).

TABLE 9.2

Length of Road in Different Periods of the Dwarkeswar Basin

Physiographic Division of Basin	Year (Road Length in km)				
	1779	1928	1945	1978	2019
Plateau proper	162.68	172.30	172.30	403.88	1,944.33
Plateau fringe	220.96	260.19	260.19	961.93	2,346.10
Alluvial plain	157.24	142.22	142.22	325.41	1,078.54

Source: Calculated by the authors from topographical map and Google earth map.

without mentioning whether it was metalled or un-metalled. At that time, total length of the road was 162.68 km in the upper part, 220.96 km. in the middle part and 157.24 km in the lower part. The length of the road network does not have increased very much up to 1945. But after 1945, length of the road has increased drastically and reached to 403.88, 961.93 and 325.41 km for the upper, middle and lower part of the study area, respectively. At present, different schemes from Government of India as well as Government of West Bengal have led to tremendous increases in the length of the road network for the three parts of the basin. Along with this transport network, several small, medium and big bridges and culvert have been constructed with improperly designed over this river basin. Several bridges were constructed on this river, such as Kashipur Bridge (near Kashipur town, Puruliya), Patakola bridge (near Bankura town, Bankura), Eklakshmi Bridge (near Rautara, Burdwan), Samroghat Bridge (near Mathuratapal, Bankura) and Abantika bridge (near Abantika, Bankura).

9.2.3.6   Dams, Water Tank, Weirs and Sluices

Plateau proper area and upper portion of plateau fringe area of this basin suffer from water scarcity during the summer season, and due to the prolonged absence of monsoon rainfall, therefore, several small water tank has been constructed to meet the requirements of local people for their daily livelihood. Table 9.3 and Figure 9.9a–c show that about 831 small water tanks were constructed till 1978. But after that, four medium-size dams have been suggested, and all of them are in operation. These are Futiary Medium Irrigation Project, Beko Medium Irrigation Project, Suvankar Dangra Medium Irrigation Project and Berai Canal Medium Irrigation Project. Except for the Berai Canal Medium Irrigation Project which is located in the middle part of the river basin, all the dams were constructed in the upper part of the basin. From its title, it is evident that these were mainly developed for the irrigational purpose, whereas the lower part of the basin has already been irrigated by the rivers Damodar and Kangsabati.

Weirs and sluices have been constructed on the river to control its flow. All the weirs and sluices mainly located in the upper parts of the river. Figure 9.10 shows the location of 154 weirs and sluices in the Dwarkeswar River basin. It has been stated earlier that this kind of construction acts as a barrier to the natural flow; therefore, it can affect the modification of flow behaviour and channel morphology of the respected streams.

9.2.3.7   River Water Extraction

The upper part of this river basin belongs to the drought-prone area. Being a drought-prone area here in this area mainly the upper portion of the middle part of the basin is associated with numerous water extraction points. These are mainly used for domestic, commercial and agricultural purposes, and we have identified 12 such stations, which are shown in Figure 9.11. These stations extract the base flow of the river as well as the underground flow of the river from the river bed. They mainly supply the entire Bankura Town, Raghunathpur Town, nearby railway stations and in

TABLE 9.3

Details of a Medium Dam Project in This Study Area

Name of the Project	River	Name of the Reservoir	Year	Submerged Area	Area (ha)	Design Gross Storage	Design Live Storage	Command Area (000 ha)	Ultimate Irrigation Potential (000 ha)
Futiary medium irrigation project	Futiary	Futiary reservoir	1997	0.079	1,871.62	2.386	1.85	0.96	1.2
Beko medium irrigation project	Beko	Beko reservoir	1990	0.142	1,759.87	2.71	2.716	1.21	2.51
Suvankar Dangra medium irrigation project	Dungrajore	Dangra reservoir	1982	0.067	5,322.01	1.525	1.42	2	2.43
Berai canal medium irrigation project	Berai	Berai weir	1986		5,142.534			3	3.63

Source: Irrigation and Waterways Department, Government of West Bengal, India, 2019.

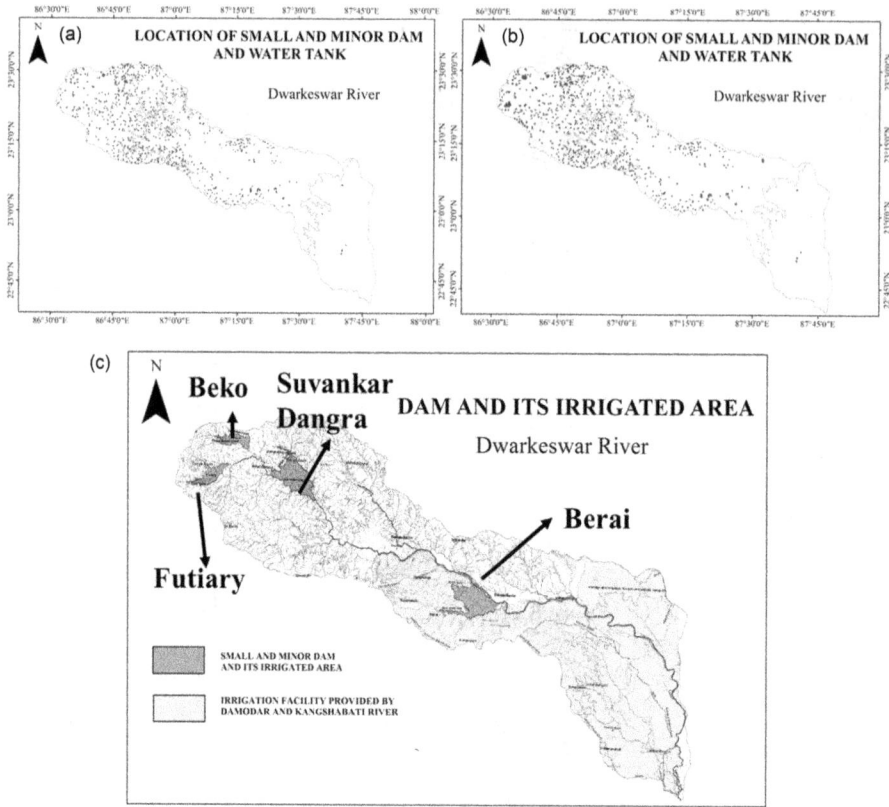

FIGURE 9.9 Location of a small water tank during 1978 (a) and present (b) and location of currently developed medium dams in the study area (c).

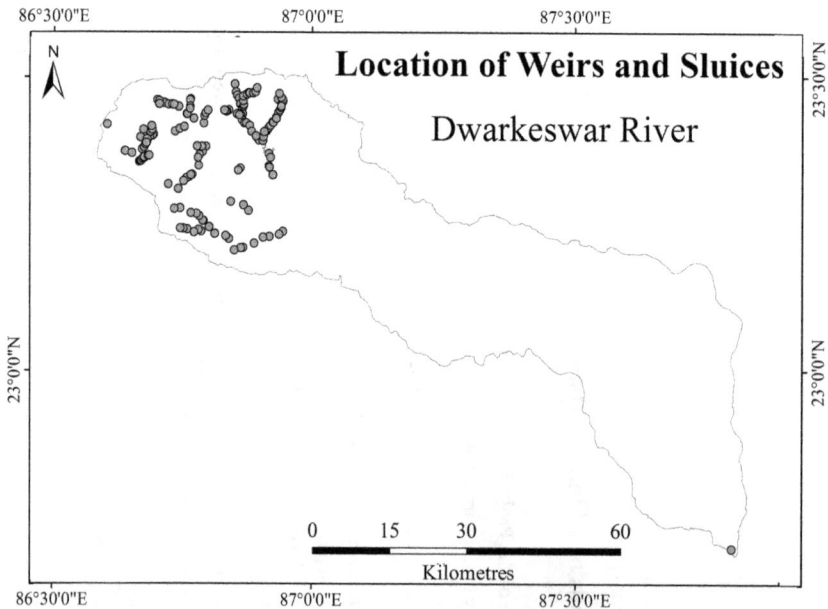

FIGURE 9.10 Location of weirs and sluices.

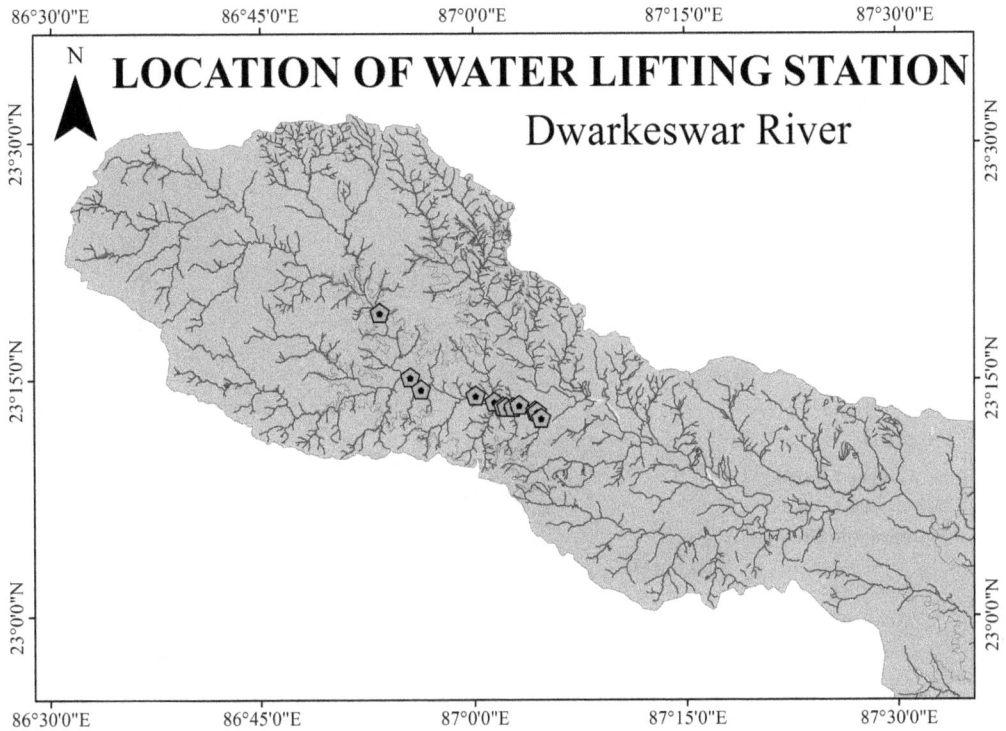

FIGURE 9.11 Location of water extraction point along Dwarkeswar River.

the local area. Water resource development is an important part of development, but we think these are the measures for the short-term development which may not be sustainable for the long term. Recently, near Patakola bridge, underground structure across the river bed has been constructed to arrest the base flow of the river to meet the requirement of Bankura town.

9.2.3.8 River Training Structures

River training structure can be defined as a structure which is constructed to improve the bank and river. In Dwarkeswar River, spar or dyke on the river bed was constructed to avoid bank erosion and channel improvement. These are structured constructed from the bank and extended transversely to the river flow direction (Zabih, 1976). These are also known by several names, the popular being spur dykes, spur and transverse dykes, retard, groin and so on (Zabih, 1976; King, 2009). Along this river, four stretches of groyne-constructed sections are shown in Figure 9.12. Among them, we have selected the groynes near Rautara Village for the ground base analysis.

9.2.3.9 Sand Mining

Sand mining is one of the important factors that can severely alter the channel morphology in a very short period of time. It has been considered that before 1985, sand mining in this river was very insignificant, but after this period, it has increased tremendously. Seventy-eight locations have been identified and shown in Figure 9.13 along the main Dwarkeswar channel, which is mainly concentrated in middle and lower parts of the river.

9.2.3.10 Morphometric Character of the Basin

Morphometric analysis of a drainage basin is one of the primary tasks to understand the basic properties of the basin (Horton, 1932; Leopold and Miller, 1956; Melton, 1958; Morisawa, 1959; Nag and Lahiri, 2011; Strahler, 1957; Doornkamp and King, 1971). Stream ordering following

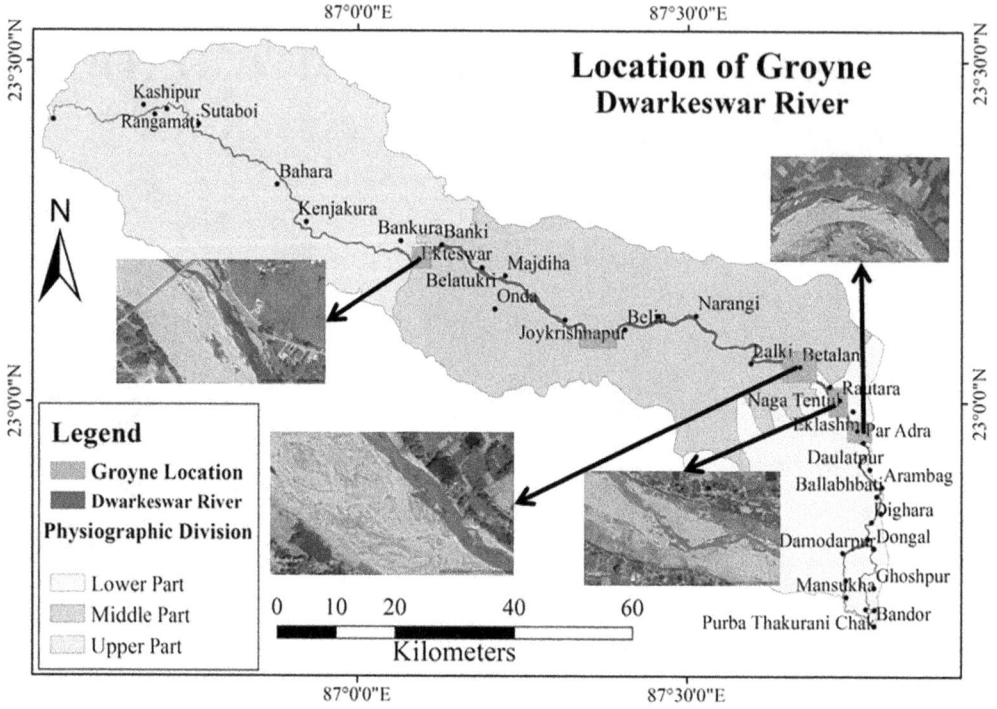

FIGURE 9.12   Location map of spar on Dwarkeswar River.

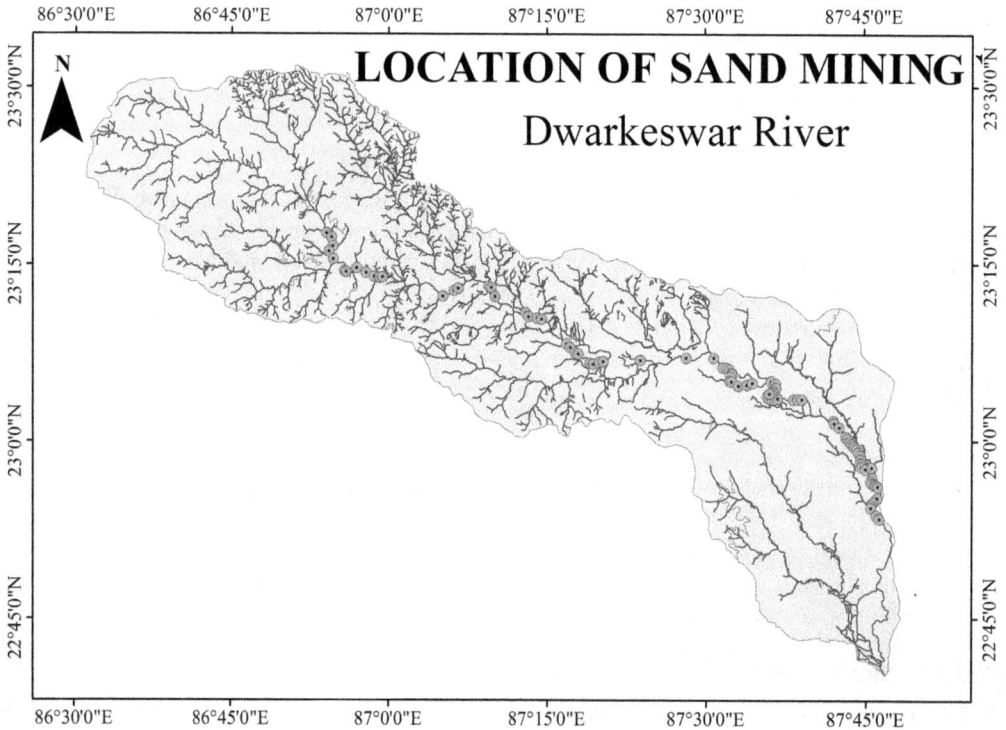

FIGURE 9.13   Location of sand mining of Dwarkeswar River.

Strahler (1957) has been calculated and shown in Figure 9.14a. Figure 9.14b is showing the nature of contour, whereas Figure 9.14c is showing the sub-basins of Dwarkeswar River. Table 9.4 and Figure 9.14a indicate that Dwarkeswar River is a sixth-order drainage basin with 152.19 km in length.

Figure 9.15a and b, Figure 9.16a and b, and Table 9.4 are showing the bivariate analysis of stream length, and stream number with stream order shows a straight line in logarithmic plotting. This demonstrated the decreasing of stream number and increases of stream length with respect to stream order logarithmic manner with r^2 value of 0.984 and 0.998, respectively, indicating homogeneous rock material subjected to prolonged weathering and erosion with characterised by lithological and

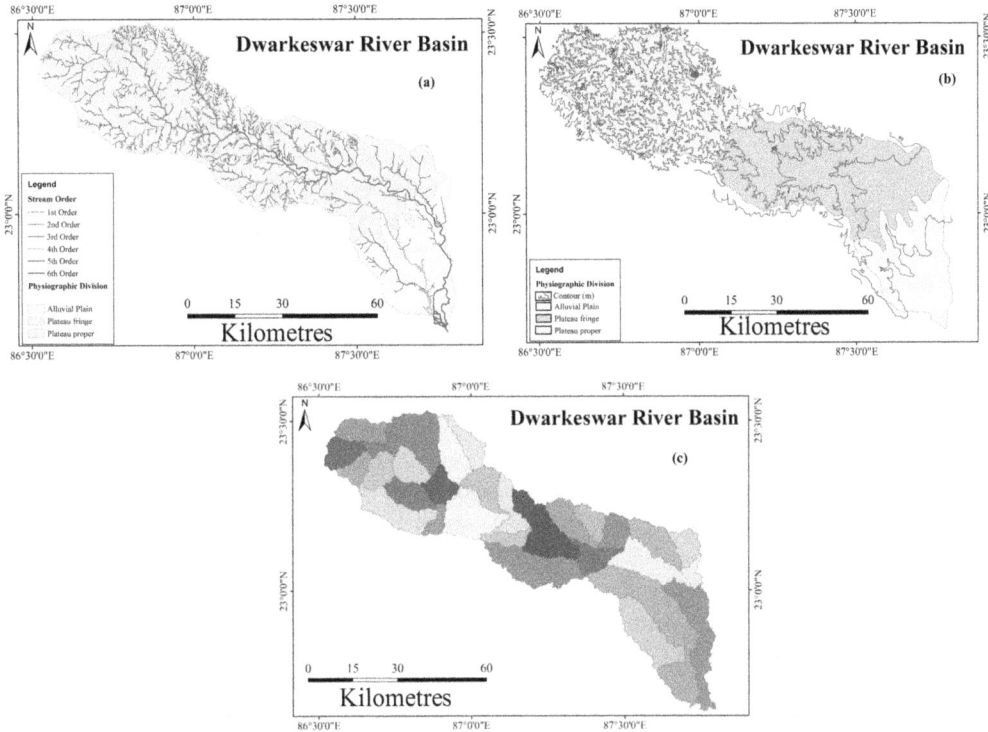

FIGURE 9.14 Map showing contour lines, channel network (a), contour line (b) and sub-basins of Dwarkeswar River (c).

TABLE 9.4

Stream Order and Its Respective Total Stream Number and the Average Length of the Stream

Stream Order	No. of Stream	Basin Area (km²)	Length of the Stream (km)	Bifurcation Ratio
1	1,073	4.35	1.11	
2	280	9.43	2.19	3.83
3	62	34.65	6.36	4.52
4	17	116.17	14.3	3.65
5	3	694.28	31.07	5.67
6	1	4,356.72	152.19	3.00

Source: Calculated by the authors from topographical map.

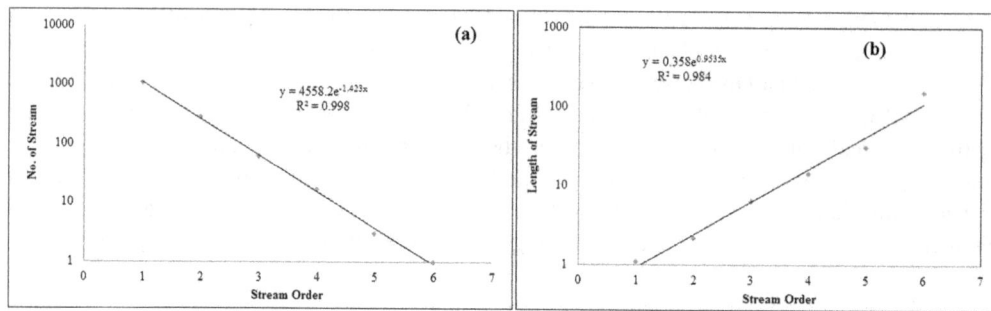

FIGURE 9.15 Bivariate relationship among stream order and stream number (a) and stream order and stream length (b).

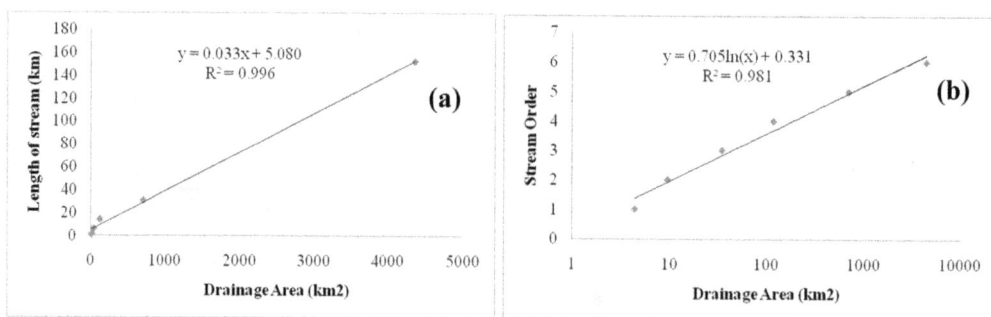

FIGURE 9.16 Bivariate relationships among the length of the stream, drainage area (a) and order of stream (b).

topographic variation (Nag and Lahiri, 2011). The bifurcation ratio of the drainage basin ranges from 3.00 to 5.67. Strahler (1964) argued that drainage basin with a high bifurcation ratio occurred where geologic structures do not have disturbed the drainage pattern. The length of the stream has been measured using GIS which reveals the characteristics of surface runoff indirectly. In this study, Table 9.4 depicts that the average length of first-order streams is about 1.11 km, and it has increased to 152.19 km for sixth-order streams. Drainage network properties of the basin have classified based on the physiographic division and are shown in Table 9.5. It has been found that the number of all order streams is very high in the middle part of the basin compared to the upper part.

TABLE 9.5
Drainage Network Properties in Different Physiographic Regions of the Basin

Stream Order	Stream Number			Average Stream Length (m)		
	Upper Part	Middle Part	Lower Part	Upper Part	Middle Part	Lower Part
1	441	680	68	946.18	1,044.35	2,010.58
2	98	201	14	1,767.17	2,069.28	5,433.05
3	22	44	2	5,640.63	5,227.34	36,707.04
4	6	15	1	6,694.89	16,917.48	86,002.03
5	1	3	-	10,089.35	31,068.97	-
6		1	1	-	1,83,887.00	1,12,184.43

Source: Calculated by the authors from Topographical map.

This kind of result might have been occurred due to the expansion of extensive agriculture in the upper part of the basin compared to the middle part.

Nature of the relationship of stream order and its mean length has been shown in respect to the drainage area, and it has been shown in Figure 9.16a. Figure 9.16b also indicates that there is a high degree of positive relationship among the length of the stream and order of the stream with a drainage area.

Stream frequency and drainage density have been calculated on the basis of 1 km² grid following Horton (1932). Higher stream frequency and drainage density have been found over the northern and middle parts of the basin as indicated in Figure 9.17a and b, whereas the upper part of the basin is devoid of high stream frequency and drainage density. The lower part of the basin is associated with low to very low stream frequency and drainage density. Along with this, Figure 9.17c and d depicts similar nature of texture ratio and drainage texture.

Relative relief, dissection index and ruggedness index have been calculated based on the 1 km² grid and is shown in Figure 9.18a–c. Figure 9.18a–c shows that the upper part is showing high to very high relative relief and dissection index, whereas the lower part is associated with very low to low values, respectively. Ruggedness index has been showing very high on the northern and middle part of the basin and very low to a low value in the lower part. Figure 9.18d shows the spatial distribution of slope which is similar to the relative relief and dissection index.

Drainage network of Dwarkeswar River is mainly dendritic in nature, although radial drainage patterns can be found on the residual hills. Sub-basin of the plateau proper area of the Dwarkeswar River basin is associated with the moderate Miller's Circularity Index (1953), and as shown in Figure 9.19, it ranges from 0.36 to 0.25. In the lower alluvial plain region, all the sub-basins are elongated in nature and their Circularity Index is very low.

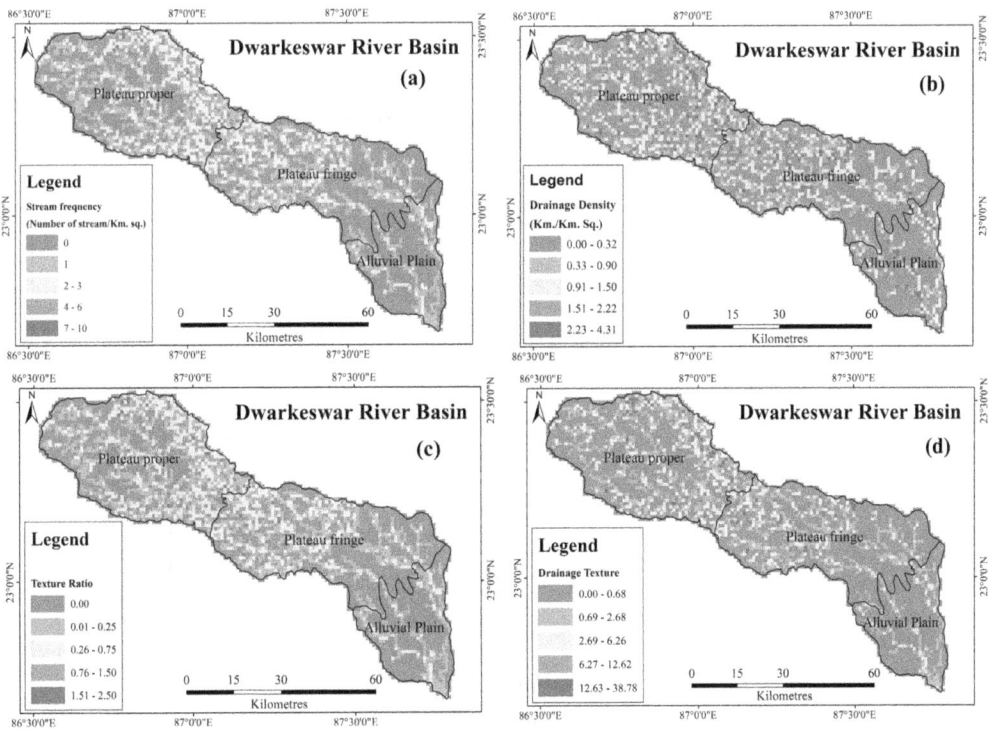

FIGURE 9.17 Stream frequency (a), drainage density (b), texture ratio (c) and drainage texture (d) of Dwarkeswar River.

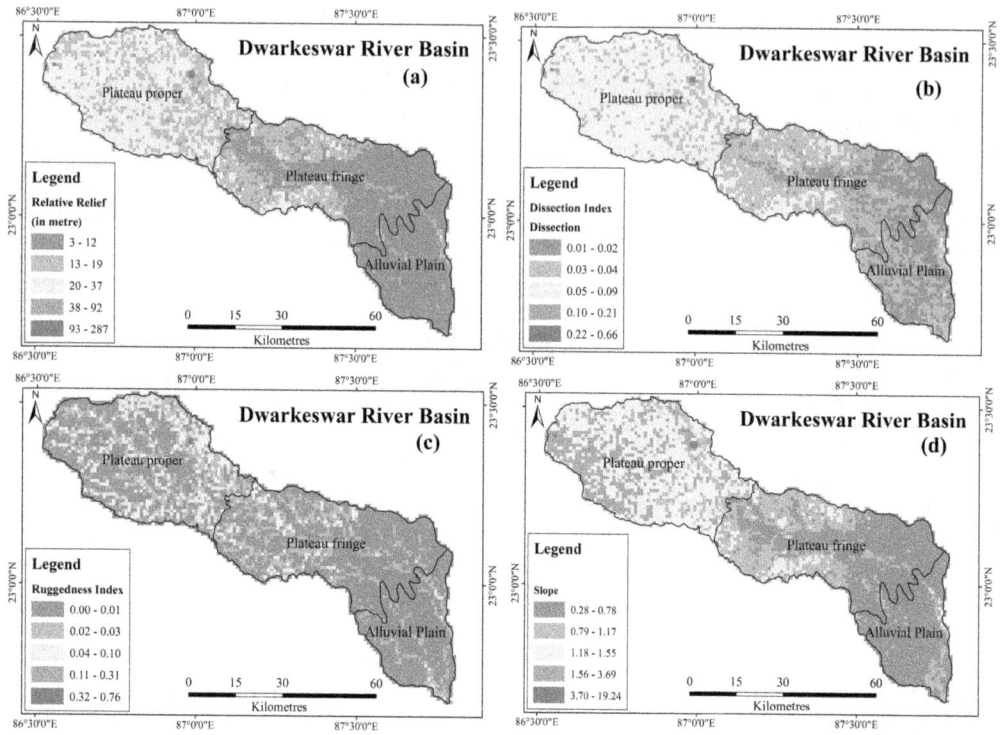

FIGURE 9.18   Relief aspect of Dwarkeswar River basin (relative relief (a), dissection index (b), ruggedness index (c) and slope (d)).

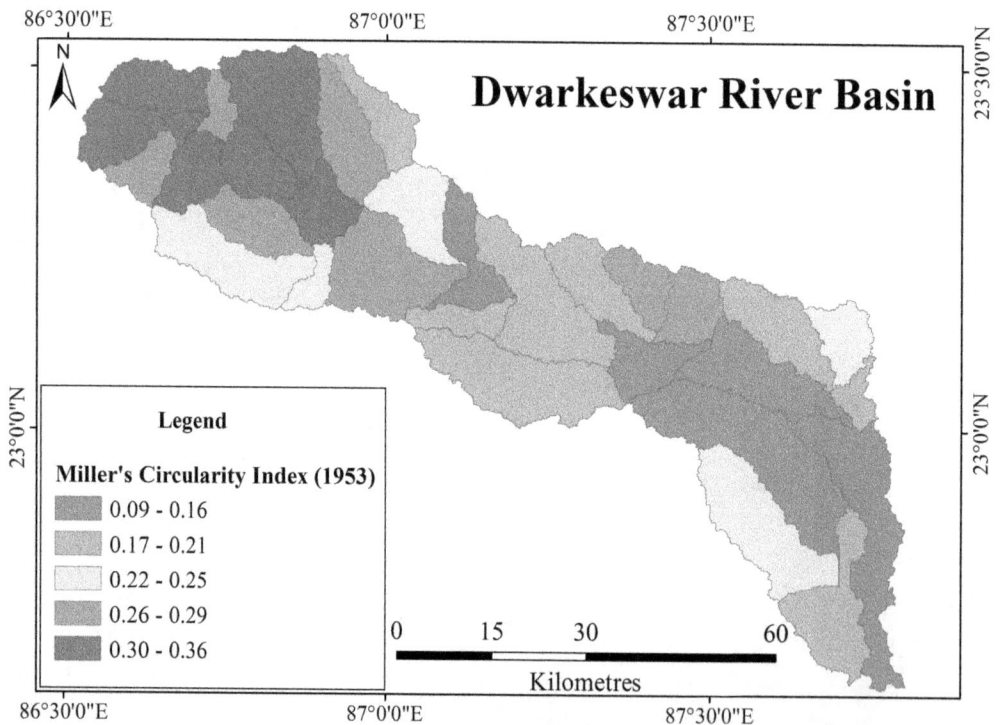

FIGURE 9.19   Sub-basin of Dwarkeswar River and its variation in Miller's Circularity Index.

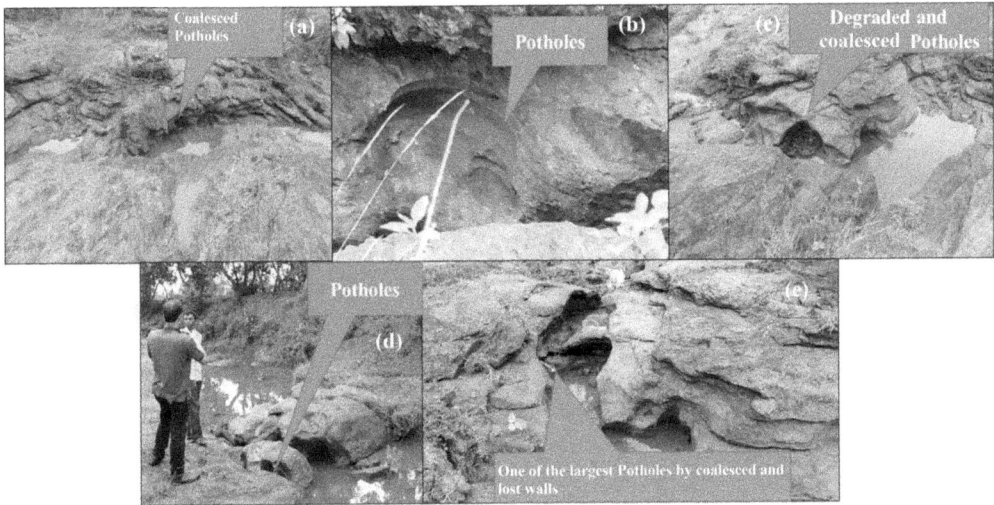

FIGURE 9.20 Presence of granitic outcrop across the channel flow and degraded and accumulated potholes (a–e).

9.2.3.11 Potholes

One of the most common features in the rapidly flowing bedrock river is potholes (Sengupta and Kale, 2011). In the upper part of the study area, several potholes have been found. The work of Elston (1917), Ängeby (1951), Lorenc et al. (1994) suggested that initiation of flow separation and differential erosion takes place due to the presence of surface irregularities in bedrock in the form of joints or bedding planes. Here, in our study area, surface irregularities can be found in the bedrock in the form of presence of granitic outcrops across the channel flow. A careful investigation from Figure 9.20a–e suggests that all the potholes have been observed in the upper part of the river course and these were degraded through time and process of accumulation significantly dominated here rather than the formation of newer potholes. Apart from this, pothole bed has become the river bed, and few potholes have a greater height than the current river bed elevation. This indicates that they might not have formed in the present fluvial regime rather than they had formed much earlier times.

9.2.3.12 Plan View of the Channel

Figure 9.21a–d shows the channel plan view of Dwarkeswar River. Figure 9.21b indicates that the upper portion of the river has been showing a straight channel, whereas Figure 9.21a depicts that plateau fringe part is associated with meandering channel pattern. Figure 9.21c demonstrates the meander scar in the plate fringe area, and anastomosing channel pattern has been found on the lower part of Arambag as shown in Figure 9.21d.

9.3 ANTHROPOGENIC SIGNATURES ON FORMS AND PROCESSES

9.3.1 Soil Erosion

Soil loss is an important aspect that can reflect all the activities that have been occurring on the river basin as well as its physical aspect altogether through its changing rate of soil erosion over time. Revised universal soil loss model has been used for the estimation of soil erosion in this basin area, which is given as follows (Renard and Laursen, 1975):

$$A = R * K * LS * C * P \tag{9.1}$$

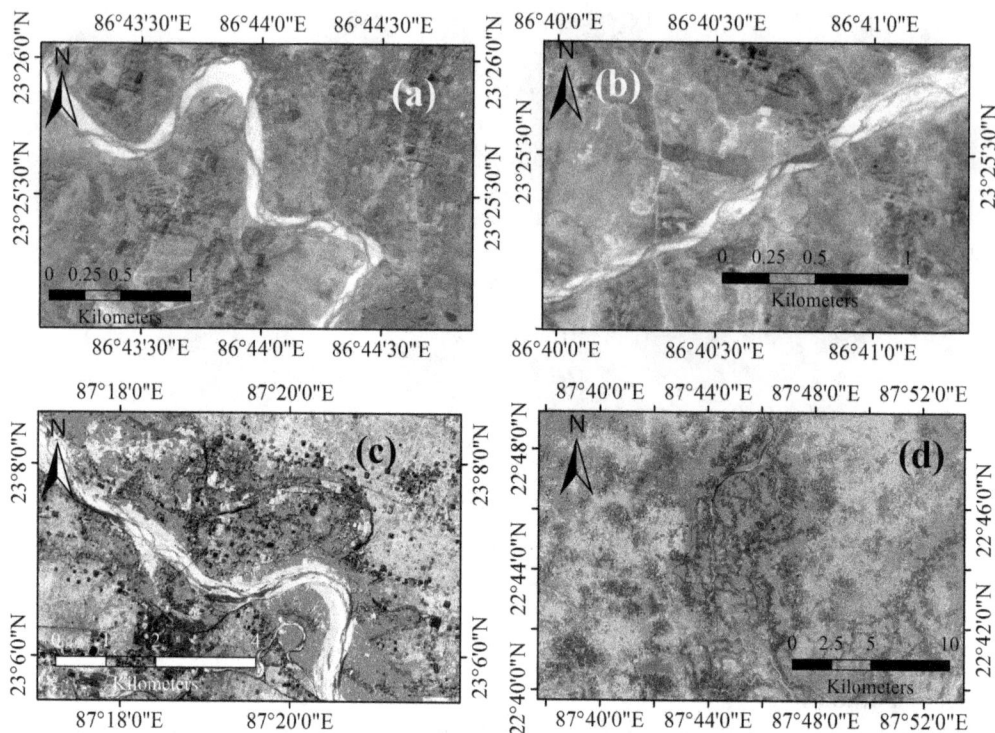

FIGURE 9.21 Channel pattern of Dwarkeswar River: (a) meandering channel pattern in the dissected plateau fringe region; (b) straight channel pattern with weirs construction and changes in channel plan-form in the plateau proper area; (c) meander scar, channel avulsion in the lower part of the plateau fringe area; (d) anabranching channel pattern in lower alluvial plain.

where A is the average annual soil erosion (tons/ha/year), R the rainfall and runoff erosivity factor (MJ mm/ha/hr/year), K the soil erodibility factor (tons/ha), LS the slope length and steepness factor, C the cover and management factor, and P the support practice factor related to slope direction.

Several scholars have used this model to estimate soil erosion. We have calculated the soil loss for the late 18th and 19th centuries and for the years 1945, 1973 and 2019. Soil erosion for the late 18th and 19th centuries has been done based on the notional map modified after Siddique (1996), and for 1945 and 1973, we have used topographical maps, and satellite images were taken into consideration for 2019.

The upper part of this region has been changed very drastically in terms of soil erosion rate. During the late 18th century, it has been found that erosion rate was very low. As shown in Table 9.6 and Figure 9.22 a, the average rate of erosion of this region during the late 18th century has been estimated as 0.34 ton ha^{-1}year^{-1}. Due to deforestation and expansion of agriculture, it has started to increase up to 2.43 tons ha^{-1}year^{-1}, which is about more than seven times greater. Table 9.6 and Figure 9.22a–h also indicate that very high degree of soil erosion found to be located in a hilly area but the high and moderate degree of soil erosion area increased from 0.09 and 1.42 ton ha^{-1}year^{-1} (late 18th century) to 16.59 and 24.7 ton ha^{-1}year^{-1} (2019), respectively. Very low to low degree of erosion decreased from 94.13 ton ha^{-1}year^{-1} (late 18th century) and 4.12 ton ha^{-1}year^{-1} (2019) to 54.7 ton ha^{-1}year^{-1} (late 18th century) and 2.08 ton ha^{-1}year^{-1} (2019). This might have occurred due to rapid deforestation and extensive expansion of agricultural land at the cost of deforestation and domination in a barren land with a single-crop agriculture system.

In the case of the middle part of the basin, the average soil erosion rate increased from 0.31 to 1.45 ton ha^{-1}year^{-1}, which is roughly 4.68 times. In this region, maximum portion of the area was

TABLE 9.6

Temporal Variations of Average Soil Loss Erosion Rate in Different Parts of the Basin

	18th Century	19th Century	1945	1973	2019
Plateau area	0.34	0.65	0.92	1.45	2.43
Plateau fringe area	0.31	0.48	0.49	0.79	1.45
Alluvial plain	0.27	0.30	0.29	0.32	0.59

Source: Calculated by the authors.

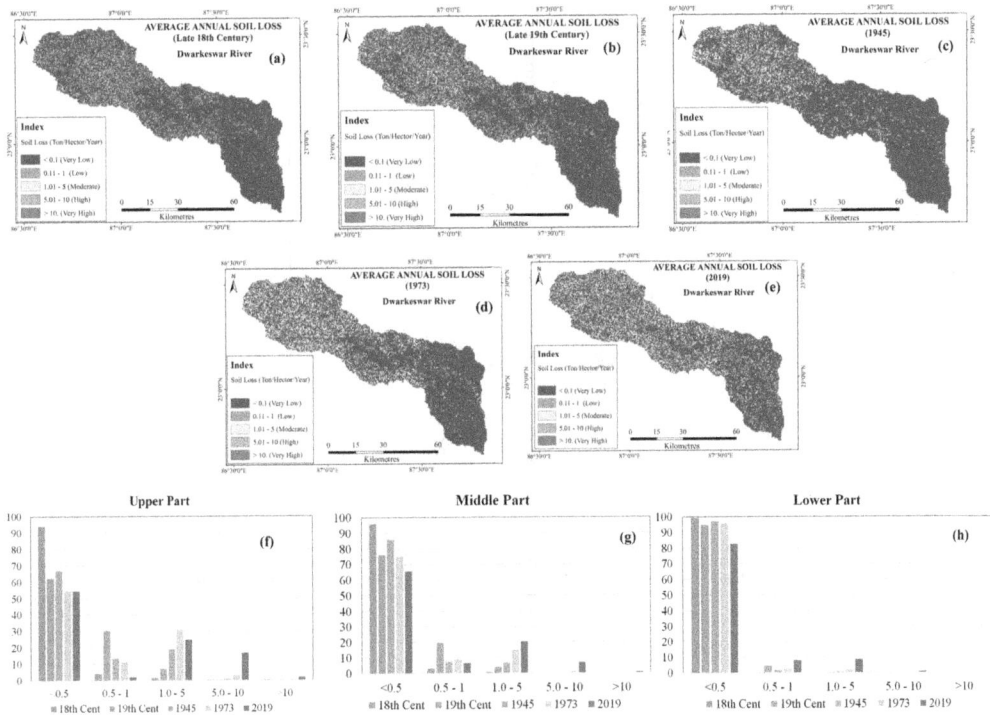

FIGURE 9.22   Spatial distribution of soil erosion for the late 18th century (a), late 19th century (b), 1945 (c), 1973 (d) and 2019 (e) using RUSEL model and a bar graph showing the spatial distribution of area in respect to its erosion rate for the upper (f), middle (g) and lower part (h) of the river basin.

under very low to a low rate of soil erosion during the late 18th century, but it has decreased to half of the middle part of the basin. The moderate and high degree of soil erosion increased its spatial coverage from nearly 1% in the 18th century to approximately 27% of the area. This phenomenon might have a relation with the encroachment of agriculture in forest cover land.

As shown in Table 9.7 and Figure 9.22a–h, average soil erosion rate (ton/ha/year) in the lower part of this region does not increase significantly compared to other parts of the basin.

9.3.2   CHANGES IN CHANNEL FLOW

Rivers around the world have been progressively transformed through dam building for human requirement (Dynesius and Nilsson, 1994; Nilsson et al., 2005), which has been well documented and acknowledged (Fuller et al., 2003; Gregory, 2006; Lane and Richards, 1997; Marsh, 1965).

TABLE 9.7

Detail Distribution of Spatial Coverage Soil Erosion Rate in Different Parts of the Dwarkeswar Basin

Erosion Rate (ton/ha/year)	Upper					Middle					Lower				
	18th Century	19th Century	1945	1973	2019	18th Century	19th Century	1945	1973	2019	18th Century	19th Century	1945	1973	2019
<0.5	94.13	62.29	66.87	54.56	54.70	95.96	76.25	85.80	75.27	65.52	99.39	94.91	97.12	95.66	82.57
0.5–1	4.21	30.27	13.17	11.00	2.08	3.06	19.52	7.33	8.91	6.53	0.26	4.52	1.80	2.47	7.99
1.0–5	1.42	6.90	18.96	30.76	24.70	0.79	3.95	6.63	14.72	20.30	0.22	0.39	0.94	1.76	8.59
5.0–10	0.09	0.33	0.83	3.19	16.59	0.06	0.17	0.22	0.92	6.94	0.09	0.07	0.13	0.07	0.82
>10	0.16	0.21	0.18	0.48	1.93	0.12	0.11	0.01	0.17	0.71	0.04	0.11	0.00	0.04	0.02

Source: Calculated by the authors.

TABLE 9.8

Temporal Variation of Annual Peak Discharge and Construction of Irrigation Project

Year	Average Gauge Height (Arambag) of 10 Years	Design Gross Storage	Cumulative Gross Storage (Cumec)
1978–1987	16.017	152.5 (2)	305
1988–1997	15.479	271.0	576
1998–2007	15.123	238.6	814.6
2008–2018	14.611	0	814.6

Source: Irrigation and Waterways Department, Government of West Bengal, India, 2019.

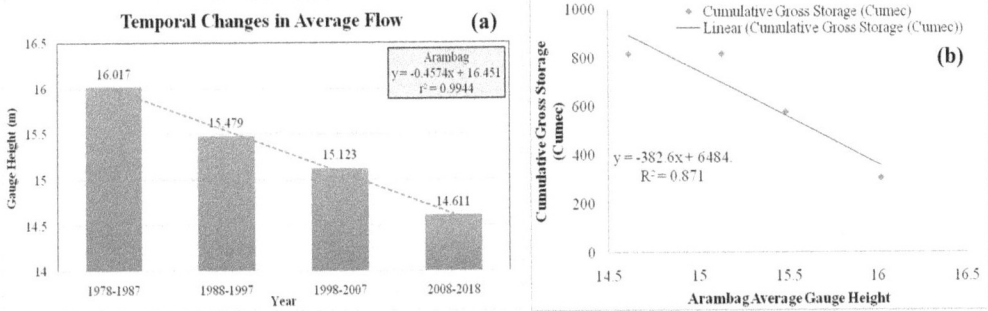

FIGURE 9.23 Temporal variation of ten-year annual peak discharge of Dwarkeswar River near Arambag station (a) and showing the relationships among temporal variation of ten-year annual peak discharge and cumulative variation of gross storage (b).

Such kind of hydrologic alteration can occasionally trigger the meteorological flood and drought (Dudley and Platania, 2007). In this Dwarkeswar River basin, four medium irrigation projects were constructed, and their details are given in Table 9.8. Thereby, the flow regime of Dwarkeswar River has been changed since 1982 with the construction of several major and minor dams for irrigation purpose especially in the upper part of the basin. Average highest gauge height in Arambag has been found from 1978 to 1987 when only two projects, namely, Suvankar Dangra Medium Irrigation Project and Berai Canal Medium Irrigation Project, were developed. Temporal variation of gauge height shows that successive construction of irrigation projects has led to altering the flow regime, which is clearly found from Table 9.8 and Figures 9.23a and b. As shown in Table 9.8 and Figure 9.23a, the lowest level of gauge height has found from 2008 to 2018. A strong ($r^2 = 0.871$) positive relationship has been found among the cumulative gross storage and changes in the river gauge height, which clearly indicates that due to the construction of the dam, river flow has been stored in this storage system that leads to a deficit in the flow. In this way, the flow regime of this river system has been altered.

9.3.3 CHANNEL PLAN-FORM CHANGES

The river Dwarkeswar rises in the extended Chhota Nagpur plateau region. The lower alluvial course of this channel is braided in nature, and the sinuosity index of every channel reach segment apart from few segments in the plateau fringe area is generally low and roughly remains constant throughout the last 80–90 years. Sinuosity index and braided index have been calculated following Friend and Sinha (1993) as shown in Figure 9.24e. Two sample sites, one from the plateau fringe area and another from the alluvial plain area as shown in Figure 9.24a–d, have taken into consideration to understand the changing nature of plan-form.

Figure 9.24f shows the temporal variation of the study reach plotting of sinuosity index against the braided channel ratio, which indicates that SI values are very low (less than 1.5). Figure 9.24f indicates that braided channel ratio was relatively high in 1979 and it has decrease into 2014. Figure 9.24b and d shows that such kind of morphological changes is occurring mainly due to sand mining in the indicated area. Uncontrolled sand mining in the river bed led to lowering of channel depth, and ultimately, the length of the channel decreases and the river becomes straight.

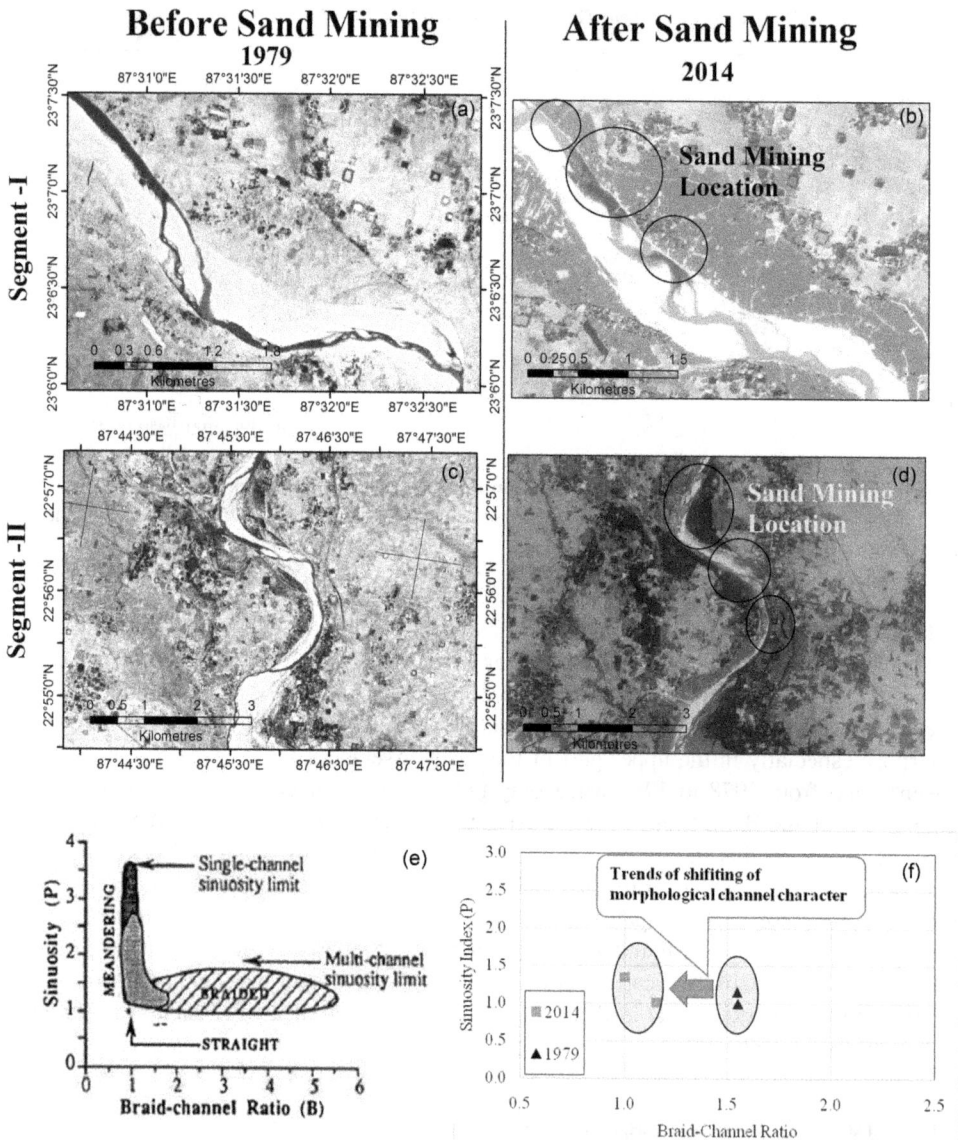

FIGURE 9.24 Reach-wise morphological changes of the channel in the plateau fringe area (segment-I (a and b)) and lower alluvial plain land (segment-II (c and d)). The year 1979 is representing the morphological character of the selected reach before sand mining and the year 2014 is representing the morphological characteristics after sand mining. (e) Nature of channel pattern (based on Friend and Sinha (1993)) and (f) channel changes from braided in 1979 to straight in 2014.

9.3.4 Variation in Channel Width

River adjusts its channel width over a long period of time. Dwarkeswar River has adjusted its channel width, showing that the width of the channel is varying over space and time. In this study, variation of channel width from several aspects of anthropogenic activities has been discussed.

The lower alluvial plain region is characterised by high population density, and sand mining (river bed sand mining) in this region initially resulted in the decrease of channel width, which had paved the way for human encroachment in this area. As shown in Figure 9.25a and b, it became quite clear that continuous sand mining in this area is limited with river bed sand mining resulting in deepening of the river bed and thereby decreases of channel width. Channel width is also being modified by the construction of river training structures. Malik and Pal (2019) studied the impact of groynes and its response on the channel morphology in Dwarkeswar River. Channel width decreases (345.57 to 283.59 m) significantly due to the construction of groyne near Rautara, Eklakshmi, in the lower alluvial part of the basin. Increases in maximum channel depth (5.94–7.77 m), stream velocity and width–depth ratio (81.20–104.63), and development of scouring (near the tip of the Groynes) have been observed due to chocking of channel capacity, thereby increasing shear stress and activation of sediment load from the tip of the groyne started through the construction of groyne (Malik and Pal, 2019; Henning and Hentschel, 2013).

Upper and lower sections of groyne construction in this area have been classified into three sections: upper reach (UR) of the groyne (UR, 750 m with 5 cross sections), middle reach (MR) or groyne-constructed area (750 m with 5 cross sections) and lower reach of the groyne (LR, 750 m with 5 cross sections) as well as Dumpy Level Survey was conducted.

As shown in Figure 9.26a–f and Table 9.9, the average width (W) of the upper part of the groyne constructed area was 338.9 m and came done to 324.8 m in MR to 306.45 m in LR. Maximum bankfull channel depth (D_{max}) of the UR was 5.77, whereas in MR and lower reach (LR), it was 7.01 and 6.52 m, respectively. Here, W is the width, D_{max} the maximum channel depth, D the mean depth, A the channel area and w/d the width–depth ratio.

So, MR of this area has been suffering from the deep scouring due to groyne construction. Mean channel depth of the MR was very high. Difference between D_{max} and D_{mean} also shows a similar result, whereas Figure 9.26a–f shows that bankfull channel cross-sectional area, W/D ratio and length of sand-bedded river decrease in the MR and increase in UR and LR.

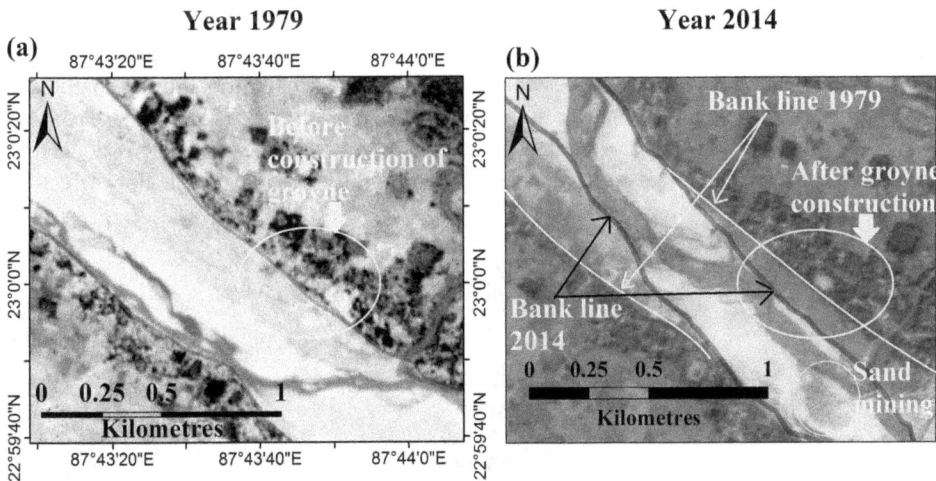

FIGURE 9.25 Channel narrowing due to the construction of groyne and sand mining; channel plan-form before the construction of groyne during 1979 (a) and channel plan-form after the construction of groyne in 2014 (b).

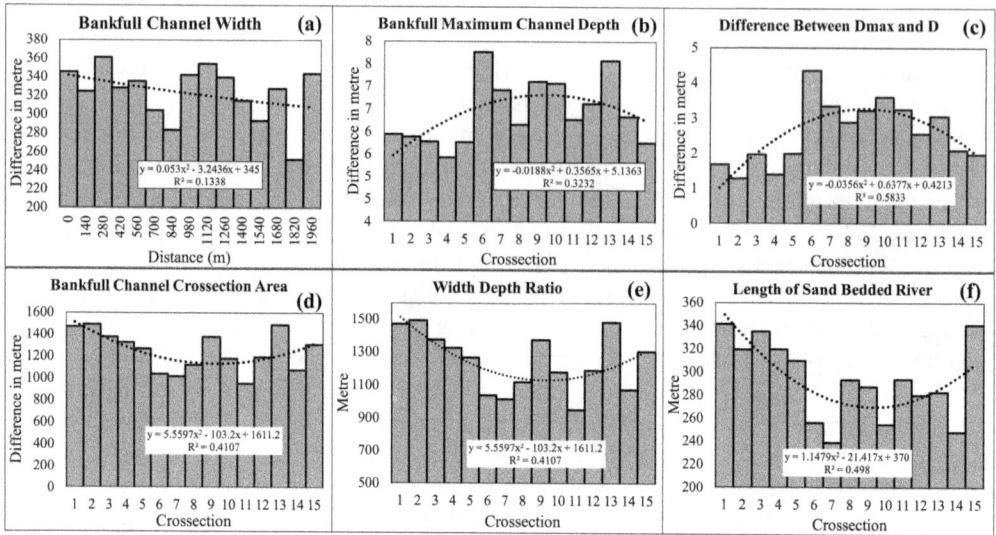

FIGURE 9.26 Downstream variations of channel morphology; variation of (a) bankfull channel width, (b) maximum channel depth, (c) the difference between maximum depth and mean depth, (d) bankfull channel area, (e) width-depth ratio and (f) sand bed.

TABLE 9.9
Reach-Wise Character of the Study Area

Reach	W	D_{max}	D	A	W/D	Stream Type
Upper	338.95	5.77	4.10	1,387.63	83.23	C5
Middle	324.88	7.01	3.52	1,146.45	92.39	B5
Lower	306.45	6.52	3.93	1,245.81	79.71	C5

Source: Field Survey with Dumpy Level from 2017 to 2019.

Simulated flow analysis has been done in HEC-RAS 5.0.7 to assess the impact of groyne constriction on the flow behaviour of Dwarkeswar River. It has been found that the flow path has been modified by the construction of groyne or spars. As shown in Figure 9.27a–c, the flow path started to deflect by the first groyne of this area due to the construction of the groyne. This deflected flow can accelerate bank erosion on its opposite side. Therefore, through changing its flow direction, the

FIGURE 9.27   Nature of flow at without groyne construction situation (a), bankfull and moderate flow condition with groyne construction (b) and lower part of the snap showing the scour formation at the tip of first groyne or spar (c).

FIGURE 9.28 Human encroachment through agriculture (a), playing field (b) and current fallow land (c) in the newly deposited area through the construction of a groyne.

deepest part of the channel will also be modified which is clearly indicated by Figure 9.27b and c. During flow deflection, deep scour formed in this area.

Construction of the groyne or spar has also resulted in the deposition of sediment at the backward side of the groyne as well as in between groyne. This has occurred due to decreases inflow velocity in this area. As a result of continuous deposition, the spur-constructed area has come under human encroachment leading to the development of playing field, agricultural land and few settlements, which is shown in Figure 9.28a–c.

Decreases of channel width due to the construction of the bridge have also been found in this river near Eklakshmi Bridge. Before the construction of the bridge, the width of the channel at this particular reach was 582 m (in 1979), but after the construction of the bridge, the channel width reduced to 232.23 m (2014), which is shown in Figure 9.29a–c. This has led to a reduction in the channel capacity in this reach, thereby increasing flow velocity as well as lowering of channel bed found along the downstream section of this bridge. In this section, sand mining is also an important contributor to lowering the channel bed simultaneously.

It has been observed that channel width varies depending on the location of sand mining. Channel width increases especially in the plateau fringe to the upper part of the lower alluvial area, where sand mining in the river bed as well as towards the flood plain area has led to increases in the channel width as well as its channel capacity. Prolonged sand mining in the river bed results in lower availability of sand from river bed; therefore, to meet the increasing demand of urbanisation, they shift their mine towards the river bank as well as towards the flood plain area, which is associated with previously deposited materials. Along with this, this section of river stretch is characterised by the low population density, which facilitates the minimum channel encroachment and thereby facilitates the sand miners to dig out the sand from the river bank. For example, before Naisarai Bazar channel width in 1979 was 322 m but continuous sand mining towards the flood plain area resulted in the increase of channel width to 365 m.

9.3.5 Changing Nature of the Pool-Riffle Sequence

Topographic high along a longitudinal course of a river is termed as riffles and pools are topographic highs. Natural pool-riffle sequence in the course of Dwarkeswar River has been altered by the huge amount of sand mining as shown in Figure 9.30a–c. Such an artificial pool increases the channel slope in the upstream section (Ghosh et al., 2016), and it results in headword erosion as shown in Figure 9.31, which can affect the channel several kilometres upstream. Thereby, these cause changes in the natural channel morphology and the pool-riffles sequence of the channel.

Excessive sand mining from river bed can cause the degradation of the river, as well as it may also lead to bank erosion, the threat to bridges, nearby structures, lowering of the groundwater system, bed coursing, channel bed degradation, riparian habitat and aquatic life through large

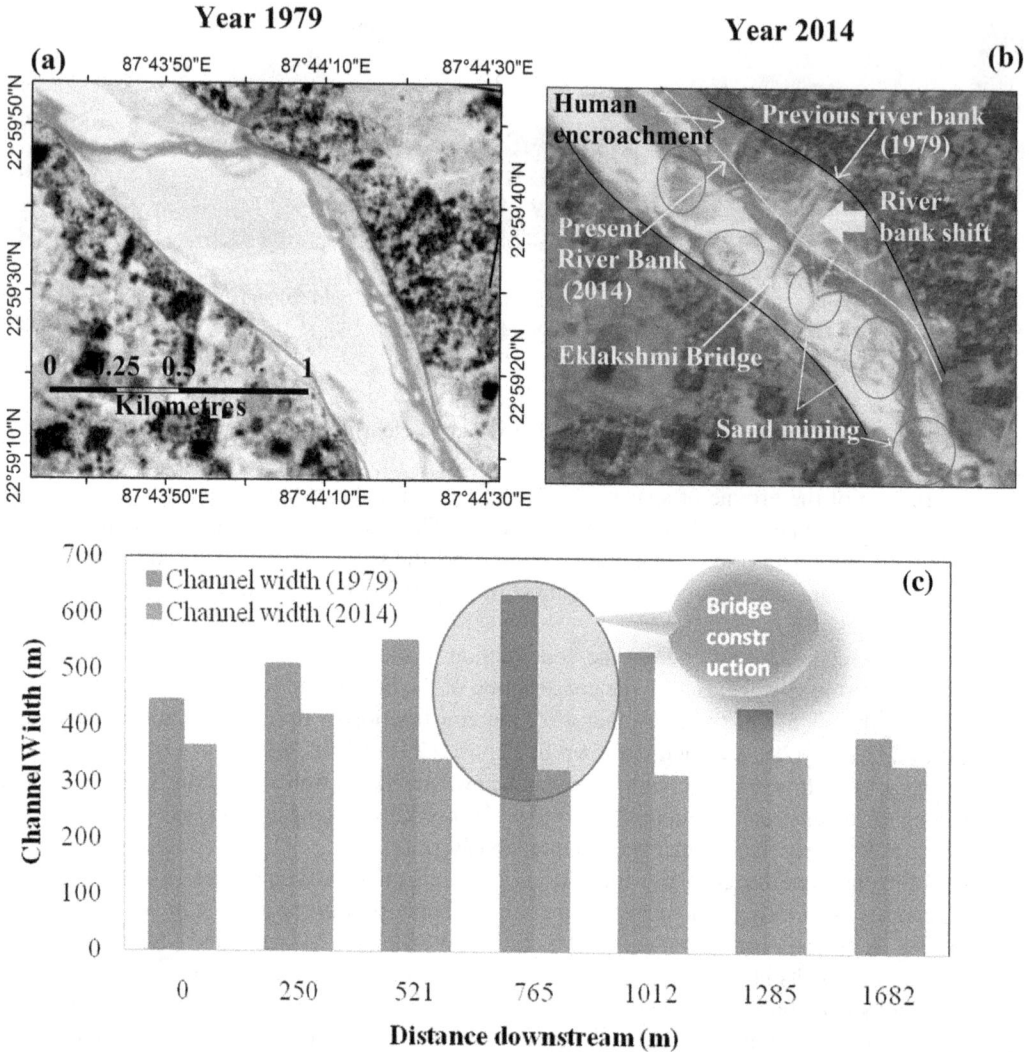

FIGURE 9.29 Channel narrowing due to bridge construction; (a) channel plan-form before the construction of the Eklakshmi Bridge (1979) and (b) after the construction of the Eklakshmi Bridge (2014); (c) bar diagram showing decreasing channel width due to the construction of Eklakshmi Bridge.

changes in the channel morphology. Apart from this, several places have been found where the sand-bedded river turned into clay-bedded river due to uncontrolled sand mining in the selected reach, as shown in Figure 9.32a–i. In some cases as shown in Figure 9.32a–c, it was found that sand mining has been so excessive that if a truck is staying over the river bed, it will not be found from the immediate distance. This has indicated the degree of channel degradation in the sand dominated area.

9.3.6   Changes in the Longitudinal Connectivity of the Flow

We have found that weirs and sluices which have been constructed over the stream have not only stored the water but also degraded the lower stretch of the stream significantly as shown in Figure 9.33a–f, which are showing several snaps of sand mining and its resultant landforms.

Channel before sand mining (1979)

(a)

Channel after sand mining (2014)

(b)

Sand mining

(c)

Expected Pool

Expected Pool

Expected Pool

0   0.5   1   2   3   4
Kilometres

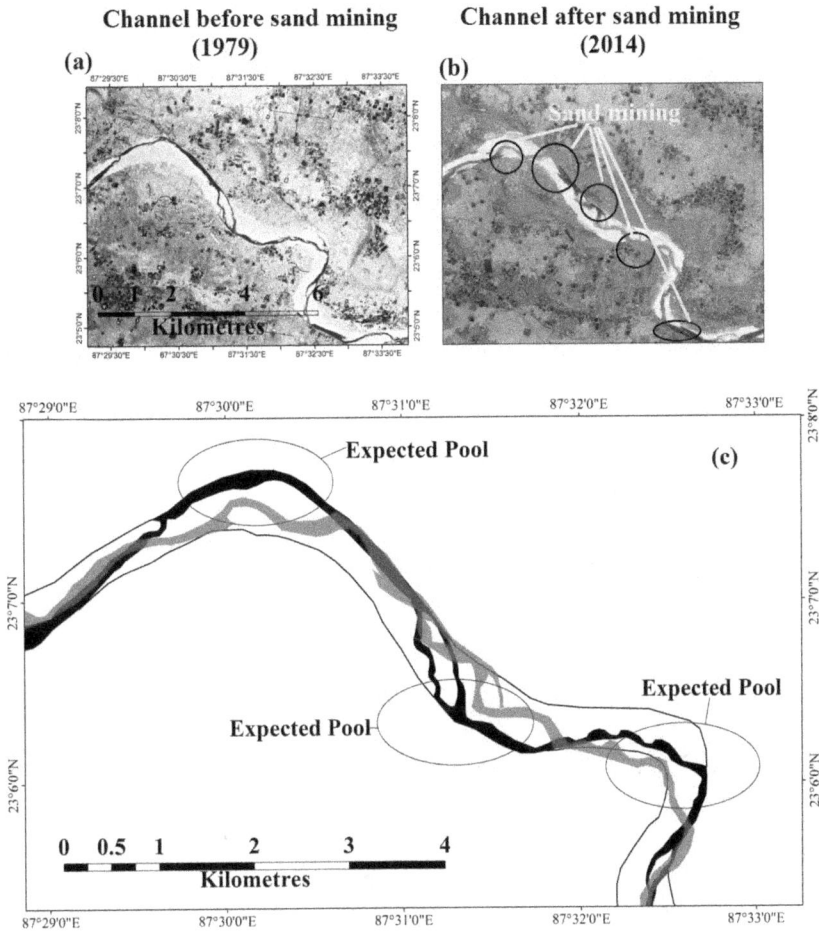

FIGURE 9.30 Channel before (a) and after sand mining (b) and variation of pool-riffles sequence of the selected site (c).

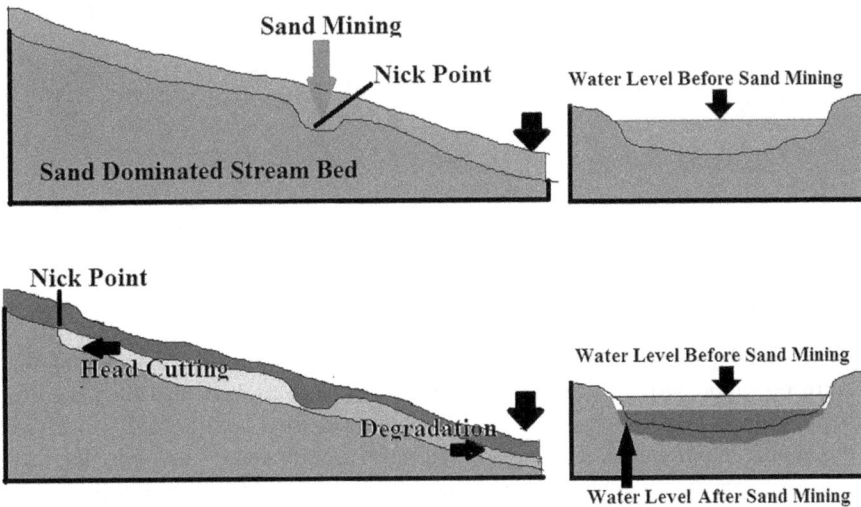

Sand Mining

Nick Point

Water Level Before Sand Mining

Sand Dominated Stream Bed

Nick Point

Head Cutting

Degradation

Water Level Before Sand Mining

Water Level After Sand Mining

FIGURE 9.31 Conceptual presentation of the impact of sand mining in Dwarkeswar River.

FIGURE 9.32 Modification of channel pattern due to sand mining and some photographic evidence of excessive sand mining; (a–c) show channel deepening through sand mining; (d and f) are the sand mining sites; (e) show the active bank erosion; (g and i) show that alteration of channel bed due to sand mining; and (h) shows the sand mining activity during rainy season.

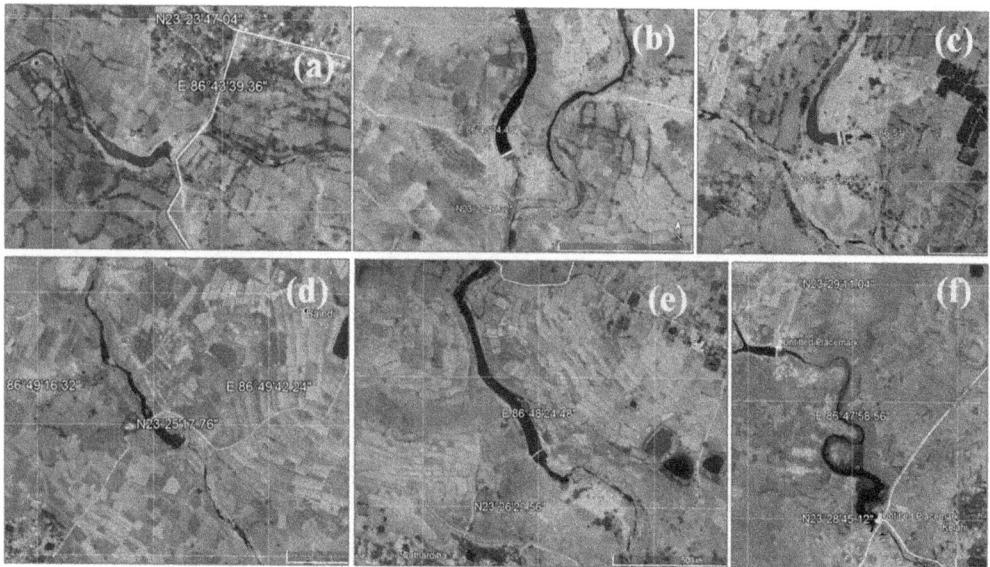

FIGURE 9.33 Impacts of weirs and sluices on the selected stretch of the river (a–f).

9.4 CONCLUSION

From the above study, we have found that the entire river basin has been affected by the several kinds of anthropogenic activity in terms of small water tank to dam, metalled road to bridge, embankment, groyne, water extraction pumping stations, sand mining and so on which has reduced the annual maximum gauge height of this river significantly. Therefore, the future prospect of this study may be associated with the degradation of river ecosystem due to the lack of sub-surface flow, and the occurrence of a small flood would be reduced which will affect the river-based ecosystem in this area. Apart from that, lowering of groundwater may be found due to loss of sub-surface flow. Lowering of bed level in the immediate future will be obvious due to ongoing excessive sand mining in this region. The lower part of this river will experience a frequent flood due to an excessive rate of soil erosion in its upper part of the basin, whereas middle part will experience more and more bank erosion, and ultimately, all together this may lead to the metamorphosis of the river system.

REFERENCES

Ängeby, O. 1951. *Pothole Erosion in Recent Water-Falls*. Royal University of Lund, Department of Geography, Lund.

Anthony, E.J., Brunier, G., Besset, M., Goichot, M., Dussouillez, P., and Nguyen, V.L. 2015. Linking rapid erosion of the Mekong River delta to human activities. *Scientific Reports*, 5, p. 14745.

Bagchi, K.K., and Majumdar, S. 2011. Dynamics of out-migration of agricultural labourers: a microlevel study in two districts of West Bengal. *Agricultural Economics Research Review*, 24(conf), pp. 568–568.

Bandyopadhyay, S., and De, S.K. 2018. Anthropogenic impacts on the morphology of the Haora River, Tripura, India. *Géomorphologie: Relief, Processus, Environnement*, 24(2), pp. 151–166.

Bandyopadhyay, S., Kar, N.S., Das, S., and Sen, J. 2014. River systems and water resources of West Bengal: a review. *Geological Society of India Special Publication*, 3(2014), pp. 63–84.

Baske, D. 1976. *Santal GanasangramerItihas* (Bengali). Pearl Publishers, Calcutta.

Bhattacharya, B.K. 1985. *West Bengal District Gazetteers: Puruliya*. Government of West Bengal, Calcutta.

Biswas, A. 1976. Temporal and spatial variations in the agriculture in the district of Birbhum in West Bengal. The University of Burdwan, Department of Geography, Burdwan [online] Available from: http://hdl.handle.net/10603/68721.

Biswas, S.S., Pal, R., Pramanik, M.K., and Mondal, B., 2015. Assessment of anthropogenic factors and floods using remote sensing and GIS on lower regimes of Kangshabati-Rupnarayan River Basin, India. *International Journal of Remote Sensing and GIS*, 4(2), pp. 77–86.

Boix-Fayos, C., Barberá, G.G., López-Bermúdez, F., and Castillo, V.M. 2007. Effects of check dams, reforestation and land-use changes on river channel morphology: case study of the Rogativa catchment (Murcia, Spain). *Geomorphology*, 91(1–2), pp. 103–123.

Brandis, D. 1869. Letter, dated November 1869. *Proceedings of Revenue Department*, Govt. of Bengal.

Bravard, J.P., Goichot, M., and Gaillot, S. 2013. Geography of sand and gravel mining in the Lower Mekong River. First survey and impact assessment. *EchoGéo*, 26, pp. 1–21.

Brunier, G., Anthony, E.J., Goichot, M., Provansal, M., and Dussouillez, P. 2014. Recent morphological changes in the Mekong and Bassac river channels, Mekong delta: the marked impact of river-bed mining and implications for delta destabilisation. *Geomorphology*, 224, pp. 177–191.

Bunn, S.E., and Arthington, A.H. 2002. Basic principles and ecological consequences of altered flow regimes for aquatic biodiversity. *Environmental Management*, 30(4), pp. 492–507.

Census of India. 2011. *District Census Hand Book of Bankura*. Government of India, Calcutta.

Census of India. 1961. *District Census Handbook of Birbhum*. Government of India, Culcutta.

Crutzen, P.J., and Stoermer, E.F. 2000. Global change newsletter. *The Anthropocene*, 41, pp. 17–18.

Doornkamp, J.C., and King, C.A.M. 1971. *Numerical Analysis in Geomorphology*. Edward Arnold, London.

Dudley, R.K., and Platania, S.P. 2007. Flow regulation and fragmentation imperil pelagic-spawning riverine fishes. *Ecological Applications*, 17, pp. 2074–2086. doi:10.1890/06-1252.1.

Duke, F.W. 1939. *Report on Denudation of Forests in Chota Nagpur and Orissa, and Afforestation of Wastelands in Western Bengal*. Revenue Department, Govt. of Bengal, Superintendent of Govt. Printing, Alipur, Calcutta.

Dutta, R. 1960. *The Economic History of India (The Victorian Age, 1837–1900)*. New Delhi: Publications Division, Ministry of Information and Broadcasting, Government of India.

Dynesius, M., and Christer N. 1994. Fragmentation and flow regulation of river systems in the northern third of the world. *Science*, 266, pp. 753–762. doi:10.1126/science.266.5186.753.

Ebisemiju, F.S. 1991. Some comments on the use of spatial interpolation techniques in studies of man-induced river channel changes. *Applied Geography*, 11(1), pp. 21–34.

Elawady, E., Michiue, M., and Hinokidani, O. 2001. Movable bed scour around submerged spur-dikes. *Proceedings of Hydraulic Engineering*, 45, pp. 373–378.

Elston, E.D. 1917. Potholes: their variety, origin and significance. *The Scientific Monthly*, 5(6), pp. 554–567.

Fetriyuna, Helmi, and Fiantis, D. 2017. Impact of land-use changes on Kuranji River basin functions chapter 8. In G. Shivakoti, U. Pradhan, & H. Helmi (Eds.), *Redefining Diversity and Dynamics of Natural Resources Management in Asia* (Vol. 4, pp. 105–114). Elsevier. Available from: https://doi.org/10.1016/B978-0-12-805451-2.00008-9 (Accessed 18 January 2019).

Friend, P.F., and Sinha, R. 1993. Braiding and meandering parameters. *Geological Society Special Publication,* 75, pp. 105–111. doi:10.1144/GSL.SP.1993.075.01.05.

Fuller, I.C., Large, A.R.G., and Milan, D.J. 2003. Quantifying channel development and sediment transfer following chute cutoff in a wandering gravel-bed river. *Geomorphology,* 54, pp. 307–323. doi:10.1016/S0169-555X(02)00374-4.

Geological Survey of India (GSI). 1999. *Geological and Mineral Map of West Bengal.* Government of India, Calcutta.

Ghosh, P.K., Bandyopadhyay, S., Jana, N.C., and Mukhopadhyay, R. 2016. Sand quarrying activities in an alluvial reach of Damodar River, Eastern India: towards a geomorphic assessment. *International Journal of River Basin Management,* 14(4), pp. 477–489.

Giglou, A.N., Mccorquodale, J.A., and Solari, L. 2018. Numerical study on the effect of the spur dikes on sedimentation pattern. *Ain Shams Engineering Journal,* 9(4), pp. 2057–2066.

Government of Bengal. (1867–1897), Letters, Procs., Rev. Dept., West Bengal State Archives, Calcutta. (i) Brandis, D., Forests, dt. 28th Nov. 1869. (ii) Dickens, C.H., P.W.D., dt. 11th May 1868. (iii) Dickens, P., Officiating Under-Secretary, dt. 20th Feb. 1869. (iv) Maguire, H.F.T., Dy. Commissioner, dt. 5th Sep. 1897. (v) Money, A., Commissioner, Bhagalpur Div., dt. 28Ut Mar. 1869. (vi) Risley, H.H., Revenue, dt. Apr. 1893. (vii) Seton-Kerr, W.S., P.W.D., dt. Apr. 1867.

Gregory, K.J. 2006. The human role in changing river channels. *Geomorphology,* 79(3–4), pp. 172–191.

Grill, G., Lehner, B., Lumsdon, A.E., MacDonald, G.K., Zarfl, C., and Liermann, C.R. 2015. An index-based framework for assessing patterns and trends in river fragmentation and flow regulation by global dams at multiple scales. *Environmental Research Letters,* 10(1), p. 015001.

Hei, P., Zhicong, C., and Xiang, D. 2009. Sediment carrying capacity spar dike open channel. In C. Zhang and H. Tang (Eds.), *Advances in Water Resources and Hydraulic Engineering* (pp. 928–932). Springer, Berlin, Heidelberg. doi:10.1007/978-3-540-89465-0_163.

Henning, M., and Bernd, H. 2013. Sedimentation and flow patterns induced by regular and modified groynes on the River Elbe, Germany. *Ecohydrology,* 6(4), pp. 598–610. doi:10.1002/eco.1398.

Horton, R.E. 1932. Drainage basin characteristics. *Eos, Transactions American Geophysical Union,* 13, pp. 350–361. doi: 10.1029/TR013i001p00350.

Humphries, P., Richardson, A., Wilson, G., and Ellison, T. 2013. River regulation and recruitment in a protracted-spawning riverine fish. *Ecological Applications,* 23(1), pp. 208–225.

Hunter, T. 1883. *Woods, Forests, and Estates of Perthshire with Sketches of the Principal Families in the County.* Henderson, Robertson and Hunter, Perth.

Hunter, W. 1877. A statistical account of Bengal [online] Available from: https://books.google.co.in/books?hl=en&lr=&id=BY8BAAAAQAAJ&oi=fnd&pg=PR9&dq=Hunter,+1877&ots=Auv_XgYdlH&sig=Iffyh13uElW8Pgxf2lxRNQhdRWQ (Accessed 15 January 2019).

Ilha, P., Rosso, S., and Schiesari, L. 2019. Effects of deforestation on headwater stream fish assemblages in the Upper Xingu River Basin, Southeastern Amazonia. *Neotropical Ichthyology,* 17(1), pp. 1–20.

Irrigation and Waterways Directorate Govt. of West Bengal (IWD). 2013. *Annual Flood Report 2013.* Government of West Bengal, India, Kolkata.

IWD. 2016. *Annual Flood Report 2016.* Government of West Bengal, India, Kolkata.

James, L.A., and Marcus, W.A. 2006. The human role in changing fluvial systems: retrospect, inventory and prospect. *Geomorphology,* 79(3–4), pp. 152–171.

Jeje, L.K., and Ikeazota, S.I. 2002. Effects of urbanisation on channel morphology: the case of Ekulu River in Enugu, South Eastern Nigeria. *Singapore Journal of Tropical Geography,* 23(1), pp. 37–51.

Khan, M.S., and Islam, A.R.M.T. 2015. Anthropogenic impact on morphology of Teesta River in Northern Bangladesh: an exploratory study. *Journal of Geosciences and Geomatics,* 3(3), pp. 50–55.

King, H., 2009. The use of groynes for riverbank erosion protection. *University of Stellenbosch CPD course "River hydraulics, stormwater & flood management".* Stellenbosch:[sn], pp. 1–21.

Kulkarni, S. 1987. Forest legislation and tribals: comments on forest policy resolution. *Economic and Political Weekly,* 22, pp. 2143–2148.

Lai, X., Shankman, D., Huber, C., Yesou, H., Huang, Q., and Jiang, J. 2014. Sand mining and increasing Poyang Lake's discharge ability: a reassessment of causes for lake decline in China. *Journal of Hydrology,* 519, pp. 1698–1706.

Lake, P.S. 2000. Disturbance, patchiness, and diversity in streams. *Journal of the North American Benthological Society*, *19*(4), pp. 573–592.

Lane, S.N., and Richards, K.S. 1997. Linking river channel form and process: time, space and causality revisited. *Earth Surface Processes and Landforms: The Journal of the British Geomorphological Group*, 22(3), pp. 249–260.

Latrubesse, E.M., Amsler, M.L., de Morais, R.P., and Aquino, S. 2009. The geomorphologic response of a large pristine alluvial river to tremendous deforestation in the South American tropics: the case of the Araguaia River. *Geomorphology*, *113*(3–4), pp. 239–252.

Lehner, B., Liermann, C.R., Revenga, C., Vörösmarty, C., Fekete, B., Crouzet, P., Döll, P., Endejan, M., Frenken, K., Magome, J., and Nilsson, C. 2011. High-resolution mapping of the world's reservoirs and dams for sustainable river-flow management. *Frontiers in Ecology and the Environment*, *9*(9), pp. 494–502.

Leopold, L.B., and Miller, J.P. 1956. *Ephemeral Streams: Hydraulic Factors and Their Relation to the Drainage Network* (Vol. 282). US Government Printing Office, Washington.

Lorenc, M.W., Mun, P., and Saavedra, J. 1994. The evolution of potholes in granite bedrock, W Spain. *Catena*, 22(4), pp. 265–274.

Maity, S.K., and Maiti, R. 2016. Understanding the sediment sources from mineral composition at the lower reach of Rupnarayan River, West Bengal, India–XRD-based analysis. *GeoResJ*, *9*, pp. 91–103.

Malik, S., and Pal, S.C. 2019. Is the construction of Groynes accelerating the degradation of channel morphology and paved the way for human encroachment in The Bengal Basin? *Advances in Space Research*, *64*, pp. 1579–1576.

Marsh, G.P. 1965. In D. Lowenthal (Ed.), Man and Nature; or, Physical Geography as Modified by Human Action. Cambridge, MA: Belknap Press of Harvard University Press.

Mei, X., Dai, Z., Du, J., and Chen, J. 2015. Linkage between Three Gorges Dam impacts and the dramatic recessions in China's largest freshwater lake, Poyang Lake. *Scientific Reports*, *5*, p. 18197.

Melton, M.A. 1958. Geometric properties of mature drainage systems and their representation in an E4 phase space. *The Journal of Geology*, *66*(1), pp. 35–54.

Mondal, B., Dolui, G., Pramanik, M., Maity, S., Biswas, S.S., and Pal, R. 2017. Urban expansion and wetland shrinkage estimation using a GIS-based model in the East Kolkata Wetland, India. *Ecological Indicators*, *83*, pp. 62–73.

Morisawa, M.E. 1959. *Relation of Morphometric Properties of Runoff In The Little Mill Creek, Ohio, Drainage Basin* (No. Cu-Tr-17). Columbia University, New York.

Nag, S.K., and Lahiri, A. 2011. Morphometric analysis of Dwarkeswar watershed, Bankura district, West Bengal, India, using spatial information technology. *International Journal of Water Resources and Environmental Engineering*, *3*(10), pp. 212–219.

Neill, C., Deegan, L.A., Thomas, S.M., Haupert, C.L., Krusche, A.V., Ballester, V.M., and Victoria, R.L. 2006. Deforestation alters the hydraulic and biogeochemical characteristics of small lowland Amazonian streams. *Hydrological Processes: An International Journal*, *20*(12), pp. 2563–2580.

Nilsson, C., Reidy, C.A., Dynesius, M., and Revenga, C. 2005. Fragmentation and flow regulation of the world's large river systems. *Science*, *308*(5720), pp. 405–408.

O'Malley, L.S. 1908. *Bengal District Gazetteers Bankura*. Kolkata [online in 1995]. Available from: http://www.bankura.gov.in/Gazet/First_seven_pages.PDF (Accessed 15 January 2019).

Ohmoto, T., Hirakawa, R., and Watanabe, K. 2009. Effects of spur dike directions on river bed forms and flow structures. In *Advances in Water Resources and Hydraulic Engineering* (pp. 957–962). Springer, Berlin, Heidelberg.

Pal, S.C., and Debnath, G.C. 2012. Morphometric analysis and associated land use study of a part of the Dwarkeswar watershed. *International Journal of Geomatics and Geosciences*, *3*(2), pp. 351–363.

Pan, S. 2013. Application of remote sensing and GIS in studying changing river course in Bankura District, West Bengal. *International Journal of Geomatics and Geosciences*, *4*(1), p. 149.

Panda, B. 1993. *Deforestation and land degradation, a case study of umran river basin Meghalaya* (Doctoral dissertation).

Renard, K.G., and Laursen, E.M. 1975. Dynamic behavior model of ephemeral stream. *Journal of Hydraulic Engineering*, *101*(5), pp. 511–528.

Rolls, R.J., Baldwin, D.S., Bond, N.R., Lester, R.E., Robson, B.J., Ryder, D.S., Thompson, R.M., and Watson, G.A. 2017. A framework for evaluating food-web responses to hydrological manipulations in riverine systems. *Journal of Environmental Management*, *203*, pp. 136–150.

Rudra, K. 2014. Changing river courses in the western part of the Ganga–Brahmaputra delta. *Geomorphology*, *227*, pp. 87–100.

Rudra, K. 2016. *State of India's Rivers for India Rivers, 2016*. Government of India, West Bengal.

Rasul, M.A. 1954. Saontal Bidroher Amar Kahini, (Bengali). Calcutta: National Book Agency.

Schlich, W. 1885. *Report on Forest Administration in the Chota Nagpore Division of Bengal*. Superintendent of Government Print, Kolkata.

Sengupta, S., and Kale, V.S. 2011. Evaluation of the role of rock properties in the development of potholes: a case study of the Indrayani knickpoint, Maharashtra. *Journal of Earth System Science*, *120*(1), pp. 157–165.

Siddique, G. 1996. *Impact of deforestation in parts of South_Western lateritic areas of West Bengal*, The University of Burdwan [online] Available from: http://hdl.handle.net/10603/65846.

Sinha, M. 2016. Gandeshwari rivulet: a geomorphic study, West Bengal, India. *Social Science Review*, *2*, pp. 78.

Strahler, A.N. 1957. Quantitative analysis of watershed geomorphology. *Eos, Transactions American Geophysical Union*, *38*(6), pp. 913–920.

Strahler, A.N. 1964. Part II. Quantitative geomorphology of drainage basins and channel networks. In *Handbook of Applied Hydrology* (pp. 4–39). McGraw-Hill, New York.

Sukhodolov, A., Uijttewaal, W.S., and Engelhardt, C. 2002. On the correspondence between morphological and hydrodynamical patterns of groyne fields. *Earth Surface Processes and Landforms*, *27*(3), pp. 289–305.

Surian, N., and Rinaldi, M. 2003. Morphological response to river engineering and management in alluvial channels in Italy. *Geomorphology*, *50*(4), pp. 307–326.

Survey of India. 1978. *Topographical Sheets (1:50,000)*. Government of India, Kolkata.

Sweeney, B.W., Bott, T.L., Jackson, J.K., Kaplan, L.A., Newbold, J.D., Standley, L.J., Hession, W.C., and Horwitz, R.J. 2004. Riparian deforestation, stream narrowing, and loss of stream ecosystem services. *Proceedings of the National Academy of Sciences*, *101*(39), pp. 14132–14137.

Szabó, J. 2010. Anthropogenic geomorphology: subject and system. In *Anthropogenic Geomorphology* (pp. 3–10). Springer, Dordrecht.

The National Bureau of Soil Survey and Land Use planning (NBSS&LUP) [online]. Available from: https://www.nbsslup.in/ (Accessed on 4 January 2019).

Thoms, M.C., and Sheldon, F. 2000. Water resource development and hydrological change in a large dryland river: the Barwon–Darling River, Australia. *Journal of Hydrology*, *228*(1–2), pp. 10–21.

Thorp, J.H., Black, A.R., Haag, K.H., and Wehr, J.D. 1994. Zooplankton assemblages in the Ohio River: seasonal, tributary, and navigation dam effects. *Canadian Journal of Fisheries and Aquatic Sciences*, *51*(7), pp. 1634–1643.

West Bengal Pollution Control Board (WBPCB). 2016. *State of Environment Report West Bengal 2016*. WBPCB, Kolkata.

Zabih, M.E. 1976. *A Study of River Bank Protection Methods*. University of Roorkee, Roorkee.

Zalasiewicz, J., Williams, M., Haywood, A., and Ellis, M. 2011. The Anthropocene: a new epoch of geological time? *Philosophical Transactions of the Royal Society A: Mathematical, Physical and Engineering Sciences*, 369, pp. 835–841. doi:10.1098/rsta.2010.0339.

Zalasiewicz, J., Williams, M., Smith, A., Barry, T.L., Coe, A.L., Bown, P.R., Brenchley, P., Cantrill, D., Gale, A., Gibbard, P., and Gregory, F.J. 2008. Are we now living in the Anthropocene? *Gsa Today*, *18*(2), p. 4.

10 Modifications of the Geomorphic Diversity by Anthropogenic Interventions in the Silabati River Basin

Priyank Pravin Patel, Sayoni Mondal and Rishikesh Prasad

CONTENTS

10.1 INTRODUCTION

Rivers have been considered as the lifeline of human civilisation since time immemorial, as they provide critical ecosystem services for human sustenance (Brierley et al., 2016). They are considered as hydrologically and ecologically connected natural systems, displaying a great degree of variation in terms of their forms and patterns, dynamics, sedimentation budgets and evolutionary history (Lewin and Ashworth, 2014). The interconnected linkages between the different forms and processes operating in river channels help to explain the various geomorphic diversities that they present (Sinha et al., 2017). Longitudinally, continuous fluvial corridors show morphological diversity as they result from a combination of physical, biological and hydrological drivers, and their varied morphological forms can be attributed much to the terrain over which the stream flows (Dufour et al., 2015; Marren et al., 2014). The hierarchy, connectivity and sensitivity of such form-process relationships help in understanding the evolutionary history of the landscape and also identify other causal factors that have led to the development of such diversity (Jain et al., 2012). The lithological structure can be taken as the starting point from which such diversities arise (Sinha et al., 2017). Varying structure and lithology, gradient-induced land use and land cover (LULC) patterns, drainage forms, bed substrate and their erosion intensity all speak about the geomorphic evolution and diversity in a channel (Pande, 2019), and this geomorphic diversity exists on a wide range of spatial and temporal scales. While at the catchment scale, regional lithology and valley confinement determine the variety of geomorphic forms created by a channel, local factors such as the presence of a rock outcrop, a tributary confluence or the deposition of large woody debris (LWD) tend to determine its reach-scale characteristics.

Simply physical factors are not solely responsible for such geomorphic diversity, and often anthropogenic influences like dam construction, gravel and sand mining and other local LULC modifications also bring about major changes to the river character (Sinha et al., 2005, 2017). Thus, the quest for more water has resulted in humans modifying fluvial ecosystems often beyond their regeneration capacity and up to a limit where much of their morphological and ecological diversity has been either altered or lost (Brierley et al., 2016). Multifaceted, complex and irreversible changes have been introduced into river channels through flow regulations, land-use modifications, floodplain fragmentation, channelisation and alteration of sediment regimes, all of which have led to the overall degradation of fluvial ecosystems around the world (Reid et al., 2008; Bertalan et al., 2018). Thus, understanding the causative factors that have created the geomorphic diversity in a channel at various spatial and temporal scales and also identifying the responsible factors for their degradation has become one of the most important tasks of river managers and policymakers (Roy and Sinha, 2017). This chapter tries to present a brief outline of the various geomorphic forms that have been created along the Silabati (or Silai) River corridor, along with identifying some of the stressors that have led to the continued degradation of its fluvial forms and processes.

For achieving the above, a number of geospatial datasets and techniques, coupled with detailed field surveys, were employed. Geological Survey of India (GSI) Quadrangle sheets were used to obtain the basin lithology attributes and the Shuttle Radar Topographic Mission (SRTM) Digital Elevation Model (DEM) dataset (spatial resolution 30 m) were used to map basin terrain attributes for landscape characterisation (Patel and Sarkar, 2010). The SRTM data was also used to extract the drainage network of the Silabati River Basin, using the common D-8 flow routing algorithm (Patel, 2013), since this would provide a fuller and more elaborate stream network for the ensuing morphometric analysis than that which could have been extracted from the topographical sheets of the area (Das et al., 2016; Patel and Sarkar, 2009). The morphometric attributes were enumerated by overlaying uniformly sized grids of 1 km × 1 km dimensions across the basin area (4,390 grids in total), with ensuing statistical analysis of the obtained values. LULC information for the basin was obtained from the Biodiversity Information System (BIS), as provided by the Indian Institute of Remote Sensing (IIRS), Dehradun. Together with this, Landsat image datasets (Table 10.1) of 1990, 2000 and 2010 (Landsat 4-5 TM) and 2017 (Landsat 8 OLI) were used for extraction of the temporal LULC coverage, enumeration of the Normalised Difference Vegetation Index (NDVI) and detecting changes therein. Furthermore, information on landscape fragmentation and disturbance caused by anthropogenic activities was also obtained from IIRS, as part of the BIS dataset (Roy et al., 2012, 2015a, 2015b), and preparation of reach-level river corridor maps to show the alterations to the natural landscape, along the most affected stretches of the trunk stream, was done using high-resolution

TABLE 10.1
Details of Image Datasets Used

Satellite/Sensor	Path/Row	Scene Date (s)
Landsat 5 TM	139/44	23 December 1990
Landsat 5 TM	139/44	18 December 2000
Landsat 5 TM	139/44	12 November 2010
Landsat 8 OLI	139/44	24 December 2017

Source: Compiled by the authors.

Google Earth images and detailed field surveys. This also involved marking the various obstructions to flow within the channel, the demarcation of significant gullied tracts within the basin area and changes to their LULC aspects from historical Google Earth image datasets. Finally, correlations were derived among the above aspects to highlight the relationships between the terrain character and its alteration by human activities.

10.2   THE RIVER AND ITS BASIN

The River Silabati (Silai) originates from the village of Chak Gopalpur in Hura Block of Purulia district in south-western West Bengal and flows through the districts of Bankura and Paschim Medinipur, to join the Darakeshwar River, with their combined flow forming the River Rupnarayan, which finally drains into the River Hugli (Figure 10.1). The river traverses a distance of 221.6 km and drains an area of 4,151 km^2. The rolling topography of the region gently slopes eastward and is covered mainly by residual granitic rocks of the Chhotanagpur Gneissic Complex in its upper catchment area. This grades into quaternary deposits to the east, consisting of the Lalgarh Formation of the Lower Pleistocene age. Younger alluvial deposits exist where the main channels drain their individual floodplains. The basin area experiences a semi-arid climate in the northern part which gradually changes to sub-humid towards the southern part of the basin. The south-west monsoon brings rain to this region, with greater precipitation in the southern basin portion, which receives an annual rainfall between 1,250 and 1,600 mm.

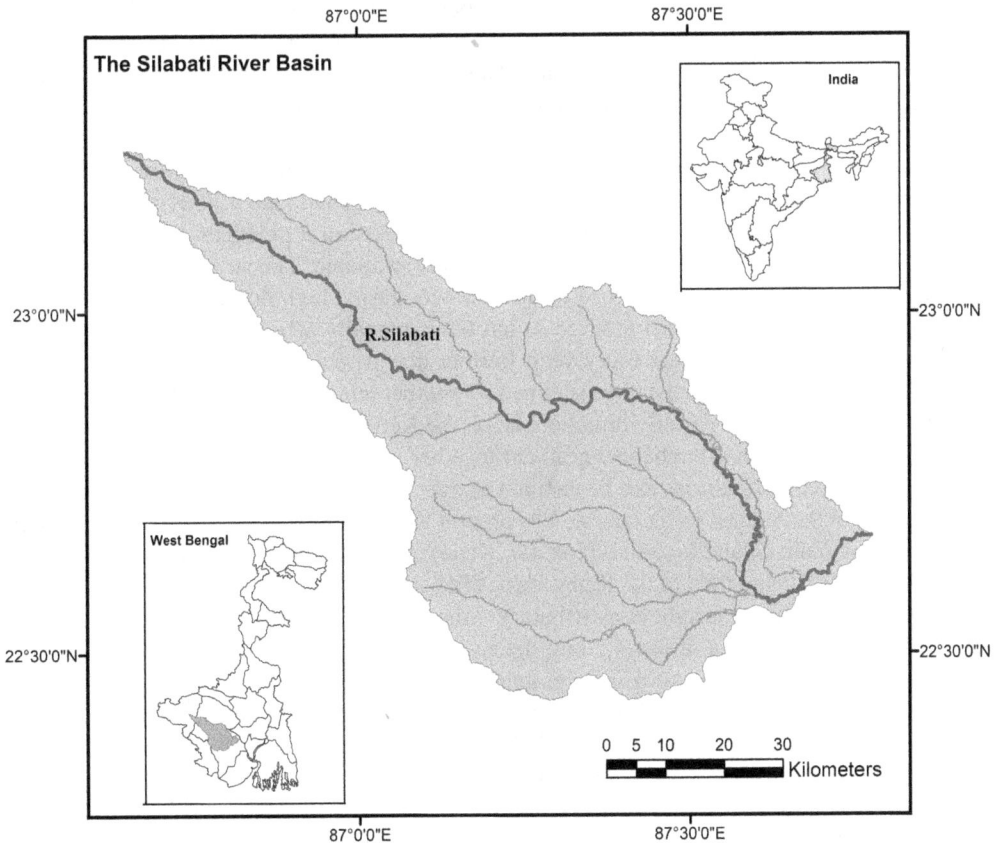

FIGURE 10.1   Location of the Silabati River Basin.

10.2.1 THE SILABATI'S COURSE

Cutting across the scattered *Sal* forest belts, which is the characteristic vegetation of the district, the Silabati River initially flows over weathered phyllitic rocks. In its upper course, the stream shows signs of entrenchment and high structural control resulting in a narrow but sinuous course, which continues till Susunia (23° 00' 13.9" N, 86° 59' 10.33" E). Thereafter, a change in the flow pattern is observed, marked by a change in bed substrate, which had earlier consisted of boulders and gravels with sand, subsequently changing to finer sand particles with a mixture of cobbles and loose rock debris. A series of five check dams within a stretch of 3.3 km and the bigger Kadam Deuli Dam (23° 06' 18.43" N, 86° 51' 39.07" E) highlight the degree of human intervention into the channel. At Chakrasol village near Lakhiapal (22° 53' 40.22" N, 87° 11' 00.35" E), the river receives its first major left bank tributary, the Jaipanda Nadi. The significant increase in discharge may be deciphered from the increased width of the active channel (60 m) with an even wider valley of 340 m. Patches of rocky boulders mixed with sandy pebbles can be seen as outcrops on the channel bed. A noticeable change in channel character and behaviour can be noticed as the river enters the lateritic lands of Paschim Medinipur through Garbeta-I Block. Here, a much widened thalweg (72 m) meanders across its wider sandy channel bed (265 m) and marks the start of the stream's middle course. Further downstream, the Silai enters the Gangani tract, locally called the *Ganganir Danga*, situated near the town of Garbeta (22° 51' 47" N, 87° 21' 13" E). Primarily composed of red soils, the area is part of the lateritic uplands occupying the north-western part of the district. The most remarkable feature of the area is the intense gullied surfaces, known as badlands, formed due to the erosive action of small rivulets originating from the slopes of the uplands (Patel and Mondal, 2019). The channel here follows a meandering course, sometimes flowing much closer to its left bank and sometimes to the right, leaving a major portion of its dried bed exposed. The channel bank characteristics show marked differences in height, with the gullied right bank reaching as high as 25–30 m, whereas the left bank displays flat slopes that are not more than 5 m in height. The bank material is comprised of unconsolidated lateritic rocks and is thus friable, which readily lends itself to both natural erosion and sand extraction for construction purposes. The morphological and pedological characteristics of these badlands on the river's right bank limit agriculture, wherein it is practised on a piece-meal basis, mainly with the help of river lift and tube well irrigation. However, the left bank, being essentially flat, favours large-scale agricultural practices, with farmers cultivating more than one crop where irrigation facilities are available. The structural composition of the valley here also influences the spread of flood waters during high flows. The high obstruction on the right side forces the channel to flood its left bank annually, where several hectares of cultivable lands are lost to floodwaters every year. Further downstream, near Panchberia village (22° 51' 58.24" N, 87° 26' 48.33" E), the river is joined by another left bank tributary, the Champa Khal, and turns to flow in a south-easterly direction. Here, the Ketia Khal distributary emerges from the main channel. The decrease in channel gradient together with the increasing water and sediment load from its upstream tributaries can be justified as the cause of origin of the Ketia Khal, and this marks the start of the Silabati lower course. The present Ketia carries almost 80% of the monsoonal discharge of the main Silai. Downstream of the Ketia emergence, the Silai flows south-eastward through the newer alluvial tract of the Kangsabati–Silabati interfluves for a considerable length, till it receives its first significant right bank tributary – the Donai Khal near Daspur-I Block. In quick succession, within a distance of just 4.5 km, the river is joined by the Buriganga Nadi (its largest right bank tributary), which in turn is a combination of three other rivers, namely, Kubai Nadi, Tamal Nadi and Parang Nadi, all of which arise from the undulating plains of the gullied lateritic terrain of Paschim Medinipur and maintain their sinuous courses to form the Buriganga Nadi (22° 34' 59.28" N, 87° 28' 49.29" E). The newer alluvium of the Kangsabati–Silabati interfluve allows the free lateral migration of these channels, and thus, they exhibit considerable widths. The main river now flows in a north-easterly direction for a length of 7.42 km and then takes a sharp bend to the north to enter Ghatal Block. It is here that the entire discharge of the Silabati accumulates and causes devastating floods every year, symptomatic of the drainage congestion that results in similarly low-lying areas (e.g. Mondal et al., 2016), with the water level rising as high as 15 m. From Ghatal, the river flows for another 6.36 km in the easterly direction to join the River Dwarakeshwar and their combined flow forms the River Rupnarayan at Bandar (22° 40′ 10.72″ N, 87° 46′ 59.05″ E).

10.2.2 BASIN PHYSIOGRAPHIC ATTRIBUTES

10.2.2.1 Surface Lithology

Locationally, the Silabati Basin lies on the eastern fringe of the Chhotanagpur Plateau, south-east of the Singhbhum Shear Zone. Therefore, the basin shows considerable variation in terms of its structure and exposes a wide array of Tertiary and Quaternary sediments. The extreme northern part of the basin lying in the Purulia district forms the last two steps of the descent from the Chhotanagpur Plateau into the Damodar Plains of West Bengal. Being a part of the Ranchi Peneplains, the physical structure of this section displays typical age old rocks of the Proterozioc Age with intrusions of Archaean rocks in between. The rolling topography of the region which gently slopes eastward is mainly covered with residual granitic rocks of the Chhotanagpur Gneissic Complex (Figure 10.2). This gneissic complex in the upper portion of the basin belongs to the Archaean Age and forms the oldest and the most extensive basement rock surface of the area, composed of granitic rocks, meta-sedimentaries and metabasics traversed by veins of epidiorite. They are, in fact, the continuation of the eastern peninsular Archaean tract of the Chhotanagpur region. The major part of the Silai's course lies over Quaternary deposits, consisting mainly of the Lalgarh Formation, which belongs to the Lower Pleistocene age. These Quaternary sediments, successively younger deposits of which

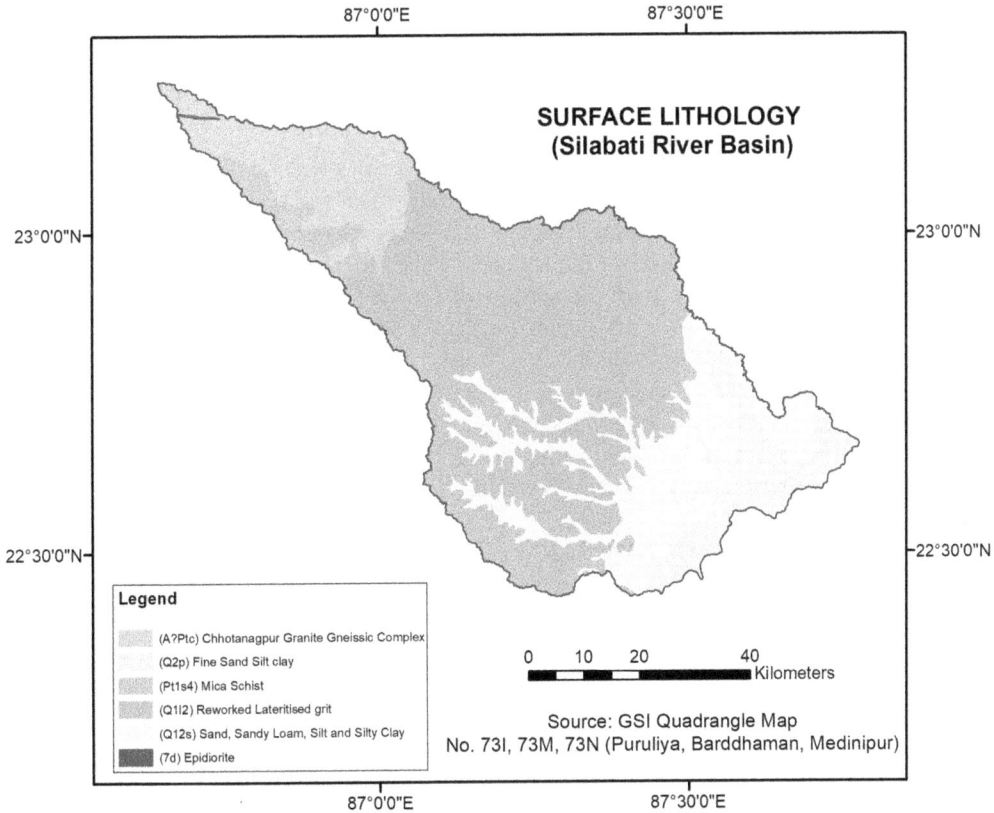

FIGURE 10.2 Surface lithology of the Silabati River Basin.

exist towards the south and east, also contain the Sijua and the Panskura Formations, are made of silty clay and fine sands that lie above the Lalgarh Formation and form the floodplains through which the river meanders. The characteristic terrain of the region is laterite which covers extensive areas in the north and west but gradually merges with deltaic alluvium in the southern and eastern parts of the basin.

10.2.2.2   Morphometric Analysis and Basin Terrain Units

The term "morphometry" was first used by De Martonne (1934) in geomorphological studies to measure the shape and geometry of any landform unit on the Earth's surface. It is the mathematical analysis and numerical measurement of the configuration of the Earth's landform features (Clarke, 1966), being the most commonly used technique in drainage basin analysis, providing a basis for "quantitative geomorphology" (Patel, 2012), and helps to evaluate the inherent characteristics of the basin and its relation to the factors that have influenced its formation. The quantitative interpretation of the Silabati River Basin using morphometric techniques was done to understand the overall character of the landscape in relation to the various terrain and lithological units over which it flows and to demarcate distinct physiographic units based on the enumerated parameters.

10.2.2.2.1   Mean Elevation and Relative Relief

The mean elevation of the basin ranges from about 210 m to almost sea level, with a steady decrease in the elevation in broad swathes from the west to the south and south-east (Figure 10.3a), following the trends of the principal valleys. The relative relief depicts the difference in height between the highest and the lowest points in the basin (Figure 10.3b). First formulated by Smith (1935), it helps in understanding the overall morphological characteristics of the basin and also its degree of dissection. The working formula is

$$\text{Relative relief (RR)} = \text{Maximum elevation} - \text{Minimum elevation} \tag{10.1}$$

The highest elevation of 217 m above mean seal level (MSL) was recorded in the extreme northern limit of the basin near the Chak Gopalpur village from which the river originates, whereas the lowest elevation of 4 m above MSL was recorded near the confluence of Dwarakeswar River with Silabati, where it joins to form the river Rupnarayan. Thus, the relative relief of the basin is 213 m. The elevation falls quite steeply from the north-western fringe towards the eastern and southern slopes which is characterised by gently sloping lowland plains drained by the main channel and its tributaries. The central portion of the basin shows zones of moderate relief through which the main Silai flows. The lower basin is characterised by zones of extremely low elevation where average height seldom reaches more than 10 m. Thus, the Silai basin is characterised by surfaces of varying elevation comprising of rolling and undulating plains in the northern and north-western flanks of the basin which slowly changes to a relatively flat terrain of minimum relief.

10.2.2.2.2   Average Slope and Slope Aspect

Slope, defined as the angular inclination of the terrain, gives a clear idea about the lithological control over the terrain and was computed following Wentworth (1930):

$$\text{Average slope} (\theta) = \tan^{-1}\left((N * CI)/636.6\right) \tag{10.2}$$

where N is the average number of contours crossing per unit length and CI is the contour interval of the map.

While relatively higher gradients can be seen in the northern part of the basin (4°–5°), the majority of the basin is covered by moderate slope representing the gently rolling plains and remnants of the Chhotanagpur Plateau, which slowly grades into plains of extreme low relief (1°–2°) at the lower

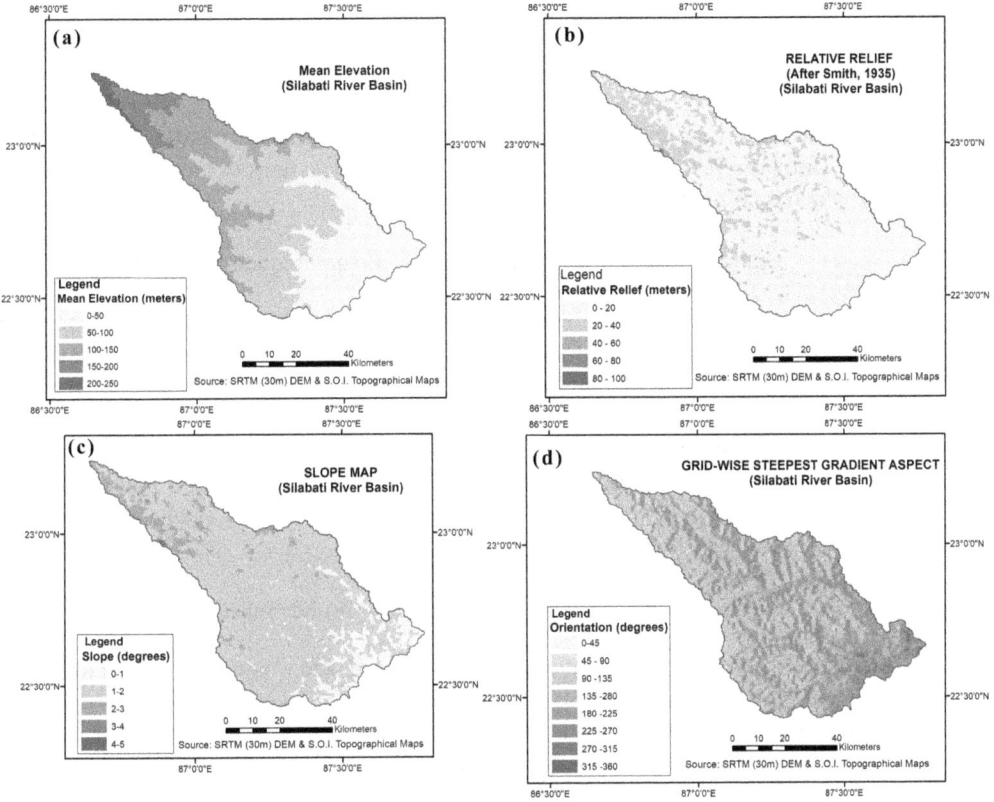

FIGURE 10.3 Morphometric attributes: (a) mean elevation, (b) relative relief, (c) slope and (d) steepest gradient.

portion of the basin near the mouth of the river (Figure 10.3c). The source area of the channel and its ensuing path through the basin can clearly be understood from the average slope map. The slope aspect map shows the local variations in the direction of the gradients (Figure 10.3d), with the overall regional slope being towards the south-east.

10.2.2.2.3 Stream Frequency, Drainage Density and Drainage Texture

Stream frequency is the number of streams per unit area (Horton, 1945), and it provides an idea of the amount of runoff generation in an area and also gives an insight into the length of overland flow (Figure 10.4a). It is computed as

$$\text{Stream frequency (SF)} = \text{No. of streams/total area} \qquad (10.3)$$

Moderate to high stream frequencies are observed throughout the basin since the topography largely consists of gently sloping plains with a large number of tributaries, distributaries and palaeochannels draining the floodplains formed by this network of streams. Moderately low values are found in the remaining stretches of the basin. Also formulated by Horton (1945), the drainage density is the ratio of the total length of all streams in a drainage basin to the total area of the basin, which is computed as

$$\text{Drainage density (DD)} = \text{Total length of all streams/total area of the basin} \qquad (10.4)$$

Zones of low to moderate drainage density can be seen to exist all over the basin within which sporadic pockets of higher densities are found, mostly due to the confluence of tributary streams with the main Silabati channel (Figure 10.4b). High drainage densities, especially in the lower portion of the basin, may be attributed to this. Drainage density is also found to be higher on the western portion of the basin due to the greater number of right bank tributaries joining the main channel therein. The Drainage Texture (DT) is the product of drainage density and stream frequency (Figure 10.4c), which is computed as follows:

$$\text{Drainage texture} = \text{Drainage density} \times \text{Stream frequency} \tag{10.5}$$

This parameter helps to discern between those areas that may have similar drainage densities but differing stream frequencies, due to a contrasting stream network arrangement (i.e., one longer sinuous channel in comparison with many short streams).

10.2.2.2.4 Dissection Index, Ruggedness Index and Hypsometric Integral

The nature of erosion across the terrain can be deciphered from the Dissection Index (De Smet, 1951), which varies from 0 (where the entire altitude has been dissected) to 1 (where no dissection has occurred). The working formula is

$$\text{Dissection index (DI)} = \text{Relative relief}/\text{Absolute relief} \tag{10.6}$$

Dissected patches can be seen along the entire length of the basin and is not concentrated to any particular zone (Figure 10.5a). The central portion of the basin is moderately dissected through which the channels have carved their valleys. High dissection in the north-western flanks can be attributed to relatively higher altitude in those regions, whereas high dissection values in the lower part may be due to the extremely flat lands situated just above MSL (i.e., both relative and absolute relief values are similarly low). Devised by Horton (1945), the Ruggedness Index gives an idea of the roughness of the landscape due to the working and reworking of surficial sediments by the channel network. It is a combined function of the general topography and the steepness of the gradient and is calculated as

$$\text{Ruggedness index (RI)} = \big(\text{Relative relief} * \text{Drainage density}\big)/1000 \tag{10.7}$$

The majority of the basin has a moderate ruggedness index (Figure 10.5b), with high ruggedness values in the north-western zone of the basin, as was also been seen for the relative relief and dissection index values. The river valleys and the lower elevation portions of the basin show low ruggedness as these grids represent typical floodplain tracts. Computed grid-wise using Pike and Wilson's (1971) formula, the hypsometric integral (HI) affirms the findings of the dissection and ruggedness index layers (Figure 10.5c).

$$\text{HI} = \big(\text{Mean elevation} - \text{minimum elevation}\big)/\big(\text{maximum elevation} - \text{minimum elevation}\big) \tag{10.8}$$

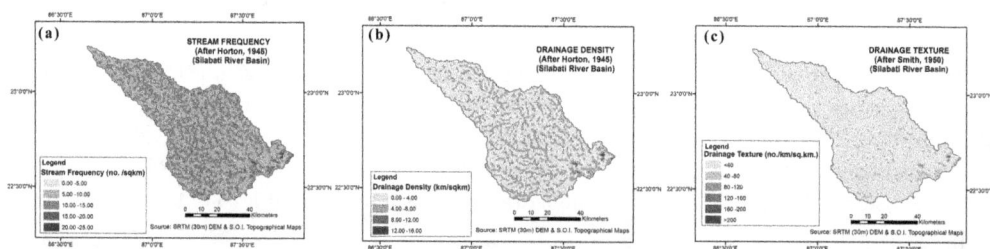

FIGURE 10.4 Morphometric attributes: (a) stream frequency, (b) drainage density and (c) DT.

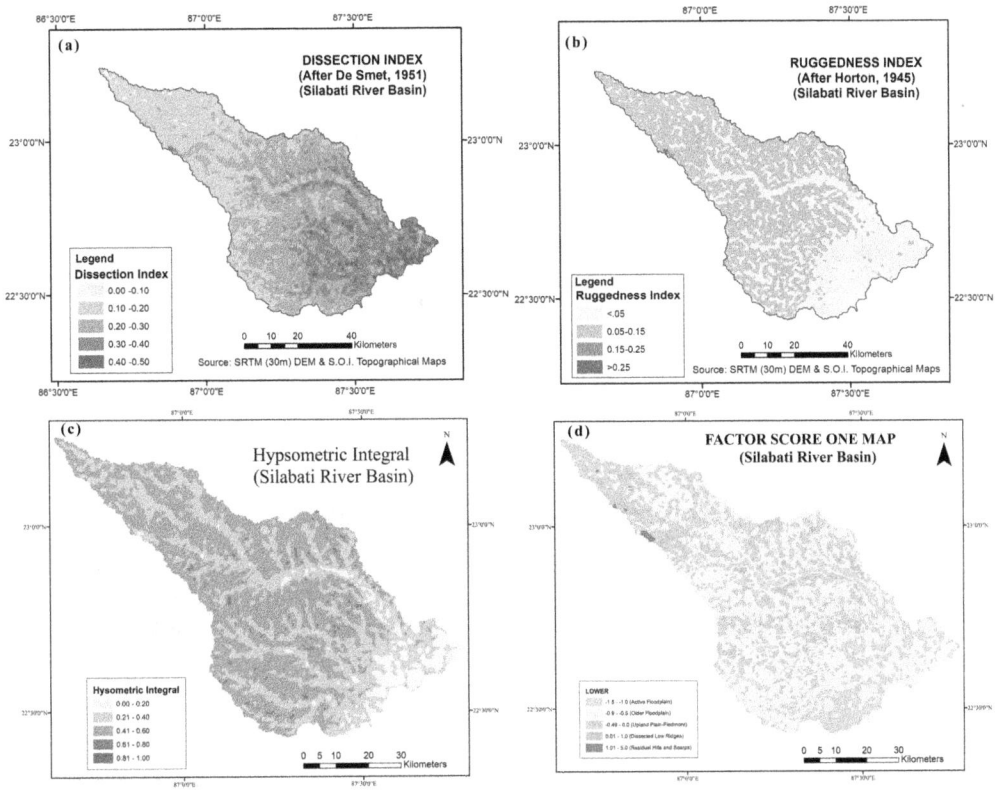

FIGURE 10.5 Morphometric attributes and landscape units: (a) dissection index, (b) ruggedness index, (c) HI and (d) factor score one map showing different physiographic units.

Ranging between 0 (almost entirely eroded) and 1 (almost entirely un-eroded), it depicts pockets of relatively lesser erosion (i.e., higher HI values), amidst overall lower to mid-range values. Thus, the basin landscape is largely in the middle to older stage of its geomorphic evolution.

10.2.2.2.5 Factor Analysis and Basin Terrain Units

The values extracted for each of the above eight parameters were reduced using a principal component analysis (PCA) transformation to derive the factor scores for each grid. The Factor Score One values were mapped across the basin surface to derive the different terrain units (Sarkar and Patel, 2011, 2012), which encapsulated the attributes of the above eight parameters (Figure 10.5d). Five distinct physiographic units were identified (Table 10.2) by grouping the Factor One scores together, after overlaying their distribution on the satellite images – active floodplain, older floodplain, upland plains – piedmont, dissected low ridges, residual hills and scarps. The active floodplains occupy a narrow strand from west to east across the central and eastern basin area, with larger patches in the south and south-east where the principal tributaries flow into the Silabati. The older floodplain is the largest unit and abuts the younger active floodplain. This unit is primarily used for agriculture, especially where it is free of lateritic hardpans. The upland plain – piedmont unit – largely denotes the low interfluves between the older floodplains of the various tributary streams. These are mostly composed of weathered lateritic cover and have sparse to dense vegetation. The other two units (dissected low ridges and residual hills and scarps) together comprise less than 5% of the basin area and are only seen in isolated patches, along the right flank of Silabati River. Some of these, like the high scarp at Garbeta, are significant for the badlands formed along their flanks. The geomorphic diversity of the basin is thus emblematic of a typical plateau-lowland drainage system, with its attendant land units (cf. Sarkar and Patel, 2016).

TABLE 10.2
Geomorphic Units within the Silabati River Basin

Geomorphic Unit	Factor Score One	Area (km²)	Area (%)
Older floodplain	−1.00 to −0.50	2,081.39	50.14
Upland plains – piedmont	−0.50 to 0.00	1,195.81	28.81
Dissected low ridges	0.00 to 0.50	148.35	3.57
Active floodplain	−1.50 to −1.00	700.42	16.87
Residual hills and scarps	1.00 to 5.50	24.94	0.60

Source: Computed by the authors.

10.3   ANTHROPOGENIC SIGNATURES ON FORMS AND PROCESSES

10.3.1   LULC Composition, Changes and Vegetation Vigour

The LULC components and pattern is a dominant controlling factor of the catchment-scale channel morphology, being itself largely determined by the present terrain units. It is also modified primarily by human activities seeking to utilise the present natural resources, and large-scale changes in land-use pattern can significantly alter surface runoff generation and sedimentation rate in a channel. Temporal changes in the LULC pattern thus need to be carefully analysed in order to ascertain the morphological changes that are being introduced into the channel. The LULC map of the basin (Figure 10.6) reveals five primary components – agricultural lands, forests (plantations, degraded, *Sal* and mixed deciduous), barren and scrub lands, orchards and settlements. Over half the basin area has been consumed by agriculture (Figure 10.7), most of which is mono-cropped and of the subsistence variety. The presence of degraded forests and barren lands point to the human-induced clear-felling and vegetation clearance in the area. Temporal analysis of the forest cover (Figure 10.8) shows a gradual decrease in the extent of the relatively dense vegetation cover classes, especially along the periphery of previous forest extents and thinning of the tree stands. This is typified by the lessening of the NDVI values across the basin area (Figure 10.9), especially in the central and eastern sections, where considerable clear-felling has occurred, as can be visually detected from the satellite image datasets.

10.3.2   Fragmentation Index

The Fragmentation Index raster layer has been obtained as part of the BIS database (Roy et al., 2012, 2015a, 2015b). As per the method formulated by the IIRS for its preparation, this index discerns the number of forest and non-forest patches in a regularly sized grid (500 m × 500 m) sampled across the entire region (IIRS, 2003a–2003d, 2011). Basically, this parameter examines the contiguity of existing forest patches, with adjacent tracts of denser vegetation cover being accorded lower scores (i.e. less fragmentation), as per the stated classification scheme. Fragmentation scores are higher (i.e., there is a greater disconnectedness of forest patches) along the fringes of former densely vegetated zones (Figure 10.10a) as these are being progressively cleared with finger-like extension of agricultural lands into them, following the courses of the small rivulets. The raster fragmentation layer was converted to vector pixels, and the mean fragmentation value for each terrain unit was thereby extracted.

10.3.3   Disturbance Index

Fragmentation of the natural landscape, from human activities, also precipitates ecological imbalances on the existent biological communities and terrain aspects. This has been quantified as the Disturbance Index in the prepared BIS (Roy et al., 2012, 2015a, 2015b), from which its raster layer was directly

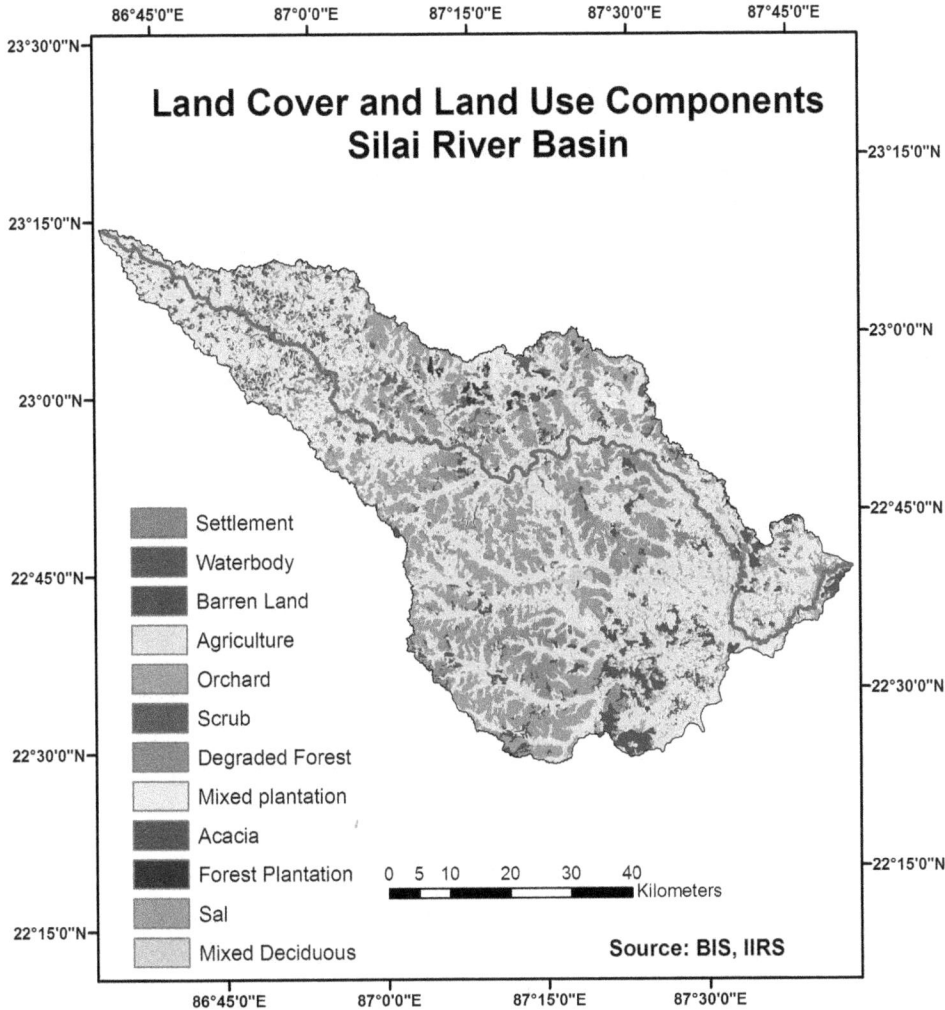

FIGURE 10.6 LULC components of the Silabati River Basin.

obtained and cropped to the basin extents. This parameter is dependent on the landscape fragmentation and patch metrics, locations of nearby villages, roads and forest clearance sites, all of which are primary disturbance drivers and the extent of possible invasiveness of other species or regeneration of the original vegetation of an area. The disturbance index is much lower along the ridges and the higher interfluves (Figure 10.10b), with the poorest scores (high values) being recorded (as per the stated IIRS classification) in the cultivated tracts that have witnessed almost total removal of the original vegetation cover of the area. Obviously, greater fragmentation of the landscapes causes a higher disturbance, with the correlation coefficient (r) of these values extracted for each grid being 0.998. Such gradual forest clearance, landscape fragmentation and ecological disturbances in plateau-lowland watersheds, along the courses of the principal stream and its rivulets, are typical characteristics of most rivers arising from the Chhotanagpur region and its extensions (Chatterjee and Patel, 2016).

10.3.4 CHANNEL CORRIDOR MODIFICATIONS

With the aforesaid fragmentation and disturbance index maps highlighting the much altered condition of the basin, further detailed insights were garnered from higher resolution land-use maps prepared along selected reaches of the river from Google Earth imagery. The channel corridor was

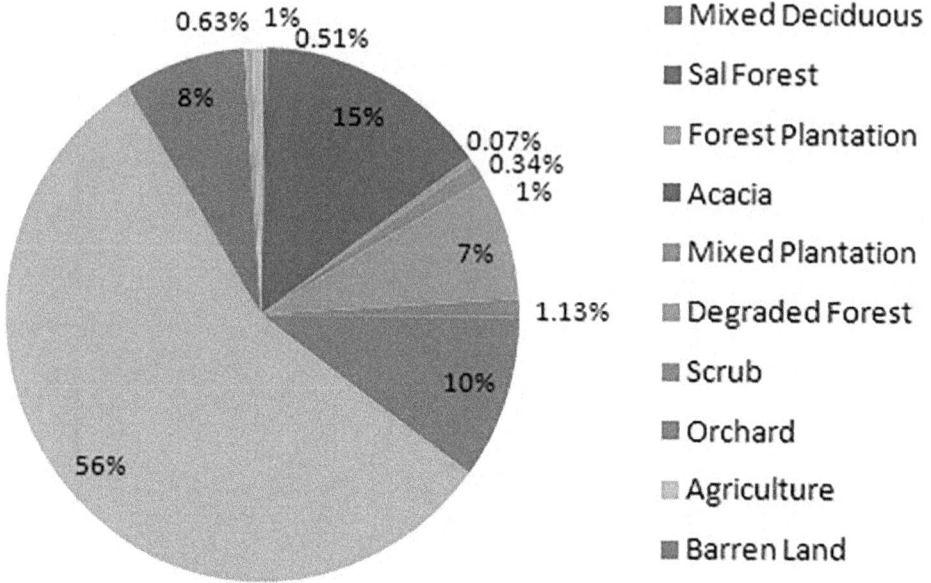

FIGURE 10.7   LULC composition of the Silabati River Basin.

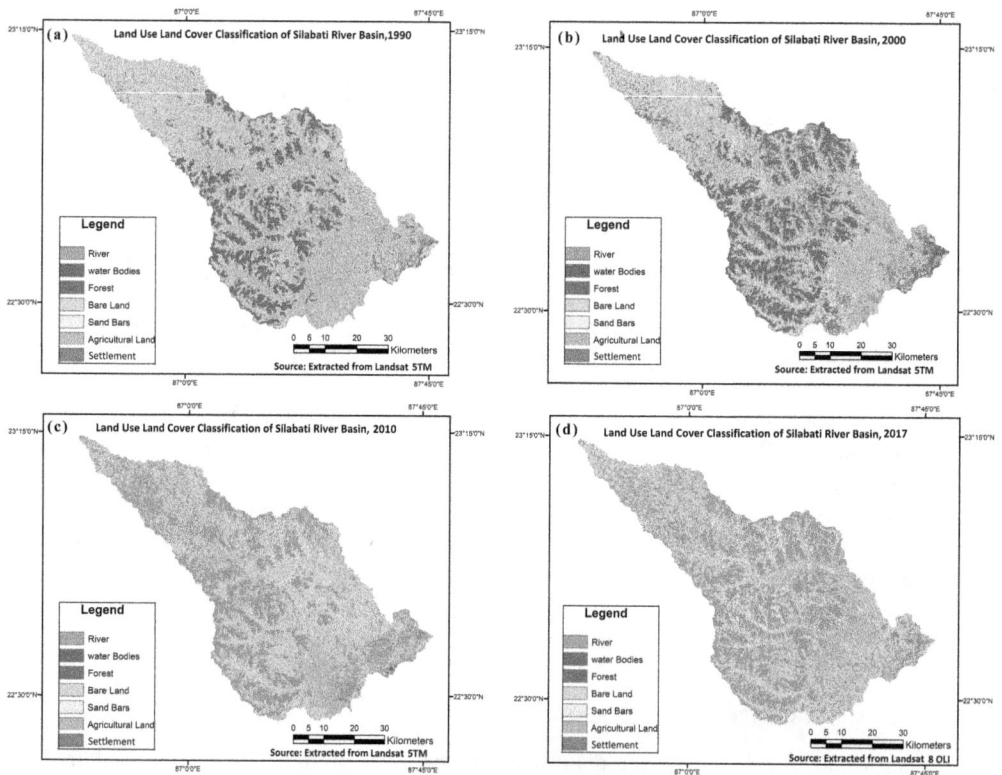

FIGURE 10.8   LULC components and changes for the years: (a) 1990, (b) 2000, (c) 2010 and (d) 2017.

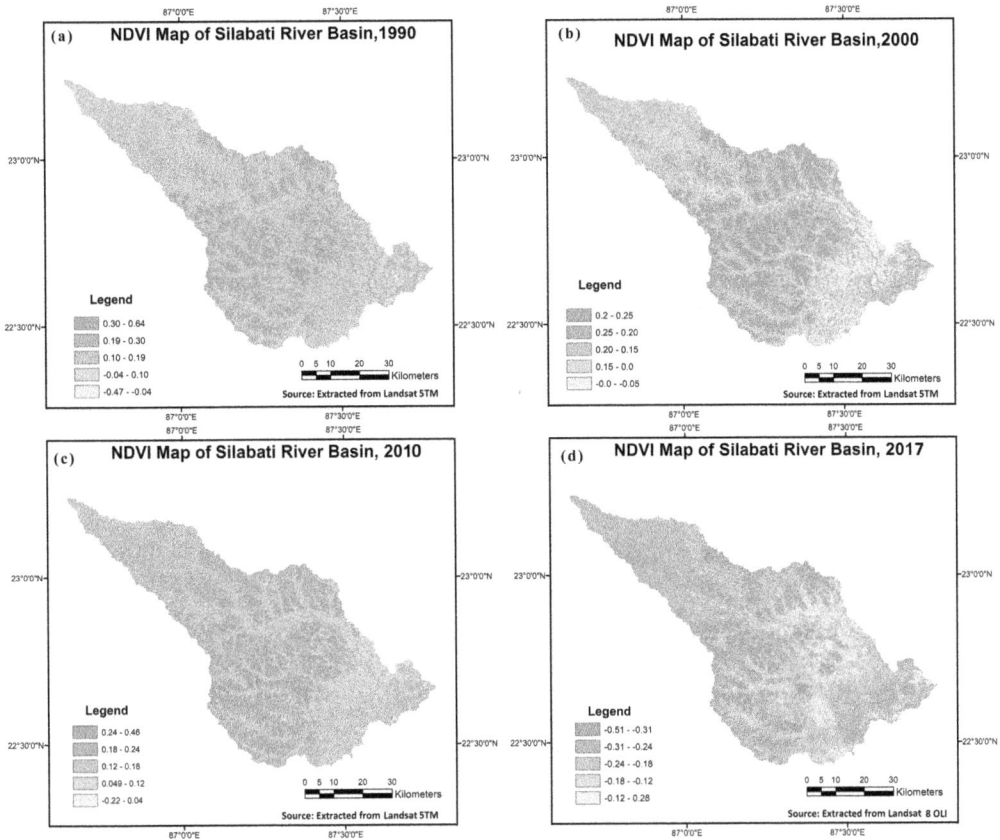

FIGURE 10.9 Normalised Difference Vegetation Index for the years: (a) 1990, (b) 2000, (c) 2010 and (d) 2017.

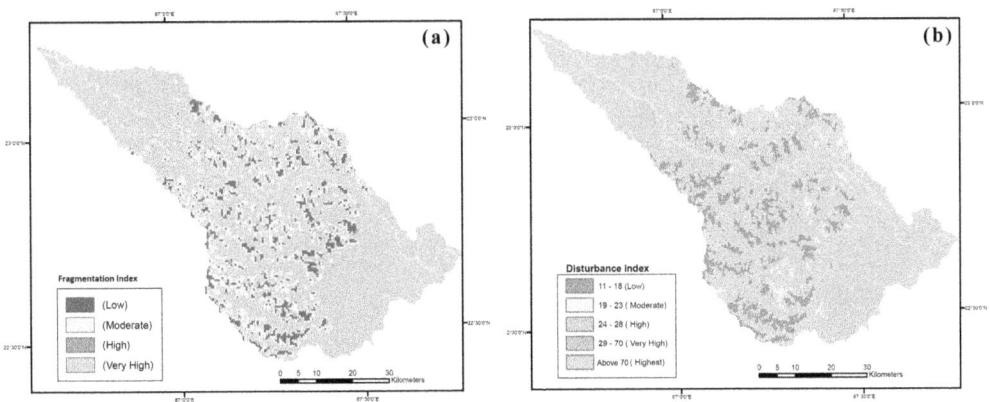

FIGURE 10.10 Biodiversity Information System database layers of (a) fragmentation index and (b) disturbance index.

demarcated, using the fixed-width method (cf. Banerji and Patel, 2019), at 500 m on either side of the left and right banklines, and 20 windows were sampled along the middle and lower courses of the Silabati (Figure 10.11), since marked variations in the fragmentation and disturbance values were recorded therein. It is along the channel corridor that intensive utilisation of the floodplain resources occurs, which have a direct bearing on overall channel morphology and health, and the

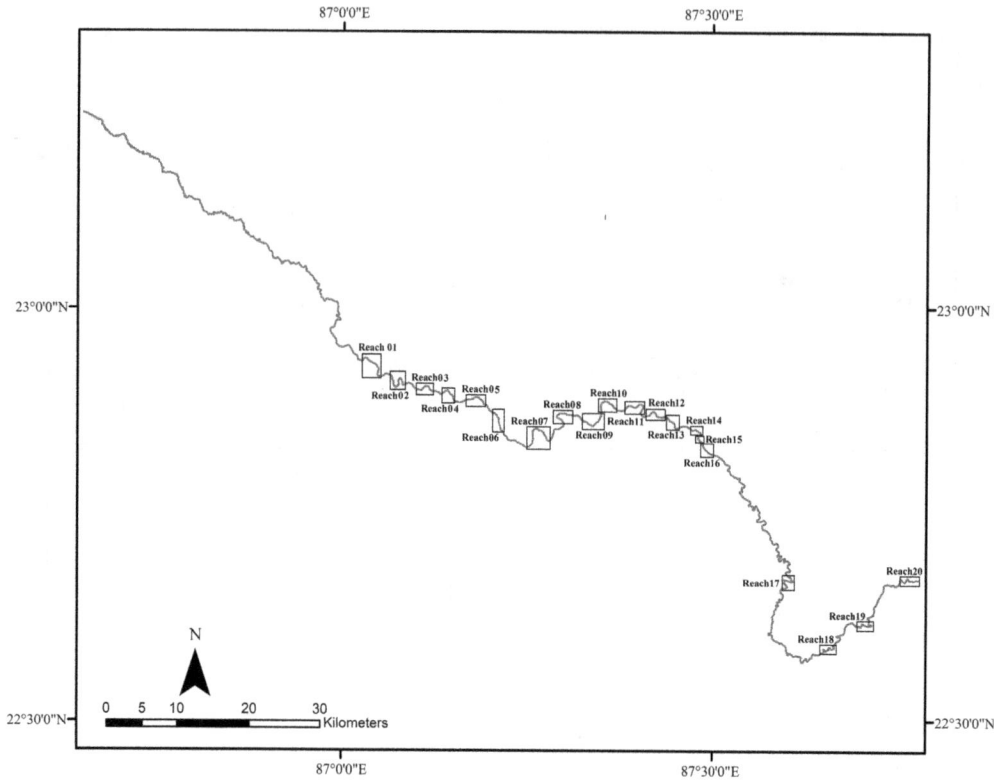

FIGURE 10.11   LULC components of sampled stream corridor windows along the Silabati River.

LULC patterns along these windows (Figures 10.12 and 10.13) symbolise the human interventions into the earlier pristine and forested riverine tracts. As the river changes from a semi-confined gravel-dominated channel into a sand-dominated one at the start of the sampled stretches, significant changes can be noticed in its channel morphology due to increase in channel width and decrease in bed gradient, and it meanders through the gentle plains in a highly sinuous course. Consequently, a change in the land-use pattern (Figures 10.14 and 10.15) has also been observed

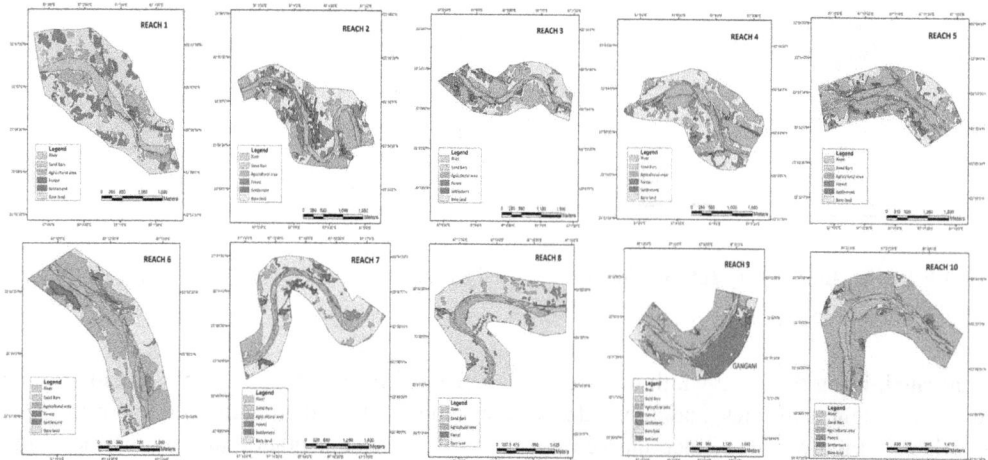

FIGURE 10.12   Reach-level LULC maps of the Silabati River corridor (reaches 1–10).

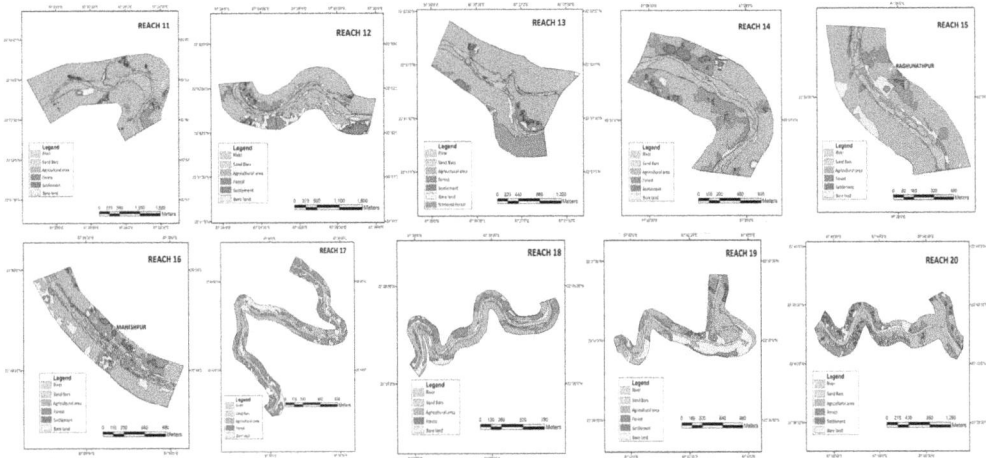

FIGURE 10.13 Reach-level LULC maps of the Silabati River corridor (reaches 11–20).

with an increased amount of cultivated lands along both the banks, supporting the settlements that thrive beside them. A steady increase in the share of agriculture occurs as the river migrates through the gentle floodplains. Intensive cultivation, with more than one crop, is practised in some stretches, producing paddy and vegetables that often cater to the needs of the nearby bigger settlements. Agriculture is carried out with the help of irrigation for the greater part of the year due to the extreme climatic regime of the region. The dominance of bare land in most of the reaches signifies the percentage of seasonally fallow lands that are intensively used for cultivation once the monsoon sets in. This automatically indicates a decline in the share of forest cover that has been readily cleared off to support the increasing needs of the small towns emerging steadily along both banks of the channel. Towns such as Simlapal, Krishnanagar, Garbeta, Chandrakona and Ghatal are the obvious results of decreasing forest cover and increasing agriculture along the river. However, a considerable presence of forest cover can be seen at few stretches in the lower course (reaches 17 and 18) due to the inaccessibility of the region in terms of transport networks, and thus, the naturalness of the channel has somewhat been maintained along these sections, from the paucity of human interventions into the immediate river environment.

10.3.5 IN-CHANNEL MODIFICATIONS

Along with changes on either bank, substantial modifications have also occurred within the main channel itself. These have been brought about by the constructions of check dams, crossing structures (mainly bridges) and, most substantially, the emergence of numerous sand-mining sites on the riverbed (Figure 10.16). The check dams are mostly present in the upper course, whereas crossing structures are present throughout. Localised sites of pollution along the river, from nearby settlements, are also seen (Figure 10.17). However, the most standout feature of the entire middle course of the river is the predominance of sandbars, arising from the huge amounts of fine sediments brought down by the channel, and these have become the prime sites for sand mining during the lean season. Reaches 1, 4, 5, 6, 7, 8, 10 and 11 in Figures 10.12 and 10.13 have considerable sand deposits that have been mined and the riverbed has been quite hollowed out, which also tends to make the channel seemingly braided at some locations, thus disrupting natural flow conditions for the greater part of the year. This intensive sand extraction has also altered the morphology and the hydrologic regime of the reaches by the formation of sand pits as deep as 2–3 m across the channel, and as a result, stagnant pools of water have formed within it (Figure 10.18). The presence of such deep pools can be noticed in reaches 15 and 16, and the river attains an apparent braided character

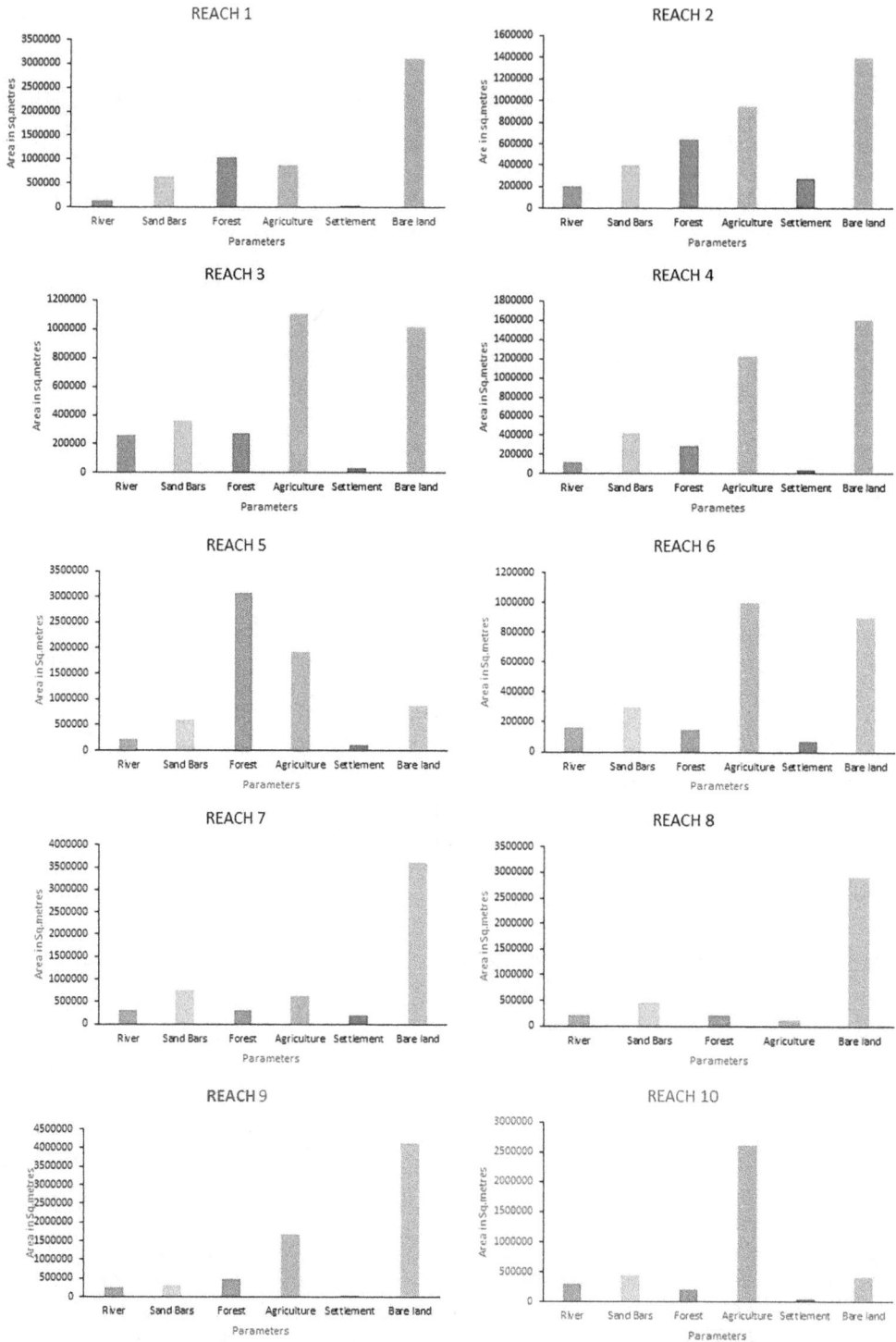

FIGURE 10.14 Reach-level LULC composition of the Silabati River corridor (reaches 1–10).

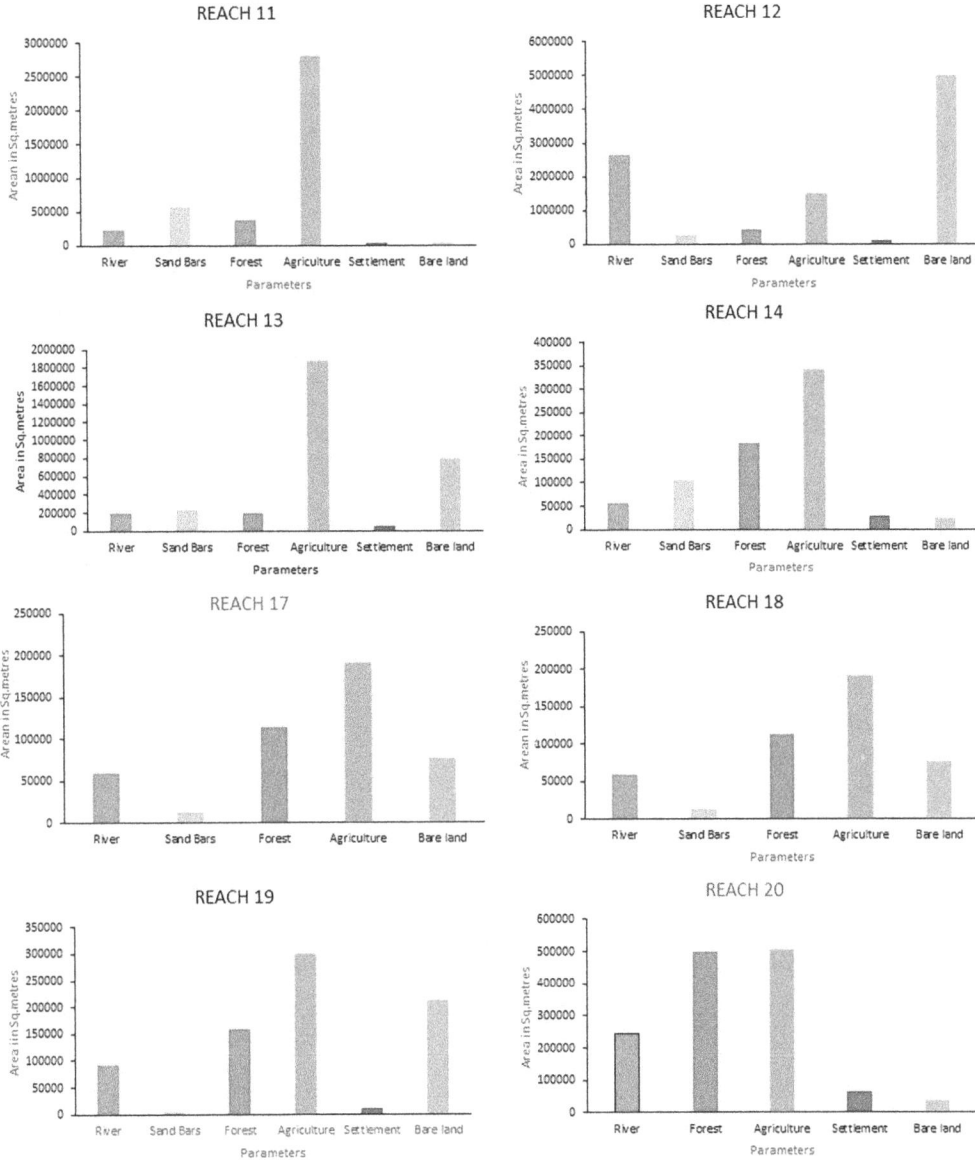

FIGURE 10.15 Reach-level LULC composition of the Silabati River corridor (reaches 11–20).

with disrupted longitudinal connectivity. Intensive sand mining activities have also resulted in alterations in channel geometry through changes in channel depth and bed slope, thereby impacting on stream power and suspended and bedload transport. Sand mining from the channel has been reported to cause bank erosion along the cliff slopes of meander bends, causing loss of valuable farmlands through channel widening at some stretches. Accelerated rates of sand mining along the middle and lower courses have led to the formation of sand pits with stagnated pools of water in between them causing marked disruption of normal flow during the lean season. These pools are characterised by shifting thalwegs, renewed channel incision due to increased velocity and often report high turbidity, thereby failing to support the fish species that normally thrive in the river. Clearing of riparian vegetation to allow the tractors to reach the channel bed has resulted in loss of connectivity between the channel and floodplain, thus adversely affecting the riparian ecosystem.

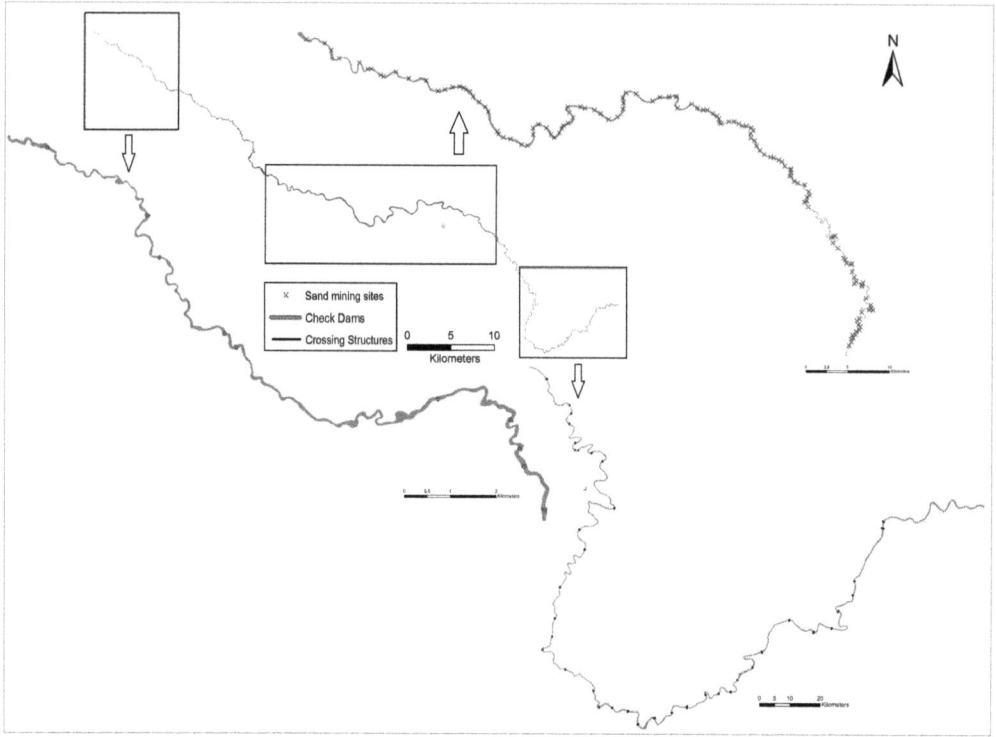

FIGURE 10.16 In-channel modification and obstruction aspects along Silabati River.

The mouzas of Shymanandapur, Keshadal, Palaschabri and Rosana are some of the worst affected pockets where mining activities have created pits of 2–3 m deep within the channel and in some parts have also exposed the bed rock beneath the channel. In the lower reaches and near its mouth, the channel narrows down considerably due to the presence of embankments on either side that have constrained its flow. Sand mining in this section is restricted due to a greater depth of the channel, which supports flow throughout the year. Agriculture is practised on the high banks on either side, and sand deposits are seldom seen in this stretch of the river.

FIGURE 10.17 Examples of river corridor and channel attributes (natural and human modified).

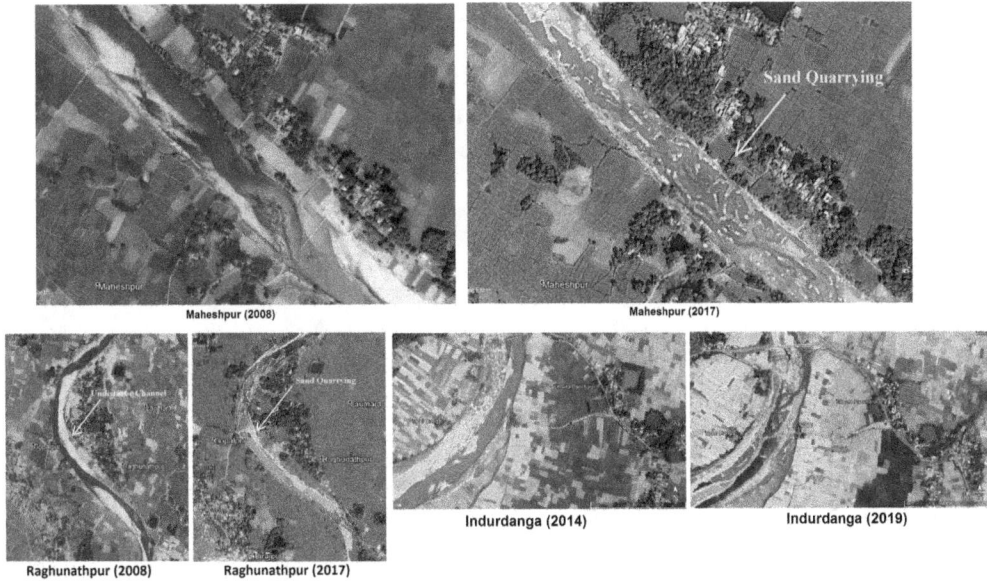

Maheshpur (2008)

Maheshpur (2017)

Raghunathpur (2008)

Raghunathpur (2017)

Indurdanga (2014)

Indurdanga (2019)

FIGURE 10.18 Examples of pre- and post-sand quarrying scenarios in select stretches along the Silabati River.

10.3.6 GULLY DEVELOPMENT

An important characteristic of reach 9 in Figure 10.12 is the presence of a lateritic badland terrain on the right bank of the river, making it prominently distinct from the other reaches. Locally known as "Ganganir Danga", this region presents a completely different picture of land use on the right bank from that on its left bank. Occupying a total area of almost 25 km², it almost renders the entire area unsuitable for agriculture with wastelands with and without scrub, barren rocky stony wastes and gullied area, mainly inhabited by pockets of rural settlements. Cashew plantations with the help of irrigation have developed in a scattered manner along the right bank. Pockets of Notified Forest Area exist, dominated by *Sal* trees, which is the characteristic vegetation of the region and also by eucalyptus plantations. The opposite left bank, with a much lower elevation, however, promotes agriculture and paddy fields are abundant. This area gets regularly inundated during the monsoon floods and as such favours intensive agriculture, with more than one crop being harvested at every season. Garbeta town is the main settlement on the right side of this lateritic tract. This area is emblematic of other such locales within the Silabati Basin, particularly in the portion of it which lies within Paschim Medinipur district that is dotted with the presence of such gullied lateritic areas. These zones are prime locations of enhanced soil loss from the region (Bandyopadhyay, 1988; Sen et al., 2004; Shit et al., 2015a, 2015b), and the expansion of the gullied landscape, together with the dynamic changes in their morphology, has given the region a distinct character (Das and Bandyopadhyay, 1995; Shit et al., 2013, 2014). These gullied lateritic landscapes within the Paschim Medinipur section of the basin have been identified using the Google Earth platform and corroborated with extensive field surveys (Figure 10.19). The primary locations for these gullies were within the deforested, open scrublands or exposed lateritic tracts, along the foothills zones of the low ridges and hills or along the banks of the principal streams, thus signifying that these develop and expand most rapidly where human activities have removed the albeit sparse, yet protective, vegetation cover over from the underlying friable lateritic surface. This has resulted in aggressive gully expansion and further degradation of the LULC components, as can be clearly discerned from historical and present Google Earth images. Particularly in the Garbeta tract, such alterations have taken the form of construction of check dams across the gullies, clearing of vegetation for tourist spots and roads to access the riverside (Figure 10.20), resulting in marked changes in the geomorphic attributes and gully morphologies (Kar and Bandyopadhyay, 1974; Patel et al., 2020).

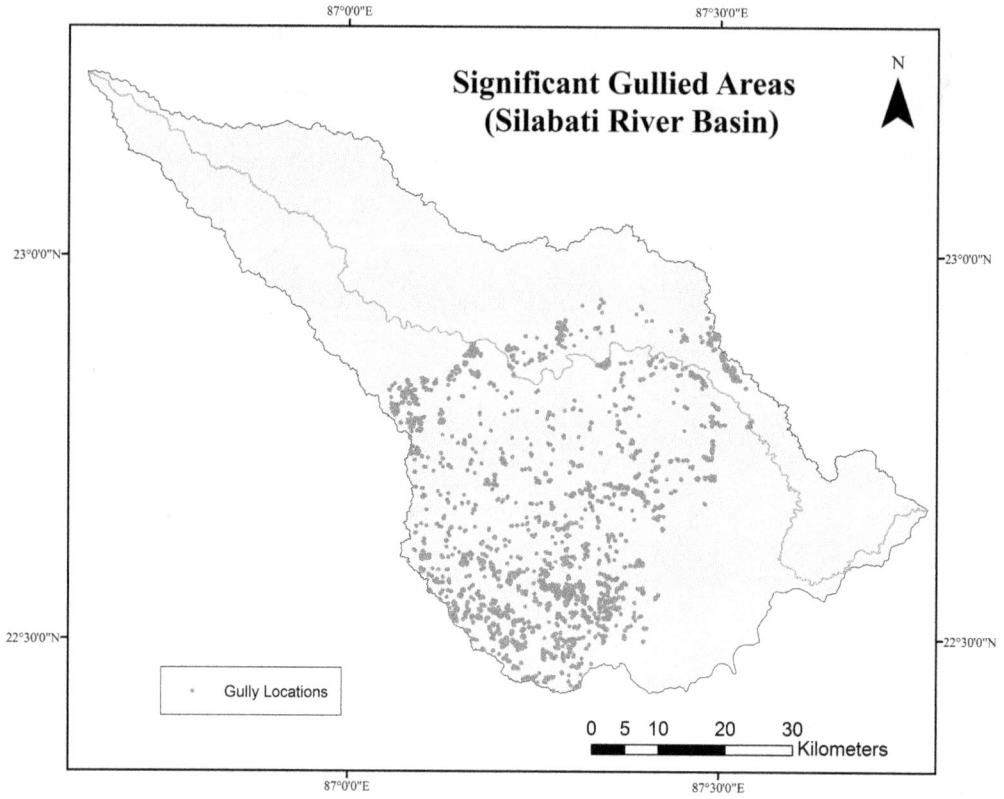

FIGURE 10.19   Significant gullied areas in Silabati Basin.

FIGURE 10.20   Landscape changes at Gangani Badlands, Garbeta.

10.3.7 GEOMORPHIC PARAMETERS AND LANDSCAPE ALTERATION CORRELATIONS

As evident from the foregoing discussions, the terrain attributes play a major role in conditioning the land modification in an area. To quantitatively assess this, a correlation analysis (Table 10.3) was performed between the constituent landscape physical variables (typified by the different morphometric parameters) and the landscape alteration variables (encapsulated by the fragmentation and disturbance indices) derived grid-wise across the entire basin area. These two landscape alteration variables correlate positively with the dissection index, stream frequency, DT and most with the drainage density, while correlating negatively with the ruggedness index, mean elevation, slope, relative relief and most with the HI. Thus, it becomes apparent that greater fragmentation and thereby disturbance into the landscape are introduced in flatter terrains traversed by multiple streams, whose floodplains and banks are easily accessible and modifiable. More rugged locales are seldom as amenable. The mean fragmentation and disturbance values have also been computed for the different Factor Score One ranges (Table 10.4), which together show the various landscape physiographic units in the basin. Their scores consistently decrease towards the more higher elevation, sloping and rugged terrain units, signifying the lower extents of human modification in these locales. Furthermore, the number of gullies (another indicator of mostly human-induced soil loss and land degradation in the area) for each of the factor one score zones/terrain units was computed. Presently, the greatest gully counts are within the piedmont zone and the dissected low ridges, with some occurring in the older floodplain tracts. Thus, as mentioned before, gullies develop along the flanks of the terraces separating the younger and older floodplains, and also originate on the flanks of the lateritic ridges and interfluves carrying on into the piedmont zone. It is this piedmont zone that shows the highest levels of loss in forest cover in recent years (as was ascertained visually from the multi-temporal classified LULC images), and the residual bare friable lateritic surface soon gets colonised by these rills and gullies. The relatively dense vegetation cover over the hills retards soil loss and thereby badland development therein. The mean fragmentation and mean disturbance values are thus highly positively correlated ($r = 0.999$), with their respective correlations with the Factor Score One values of these terrain units are 0.843 and 0.834 (both negative relations).

10.4 CONCLUSION

The Silabati River Basin represents a typical example of a Plateau fringe – *Rarh* landscape – alluvial tract drainage network. The topography of the basin remains largely controlled by its lithological aspects, with higher relief and slope aspects spatially correlating with locations of metamorphic to lateritic rocks. Unconsolidated alluvial deposits fill the intervening valleys and allow stream meandering and floodplain formation. Agriculture here, as elsewhere, has been the main driver of natural landscape and vegetation modification, resulting in fragmentation of the tree stands and replacement of the original forest cover with orchards and plantations in some areas. Expectedly, the greatest landscape alterations have occurred in the floodplain (newer and older) units, with these representing the interlinked dual physical-human entities (Dasgupta and Patel, 2017) crafted herein over centuries, from river deposition and anthropo-geomorphic modifications.

The basin area is typified by two main physical aspects. One is the lateritic cover over its upper plains and piedmont zones, as is emblematic of the *Rarh* region (Ghosh and Guchhait, 2015, 2019), over which gullies have developed abundantly in some locations and cause substantial soil loss from the region, accentuated by human enhanced process erosivity and surface erodibility. Some measures have been sought to control the soil loss from such gullied tracts using vegetation and other geotechnical means (Shit and Maiti, 2014), but this remains an ensuing challenge, as tourism in locations like Gangani, has increased the human footprint across the natural scape.

The other phenomenon that regularly affects the basin is annual floods, particularly in its lower reaches in Ghatal Block, which arise from drainage congestion and water-logging due to the high influx of tributaries into the main channel over a short distance, in an area of very low relief and slope.

TABLE 10.3

Correlations between Landscape Morphometric and Alteration Indices

	MEAN_ELEV	RR	DI	RI	SF	DD	DT	SLOPE	HI	FRAG_MEAN	DIST_MEAN
MEAN_ELEV	1.00										
RR	0.71	1.00									
DI	−0.73	−0.22	1.00								
RI	0.44	0.72	−0.13	1.00							
SF	−0.09	−0.12	−0.06	0.37	1.00						
DD	−0.26	−0.25	0.09	0.40	0.73	1.00					
DT	−0.19	−0.21	0.03	0.35	0.85	0.92	1.00				
SLOPE	0.73	0.88	−0.30	0.55	−0.19	−0.38	−0.31	1.00			
HI	0.46	0.49	−0.29	0.12	−0.31	−0.53	−0.44	0.54	1.00		
FRAG_MEAN	−0.11	−0.31	0.04	−0.10	0.18	0.28	0.22	−0.24	−0.57	1.00	
DIST_MEAN	−0.12	−0.31	0.05	−0.10	0.18	0.28	0.23	−0.25	−0.57	0.99	1.00

Source: Computed by the authors.

TABLE 10.4

Landscape Geomorphic Unit Relations with Landscape Alteration

Factor Score One	Landscape Unit	Mean Fragmentation	Mean Disturbance	No. of Gullies
−1.50 to −1.00	Active floodplain	172.99	173.00	1
−1.00 to −0.50	Older floodplain	150.93	152.56	310
−0.50 to 0.00	Upland plains – piedmont	121.88	125.68	2,676
0.00 to 0.50	Dissected low ridges	124.00	127.46	2,075
0.50 to 1.00		125.51	129.68	615
1.00 to 1.50	Residual hills and scarps	142.72	142.52	26
1.50 to 2.00		113.16	122.33	0
2.00 to 2.50		114.39	122.99	0
2.50 to 3.00		97.83	108.48	0
3.00 to 3.50		108.63	117.11	0
3.50 to 4.00		118.79	125.91	0
4.00 to 4.50		90.54	102.03	0
4.50 to 5.00		38.36	58.01	0
5.00 to 5.50		2.00	26.00	0

Source: Computed by the authors.

Further research for gauging the flood susceptibility of the region (e.g. Sahana and Patel, 2019) needs to be undertaken, along with ascertaining the socioeconomic vulnerability of the residents that may arise from such hazards. Such floods have themselves been drivers of land-use changes in the basin area, as is also seen in other basins of the *Rarh* region (Patel and Dasgupta, 2009), with the construction of high embankments along the lower courses of the Silabati signifying the constant threat posed by them. Riverbank erosion accompanies such flood hazards, and the installation of riparian buffers to abate this and the flood effects (Mondal and Patel, 2018, 2020) has been attempted at certain locations along the river's lower course.

Possibly, the greatest threat to the stream stability and ecological condition comes from the rampant extraction of sand from the riverbed, which was identified during the field surveys as one of the most important causes of channel degradation and bankline shifts, especially in the middle and the lower courses. The direct impacts include alteration of species habitats and ecosystems, changes in hydraulic regimes and in riffle-pool sequences, and loss of floodplain-channel linkages (Bhattacharya et al., 2019), whereas the indirect changes are linked to the long-term changes that are incurred in the fluvial ecosystem due to changes in the physical processes operating in the channel and also water quality degradation due to changes in the physico-chemical composition of such lotic ecosystems.

NOTE

The authors acknowledge that this research has been funded by a Department of Science and Technology, Government of West Bengal Grant and a University Grants Commission Start-Up Grant under its Faculty Research Promotion Scheme for early career researchers awarded to Priyank Pravin Patel and also aided by the FRPDF allotment by Presidency University to him. The grant allotment from the UGC-SRF Award of Sayoni Mondal is also acknowledged.

REFERENCES

Bandyopadhyay, S. (1988) Drainage evolution in a badland terrain at Gangani in Medinipur District, West Bengal. *Geographical Review of India*, 50(3), 10–20.

Banerji, D., Patel, P.P. (2019) Morphological aspects of the Bakreshwar River Corridor, West Bengal, India. In: B. Das, S. Ghosh, A. Islam (eds.) *Advances in Micro Geomorphology of Lower Ganga Basin – Part I: Fluvial Geomorphology*, Springer International Publishing, Cham, pp. 155–189. doi: 10.1007/978-3-319-90427-6_9.

Bertalan, L., Novak, T.J., Nemeth, Z., Rodrigo-Comino, J., Kertesz, A., Szabo, S. (2018) Issues of meander development: land degradation or ecological value? Example of Sajo River, Hungary. *Water*, 10(11), 1613. doi: 10.3390/w10111613.

Bhattacharya, R., Dolui, G., Chatterjee, N.D. (2019) Effect of instream sand mining on hydraulic variables of bedload transport and channel planform: an alluvial stream in South Bengal Basin, India. *Environmental Earth Sciences*, 78, 303. doi: 10.1007/s12665-019-8267-3.

Brierley, G.J., Yu, G., Li, Z. (2016) Geomorphic diversity of rivers in the upper Yellow River basin. In: Brierley, G.J., et al. (eds.) *Landscape and Ecosystem Diversity, Dynamics and Management in the Yellow River Source Zone*, Springer International Publishing, Cham, pp. 59–77.

Chatterjee, S., Patel, P.P. (2016) Quantifying landscape structure and ecological risk analysis in Subarnarekha sub-watershed, Ranchi. In: Mondol, D.K. (ed.) *Application of Geospatial Technology for Sustainable Development, University of North Bengal, India*, North Bengal University Press, Raja Rammohunpur, pp. 54–76.

Clarke, J.L. (1966) Morphometry from Maps. In: Dury, G.H. (ed.), *Essays in Geomorphology*, Heinemann, New York, pp. 235–274.

Das, K., Bandyopadhyay, S. (1995) Badland development over laterite duricrusts. In: Jog, S.R. (ed.), *Indian Geomorphology, Volume I: Erosional Landforms and Processes*, Rawat Publications, New Delhi, pp. 31–41.

Das, S., Patel, P.P., Sengupta, S. (2016) Evaluation of different digital elevation models for analyzing drainage morphometric parameters in a mountainous terrain: a case study of the Supin–Upper Tons Basin, Indian Himalayas. *SpringerPlus*, 5, 1544. doi: 10.1186/s40064-016-3207-0.

Dasgupta, R., Patel, P.P. (2017) Examining the physical and human dichotomy in geography: existing divisions and possible mergers in pedagogic outlooks. *Geographical Research*, 55(1), 100–120. doi: 10.1111/1745-5871.12220.

De Martonne, E. (1934) Les régions arides du nord argentin et chilien. *Bulletin de Association de Géographes Francais*, 79, 58–62.

De Smet, R. (1951) Principles elementaires de morphometrie. *Rev. Cercle des Sciences*, 1, 13–16.

Dufour, S., Rinaldi, M., Piegay, H., Michalon, A. (2015) How do river dynamics and human influences affect the landscape pattern of fluvial corridors? Lessons from the Marga River, Central-Northern Italy. *Landscape and Urban Planning*, 134, 107–118. doi: 10.1016/j.landurbplan.2014.10.007.

Ghosh, S., Guchhait, S.K. (2015) Characterization and evolution of primary and secondary laterites in north-western Bengal Basin, West Bengal, India. *Journal of Palaeogeography*, 4(2), 203–230. doi: 10.3724/SP.J.1261.2015.00074.

Ghosh, S., Guchhait, S.K. (2019) *Laterites of the Bengal Basin: Characterization, Geochronology and Evolution.* Springer Nature, Cham.

Horton, R.E. (1945) Erosional development of streams and their drainage basins: hydrophysical approach to quantitative morphology. *Bulletin of the Geological Society of America,* 56, 275–370. doi: 10.1130/0016-7606(1945)56[275:EDOSAT]2.0.CO;2.

IIRS (2003a) Biodiversity Characterization at Landscape Level in Andaman and Nicobar Islands India using satellite remote sensing and Geographical Information System. Joint Project Report of Department of Biotechnology and Department of Space, Government of India, Indian Institute of Remote Sensing, Dehradun, India.

IIRS (2003b) Biodiversity Characterization at Landscape Level in Western Ghats India using satellite remote sensing and Geographical Information System. Joint Project Report of Department of Biotechnology and Department of Space, Government of India, Indian Institute of Remote Sensing, Dehradun, India.

IIRS (2003c) Biodiversity Characterization at Landscape Level in North East India using satellite remote sensing and Geographical Information System. Joint Project Report of Department of Biotechnology and Department of Space, Government of India, Indian Institute of Remote Sensing, Dehradun, India.

IIRS (2003d) Biodiversity Characterization at Landscape Level in Western Himalayas India using satellite remote sensing and Geographical Information System. Joint Project Report of Department of Biotechnology and Department of Space, Government of India, Indian Institute of Remote Sensing, Dehradun, India.

IIRS (2011) Biodiversity Characterization at Landscape Level in Northern Plains using satellite remote sensing and Geographical Information System. Joint Project Report of Department of Biotechnology and Department of Space, Government of India, Indian Institute of Remote Sensing, Dehradun, India.

Jain, V., Tandon, S.K., Sinha, R. (2012) Application of modern geomorphic concepts for understanding the spatio-temporal complexity of the large Ganga River dispersal system. *Current Science,* 103(11), 1300–1316.

Kar, A., Bandyopadhyay, M.K. (1974) Mechanism of rills: an investigation in microgeomorphology. *Geographical Review of India,* 36, 204–215.

Lewin, J., Ashworth, P.J. (2014) Defining large river channel patterns: alluvial exchange and plurality. *Geomorphology,* 215, 83–98. doi: 10.1016/j.geomorph.2013.02.024.

Marren, P.M., Grove, J.R., Webb, J.A., Stewardson, M.J. (2014) The potential for dams to impact lowland meandering river floodplain geomorphology. *The Scientific World Journal,* 1–24. doi: 10.1155/2014/309673.

Mondal, S., Patel, P.P. (2018) Examining the utility of river restoration approaches for flood mitigation and channel stability enhancement: a recent review. *Environmental Earth Sciences,* 77, 195. doi: 10.1007/s12665-018-7381-y.

Mondal, S., Patel, P.P. (2020) Implementing Vetiver grass-based riverbank protection programmes in rural West Bengal, India. *Natural Hazards,* https://doi.org/10.1007/s11069-020-04025-5

Mondal, S., Sarkar, A., Patel, P.P. (2016) Causes of drainage congestion in the Moyna Block, Purba Medinipur, West Bengal. In: Mondol, D.K. (ed.) *Application of Geospatial Technology for Sustainable Development,* University of North Bengal, India, North Bengal University Press, Raja Rammohunpur, pp. 1–9.

Pande, A. (2019) Appraisal of geomorphic diversity with special reference to basin-area extremity in central-lesser Himalaya. *Journal of the Geological Society of India,* 94, 375–386. doi: 10.1007/s12594-019-1325-3.

Patel, P.P. (2012) An exploratory geomorphological analysis using modern techniques for sustainable development of the Dulung river basin. Unpublished PhD Thesis, University of Calcutta, Kolkata. Available at https://sg.inflibnet.ac. in/handle/10603/156681.

Patel, P.P. (2013) GIS techniques for landscape analysis – case study of the Chel River Basin, West Bengal. Proceedings of State Level Seminar on Geographical Methods in the Appraisal of Landscape, held at Dept. of Geography, Dum Dum MotijheelMahavidyalaya, Kolkata, on 20th March, 2012, pp. 1–14.

Patel, P.P., Dasgupta, R. (2009) Flood induced land use change in the Dulung River Valley, West Bengal. In: Singh, R.B., Roy, S.D.D., Samuel, H.D.D.K., Singh, V.D., Biji, G.D. (eds.) *Geoinformatics for Monitoring and Modelling Land-Use, Bio-diversity and Climate Change – Contribution Towards International Year of Planet Earth,* Vol. 1, NMCC Publication, Marthandam, pp. 103–123.

Patel, P.P., Dasgupta, R., Mondal, S. (2020) Using ground-based photogrammetry for fine-scale gully morphology studies: some examples. In: Shit, P.K., Pourghasemi, H.R., Sankar, G. (eds.) *Gully Erosion Studies from India and Surrounding Regions,* Springer International Publishing, Cham, pp. 207–220. doi: 10.1007/978-3-030-23243-6_12.

Patel, P.P., Mondal, S. (2019) Terrain – landuse relation in Garbeta-I block, Paschim Medinipur District, West Bengal. In: Mukherjee, S. (ed.) *Importance and Utilities of GIS,* Avenel Press, Burdwan, pp. 82–10. ISBN: 978-81-939776-6-8.

Patel, P.P., Sarkar, A. (2009) Application of SRTM data in evaluating the morphometric attributes: a case study of the Dulung River Basin. *Practicing Geographer*, 13(2), 249–265.

Patel, P.P., Sarkar, A. (2010) Terrain characterization using SRTM data. *Journal of the Indian Society of Remote Sensing*, 38(1), 11–24. doi: 10.1007/s12524-010-0008-8.

Pike, R.J., Wilson, S.E. (1971) Elevation – relief ratio, hypsometric integral and geomorphic area – altitude analysis. *Bulletin of the Geological Society of America*, 82, 1079–1084. doi: 10.1130/0016-7606(1971)82[1079:ERHIAG]2.0.CO;2.

Reid, H.E., Gregory, C.E., Brierley, G.J. (2008) Measures of physical heterogeneity in appraisal of geomorphic river condition for urban streams: twins stream catchment, Auckland, New Zealand. *Physical Geography*, 29(3), 247–274. doi: 10.2747/0272–3646.29.3.247.

Roy, N.G., Sinha, R. (2017) Integrating channel form and processes in the Gangetic plains rivers: implications for geomorphic diversity. *Geomorphology*, 302, 46–61. doi: 10.1016/j.geomorph.2017.09.031.

Roy, P.S., Behera, M.D., Murthy, M.S.R., Roy, A., Singh, S. et al. (2015a) New vegetation type map of India prepared using satellite remote sensing: comparison with global vegetation maps and utilities. *International Journal of Applied Earth Observation and Geoinformation*, 39, 142–159. doi: 10.1016/j.jag.2015.03.003.

Roy, P.S., Kushwaha, S.P.S., Murthy, M.S.R., Roy, A., Kushwaha, D., Reddy, C.S. et al. (2012) *Biodiversity Characterization at Landscape Level: National Assessment*. Indian Institute of Remote Sensing, ISRO, Dehradun. ISBN: 81-901418-8-0.

Roy, P.S., Roy, A., Joshi, P.K., Kale, M.P., Srivastava, V.K., Srivastava, S.K. et al. (2015b) Development of decadal (1985–1995–2005) land use and land cover database for India. *Remote Sensing*, 7, 2401–2430. doi: 10.3390/rs70302401.

Sahana, M., Patel, P.P. (2019) A comparison of frequency ratio and fuzzy logic models for flood susceptibility assessment of the lower Kosi River Basin in India. *Environmental Earth Sciences*, 78, 289. doi: 10.1007/s12665-019-8285-1.

Sarkar, A., Patel, P.P. (2011) Topographic analysis of the Dulung R. Basin. *The Indian Journal of Spatial Science*, 2. Article 2.

Sarkar, A., Patel, P.P. (2012) Terrain classification of the Dulung Drainage Basin. *The Indian Journal of Spatial Science*, 2, 1–8.

Sarkar, A., Patel, P.P. (2016) Land use – terrain correlations in the piedmont tract of Eastern India: a case study of the Dulung River Basin. In: Santra, A., Mitra, S. (eds.) *Handbook of Research on Remote Sensing Applications in Earth and Environmental Studies*, IGI Global, Hershey, pp. 147–193. doi: 10.4018/978-1-5225-1814-3.ch008.

Sen, J., Sen, S., Bandyopadhyay, S. (2004) Geomorphological investigation of badlands: a case study at Garhbeta, West Medinipur District, West Bengal, India. In: Singh, S., Sharma, H.S., De, S.K. (eds.) *Geomorphology and Environment*, ACB Publication, Kolkata, pp. 204–234.

Shit, P.K., Bhunia, G.S., Maiti, R. (2013) Assessment of factors affecting ephemeral gully development in badland topography: a case study at GarhbetaBadland (PaschimMedinipur). *International Journal of Geosciences*, 4(2), 461. doi: 10.4236/ijg.2013.42043.

Shit, P.K., Bhunia, G.S., Maiti, R. (2014) Morphology and development of selected badlands in South Bengal (India). *Indian Journal of Geography and Environment*, 13(1), 161–171.

Shit, P.K., Maiti, R. (2014) Gully erosion control-lateritic soil region of West Bengal. *Indian Science Cruiser*, 28(3), 54–61.

Shit, P.K., Nandi, A.S., Bhunia, G.S. (2015a) Soil erosion risk mapping using RUSLE model on Jhargram sub-division at West Bengal in India. *Modeling Earth Systems and Environment*, 1, 28. doi: 10.1007/s40808-015-0032-3.

Shit, P.K., Paira, R., Bhunia, G., Maiti, R. (2015b) Modeling of potential gully erosion hazard using geo-spatial technology at Garbheta block, West Bengal in India. *Modeling Earth Systems and Environment*, 1, 2. doi: 10.1007/s40808-015-0001-x.

Sinha, R., Jain, V., Babu, G.P., Ghosh, S. (2005) Geomorphic characterization and diversity of the fluvial systems of the Gangetic Plains. *Geomorphology*, 70, 207–225. doi: 10.1016/j.geomorph.2005.02.006.

Sinha, R., Mohanta, H., Jain, V., Tandon, S.K. (2017) Geomorphic diversity as a river management tool and its application to the Ganga River, India. *River Research and Applications*, 1–21. doi: 10.1002/rra.3154.

Smith, G.H. (1935) The relative relief of Ohio. *Geographical Review*, 25, 272–284.

Wentworth, C.K. (1930) A simplified method of determining the average slope of land surfaces. *American Journal of Science*, 117, 184–194.

11 Tidal Morphology and Environmental Consequences of Rasulpur River in the Era of Anthropocene

Pravat Kumar Shit, Gouri Sankar Bhunia,
Manojit Bhattacharya, Avijit Kar
and Bidhan Chandra Patra

CONTENTS

11.1 INTRODUCTION

Estuaries are fringed via tidal marshes, which are inimitable environments with very high biomass that adapt the confined hydro-morphodynamic circumstances (Lokhorst et al., 2018; Friedrichs, 2010). Morphological procedures of the fluvial channel are measured by fluid and compact fluxes through hydraulic armies exercised by the tide and sediment conveyance, attrition and aggradation, particularly in flood management aspect. The climatic variations and the man-induced pressure in preceding 100 years have transmuted most of the river hydro-systems (Duțu et al., 2014). Vegetation distresses hydro-morphodynamics in watercourses (Corenblit et al., 2009), and this consequence has been revealed on the scale of discrete tidal marshlands (Temmerman et al., 2007). Tidal wetland improves aggradation both through shorten flow and particle detention, slightly analogous to ensues on river deltas, but tidal marsh is not deliberated a predominantly operative stream as well as bank stabiliser (Lokhorst et al., 2018).

The existing coastline of Digha, West Bengal, is a vibrant nature and present in unbalanced condition (Pitchaikani et al., 2016). It is a very challenging aspect to forecast about the nature of the coastline fluctuating of this specified region. A certain part of the Digha coast has been retrograding, and other sections have been progressing in environmental nature (Pitchaikani and Mukherjee, 2017). The embankments have clogged with the rhythm of tidal sequences that may show ongoing accretionary phases in the adjacent region (Shuai et al., 2017; Dhara and Paul, 2016). Moreover, the native people of the region directly affected by the accumulative siltation rate in the river beds in the past three to four decades (Laha et al., 2014).

Recently, Rasulpur adjoining areas are being experienced by the newer techno-agricultural system from its past trend, more profitable brick manufacturing, fish farming techniques and efforts towards coastal tourism. As a result, the last decade reflects a large scale of land-use change patterns in Rasulpur riverside. In fact, more than 80% of emerging brick fields and 85% of fish farms has been developed between 2001 and 2011 along with the bank of rivers, linked canals encroaching most of the agricultural lands. Now 17.8% and 4.7% of the land cover of the area is used for fish culture and brick manufacturing, respectively, while about 30% of 58% agricultural lands are characterised by the multi-agricultural system. As a result, most of the typical and periodic wetlands are the feeding and breeding ground of all indigenous fish species, and other aquatic lives have loosed its natural existence quickly. Due to this rapid habitat loss; unregulated or illegal killing and collection of those indigenous species; unplanned application of chemical fertilisers, pesticides, fungicides and insecticides; soil, land and water pollution from foresaid activities; competition of those species with other species; evolving newer diseases; and also predation about 36 indigenous fish species have been threatened in different ways. According to global threatened categories of IUCN, in this area, eight species have already extinct, eight species are critically endangered, about ten species are threatened and ten species are vulnerable. This is one of the typical scenarios of loss of fish diversity responsible for the exploitation of local fish resources and also irreversible destruction of entire aquatic ecosystem.

Natural morphological structure of river was partially amended as embankment barricades with respect to regular tidal flow of rivers (Prandle and Lane, 2015). Therefore, the rapid changing of population expansion and succeeding proliferation of resource exploitation destabilise the ecological stability of the natural ecosystem (Mitra, 1984). This fast changing aspect is exercising persistent ecological imprint, wherein stability between management and exploitation has been endangered recurrently (Milligan et al., 2009). Local inhabitants transfigure the creeks into numerous freshwater ponds for their daily usage along with little irrigation tenacities (Yennawar and Tudu, 2014). The marginal people of this region are dependents on the agriculture and fishing practices to a greater extent (Sen et al., 2017). Consequently, the agricultural lands were converted into pisciculture by consenting saline water *via* creeks into the cultivated fields (Endo et al., 2011). This alteration creates the paddy field saline in one side and alternatively reformed the natural flow of creeks. Such variation in long run distressed the symmetry of rivers, as creeks are the imperative element of estuarine geomorphology and BioNetwork. Present investigations provide recommendation for future improving the planning and management supports of river course to sustain the underlying anthropogenic process with considering the ecosystem values and the needs of the users. This study also figures out the threatened indigenous freshwater fish species, and their natural habitat leads to an in-depth analysis of the actual reasons for threatening of fish species related to the changes in LULC patterns.

11.2 THE RIVER AND ITS BASIN

Rasulpur River rises in the southwest of Purba Medinipur district (West Bengal) as a Bagda river, flows in east and southeast directions, and falls into the river Hooghly as shown in Figure 11.1. Geologically, Rasulpur River was a distributary link channel of Subarnarekha River within the Contai coastal plain in West Bengal (India). The river is stressed over the lowland basin between

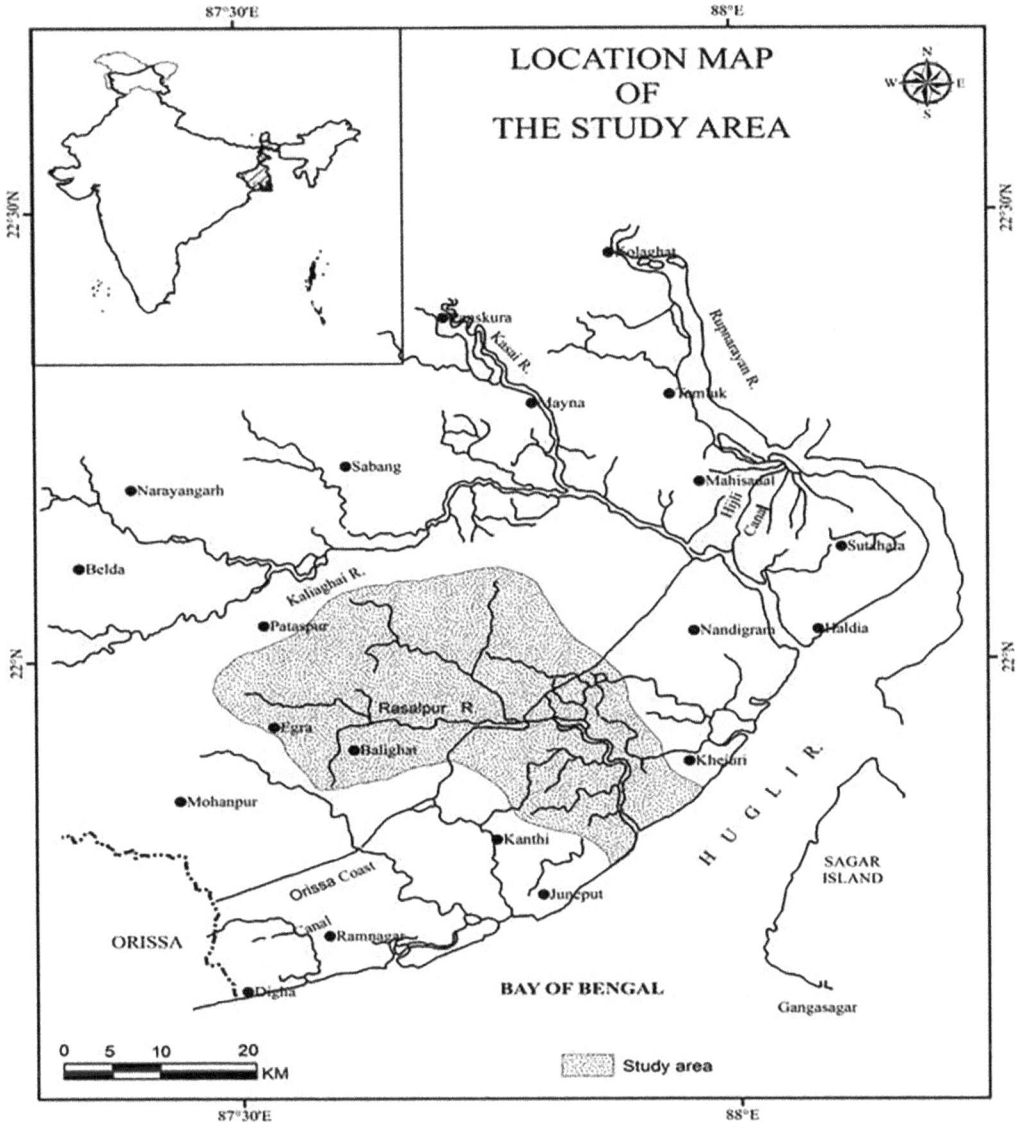

FIGURE 11.1 Catchment area of Rasulpur tidal river.

Pataspur and Bhagwanpur areas with small stream networks (Figure 11.2). Since the last decades, this river is transmuted into a beheaded stream and is not bringing freshwater. Currently, Rasulpur River is now a major tidal pass which used to trap shoreline sediments at the surface by tidal inflows and outflows. The depositional flats and bars extended rapidly around the channel mouth with the interaction of sea waves and tidal waves along with sediment supply in the region. Tidal water spread over the inland basin through the tidal pass from the month of July to November, and the lowland basin around the channel margins are used to develop into marshy tracts with high levels of tidewater and rainwater storage in the river basin floor.

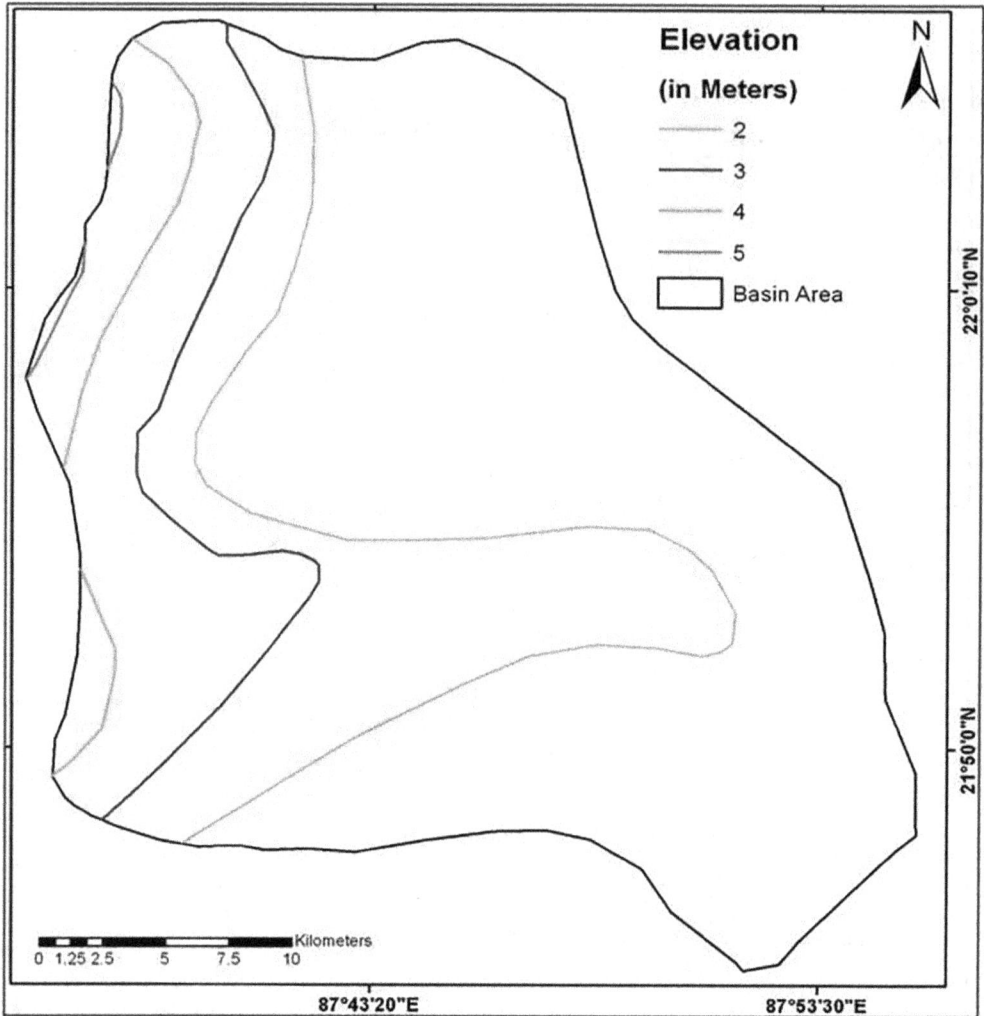

FIGURE 11.2   Contour map of the Rasulpur River basin.

11.2.1   Existed Types of Niches in the Study Area

Since the study area is one of the segments of coastal fluvial basins of West Bengal and included Rasulpur–Pichhaboni watershed, a variety of niches as well as habitats are found. Mainly five types of niches for freshwater indigenous fish species are found:

1. Small domestic ponds: existed throughout the study area.
2. Big ponds: either owned by individuals or few families, maybe as a social resource also, existed throughout the study area.
3. Rain-fed canals/drains/irrigation canals: frequently observed at the roadside position and somewhere connected with main or base rivers or channels. It should be notified that Orissa Coast Canal has passed through Rasulpur Riverside-I CD Block.
4. Land shaping ponds: mainly excavated for agricultural irrigation purpose. The magnitude of such type of niches has been improved through different kinds of governmental schemes and projects during different planning periods.

5. Low-lying inundated paddy fields: observed throughout the study area. Without the above, there are observed much of coastal wetlands, tidal canals, river channels, creeks and basin lowlands on and along the coastal belt. But, those are affected and influenced by saline water and saline water fishes.

11.2.2   MORPHOLOGY OF LANDSCAPE AND TIDAL RIVER PROCESSES

Table 11.1 illustrates the details description of basin morphology. Figures 11.3 and 11.4 represent the morphological characteristics of Rasulpur tidal river. At present, the river is transformed into a beheaded stream and are not supplying freshwater other than the rainwater of catchment area into

TABLE 11.1

Basin Characteristics of Study Area

Area (km²)	896.456
Perimeter (km)	124.0931
Length (km)	44.94
Form factor	0.44388
Circulatory ratio	0.73155
Elongation ratio	0.87895
Laminscate ratio	1.76940

Source: Satellite images.

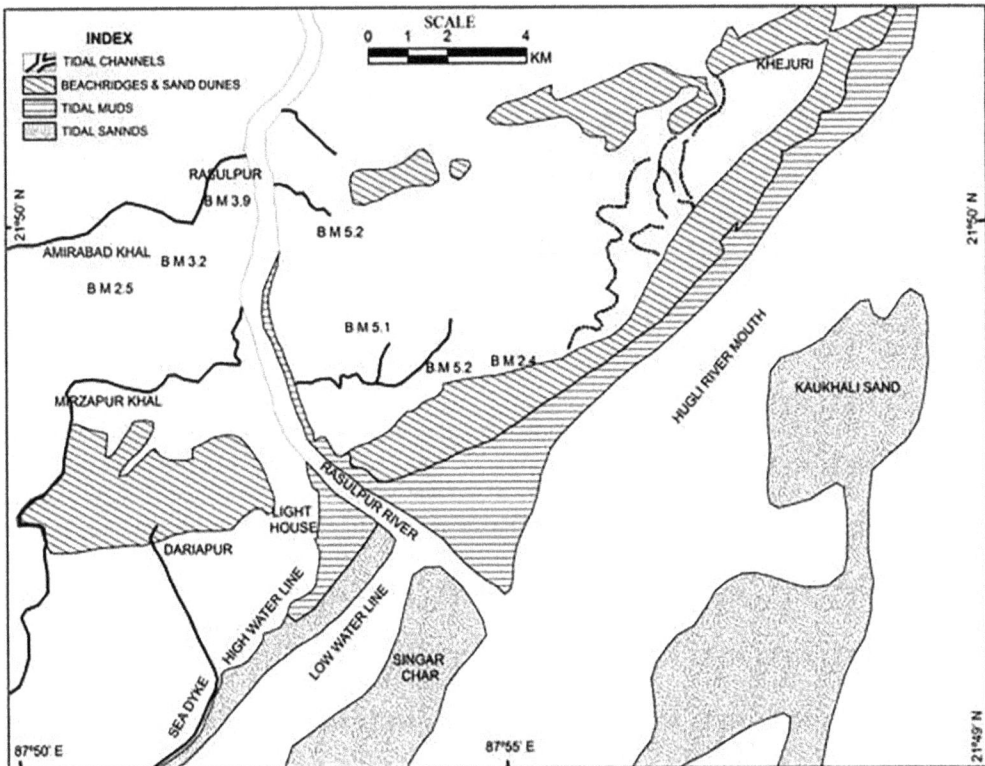

FIGURE 11.3   Morphological landscape of Rasulpur tidal river. (Modified after Paul (2002).)

FIGURE 11.4 Tidal characteristics of Rasulpur River: (a and b) bank erosion with scouring, (c and d) subsidence due to basal erosion, and (e and f) primary wave acts.

the Bay of Bengal. Thus, Rasulpur River is now a major tidal pass which used to trap shoreline sediments at the sea face by tidal inflows and outflows. Depositional flats and bars are extended rapidly around the channel mouth with the interaction of sea waves and tidal waves along with sediment supply in the region, as explored in Table 11.2. Marine sediments are also transported into the channel bed by active tidal bores. Tidewater spread over the inland basin through tidal pass from the month of July to November. This time, the lowland basins around the channel margins are used to develop into a marshy tract with levels of tidewater and rainwater storage in the basinal floor.

Channel margin mudflats are extended along the intertidal region. The lower course is protected from both sides by long earthen embankments down the Hijli Tidal Canal. Table 11.3 explains that there are many feeder channels of tidal flows along the lower course of Rasulpur River to spill the tidal flood waters over the lowland areas. Figure 11.5 shows river courses change from 1987 to 2019. Minor course changes only lower part of basin area due to the tidal bore and anthropogenic activities.

TABLE 11.2

Geo-Environmental Characteristics, Processes and Particle Size Distribution

Sub-Environment	Bio-Geomorphological Characteristics	Sediment Size Distribution (%)		
		Sand	Silt	Clay
Sub-tidal region	Strong tidal currents	24.50	47.40	28.50
Lower mud flat	Abundant bioturbation	28.30	45.70	24.00
Upper mud flat	Algal binding of sediment	24.40	44.20	31.40
Swamp floor	Mangroves and slow accumulation of sediment	30.90	40.10	29.00

Sources: Field survey-2018 December.

TABLE 11.3

Human Intervention of Sediment Distribution and Associated Problems

Zone of Sediment Sinks	Obstruction Caused by Human Activities	Results	Future Risks
Estuary channels	Channelisation with the construction of great levees on the estuaries; guide walls and cross spurs; other tidal river training structures	In-channel siltation, reduced overbank deposition of silt onto swamps, increased velocity, changes in the course	Coastal flooding, erosion, change in the course, inundation
Riverine wetlands	Reclamation and development intervention; fish farm plots with earthen banks	Reduced over bank deposition of silts onto the wetlands and change the salinity conditions of marshland plants	Flooding, inundation
Coastal wetland swamps; salt marshes; mangrove swamps	Reclamation with protective embankments; cultivation on deltaic alluviums	No sediment stockpiling function, loss of drainage capacity	Remain as lowland vulnerable to flooding and inundation

Sources: Field su1rvey-2018 and Paul (2002).

11.3 ANTHROPOGENIC SIGNATURES ON FORMS AND PROCESSES

11.3.1 CHANGES IN LAND USE/LAND COVER

The LULC characteristics of specified study area were generated through supervised classification technique with maximum-likelihood algorithm. This decision rule is reliant upon the likelihood that a pixel belongs to a specific class with the maximum possibility among numerous potentials (Deb and Nathr, 2012). This classification citation reproduces the comprehensive identification conceivable in portraying the LULC and landforms. For each of the LULC classes, a total of five sites were chosen. The LULC map of the study area has been prepared for 1987 as shown in Figure 11.6, 2005 as in Figure 11.7 and 2019 as in Figure 11.8.

The entire basin area has been divided into eight LULC classes, namely, mixed settlement, sand, dense mangrove forest, marsh, cultivated land, agricultural fallow, river and wet sand zones. The areal distribution of LULC characteristics of the river basin is represented in Table 11.4. In 1987, the area of mixed settlement was classified as 83.61 km² (9.31%) which is increased up to 118.89 km² (13.24%). However, the area of mixed settlement in the study area is reflected in an increasing trend,

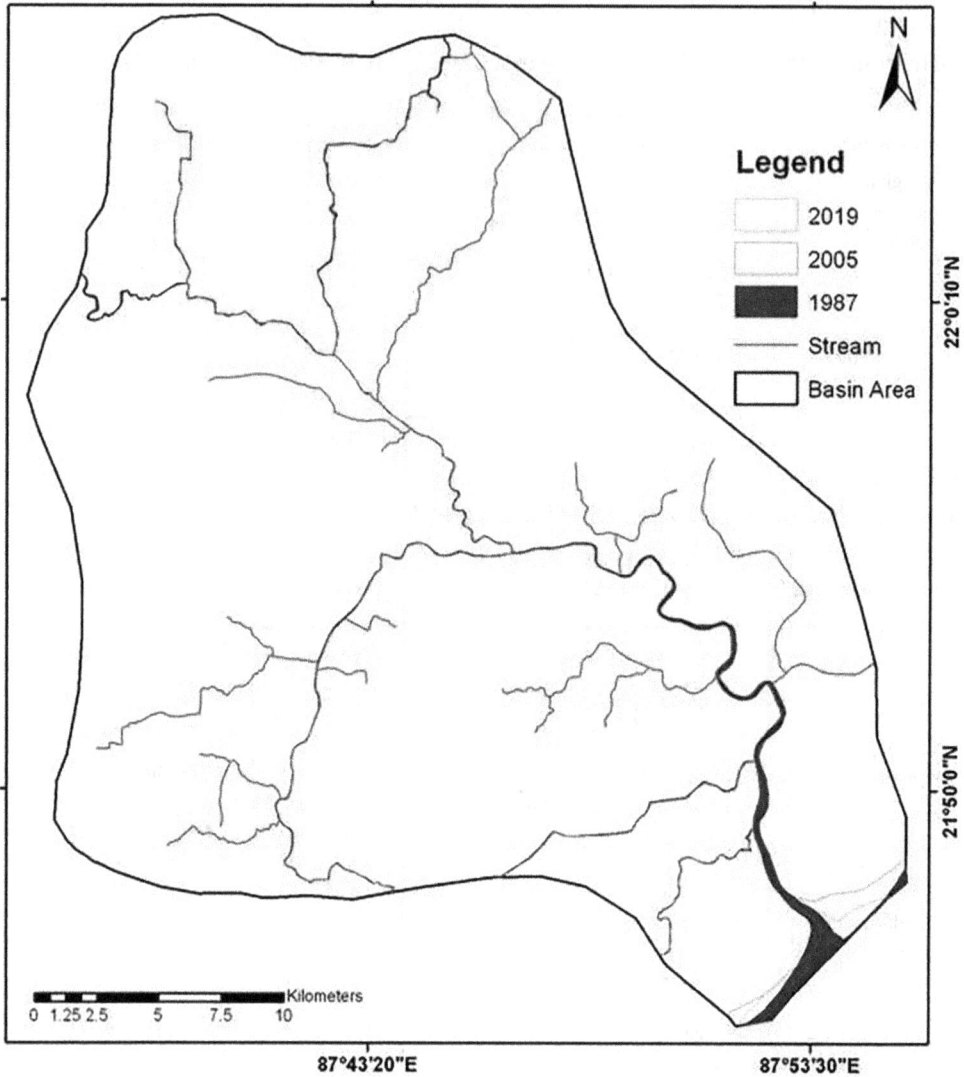

FIGURE 11.5 Rasulpur River basin map.

while the dry sand-covered area in the study site is slightly increasing in trend. In 1978, the classified dry sand was 1.21%, followed by 1.26% in 2005 and 1.35% in 2019. River water in the study area is slightly increasing in trend, covered about approximately 12.95 km^2 (1.44%) in 1987, whereas it has been increased by 17.05 km^2 (1.90%) in 2019. Most of the study area is covered by cultivated land and agricultural fallow. The agricultural fallow land within the study area is decreasing in continuous trend covered by 402.75 km^2 in 1987 and 225.18 km^2 in 2019. The area of cultivated land in the study area is an increasing trend. In 1987, 267.83 km^2 areas covered by cultivated land, whereas it has been increased to 310.07 km^2 area in 2019 (Table 11.5).

11.3.2 Pisciculture Development

A total of 62.57 km^2 areas were delineated as pisciculture zone in Rasulpur River basin derived from the Sentinel-2 data with a spatial resolution of 10 m. To identify the pisciculture zone,

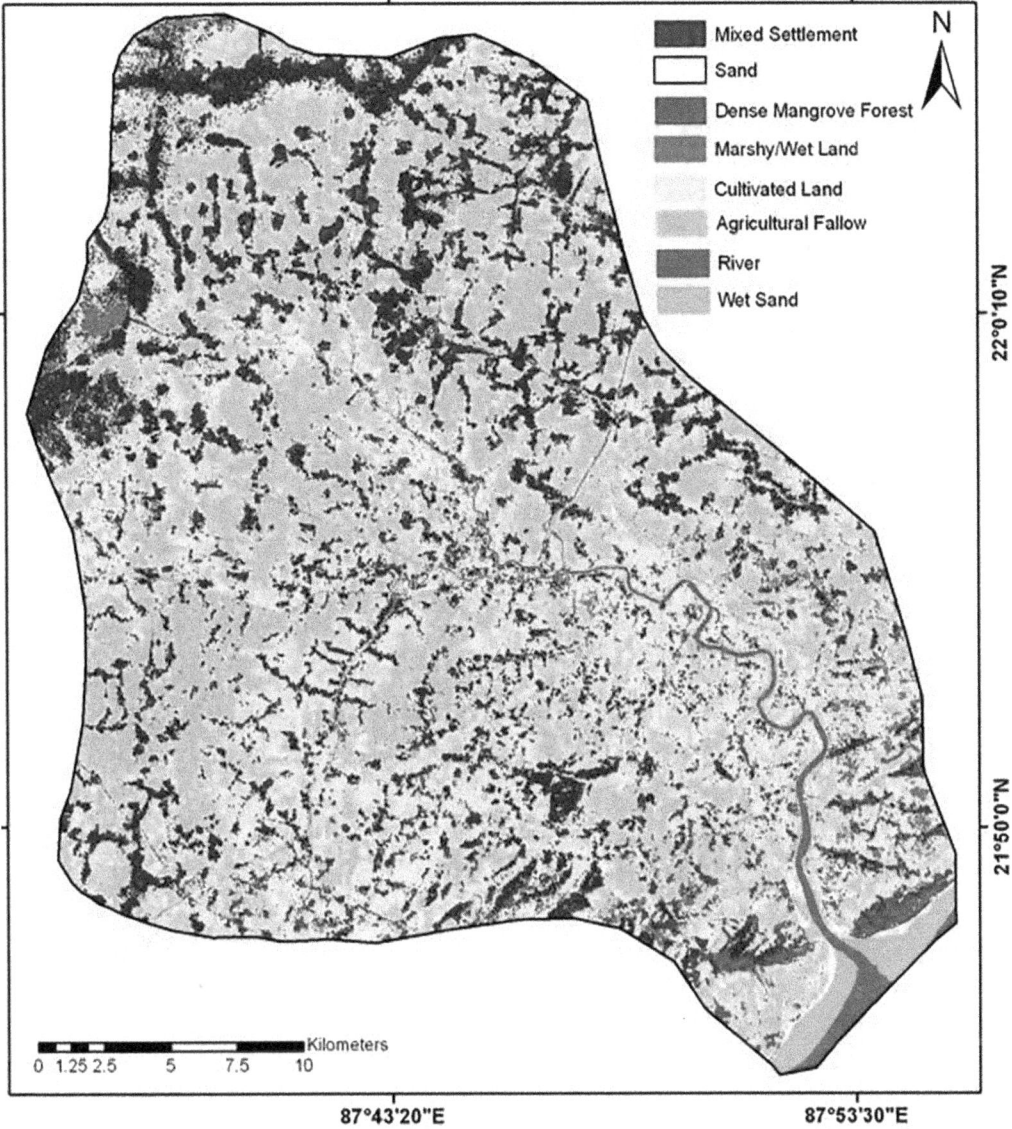

FIGURE 11.6 LULC map of Rasulpur River basin in 1987.

normalised difference water index (NDWI) was used following scientific protocol (Xu, 2006). Fishing cultivation has appeared the most lucrative economic growth of this region (Figures 11.9 and 11.10). A significant amount of agrarian land has been converted into pisciculture at the southern and western ends, wherein river, canal and creeks have been silted promptly. This modification amends the performances of contiguous agricultural lands and impedes the normal flow of the tide. Conversely, various creeks of Rasulpur River have been altered into freshwater ponds and Bricklin Industry, and accordingly, creeks are deteriorating to appear the morphological symmetry.

Though scrutinising the fishing procedures, fisherman consumed transient life in the deep ocean for a week or more, and during this time, they assemble firewood for their everyday culinary from the mangrove forests.

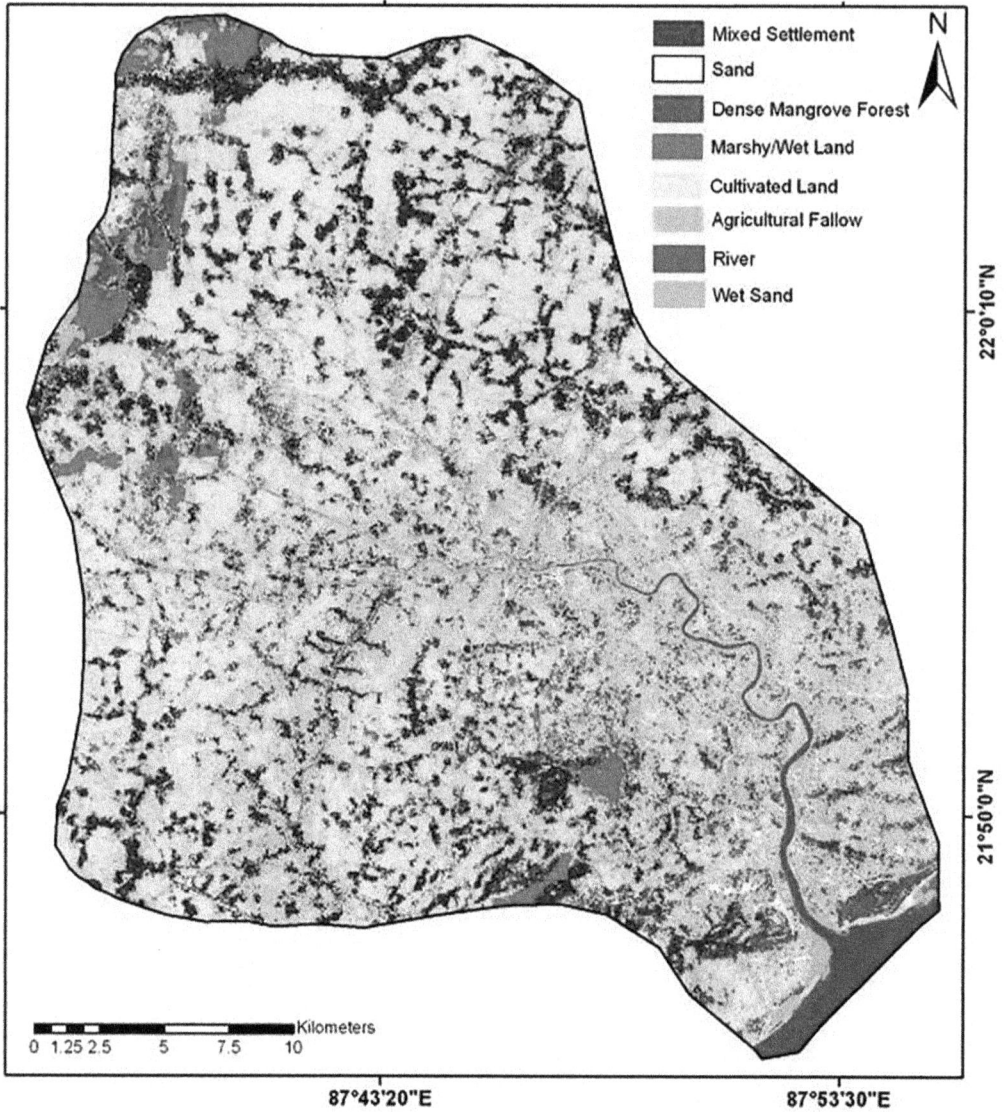

FIGURE 11.7   LULC map of Rasulpur River basin in 2005.

11.3.3 Crisis of Indigenous Freshwater Fish Species

In biogeography, a species is defined as indigenous to a given region or ecosystem if its presence in that region is the result of an only natural process, with no human intervention. The term is equivalent to "native" in less scientific usage. Every natural organism (as opposed to a domesticated organism) has its natural range of distribution where it is regarded as indigenous. Outside this native range, a species may be introduced by human activity; it is then referred to as an introduced species within the regions, whereas it was anthropogenically introduced. Indigenous fish species are found generally in freshwater systems. The indigenous freshwater fish species (IFFS) form a major component of food consumed by families, especially those living closer to freshwater resources. IFFS found in the vast inland water resources provide not only nutrition but also livelihood opportunities and income to many fishers.

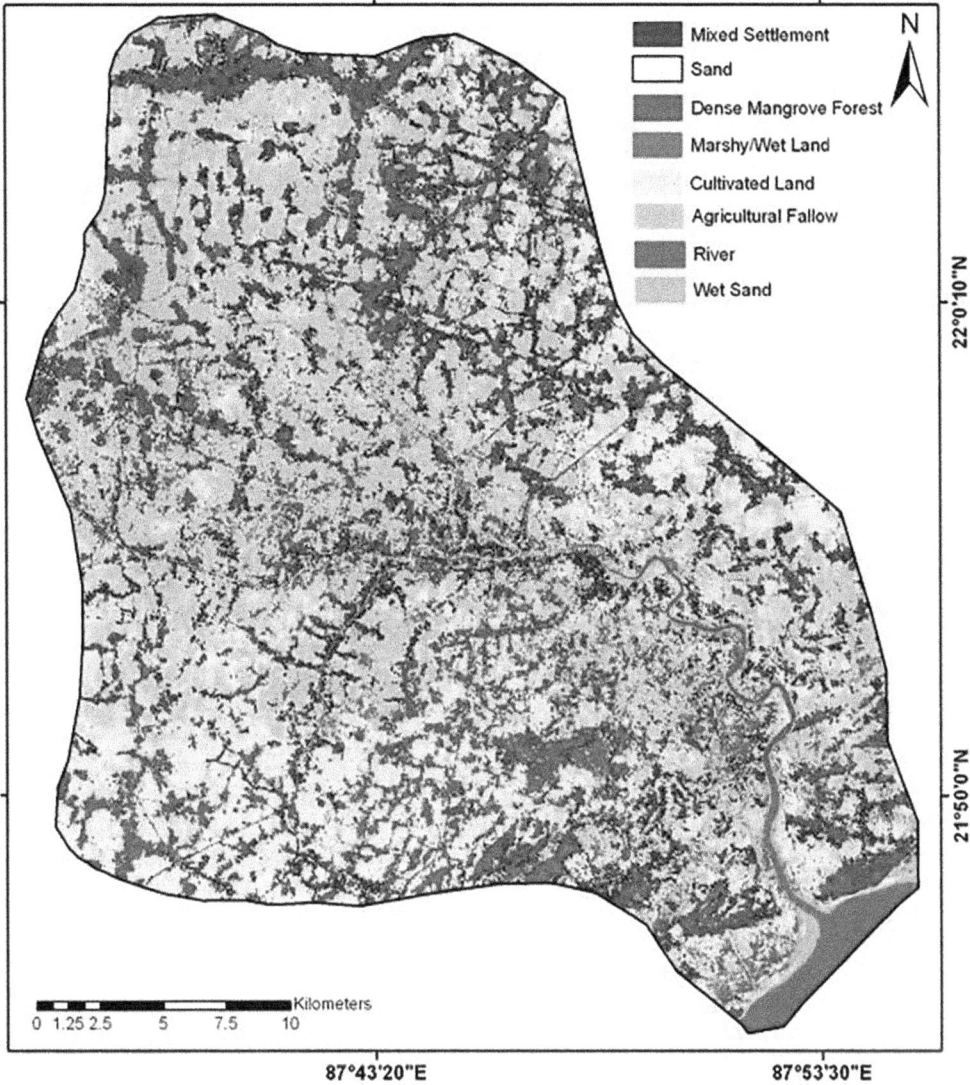

FIGURE 11.8 LULC map of Rasulpur River basin in 2019.

To explain the present status of fish diversity in Rasulpur River and adjoining areas, several intensive questionnaires surveys had been conducted as a source of secondary data from different groups, Fisheries Development Office, Animal Resource Development Office and District Brickfield Association besides the real-time sample collection from different parts of the study areas in regular mode of interval as a primary data source. The present investigation was focused on the ecological meltdown connected to the natural environment. The nutritional values of IFFS are illustrated in Table 11.6 in detail.

11.3.4 PRESENT STATUS OF INDIGENOUS FRESHWATER FISH SPECIES

From the survey and study on the selected area, 114 fish species have been identified and considered to justify the status of IFFS. Among those, 51 (44.7%) were IFFS and 63 (55.3%) were others. Survey as well as this study shows the tremendous result that out of 51 identified and selected

TABLE 11.4

Areal Distribution of LULC Characteristics of Rasulpur River Basin during Period from 1987 to 2019

Land Use/Land Cover Type	1987 (km²)	Percent	2005 (km²)	Percent	2019 (km²)	Percent
Mixed settlement	83.61	9.31	100.75	11.22	118.89	13.24
Sand	10.85	1.21	11.28	1.26	12.13	1.35
River	12.95	1.44	15.04	1.67	17.05	1.90
Wet sand	13.55	1.51	16.09	1.79	20.00	2.23
Dense/mangrove forest	68.08	7.58	120.21	13.38	137.76	15.34
Marshy/wet land	38.60	4.30	55.42	6.17	57.11	6.36
Cultivated land	267.83	29.82	290.34	32.32	310.07	34.52
Agricultural fallow	402.75	44.84	289.09	32.18	225.18	25.07

TABLE 11.5

LULC Change Dynamics of Rasulpur River Basin during Period from 1987 to 2019

LULC Type	1987–2005		2005–2019		1987–2019	
	Areal Change (km²)	Change Rate	Areal Change (km²)	Change Rate	Areal Change (km²)	Change Rate
Mixed settlement	17.13	1.14	18.15	1.00	35.28	2.34
Sand	0.44	0.22	0.85	0.42	1.29	0.66
River	2.09	0.90	2.01	0.74	4.11	1.76
Wet sand	2.54	1.04	3.92	1.35	6.46	2.65
Dense/mangrove forest	52.13	4.25	17.55	0.81	69.68	5.69
Marshy/wet land	16.82	2.42	1.70	0.17	18.51	2.66
Cultivated land	22.52	0.47	19.73	0.38	42.24	0.88
Agricultural fallow	−113.66	−1.57	−63.90	−1.23	−177.56	−2.45

species (Table 11.7), about 70.6% is threatened more or less where only 29.4% is out of danger in recent. According to the IUCN Red Data Book List, all of the threatening species have been divided into different intimidating categories where it is observed that among the threatening species in the study area, Rasulpur riverside, about 33% is under extinct level unluckily, about 28% is going to critically endangered position, about 22% is included endangered category, and approximately 17% is at vulnerable level. The following figures and tables reflect the recent status of major freshwater fish species existed in the study area.

The species diversity of fish indicated a positive association of species richness across the Rasulpur River sites and could be developed by the biodiversity conservation managers for prioritisation of preservation sites and habitat reinstatement. During the study period, it was observed that the variety of species were high in accordance to with the extent of water body. Most of the fishes were collected from the low-laying area. Shaikh et al. (2011) also suggested that low-lying areas are suitable for the freshwater fish diversity. Table 11.8 shows the abundance of fish species in relation to physio-chemical parameters of water. The maximum level of water is decreased during summer due to hot air and water temperature; therefore, most of the fishes are migrated towards lowland for survival. Moreover, fish species were abundant during winter due to the preference of clear water, maximum amount of phytoplankton and zooplankton. Present results also corroborated with the earlier study (Prakash et al., 2007). Water pH is also an important indicator of overall productivity that causes habitat diversity. In the study area,

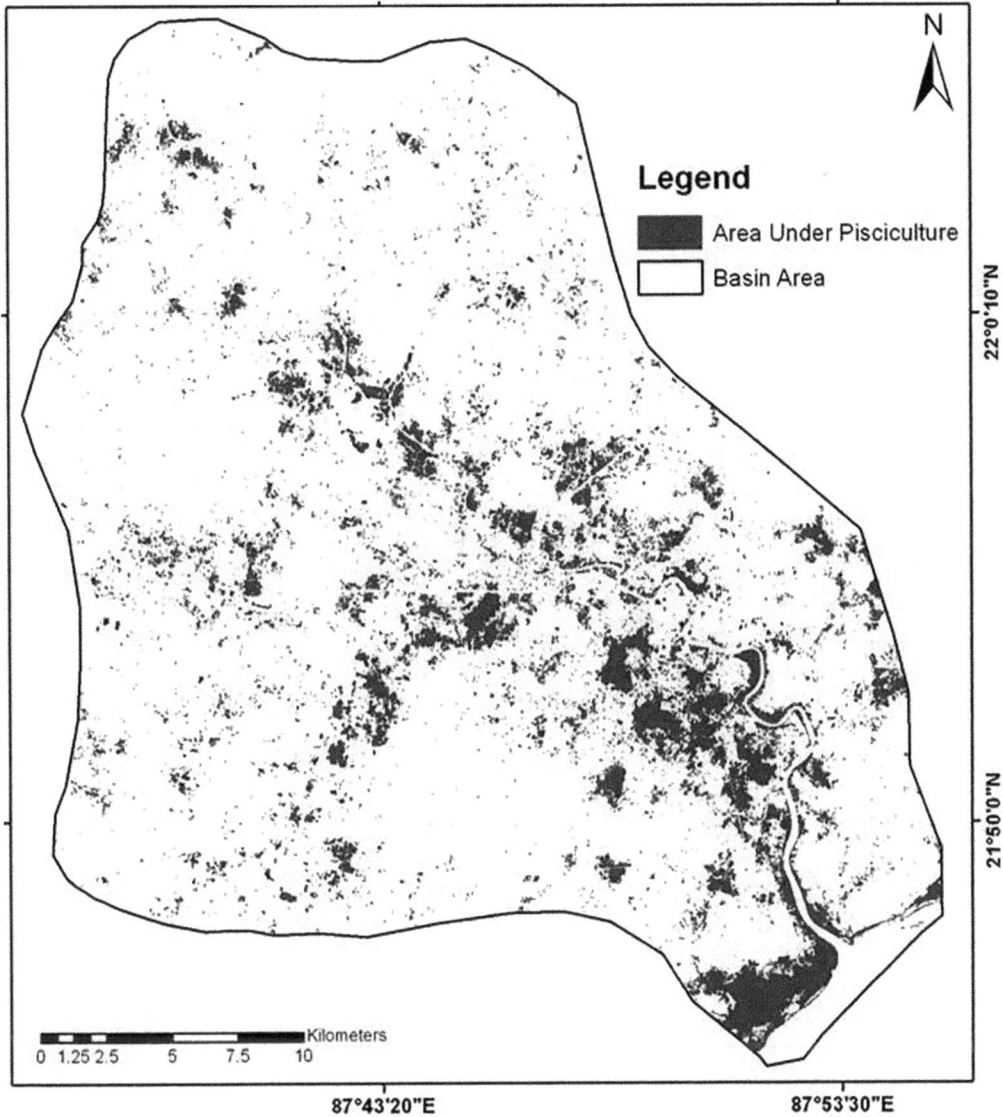

FIGURE 11.9 Spatial distribution of pisciculture zone in Rasulpur River basin derived from Sentinel-2 Satellite data.

pH value is varied between 6.1 and 7.8 and the seasonal mean pH was close to neutral. Goldman and Horne (1983) suggested that pH < 5.0 can sternly decrease aquatic species diversity. Dissolve oxygen (DO) is a noteworthy sign of water quality, and turn down of DO has a severe insinuation of aquatic organisation. This permits admittance to the energy that fuels aerobic tricks. DO value was more than 5 mg l^{-1} at all the study locations. OD value was also calculated to measure the degree of water pollution and useful for evaluating the self-purification the capacity of a water body. The OD value was slightly higher in summer season due to higher microbial activity and temperature, whereas it was low in the winter season. Turbidity values were slightly higher in the study area that may be due to severe bank erosion. However, the study reveals that physio-chemical characteristics and habitat variables influence fish assemblage structuring in the river Rasulpur (Table 11.8).

FIGURE 11.10   Pisciculture development along the Rasulpur River basin.

11.3.5   DEVELOPMENT OF BRICK INDUSTRY IN THE RASULPUR RIVER BASIN

In the study area, a total of 39 brick fields were demarcated (Figure 11.11). Most of the brick fields were located on the left side of the river bank. Before 2004, only twelve brick fields were established in the whole basin. Twenty-seven new brick fields were constructed between 2005 and 2018 (Figure 11.12). The condition of the brick fields can be well understood with the aid of the area they inhabit, and labour and raw materials used for increasing their production levels. The area of the brick fields within the basin areas varies between 0.0015 and 0.18 km². Out of the 39 brick field, 18 brick fields are small in size (i.e., between 0.02 and 0.05 km²). Among them, 13 brick fields are considered as medium sized, ranging between 0.06 and 0.098 km². The remaining eight brick fields are considered as a large brick industry, having an area of 0.10 km² within the basin area. Several factors such as habitat availability, water quality, flow variability and nutrient supplies from riparian habitats organise the profusion and proliferation of river fishes (Pouilly et al., 2006). Distribution of freshwater fish depends on the climate, stream morphology, and biotic and chemical factors.

TABLE 11.6
Nutritional Value of Indigenous Freshwater Fish Species along the Rasulpur River

Fish Species (Per 100 g Raw, Edible Parts)	Scientific Names	Vitamin (mg)	Calcium (mg)	Iron (mg)
	Small Indigenous Species			
Mourala	*Amblypharyngodon mola*	1961	1070	11
Dhela	*Rohtee catio*	935	1260	-
Darkina	*Esomus danricus*	1459	-	
Chanda	*Parambassis* sp.	343	1161	-
Punti	*Puntius* sp.	39	1058	-
	Edible Big Fishes			
Hilsa	*Hilsa hilsa*	72	126	6
Silver carp adult	*Hypophthalmichthys molitrix*	19	268	-
Rohu	*Labeo roheta*	25	319	-
Silver carp juvenile	*H. molitrix*	15	-	-
Tilapia	*Oreochromis niloticus*	21	-	7

TABLE 11.7
Available Fish Species and Their Common Name, IUCN (ver. 2017–2013) Status and Occurrence Sites in Rasulpur River, West Bengal Coast, India

Scientific Name	Common Name	IUCN Status	Site of Occurrence	Availability of Fishes in Different Time Scale		
				Year 1985–1986	Year 2000–2001	Year 2016–2017
Anguilla bengalensis (Gray, 1831)	Bam	NT	U,M,L	++	++	+
Strongylura strongylura (van Hasselt, 1823)	Gar fish	NE	U,M,L	++	+	-
Hyporhamphus limbatus (Valenciennes, 1847)		LC	U,M	++	+	-
Hemiramphus far (Forsskål, 1775)	Bak	NE	U,M,L	+++	+++	++
Anodontostoma chacunda (Hamilton, 1822)	Koi puti	NE	U,M,	+++	++	++
Tenualosa ilisha (Hamilton, 1822)	Hilsa	LC	L	+++	++	+
Tenualosa toli (Valenciennes, 1847)	Kokila	NE	L	++	+	+
Gudusia chapra (Hamilton, 1822)		LC	M,L	+++	++	++
Chirocentrus dorab (Forsskål, 1775)	Khanda	NE	M,L	+++	+++	+++
Coilia ramcarati (Hamilton, 1822)	Tapertail anchovy	NE	M,L	+++	++	+
Coilia reynaldi (Valenciennes, 1848)	Amadi	NE	L	+++	++	+
Coilia neglecta (Whitehead, 1967)		LC	M,L	++	+++	+++

(Continued)

TABLE 11.7 (*Continued*)

Available Fish Species and Their Common Name, IUCN (ver. 2017–2013) Status and Occurrence Sites in Rasulpur River, West Bengal Coast, India

Scientific Name	Common Name	IUCN Status	Site of Occurrence	Year 1985–1986	Year 2000–2001	Year 2016–2017
Coilia dussumieri (Valenciennes, 1848)	Amude	NE	L	+++	++	+
Setipinna phasa (Hamilton, 1822)	Phasa	LC	U,M,L	+++	+	++
Setipinna taty (Valenciennes, 1848)	Phasa	NE	U,M	+	+	-
Chela laubuca (Hamilton,1822)	Dankena	NE	U,M	+++	+++	++
Systomus sarana (Hamilton, 1822)	Sar Punti	LC	U,M	++	+++	+
Puntius sophore (Hamilton, 1822)	Deshi Punti	LC	U,M	+++	+	++
Esomus danrica (Hamilton, 1822)	Darikhana	LC	U,M	+++	++	+
Pethia ticto (Hamilton, 1822)	Tit Punti	LC	U,M	+++	+++	+++
Salmostoma bacaila (Hamilton, 1822)		LC	M,L	+++	++	++
Labeo calbasu (Hamilton, 1822)	Kalbose	LC	U	++	++	-
Labeo bata (Hamilton, 1822)	Bata	LC	U	+++	+	-
Labeo rohita (Hamilton, 1822)	Rui	LC	U,M	+++	++	+
Cirrhinus mrigala (Hamilton, 1822)	Mrigal	LC	U	++	+	+
Labeo catla (Hamilton, 1822)	Catla	LC	U	++	+	-
Danio rerio (Hamilton, 1822)		LC	U,M	+++	+++	+++
Lepidocephalichthys guntea (Hamilton, 1822)	Gunte	LC	U	+++	+	+
Aplocheilus panchax (Hamilton, 1822)	Trichokha	LC	U,M,L	+++	+++	+++
Chelon parsia (Hamilton, 1822)	Goldspot mullet	NE	L	+++	++	++
Moolgarda seheli (Forsskål, 1775)	Grey mullet	NE	M,L	+++	++	+
Rhinomugil corsula (Hamilton, 1822)	Corsula mullet	LC	L	+++	++	+
Mugil cephalus (Linnaeus, 1758)	Ain	LC	L	++	+	+
Drepane punctata (Linnaeus, 1758)	Moonfish	LC	U,M,L	++	++	+
Glossogobius giuris (Hamilton, 1822)	Balia	LC	U,M,L	++	+	+
Pseudapocryptes elongates (Cuvier, 1816)	Chewa	NE	U,M	++	++	+
Brachygobius nunus (Hamilton, 1822)	Nona Bele	NE	U	++	+	+
Lates calcarifer (Bloch, 1790)	Bhekti	NE	M,L	+	+	-
Lutjanus russelli (Bleeker, 1849)		NE	M,L	++	+	+
Mene maculata (Bloch & Schneider, 18-1)	Chanda	NE	L	+++	+	+

(Continued)

TABLE 11.7 (*Continued*)
Available Fish Species and Their Common Name, IUCN (ver. 2017–2013) Status and Occurrence Sites in Rasulpur River, West Bengal Coast, India

Scientific Name	Common Name	IUCN Status	Site of Occurrence	Availability of Fishes in Different Time Scale		
				Year 1985–1986	Year 2000–2001	Year 2016–2017
Monodactylus argenteus (Linnaeus, 1758)	Chanda	NE	L	+++	++	++
Polynemus paradiseus (Linnaeus, 1758)		NE	L	+++	++	+
Johnius amblycephalus (Bleeker, 1855)		NE	M,L	++	++	+
Johnius coitor (Hamilton, 1822)		LC	L	+++	+++	+++
Macrospinosa cuja (Hamilton, 1822)	Cuja bola	NE	U,M,L	+++	++	+
Pampus argenteus (Euphrasen, 1788)	Pomfret	NE	L	++	+	-
Terapon jarbua (Forsskål, 1775)		LC	U,M,L	+++	++	+
Arius arius (Hamilton, 1822)		LC	U,M,L	++	+	+
Mystus gulio (Hamilton, 1822)		LC	U,M	+++	++	+
Sperata seenghala (Sykes, 1839)		LC	U,M	++	++	+
Pangasius pangasius (Hamilton, 1822)	Pungas	LC	U,M	++	+	+++
Ompok pabda (Hamilton,1822)	Pabda	NT	U,M,L	+++	+	+
Ompok bimaculatus (Bloch,1787)	Pabda	NT	U,M,L	+++	+	-
Wallago attu (Schneider, 1801)	Boal	NT	U,M,L	+++	-	-
Platycephalus indicus (Linnaeus, 1758)	Bartail flathead	DD	U,M,L	+++	++	+
Oreochromis niloticus (Linnaeus, 1758)	Tilapia	LC	U	+	+	+++

Habitat destruction and environmental dilapidation have gravely affected the fish in the study area. However, linkages with physiological and behavioural characteristics are poorly understood for freshwater fishes (Tongnunui et al., 2016). Most of the species were below average in abundance and are related to the physical habitat and some chemical factors, such as DO, pH and alkalinity (Suvarnaraksha et al., 2012). Table 11.9 shows the main responsible causes for threatening of fish species in the Rasulpur River basin.

11.3.6 OVERALL EFFECTS ON ENVIRONMENTAL GEOMORPHOLOGY

Man-induced strife is poses a severe risk to the functioning of estuarine ecosystem. Natural morphological system of rivers was partially modified as embankment prevents the natural tidal flow of river (Danda, 2007; Bandyopadhyay, 2000). Fluvial geomorphic systems lean to sustain a dynamic stability among the ambient sediment load, water discharge and channel geometry (Al Masud et al., 2018). Sediment grain size, channel pattern and stream gradient may regulate along the intangible

TABLE 11.8

Abundance of Fish Species in Relation to the Water Parameters in the Study Areas

	PH	Temp	Salinity	TDS	Turbidity	Cond.	OD	DO	Species Abundance
PH	1								
Temp	0.61055	1							
Salinity	0.59191	0.75637	1						
TDS	−0.2337	0.2522	0.08539	1					
Turbidity	−0.0681	0.10409	0.13502	0.13773	1				
Cond.	−0.1651	0.16445	0.05666	0.78015	0.03296	1			
OD	0.26069	−0.1621	0.13271	−0.0348	0.09864	0.27923	1		
DO	0.07653	−0.1185	−0.2159	−0.0003	−0.2773	−0.1378	−0.1037	1	
Species abundance	0.07109	−0.1967	−0.2036	−0.0652	−0.3933	−0.0174	0.06115	0.70329	1

gradient to preserve by equilibrium circumstances. Since the 20th century, LULC has been thoroughly modified and afterwards distress the morphological in addition to the estuarine ecosystem of the region. Tidal river floodplain system and its ecological environment and human well-being thereafter are closely interlinked which has been aggravated by structural intervention processes (Hoitink and Jay, 2016). The appropriate interface between land and water for assigning rights to use land and water resources concerned with protection crops, settlements, industrial development and other resources makes coastal people with high magnitude and other income generating activities. All these facts axiomatically bear out the constant environmental dilapidation of the region in the form of fishery production at the mouth of Rasulpur River and north of the river basin and Brick kiln Industry growth at the south. After all, such extreme developments will absolutely the normal accessory of man-nature reciprocal rapport.

Likewise, fishes from upper freshwater zone, e.g. *Salmostoma bacaila*, *Wallago attu*, *Glossogobius giuris*, and *Xenentodon cancila*, also get entrapped in this deeper pool. This fairly long and deeper area supports a commercial fishery substantially, all through the year including summer season. *Hyporhamphus limbatus* is a coastal fish species that enter into the estuary and sometimes even reaching transitional zone. Reportedly, it may even be found in the freshwater area of some estuaries. During the present survey, this species of half-beak was, however, not encountered in the freshwater zone of the Rasulpur River part. All the members of order Cypriniformes are restricted to the freshwater stretch from upper division of the Rasulpur riverside to about 5–6 km downstream ways. *Notopterus notopterus*, *Chitala chitala*, *Sperata aor*, *Wallago attu*, *Ompok pabda*, *Gagata cenia*, *Mystus vittatus*, *Eutropichthys vacha*, *Pangasius pangasius*, *Ailia coila*, etc. were also found in the freshwater zone. Although *G. giuris* demonstrated a wide range of distribution, *P. lanceolatus* was encountered only from low saline to transitional zone. *Lates calcarifer* and *Tenualosa ilisha*, though not frequent, are found in the transitional zone. Differences in the salinity regime in different areas influence habitat conditions and therefore the fish community. Although salinity is considered a prime factor, there may be other important factors that may operate to influence the fish diversity, e.g. silt deposition, thermal variation in the estuarine ecosystem, irrational fishing activities, scarcity of freshwater discharge, and the impact of pollutants components and accumulated non-biodegradable substances in a given area. All these may affect species composition either directly or indirectly by affecting fish food abundance of a given area in the estuary. Another important aspect affecting fish diversity is the indiscriminate release/casual entry of fish species not only by way of ranching, but for other reasons as well. Flooding of some areas during monsoon, when

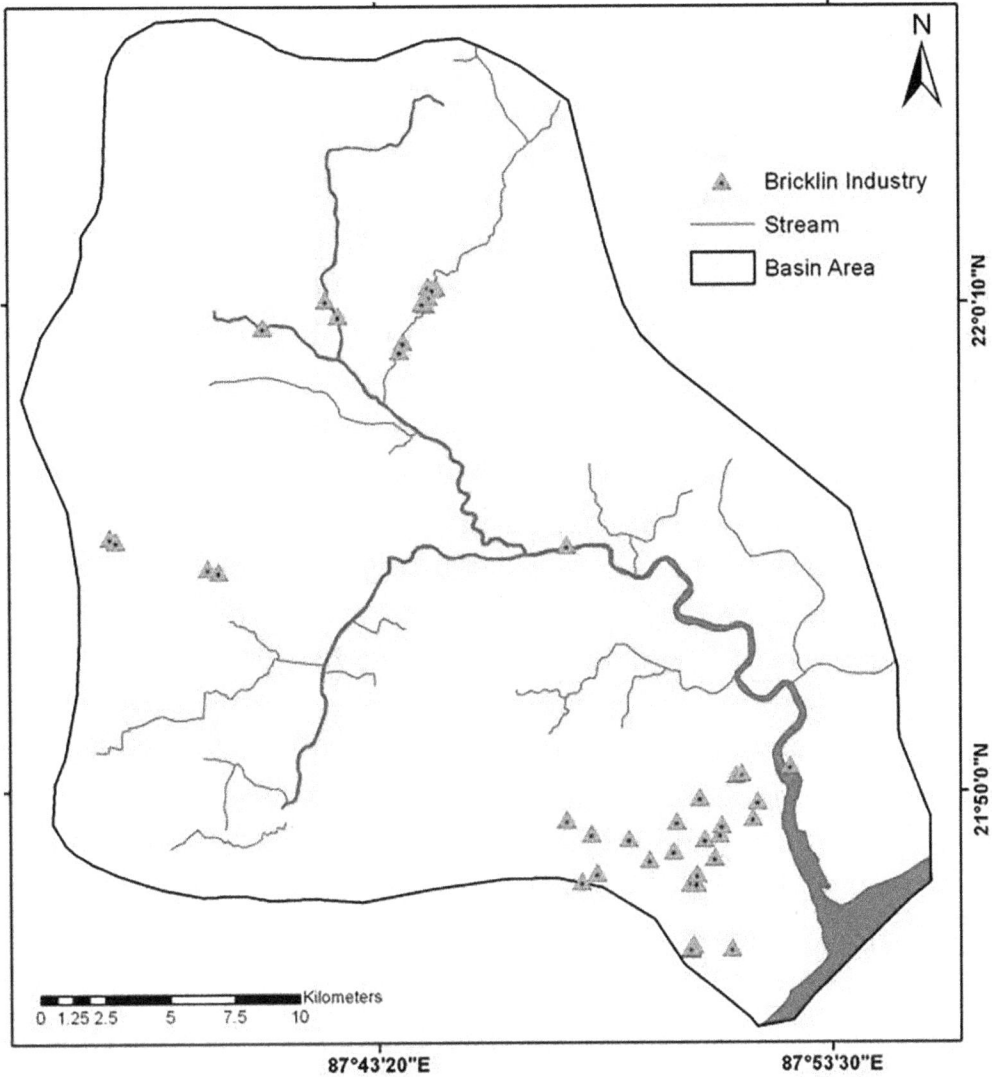

FIGURE 11.11 Spatial distribution of brick industry in Rasulpur River basin.

huge areas alongside the estuary get inundated, may also result in the entry of fish species including exotic ones, altering biodiversity. In Rasulpur River and adjoining areas, silting of the mouth has resulted in large sand bars, mudflats and raised estuary bed, which virtually restricts tidal ingress long enough landwards. This has led to almost freshwater conditions to prevail even during winter and summer months in the upper reaches of the estuary from above Ramnagar–Rasulpur. In an adverse situation like this, the movement of migratory freshwater fish species further upstream is absolutely hindered.

Brick fields are well thought-out as the sign of increased urbanisation with the growing economy. Brick industry is a labour-intensive sector, and a large number of the efficient work forces are needed for this industry (Bandyopadhyay et al., 2013). Approximately, 130–200 labourers are used in major of the brick fields. Most of the brick fields absorb the migrant labour of this basin area (Roy and Kunduri, 2018). However, these brick fields have major impact on the pollution of

FIGURE 11.12   Brick industry in Rasulpur River basin.

the river water. Both total suspend solid and turbidity is high in the Rasulpur river water that may be owing to the existence of minute particles of bricks in addition to the piece of ashes coming from those brick field (Bandyopadhyay et al., 2013). Moreover, the high absorption of chloride and alkalinity in the river water is mainly because of the brick field's effluents and also imminent transportation system within the basin area. In the study area, utilising the river bed as a road for movement of the heavy loaded truck is disturbing the river bed to a greater extent. The recurrent movement of trucks along the river bed makes river bed strangely hard and also impedes the usual flow of the river (Patra et al., 2015). Moreover, the intuitive quarrying of sand from the river bed is increasing unevenness of the river bed. The united effect of sand compilation and deposition of waste bricks, and ashes within the river guides to alter in the channel course by generating pools, ripples and bars along Rasulpur River and its tributaries. Therefore, the government should take some compulsory actions for preventing and scheming all these unempirical and unlawful actions to protect the riverine environment.

TABLE 11.9

Responsible Causes due to Threatening of Indigenous Freshwater Fish Species

Responsible Factors/Causes for Threatening the Species	Sub-Causes
Habitat loss, modification and fragmentation	Settlement Expansion and encroachment of aquatic habitats • Extension and development of commercial construction and activities and capturing the freshwater habitats • Fish Farming and shrimp cultivation replacing the natural habitats • Brick manufacturing covering the natural habitats as well as riverine or roadside wetlands • Double cropped agricultural land expansion loosening the opportunity of seasonal wetlands • Shrinkage of natural wetlands by changing natural process and also uncivilised works of civilised inhabitants • Shrinkage of seasonal wetlands (specifically crop lands) due to different anthropogenic activities • Damaging the river or drainage system by different natural and anthropogenic activities or changes
Habitat pollution and degradation	• Pesticides and pollution from modern agriculture • Pollution from fish farming and shrimp cultivation • Pollution from brick manufacturing • Pollution from settlement or constructing pockets
Unregulated collection and killing	• Over-collection and exploitation for fulfilment the huge food demands for explosive population • Over-collection and exploitation of immature species at their initial and early young stage
Diseases	• Disease from pollution/toxic elements from different sources • Microbial disease
Competition with other species	• Competition with other natural indigenous fish species • Competition with other natural aquatic species • Competition from exotic species • Competition with other artificial/cultivated/imported fish species
Predation	• Intensive subsistence/livelihood predation • Extensive/commercial/economic predation
Declining the growing and breeding space and environment	• Habitat loss due to above causes • Polluting and degrading the environment due to above causes • Changing land use and environment due to both physical and anthropogenic changes • Drastically land use change through the development of the region

11.4 CONCLUSION

The above discussion reveals that some relevant incorporation may be helpful to the future researchers:

- The bathymetric prediction in the estuaries may be done through high-resolution satellite data and ground based observation.
- Sediment transport model with a different fraction of bed layer may be developed to understand the morphological system of the tidal river.
- In future, the hydro-dynamic model may be developed to simulate the flow in the river including tidal zone.

- As such, investigations need to be carried out to understand the impacts of climate change on tidal river system.
- The relationship between species numbers, total abundance of all fish and environmental parameters need to be examined. Sustainable strategies require investigating more fish, consumption and saving fish the community of this riverine ecology.

REFERENCES

Al Masud M, Moni NN, Azad AK, Swarnokar SC. 2018. The impact of tidal river management on livestock in the Ganges-Brahmaputra Basin. *Journal of Dairy & Veterinary Sciences*, 6(5). JDVS.MS.ID.555696.

Bandyopadhyay S. 2000. Coastal changes in the perspective long term evolution of an estuary: Hugli, West Bengal, India. Proceeding International Quaternary Seminar on INQUA shoreline. (pp. 103–115). Indian Ocean Sub Commission.

Bandyopadhyay S, Ghosh K, Saha S, Chakravorti S, De SK. 2013. Status and impact of brick fields on the river Haora, West Tripura. *Transactions*, 35(2), 275–285.

Corenblit D, Steiger J, Gurnell AM, Naiman RJ. 2009. Plants intertwine fluvial landform dynamics with ecological succession and natural selection: a niche construction perspective for riparian systems. *Global Ecology and Biogeography*, 18, 507–520.

Danda AA. 2007. *Surviving in the Sundarbans: Threat and responses.* Enschede: University of Twente.

Deb SK, Nathr RK. 2012. Land use/cover classification- an introduction review and comparison. *Global Journal of Researches in Engineering Civil and Structural Engineering*, 12(1), 5–16.

Dhara S, Paul AK. 2016. Embankment Breaching and its impact on local Community in Indian Sundarban – a case study of some blocks of South West Sundarban. *International Journal of Innovative Science, Engineering & Technology*, 3(2), 23–32.

Duţu LT, Provansalb M, Cozc JL, Duţu F. 2014. Contrasted sediment processes and morphological adjustments in three successive cutoff meanders of the Danube delta. *Geomorphology*, 204, 154–164.

Endo T, Yamamoto S, Larrinaga JA, Fujiyama H, Honna T. 2011. Status and causes of soil salinization of irrigated agricultural lands in Southern Baja California, Mexico. *Applied and Environmental Soil Science*, 2011, 12. Article ID 873625. doi: 10.1155/2011/873625.

Friedrichs CT. 2010. Barotropic tides in channelized estuaries. *Contemporary Issues in Estuarine Physics*, 27–61.

Goldman CR, Horne AJ. 1983. *Limnology.* New York: McGraw-Hill.

Hoitink AJF, Jay DA. 2016. Tidal river dynamics: implications for deltas. *Reviews of Geophysics*, 54. doi: 10.1002/2015RG000507.

Laha AK, Chatterjee S, Bera K. 2014. Flood Hazard cause assessment and their mitigation option using geoinformatics technology. *International Journal of Scientific and Research Publications*, 4(8), 1–7.

Lokhorst IR, Braat L, Leuven JRFW, Baar AW, Oorschot MV, Selaković S, Kleinhans MG. 2018. Morphological effects of vegetation on the tidal–fluvial transition in Holocene estuaries. *Earth Surface Dynamics*, 6, 883–901.

Mitra A. 1984. Rising population and environmental degradation. *Yojana*, 28(18), 4–8.

Milligan SR, Holt WV, Lloyd R. 2009. Impacts of climate change and environmental factors on reproduction and development in wildlife. *Philosophical Transactions of the Royal Society of London. Series B, Biological Sciences*, 364(1534), 3313–3319. doi: 10.1098/rstb.2009.0175.

Paul AK. 2002. Coastal Geomorphology and Environment: Sundarban Coastal Plain, Kanthi Coastal Plain, Subarnarekha Delta Plain. Kolkata: ACB Publications.

Patra P, Guray A, Ganguly S. 2015. A study on Brick Kiln Industry in Pursura Block of Hooghly District, West Bengal. *International Journal of Applied Research*, 1(9), 95–99.

Pitchaikani JS, Kadharsha K, Mukherjee S. 2016 July. Current status of seawater quality in Digha (India), Northwestern coast of the Bay of Bengal. *Environmental Monitoring and Assessment*, 188(7), 385. doi: 10.1007/s10661-016-5383-3.

Pitchaikani JS, Mukherjee S. 2017. A wake up call for protecting Digha coast, West Bengal, Northeast coast of India. *Indian Journal of Geo Marine Sciences*, 46(04), 771–773.

Pouilly M, Barrera S, Rosales C. 2006. Changes of taxonomic and trophic structure of fish assemblages along an environmental gradient in the Upper Beniwatershed (Bolivia). *Journal of Fish Biology*, 68, 37–156

Prakash JW, Asmon J, Regint GS. 2007. Water quality analysis of Thirparappu reservoir, Kanyakumari. *Indian Journal of Environmental Protection*, 27(8), 733–736.

Prandle D, Lane A. 2015. Sensitivity of estuaries to sea level rise: vulnerability Indices Estuarine. *Coastal & Shelf Sciences*, 160, 60–68.

Roy SN, Kunduri E. Migration to brick kilns in India: an appraisal. Available at: http://www.indiaenvironmentportal.org.in/files/file/Migration%20to%20Brick%20Kilns%20in%20India.pdf.

Sen S, Sen G, Chatterjee M. 2017. Coastal vulnerability: a case study along Digha Sankarpur coast West Bengal, India. *Indian Journal of Geo Marine Sciences*, 46(02), 259–265.

Shaikh HM, Kamble SM, Renge AB. 2011. The study of ichthyofauna diversity in Upper Dudhna project water reservoir near Somthana in Jalna district (MS) India. *Journal of Fisheries and Aquaculture*, 2(1), 8–10.

Shuai P, Myers K, Knappett P, Cardenas MB. 2017. Tidal and seasonal river stage fluctuations impact the formation of permeable natural reactive Barriers in Riverbank Sediments. *American Geophysical Union*. Abstract #H12E-03.

Suvarnaraksha A, Lek S, Lek-Ann S, Jutagate T. 2012. Fish diversity and assemblage patterns along the longitudinal gradient of a tropical river in the Indo-Burma hotspot region (Ping-Wang River Basin). *Hydrobiologia*, 694, 153–169

Temmerman S, Bouma T, Van de Koppel J, Van der Wal D, De Vries M, Herman P. 2007. Vegetation causes channel erosion in a tidal landscape. *Geology*, 35, 631–634.

Tongnunui S, Beamish FWH, Kongchaiy C. 2016. Fish species, relative abundances and environmental associations in small rivers of the Mae Klong River basin in Thailand. *Agriculture and Natural Resources*, 50(5), 408–415.

Xu H. 2006. Modification of normalised difference water index (NDWI) to enhance open water features in remotely sensed imagery. *International Journal of Remote Sensing*, 27(14), 3025–3033.

Yennawar P, Tudu P. 2014. Study of macro-benthic (invertebrate) fauna around Digha coast. *Records of the Zoological Survey of India*, 114(Part-2), 341–356.

12 The Jalangi
A Story of Killing of a Dying River

Balai Chandra Das and Soma Bhattacharya

CONTENTS

12.1 INTRODUCTION

The moment when the first able man [Homo habilis, who lived during the Gelasian and early Calabrian stages of the Pleistocene geological epoch (Schrenk et al., 2007)] on the earth surface picked up a piece of stone in his hand and threw it to hit his prey, the first anthropogenic signature was put on the earth forms in the Palaeolithic age (2.4–1.5 million years ago). But that act of picking up a piece of stone and alteration of the natural earth form by the first man is not considered as anthropogenic imprints on landforms or anthropogeomorphological event. Because at that time, man's effort to alter earth's natural form was not only unconscious but also its magnitude was no way bigger than any other animals and it could not give the earth surface any remarkable perpetual change in natural form. His effort was as if a part of natural processes. That change of earth surface by human effort in comparison with the total area of the earth's surface was very negligible.

Here, the question arises: what is the definition of artificial geo-forms? What magnitude of alteration of land by human beings can be considered as a threshold point to designate a landform "artificial"? Several methods of quantifying the anthropogenic transformation of the earth's surface were formulated by Lóczy and Pirkhoffer (2009), Hooke (2000), Erlich and Erlich (1990), and Rózsa and Novák, (2011), all of which are referred in different chapters of this volume. Price et al. (2011) suggested the measurement of the rate of anthropogenic processes in terms of the volume of materials moved per unit time. This could be further extended as volume of materials moved over a unit area per unit time. However, they also found the

> magnitude of impact (quantity and spatial extent) of material moved and its rate correlates with increasing population and plotted possible stratigraphical markers in the Anthropocene based on the use of UK-produced natural resources over time compared with population growth, urbanization, and some major events and developments.

A small group of shifting nomadic people cannot put any measurable and enduring prints on the earth's surface. Whatever they do, do as a part of the natural system on nature and for nature. Reshaping and reworking on the earth's surface of significant magnitude became only possible when our ancient forefathers felt the necessity of making a permanent settlement. So, the sedentary mode of living in the evolutionary history of the human species on the earth surface may be considered as the point of revolution not only because of his acquiring knowledge of agriculture and animal rearing as a compulsory association of permanent settlement but also his role as anthropogeomorphic agent. At this very point in time, he had to make a terrace on the undulated hill slope to prepare a piece of land for producing grains; he had to clear forest not only for that piece of agricultural land but also to make his hut and agricultural tools. Thus, the man put the first significant signature on reshaping the earth's surface. During the dry season, he felt the need of irrigation for his crops. He diverts the partial flow of the spring towards his field by not only reshaping the earth form making ditch to supply his dry land with water artificially but also interfering with the natural fluvial processes of shaping geo-forms. Time taught him the techniques of storing and preserving resources whatever he collects. As a result of this technical achievement coupled with his growing mental quality of more demand, he gathered more and more resources superseding manifold his need. He cleared more and more areas of forest and terraced them for expansion of his agricultural land, made longer ditches to irrigate those fields and mined more and more stones to enlarge his dwelling into a big house. Up to this point in time, the collective energy of human society was not so large to reshape the earth's surface of a significant magnitude. But his energy was compounded after the invention of the wheel. During the 18th century, inanimate energy replaced animate energy in moving wheels and machines and his greed rolled to the farther and started major anthropogenic atmospheric and landscape influence (Figure 12.1).

Automation and industrial revolution multiplied the human energy manifold in reworking with the earth materials, and he reached the point when his glorious and proudly deeds of reshaping earth's surface felt acting against him. He became geological and geomorphological agents in the field of anthropogeomorphology and created the era of Anthropocene. In the last few decades, during the period of Great Acceleration (Goudie, 2018), human society employed tremendous energy to interfere with natural geomorphic processes and to rework on forms. He built giant dams on the river to harvest electric power, made millions of kilometres of canals to irrigate millions of acres of land, levelled the huge terrain to lay railway tracks and roads, excavated tunnels of miles of length to overcome the impassable barrier of mountains, and dug crores of tones of earth materials to mine ores. The energy released by human society is negligible in comparison with the tectonic processes, but it is not incomparable with the exogenic processes operating on reshaping the earth's surface (Szabó, 2010). Rózsa (2007) has estimated that about one-third of the earth's continental surface ($149 \times 10^6 km^2$) was affected by the activity of man. Of these, arable land and plantations were $15 \times 10^6 km^2$, grazing land $35 \times 10^6 km^2$, forests $38 \times 10^6 km^2$, built-up areas $2 \times 10^6 km^2$ and other types of land use $10 \times 10^6 km^2$ (Loh and Wackernagel, 2004, Rózsa, 2007). Douglas and

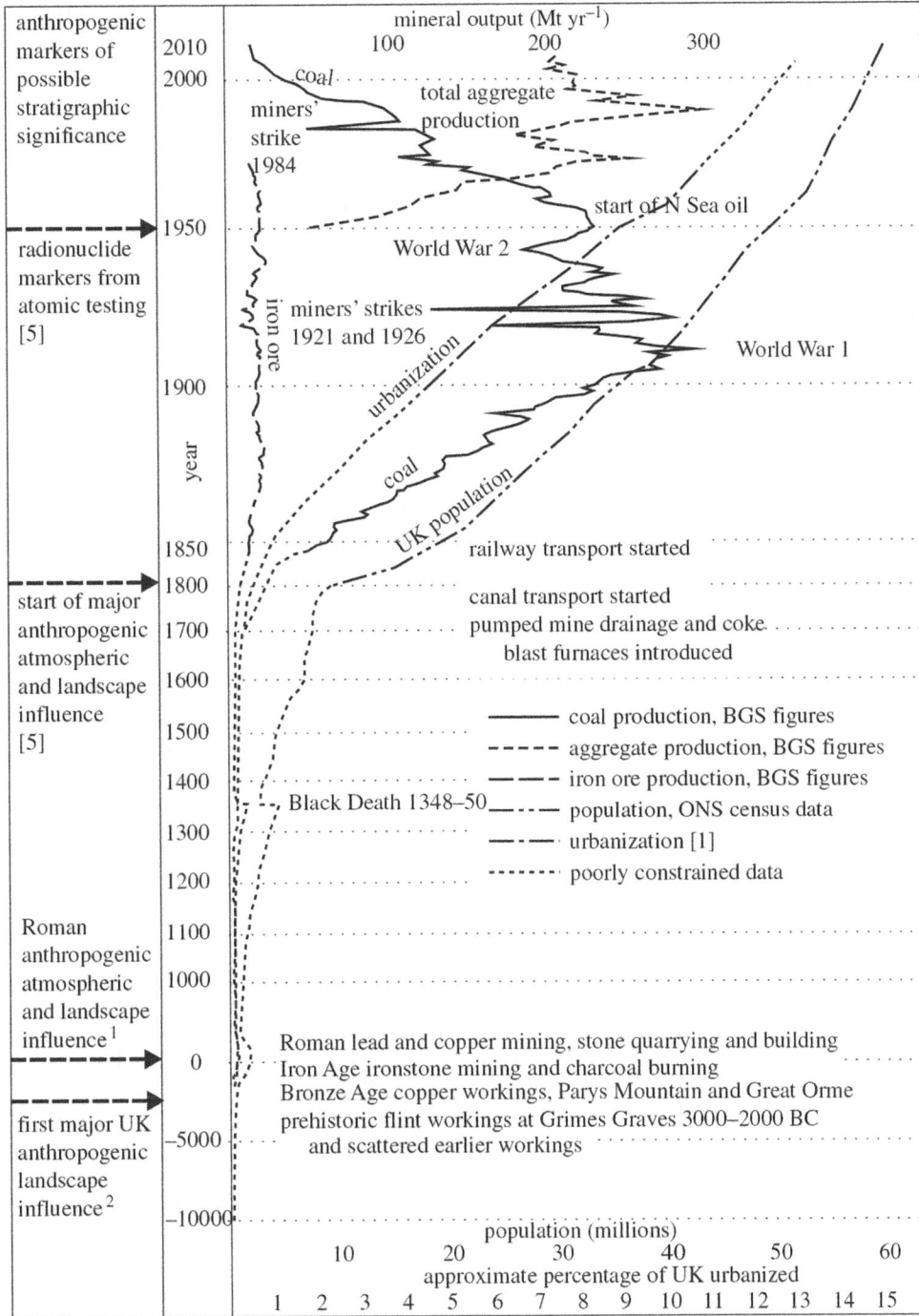

FIGURE 12.1 Possible stratigraphical markers in the Anthropocene. (After Price et al. (2011).)

Lawson (2001) estimated 972 million tonnes of earth material deliberately moved by humans in Great Britain each year. Twenty million tonnes of rock spoil produced from Channel Tunnel, linking Great Britain and mainland Europe, formed 40 ha of new land. The Crossrail railway, linking east and west London, produced 5 million tonnes of material, which has formed 600 ha of the made ground at Wallasea Island, Essex, eastern England (Price et al., 2011). It is estimated by Price et al. (2011) that the worldwide deliberate annual shift of sediment by human activity is 57,000 million tonnes, which is 2.59 times the transport of sediments by rivers to the oceans (22,000 million tonnes).

Different types of anthropogeomorphic processes are classified by Price et al. (2011) into two groups: intentional and unintentional. Both these groups were further divided into two sub-groups (Figure 12.2).

In the era of Anthropocene, the Jalangi River basin also experienced a lasting signature on its geo-forms which has been multiplied in the period of Great Acceleration. The present chapter will illuminate light on those artificial grounds and anthropo-processes reshaping or reworking on those *anthropogeoforms* within the Jalangi River basin.

Jalangi River basin is a deltaic monotonous plain interspersed with the river and its spill channels. Changes in river courses, avulsions, channel shifting, bank erosion, meandering and formation of oxbow lakes, breaching of levees and flooding, braiding, anabranching and channel decaying are obvious inseparable natural processes of delta building mechanism. Yet human efforts, either deliberately or unintentionally, through the processes of construction of Farakka Barrage, agricultural practices, embankment making, ferry services, fishing practices, soil cutting from river bank and bed, making of earthen dam across the river, road crossings, urbanisation, brick kilns, etc. have put significant marks on deltaic landforms of the Jalangi River basin and intervened significantly the natural fluvial processes. Human livelihood practices such as agriculture, fishing, transport and housing have erased several meso- and micro-landforms within the basin as well, and he has created many artificial landforms such as earthen embankments, roads and rails, terracing, levelling of undulated terrain for agriculture, filling of natural low-lying marshes and water bodies, excavating ditches along roadside and embankment side, canals and ponds. By changing landforms, man has indirectly altered the fluvial processes that are shaping and reshaping those forms. After the green revolution in India, new agricultural practices in the Jalangi River basin have broken all past records of the human effort of reworking with earth surface forms and processes not only by multiplying cropping intensity but also by expanding agriculture in more and more virgin areas. Keeping pace with the agricultural revolution and the need for all other infrastructural facilities such as housing,

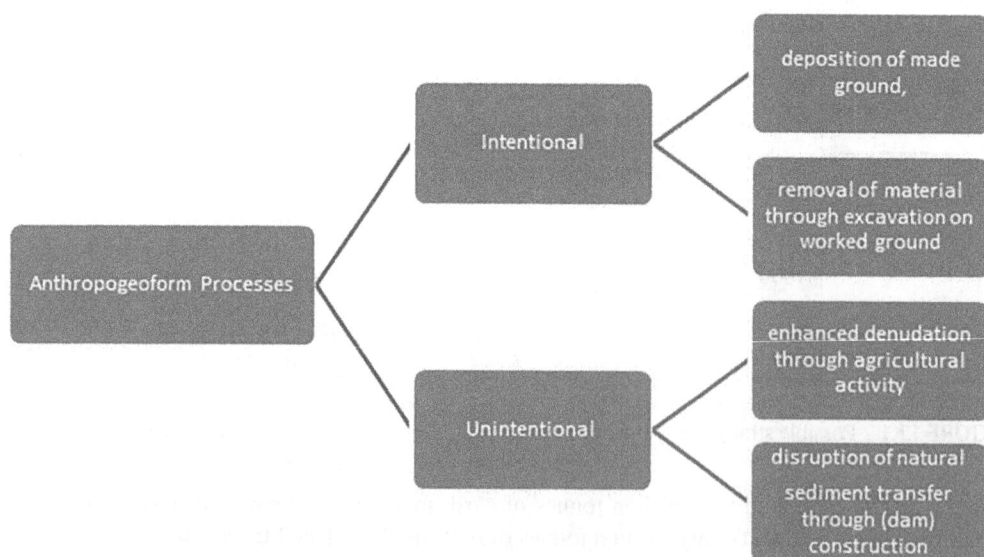

FIGURE 12.2   Anthropogeomorphic processes. (After Price et al. (2011).)

transport, education, business and commerce, he altered the land at a manifold higher rate than the natural processes. Jalangi River was directed several times to channelise its flow at expected volume in desired ways during British rule.

In this chapter, an effort is given to illuminate these entire human signatures on fluvial forms and processes in the Jalangi river basin. Finally, this chapter will shed its search on the processes of killing of a dying river—the Jalangi, by conscious and unconscious efforts of man.

12.2   THE RIVER AND ITS BASIN

"Jalangi" is a Bengali word meaning "watery body" or "the body is made of water". So it was (yes was, a past form of tense is used consciously here) a river of water. But now, it is a 233 km long moribund river in the Bhagirathi-Hooghly Basin. The river is simultaneously a distributary of the river Padma and tributary of the river Bhagirathi. The reach of 9.1 km (Table 12.1) from abandoned offtake at Char Madhubona (Figure 12.3) near Gopalpur (Ghat) to Sialmari River–Jalangi River confluence at Kupila is erased out from the map of the region and 41.9 km from Sialmari River–Jalangi River confluence to Bhairab River–Jalangi River confluence at Moktarpur, though traceable but dead at present. The reach downstream to the Bhairab river–Jalangi River confluence up to Jalangi–Bhagirathi confluence at Swarupganj (182 km) is being maintained by the base flow of seepage water, contribution of Bhairab and other spills during two months of the rainy season.

Up to the late 19th or early 20th century, the river was one of the three (Bhagirathi, Mathabhanga, and Jalangi – three *Nadia Rivers*) main waterways of south Bengal. Sometimes, the river Jalangi was more suitable as a navigation route than Bhagirathi and Mathabhanga (Garrett, 1910; Ferguson, 1912). The first steamer record to pass through the river Jalangi dates back to 21 October 1830 and mentioned as

> .... we left Calcutta on the 14th October 1830 in the steamer 'Hooghly' towing the 'Soonamokee' with Lord William Bentinck and suite; the steamer drew 4 feet 6 inches. On the 21st of October, we passed through the Jellinghy into the Ganges with nothing less than 6 feet.

(Reaks, 1919)

Since then, the river Jalangi allowed hundreds of steamers and boats of considerable sizes to ply through till 1930, a time span of 100 years. But now the offtake of the river is completely closed to allow any boat to pass into the river Padma.

TABLE 12.1
Reaches of the River Jalangi and Present State

Sl. No.	Reach Name		Length (km)	Comments
	From	**To**		
1	Offtake at Char Madhubona	Sialmari confluence at Kupila	9.10	Untraceable
2	Sialmari confluence at Kupila	Bhairab confluence at Moktarpur	41.9	Dead
3	Bhairab confluence at Moktarpur	Suti confluence	42.7	Rapidly dying + killing processes by man are in operation
4	Suti confluence	Kalma *Khal* confluence at Radhanagar	55.2	Rapidly dying + killing processes by man are in operation
5	Kalma *Khal* confluence at Radhanagar	Bhagirathi–Jalangi confluence	84.1	Dying + killing processes by man are in operation
Total	Offtake at Charmadhubona	Bhagirathi–Jalangi confluence	233	

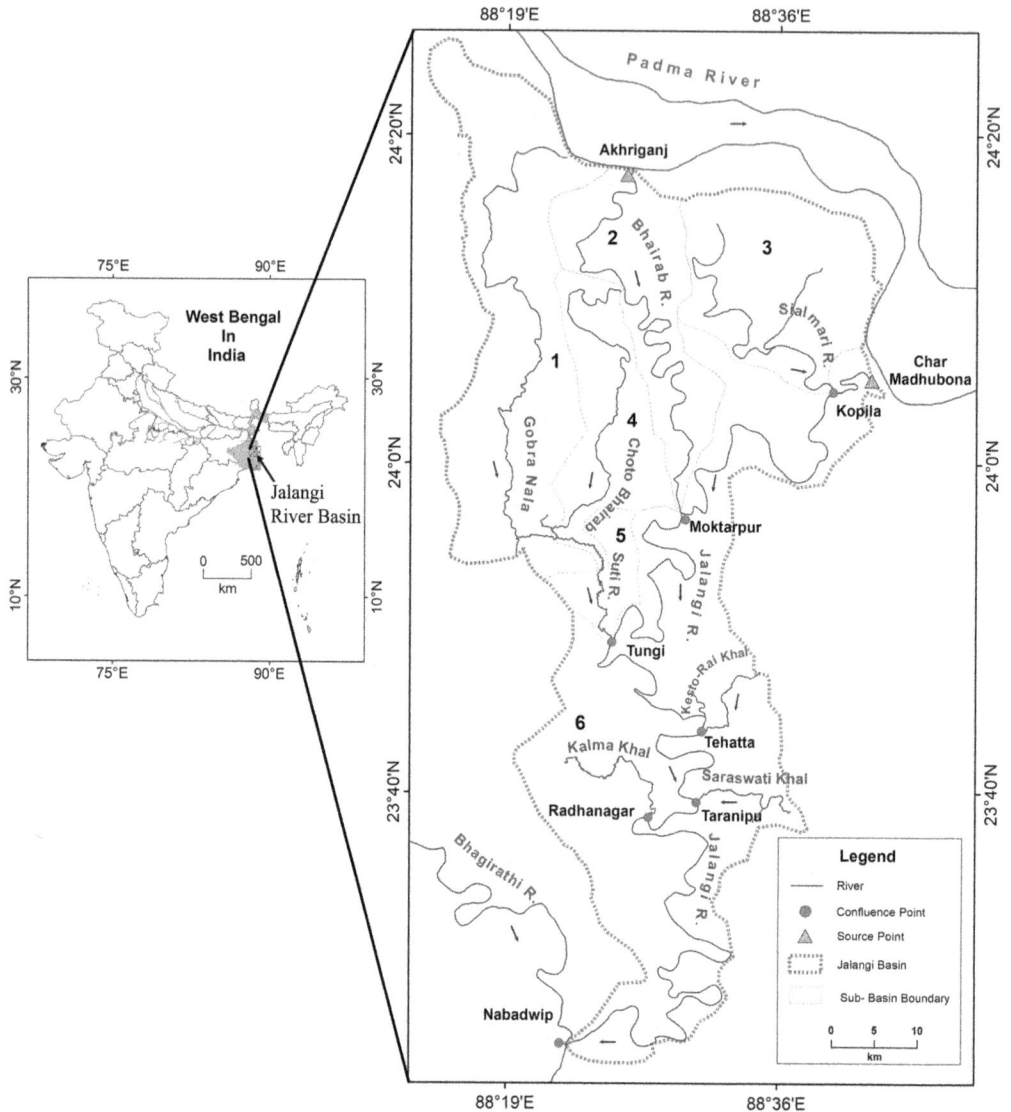

FIGURE 12.3   Jalangi River and its tributaries.

The Jalangi leaves the Ganges or Padma at the extreme north of the Nadia district at 24°05′26″ N and 88°41′53″ E. After leaving offtake, the river pursues a meandering course along the north-west border of Nadia district for 51 km up to Moktarpur village. This length is dead because there is no connection with the river Padma from where formerly the river would get its supply. From Moktarpur, the river runs 45 km up to Gopinathpur and Sahebnagar villages of C.D. Block Tehatta-I and C.D. Block Tehatta-II, respectively, separating the district of Murshidabad from the district of Nadia. Thence, the river enters the district of Nadia and never left the district before falling into the river Bhagirathi. After Gopinathpur and Sahebnagar villages, the river follows a tortuous course in a southerly direction by Palashipara, Tehatta, Bara Andulia, Chapra and Doier Bazar until it reaches Krishnanagar, the headquarter of the district of Nadia. From Krishnanagar, the river proceeds due west, and after a total course of 233 km, it debouches into the river Bhagirathi at Swarupganj, opposite the town of Nabadwip, the birthplace of Sri Chaitanya, at 23°29′23″ N and 88°28′57″ E. The united flow of the river Bhagirathi and the river Jalangi, downstream to Nabadwip or Jalangi–Bhagirathi confluence is known as the river Hooghly. Out of a total course of 233 km, 96 km (51 + 45 km) is running along the border of Nadia and Murshidabad districts and rest 137 km through the Nadia district separating different villages of seven C.D. Blocks (except C.D. Blocks Karimpur-I and II where the river makes natural boundary between Murshidabad and Nadia) on left and right banks (Table 12.2). Jalangi is a "river of Nadia" in true sense because, unlike Bhagirathi and Mathabhanga–Churni, the river is flowing its whole length either along the border or through the district of Nadia.

It is rather impossible than difficult to the precise delineation of the drainage basin of a deltaic river like Jalangi. Yet an attempt of approximation in this regard was taken. For the purpose, Google earth image of the inundated area by flood during 2000, contour map (Bagchi, 1978), flood slope map (Biswas, 2001, plate-19) and transport networks were taken into consideration. Estimated basin area is 2,815.33 km².

It is important to mention here that APPE & M Cell, I & WD, Government of West Bengal published Annual Flood Report-2016, where the flow path of the river Jalangi seems not clear. It was mentioned there –

> On the left bank of the Bhagirathi river system the Bhairab-Jalangi-Sealmari group of rivers originate from Ganga-Padma at Akherigunj in Murshidabad district and meet the Bhagirathi at Swarupgunj in Nadia District.

But Bhairab, Jalangi, and Sealmari (Sialmari) are not a single river. Rather Bhairab and Sealmari are tributaries of the river Jalangi. Therefore, their points of origin are not at the same place "Akherigunj", actually from where the river Bhairab was taking off from the river Padma.

Again, it was mentioned –

> The river Jalangi originates from the right bank of the river Padma in Murshidabad district, 165 km. downstream of Farakka. Jalangi is dead for all purposes except during the periods of heavy rain, when it receives water from Padma. The river ends its journey by finally outfalling into the river Bhagirathi near Nabadwip town of Nadia district. The major tributary of Jalangi is river Bhairab which starts its journey from the river Ganga near Lalbag of Murshidabad district. It is now almost a dead channel but during rainy season it receives water from Padma.

First, offtake of the river Jalangi at Char Madhubona near Bausmari in C.D. Block Karimpur-I (formerly from near the village Jalangi of Murshidabad from where the river got its name "Jalangi") has lost its existence and erased out from the landscape permanently and no way it receives any supply from the river Padma. Offtake of the river Jalangi is ~132 km downstream from Farakka. Second, Akherigunj or Ankhriganj is the offtake of Bhairab, not of Jalangi.

12.2.1 TRIBUTARIES

Tributaries play the key role in the flow of the river Jalangi because Jalangi's own offtake from the river Padma is literally erased out. They contribute to the river Jalangi not only by draining back Marshes but also by seepage water.

River Sialmari: Sialmari is the first contributing channel to the river Jalangi. The river takes off from the river Padma, about 5 km east of Hursi in C.D. Block Raninagar-I or 7 km north of village Malibari in C.D. Block Raninagar-II of the district of Murshidabad. There is a village a few kilometres south of the offtake of the river named "Sialmari" from which the river had got her name. After taking off, the river pursues a tortuous course towards the south via Raninagar, Kasba Goas, Domkal and Raipur to fall into the river Jalangi at *Kupila* on the right bank (Figure 12.3). The offtake of the river Sialmari from the river Padma is also dried up, and the river no more gets any supply from the Padma. As a result, Sialmari and the upper reach of the river Jalangi have been deteriorated badly. Now it contributes to the river Jalangi by draining the Bhairab–Sialmari interfluvial tract in the district of Murshidabad (Table 12.3).

TABLE 12.2

C.D. Blocks and Villages on the Banks of the River Jalangi in the District of Nadia

C.D. Blocks & Villages on Left Bank			C.D. Blocks & Villages on Right Bank		
1. KARIMPUR-I	3. TEHATTA-I	5. KRISHNANAGAR-I	1. TEHATTA-II	3. KRISHNANAGAR-II	4. NABADWIP
Kuchaidanga	Kanainagar	Naldaha	Sahebnagar	Rupdaha	Sardanga
Takipur	Rajapur	Bhandarkhola	Panchdara Abhaynagar	Pathradaha	Ghasighata
Uttar Krishnapur	Baksipur	Jaba	Palashipara	Sonatala	Mollapara
Durlabhpur	Sardanga	Haranagar	Ramnagar	Char Mahatpur	Ballaldighi
Dhanerpara	Shrirampur	ChakGokulnagar	Rudranagar	Debipur	Rudrapara (P)
Jayrampur	Gopinathpur	Ghurni	Chak Rudranagar	Uttar Jhitkipota	
Gopalpur	Raninagar	Krishnanagar	Hanspukuria	Paschim Panditpur	
ChakMadhubona	Radhanagar	Ruipukur	Natipota	Purba Panditpur	
Abhaypur	Krishnanagar	6. NABADWIP	Iswarchandrapur	Krishnachandrapur	
Uttampur	Syamnagar	Tiorkhali	Chanderghat	Sahebnagar	
Karimpur (CT)	Tarangar	Maheshganj	Chak Hanspukuria	Mayakol	
2. KARIMPUR-II	Swaruppur	Gadigachha	Chak Ramnagar	Par Media	
Dogachhi	Nishchintapur		Kustia	Bahadurpur	
Dhoradaha	Raghunathpur		2. NAKASHIPARA	Bishnunagar	
Jayghata	Krishnachandrapur		Shibpur		
Monoharpur	Jitpur		Radhanagar		
Saguna	Tehatta		Patikabari		
Natidanga	Kulgachhi		Birpur		
Char Moktarpur	Khaspur		Petuabhanga		
Fazil Nagar	Paschim Chakjaliapara		Teghari		
Lal Nagar	Taranipur		Mota		
Narayanpur	Puthimari		Dogachia		
Char Brindabanpur	4. CHAPRA		Saligram		
Sadipur	Hatisala				
CharJagaipur	Mahesnagar				
	Bara Andulia				
	Gopinathpur				
	Hatra				
	Talukhuda				
	Brittihuda				
	Gokhurapota				
	Pitambarpur				
	Dwip Chandrapur				
	Bangaljhi				
	Mahatpur				
	Dakshin Sonatala				

TABLE 12.3

Tributaries, Their Source Region and Location of the Confluence with the River Jalangi

Name of the Channel	Source Region	Location of the Confluence with Jalangi
Sialmari River	24°13′10.22″ N and 88°29′31.85″ E. Taking-off from the river Padma, north to Malibari in P.S. Raninagar-II, Murshidabad.	24°04′23.64″ N and 88°39′25.70″ E. Kupila in P.S. Domkal, Murshidabad
Bhairab River	24°18′02″ N and 88°26′43.80″ E. Receiving supply from Padma at Hursi in P.S. Raghunathganj-II, Murshidabad.	23°51′01″ N and 88°29′51.25″ E. Moktarpur in P.S. Karimpur-II, Nadia.
Choto Bhairab	Drains low-lying areas of Mirpur, Dhanaipur and Gangaprasad and runs through *Kālāntar* tract of Hariharpara	It does not meet the river Jalangi. Rather it meets the Suti river at Trimohini (23°55′26″ N and 88°22′06″ E)
Gobra-Suti Nala	Taking-off from the river Padma, about 3 km upstream of Bhairab offtake and draining *Kālāntar* tract to fall in Jalangi at Tungi.	23°49′10.26″ N and 88°25′31.85″ E. Tungi in P. S. Nawda, Murshidabad.
Kalma *Khal*	Draining Bil Kumari and Pon bil of P.S. Tehatta-II	23°47′13.26″ N and 88°29′31.75″ E. Radhanagar in P.S. Nakashipara, Nadia

River Bhairab: It is the second (in terms of position from offtake) contributing channel to the river Jalangi but contributes the largest volume of water and therefore the main tributary of the river. It is also a distributary of the river Padma and at the same time a contributor to the river Jalangi. At present, the river Bhairab, being fed by the river Padma, is maintaining the flow of the river Jalangi, for one or two monsoon months only. The river takes off from the river Padma at *Ankhriganj* (Figure 12.3), about 5 km upstream from Hursi, just following the border of C.D. Block Bhagwangola-II and C.D. Block Raninagar-I in the district of Murshidabad. Thence, it pursues a southward course up to tri-junction point of C.D. Block Raninagar-I, C.D. Block Jalangi and C.D. Block Berhampore of the district of Murshidabad, from where it bifurcates. The eastern branch is the main Bhairab and pursues a southerly course via Islampur, Bhagirathpur, Swaruppur, Garibpur and Dakshin Jitpur to fall into the river Jalangi at *Moktarpur* in C.D. Block Karimpur-II of Nadia district. The river makes the natural border of C.D. Blocks of Berhampore–Raninagar-I, Berhampore-Domkal, Hariharpara–Domkal and Nawda – Domkal of the district of Murshidabad. After bifurcation from the tri-junction point of Jalangi–Raninagar-I–Berhampore, the right branch pursues a westerly course for about 7 km along the border of C.D. Blocks of Jiaganj and Berhampore and thence turns southwards to traverse Berhampore and Hariharpara C.D. Blocks. From 1–2 km north of Protappur in the southern part of Hariharpara C.D. Block, this branch bifurcates again, the left branch to join original (main) Bhairab near its confluence to Jalangi and the right branch traverse southwards to meet *Gobra Nala* near Goghatia in C.D. Block Nawda.

Choto Bhairab: This third tributary drains low-lying areas of Mirpur, Dhanaipur and Gangaprasad and runs through Hariharpara to fall into the Gobra–Suti Nala at (23°55′26″ N and 88°22′06″ E) Trimohini near Jhowbona on Beldanga–Amtala road.

Gobra-Suti Nala: This tributary of the river Jalangi would formerly receive its supply from the river Padma, 2 km upstream from the offtake of river Bhairab, and pursues a westerly course up to Bhander–Kismat, south of Bhagwangola. Then, it crosses the Bhagwangola–Hursi–Raninagar road and flows southwards through the C.D. Blocks of Bhagwangola-II, Jiaganj, and Berhampore up to 2 km north of Tarakpur, via Mrigi, Jiaganj, Lalbagh and Begamnagar. From 2 km north of Tarakpur, it follows the border of C.D. Blocks of Berhampore–Hariharpara and Beldanga-I–Hariharpara, till it branches off into many channels in *Kālāntar* (the low land, especially marshy land in between

Bhagirathi and Jalangi rivers) depression. Now, collecting water through different channels from *Kālāntar* back swamp, it reunites to be *Suti Nala* to fall into the river Jalangi at Tungi (Figure 12.3).

Kalma Khal: This minor contributing channel drains *beel* Kumari and Pon *beel* ('*beel*' sounds as bil means a marsh) of C.D. Block Tehatta-II to contribute to the river Jalangi during rainy months. The location of the confluence of the channel with the river Jalangi is 23°47′13.26″ N and 88°29′31.75″ E at the village Radhanagar in C.D. Block Nakashipara, Nadia.

12.2.2 Spills

There are four mention-worthy spills of the river from its left bank. These spills have important roles in maintaining the channel of the river. They carry silt-laden water during flood into the back Marshes or into the lower flood plains and deposit to raise the level. But when the water level of the river goes down, reverse flow of silt-free clear water through these spills maintains the lean season flow of the river facilitating the scour of the loose slurry bed deposits. Namable spill channels of the river are discussed as follows.

Bhairab Khal – It is not the Bhairab river discussed under the head TRIBUTARIES, and obviously a different channel. It leaves the river Jalangi at 23°59′10.16″ N and 88°37′31.85″ E at Uttampur, near Karimpur bazaar in C.D. Block Karimpur-I, Nadia, and runs through C.D. Block of Karimpur-I and passes into the Meherpur of Bangladesh (Table 12.4).

Kesto-Rai Khal – The spill leaves the river Jalangi at 23°44′19.20″ N and 88°31′31.55″ E near Tehatta ferry ghat, C.D. Block Tehatta-I and runs north-west towards Gobri *beel* and Chand *beel*, in C.D. Block Tehatta-I and Margangni in C.D. Block Karimpur II, and Meherpur, Bangladesh. The spill is also called as Gobri and Chand *beel* in draw channel.

Saraswati Khal – The spill drains the low lands of Meherpur district in Bangladesh and meets the river Jalangi at 23°42′29.20″ N and 88°31′11.55″ E near Putimari, C.D. Block Tehatta-I.

River Anjana – This is a distributary of the river Jalangi takes off from the feeder river Jalangi at Krishnanagar. It connects the river Churni with the river Jalangi. The geomorphology of the river Anjana and its basin has also been reworked and reshaped by man at a significant magnitude which is illustrated in the 16th chapter of this volume.

12.2.3 Marshes/*Beels* and Relict Channels

There are numerous *beels* or Marshes distributed all over the Bhairab–Jalangi river basin in the district of Nadia and Murshidabad. Most of them are now disappeared mainly by human interventions

TABLE 12.4
Spills, Their Source Region and Location of the Confluence with the River Jalangi

Name of the Spill Channel	Connecting/Source Region	Location of the Confluence with Jalangi
1. Bhairab *Khal*	Meherpur, Bangladesh	23°59′10.16″ N and 88°37′31.85″ E. Uttampur, near Karimpur bazaar in P.S. Karimpur-II, Nadia.
2. Kesto-Rai Khal	Gobri bil, Chand bil, in P.S. Tehatta-I and Margangni in P.S. Karimpur II, and P. S. Meherpur, Bangladesh.	23°44′19.20″ N and 88°31′31.55″ E. Near Tehatta ferry ghat, P.S. Tehatta-I.
3. Saraswati *Khal*	Meherpur, Bangladesh.	23°42′29.20″ N and 88°31′11.55″ E. Putimari, P.S. Tehatta-I
4. River Anjana	River Churni at Majdia P.S. Krishnaganj and Hazrapur P.S. Ranaghat-II	23°24′44.30″ N and 88°22′49.07″ E. Nagendranagar (ward No. 24), Krishnanagar Municipality,

TABLE 12.5

Beels or Swamps Located along the Course of the River Jalangi

C. D. Block	Sl. No.	Name of *Beel* (Swamp)	The shape of *Beel* (Swamp)	Location of Central Point	Area (km²)
Karimpur-I	1	Karua *beel*	Irregular	23°58′00″ N 88°38′00″ E	1.45
	2	Tengramari *beel*	,,	23°57′30″ N 88°37′00″ E	1.05
Karimpur-II	3	Topla *beel*	Ox-bow lake	23°57′00″ N 88°30′30″ E	0.70
	4	Margangni-Paikpara *khal*	Relict channel	_____	1.00
Nakashipara	5	Patpukur *beel*	Irregular	23°36′00″ N 88°31′30″ E	–
Chapra	6	Gokhurapota *beel*	Irregular	23°32′00″ N 88°30′30″ E	–
Tehatta-I	7	Morakodi *beel*	Ox-bow lake	23°50′00″ N 88°29′30″ E	1.10
	8	Damuk *beel*	,,	23°49′30″ N 88°30′00″ E	0.80
	9	Argarini *beel*	Relict channel	23°48′00″ N 88°29′00″ E	0.15
	10	Banur *beel*	Ox-bow lake	23°47′00″ N 88°29′15″ E	0.75
Tehatta-II	11	Baragadi *beel*	Irregular	23°48′30″ N 88°25′00″ E	0.25
	12	Sati*khali beel*	Relict channel	23°45′30″ N 88°27′00″ E	0.20
	13	Margangni *beel*	Ox-bow lake	23°42′00″ N 88°30′00″ E	1.25
Krishnanagar-II	14	Nowapara *beel*	,, (Das 2017)	23°30′30″ N 88°29′00″ E	1.00
	15	Hasadanga *beel*	,,	23°27′30″ N 88°27′30″ E	1.20
Krishnanagar-I	16	Shyamnagar *beel*	Relict channel	23°25′00″ N 88°27′00″ E	2.00
Nabadwip	17	Alokananda	,,	23°24′00″ N 88°24′30″ E	1.5
	18	Kalatala *beel*	,,	23°23′30″ N 88°23′30″ E	1.25
	19	Bhairab R.		24°02′52″ N 88°30′33″ E	

through the processes of agriculture, road and embankment construction, building constructions, etc. or by natural processes. Some of them located along the course of the river Jalangi in the district of Nadia are listed in Table 12.5.

12.2.4 MAJOR LANDFORMS

12.2.4.1 Extra-Channel Geomorphology

Relief – Jalangi basin including basins of contributing spills covers an area of 2,815.33 km² (Figure 12.4, Table 12.6) of the district of Nadia and Murshidabad. More than 63% of the total basin area has an altitude between 15.1 m and 27.00 m. Only 1.91 km² area (0.07%) of the lower part of the Suti and Jalangi basin lies below the altitude of 5 m. Medium altitude of 5.1–15.0 m covers an area of 1,035.46 km² (36.78%) of the northern part of Gobra basin, Bhairab basin, and Sialmari basin. Altitude decreases from north to the south.

Slope – The general slope of the Jalangi basin is towards the south which is followed by tributary channels and the river Jalangi itself. Slope map (Figure 12.5) and area under different slope categories (Table 12.6) derived from satellite imagery under ArcGIS software show that 93.83% of the total basin area, i.e., 2,641.79 km², has a slope lower than 4.385°. Only 6.17% area has a slope >4.385°. The slope of the basin is very low with the monotonous landscape. The lowest slope of <1.67° covers 56.97% of the total basin area.

Flood slope is the general slope of the Jalangi basin determined by the pattern of contours which is followed by flood water to move downward during floods or rains. For identification of flood slope, maps from Biswas (2001) and topographical maps were used as tools. The general slope of the Jalangi–Bhagirathi interfluvial tract of C.D. Blocks of Beldanga-II in Murshidabad and C.D.

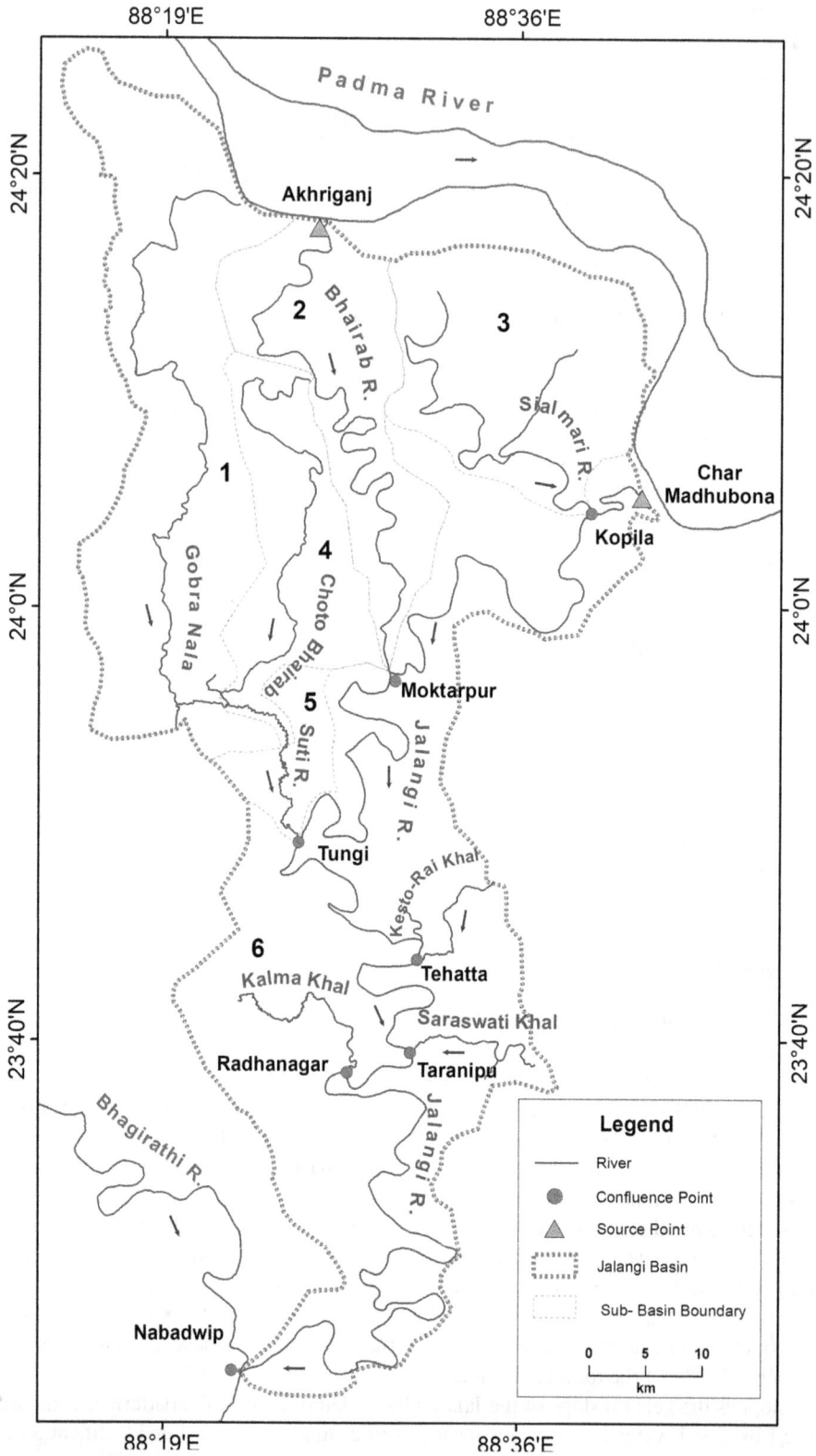

FIGURE 12.4 Jalangi and its tributaries.

TABLE 12.6
Relief and Slope Distribution of the Jalangi River Basin

Elevation in m	Area in km²	%	Slope in Degrees	Area in km²	%
0–5	1.91	0.07	0–1.670	1,603.96	56.97
5.1–15	1,035.46	36.78	1.671–4.384	1,037.83	36.86
15.1–27	1,777.96	63.15	4.385–19.12	173.54	6.16
Total	2,815.33	100.00	Total	2,815.33	100.00

Blocks of Tehatta-II, Kaliganj, Nakashipara, Krishnanagar-II and part of Nabadwip in Nadia district is towards south and overland flow during floods converges into the lower reach of the river Jalangi from Krishnanagar to Swarupganj causing yearly inundation in C.D. Blocks of Nabadwip and Krishnanagar-II.

Catchments area (2,815.33 km²) of the river Jalangi slopes towards the south and is classified and delineated into six sub-catchments, which are given as follows:

Padma–Jalangi-Sialmari interfluvial tract – Slopes of Padma-Jalangi-Sialmari interfluvial tract are from north to south and a portion of surface runoff comes to the river Jalangi directly and portion via river Sialmari (Biswas, 2001, plate-18 & 19).

Sialmari–Bhairab interfluvial tract – Also slopes southward and water from the region come largely through river Sialmari and Bhairab, and partly direct to the river Jalangi. During 1 or 2 monsoon months, river Bhairab gets supply from the river Padma and debouches into the river Jalangi at Moktarpur. Like Jalangi, the offtake of the river Sialmari is also closed, and it collects water mainly from the Ganges–Sialmari and Sialmari–Bhairab interfluvial tracts (Figure 12.6).

Bhairab–Suti interfluvial tract – Suti Nala supplies considerable cusecs during monsoon months. Draining the *Kālāntar* tract of CD Block Hariharpara and CD Block Beldanga-I of Murshidabad district during and after rains, Suti falls in Jalangi at Tungi. There was a gauge station equipped by only a piece of graduated bamboo on the river Suti at Tungi which was looked after by very poorly paid Kalu Dafader, a fisherman, who also took the reading of the gauge for several years. Now it is closed.

Suti–Bhagirathi interfluvial tract – This natural interfluvial tract is divided into two parts by National Highway-34 (NH34):

Suti–NH34 intermediate tract – Flood slope of this region is from north to south and southeast. The tract between the river Suti and NH34 is known as *Kālāntar.* Part of the overland flow moves south and falls directly into the river Jalangi. A portion of the overland flow passes through the bridges under NH34 and Krishnanagar–Lalgola railway line and concentrate to fall into the lower reach of the river Jalangi from Krishnanagar to Swarupganj causing yearly inundation in C.D. Blocks of Nabadwip and Krishnanagar-II.

NH34–Bhagirathi intermediate tract – Flood slope of this region is also from north to south and southeast and the collective volume of Suti–NH34 intermediate tract and NH34–Bhagirathi intermediate track falls jointly into the river Jalangi and the river Bhagirathi through C.D. Block of Nabadwip and Krishnanagar-II causing routine annual inundation, especially at Mayapur and Swarupganj (Figure 12.7).

Left-bank side of the river Jalangi (south bank) also slopes southward and south-eastward and surface runoff or overland flow do not go into the river Jalangi, rather overbank flow from Jalangi spills towards south or south-east. Only a portion of surface runoff over the bank (left) comes again into the river through Kesto-Rai Khal and Saraswati Khal, and debouches at Tehatta and Shibpur, respectively, during post-monsoon months.

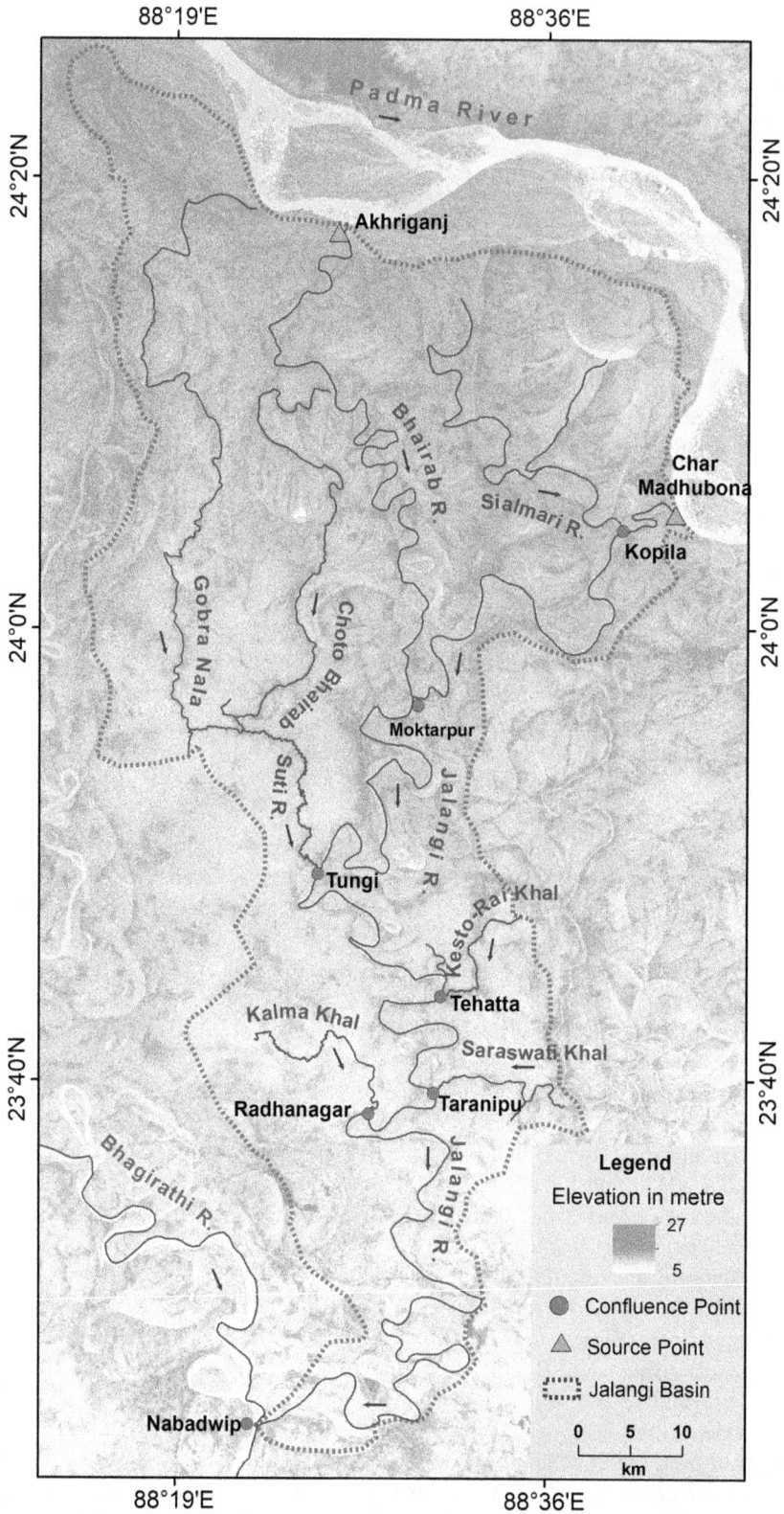

FIGURE 12.5　Relief variation of the Jalangi river basin.

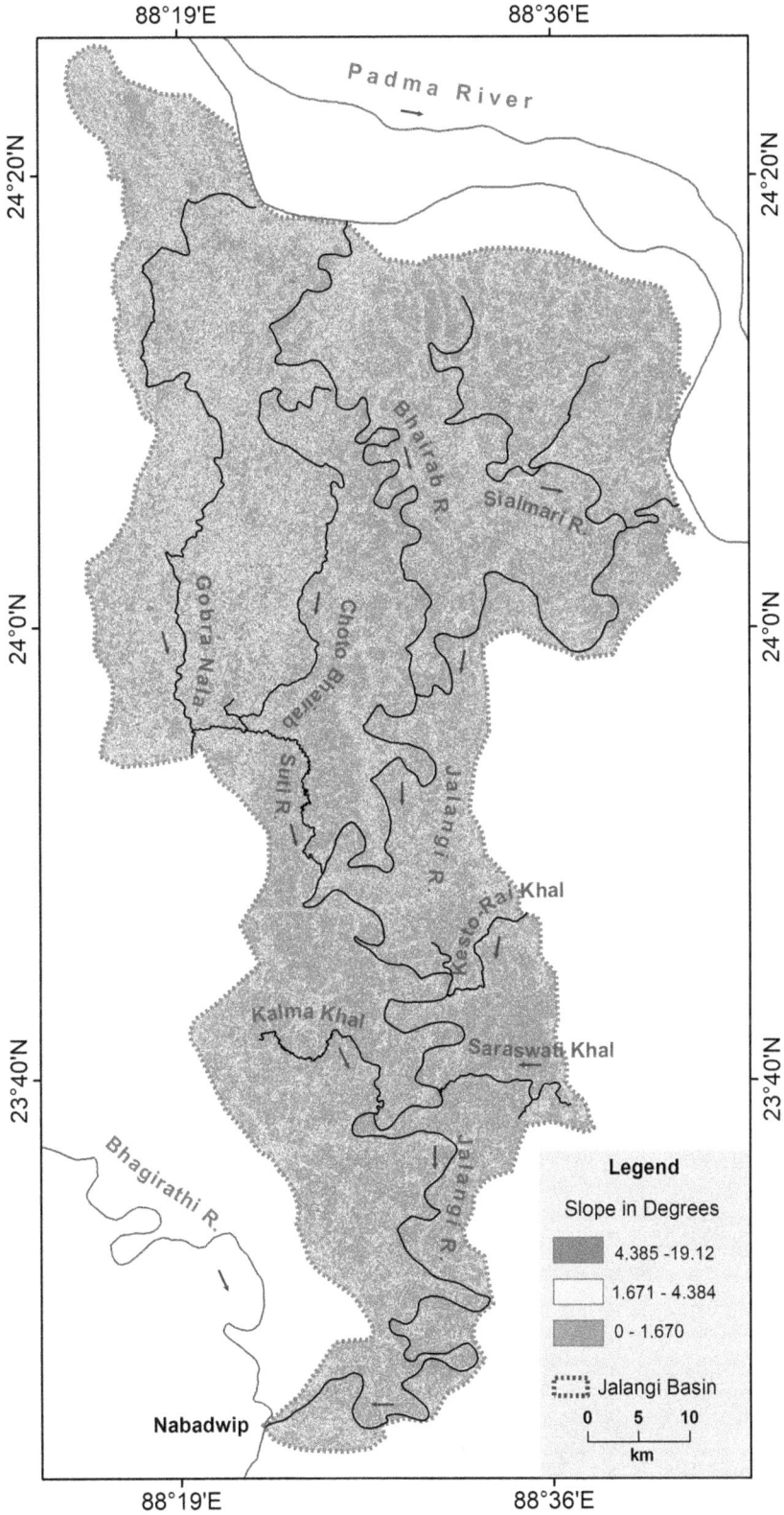

FIGURE 12.6   Slope distribution of the Jalangi river basin.

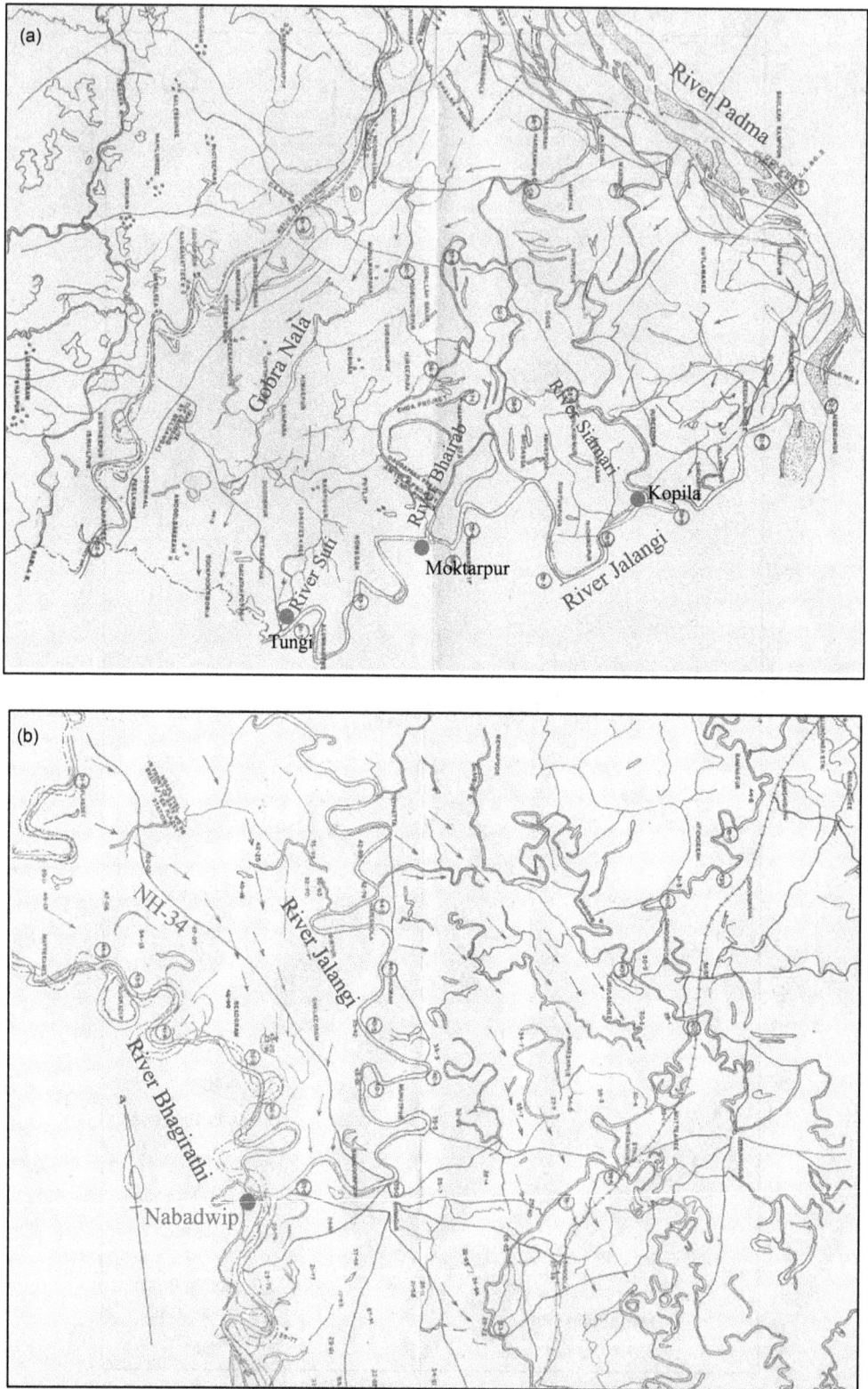

FIGURE 12.7 Flood slope directions of the (a) upper and (b) lower Jalangi basin. (Biswas (2001, plate-18).)

12.2.4.2 Channel Geomorphology

Channel morphology includes two-dimensional and three-dimensional channel geometries which can largely be characterised in terms of adjustable variables – width, depth, slope and meander form (Knighton, 1984).

Cross-profiles – To study the cross-sectional morphometry of the river Jalangi, the authors had made several cross sections on the river from its offtake to confluence. Based on those cross sections, the following morphometric information is derived. For cross-sectional morphometry, the bank-full stage was considered.

a. Width (w)

Richards (1976) found the bank-to-bank distance or width of a channel varies between riffles $\left(w = 4.54Q^{0.33} \right)$ and pools $\left(w = 3.85Q^{0.33} \right)$. The average channel width of the river Jalangi is 162.58 m, the highest being 224.21 m at Moktarpur–Velanagar and the lowest being 75.89 m at Moktarpur–Chandpur. At Moktarpur–Velanagar, flat slope and the confluence of the river Bhairab, the river is shallow, and lateral erosion has widened the channel maximum. But at Moktarpur–Chandpur, immediately upstream of the maximum width, the river has the lowest width. This is because first, from Moktarpur to offtake at Char Madhubona near *Gopalpur Ghat*, this reach of the river Jalangi has been deteriorated since the first decade of the 20th century. Second, the anthropogenic processes like the closing of the offtake of the river to construct State Highway-11, ploughing within river bed, making tanks by diking the moribund channel of the river have decayed the river to such an extent that it is worthy to say the reach is dead. This is why the width of the river is least at this section.

b. Mean depth ($\bar{d}$)

Mean depth is expressed as A/w where A = cross-sectional area and w = width. The mean depth of 20 cross sections of sample study is shown in Table 12.7. The highest mean channel depth of 7.74 m was recorded at Shibpur–Hatisala cross section, and the lowest was measured (0.82 m) at Jitpur–Saguna (Figure 12.8). The depth of the river is literally untraceable at Bausmari, Char Madhubona, and Gopalpur. All the pools like Shibpur–Hatisala cross section have higher depth than those reach of riffles. Jitpur–Saguna cross section is on the dead reach (mentioned above) of the river. Hence, its depth was least.

c. Cross-sectional area (A)

It is the product of channel width (w) and mean depth ($\bar{d}$) (Richards, 1982). The largest channel *cross-sectional area* measured at Krishnanagar–Mayakol was 265,042.70 m², which is nearly 125 times greater than the smallest one at Jitpur–Saguna (2,106.36 m²). Causes of such great differences in the channel *cross-sectional area* at two cross sections are already explained under sub-headings "Width" and "Mean depth" as channel *cross-sectional area* is the product of those two.

d. w/$\bar{d}$ ratio

It is the ratio between channel width (w) and the mean channel depth ($\bar{d}$) and expressed as w/$\bar{d}$ (Schumm, 1960). The highest w/$\bar{d}$ ratio (136.16) or very wide shape of the channel was observed at Jayrampur–Chanderpara, and the lowest was 17.346 at Shibpur–Hatisala. Jayrampur–Chanderpara cross section of the river is on dead reach. So, the depth is very low. But as it is immediately above the Sialmari–Jalangi confluence, the river is very wide. Moreover, the channel is extremely silted by washed down soil from ploughed banks.

Meander properties: In this section, channel's areal pattern is discussed under following sub-heads.

a. Wavelength (λ)

Wavelength (λ) of a meander is the distance between troughs or crests of corresponding bends of meander (Richards, 1982). Wavelength varies from 1,500 m at Jayrampur to 10,000 m at Karimpur (Table 12.8), and the average wavelength is 3,646.74 m. Although

TABLE 12.7

Downstream Variations of Channel Width, Mean Depth, Lowest Bed Level, Bank Height, Meander Amplitude and w/d Ratio

Villages	Distance from Offtake (m)	Width (m)	Mean Depth (m)	Lowest Bed Level in m (a.m.s.l.)	w/d Ratio	Bank Height (a.m.s.l.)
Madhubona	0.00	[a]	[a]	19.00	[a]	19.00
Gopalpur Ghat	2.25	[a]	[a]	18.50	[a]	18.50
Jayrampur–Chanderpara	12.50	217.85	1.60	15.66	136.16	18.00
Karimpur–Bakshipur	22.00	204.55	3.49	10.65	58.67	17.00
Jitpur–Saguna	30.00	84.82	0.83	14.47	102.47	16.50
Moktarpur–Chandpur	48.00	75.89	1.60	11.91	47.42	18.00
Moktarpur–Velanagar	48.10	224.21	3.16	12.40	70.88	18.00
Rajapur–Patikabari	69.00	157.00	4.19	7.53	37.45	14.30
Tungi–Gopinathpur	90.50	160.00	4.23	7.01	37.85	14.00
Bali–Gopinathpur	90.60	164.06	4.22	7.03	38.92	14.00
Tehatta–Natipota	110.50	114.33	5.98	−0.59	19.13	12.00
Shibpur–Hatisala	135.50	134.36	7.75	−0.81	17.35	12.00
Maheshnaga–Radhanagar	141.75	124.67	7.08	0.91	17.62	11.50
Bara Andulia–Teghari	153.00	216.50	3.57	6.00	65.80	12.50
Brittihuda–Dogachhia	160.75	160.00	3.08	5.50	51.90	11.25
Char Mahatpur–Sonatala	175.50	100.65	3.95	4.00	25.51	12.25
Char Mahatpur–Pukuria	183.00	182.10	4.91	5.19	37.11	12.50
Debipur–Tilakpur	196.50	158.65	6.26	1.55	25.34	12.75
Krishnanagar–Mayakol	206.75	193.56	6.62	−2.90	29.23	10.00
Swarupganj–Mayapur	220.50	130.79	7.34	−2.69	17.81	9.50
Max.	220.50	224.21	7.75	19.00	136.16	19.00
Min.	0.00	75.89	0.83	−2.90	17.35	9.50
Average	104.84	155.78	4.44	7.02	46.48	14.18
SD	73.16	45.41	2.05	6.75	31.68	3.04
CV	0.70	0.29	0.46	0.96	0.68	0.21

[a] River channel is not traceable at these stations

both these maximum and minimum wavelengths of meanders are in the dead reach of the river, why such an acute bend has been formed at Jayrampur and a broad one at Karimpur is not known. Moreover, the reach is dead now .and hence. the dynamics of the river which might be responsible for such bending are also absent. Wavelength (λ) and channel width (w) vary proportionately, and it was found that $\lambda = 23.73\,w$. The wavelength of different magnitude shows relation with other meander properties like meander amplitude (a_m) and radius of curvature of meander (r_c). Leopold and Wolman (1960) found $\lambda = 4.7 r_c^{0.98}$. For the river Jalangi, it was $\lambda = 3.45 r_c$ or $r_c = 0.29\,\lambda$ (Figure 12.9).

b. Bend amplitude (a_m)

　　Perpendicular height of the apex of meander bends from meander axis is termed as *bend amplitude* or *meander amplitude* (a_m). The average amplitude of the bend of Jalangi is 5,107.14 m. The highest value of 8,500 m is recorded at Taranagar near the Radhanagar

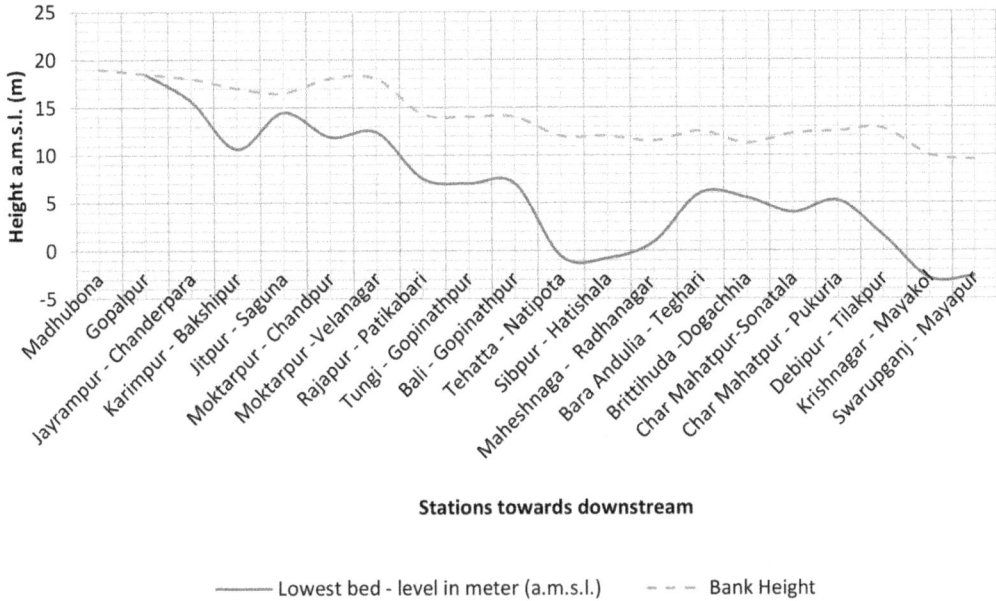

FIGURE 12.8 Variation of bank height and channel depth towards downstream. (Field survey during 2007–2010.)

loop, and the lowest amplitude of meander is found at Jayrampur, which is 1,250 m only. The overall meander amplitude of the river is very high which is maybe caused by positive feedback mechanism between the *amplitude of meander* (λ) and *reduction of slope* (s). Meander amplitude increases with increasing channel width towards downstream (Figure 12.10), and we found $a_m = 29.08$ w. Meander amplitude, wavelengths (λ) and radius of curvatures (r_c) of the river Jalangi are related as $a_m = 1.04\lambda$ and $a_m = 2.79 r_c$.

c. The radius of curvature (r_c)

   The average radius of curvature of meander bends of the river Jalangi is 1,113.10 m (Table 12.8). The largest radius of curvature of the meander of 4,250 m was recorded at Karimpur, and the smallest of 250 m was recorded at Jayrampur. The radius of the curvature of a meander bend is the scaled properties of a river (Das, 2014b) and increases with increasing channel width. Channel width increases gradually towards downstream. Although Karimpur is located to the downstream of Jayrampur, the largest r_c ought to be recorded far downstream near Krishnanagar. *Radius of curvature/width ratio* (r_c/w) is a dimensionless measure of the relationship between r_c and w. The average value of the ratio 6.88 implies that $r_c \sim 7w$ (Figure 12.11) which is significantly different from finding of earlier studies ($2w < r_c < 4w$ by Knighton, 1977; $r_c = {}^{0.98}\!\sqrt{2.32 w^{1.01}}$ by Leopold and Wolman, 1960). We found $a_m = 2.79 r_c$ and $\lambda = 2.88 r_c$. Downstream variation in r_c represents no such significant trend at all (Figure 12.10).

d. Sinuosity index (SI)

   Ratio between channel length and meander axis length (L/m_{ax}) was used for calculating the SI to illuminate the nature of the meander of the river Jalangi. The average SI of the river is 2.67. The sinuosity of the river is very high, and some meander necks like Radhanagar loop and Panditpur loop are waiting for cutoff to be ox-bow lakes. But due to detachment from the feeder river Padma, the river is too feeble to cut the necks or chutes of the meander. So Panditpur loop with 220 m neck width (equal to 2 w) and Radhanagar loop with 432 m neck width are there for decades.

TABLE 12.8

Downstream Variations of Meander Properties

Stations	r_c	λ	a_m	w	r_c/w	λ/w	S_mI	F_mI
Madhubona	0.00	0.00	0.00	0.00	0.00	0.00	0.00	0.00
Gopalpur	0.00	0.00	0.00	0.00	0.00	0.00	0.00	0.00
Jayrampur	250.00	1,500.00	1,250.00	217.85	1.15	6.89	0.20	0.83
Dogachhi	2,250.00	7,000.00	4,750.00	198.65	11.33	35.24	0.47	0.68
Karimpur	4,250.00	10,000.00	4,750.00	204.55	20.78	48.89	0.89	0.48
Saguna	500.00	5,000.00	5,000.00	84.82	5.89	58.95	0.10	1.00
Fazilnagar	1,000.00	5,875.00	3,750.00	224.21	4.46	26.20	0.27	0.64
Char Brindabanpur	1,250.00	3,750.00	3,750.00	201.50	6.20	18.61	0.33	1.00
Raninagar	250.00	2,500.00	3,500.00	189.35	1.32	13.20	0.07	1.40
Radhanagar	500.00	3,000.00	3,500.00	164.06	3.05	18.29	0.14	1.17
Tarangar	750.00	2,000.00	8,500.00	169.10	4.44	11.83	0.09	4.25
Chak Rudranagar	250.00	1,750.00	8,000.00	158.95	1.57	11.01	0.03	4.57
Natipota	1,250.00	4,250.00	6,000.00	110.50	11.31	38.46	0.21	1.41
Khaspur	1,125.00	4,750.00	6,000.00	126.57	8.89	37.53	0.19	1.26
Chanderghat	750.00	4,750.00	7,500.00	152.05	4.93	31.24	0.10	1.58
Mahesnagar	750.00	5,250.00	7,500.00	131.60	5.70	39.89	0.10	1.43
Brittihuda	1,375.00	3,250.00	3,750.00	129.65	10.61	25.07	0.37	1.15
Bangaljhi	625.00	3,750.00	6,250.00	153.20	4.08	24.48	0.10	1.67
Mahatpur	2,000.00	3,750.00	6,000.00	182.10	10.98	20.59	0.33	1.60
Panditpur	875.00	1,750.00	6,500.00	158.65	5.52	11.03	0.13	3.71
Ghurni	500.00	2,500.00	3,500.00	132.50	3.77	18.87	0.14	1.40
D.L. Roy Bridge	1,375.00	3,750.00	3,750.00	193.56	7.10	19.37	0.37	1.00
Chakgokul Nagar	1,500.00	3,750.00	3,750.00	130.79	11.47	28.67	0.40	1.00
Max	4,250.00	10,000.00	8,500.00	224.21	20.78	58.95	0.89	4.57
Min	0.00	0.00	0.00	0.00	0.00	0.00	0.00	0.00
Average	1,016.30	3,646.74	4,663.04	148.44	6.28	23.67	0.22	1.44
SD	918.79	2,217.14	2,303.17	58.79	4.86	14.86	0.20	1.18
CV	0.90	0.61	0.49	0.40	0.77	0.63	0.91	0.82

e. Meander shape index (S_mI)

This was defined as the ratio between the amplitude (a_m) and radius of curvature (r_c) and formulated as

$$S_mI = \frac{r_c}{a_m} \left(Das, 2013, 2014\right).$$

The smaller the value of S_mI is, the more intense the meander is. It was found that maximum S_mI and minimum S_mI were 0.89 and 0.03, respectively. Median S_mI was 0.14. Therefore, 100% meanders have $a_m > r_c$, and those are more sinuous than a regular sine curve. More than 50% meanders were acute meander (median $S_mI = 0.14$). The average S_mI was 0.22. Downstream variation in S_mI shows no significant trend at all (Figure 12.12).

f. Meander form index (F_mI)

Meander form index is the measure of the degree of intensity of curvature of the meander. This was defined as the ratio between the amplitude (a_m) and wavelength of meander (λ) and formulated as

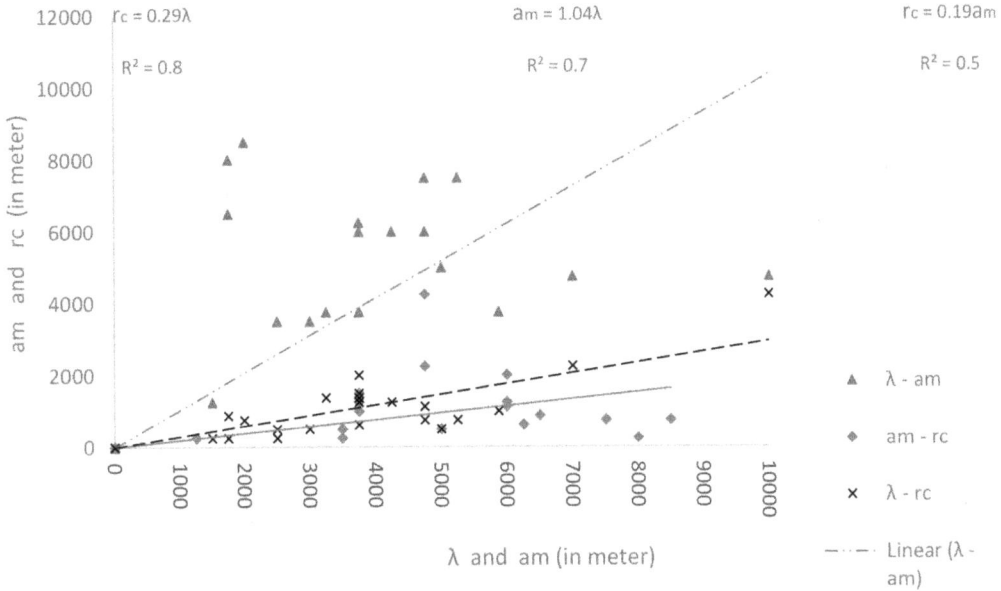

FIGURE 12.9   Relation amongst a_m, r_c and λ.

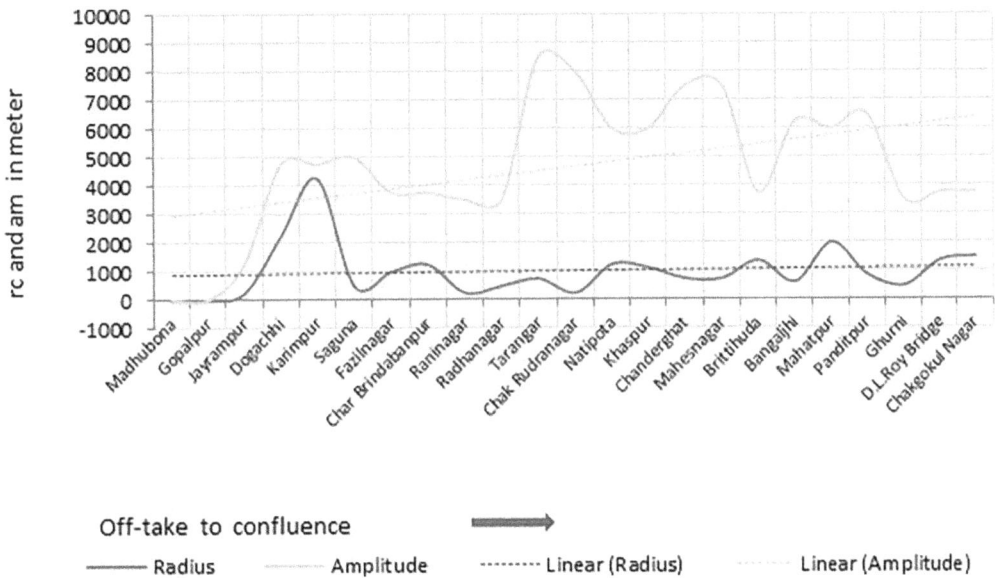

FIGURE 12.10   Downstream variations of the radius of curvature (r_c) and amplitude (a_m) of meanders. (Topographical map Nos. 78D/7, 8, 9, 12, 79A/5, 6, 7, 9.)

$$F_m I = \frac{a_m}{\lambda} \left(\text{Das}, 2013, 2014 \right).$$

The higher the value of $F_m I$ is, the more intense the meander is. Maximum $F_m I$ and minimum $F_m I$ were 4.57 and 0.48, respectively. Average $F_m I$ was 1.44, and the median value was 1.17. This implies that more than 50% of meanders were acute in nature.

Cross-sectional and Meander Properties – Meander properties like wavelength (λ), amplitude (a_m) and radius of curvature (r_c) all increase with increasing channel width (w).

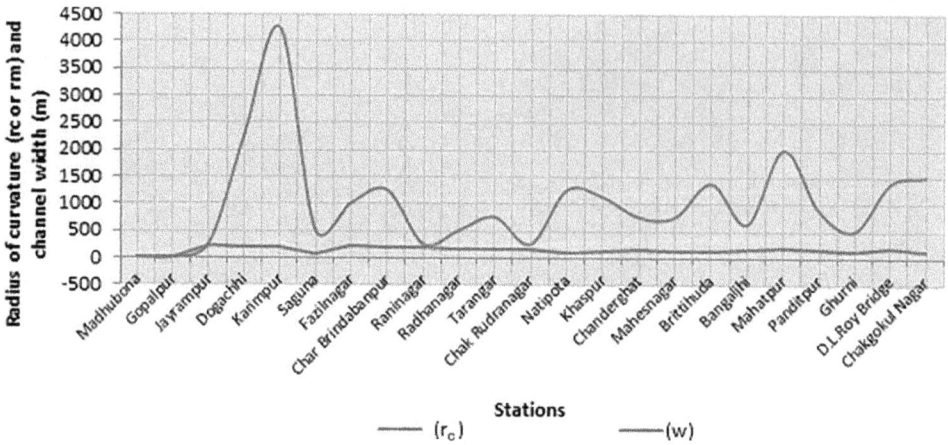

FIGURE 12.11 Variation in the radius of curvature (r_c) and channel width (w) towards downstream (Tables 12.7 and 12.8). (Field survey and Topographical map Nos. 78D/7, 8, 9, 12, 79A/5, 6, 7, 9.)

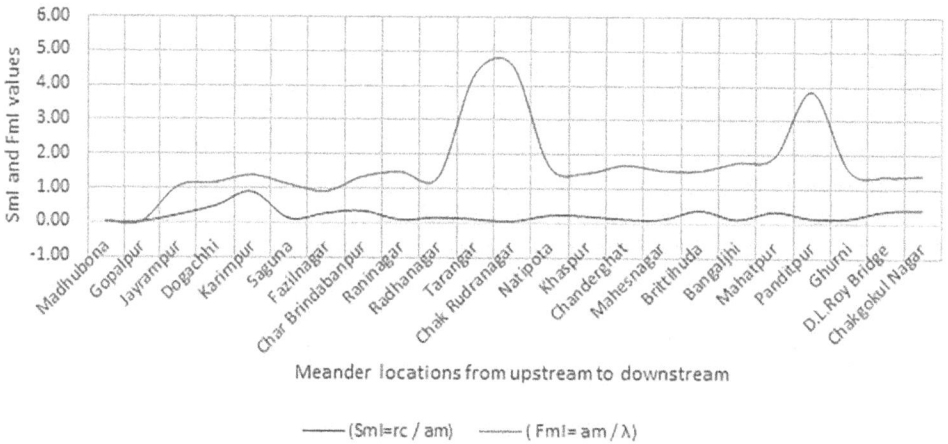

FIGURE 12.12 Downstream variations of S_mI and F_mI. (Topographical map Nos. 78D/7, 8, 9, 12, 79A/5, 6, 7, 9.)

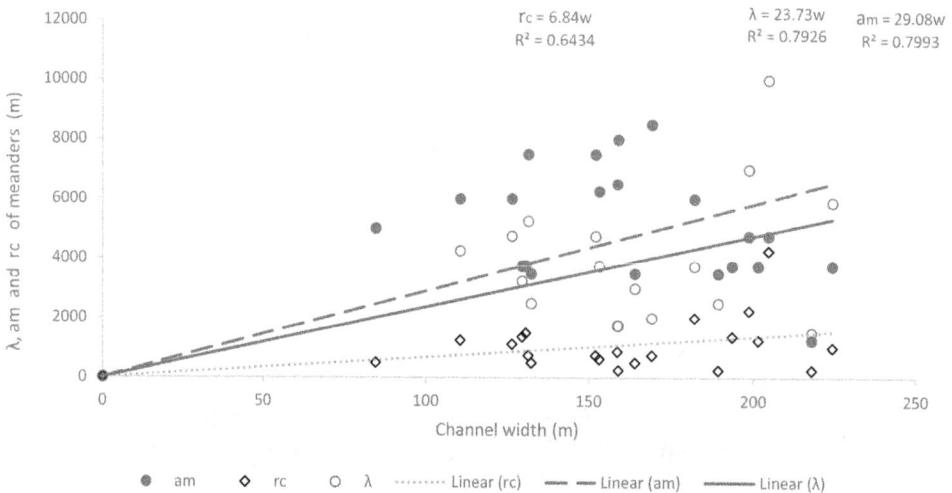

FIGURE 12.13 Relation amongst channel width and meander properties.

The wavelength-channel width relation of the river Jalangi is represented as $\lambda = 23.73\,w$. It implies that the wavelength (λ) of the meander bend of the river is about 24 times of channel width (w). The amplitude of meander bend and the channel width is related as $a_m = 29.08\,w$ (Figure 12.13). Therefore, the average amplitude of meander bends is larger than wavelength which explains the highly tortuous nature of the river. The radius of curvatures of meanders and channel width is related as $r_c = 6.84\,w$.

12.3 ANTHROPOGENIC SIGNATURES ON FORMS AND PROCESSES

12.3.1 IN-CHANNEL ANTHROPOGEOMORPHOLOGY

12.3.1.1 Rail and Road Crossings

At present, there are three complete road-bridges and two rail-bridges on the river Jalangi. Three incomplete road bridges are also waiting for completion. Information of daises and piers of all these eight bridges is listed in Table 12.9. From the holistic viewpoint, an engineer not only joins two roads/rails on opposite banks of a river but also makes the best match between the maximum strength of the bridge and minimum interference with the river. But when the river is of a dying kind, intentionally or unintentionally, the design of bridges focuses only on longevity and load capacity at the cost of the river's health. The dais at base of piers of all the bridges on Jalangi, Churni, Bhairab, Sialmari and other dying rivers of deltaic West Bengal constructed after the post-Farakka Barrage period (1975) proves this statement. Dais and piers designed neglecting the river's health not only interfere with the flow pattern of the river but also modify channel bed and bank morphology significantly.

Out of eight bridges (Figure 12.14), four is sited at Krishnanagar within a reach of 500 m only. Two of these, one railway bridge on the Sealdah–Lalgola Railway and another road bridge on NH34, were constructed before the Farakka Barrage project in 1905 and the 1960s, respectively (Das, 2019).

A New Railway Bridge (NRB) was constructed in 2012 to host the second track of the Sealdah–Lalgola Railway. But unlike Old Railway Bridge (ORB) and *Dwijendra Setu* (DLRB) on NH34, NRB was designed in such a way that it created a huge obstacle to the flow of the river and made a great change in the channel morphometry (Table 12.10). To assess the impact of the bridge construction on river channel morphometry and hydraulics, first-hand data were collected before and after the construction of the bridge. Velocity distributions across the channel were measured using a submerged-float method and found a directional change as well as a change in magnitude also. Channel asymmetries were calculated using formula $A^* = \dfrac{A_r - A_l}{A}$ (Knighton 1981), and flow asymmetry was determined using formula $A^Q = \dfrac{Q_r - Q_l}{Q}$ (Das 2019). It was found that because of bridge construction, channel asymmetry and flow asymmetry have increased considerably. Due to the construction of the NRB, mainly because of its midchannel dais, the left bank line at downstream has retreated 7 m swallowing 1,125 m^3 of soil.

Construction of a pier (and a dais of significant dimensions at its base) within midchannel had greatly changed the processes of flow dynamics of the reach. This change in processes, in turn, shaped the channel forms to set equilibrium between forms and processes.

A. Adjustment in hydraulics:

Impulsive construction of NRB made the main current of the river flow along the left bank at the reach bringing a significant change in flow asymmetry. A similar finding was recorded in the study of Keeley in 1971. Flow asymmetry (A^Q) at NRB station was calculated before and after construction, which was −0.03 and −0.65, respectively, a change of

TABLE 12.9

Features of Bridges on the River Jalangi

Sl. Number	Name of the Bridge	Location of Road Crossings	Latitude/ Longitude	No. of Piers	Width of Piers	Dais at the Base of Piers
1.	Karimpur–Bakshipur	On Karimpur–Domkal Road	23°58′58″ N and 88°37′20″ E	5 (2 piers within the wetted channel)	2 m	Absent
2.	Fazilnagar–Amtala	On proposed Fazilnagar–Amtala Road	23°55′48″ N and 88°27′32″ E	4, (3 completed from the right bank)	Round piers with diameter 2.5 m	Round dais with 7 m diameter
3.	Radhanagar–Patikabari	On Radhanagar–Patikabari–Nawda–Amtala Road	23°49′20″ N and 88°28′14″ E	6, (2 on the left bank, beyond channel)	Paired piers on the round dais. Diameter 1.5 m each	Round dais with 7.5 m diameter
4.	Dwijendralal Setu, inaugurated by Honorable Pūrta & Ābāsana Minister Sri Jatin Chakraborty on 1st July 1979	On Palashipara–Tehatta Road	23°47′48″ N and 88°27′09″ E	3, (2 piers within the wetted channel)	Squared piers, one side 2.5 m	-
5.	*Dwijendra Setu*, named by Honorable Pūrta & Sarak minister Sri Kshiti Goswami on 21st July 1995	On NH34 at Krishnanagar	23°24′48″ N and 88°28′16″ E	6 (3 piers within the wetted channel)	Oval piers, width 2.5 m	Absent
6.	Proposed Road Bridge-II Krishnanagar	On NH34 at Krishnanagar	23°24′48″ N and 88°28′17″ E	Under construction	NA	NA
7.	Old Rail Bridge at Krishnanagar	On Sealdah–Lalgola Railway at Krishnanagar	23°24′48″ N and 88°28′11″ E	5 (3 within the channel)	Oval piers 3.95 m	Absent
8.	New Rail Bridge at Krishnanagar	On Sealdah–Lalgola Railway at Krishnanagar	23°24′48″ N and 88°28′12″ E	5 (2 within the channel)	2 m	Rectangular dais with dimension 15.25 m ×10.65 m

23.89 times. This happened because before the construction of NRB, the velocity of the reach was $0.60 \, m^{-s}$ along left bank (as on 16.08.2009), $0.48 \, m^{-s}$ in midchannel and $0.36 \, m^{-s}$ along the right bank. The average velocity of the reach was $0.48 \, m^{-s}$.

During the construction of NRB, about half of the wetted channel along the right bank was spurred by earth to facilitate the access to the midchannel pier for constructional work of the bridge. This was done by putting $130,000 \, ft^3$ or $3,681.20 \, m^3$ of soils (Figure 12.15) in the river bed which has not yet been removed even after completion of the construction. This earthen spur across the river has forced the lion share of the flow towards the left bank. Moreover, the dais of the midchannel pier has shortened the width of the channel resulting in higher velocity along the left bank in

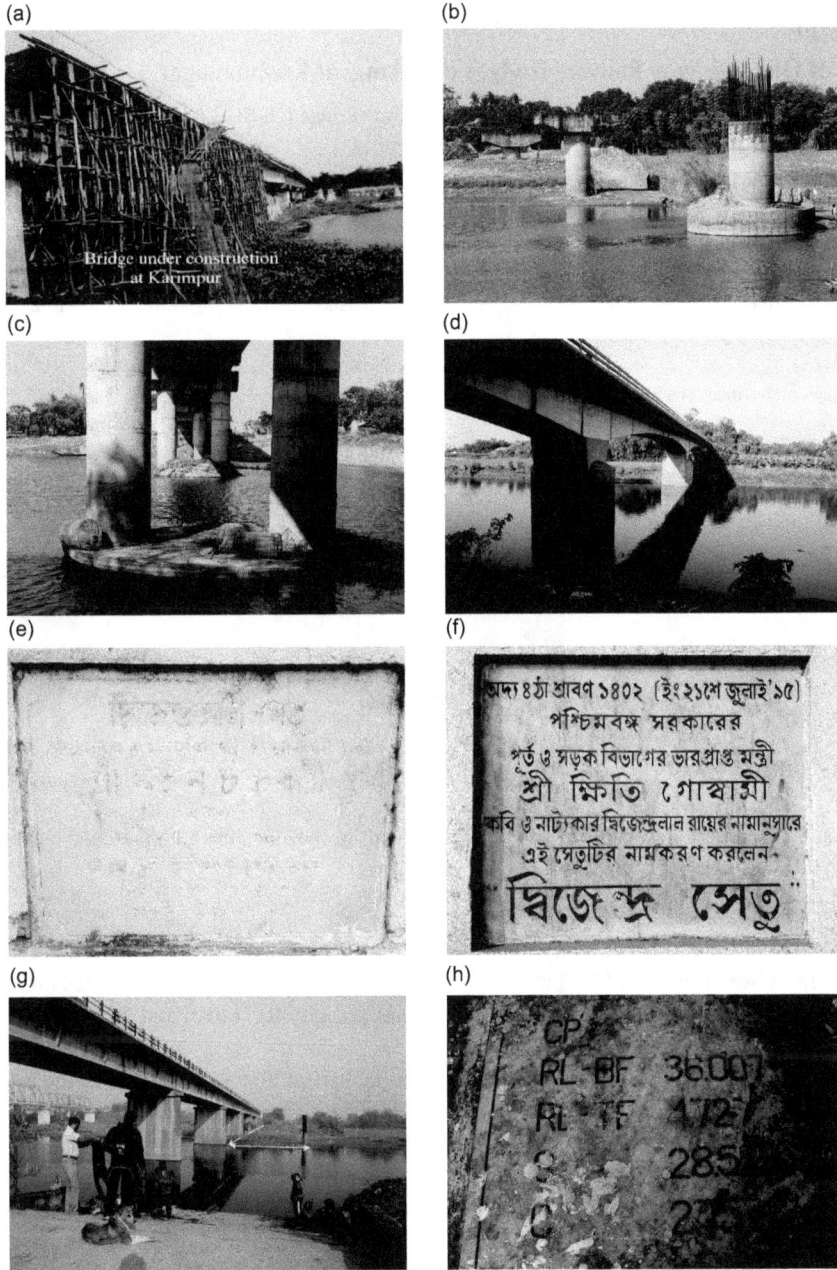

FIGURE 12.14 Bridges on the river Jalangi. (a) bridge under construction at Karimpur on the dead reach of the river where earthen *bāndh* was made from one bank to another bank to facilitate construction of the bridge. (b) Mid-Channel dais of Amtala bridge (under construction) and (c) Radhanagar–Patikabari bridges interfere more with the flow process than piers with a sharp nose. (d) Square-nose piers of Dwijendralal Setu at Palashipara make much more obstacle to flow than sharp- and round-nose pier. (e) Name of the bridge sculpted on Dwijendralal Setu and (f) *Dwijendra Setu*. (g) For construction purpose of the proposed second bridge of NH34 at Krishnanagar, again half of the channel width was closed by putting soil (marked with arrows). (h) the marble-encrypt at the base (left bank) of the NRB at Krishnanagar (CP = Concrete pile foundation; RL-BF = Depth from rail level to bottom of foundation; RL-TF = Depth from rail level to top of foundation; S = Sand; C = Clay).

TABLE 12.10

Features of Old and New Railway Bridges on Jalangi at Krishnanagar

Item	Old Railway Bridge (ORB)	New Railway Bridge (NRB)
Number of piers	5	5
Number of piers within the river channel	3	2
Width of piers	3.95 m	2 m
Length of piers (horizontal)	9.30 m	9.15 m
Cross sectional area of piers	37.2 m^2	18.3 m^2
Length of the *dais* (the platform on which pier is based)	No dais	15.25 m
Width of dais	No dais	10.65 m
Height of daises above mean sea level (MSL)	No dais	Variable (5.55 m for midchannel dais)
Lowest bed level of the river		3.87 m below MSL

FIGURE 12.15 Spur/*bāndh* was made up to more than half of the channel width putting a huge volume of soil (130000cft) into the river bed.

between piers of ORB and NRB. It is found from Table 12.12 that after construction, the velocity of the current in between piers of ORB and NRB along the left bank was 0.995 m^{-s} (as on 24 August 2013) which was 1.4 times higher than the midchannel velocity (0.71 m^{-s}) and 2.1 times higher than velocity (0.47 m^{-s}) along the right bank (Figure 12.16).

The most decisive structure for the interference of the hydraulics of the river was the midchannel dais of the pier. Its length, breadth, and height (Table 12.10; Figure 12.17) are so designed, which makes obstacles to about 4.95% of total flow (calculated in respect of mean water level (MWL), the average of the highest flood level (HFL) and the lowest water level (LWL)) of the river. The height of the dais corresponds to 97.30% (Table 12.11) of the LWL. This is very crucial so far as the lean season flow and self-maintenance of the channel are concerned. Similar observations were found in the study of Simon and Downs (1995) and Musy and Higy (2011).

  B. Adjustment of channel morphometry:

     Cross section at NRB station before and after the construction of the bridge (Figure 12.18) shows that there was a great change in the morphometry of the channel. Before the construction of NRB, the right bank (7.545 m) was 1.205 m higher than after the construction of NRB (6.34 m). Soil cutting from the right bank to facilitate the movement of lorries carrying construction materials was responsible for that change in bank height.

Onsite channel deepening due to constriction and consequent scouring of channel bed was noted by Schumm (2005) and Inglis (1949). NRB made the channel constriction and flow concentration. Concentrated flow in the constricted channel at NRB station stimulated bed to scour as a result of

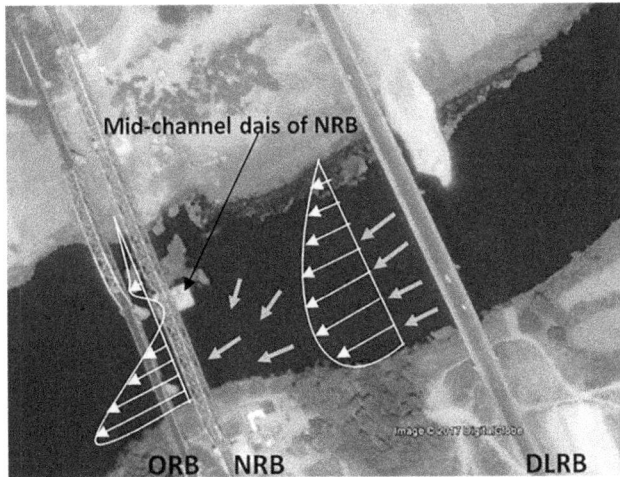

FIGURE 12.16 Flow direction (Thick arrows) immediately downstream of DLRB is slightly diverted towards the left bank. But in between NRB and ORB, it is converging towards left bank because of the mid-channel dais (pointed by black arrow). Velocity distribution across the channel is proportional to white thin arrows.

which thalweg level (d_{max}) before NRB was −2.94 m which was −3.87 m after the flood of 2012. To keep pace with the volume, the river scoured its bed 0.93 m deeper towards the left bank. Due to the construction of earthen spur up to midchannel from the right bank and location of the midchannel dais, the flow was diverted towards the left bank. This diverted flow pushed thalweg point 12 m (34 m from left bank before construction of NRB 22 m from left bank after construction of NRB) towards the left bank. After NRB, midchannel dais reduced the width of the channel.

Diverted and concentrated flow triggers bank erosion (Keeley, 1971) and onsite channel deepening (Schumm, 2005; Inglis, 1949). At NRB station, channel asymmetry before and after NRB was −0.03 and −0.57, respectively, with an increase of 21.05 times (Table 12.12). Channel asymmetry at 107.5 m downstream before NRB was −0.15 which became −0.36 after NRB, with an increase of 2.40 times.

The dais at the base of the midchannel pier created a considerable obstruction and created diversion of flow (Figure 12.19). The flow was diverted and concentrated towards the left bank. The midchannel dais of the pier was completed before the flood of 2012, and the single flood of 2012 engulfed about 500 m² of agricultural lands by the river (Figure 12.20). About 15 trees on the bank have been swallowed by Jalangi. Here, the river has widened its width by 5.93% cutting 7 m of the left bank. The volume of eroded bank materials was estimated to be at least 1,125 m³. A whirl of a back current of 72 m length and 7 m width was developed which scoured the left bank immediately downstream of the construction work. Indian Railway, as reported by the then Executive Engineer, did not take any consent regarding the design of the NRB across the river Jalangi from Executive Engineer, Irrigation and Water Ways Department, Government of West Bengal. Some bamboo porcupines with sandbags, financed by Eastern Railway and executed by the "Irrigation and Water Ways Department", have been introduced to protect further erosion of the bank. A shoal along the right bank has been developed 250 m downstream of the NRB site.

Recently, the construction of another Road Bridge-II on NH34 to the upstream of the *Dwijendra Setu* has started. To facilitate construction of piers, in 2016 again ~3,600 m³ or ~123,600 ft³ of soil was put into the river bed to make a spur up to midchannel from the right bank modifying channel cross-sectional geometry (Figure 12.21). However, although the construction process is now withheld, that enormous volume of soil remained in the river bed as was during the construction of NRB (Figure 12.14g).

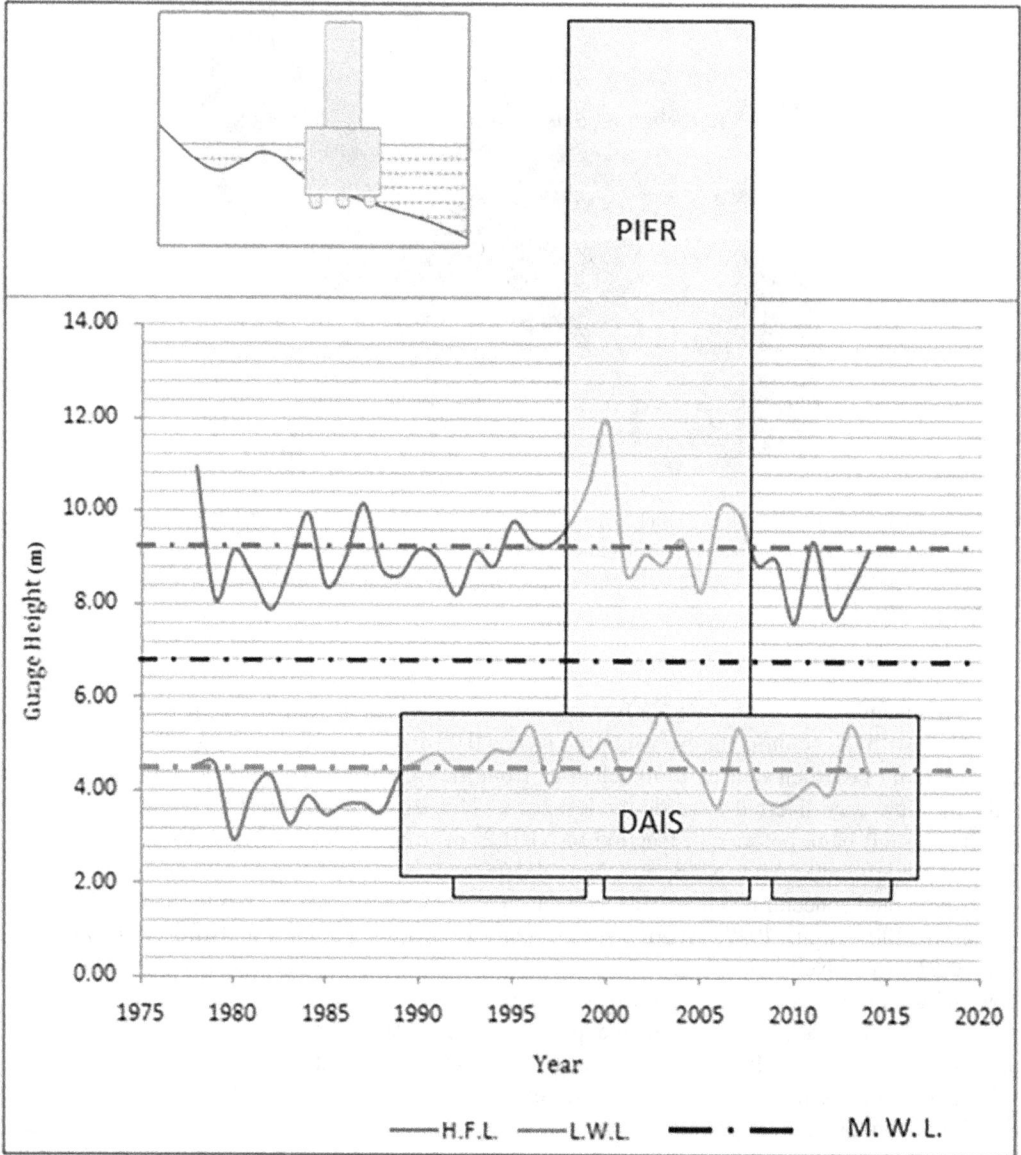

FIGURE 12.17   Square nose of midchannel dais creates an obstacle to the flow many times than the round-nose pier does.

TABLE 12.11

Obstacle of Flow by Midchannel Dais and Pier in the River Jalangi at Krishnanagar

The cross-sectional area of channel at a MWL of 6.77 m	585.83 (m²)
The cross-sectional area of the dais	33.44 (m²)
The cross-sectional area of piers making obstacle to flow regime	2.44 (m²)
The cross-sectional area of base-piers making obstacle to flow regime	6.58 (m²)
Total obstacle made to flow regime	42.46 (m²)
% of obstacle made to flow regime	4.95

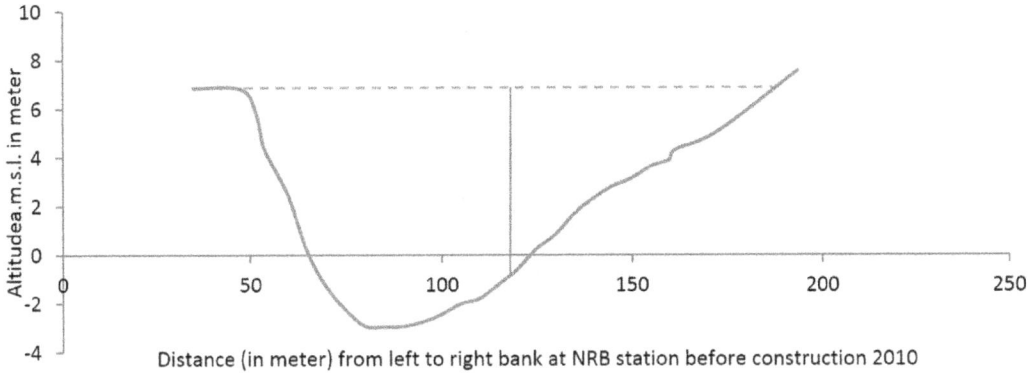

FIGURE 12.18 Cross section of the river at Krishnanagar in 2009 before construction of NRB.

12.3.1.2 Channel Training

River training in moribund (Bagchi, 1978) deltaic Bengal during British rule which caused changes in channel pattern and the course of the rivers were mainly

i. Straightening of river course by cutting meander neck
ii. Closing offtake to divert flow through desired channel
iii. Opening of new offtake
iv. Clearing of snags and presently
v. Bank protection.

There is no such history of artificially closing and opening of offtake of the river Jalangi. In 1830–1831, Mr. May's attention was called to the encroachment on a village near Krishnanagar and in considering the means for arresting its further advance advocated a plan for improving the navigation by cutting a series of cuts through necks of larger bends, thereby shortening its course by increased current obtaining greater depth. The proposed cut was not made, for what reason does not appear (Hunter, 1877). Straightening of meandering rivers by cutting necks is not a permanent solution felt repeatedly all over the world. The very mechanism of delta building inherits the property of the oscillation of the river course within the meandering belt. So, any effort to straighten meanders will lead meandering soon (Figure 12.22).

However, some measures taken to train the river by the then Nadia River Division of Public Works Department authority is listed in Table 12.13 (see also Chapter 13).

12.3.1.3 Ferry Service and *Bāndh* across the River

At present, the most decisive anthropogenic intervention within the river bed that altered channel forms significantly and interfered flow and river processes is the activities related to ferry service. For the purpose of reducing the length of ferry services, the earthen dam across the river is very common in the river Jalangi particularly upstream to the Suti confluence. From both sides of the river, soils are put into the river bed leaving a very narrow channel where boatmen keep three or four boats tied together to make faster, easier and heavier ferry service. Even four-wheeler motor vehicles, lorries, etc. also move across these earthen dams and boat-made bridges. These earthen dams reduce the span for free flow of the river accelerating bed scour.

There are 37 (Table 12.14) major *ferry ghats* over the river Jalangi out of which sixteen *Ghats* put soil in the river bed from both banks to make the channel narrow for the sake of easier, faster and heavier ferry service during the lean season. On average, every year at each ferry *Ghat* boatman charges a huge quantity of sediment to the river putting approximately ~15,000 ft^3 or ~425 m^3 of soil during the lean season. Therefore, the total volume of soil put into the river per year is

TABLE 12.12

Increased Asymmetries of Flow and Morphometry of Channel

| Year | Channel Asymmetry | | Bed Asymmetry | | Flow Asymmetry | | Average Velocity | | | |
| | Bridge Site Station | Downstream Station | Bridge Site Station | Downstream Station | Bridge Site Station | Downstream Station | In between DLRB and ORB (m min⁻¹) | | In between NRB and ORB (m min⁻¹) | |
							The Left Side of the Midchannel Pier	The Right Side of the Midchannel Pier	Left Part of the Channel	The Right Part of the Channel
2010	−0.03	−0.15	−0.20	−0.20	−0.03	−0.02	34.95	32.83	–	–
2013	−0.57	−0.36	−0.55	−0.29	−0.65	−0.053	35.9	29.05	59.7	28.4

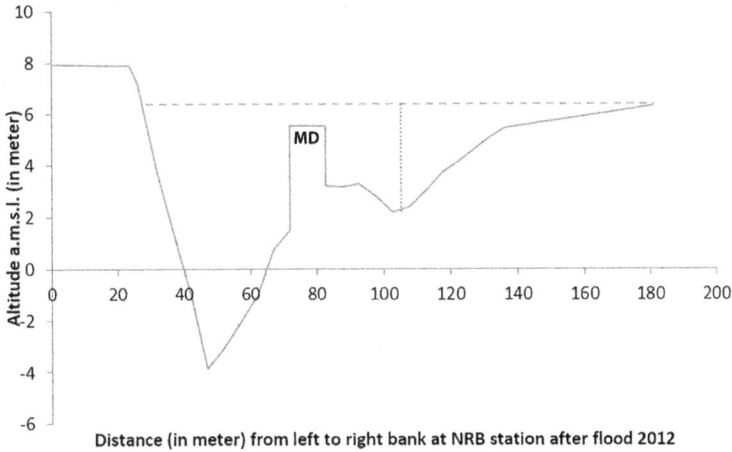

FIGURE 12.19 Cross section at NRB station after the flood of 2012 has been changed abruptly due to the construction of the midchannel dais MD.

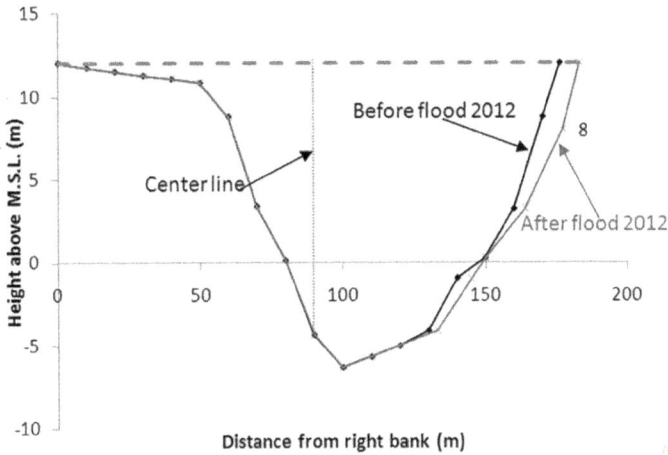

FIGURE 12.20 Superimposed cross sections at downstream of the bridge showing bank shift after the flood of 2012.

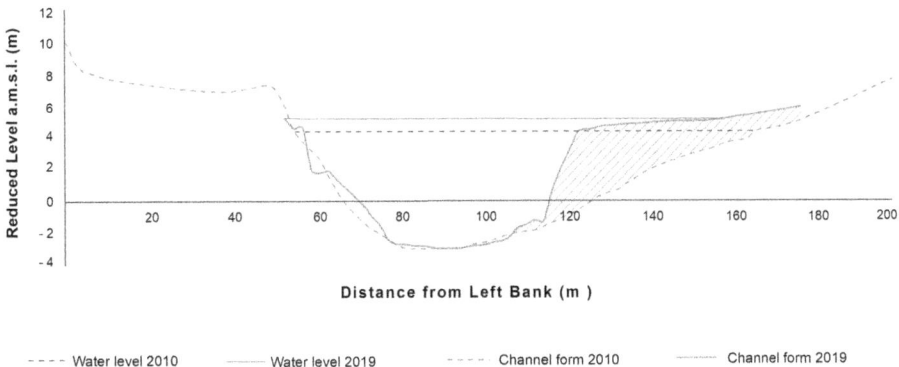

FIGURE 12.21 Change in cross-sectional geometry of the channel of the river at construction site of the proposed bridge at Krishnanagar, immediately upstream of *Dwijendra Setu*. Portion of the channel was filled by soil (shown as shaded portion to the right side) to facilitate construction of pier within the channel.

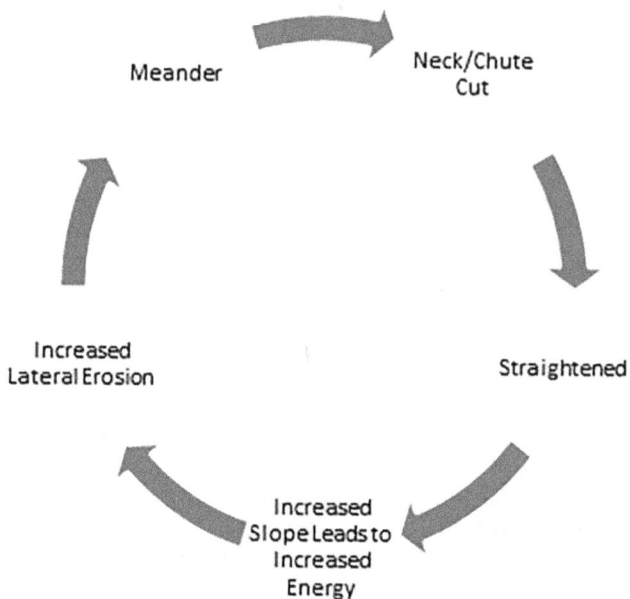

FIGURE 12.22 Artificial channel straightening leads meandering soon after it gets extra energy because of the increased slope.

TABLE 12.13
Measures Taken to Train the River Jalangi

Year	Nature of River Training	Source of Information	Remarks
1830–1831	Mr. May's attention was called to the encroachment on a village near Krishnanagar and, in considering the means for arresting its further advance, advocated a plan for improving the navigation by cutting a series of cuts through necks of larger bends, thereby shortening its course by increased current obtaining greater depth	Hunter (1877, p. 25)	The proposed cut was not made, for what reason does not appear.
1835	Work on Nadia Rivers stopped temporarily as the government thought that the result obtained was not worth the cost.	Hunter (1877, p. 220)	
1905–1906	"Nemotha" a converted dredger with 14-inch suction was employed to work on Maricha shoal.	Reaks (1919)	Entrance remained open in 1905–1906
1906–1907	Two dredgers "Alpha" with 18-inch suction and "Rescue" with 12-inch suction worked on the entrance of the Bhairab and on an average lifted 1,129–2,839 cft of soil per hour.	Reaks (1919)	Entrance remained open in 1906–1907
December 1906–March 1907	Two small dredgers were hired and worked on head shoals of the river Jalangi costing Rs. 38,000	Garrett (1910)	Effort was satisfactory

(Continued)

TABLE 12.13 (*Continued*)
Measures Taken to Train the River Jalangi

Year	Nature of River Training	Source of Information	Remarks
1908	Two dredgers were again employed at a cost of Rs. 12,000 during cold weather dredging 6.25 miles or 10.06 km of the river.	Do	Effort was satisfactory
1916–1917	Important shoals in the river with a length of 21,485 ft were trained	Reaks (1919)	Effect not mentioned
1929	A cut was made by Nadia river division from the old Jalangi to flush the Bhairab *Khal*, an almost dead sluggish river full of water hyacinths and weeds.	Majumder (1978)	The effort was not satisfactory
1930	At Narayanpur, 45 miles below Moktarpur, a dried-up *khal* which falls into the Bhairab *Khal* near Kamdevpur, Morgangnee project to excavate this *khal* connecting Jalangi was under consideration	Majumder (1978)	Whether implemented or not, unknown
2004	Bahadurpur, Bank protection scheme at Bahadurpur on the right bank, the average annual rate of bank erosion was 400 m² for which average annual loss estimated (2004) to be Rs.15.68 lakhs.	Department of Irrigation & Water Ways, 2004, Jalangi Bhawan,	-
2008	Natun Shambhunagar, Bank protection scheme, At Ruipukur–Natun Shambhunagar, the village road, agricultural lands (3.5 acres) and houses (5 nos.) have been swallowed by left bank erosion.	Department of Irrigation & Water Ways, 2004, Jalangi Bhawan,	Project complemented
2010	A length from Bhairab confluence upward of the channel of the river was excavated under 100 days' work program.	Offices of the B.D.O., Karimpur-II, Nadia, and Domkal, Murshidabad 2010	Done
2010	Bank protection scheme, near *Dwijendra Setu*, Krishnanagar	Department of Irrigation & Water Ways, 2010, Jalangi Bhawan	Project complemented
2011	Bank protection with piers and sandbags piling at Radhanagar, immediately downstream to the Kalma *Khal*–Jalangi confluence (Figure 12.23).	Department of Irrigation & Water Ways, 2011, Jalangi Bhawan	Project complemented

$6,800 \text{ m}^3 = \left(425 \text{ m}^3 \times 16 \text{ ferry Ghats}\right)$. During the rainy season, the river swept away a part of this huge quantity of soils and deposits a lion share into its bed or along banks, because the river has no sufficient energy to carry away the whole volume of soil introduced into it every year.

At Moktarpur–Bhelanagar ferry ghat, immediate downstream to the Bhairab–Jalangi confluence, in 2019, the spur reducing the width of the channel has been made by bricks so that the river cannot wash away during floods (Figure 12.24). The brick spur is made from the right bank closing 66.5 m of the channel width and leaving the only 8.25 m (Figure 12.24, from the right side of the photo 2.20 + 2.60 + 1.70 + 1.75 m) with intermittent piers made of sandbags for passing the flow along the left bank. The spur from the left bank is earthen. I wonder, after a heavy flood, the river

TABLE 12.14

List of Ferry Ghats and Their Location, and Mode of Operation on the River Jalangi

Reach		Length	Sl.	Location of Ferry Ghat	
From	To	(km)	No.	with Earthen *Bāndh*	Comments
Offtake at Char Madhubona	Sialmari confluence at Kupila	9.10	1.	Nil	The reach is erased out and no need to make any *bāndh* to cross the river
Sialmari Confluence at Kupila	Bhairab Confluence at Moktarpur	41.9	2.	Jayrampur–Kupila ferry ghat	Bamboo bridge
			3.	Durlabhpur–Rajapur ferry ghat	Earthen *bāndh* and Bamboo bridge
			4.	Kuchaidanga ferry ghat	Boat/Bamboo bridge
			5.	Karimpur Pattabuka ferry ghat	Boat/Bamboo bridge
			6.	Karimpur–Boxipur ferry ghat	Channel width narrowing by earthen *bāndh* and bridging the gap by boats tightened together
			7.	Karimpur Shisa Kadamtala ferry ghat	Boat/Bamboo bridge
			8.	Baishnabpara ferry ghat	Bamboo bridge
			9.	Sahebpara ferry ghat (Simultola)	Bamboo bridge
			10.	Dogachi Kushaberia (Ferry) Ghat	Earthen *bāndh* and Bamboo bridge
			11.	Char Moktarpur Ghat	Bamboo bridge
Bhairab confluence at Moktarpur	Suti confluence at Bali-Tungi	42.7	12.	Bhelanagar ferry ghat	Channel width Narrowing by earthen *bāndh* and bridging the gap by boats tightened together
			13.	Amtala–Fazilnagar ferry ghat	Wide bamboo bridge on the pier of sandbags with the intermittent passage of water flow (Figure 12.25)
			14.	Bonchadanga ferry ghat	Do
			15.	Gopalnaga–Paschim Char Brindabanpur ferry ghat	Bamboo bridge
			16.	Paranpur ferry ghat	Wide bamboo bridge on the pier of sandbags with the intermittent passage of water flow
			17.	Char Jagaipur ferry ghat	Wide bamboo bridge on the pier of sandbags with the intermittent passage of water flow. A brick kiln on the right bank is near the ferry ghat
			18.	Kanainagar ferry ghat	Wide bamboo bridge on the pier of sandbags with the intermittent passage of water flow.
			19.	Kanainagar second ferry ghat	Earthen *bāndh* and Bamboo bridge
			20.	Uttar Char Chandpur ferry ghat	Wide bamboo bridge on the pier of sandbags with the intermittent passage of water flow.

(Continued)

TABLE 12.14 (*Continued*)
List of Ferry Ghats and Their Location, and Mode of Operation on the River Jalangi

From	To	Length (km)	Sl. No.	Location of Ferry Ghat with Earthen *Bāndh*	Comments
			21.	Patikabari ferry ghat downstream to Bridge	Wide bamboo bridge on the pier of sandbags with the intermittent passage of water flow.
			22.	Raninagar–Patikabari ferry ghat	Bamboo bridge
			23.	Raninagar–Tungi ferry ghat	Wide bamboo bridge on the pier of sandbags with the intermittent passage of water flow.
Suti confluence	Kalma *Khal* confluence at Radhanagar	55.2	24.	Gopinathpur–Bali ferry ghat	Bamboo bridge
			25.	Gopinathpur ferry ghat	Bamboo bridge
			26.	Raghunathpur-Iswarchandrapur ferry ghat	Bamboo bridge
			27.	Krishnachandrapur ferry ghat	Bamboo bridge
			28.	Tehatta ferry ghat	Earthen *bāndh* and Bamboo bridge
			29.	Hanspukuria ferry ghat	Bamboo bridge
			30.	Kulgachi–Chanderghat Ferry	Bamboo bridge
			31.	Nimtala–Chanderghat ferry	Do
			32.	Putimari–Shibpur Ghat	Do
Kalma *khal* Confluence at Radhanagar	Bhagirathi–Jalangi Confluence	84.1	33.	Birpur ghat ferry	Wide bamboo bridge
			34.	Bara Abdulia ferry ghat	Bamboo bridge
			35.	Gopinathpur Teghari ferry ghat	Narrowing of the channel by earth filling and bamboo bridge
			36.	Brittihuda ferry ghat	Narrowing of the channel by earth filling and bamboo bridge
			37.	Gokhurapota ferry ghat	Do
			38.	Sonatala ferry ghat	Do

may engulf the left bank because the flow will not be able to wash semi-permanent brick-made spur of the right side and will find its path eroding left bank.

12.3.1.4 Brickfields

According to District Land and Land Reform Office (DL & LRO), Nadia, Government of West Bengal (2009) record, the average soil cut by each brickfield is 133,110 ft³ per annum. There are 455 brickfields (Figure 12.26) in the basin area of 2,815.33 km² of the river Jalangi. Therefore, total soil moved by brickfields is 60,565,050 ft³ per annum or 1,715,011.23 m³ per annum. The total dry weight of the volume is 1,395,418,752 kg. On the other hand, total sediment contributed by Jalangi–Churni system to the river Bhagirathi was estimated (Guchhait et al. 2016) to be 3,710,000,000 kg year⁻¹. Jalangi and Churni share basin area of 2,815.33 km² and 2,585.61 km², respectively. So, out of 3,710,000,000 kg year⁻¹ of sediment contributed by Jalangi–Churni system, Churni alone contributes 1,770,478,565.33 kg year⁻¹, and Jalangi alone contributes

FIGURE 12.23 Use of pilling and sandbags for bank protection (in between two arrows) at Radhanagar, immediately downstream of Kalma khal and Jalangi confluence.

FIGURE 12.24 Moktarpur ferry ghat (a) in 2009. 1 to 6 numbers are of boats used to support bamboo structure of river crossing. Use of boats neither made significant obstacle to the flow of the river nor changed channel cross-sectional form. (b) After ten years, in 2019, the brick spur (marked with large arrow) is made from right bank closing 60 m of the channel width and leaving the only 8.25 m (from the right side of the photo 2.20 + 2.60 + 1.70 + 1.75 m) with three intermittent piers made of sandbags for passing the flow along left bank. The small arrow shows earthen spur from the left bank.

1,939,521,434.66 kg year^{-1}. Therefore, brickfield as an anthropogeomorphic agent translocates

$$71.95\% \left(\frac{\text{Soil translocated by brick-fields}}{\text{Sediment transported by Jalangi system}} = \frac{1,395,418,752 \text{ kg year}^{-1}}{1,939,521,434.66 \text{ kg year}^{-1}} \times 100 \right) \text{ of earth}$$

materials as compared to the sediment translocated by the Jalangi system (Table 12.15).

Soil cutting from river banks and river beds is a problem exclusively related to the brickfield. Invariably all the brickfields are sited on the higher concave bank of meander turn and every brick-field makes *Khadan* (pond-like water body attached to the river made by cutting soil on the bank)

FIGURE 12.25 Two arrows are pointing towards upstream, and the river is flowing in between arrows intermittent by piers of bamboo bridge. To the right side of the right arrow, an earthen spur is extended more than half of the channel width. During the last 10–12 years, washed soil from man-made spur silted downstream to the lee side of the spur, and a mid-channel bar was created on which boys are walking.

to trap silt in those *Khadans* (Figure 12.27) during floods. But these *Khadans* become further extended by the attack of helical secondary current on concave bank causing bank erosion and gradual shifting of the river (Das, 2014b).

Out of 17 C. D. Blocks of the district of Nadia, 9 C. D. Blocks are along the course of river Jalangi within which there were 159 brick fields in 2009 (Das, 2014a). Out of 159 brickfields, 71 (44.65%) were on Jalangi banks. As per government record, 159 brick fields cut 19,835,669 ft^3 of soils per year. Seventy-one brickfields cut 9,559,995 ft^3 of soils per year out of which 2,867,999 ft^3 (30%) soils are cut from banks and the bed of river Jalangi (Figure 12.28).

Some brickfields cut silt and soil from the river bed. This practice not only instantly changed the channel form but also set a stimulant to the river processes which might have stirred the hydrodynamic stability of the river (Gailot and Piegy, 1999). Excavating *Khadans* and dumping thousands of cubic meters of soil change local geo-forms significantly. *Khadans* are significant anthropogeomorphological imprints on the earth's surface and last for several decades. On the other hand, although it is a temporary phenomenon, soil dumping for making bricks, sometimes, takes the size and shape of a hill bringing a significant change in the landscape (Figure 12.29).

12.3.1.5 Fishing

Different fishing practices, very common all along the course of the river Jalangi, obstruct the free flow of the river. Fishermen put Kāntā (branches of bushes and trees) into the river channel to make *komot* (cluster of varying sizes from few hundreds square metres to few thousand square metres of branches of bushes and trees submerged into the river) (Figure 12.30) so that fishes gather into that *komot*. These *komots* not only interfere with the flow process of the river but also stimulate the huge growth of submerged aquatic plants. These submerged aquatic plants prohibit bed scouring and trap suspended and bedload sediments to settle rapidly and raise the river beds. *Komot* practice is found all along the course of the river, and it deprives the poor fishermen's community of earning their livelihood by fishing, because, to put kāntā into the river and to keep them in situ in a flowing river,

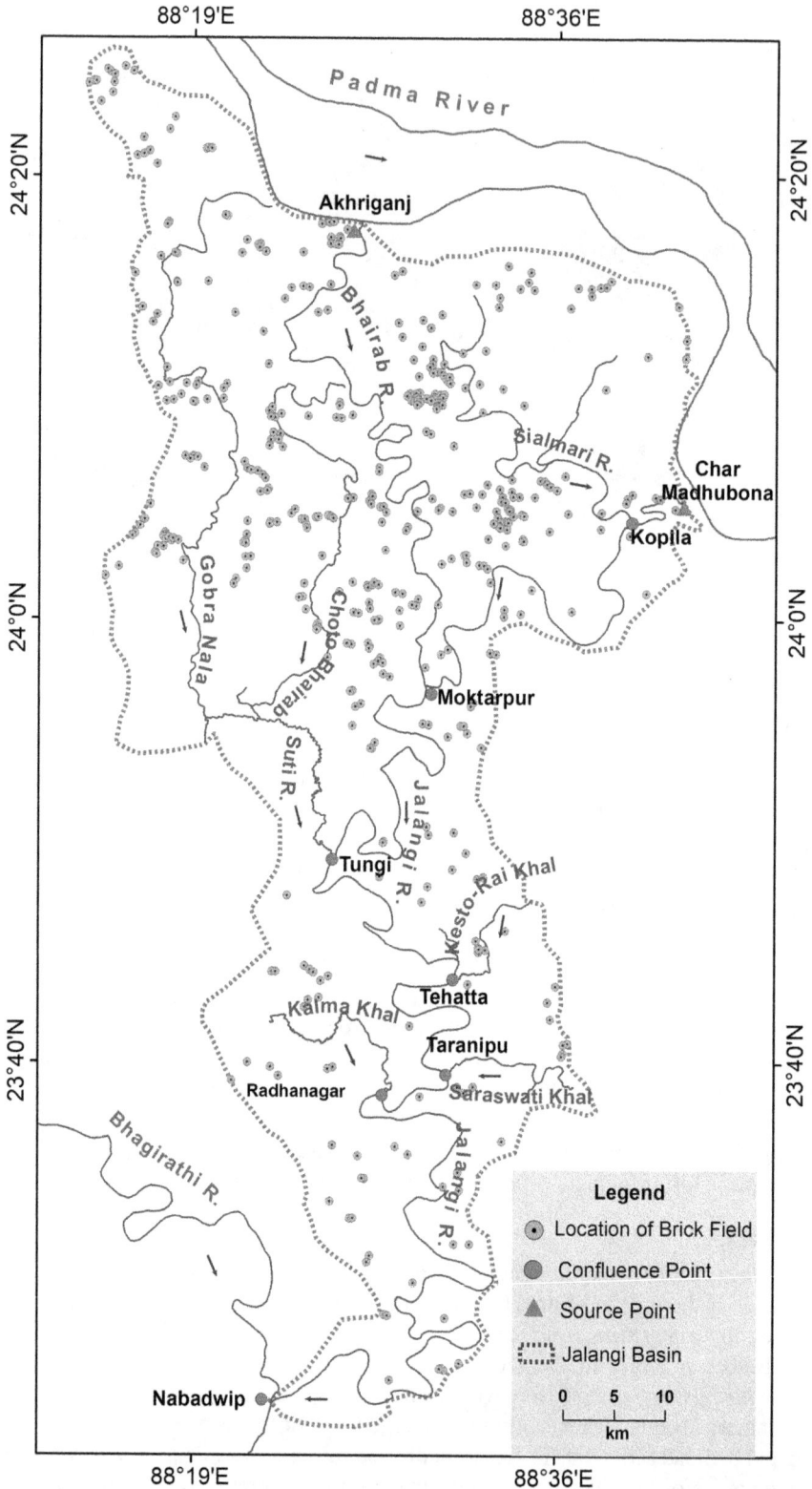

FIGURE 12.26 Location of brick kilns in the Jalangi river basin.

TABLE 12.15

Comparison of Earth Material Translocation by the River Jalangi and Ferry Ghats, Brickfields, Embankments, Bridges

Item of Geomorphology	Amount of Earth Material Moved by Volume	Amount of Earth Material Moved by the Weight
Sediment contributed to the Bhagirathi-Hooghly system by the river Jalangi	2,526,668.73 m^3 year^{-1}	2,055,823,812 kg year^{-1}
Ferry ghats	6,800 m^3 year^{-1}	5,532,819.46 kg year^{-1}
Brickfields	1,715,011.23 m^3 year^{-1}	1,395,418,752 kg year^{-1}
Embankment/road	17,758,600 m^3	14,449,284,890 kg
Bridge (three)	3×3,681.20 m^3	8,985,625.14 kg

FIGURE 12.27   (a) Soil cutting from bank and (b) *Khadan* reshaping river bank at Raninagar of C.D. Block Tehatta-I.

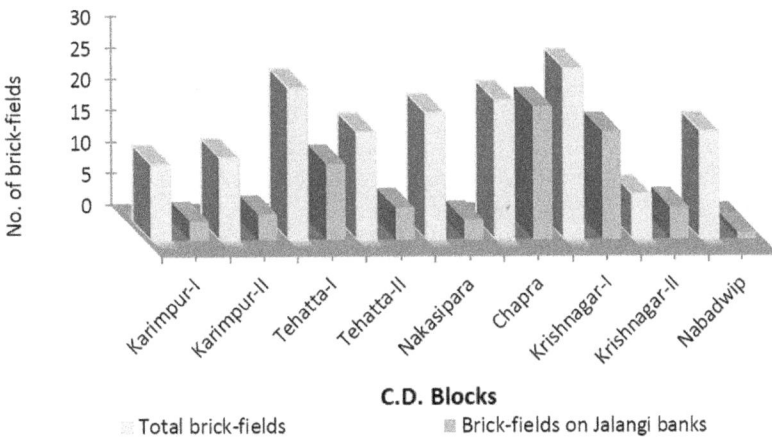

FIGURE 12.28   Distribution of brickfields on Jalangi banks and in respective C.D. Block in the district of Nadia.

FIGURE 12.29 (a) Soil-made hills at brickfields are the most temporary anthropogenic landforms. (b) soil cut deformed banks and made many *Khadans*.

FIGURE 12.30 *Komots* (4 to the left and 3 to the right of the channel) appear as green patches within the channel near Krishnanagar makes obstacle to the flow and encourage the growth of submerged aquatic plants. The arrow shows downstream direction.

it requires huge capital and labours which poor fishermen cannot afford. Moreover, *komot* constricts the space for fishing of poor fishermen not only because of its physical occupation of space in to the river but also the owners of *komots*, empowered by muscle and political powers, restricts fishing of poor fishermen within reach of kilometres of the river where they have put the *komot*. Giving lease of "river channel" to the fishermen cooperative society has widened the misery of these poor fishermen because their movement in the river becomes restricted again by a rich member of the society whose occupation is not fishing and who does not have any fishing instruments at all (Das, 2013).

Fishing instruments such as *Jhānp* (wall-like mat, made of the bamboo splinter) and *bānā* (bamboo-stick made mat to pass water but not fish through it) across the river bed create a significant obstacle to the free flow of the river modifying river-processes (Figure 12.31). This, in turn, due to retarded velocity, causes in-bed siltation and gradual deterioration of the river.

12.3.1.6 Immersion of Idols

The majority of the major settlements of Jalangi river basin are sited on river banks. And there is a common practice of idol worship amongst the inhabitants. It was estimated that the approximate number of Jagaddhatri idols worshiped in the town of Krishnanagar in 2019 was 160. The average volume of soil and weight of paddy straw needed for the making of a Jagaddhatri idol are 75 cft and

FIGURE 12.31 Use of *bānā* for fishing obstacle flow of the river significantly.

58 kg, respectively. The volume of soil and weight of paddy straw needed for the making of different idols and the approximate number of idols worshiped each year were estimated (Table 12.16) on the basis of the interview conducted among idol makers.

Almost all these idols are immersed in the river Jalangi after worship. So, all the soil and paddy straw along with wooden and bamboo structures interfere channel's flow processes and add huge sediments into the river reshaping channel geometry and bedforms (shoals). It was estimated that the total volume of soil added each year from idols to the river adjacent to the town of Krishnanagar is 32,687 cft. Straw added to the river per year was 23681.5kg. In 2019, some idols were removed from the river by municipality authority, but most remained in the river. At local scale, these practices of idol immersion are playing a significant role in reshaping the channel geometry and interfering flow processes of a moribund river like Jalangi.

12.3.1.7 Withdrawal of Water

Excessive withdrawal of water and consequent decrease in volume and velocity is one of the prime cause of decaying of the rivers. The river Jalangi is not being able to be fed by the river Padma due to post-Farakka lowering of water level of the river Padma than that of Jalangi bed. Along with the lowering of water level of the river Padma, water from the river Jalangi and its feeder rivers is increasingly being pumped out to irrigate agricultural lands. There are 124 River Lift Irrigation

TABLE 12.16

Soil and Paddy Straw from Immersed Idols Added into the River Jalangi at Krishnanagar

Sl. No.	Idol	No. of Idols Worshiped in Krishnanagar Town	Needed Soil per Idol (cft)	Needed Straw per Idol (kg)	Total Soil Needed (cft)	Total Straw Needed (kg)
1	Jagaddhatri	160	75	58	12,000	9,280
2	Durga	61	140	105	8,540	6,405
3	Saraswati	712	3	2.5	2,136	1,780
4	Kali	278	12	5.5	3,336	1,529
5	Laxmi	975	3	2.5	2,925	2,437.5
6	Others	750	5	3	3,750	2,250
Total					32,687	23,681.5

(RLI) stations on the banks of the river Jalangi within the district of Nadia to irrigate more than 10% of irrigated land (Das, 2013). The decrease in volume leads to a decrease in velocity which in turn compels the river to deposit its silt load within the river bed causing rapid deterioration.

12.3.2 Signature on Long-Profile Fragmentation

The more the river deteriorated, the more is the anthropogenic modification of the river channel. The upper reach of the river Jalangi from offtake to Bhairab–Jalangi confluence is deteriorated the worst, and men have fragmented the river into ponds as private property (Figures 12.32 and 12.33). This practice of fragmentation of the river channel is also found in many other rivers of Bhagirathi-Hooghly basin like Anjana (see Chapter 16), Kana–Ghia–Kunti (see Chapter 15), Saraswati and Jamuna (see Chapter 14).

From offtake to Sialmari confluence, about 10 km reach of the river Jalangi is fragmented into ponds and crop fields by full or partial encroachment of the river bed (Figure 12.33). This scenario is also common for Sialmari, Bhairab, Suti, Gobra and other channels of the Jalangi basin. About 800,000 m² of the river bed of upper reach has been either converted into ponds or crop fields of private ownership. Although they were not ready to show the papers, yet most of the owner said that they have purchase deeds of their occupation!

12.3.3 Extra-Channel Anthropogeomorphology

12.3.3.1 Farakka Barrage

Farakka Barrage on the river Ganges was planned to feed the river Bhagirathi to save the Port of Calcutta, the economic hearth of the eastern and north-eastern India. It was inaugurated in 1975. But the post-Farakka lowering of the water level of Padma has deteriorated all its branches such as Bhairab, Jalangi and Mathabhanga. Because due to the lowering of Padma water level, flow does not enter these branch channels. At Jayrampur, the bed level of the river Jalangi is 15.66 m, and at

FIGURE 12.32 Jalangi is fragmented into privately-owned ponds (marked with boundary) at Mathpara, west of Gopalpur Ghat, Karimpur.

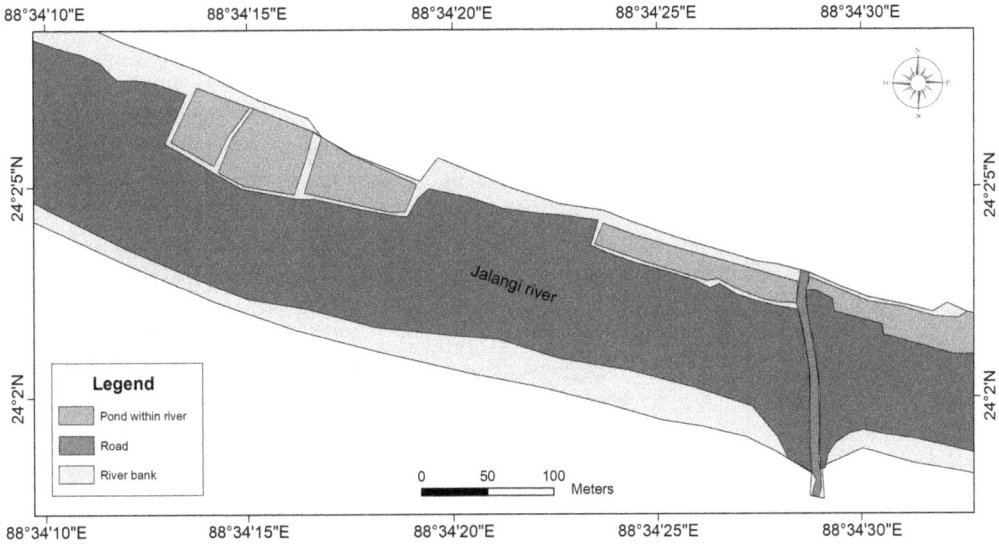

FIGURE 12.33 Encroachment within the river channel to make private ponds near Sahebpara ferry *ghat*, Karimpur.

Gopalpur Ghat, the river is almost disappeared with a bed level of 17.76 m. But the average (during 2007–2009) highest water surface level of the river Padma during monsoon months (July–October) is only 16.75 m which is 1.01 m below the Jalangi bed level to enter into the river. That is why the water from the Padma cannot feed the river Jalangi. During December 2016, the water surface level of the river Padma at Ankhriganj was 15 m, and thalweg level of the river Bhairab was 17 m. So, thalweg at Bhairab offtake is 2 m higher than the water level of Padma, and flow cannot enter into the Bhairab to feed the river Jalangi.

12.3.3.2 Embankments and Road-Networks

Flushing of land during floods is not only necessary to raise and fertilise the land and, in the interest of public health, but also essential for the conservancy of the river itself (Majumdar, 1941). Owing to flatter gradient and consequently comparatively less velocity, the river is normally unable to transport its silt. During floods, rivers require certain spill areas where it could relieve itself of a portion of its silt which would otherwise deposit in its bed and would gradually deteriorate it. But embankments do not allow the river to spill.

At present, there are 10,550 km long embankments along the banks of rivers (Rudra, 2010) of West Bengal out of which more than 60% were inherited from the Zamindars or landlords (Basu, 2002). Doors to spill areas for the river Jalangi are effectively closed from all sides by numerous embankments. State Highway-11 from Krishnanagar to Jalangi via Karimpur and Krishnanagar–Swarupganj road is the *second-front* protection on the left bank. The first-front protection against flood is the village roads (small-scale embankments) now converted into metalled roads under Pradhan Mantri Gram Sadak Yojana. Karimpur–Domkal, Palashipara–Domkal, Krishnanagar–Nabadwip, Bethuadahari–Birpur and Chowgacha–Hulorghat state highways and Krishnanagar–Bethuadahari NH-34 are some embankments converted into metalled roads on the right bank. There are many *first-front* embankments such as Gopalpur Ghat to Jayrampur in C.D. Block Karimpur-I; Moktarpur–Natipota road in C.D. Block Karimpur-II; Kalabagha–Birpur road in C.D. Block Nakashipara; and Gait road in Krishnanagar. Except for NH34, all most all the state highways or other metalled roads mentioned above are on previously existing embankments built by local Zamindars or landlords and renovated later on by colonial rulers.

During the 1950s, State Highway-11 (SH-11) was constructed erasing the last line of the river Jalangi. It runs from Karimpur to Berhampore via Gopalpur Ghat–Jalangi. A ferry ghat was there at Gopalpur Ghat even during the last part of the 1930s, and present-day SH-11 was an unmetalled road discontinued by the river Jalangi at Gopalpur ghat. Therefore, it appears that after the construction of this SH-11, the last sign of the offtake of the river Jalangi was permanently closed. The present-day landform of the source region of the river Jalangi is much more anthropogenic than riverine.

Total length of the roads within the basin is 887.9 km. The average width of a two-lane national highway in India is 14 m. If we consider roughly the width of all the roads within the basin is 10 m and an average height of 2 m from the general ground surface, then total earth material moved by anthropogenic processes of road construction in the Jalangi river basin is 17,758,600 m³ (Figure 12.34; Table 12.17).

In 1972, an embankment of 1,700 m length was constructed along the left bank of the river Jalangi to arrest flood water otherwise which could inundate the town of Krishnanagar (Figure 12.35). Man has reshaped the earth equal to twice the volume (240,478.6 m³) of this embankment (see section "Made and Worked Ground" of this chapter and Chapter 16).

12.3.3.3   Agriculture

On-bank and in-bed agriculture have two dimensions in the anthropogeomorphology of the river channel. One is reshaping of channel forms, and the other is controlling of sediment concentration of flow. Due to the on-bank (Moktarpur, Putimari) and in-bed (Jayrampur, Saguna, Bakshipur) ploughing, not only the channel morphology of the river has been changed significantly but also the

(a)

(b)

Basin Boundary (JALANGI)
Road Networks

N

0   12.5   25
      Kilometres

Earthen Embankment to Protect Flood or to Arrest Spill

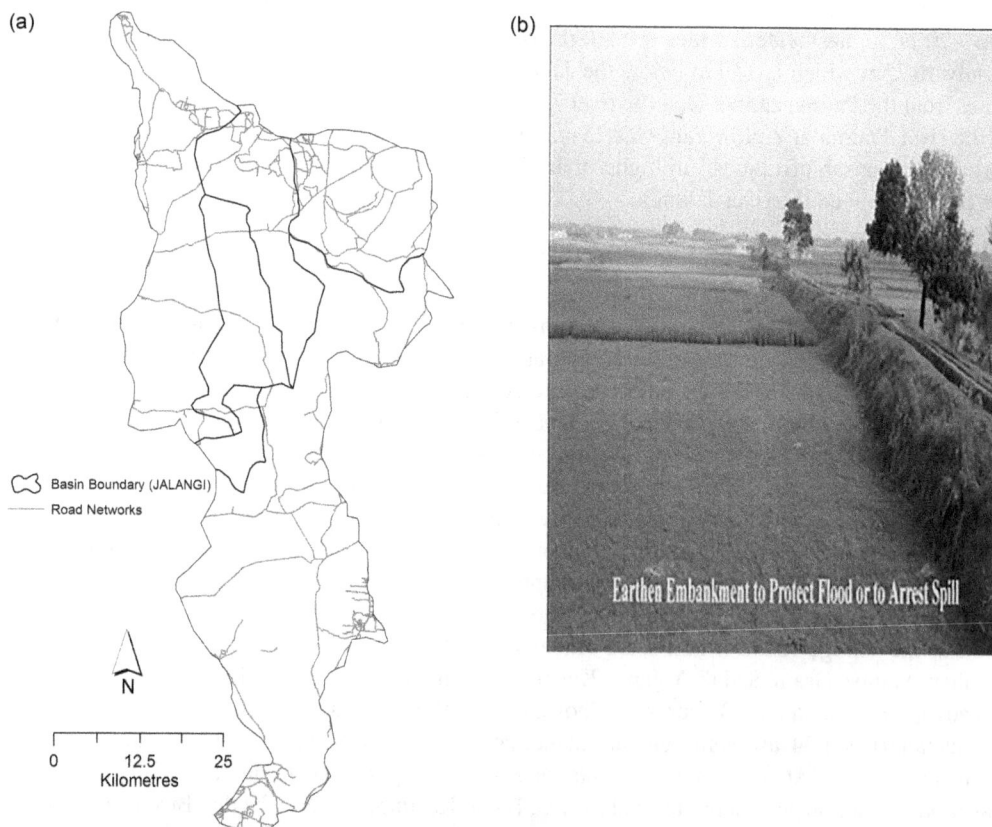

FIGURE 12.34   (a) Road networks in the Jalangi basin. (b) the earthen embankment on the bank of the river Jalangi at Saguna near Karimpur.

TABLE 12.17

Sub-Basin Wise Account of Road Length, Density and Volume of Earth Material Moved

Name	Area (m²)	Area (km²)	Perimeter (m)	Road Length (m)	Road Density (km/km²)	The Volume of Earth Translocation (m³)
Bhairab River	287,896,885.2	287.9	100,681.9877	94,925	0.33	1,898,500
Chhoto Bhairab River	235,852,263.9	235.85	93,657.11299	28,401	0.12	568,020
Gobra Nadi	623,577,751.6	623.58	165,199.548	180,297	0.29	3,605,940
Jalangi River	1,249,681,991	1,249.68	269,599.8833	384,392	0.31	7,687,840
Sialmari River	361,010,185.8	361.01	77,370.73852	194,750	0.54	3,895,000
Suti River	74,345,769.95	74.35	46,708.17936	5,165	0.07	103,300
Total	2,832,364,847	2,832	75,3217	88,7930	2	17,758,600

Source: Google Street Map 2019.

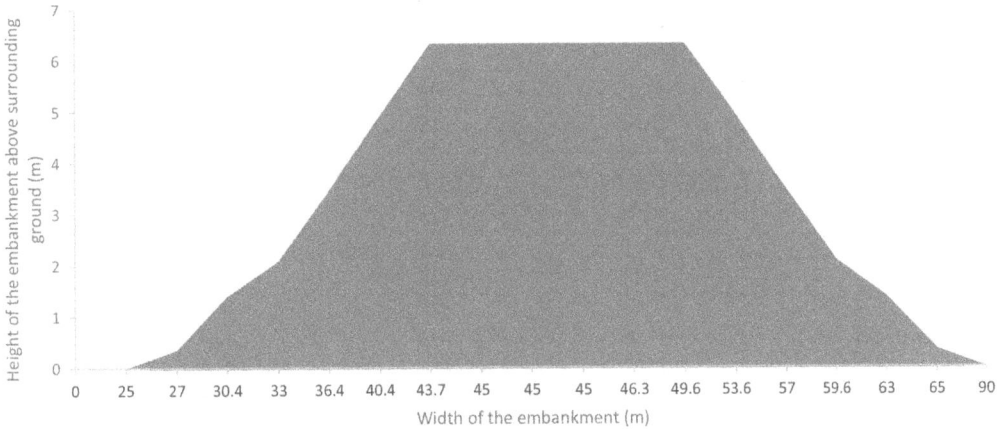

FIGURE 12.35 Cross-section of the earthen embankment along the left bank of the river Jalangi and closing the entrance of the river Anjana to protect the town Krishnanagar from flood water of Jalangi.

soil became vulnerable to be washed away by surface runoff affecting sediment concentration of flow. Terracing and flattening of the river bed for agriculture are very common in the upper reach of the river from offtake to Bhairab–Jalangi confluence at Moktarpur (Figure 12.36). Excessive silt charge to the river is mainly caused by the addition of soil to the river water from bank erosion and soil washed down by surface runoff from ploughed field.

12.3.3.4   Land Use Land Cover (LULC) Change

Land use land cover change (LULC) is one of the most important indicators of the anthropogenic signature on landforms. Virgin land of deltaic Jalangi basin was a monotonous plain land with numerous scares of palaeochannels. With the advance of time, the basin was gradually inhabited by man which was accelerated during the period of Great Acceleration. Cropland and rural land use cover >95% of total land (Table 12.18), which was 95.23% in 2005–2006 and 95.34% in 2015–2016, a nominal increase in rural land use and a decrease in cropland (Figures 12.37 and 12.38).

All the water bodies including rivers, lakes and ponds and inland water bodies have been decreased. Deciduous vegetation and plantation have decreased in this decade. On the other hand, fallow land, scrub forest and forest plantation have been increased. All the human efforts like mining and urbanisation, responsible for reshaping the landforms, have been increased.

FIGURE 12.36 (a) Anthropogenic effort making the river bed flat, suitable for cropping. (b) paddy is planted on the flatbed of the river Jalangi at Moktarpur.

TABLE 12.18
Decadal Changes in LULC from 2005–2006 to 2015–2016

Land Use Type	LULC 2005–2006	%	LULC 2015–2016	%
Deciduous	5.82	0.21	3.62	0.13
Inland water body	20.64	0.73	19.53	0.69
Mining	5.96	0.21	7.89	0.28
Lake/ponds/reservoir	11.43	0.41	10.82	0.38
River	70.59	2.51	64.64	2.30
Fallow land	0.11	0.00	0.19	0.01
Scrub forest	0.27	0.01	0.34	0.01
Plantation	0.74	0.03	0.19	0.01
Forest plantation	1.18	0.04	1.40	0.05
Urban	17.72	0.63	22.45	0.80
Rural	608.35	21.61	624.85	22.19
Cropland	2,072.52	73.62	2,059.41	73.15
Total	2,815.33	100.00	2,815.33	100.00

12.3.3.5 Made and Worked Ground

All the intentional anthropogeomorphic forms are classified as either made or worked ground. Embankments, roads, spurs, earth filling for settlements, etc. are an example of made ground discussed in earlier sections. Again, in this section, we will focus on a special type of made ground which is made of an inevitable waste of modern civilisation. Plastics are the most voluminous waste in present-day India. India generates close to 26,000 tons of plastic a day, according to a **Central Pollution Control Board** estimate from 2012. At less than 11 kg, India's per capita plastic consumption is nearly a tenth of the US, at 109 kg. It is estimated that the total population in the Jalangi basin is 3,728,877. Therefore, the average annual plastic consumption in the basin is 41,017,650 kg. A significant part of this huge plastic waste along with other solid wastes is being used for landfill. A considerable portion of plastic waste from Krishnanagar municipality goes to the Jalangi river and affects badly not only the aquatic life but also the flow process of the river. Another significant portion is disposed of at Godadanga area (about 14 acres of land), the outskirt of the municipality where the made ground is raised by meters. Another landfill is located near *Dwijendra Setu* on Jalangi river where 5,243 m^2 (4,217 m^2 + 1,026 m^2) of land is filled with plastics and other solid waste of Krishnanagar municipality (Figure 12.39).

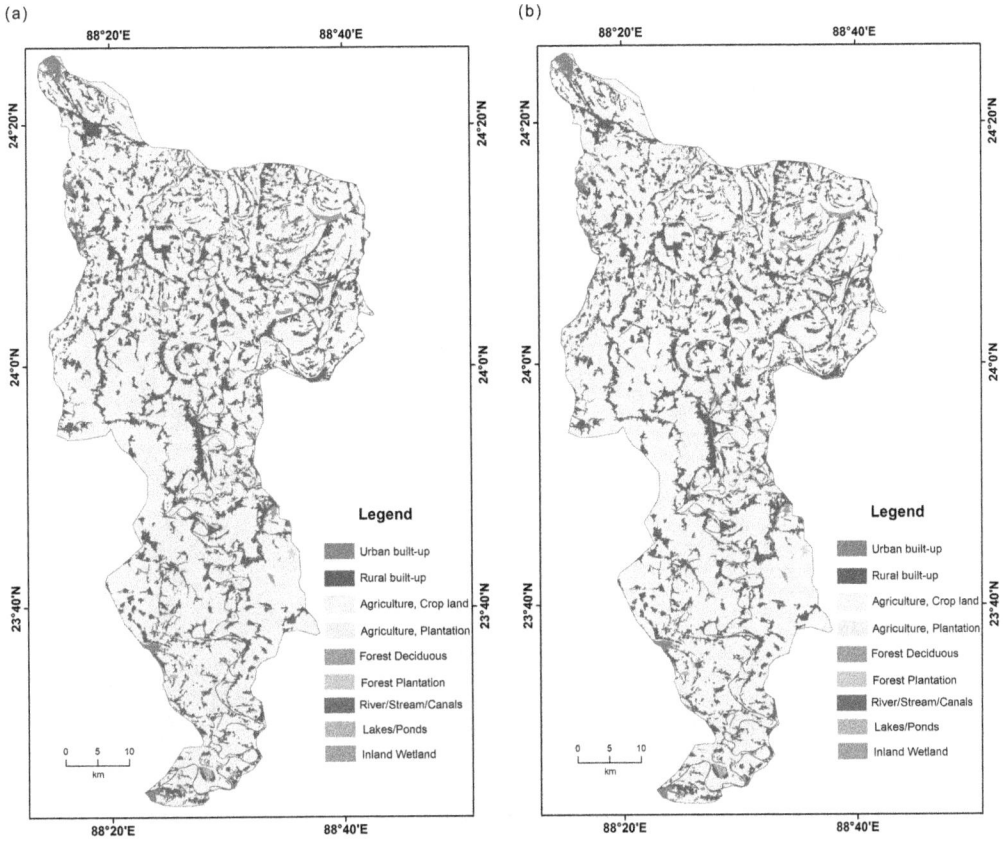

FIGURE 12.37 Decadal changes in LULC indicate anthropogenic signature on landforms of (a) 2005–2006 and (b) 2015–2016.

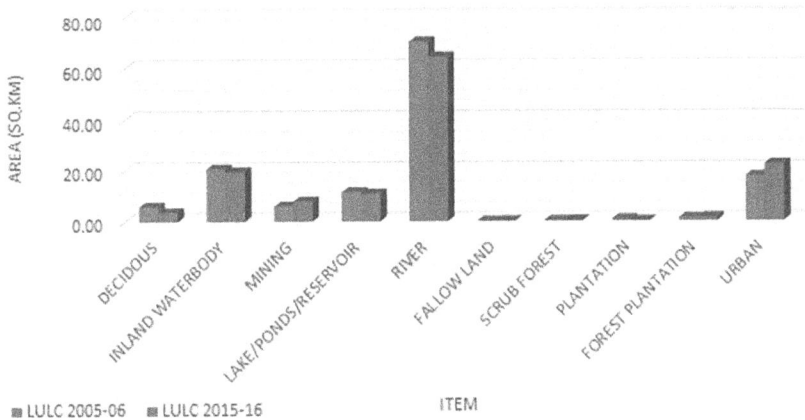

FIGURE 12.38 Changes in LULC during the decade from 2005–2006 to 2015–2016.

Worked ground and made ground are found together all along the roads and embankments. The earth materials accumulated to made grounds like embankments and roads are equal to the earth materials removed from the worked ground like ditches or pits along roads and embankments (Figure 12.39). Therefore, landforms affected by anthropo-processes are equal to twice the made ground or worked ground (in volume).

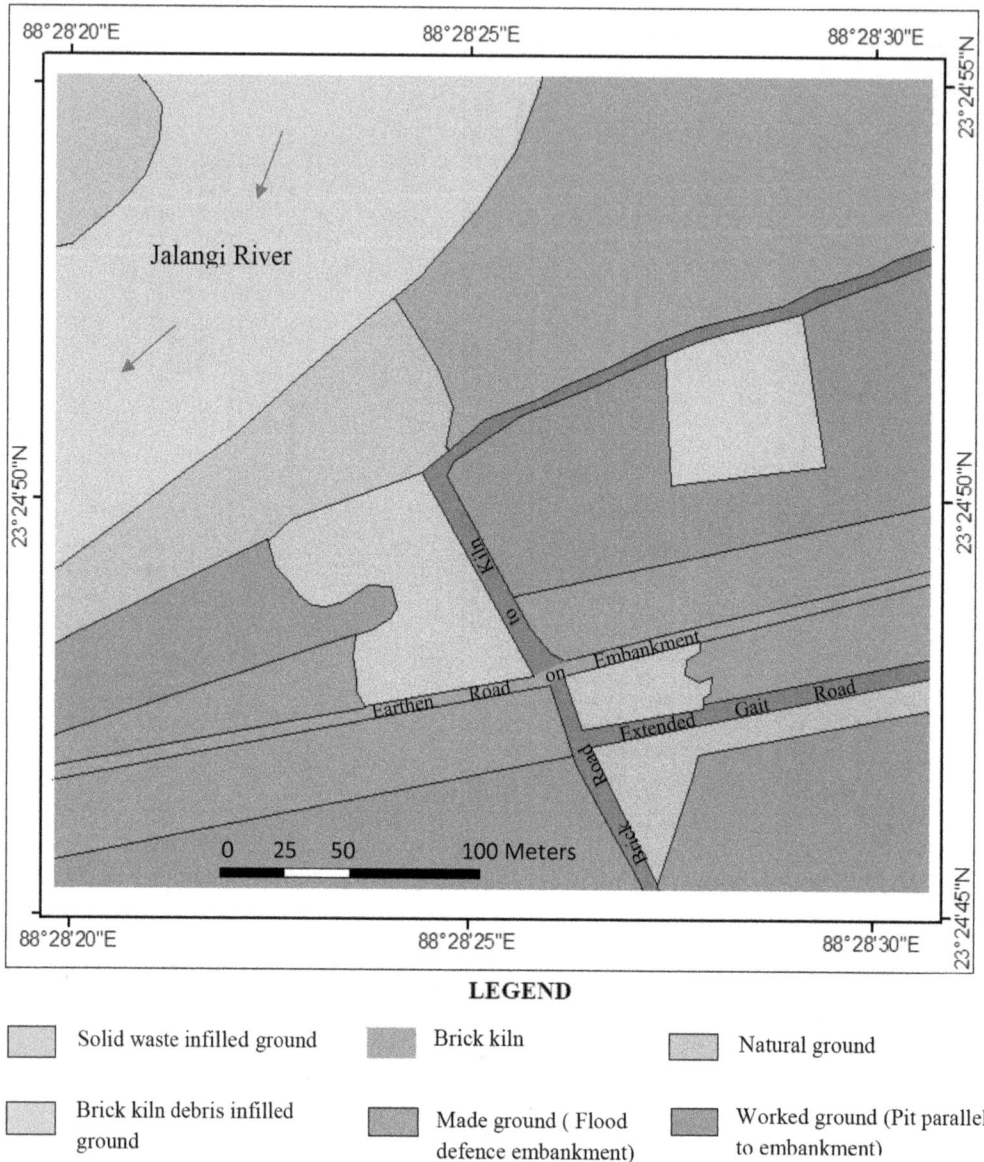

FIGURE 12.39   Made and worked ground near *Dwijendra Setu* on the river Jalangi at Krishnanagar.

12.4   CONCLUSION

The anthropogeomorphology of the Jalangi river basin may be divided into two categories – in-channel and extra-channel. Significant and considerable forms under both these types are either worked or made ground. These anthropogenic forms and processes which have made a substantial alteration in the natural geomorphology of the basin are related to cut/dredging in the channel, *bāndh* across the channel, fragmentation of channel into ponds, road and rail crossings and changes in channel form and processes, fishing practices and interventions with channel processes, etc. This chapter extends the front line of further study of anthropogeomorphology related to all the aspects mentioned in the basin. There is ample scope of research on soil cut by brickfields and ferry ghats from banks and how they interfere with the natural environmental system for finding a line towards sustainable management of the channel and its basin. The present study has its limitation of lack of sufficient data and information. Intensive study and gathering of sufficient quantitative and qualitative data/information on earth materials moved by anthropogenic activities and on *anthropogeomorphic* processes related to *bāndh* across the channel, fragmentation of channel into ponds, road and rail crossings, fishing practices intervening channel processes, agricultural practices controlling sediment charge in the channel, etc. will extend the frontline of research in several directions.

Besides pointing towards the different avenue of research directions, this study complies with the pros and cons of the different applied fields of knowledge related to the construction of bridges, houses, roads and rails, and embankments. Future projects of all these types of constructional works may incorporate ideas of this study for implementing coming projects within the area of the Jalangi basin. All the bridges constructed during the 21st century have dais at the base of piers and put thousands of cubic meters of soil into the river channel. These constructional designs and processes badly affected river health. Reshaping of the piers and dais at the base of piers, which have already constructed, is not practicable. So, at least removal of soil from the river channel is suggested. The bridge at Fazilnagar–Amtala is not yet complete. More piers are to be constructed on Nadia's side. Future construction of a pier(s) within the channel of that bridge may exclude dais at the base of the pier. Public Works Department, local government authority, environmental engineers, planners at the local and regional levels, and others concerned may use this knowledge base for implementing their future projects.

The basin has a wide range of anthropogeomorphic forms which are altering the in-channel and extra-channel geomorphic processes. Although these human interventions with the natural landscape of the Jalangi basin started long back, yet have been accelerated during the last three or four decades of the 20th century and 19 years of the 21st century. LULC of the basin changed abruptly bringing a substantive change in the landscape of the catchment area of the basin and channel processes. Road and railway crossings, embankments, ferry services, brickfields, fishing practices, channel encroachments, etc. play a significant role in anthropogeomorphology of the basin.

Road and rail crossings, especially of NH34 and Sealdah–Lalgola Railway at Krishnanagar, reshaped channel geometry and hydraulics of the channel significantly. *Bāndh*s and spurs across the river for facilitating easier and heavier ferry services are acting as slow poisons for certain death of the river Jalangi. Moktarpur and Amtala ferry ghats are of special concern in this regard because more than half of the wetted channel is closed by bricked *bāndh*. A significant amount (17,758,600 m³) of earth materials are translocated by construction of roads and embankments. Brickfields also have a considerable role in anthropogeomorphology of the basin. It alone translocates 1,715,011.23 m³ of soil per annum. River channel encroachments for construction and fragmentation of river channel into ponds have changed the channel geomorphology substantially. Hydraulics of the flow of the river Jalangi is also to some extent controlled by different fishing practices. So, the anthropogeomorphology of the Jalangi river basin has a significant share in reshaping the landforms and interfering with the landform processes and got accelerated to outperform the share of natural forms and processes in the near future.

REFERENCES

Bagchi, K.G. (1978). *Diagnostic Survey of Deltaic West Bengal*, Department of Geography, C.U, Calcutta, fig-3.

Basu, P.K. (2002). *Seminar on Flood & Drainage Problem of Rivers of West Bengal*, Calcutta Mathematical Society, Kolkata, pp. 1–5.

Biswas, K.R. (2001). *Rivers of Bengal* (Vol-I), Government of West Bengal, pp. xviii, xxix, 87, plate-18 and 19.

Das, B.C. (2013). Changes and deterioration of the course of river Jalangi and its impact on the people living on its banks Nadia West Bengal. Ph.D. Thesis. The University of Calcutta.

Das, B.C. (2014a July–December). Impact of in-bed and on-bank soil cutting by brick fields on Moribund deltaic rivers: a study of Nadia River in West Bengal. *The NEHU Journal*, 12(2), 101–111.

Das, B.C. (2014b). Two indices to measure the intensity of meander. In M. Singh, et al. (eds.), *Landscape Ecology and Water Management: Proceedings of IGU Rohtak Conference* (Vol. 2), Advances in Geographical and Environmental Sciences, pp. 233–245. doi: 10.1007/978-4-431-54871-3_17.

Das B.C. (2017). Bathymetric and chemical analysis of an ox-bow lake in view of aquaculture. Journal of Aquaculture & Marine Biology, 6(6), 1–5. ISSN-2378-3184, https://medcraveonline.com/JAMB/bathy-metric-and-chemical-analysis-of-an-ox-bow-lake-in-view-of-aquaculture.html

Das B.C. (2019). A study on impact of bridge construction on channel dynamics, West Bengal, India. *Scientific Journal of K F U (Humanities and Management Sciences). Saudi Arabia.* 20(1), 265–279. ISSN-1319-6944, https://services.kfu.edu.sa/scientificjournal/Handlers/FileHandler.ashx?file=h20114.pdf&Folder=UploadFiles

Douglas, I. and Lawson, N. (2001). The human dimensions of geomorphological work in Britain. *Journal of Industrial Ecology*, 4, 9–33. doi: 10.1162/108819800569771.

Erlich, P.R. and Erlich, A.H. (1990). *The Population Explosion.* Simon and Schuster, New York, 320 p.

Ferguson, J. (1912). On recent changes in the delta of the Ganges, Pub-Calcutta, Bengal Secretariat Press Reprinted from the *Quarterly Journal of the Geological Society of London*, Vol-XIX, 1863. Edited by Biswas K.R. 2001, Rivers of Bengal, pp. 184–205.

Gailot, S. and Piegy, H. (1999). Impact of gravel mining on stream channel and coastal sediment supply: example of the Clavi bay in Corsica (France). *Journal of Coastal Research*, 15 (3), 774–788.

Garrett, J.H.E. (1910). *Bengal District Gazetteer.* Bengal Secretariat Book Depot, Nadia, reprinted in 2001, pp. 14, 15, 26, 92–134.

Goudie, A. (2018). The human impact in geomorphology – 50 years of change. *Geomor.* doi: 10.1016/j.geomorph.2018.12.002.

Guchhait, S.K., Islam, A., Ghosh, S., Das, B.C. and Maji, N.K. (2016). Role of hydrological regime and floodplain sediments in channel instability of the Bhagirathi River, Ganga-Brahmaputra Delta, India. *Physical Geography.* doi: 10.1080/02723646.2016.1230986.

Hooke, R.L. (2000). On the history of humans as geomorphic agents. *Geology*, 28(9), 843–846.

Hunter, W.W. (1877). *A Statistical Account of Bengal.* Vol-II, Trubner & Co, London, p. 19, 25, 29, 33, 140, 220.

Inglis, C.C. (1949). *The Behavior and Control of Rivers and Canals.* Central Water Power Irrigation and Navigation Research Station, Poona. Research Publication 13.

Keeley, J.W. (1971). *Bank Protection and River Control in Oklahoma.* Federal Highway Administration, Oklahoma Division, Oklahoma City.

Knighton, A.D. (1977). The meander problem. *Geography*, 62(2), 106. Geographical Association.

Knighton, A.D. (1981). Asymmetry of river channel cross-sections: part Quantitative indices. Earth Surf Process Landf. VI: 581–588.

Knighton, D. (1984). *Fluvial Forms and Processes, A New Perspective.* Edward Arnold Pvt. Ltd, London, pp. 1, 2, 164.

Leopold, L.B. and Wolman, M.G. (1960). River meanders. *Geological Society of America Bulletin*, 71, 769–794.

Lóczy, D. and Pirkhoffer, E. (2009). Mapping direct human impact on the topography of Hungary. *Zeitschrift für Geomorphologie*, 53(3), 145–155.

Loh, J. and Wackernagel, M. (eds.) (2004). *Living Planet Report 2004.* World Wildlife Fund International, Gland.

Majumder S.C. (1941). Rivers of the Bengal Delta, In: K.R. Biswas, *Rivers of Bengal*, Vol. I. Department of Higher Education, Govt. of West Bengal, p. 17, 18, 54.

Majumder, D. (1978). *West Bengal District Gazetters Nadia.* Government of West Bengal, pp. 5, 7, 16.

Musy, A. and Higy, C. (2011). *Hydrologya Science of Nature.* Science Publishers Enfield, New Hampshire, p. 265.

Price, S.J., Ford, J.R., Cooper, A.H. and Neal, C. (2011). Humans as major geological and geomorphological agents in the Anthropocene: the significance of artificial ground in Great Britain. *Philosophical Transactions of the Royal Society A*, 369, 1056–1084. doi: 10.1098/rsta.2010.0296.

Reaks, H.G. (1919). Report on the physical and hydraulic characteristics of the rivers of the delta. In *Report on the Hooghly River and Its Head-Waters* (Vol-I). The Bengal Secretariat Book Depot, Calcutta, In Biswas (2001), Rivers Of Bengal, Vol-II, Government of West Bengal, pp. 87, 107.

Richards, K.S. (1976). The morphology of riffle-pool sequences. *Earth Surface Processes*, 1, 71–88.

Richards, K.S. (1982). *Rivers Forms and Processes in Alluvial Channels.* The Blackburn Press, Caldwell, NJ, p. 192.

Rózsa, P. (2007). Attempts at qualitative and quantitative assessment of human impact on the landscape. *Geografia Fisica e Dinamica Quaternaria*, 30, 233–238.

Rózsa, P. and Novák, T. (2011). Mapping anthropogenic geomorphological sensitivity on a global scale. *Zeitschrift für Geomorphologie*, 55(1), 109–117.

Rudra, K. (2010). Banglar Nadikatha. In: *Bengali*, Sishu Sahitya Sansad Pvt. Ltd, Kolkata. pp. 36–40.

Schrenk, F., Kullmer, O. and Bromage, T. (2007). The earliest putative homo Fossils. Chapter 9. In W. Henke, I. Tattersall (eds.), *Handbook of Paleoanthropology*, pp. 1611–1631. doi: 10.1007/978-3-540-33761-4_52.

Schumm, S. A. (2005). *River Variability and Complexity*. Cambridge University Press, New York, p. 165.

Simon, A. and Downs, P.W. (1995). An interdisciplinary approach to evaluation of potential instability in alluvial channels. *Geomorphology*, 12, 215–232.

Szabó, J. (2010). Anthropogenic geomorphology: subject and system. In Szabó, J., Dávid, L. and Lóczy, D. (eds.) (2010). *Anthropogenic Geomorphology: A Guide to Man-Made Landforms*, pp. 3–10. Springer, Dordrecht Heidelberg London New York, doi: 10.1007/978-90-481-3058-0.

13 Anthropo-Footprints on Churni River
A River of Stolen Water

Biplab Sarkar, Aznarul Islam and Balai Chandra Das

CONTENTS

13.1 INTRODUCTION

In the era of Anthropocene, modifications of river channel and its flow have become both widespread and intensive through the execution of national dam-building programmes (Guo et al., 2020; Kong et al., 2020; Li et al., 2018), inter-basin transfers of water (Zhou et al., 2017; Wang et al., 2013), direct abstraction of river flow, catchment land-use practices that affect the movement of water into the river (Boix-Fayos et al., 2007), etc. The hydro-geomorphological responses of river to extensive human interventions in the era of Anthropocene, especially during the 20th century, have become diversified and that have been investigated by numerous researchers from the standpoint of different scientific disciplines (Gregory and Chin, 2002; Hooke, 2006; Knox, 1977; Gregory, 2006; Arnaud et al., 2019; Downs and Piégay, 2019; Gergel et al., 2002; Surian and Rinaldi, 2003; Tarolli and Sofia, 2016; Wyzga, 1993). Charlton (2008) mentioned two types of human interventions: a. direct modification of channel and b. changes within the drainage basin. Regarding the direct channel intervention or modi-fication, execution of channelisation or channel regulation has been observed in many large rivers across the world, which indicates some engineering actions like straightening, embanking, enlarg-ing of channel or protecting the existing channel for multiple developmental purposes such as flood control, improvement of drainage and maintenance of channel navigation (Crickmay, 1974; Charlton, 2008; Huggett, 2007; Wohl, 2018; Gibling, 2018). In addition, these channelised rivers have also undergone further modification by other engineering actions such as installation of dams, construc-tion of stream crossing, and sand and gravel mining from river bed. Therefore, these interventions have had immense effect on macro- and microlevel hydro-geomorphological changes of river. On the other hand, long-term land-use changes within the basin alter the amount of water and sediment supplied to river, which, in response, alter the geometry and morphology of channel (Botter, 2019; Fuller et al., 2015; James and Marcus, 2006; Ortega et al., 2014). The sediments produced by land-use land-cover (LULC) change are termed as anthropogenic alluvium. So, the wide variety of land uses marked and their share measured in yielding of the sediments by researchers include mining opera-tions (Gilbert, 1917; Pickup and Warner, 1984), timber harvesting (Varnum and Ozaki, 1986; Madej and Ozaki, 1996), agriculture (Happ et al., 1940; Costa, 1975; Meade and Trimble, 1974;Trimble and Lund, 1982; Jacobson and Coleman, 1986; Jacobson and Pugh, 1992; Islam et al., 2020; Schottler et al., 2014), urbanisation (Wolman, 1967; Wolman and Schick, 1967; Hammer, 1972; Ebisemiju, 1989) and mixtures of different land uses.

All the rivers of Bengal delta have strangely been subjected to continuous modification by human activities since long time (Islam and Guchhait, 2017, 2018). Many rivers (Bhagirathi, Jalangi and Churni) have been observed to be embanked for flood mitigation, dammed for hydropower produc-tion, diverted for irrigation of agricultural land and channelised for good navigation (Allison, 1997; Allison and Kepple, 2001; Goodbred, 2003; Kuehl et al., 2005; Parua, 2009; Parua, 2010; Rudra, 2010). The first human intervention was prominently found to build embankments along the major rivers of Bengal delta to protect the densely populated areas from flood and to keep transportation system (roads and railways) uninterrupted. But during the British rule in India, human interventions had got its diversity and the acceleration where apart from the extension embankments, installation of dams and numerous culverts and bridges over river; deforestation, on-bed agricultural practice and sand mining from river bad had become prominent (Rudra, 2011, 2014). Recently, this accel-eration has got its highest degree by the advancement of new technology. Churni is an important river of Bengal delta and also an international river between India and Bangladesh, and a consider-able length of it serves as international border between two countries. Human efforts, through the processes of construction of agricultural practices, embankments, ferry services, fishing practices, soil cutting from river banks, road crossings, urbanisation, brick industry, etc. have changed fluvial landforms of the river basin. The chapter will focus on the human efforts influencing the processes involved in evolution of this deltaic basin.

13.2   THE RIVER AND ITS BASIN

13.2.1   COURSE OF THE MATHABHANGA–CHURNI

During the British rule in Bengal, Bhagirathi, Jalangi and Mathabhanga rivers were collectively known as "Nadia Rivers" (Hunter, 1912; Garrett, 1910) or "Kishnaghur Group of Rivers" (Fergusson, 2012). Among these three rivers, though grouped under the name of cities of a province of India, Mathabhanga is the only river which has found its path crossing and re-crossing the international boundary between India and Bangladesh. The river takes off from the river Padma (24°03′43″ N and 88°44′21″ E) in Bangladesh (Figure 13.1). Char Mahiskundi to the east of Char Madhugari village is located to the extreme north-east of the CD Block Karimpur-I. Char Mahiskundi is to the east of border fencing, and a river is taking off from a thin bypass channel of the river Padma. That river is Mathabhanga. Yes, it is Mathabhanga (beheaded) because its offtake is detached by a big sand bar

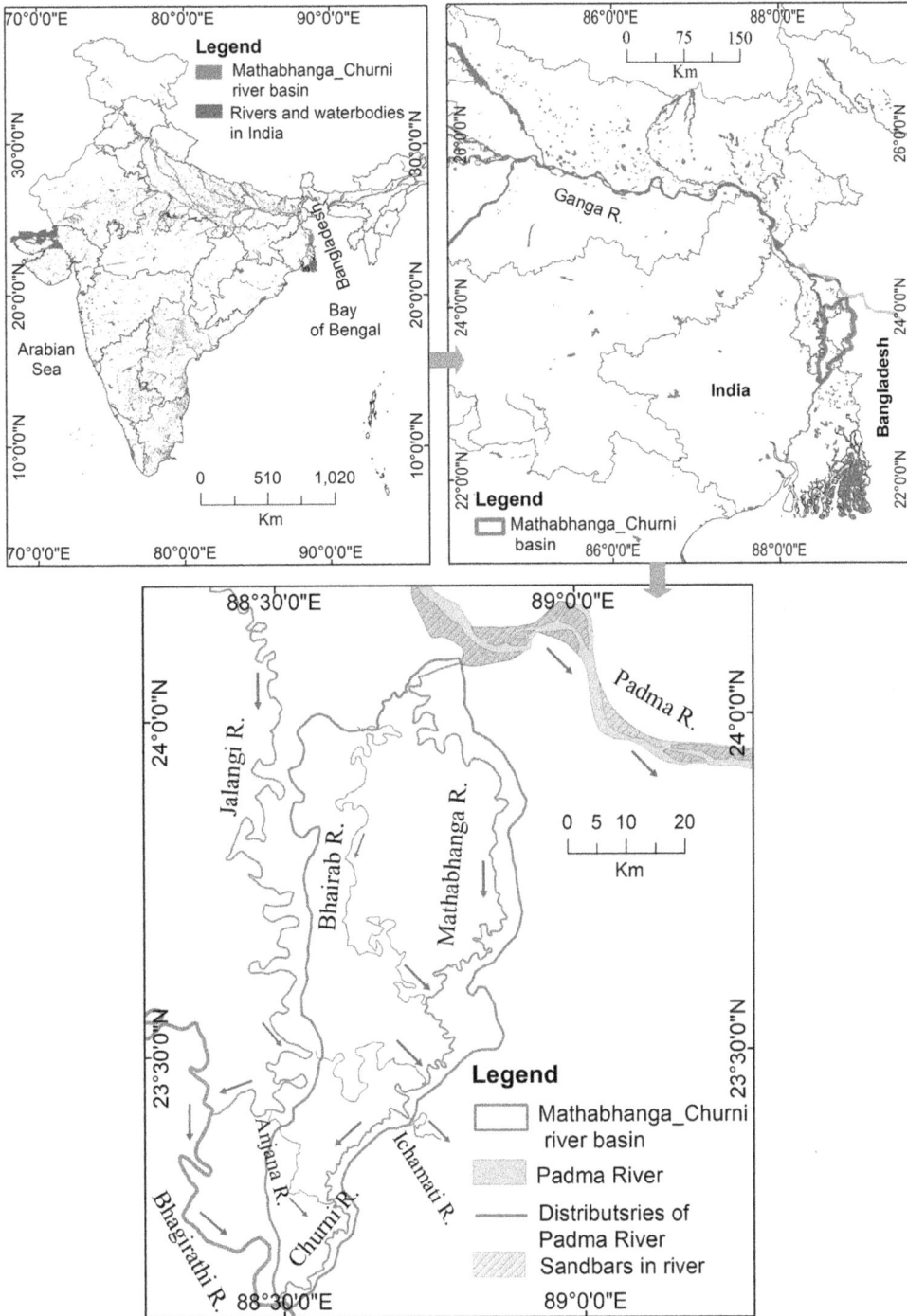

FIGURE 13.1 Location of the study area.

(char) from the feeder channel except for monsoon months. The village to the opposite bank (east or left bank) is Jamalpur in Bangladesh. Vagjot–Munsiganj road and Jamalpur road converge in this village. To the west of this village, after taking off from Padma, the river Mathabhanga runs south by Madhugari–Char Meghna–Ramnagar–Shikarpur–Darpanarayan villages of CD Block Karimpur-I.

Then, the river turns east and enters Bangladesh to run through Kajipara–Mamudpur–Shyampur–Moubariya–Khalisakundi–Rajapur–Rajnagar   Naodapara–Voladanga–Hatboyalia–Bamanagar–Munsiganj Bazar–Mominpur–Boyalmari–Sialmari–Akundabariya–Hatikata–Taltala–Islampara and Chuadanga. Chuadanga–Meherpur Highway bridge is here on the river. After Chuadanga, the river Mathabhanga flows by Hatkaluganj–Vimrulla–Pirpur–Bishnupur–Juranpur–Damurdaha–Puratan Bastupur–Patachora–Raghunathpur–Subalpur–Madna–Shyampur–Darshana of Bangladesh. Thence, the river runs a tortuous path towards south-west and re-enters India at Gede. Thence, the river flows by Bijaypur–Gobindapur–Banpur–Teghari–Gopipur–Swarnakhali–Chak Gutra–Durgapur–Mathurapur–Majdia. The name "Mathabhanga" ends at Majdia, and the river bifurcates into the rivers Ichamati and Churni.

Churni the right branch turns north-west from Majdia up to Shibnibas. From Shibnibas, the river turns its path towards south-east by Chandannagar–Kastopur–Natungram–Raypur–Benali–Kalipara–Batna–Hanskhali–Gayesh–Pichpur–Bapujinagar–Takshali–Byaspur–Barhatta–Radhakantapur–Anandanagar–Aranghata–Paharpur–Kalinarayanpur–Aistala–Ranaghat–Anulia–Nandighat–Masunda–Anandadham–Majdia–Paschim Shambhupur–Shibpur to debouch into the river Hooghly.

13.2.2 Basin Geometry and Morphometry

Morphometric attributes of drainage basin are the key factors controlling the hydrological behaviour of a basin. Therefore, consideration and quantification of morphometric attributes are essential for investigation many aspects of hydrological modelling. The basic morphometric attributes dealt for the Mathabhanga–Churni River Basin are area, length, perimeter, shape relative relief and slope. Mathabhanga–Churni River covers the basin area of about $2,585\,km^2$ having a perimeter of about $295.30\,km$. The linear length of the basin along the principal river is measured to be ~110 km, whereas the widest length of the basin is found to be ~40 km (Figure 13.2). Many geomorphologists have propounded their own indices for qualifying the basin shape, but the most prevalently used indices are compactness index of Gravelius (1914), form factor of Horton (1932) basin circularity of Miller (1953), basin elongation of Schumm (1956) and lemniscate ratio of Chorley et al. (1957) as presented in Table 13.1.

The value of shape factor ($F = 0.21$) of Horton suggests that the shape of the Mathabhanga–Churni River is elongated in nature. Similarly, values of basin circularity ($c = 0.37$), basin elongation ($E = 0.52$), compactness index ($C = 1.63$) and Lemniscate ratio ($K = 3.67$) also support the elongation character of the basin shape. This shape is an indication of lower peak flow for longer duration which has been prominantly observed in the river. Besides, this shape of the basin helps in managing the flood flow comparatively easier.

13.2.3 Fluvio-Origin Features in Basin: Palaeochannel and Oxbow Lake

The entire basin is remarkably characterised by the floodplain which experiences flood at least once in every decade. The river Mathabhanga–Churni in its historical past was hyperdynamic and left numerous abandoned channels as fluvio-origin features, though the lower part these features are less prominent in number. Now, Mathabhanga–Churni has become less dynamic or almost stable because of the fact that the river receives water from the river Ganga only in monsoon period and rest of the year remains disconnected by high siltation in its source. Due to this fact, no cutoff has occurred in the last 50 years. Throughout the course of the river, ten abandoned channels have been identified prominently (Figure 13.3). As these abandoned channels are helpful for the historical reconstruction of the present river course, width of each of these eleven abandoned channels has been compared to the width of present channel adjacent to abandoned channel. ANOVA reveals the fact that there is a significant difference of width between abandoned channel and its adjacent present course as presented in Table 13.2.

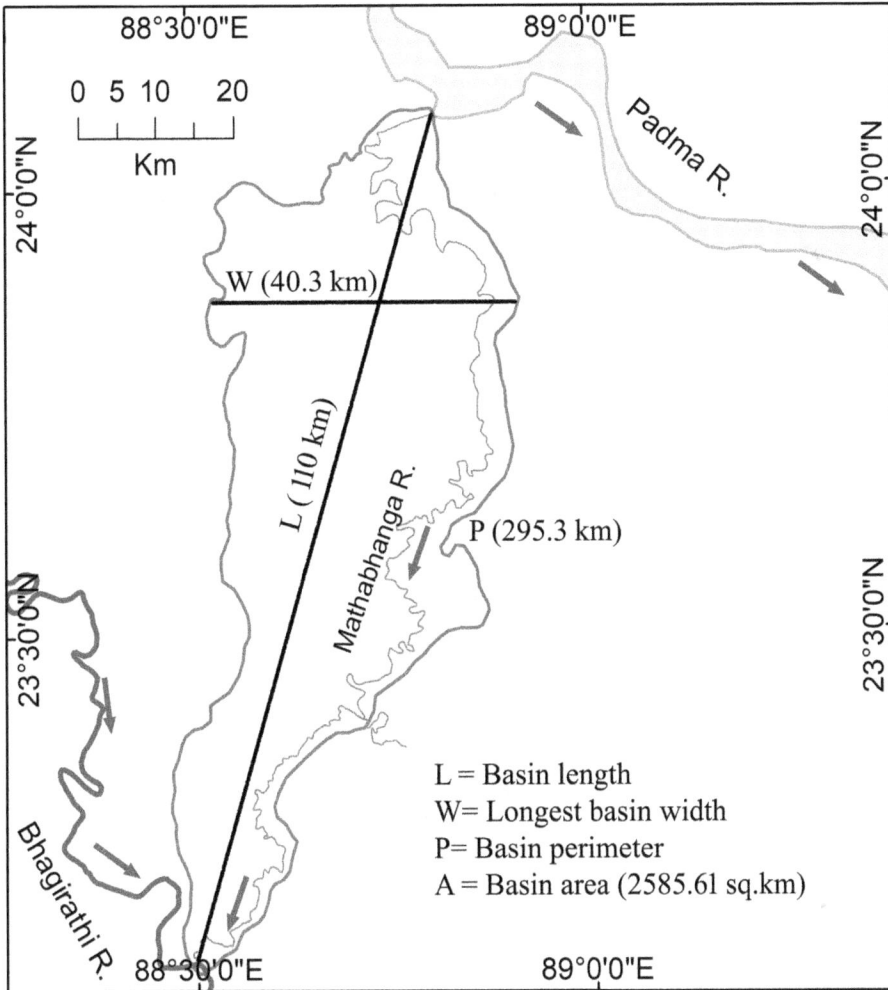

FIGURE 13.2 Geometry of the Mathabhanga–Churni River Basin.

TABLE 13.1

Shape Indices of Drainage Basin

Shape Indices	Equations	Scale Identifying Basin Shape
Compactness index (C)	$C = \dfrac{P}{2\sqrt{\pi A}}$	C close to 1 indicates circular shape and more compactness, and the higher the value of C, the lesser the compactness.
Form factor (F)	$F = \dfrac{A}{L^2}$	C ranges from 0 (representing highly elongated) to 1 (representing perfect circular).
Basin circularity (c)	$c = \dfrac{4\pi A}{P^2}$	C ranges from 0 (representing a line) to 1 (representing a circle)
Basin elongation (E)	$E = \dfrac{2\sqrt{A}}{L\sqrt{\pi}}$	E varies from 0 to 1 where <0.5, 0.5–0.7, 0.7–0.8, 0.8–0.9 and >0.9 represent highly elongated, moderately elongated, less elongated, oval and circular, respectively.
Lemniscate ratio (K)	$K = \dfrac{L^2\pi}{4A}$	K < 0.6, 0.6 > K < 0.9 and K > 0.9 represent basin shape as circular, oval, and elongated, respectively.

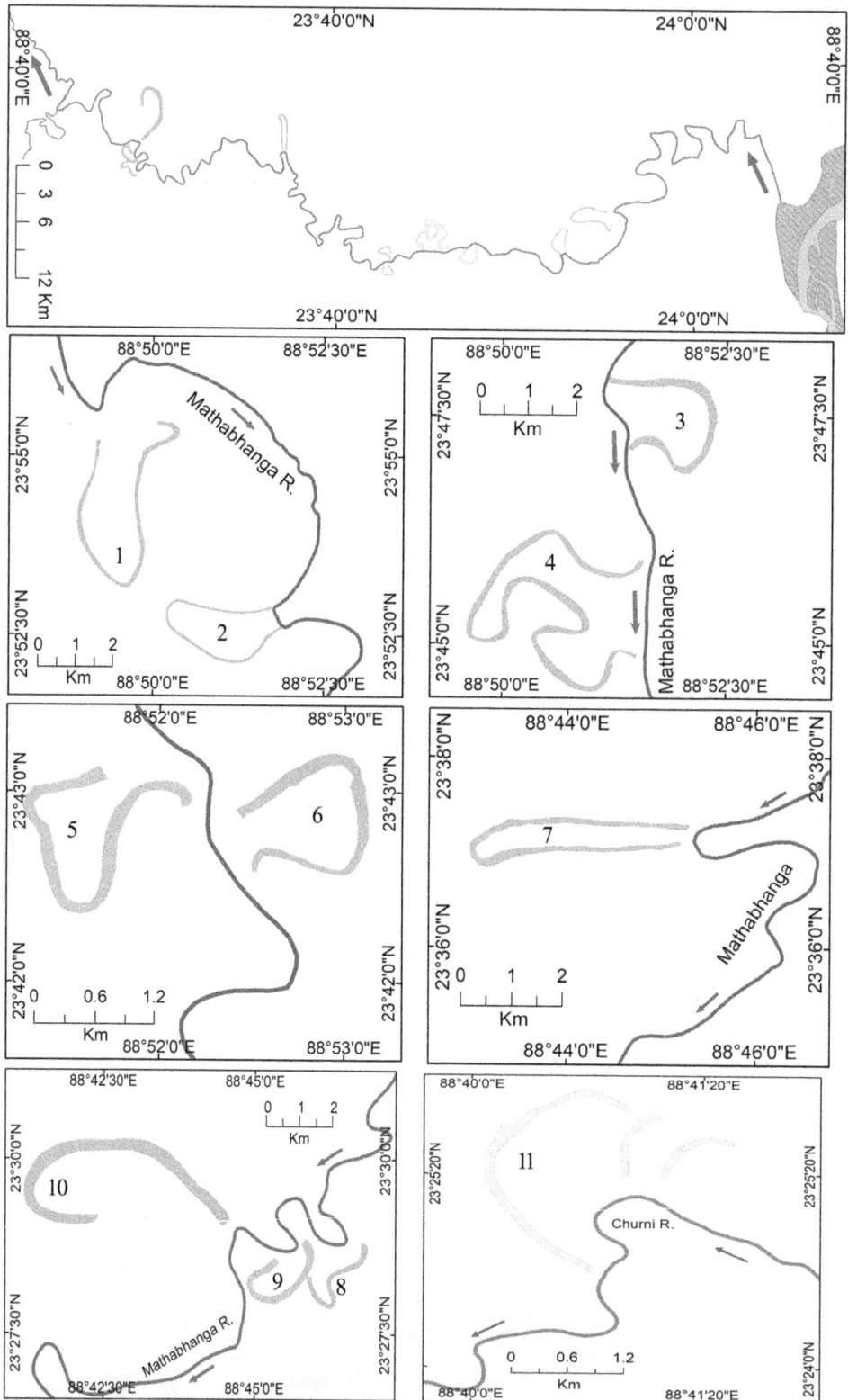

FIGURE 13.3 Abandoned channels of the river Mathabhanga–Churni.

TABLE 13.2
ANOVA: Comparing the Significant Difference of Width between Present Channel and Adjacent Abandoned Channels

Channel Id	F	P-value	F crit	Remarks
1	103.7374248	6.72911E-09	4.413873419	Alternative
2	36.28521486	1.07359E-05	4.413873419	Alternative
3	52.07578065	1.03002E-06	4.413873419	Alternative
4	23.06359283	0.000142631	4.413873419	Alternative
5	36.01074078	1.12493E-05	4.413873419	Alternative
6	26.61879071	6.58455E-05	4.413873419	Alternative
7	211.5499232	2.15643E-11	4.413873419	Alternative
8	37.8088833	8.31986E-06	4.413873419	Alternative
9	85.96113208	2.82311E-08	4.413873419	Alternative
10	30.32183623	3.14416E-05	4.413873419	Alternative
11	38.01074078	1.11493E-05	4.413873419	Alternative

Source: Computed by authors, 2019.

13.2.4 NATURAL VALLEY CONFIGURATION

Natural valley of the river Churni, in most cases, is characterised by the presence of natural vegetation on both the banks and is steep in nature. Moreover, most of the valleys are "U" shaped, and the deepest point (thalweg) locates very close to middle of the channel representing the ideal alluvial channel (Figure 13.4). For representing the present character of the natural valley of the river Churni, nine cross-sections have been taken (three from each stretch: upstream, middle stream and downstream). The average bankfull channel width and depth are recorded to be 57 and 4.5 m for upstream, 70 and 5 m for middle stream, and 66 and 4.66 m for downstream. Therefore, the average values of width–depth ratio for upstream, middle stream and downstream have become 13.77, 14.63 and 14.19, respectively. Moreover, the average values of cross-sectional area for upstream, middle stream and downstream recorded are 236, 351 and 323 m², respectively. However, the study of a real asymmetry of cross-sections ($A^* = 0.01$, $A_1 = 0.31$ and $A_2 = 0.14$ for upstream; $A^* = 0.10$, $A_1 = 0.23$ and $A_2 = 0.09$ for middle stream; and $A^* = 0.13$, $A_1 = 0.23$ and $A_2 = 0.09$ for downstream) indicates that channels are relatively symmetric due to the fact that the flow characteristics such as flow depth, discharge, velocity and energy are not in favour of altering the valley configuration and lateral migration of channel.

13.2.5 FLOODPLAIN SEDIMENT

As the Mathabhaanga–Churni River Basin is an important part of the Ganga–Brahmaputra Delta, sand of different grain size is the principal component of floodplain sediment stratigraphy. The sand-dominated stratigraphy most commonly found is characterised by a fining-up sequence. Datta and Subramanian (1997) reported that the lion's share (~80%) of bed sediments in the Ganga–Brahmaputra Delta comes into the category of fine to very fine sand having the mean grain size varying from 2.5 to 4 Φ (1/6–1/16 mm). Moreover, in the Ganga–Brahmaputra Delta surface sediments are generally characterised by silty to clayey silts, whereas the sediments of lower part are dominated by sand of different sizes (Garzanti, 2019; Guchhait et al., 2016; Jha and Bairagya, 2011; Mukherjee et al., 2009; Weinman et al., 2008). BH1 in the study area portrays that ~35 m depth from the surface is dominated by clay of different types, and from ~35 to ~55 m, fine sand is the principal

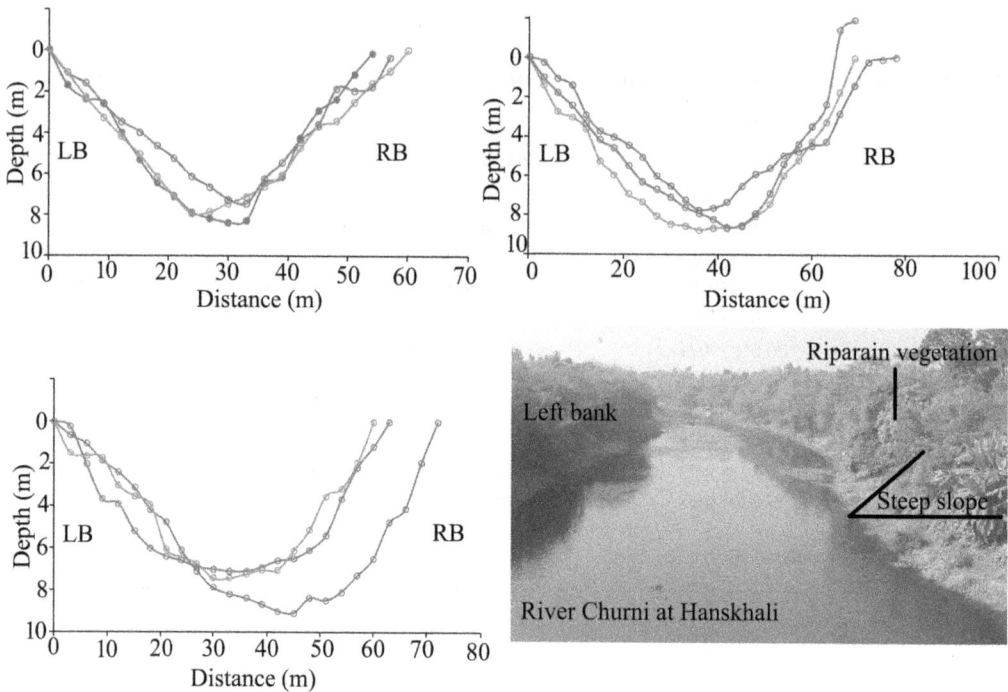

FIGURE 13.4 Pattern of natural valley in the river Churni.

constituent, below which is the layer of fine to medium sand (Figure 13.5). BH2 is sand-dominated layer where fine sand covers from surface to 55 m depth. Five thin clay layers have also found within the fine sandy layers at different depths. Below 55 m, another two layers of 5 m thickness; that is, fine to medium sand and medium to coarse sand exist (Figure 13.5). In BH3 located in the upper part of the basin, alternating layers of clay and sand are found up to the depth of 35 m, and below then, fining-up sequence of sandy layers is present (Figure 13.5). Furthermore, the mineralogical study of these sediments reveals the fact that among the clay minerals, illite and kaolinite are dominant figures (40%–65%), whereas smectite and chlorite are found at lesser amount. The trace metals such as arsenic, lead and mercury are also found with alarming concentration in the sediments of the study area (Koenig, 2011; Mukhopadhyay et al., 2006; Bhattacharya et al., 2011; Mukherjee and Fryar, 2008; Ghosal et al., 2015).

13.3 ANTHROPOGENIC SIGNATURES ON FORMS AND PROCESSES

13.3.1 CHANNEL REGULATION DURING HISTORICAL PAST: ARTIFICIAL CANAL VS. NATURAL RIVER

The river Churni is mentioned as an artificial canal in much local literature (Bandyopadhaya, 2007; Chatterjee, 2017; Chattopadhyay, 2007). It is also said that Maharaja Krishnachandra made this artificial canal usurping a part of the flow of river Ichamati. As the artificial canal seized water from the river Ichamati, it is named as "Churni" meaning the "female thief" (Sahitya Samsad, 2000; Das, 2003). If it is true that the river is an artificial canal, then it is the most significant and biggest anthropogenic signature on the geomorphology of the area. The present section will illuminate light on the discourse to search the truth (Figure 13.6).

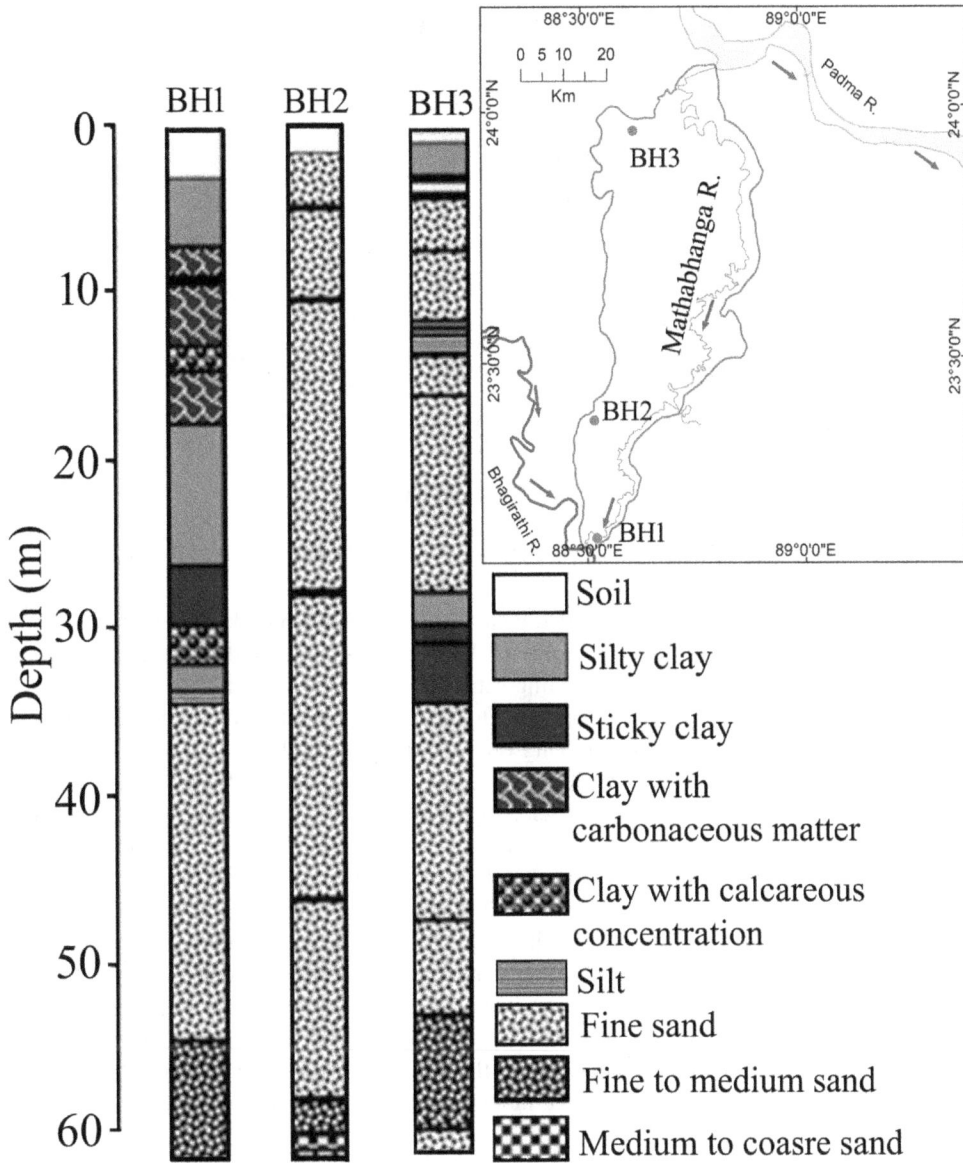

FIGURE 13.5 Sediment stratigraphy in the floodplain of the river Mathabhanga–Churni – BH1 (at Ranaghat), BH2 (at Krishnanagar), BH3 at (Karimpur).

13.3.1.1 Stories as Talked

Being afraid of an invasion by *Bargis* (the Maratha soldiers who indulged in large-scale plundering of the countryside of western part of Bengal for about ten years from 1741 to 1751) Maharaja Krishnachandra of Nadia was forced to shift his residence for a time and made a new palace at Shibnibas (Garrett, 1910; Mallick, 1911; Dutta, 1996; Bandyopadhaya, 2007). Nasrat Khan, a fakir, would live there and that is why Shibnibas was known as "Nasrat Khan's Ber" before Nadia palace established in 1742. Mallick (1911) wrote that Nasrat Khan was a notorious dacoit, and to defeat him, Maharaja Krishna Chandra raided "Nasrat Khan's Ber" and he chose that safer place to build his new residence "Shibnibas". Ray (1875) wrote that to escape from attack by Bargis Maharaja Krishnachandra had chosen a place near Ichamati River, 19.31 km away from Krishnagar.

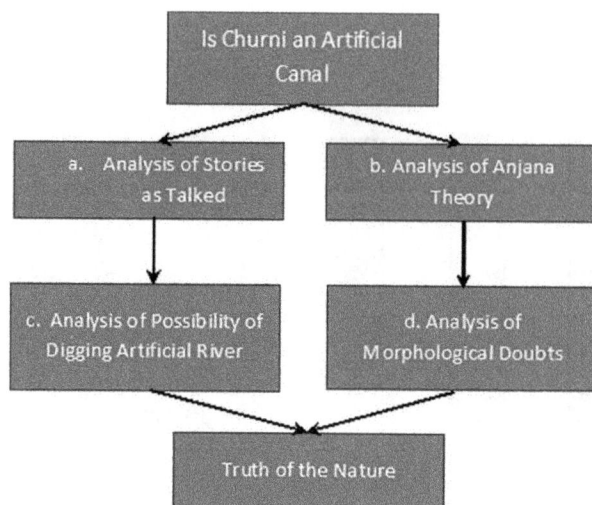

FIGURE 13.6 Flow chart showing natural channel vs. artificial canal debate.

That place was full of jungle and surrounded by a waterbody. A ditch of 457.2 m was dug from the east of that surrounding waterbody to join it with the river Ichamati. Another ditch of 9.66 km was also excavated from the west of that encircling waterbody to join it with the confluence of Anjana River at the north of Hanskhali. Being connected with river Ichamati and Anjana, the waterbody surrounding that place turned into a flowing one. As the waterbody was circular in plan-view, Maharaja Krishnachandra named it 'Kankana' and the city was named 'Shibnibas'. Whatever be the story, the common and prime factor for selecting Shibnibas for new residency was "safety for his family", afraid of *bargi* attack. Now the question arises, why Shibnibas was safer than Krishnagar?

Actually, it seems logical that as Maharaja Krishnachandra was not capable of defeating Marhatta Bargis, he just was looking for a place to escape *bargi* raids. So, he chose (not found by chance or by heavenly wish) "Nasrat Khan's Ber", a dense jungle surrounded by water bodies. Maybe he was well aware of this place, because their ancestral residence was at Matiari, very near to this Shibnibas. The dense jungle was inaccessible to foreign invaders. *Bargis* will not dare to penetrate such an unknown jungle. Moreover, during an attack, if it happens at all, members of his family could get ample scope of camouflaging themselves in the jungle. All these made Shibnibas safer than Krishnagar. Neither more defensive fort nor more impassable moat, nor larger and more efficient troop of soldiers made Shibnibas safer.

Maharaja Krishnachandra established a Siva Temple near the palace and named the new capital "Shibnibas" adobe of god Siva (Mallick, 1911; Dutta, 1996; Bandyopadhaya, 2007). To protect the "Shibnibas" from the plundering of *Bargis*, a moat was dug around the palace, and it was named *Kankana* a ring-like ditch (Bandyopadhaya, 2007). To make the moat impassable to the invaders, it was filled with water from the river Ichamati. For this purpose, a canal of 500 yards was dug to the east of the moat connecting the river Ichamati. But for the sake of longevity and to enable scouring of loose deposits in its bed, the moat needs to be flowing. So, it was connected with the "confluence" of the river Anjana to the north of Hanskhali (Dutta, 1996; Mallick, 1911). Maharaja Krishnachandra gave birth to the river Churni thieving water from Ichamati and that is why the river was named as "Churni" female thief. But a few questions arise from these stories.

First, the Bengali word "Ber" is associated with the "Nasrat Khan's Ber", meaning "Gher" or boundary. Most probably, the place where Nasrat Khan would live was encircled by a circular ditch full of water. Maharaja Krishnachandra, very naturally, chose that naturally protected place for his palace. Therefore, Shibnibas, before its occupation by Nadia Raj, was surrounded by a water body which was dredged and renovated to a moat by Maharaja Krishnachandra.

Second, from the moat, a canal of 500 yards was dug towards the east to connect with the river Ichamati (Mallick, 1911). But from the present-day ruins of palace or Siva temple distance of Ichamati–Churni bifurcation point is 2,000 yards. Therefore, if only 500-yard excavation was enough to connect with the river Ichamati, then the course of the river Churni was there before the occupation. Maybe it was badly decayed, and Maharaja Krishnachandra dredged and renovated a very small length (500 yards) of the reach to get water into it from the river Mathabhanga-Churni.

Third, from the moat, another canal of 3 krosh or 6 miles was dug towards the west to connect with Anjana Mohana (confluence) to the north of Hanskhali (Mallick, 1911). Here the Bengali word "mohana" is worthy for special consideration. "mohana" means the place where a river debouches into the sea. There is another word "sangam" meaning the place where one river meets another river. Perhaps, Mallick (1911) did not make the difference between "mohana" and "sangam", because during the period of Krishnachandra, no sea was there at Hanskhali. Therefore, the only possibility was that river Anjana would fall into another river at Hanskhali. And the canal dug up to that meeting point (Sangam) of two rivers. Then, what was the name of that second river in which the river Anjana would fall?

During the period of Rennel, the river Churni was not connected with the river Mathabhanga (Rudra, 2010). That is why Rennel drew his map showing the river Churni detached from the river Mathabhanga. Below Shibnibas in Rennel's time, Mathabhanga continued along the present Ichhamati. Churni was a mere nullah and was flowing along the north of Aranghata and Ranaghat (Reaks, 1919). In the last years of the 18th century, greater discharge into the Mathabhanga opened out the modern Churni from Shibnibas towards south-west through a tract where Rennel shows no stream and then joined the old course near Ranaghat (Reaks, 1919). These statements reveal that during the 1760s, an old deteriorated course of a river was there through which the present-day Churni had established its course. Moreover, that old nullah along Aranghata and Ranaghat was in a deteriorated state in Rennels's time. Whatever the name (although no other name except Churni is found in any literature/record) of that river was, it was present-day Churni. What Maharaja Krishnachandra did was nothing but a renovation of the upper part of that deteriorated channel of the river Churni for the sake of the moat protecting the Royal Palace.

In Rennel's time, the river Kumar (upper reach of the river Mathabhanga was also named so during that time) was taking off from near the offtake of the river Jalangi. That river Kumar became popular as Mathabhanga in the later period (Rudra, 2010). For two probable reasons, Rennel might have labelled that river as Kumar. First, during that time, flow through the river Kumar was much higher than that through the lower reach (reach after bifurcation into Kumar and Mathabhanga) of the river Mathabhanga which was witnessed in many years during the first half of 19th century in 1820–1821, 1825–1830, and 1852 (Reaks, 1919). So, Rennel named the river after the name of the comparatively larger branch "Kumar". Second, Kumar, Churni and Ichamati are three separate branches of a single river taking off from river Padma, about 5.5 km downstream of the offtake of the river Jalangi. Presently, that river is named "Mathabhanga". Therefore, during the time of Maharaja Krishnachandra (~30 years before Rennel's survey, when he drew the map of the river Kumar), the river Mathabhanga was there. In 1751, Nawab Alivardi Khan (1740–1756) made a treaty with the Marathas and the *Bargis* left Bengal forever. Maharaja Krishnachandra came to Shibnibas in 1742. Rennel surveyed the province during the 1770s. Therefore, if the river Churni was there in Rennel's time, obviously it was there in the time of Maharaja Krishnachandra also. It is mentioned about pre-occupied Shibnibas in "Kshitish–Vanshabali–Charit" (Ray, 1875) that the place was 'full of jungle and surrounded by water body'. Probably, the place was like an island encircled by an oxbow lake of the river Churni. Again, in the work of Mukhopadhaya (1805), we found proof of the existence of the river Churni. He mentioned that "one day … they reached a place and saw a very pleasant piece of island encircled by the river and there are numerous birds here and there". This statement reveals that the then Shibnibas was encircled by an oxbow lake and naturally it was a rejected reach of the river Churni.

Then, what about the canals to the east and west of the moat of Royal Palace dug by Maharaja Krishnachandra? It seems logical that the oxbow lake encircling Royal Palace was partially dug and renovated to convert it into the moat. And to make the moat impassable by the enemy soldiers, the decayed channel of the river Churni was dug to the east and west of the moat so that the moat became full of moving water from Churni.

13.3.1.2 Anjana Theory

It is said (Ray, 1875; Mallick, 1911; Dutta, 1996) that the river Anjana after bifurcation at Jatrapur left branch flows towards north passing by Mahishnengra village (J.L. No. 108; PLCN. 01550500) wherefrom the river turns southward and runs by Beraberia, Tegharia, Chitrashali, Hematpur, Betna Kuthi Para, Jhinuk Ghata and Hanskhali (J.L. No. 53; PLCN. 01564100) wherefrom it turns southeast by Bapujinagar, Goruapota, Mamjoyan and Byaspur where it unites with the right branch. The united branch runs by Aranghata, Kalinarayanpur, Ranaghat, Anulia, Ananda Dham and Shibpur to debouch into the river Bhagirathi. Therefore, according to this theory, before Maharaja Krishnachandra, the course of the present-day Churni River after Hanskhali was actually the course of the river Anjana. Maharaja Krishnachandra dug a canal of 500 yards to the east of the moat around Royal Palace connecting the river Ichamati and another canal of 3 krosh or 6 miles towards the west to connect with Anjana Mohana at Hanskhali.

But the word "Anjana Mohana" surely points towards the existence of a second river with which river Anjana would meet. Moreover, the general slope of the region concerned is towards the south or south-west. Then, why a small channel like the left branch of the river Anjana moved towards the north (after bifurcation at Jatrapur) and north-east up to Hanskhali? Moreover, fundamental variables of channel morphology, i.e., width and depth of the river Anjana (both of right and left branch) significantly mismatches the width and depth of the channel of the river Churni. So, there is an ample scope of the doubt to the theory that the course of the river Churni from Hanskhali onwards is the old course of the river Anjana. Finally, no part of the river Churni is known as Anjana.

If Maharaja Krishnachandra wanted to make Shibnibas safer than Krishnagar by excavating gigantic moat and creating an artificial river like Churni mightier than Anjana, then

i. Why not he widened and deepened the moat around Krishnagar palace making it more difficult to pass? This could be cheaper and more effective against the enemy soldier.
ii. As per this theory, an artificial canal to the west of the moat of Shibnibas residence meets the then Anjana at Hanskhali. Therefore, Anjana was reasonably mightier than the artificial canal (*Kankana* or Churni) excavated, and Maharaja Krishnachandra could have a safer place at Krishnagar on Anjana bank than Shibnibas on the bank of that artificial canal. He needed not to excavate an artificial river like Churni to give adequate protection to his palace. Rather he could take the locational opportunity of the safer Krishnagar palace on the bank of the river Anjana. Therefore, Anjana Theory is discarded.

13.3.1.3 Possibility of Digging Artificial River

It is popular that the river Churni was artificially dug by Maharaja Krishnachandra thieving water from the river Ichamati, so it was named "Churni". It is not uncommon that the then Maharajas would take the initiative of such a behemoth task. But if it happened, then some questions arise:

i. To make it economic, reasonably the canal should be straight, not sinuous as the river Churni is.
ii. Generally, this kind of initiative would aim to facilitate irrigation. But there is no such historical evidence that Maharaja Krishnachandra cut irrigation canal to facilitate irrigation of agricultural lands of his kingdom. Therefore, it seems unbelievable that only to protect the Royal Palace, he dug a whole river! Rather, being a wise ruler, he always had chosen

such a place that is naturally protected. As was the case of Krishnagar Rajbari, the palace was built on the bank of the river Anjana so that the course of the river could be used for making the moat. He did not cut the whole Anjana River to protect the Rajbari, the Royal Palace.

iii. It would be irrational to think that the then engineers or landscape developers would not consider the cost-benefit of any project. They were much aware of local landforms. So, it seems probable that they renovated the decayed channel of the river Churni to protect Shibnibas. It is unbelievable that a river was made artificially.

iv. Many said that Maharaja Krishnachandra not only named "Shibnibas" but also named the river "Churni". However, no authentic source about this was found. In *Maharaja Krishnachandra Rayasyo Charitrang*, Mukhopadhaya (1805) wrote that Krishnachandra named the city as "Shibnibas" and the river as *Kankana*. But from offtake to confluence, no part of the river is known as *Kankana*. In Bengali, *Kankana* means bangle. Krishnachandra named the moat, which was the renovated oxbow lake of the river Churni and which was protecting the palace at Shibnibas encircling it like a bangle, *Kankana*. Mukhopadhaya (1805) wrote in another line that the place, where Shibnibas sited is, was surrounded by river. That river was nothing but an oxbow lake of the river Churni. In *Kshitish Vansabali Charita*, Ray (1875) wrote "whether the name 'Churni' was given by Maharaja Krishnachandra to the river from Shibnibas to Shibpur or the part of the river which existed in earlier period owed the name 'Churni' is not known". In this sentence, two points are significantly worth mentioning. First, a clear doubt was there about whether Maharaja Krishnachandra named the river "Churni". But 70 years before Ray (1875), Mukhopadhaya (1805) wrote that Maharaja Krishnachandra named the river *Kankana*. If older is considered as certifiable, then the name of the river ought to be *Kankana*, not Churni. But no part of the river is known as *Kankana*. It seems obvious that Maharaja Krishnachandra named that oxbow lake (converted into moat) *Kankana* not the whole river.

13.3.1.4 What Nature Inscribed

Three morphological points of inclination are there to believe the river Churni as an artificial channel. First is the notable absence/scarcity of oxbow lakes along its course. In this deltaic part, all other distributaries (rivers Jalangi, Bhagirathi, and Ichamati) have their many oxbow lakes along their courses. Is this because of the artificial origin of the river? Second is its lower sinuosity index. Third is the low width/depth ratio. Most of the deltaic rivers have a very large width in comparison with the depth resulting larger width/depth ratio (Figure 13.6).

But both these lower width/depth ratio and scarcity of oxbow lakes may be explained as natural phenomena. Bank materials of the river Jalangi are coarser and prone to erosion. As a result, the river widens its channel by wandering in easily erodible bank materials. So, its width/depth ratio is higher. On the contrary, the bank materials of the river Churni are much finer and resistant to erosion and therefore are relatively stable, less oscillating and lacking oxbow lakes. So, the width/depth ratio is smaller. Moreover, the river Mathabhanga–Churni is connected with the river Padma and gets its flow for several months in rainy season. So, the river can maintain its depth by scouring loose sediments from its bed which the river Jalangi lacks. Therefore, deeper channel and absence/scarcity of oxbow lakes do not prove its artificial origin.

13.3.1.5 Near the Truth

Kankana is not the name of the river Churni. The moat encircling Shibnibas which was actually a decayed oxbow lake of the river Churni was named *Kankana* by Maharaja Krishnachandra.

Churni is not an artificial canal. Before the establishment of the palace at Shibnibas, Churni was there although it was badly decayed. Maharaja Krishnachandra renovated a part of that decayed channel and joined it with the *Kankana*.

Maharaja Krishnachandra might have not named the river Churni.

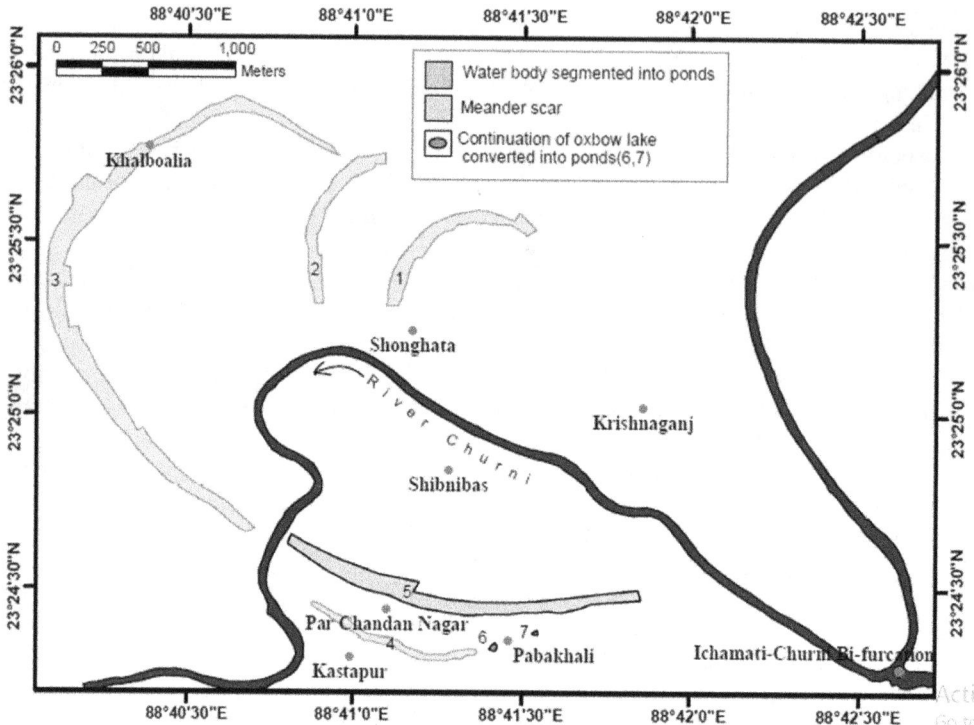

FIGURE 13.7 Presence of these oxbow lakes and paleochannels reject the artificial origin hypothesis of the river Churni.

Therefore, the Churni, the daughter of the river Mathabhanga is not an artificial river thieving water of the Ichamati. Eventually, it became mightier than the river Ichamati, and the river Mathabhanga found its path through the river Churni to relieve its lion share of volume. Churni became the main channel and Ichamati the secondary one.

13.3.2 ARTIFICIAL CHANNELISATION DURING BRITISH RAJ

Since the first quarter of the 19th century, Nadia rivers (Bhagirathi, Jalangi and Mathabhanga) were of prime concern of British Raj and different artificial efforts were adopted to keep them navigable (Table 13.3). Abnormally winding offtakes of these rivers and huge sand deposit closing offtakes made Bhagirathi and Jalangi less suitable in comparison to Mathabhanga for navigation and channel training.

Several cuts across the winding channel of the river to facilitate navigation through the straightened course were made during the first half of the 19th century. The navigation route of the river Mathabhanga–Churni was shortened by 10.46 km (12.3–1.84 km) by a single cut of 1.84 km just above Pangasi offtake during this period (Figure 13.7).

13.3.3 ROAD-STREAM CROSSING

Stream crossings such as culvert and bridge (foot road and rail) installed for augmenting the accessibility and connectivity of transportation in the Mathabhanga–Churni River Basin are different from their dimension, structure and orientation. The intensity of stream crossing depends upon the regional topography and terrain, planning for transportation. The river Mathabhanga–Churni passing through highly dense urban area is generally found to have many numbers of crossings.

TABLE 13.3
Chronological Arrangement of Anthropogenic Signatures on Mathabhanga–Churni River Channel during the 19th Century

Year	Superintending Engineer of Nadia River Division	River Training Scheme Had Taken	Result	Source of Information
1819	Mr. C. K. Robinson was appointed Superintendent and Collector of Mathabhanga River. With this appointment, Nadia River Division of Public Works Department came into existence	An effort to close the offtake of River Kumar by "bandals" (a branch of the river Mathabhanga was diverting 5/6th of the water to the south-east) was adopted.	Unsuccessful attempt	Reaks (1919)
1820 June	Mr. May	Attention was given to clear snags	-	Reaks (1919)
1821	Mr. May	Cleared many snags including 300 timbers sunken and many wrecked boats, and cut down leaning trees on banks likely fall into the river	Cleared	Garrett (1910) and Reaks (1919)
1821	Mr. May	To close the offtake of the river Kumar which was still taking away 3/4th of the Mathabhanga supply	A cassion and some old boats were sunk across the offtake of the river Kumar. Kumar offtake was closed.	Reaks (1919)
1821	Mr. May	Straightening of the channel of the river Mathabhanga to attract water into the main channel.	A cut of 1,540 yards (1,408.176m) was made across a bend of the river Mathabhanga to attract the water. Mathabhanga was navigable throughout with 3 ft draft.	Reaks (1919)
1821	Mr. May	Closing of the offtake River Pangasi, the second branch of the river Mathabhanga takes off at 23°44′09″ N and 88°52′02″ E.	Unsuccessful attempt	Reaks (1919)
1823	Mr. May	Dredging of Mathabhanga offtake	A dredging machine drawn by bullocks was employed, but owing to deteriorated channel and severely closed offtake by sand deposit, the attempt was unsuccessful.	Reaks (1919)

(Continued)

TABLE 13.3 (Continued)

Chronological Arrangement of Anthropogenic Signatures on Mathabhanga–Churni River Channel during the 19th Century

Year	Superintending Engineer of Nadia River Division	River Training Scheme Had Taken	Result	Source of Information
1824	Mr. May	Due to better condition of Jalangi and Bhagirathi, attention was given to those two rivers also and Mr. May was given additional charges	—	Reaks (1919)
1825–1830	Mr. May	Two attempts were taken to divert the water flowing away through the Pangasi.	Attempts were unsuccessful	Reaks (1919)
1828–1829	Mr. May	Maintaining minimum depth for navigation	A steam dredge was employed below Kumar entrance without much result due to unsuitable machinery and heavy draft (6 ft) of the dredging boat.	Reaks (1919)
1831–34	Mr. May	To have a satisfactory solution of Pangasi problem, a proposal was made to divert the current into a loop taking off the right bank of the Mathabhanga above the Pangasi offtake.	—	Reaks (1919)
1840	Captain Smyth	—	—	Reaks (1919)
1852	Captain Smyth	Construction of a channel for 10 miles below the river Pangasi by artificial banks with the use of kodalis	A passage was maintained for small boats	Reaks (1919)
1913	Captain Smyth; Mr. F. A. A. Cowley, Superintending Engineer, Central Circle and Deputy Secretary, Irrigation Department	Opined that only Mathabhanga out of all the Nadia rivers is worthy to be kept artificially navigable because its offtake is just above the Sara bridge and not abnormally winding as is common for two other Nadia rivers.	—	Reaks (1919); Reaks (1919)

FIGURE 13.8 Cut (the length shown by second bracket) and straightening of tortuous path of the river Mathabhanga just above the offtake of the river Pangasi.

Channel and stream crossing are inversely related because if channel migrates laterally, destruction of bridge is inevitable, and on the other hand, installed bridge has strong influence on downstream hydro-morphology (Crickmay, 1974; Gupta, 1982; Chin and Gregory, 2001; Roy and Sahu, 2017; Roy, 2013; Surian and Rinaldi, 2003; Wang et al., 2014). Hence, understanding the relationship between channel morphology and stream crossing helps in reducing channel instability, improving stream habitat and avoiding or minimising the adverse impact on environment (Blanton and Marcus, 2009, 2014; Bouska et al., 2010; Burchsted et al., 2014; Chin et al., 2017; Johnson et al., 1999; Marion et al., 2014). So, construction must be preceded by in-depth investigation on the morphological, hydrological and ecological impact supposed to occur in different spatial and temporal scales.

13.3.3.1 Distribution, Types and Geometry of Road-Stream Crossing

The bridges constructed over the river Churni are of three types: road, rail and foot. Although the foot bridge has less control on channel morphology, it is also taken into consideration as the anthropogenic signature on river channel. The road bridge at Shiblibas (Figure 13.9b) is the perfect example of grouped piers (consisting of ten piers in each group). Their piers are cylindrical in shape having a perimeter of 5 m. Among the four grouped piers, two are bank attracted, not in contact with water in normal flow, whereas two are very proximate to middle of the river channel and stand a portion of 0.7–1.4 m below the water level during pre-monsoon and post-monsoon season. The river at Kalinarayanpur has been intensively modifying since last 5 years (2014–2019) due to newly constructed railway bridge by destroying the existing old bridge. Now, three bridges (one foot and two rail) are existing very proximate to each other and playing a crucial role in contracting river channel. Each rail bridge stands on four piers (two are circular in shape having a diameter of 32 m and located at river bed where flow depth maintains at least 1.15 m in dry period, and another two are rectangular in shape and located at river on the bank of the river (Figure 13.9b)).

FIGURE 13.9   Geometry of road-stream crossing (a) bridge at Kalinarayanpur, near Ranaghat, (b) bridge at Shibnibas, (c) development of scouring at Kalinarayanpur Bridge.

13.3.3.2   Effect

a. Channel morphology:

The effects of bridge on the channel morphology of the river Churni have been clearly noticed. Regarding Shibnibas bridge, the changes in cross-sectional area and average depth with increasing distance from the bridge piers follow a general trend. The values of cross-sectional area increase with increasing distance from the bridge for both upstream and downstream parts of the bridge. In downstream at every distance of 30 m, the values of cross-sectional area have been recorded as 107, 141 and 160 m², whereas in case of upstream part, these are 118, 165 and 208 m². Another observation has also been noticed that the values of cross-sectional area are high in upstream part compared to that of the downstream part. This is due to the fact that in general, the downstream of the bridge is morphologically the most susceptible part to change by the construction of bridge than that of upstream part. Besides, the average depth appears in the same way as observed for the cross-sectional area. The average depth (downstream part) increases with increasing distance (d = 2.1 m at 30 m, d = 2.8 m at 60 m and d = 2.95 m at 90 m down of the bridge), and the same observation has also been found for upstream part (d = 2.3 m at 30 m, d = 3.3 m at 60 m and d = 2.95 m at 90 m down of the bridge).

On the other hand, the morphological changes near Kalinarayanpur Bridge (near Ranaghat) appear differently from those of the bridge at Shibnibas. The value of cross-sectional area (downstream of the bridge) decreases with increasing distance (A = 569 m² at immediately below the bridge, A = 306 m² at 30 m down and A = 273 m² at 60 m down). This decreasing trend appears due to the fact that at bridge site, the formation of local and contraction scouring plays a crucial role for increasing the average depth of the channel, and thereby, the value of cross-sectional area has been recorded higher than that of the distant cross-sections (Richardson and Davis, 1995). The similar observation has also been found for upstream stretches of the bridge (A = 286 m² at immediately up the bridge, A = 193 m² at 30 m distance). In addition, the average depth at bridge site (downstream) has been recorded as 6.84 m, whereas it appears as 4.32 m at a distance of 50 m. The high value of average depth at bridge site has been recorded because of scouring factor. Besides, in the upstream part, naturally, with increasing distance, the value of average depth has been found increasing (d = 3.36 m at 30 m, d = 3.96 m at 60 m, d = 4.28 m at 150 m distance).

b. Channel asymmetry

Natural river is generally found asymmetric throughout its course, and it has in most cases been found that the higher the order, the higher the value of asymmetry. Moreover, the direct modifications of river channel by numerous anthropogenic activities have made the river channel more asymmetric. The river Churni has also experienced the same thing. Thus, the indices of channel asymmetry (A^*, A_1 and A_2) developed by Knighton (1981) have been applied to compare how different human activities play an important role in increasing the channel asymmetry of the river:

$$A^* = \frac{(A_r - A_l)}{A}$$

$$A_1 = (2X \times d_{max})/A$$

$$A_2 = 2X (d_{max} - d)/A$$

where A_r is indicated as the cross-sectional area of the right of the channel centreline and A_l is cross-sectional area of the left of the channel centreline; A ($= A_r + A_l$) is the total cross-sectional area of the channel; X is the distance between channel central line (Lc) and d_{max} (measured as positive sign when Lc is to the right and vice versa); d_{max} is the maximum depth; and d is the average channel depth. The value of A^* being zero represents no asymmetry present in cross section. Moreover, the exclusive limits of A^* are −1 and +1 representing extreme left and right asymmetry, respectively.

The nature of asymmetry reveals the fact that the asymmetry is high at the immediately up and below the bridge and with increasing distance the asymmetry decreases. Concerning the bridge at Shibnibas, all the asymmetry values appear as positive. Immediately below the bridge, A^*, A_1 and A_2 have been found as 0.27, 0.67, and 0.37, respectively, whereas at 90 m below the bridge, they appear as 0.23, 0.61 and 0.29, respectively. Similarly, at immediately up of the bridge, A^*, A_1 and A_2 have been found as 0.28, 0.70 and 0.36, respectively, whereas at 90 m up of the bridge, they appear as 0.03, 0.35 and 0.17, respectively (Figure 13.10a). Moreover, it has been clearly noticed that the reduction of asymmetry with increasing distance is high for the upstream part compared to the downstream part. This is due to the fact that channel below the bridge has been more effected for a long distance compared to the upstream part. On the other hand, concerning the bridge at Kalinarayanpur (Ranaghat), both the positive asymmetry and negative asymmetry have been recorded. Similar to the Shibnibas bridge, reduction of asymmetry with increasing distance is found high for the upstream part compared to the downstream part. At immediately up of the

(a)

(b)

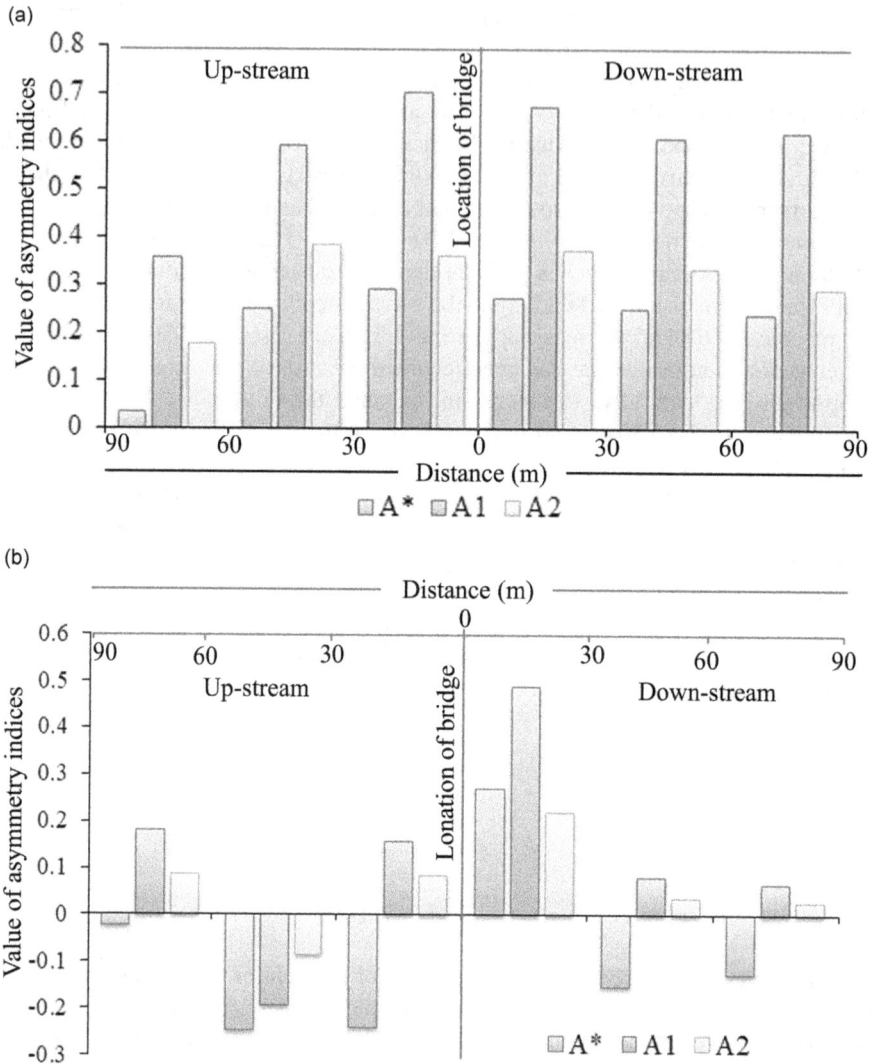

FIGURE 13.10   Nature of channel asymmetry at bridge sites: (a) at Shibnibas and (b) at Kalinarayanpur.

bridge, A^*, A_1 and A_2 have been found as −0.23, 0.15 and 0.09, respectively, whereas 90 m up of the bridge, they appear as −0.02, 0.18 and 0.08, respectively. Moreover, at immediately below the bridge, A^*, A_1 and A_2 have been found as 0.27, 0.48 and 0.21, respectively, 90 m below the bridge, they appear as −0.12, 0.06 and 0.02, respectively (Figure 13.10b).

c. Bed scouring at bridge piers:

The important morphological change that occurs by the installation of piers of bridge is the development of scouring. Generally, two types of scouring are observed near bridge site: contraction and local scouring.

The contraction scouring appears when flow area of a river is reduced by the bridge and embankments (Figure 13.11a and b). The embankments force the overbank flow to run through the channel, and as a result, the average velocity and bed shear stress increase to remove more bed material from the constricted reach (Abid, 2017; Al-Shukur and Obeid, 2016; Gampathi, 2010; Ghorbani, 2008; Govindasamy, 2009; Kothyari, 2007; Khan et al, 2016; Laursen and Toch, 1956; Rahman and Haque, 2003; Saha et al., 2018). On the other

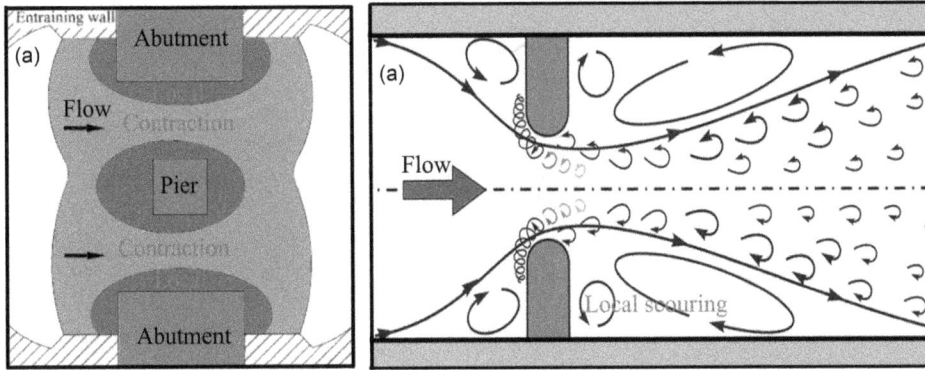

FIGURE 13.11 Flow interruption by bridge and development of contraction and local scouring.

side, Hoffmans and Verheji (1997) observed that the development of scouring around the bridge piers (local scouring) is one of the principal causes of bridge failure. When flow is interrupted by the intrusion of piers, velocity increases and vortices occur around the piers which, as a result, remove the bed material with the flow developing the local scouring. In the river Churni, the role of hydraulics and bridge in scouring development has been analysed thoroughly.

13.3.3.2.1 Methods for Estimating the Local Scouring Depth

CSU method – Estimating the local scouring depth helps for understanding the changing pattern of river bed configuration induced by installed piers. The well-established equations developed by Colorado State University, also known as CSU equations, have been used to estimate the maximum depth of scouring, which is given by

$$\frac{y_s}{a} = 2.0 \; K_1 K_2 K_3 K_4 \left(\frac{y_1}{a}\right)^{0.35} Fr_1^{0.43}$$

where y_s represents the scour depth (m), y_1 is the flow depth directly upstream of the pier(m), K_1 is the correction factor for pier nose shape from Table 13.4 and Figure 13.12, K_2 is the correction factor for angle of attack of flow from Table 13.4, K_3 is the correction factor for bed condition from Table 13.4, K_4 is the correction factor for armouring by bed material size ($K_4 = 1.0$ for sand bed

TABLE 13.4

Correction Factor of Different Parameters

Correction Factor, K_1, for Pier Nose Shape		Correction Factor, K_2, for Angle of Attack, θ, of the Flow				K_3, for Bed Condition		
Shape of Pier Nose	K_1	Angle (Skew Angle of Flow)	L/a = 4 (L = Length of Pier, m)	L/a = 8	L/a = 12	Bed Condition	Dune Height (m)	K_3
Square nose	1.1	0	1.0	1.0	1.0	Clear-water scour	N/A	1.1
Round nose	1.0	15	1.5	2.0	1.5	Plane bed and antidune flow	N/A	1.1
Circular cylinder	1.0	30	2.0	2.75	3.5	Small dunes	$3 > H \geq 0.6$	1.1
Group of cylinders	1.0	45	2.3	3.3	4.3	Medium dunes	$9 > H \geq 3$	1.2 to 1.1
Sharp nose	0.9	90	2.5	3.9	5.0	Large dunes	$H \geq 9$	1.3

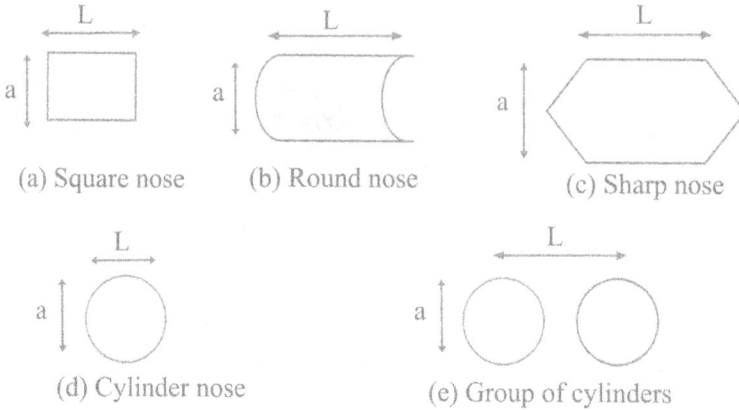

FIGURE 13.12 Common shapes of piers.

material), a is the Pier width (m), Las is the length of pier (m), Fr_1 is the Froude Number directly upstream of the piers, and g is the acceleration of gravity (9.81 m s^{-2}) (32.2 ft s^{-2}).

The correction factor, K_2, for angle of attack of the flow, θ, is calculated using the following equation:

$$K_2 = \left(Cos\theta + \frac{L}{a} \, Sin\theta \right)^{0.65}$$

Laursen's equation: local and contraction scouring – Given a pier in a flume with different flow and sediment conditions, Laursen observed that equilibrium scour depth mainly increases with approach flow depth. Considering variations in pier diameter and sediment size, Laursen developed the relationship for local scouring depth shown in the following equation:

$$\frac{b}{y_s} = 5.5 \left[\frac{\left(\dfrac{y_s}{11.5h} = 1 \right)^{\frac{7}{6}}}{\sqrt{\dfrac{\tau_1}{\tau_c}}} - 1 \right]$$

where y_s is the scour depth (m), h the flow depth (m), b the pier diameter (m), τ_1 the grain bed shear, lb m^{-2}, and τ_c the critical shear at sediment threshold, lb m^{-2}. In Laursen's equation, scour depends on the characteristics of the flow field (depth and bed shear), the structure (pier diameter) and the sediment (critical shear at sediment threshold). For the maximum potential scour depth $(y)_s$, the following equation will be applicable when $\tau_1 = \tau_c$.

$$\frac{b}{y_s} = 5.5 \left\{ \left[\left(\frac{y_s}{11.5} + 1 \right)^{\frac{6}{7}} - 1 \right] \right\}$$

Laursen has also formulated an equation for estimating the depth of contraction scour considering the basic parameters of hydraulics and channel morphology. The equation for estimating the depth of contraction scour is shown as follows:

$$\frac{y_2}{y_1} = \left(\frac{Q_t}{Q_c} \right)^{6/7} \left(\frac{W_1}{W_2} \right)^{K_1} \left(\frac{n_2}{n_1} \right)^{K_2}$$

TABLE 13.5
Scouring Depth at Bridge Site

Location	Discharge (Q) m3 s−1	Velocity (V) m s−1	Types of Pier Shapes	Measured Scour Depth (m) Contraction Scour	Measured Scour Depth (m) Local Scour	Theoretical Value (m) Contraction Scour Laursen's	Theoretical Value (m) Local Scour CSU	Theoretical Value (m) Local Scour Laursen's
A	50.98	0.35	Grouped Circular	-	2.72	-	2.85	2.95
B	70.21	0.42	Circular	2.58	3.2	2.95	2.85	3.10

Source: Computed by authors, 2019.
A, Shibnibas; B, Hanskhali; C, Kalinarayanpur; CSU, Colorado State University.

where y_1 is the average depth in the main channel, y_2 the average depth in the contracted section, W_1 the width of the main channel, W_2 the width of the contracted section, Q_c the discharge in the main channel upstream of the bridge, Q_t the discharge in the contracted section; n_1 the manning n in the main channel and n_2 the manning n in the contracted section

Measurement of scouring depth at Shibnibas and Kalinarayanpur Bridge – Both the contraction and local scouring have been observed in the river Churni where both are prominently observed for Kalinarayanpur Bridge, whereas only local scouring has been prominently observed for Shibnibas bridge. The scouring depth (both the observed and theoretical) due to the effect of bridge piers is shown in Table 13.5. The table values portray that the contraction scouring is more prominent for Kalinarayanpur Bridge. Here, the true channel has been contracted from both sides by 20 m, and the narrow channel produces the high velocity (0.47 m s⁻¹) and shear stress (160 Kg m⁻²) compared to the other site of the channel. The average depth with respect to wetted perimeter in the contracted channel has been recorded as 2.58 m where the deepest point has been recorded as 3.4 m. The depth and velocity discharge shear stress all have been increased at the immediately lower part of the bridge which promote to develop local scouring. In the immediate lower part of the bridge the local scouring has been observed in the right bank of the river. The average scouring has been recorded to be 3.2 m where the highest depth has been recorded to be 4.85 m. Similarly, in case of Shibnibas, the local scouring has been prominently around pillars located near the right bank of the river. The average scouring depth has been recorded as 2.72 m (Table 13.4).

13.3.4 AGRICULTURE

In many scholastics works, the modern agricultural practice (heavy inputs of pesticides, fertiliser and insecticide) has been proved to be the key factor for augmenting the nutritional concentration in adjacent river. Therefore, agriculture has been treated as a dominated nonpoint source of pollution in a river basin (Adusumilli et al., 2010; Chimwanza et al., 2006; Dunca, 2018; Hans et al., 1999; Moss, 2007; Mutisya and Tole, 2010; Toner, 1986). In the river Churni, ~5.8% (3.8% left bank and 2.0% right bank) of the total valley area is used for agriculture (on-bed agriculture) mostly in pre-monsoon and post-monsoon, but the share is significantly less during monsoon because of valley inundation (Figure 13.12a–c). During pre-monsoon and post-monsoon, when the discharge becomes minimum in the river Churni due to disconnection from the river Ganga, runoff from on-bed agriculture significantly controls the status of water. Moreover, in the monsoon period when the amount and intensity of rainfall are high, pollutants from off-bed agriculture are the major controlling factors of pollution.

Distribution and pattern of agriculture – In the winter season, almost all variety of vegetables are cultivated in the sloppy bed of the Churni River. The main variety includes mustard, onion, brinjal, pumpkin and carrot. It has been observed during the field verification that there are tendencies to cultivate more than one crop in a single farm. River lift is the main form of irrigation (Sarkar and Islam, 2019). Traditional manual methods as well as small water pump sets are being used in the river lift irrigation method to irrigate the on-bed agricultural lands.

Change in valley configuration – In agriculture-dominated channel of the river Churni, one bank where agricultural practice is performed is less steep compared to antipodal bank of the river (Figures 13.13a and 13.14b). With slight modification of bank when agricultural tracts are prepared, the character of bank deviates from the natural bank. Moreover, when stepped cultivation is practiced by huge modification of the bank, the slope appears differently compared to slope of the antipodal bank. Besides, these changes may also induce the other attributes of the channel such as width, average depth, cross-sectional area and channel asymmetry (Figures 13.13b and 13.14a). For investigating the effect of anthropogenic intervention on the channel morphology, six cross sections have taken out. The average values of width and depth have been recorded as 73.66 and 3.99 m, respectively, and therefore, the W/d is 18.46. Moreover, the average value asymmetry of agriculture-dominated valley ($A^* = 0.13$, $A_1 = 0.34$, $A_2 = 0.14$) reveals the fact that the asymmetry in agricultural dominated valley is slightly more than that of the natural valley (Figure 13.15).

FIGURE 13.13 Transformation of river bed into agriculture: (a) intensive crop cultivation at sloppy bank near Shibnibas, (b) cutting of slope and stepped cultivation, (c) cultivation of crops on gentle valley slope.

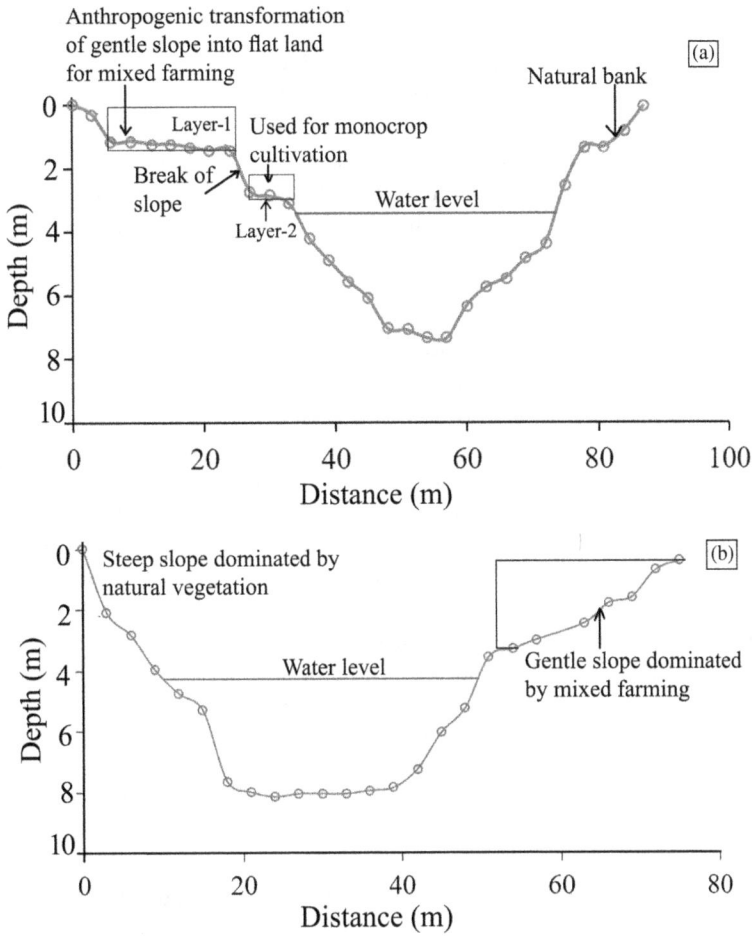

FIGURE 13.14 Cross-sections over agriculture dominated valley: (a) stepped agricultural practice modifying the natural slope near Shibnibas, (b) slight modification of valley slope and practice of agriculture.

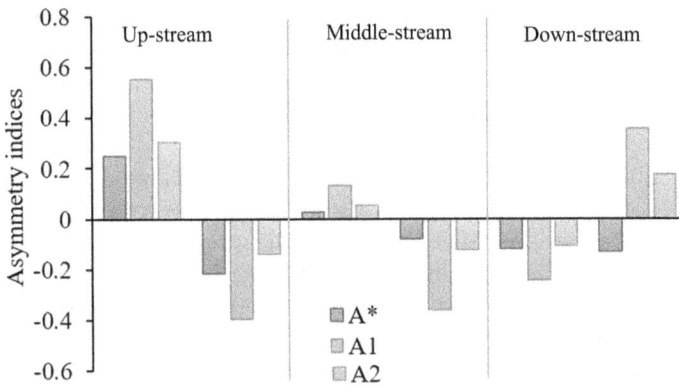

FIGURE 13.15 Channel asymmetry in agricultural dominated area.

13.3.5 Nutritional Imbalances in Water by Surface Runoff from Agricultural Land

Agricultural activities in the river basin, characterised by the application of excessive fertilisers and pesticides, are the important nonpoint sources of pollution which, unlike point sources that can be observed and measured with the highest accuracy, is basically the proximate estimation of non-observable nonpoint discharges. In most cases, the high concentration of phosphate, nitrate and ammonia is the result of agricultural runoff (Sarkar and Islam, 2020). In this study, the contribution of on-bed agricultural practice in increasing the concentration of these nutrients in the river Churni has been assessed in detail.

Phosphate – The concentration of phosphate in uncontaminated water is expected to be in the range of 0.01–0.03 mg L^{-1}, but when it ranges from 0.03 to 0.08, plant growth in water is highly accelerated and when it exceeds the level of 0.1 mg L^{-1}, high eutrophication is inevitable by accelerated growth of aquatic macrophytes. Moreover, phosphate concentration in water more than 0.1 mg L^{-1} also signals that industrial waste water, fertiliser used in agriculture and sewage from rural and urban areas are the major sources of phosphate pollution. At Majhdia, the high concentration is recorded to be 0.109 mg L^{-1} in post-monsoon season followed by 0.096 mg L^{-1} in monsoon and 0.084 mg L^{-1} in pre-monsoon season. Similarly, at Ranaghat, post-monsoon has recorded the highest level of phosphate as 0.119 mg L^{-1} followed by pre-monsoon (0.112 mg L^{-1}) and monsoon (0.110 mg L^{-1}) (Figure 13.16a). Thus, in both the upstream and downstream stretches of the river, the concentration of phosphate exceeds the desirable limit of 0.03 mg L^{-1}. Consequently, the accelerated growth of rooted aquatic macrophytes has been found in many parts of the river. In addition, when these macrophytes die, the microbial breakdown processes start and consequently dissolved oxygen depletes which creates harmful environment for other aquatic lives also. Sarkar and Islam (2020) observed very low concentration of dissolved oxygen (2.22 ± 2.80, 3.49 ± 1.79 and 1.61 ± 1.35 at Majhdia, and 3.80 ± 1.73, 3.41 ± 1.86 and 3.53 ± 0.90 at Ranaghat for pre-monsoon, monsoon and post-monsoon, respectively) due to which the diversity and production of fish species fall at alarming rate.

Nitrate – Neill (1989) observed that there is a direct relationship between the percentage of ploughed land with higher application of nitrogen fertiliser and the higher nitrate loading in river water. Moreover, a significant amount of anthropogenic nitrogen inputs which remain unused by crops are added to adjacent waterbodies. And transportation of this nitrogen is largely dependent on many factors such as timing and type of fertiliser application, amount of fertiliser, soil properties, crop types and amount of rainfall. On the banks of the river Churni, a variety of crops are cultivated with intensive application of N-containing fertilisers such as urea and ammonia, and they are transported easily from the sloppy land into river being an important constituent of agricultural runoff. The reference loading of nitrate in river for physiological activities of aquatic lives is 0.2 mg L^{-1}, but at Majhdia, post-monsoon has recorded the highest nitrate concentration as 1.21 mg L^{-1} followed by monsoon (0.59 mg L^{-1}) and pre-monsoon (0.56 mg L^{-1}). Similarly, at Ranaghat, 1.14 mg L^{-1} is recorded in post-monsoon followed by pre-monsoon (0.94 mg L^{-1}) and monsoon (0.85 mg L^{-1}) (Figure 13.16b). Therefore, in every season and both the stations, nitrate concentration exceeds the desirable range, and consequently, the fish species have been suffering from respiratory problems and damages of the nervous system.

Ammonia – The addition of ammonia into the river water follows the same processes as discussed for nitrate. The concentration of ammonia in Churni River has crossed the desirable limit of 0.1 mg L^{-1} for all the seasons. At Majhdia, the highest average concentration has been recorded in pre-monsoon (5.33 mg L^{-1}) followed by monsoon (032 mg L^{-1}) and post-monsoon (0.31 mg L^{-1}). Similarly, at Ranaghat pre-monsoon has recorded the highest concentration (0.50) followed by post-monsoon (0.40 mg L^{-1}) and monsoon (0.34 mg L^{-1}) (Figure 13.16c). This undesirable concentration of ammonia causes the dropping of disease resistance of fish species in Churni River. Therefore, suffering from different diseases such as gill damage, osmoregulatory imbalance and slow destruction of mucous-producing membranes by ammonia poisoning is inevitable (Sarkar and Islam, 2020).

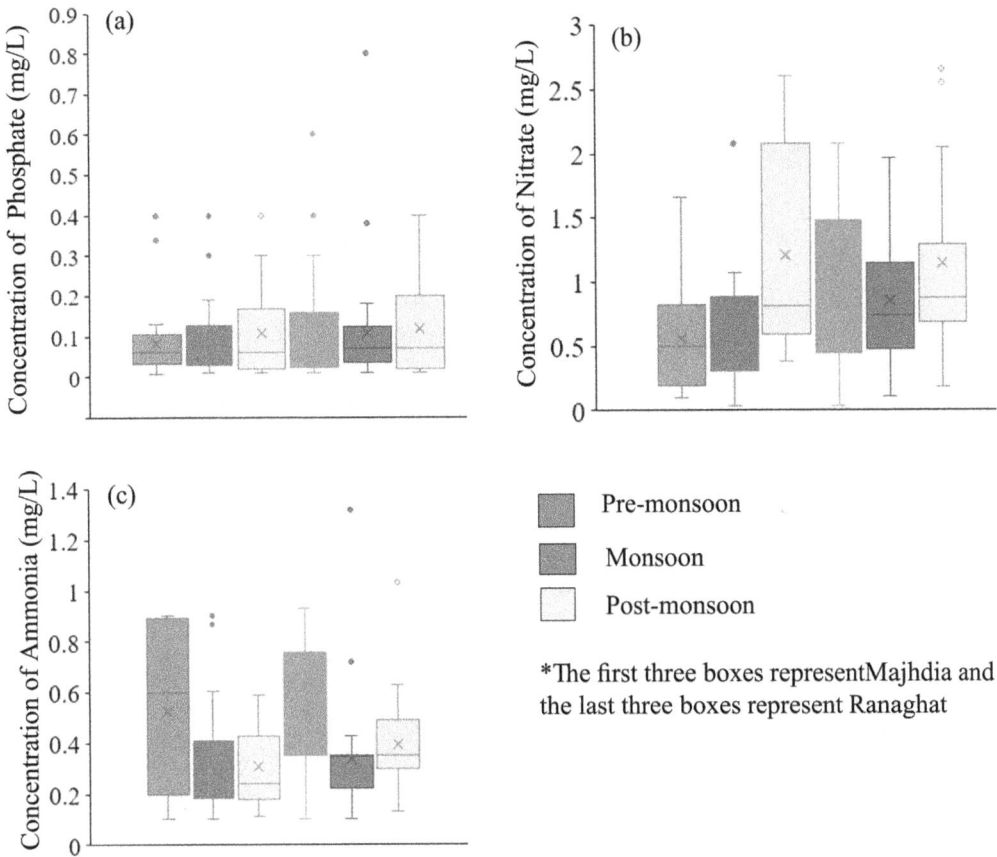

FIGURE 13.16 Concentration of (a) phosphate, (b) nitrate and (c) ammonia in the river Churni.

13.3.6 BRICK KILN

Thirty-four major brick kiln industries have been emerged prominently at both the banks of Mathabhanga–Churni River Basin (Figure 13.17e). Among thirty-four industries, eighteen are located along the river Mathabhanga (from the source to the bifurcation point of Ichamati and Churni Rivers at Shibnibas), and another sixteen are on the bank of Churni River (from the bifurcation point to the confluence point with Hooghly River at Payradanga). Majority of the industries have been continuing their brick production depending on the collection of sand from the bank of the river. The illegal and unscientific practice of bank excavation has strongly altered the morphology and hydraulics of channel. Moreover, the excavation is generally performed at the natural bank which is covered by natural vegetation, and therefore, vegetation clearance is also inevitable which further leads to the bank more vulnerable to soil erosion (Figure 13.17d). Consequently, sediment load in the river is increased at alarming rate. Further, to protect the bank from soil erosion during monsoon an attempt has been taken by planting the vetiver grass on the bank (Figure 13.17c). In the upper part of the river basin, the striking feature of channel modification has been found at the brick kiln of Betbaria (Khulna division, Bangladesh). About 10,000 m² area on the bank has been identified as the bank excavation area and the continuous process of bank excavation gives birth to a pond of ~3,000 m². This pond is connected with the flow of the river, even in the dry season of the year (Figure 13.17a). Another example of bed excavation by brick kiln has been found near Goalgram (Daulatpur police station, Bangladesh). The industry is located ~200 m away from the river, and 1,000 m² has gone under excavation and due to which the width of the valley has been widened from

FIGURE 13.17 Collection of sediments from bank for brick production: (a) formation of pond by bed excavation, (b) massive destruction of river bank by sediment collection for brick production near Ranaghat, (c) protection of bank by planting vetiver grass near Hanskhali.

30 to 55 m. In the Churni River Basin, the brick kiln located at Hanskhali (middle course the river) has drastically changed the valley configuration. The natural valley width in this area ranges from 40 to 50 m, but at the brick kiln, the valley extends up the 99 m. The total cross-sectional area has also been increased from ~300 to ~430 m² and the average depth from ~3.3 to ~4 m. Moreover, as

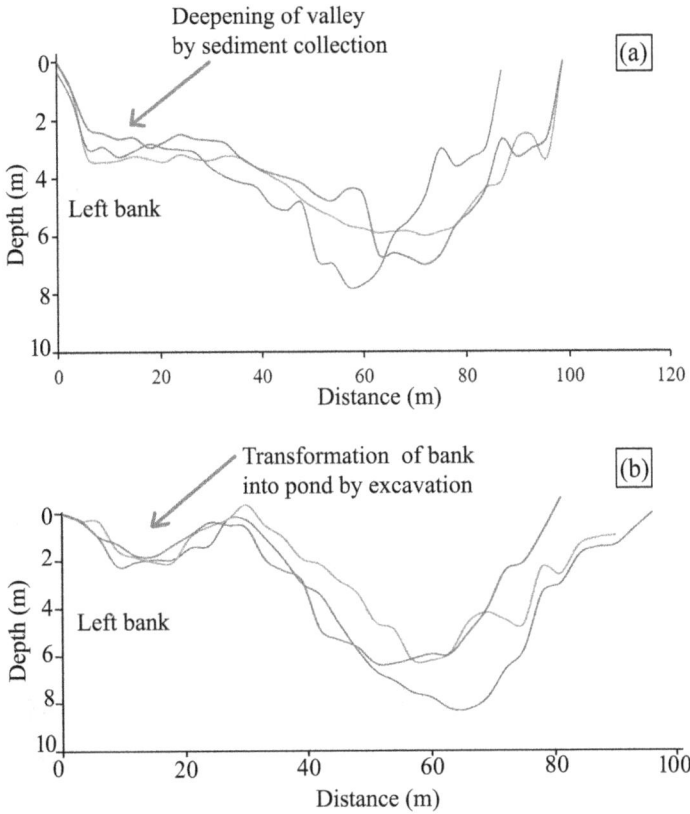

FIGURE 13.18 Modification of cross-section near brick kiln industry: (a) near Hanskhali and (b) near Ranaghat.

the excavation is performed at the left bank of the river, the cross-sectional area of the left side from the valley centre line is more than that of the right side. Therefore, the channel has become more asymmetric in nature. The asymmetry study reveals that the average value of A*, A_1 and A_2 is 0.230, 0.204 and 0.102, respectively, for Hanskhali (Figure 13.18a). Finally, the river bank near Ranaghat Municipality (downstream stretch of the river) has been extremely modified than the upper and middle course of the river because eleven brick kilns are existing very close to each other and all the industries have made the bank ponds by excavation and thereby the cumulative effect on channel is astonishing (Figure 13.17b). The average asymmetry value of A*, A_1 and A_2 at Ranaghat is 0.238, 0.431 and 0.262, respectively (Figure 13.18b).

13.3.7 ANTHROPOGENIC SIGNATURE OVER THE BASIN: LAND-USE AND LAND COVER CHANGE

The observation in the last three decades on the LULC reveals the fact that the changes in the share of each LULC type have been eye-catching (Figure 13.19a and b). In 1991, the dominant share belonged to vegetation (51.02%) which has been reduced to 21.80% in 2018 (Figure 13.19c). Besides, waterbodies in this area have also been reduced from 7.56% in 1991 to 1.18%. This massive reduction, therefore, is the result of expansion of agricultural and settled area at faster rate. Now in 2018 the dominant LULC type is agriculture which has been raised up to 53.37% from 23.22% in 1991. Similarly, the settlement area has also been expanded up to 22.95% in 2018 from 18.18% in 1991 (Figure 13.19c). Thus, the change of LULC type reveals that in the period of three decades, the areas of vegetation and waterbodies have been reduced at alarming rate, whereas the areas of

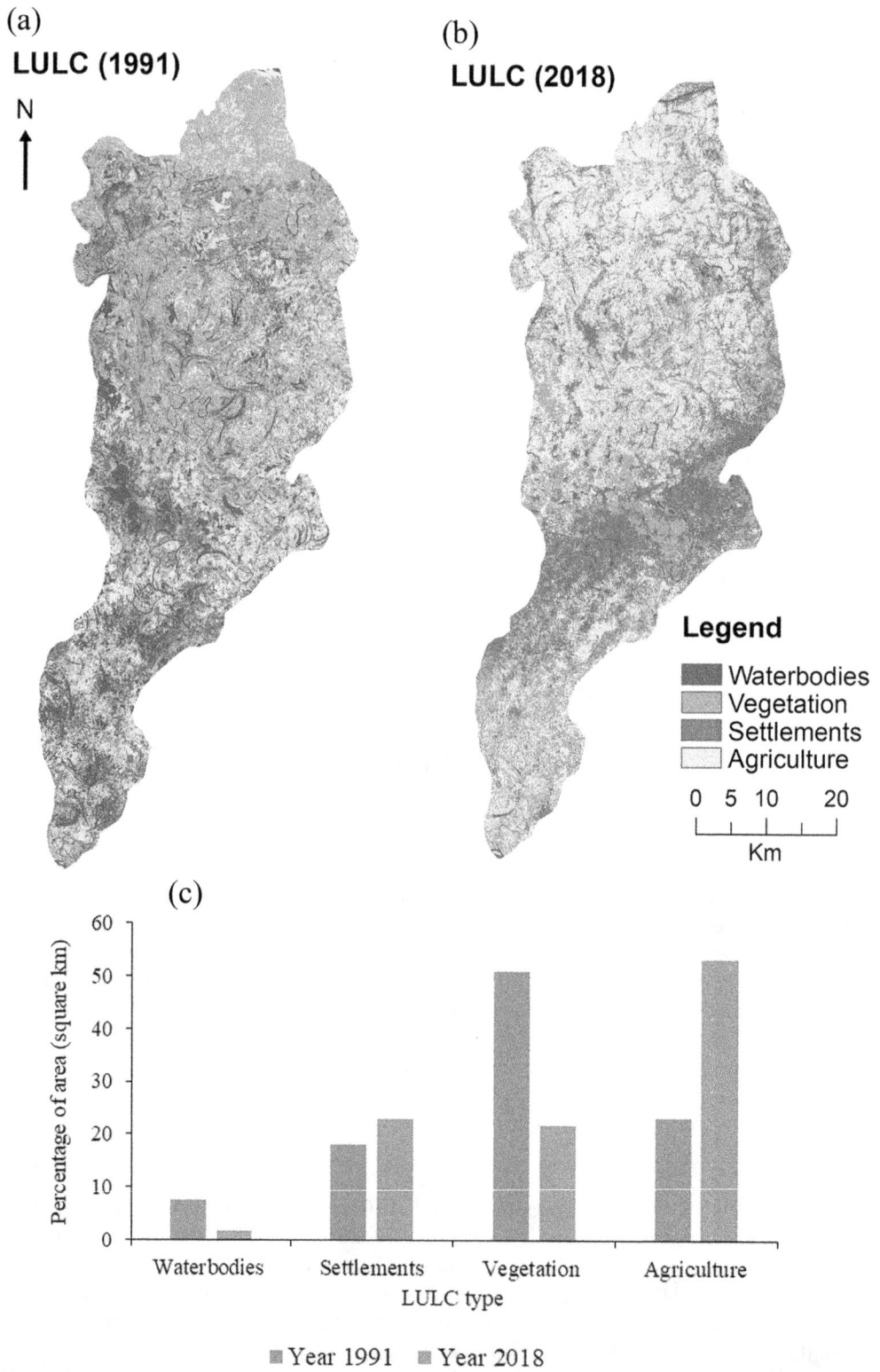

FIGURE 13.19 Changing pattern of LULC in the Mathabhanga–Churni River Basin: (a) LULC in 1991, (b) LULC in 2018, (c) share of LULC types.

TABLE 13.6
Comparison of Channel Morphology

Channel Type		Mean and SD	CV	Range	IQR	Skewness
Natural	A*	0.08 ± 0.09	116.71	0.29	0.13	0.15
	A_1	0.26 ± 0.13	53.12	0.43	0.19	0.35
	A_2	0.11 ± 0.05	50.47	0.15	0.09	−0.12
	w	64.44 ± 7.65	11.87	24	9	0.46
	$D_{average}$	4.61 ± 0.67	14.59	1.81	1.03	0.48
	CS area	303.81 ± 66.27	21.81	187.94	74.05	0.17
Agricultural	A*	0.13 ± 0.08	61.28	0.22	0.11	0.21
	A_1	0.34 ± 0.14	42.04	0.42	0.12	−0.01
	A_2	0.14 ± 0.08	57.44	0.25	0.06	0.27
	w	73.66 ± 10.26	13.94	26	15	−0.11
	$D_{average}$	3.99 ± 0.67	16.86	1.84	0.48	1.73
	CS area	292.23 ± 33.68	11.52	86.85	12.24	−2.29
Road-stream crossing	A*	0.19 ± 0.09	46.82	0.26	0.10	−1.16
	A_1	0.39 ± 0.24	62.50	0.63	0.43	−0.10
	A_2	0.20 ± 0.13	67.95	0.35	0.25	0.06
	w	59.66 ± 8.84	14.82	27	7.50	1.19
	$D_{average}$	3.57 ± 1.37	38.61	5.45	1.52	0.92
	W/D	18.77 ± 7.09	37.81	25.72	7.67	1.51
	CS area	231.45 ± 127.14	55.04	462	137.61	1.80
Brick field	A*	0.22 ± 0.11	49.96	0.320	0.10	−0.41
	A_1	0.34 ± 0.18	53.43	0.420	0.29	0.47
	A_2	0.18 ± 0.14	78.86	0.312	0.23	0.66
	W	87 ± 66.38	17.31	39	15.75	−1.38
	$D_{average}$	3.33 ± 0.67	20.31	1.63	1.02	−0.12
	CS area	338.75 ± 15.05	19.59	177.37	74.05	−0.82

Source: Computed by authors, 2019.

settlement and agriculture have been expended at faster rate. These changes have greatly increased the amount of basin runoff and sediment discharge into the river. Moreover, the runoff from the agricultural land has also increased the concentration of phosphate, nitrate, ammonia, etc. which lead to eutrophication in river, thereby increasing the level of BOD and COD and decreasing DO in water. Moreover, the runoff from rural and urban areas has also increased the dimension and level of pollution (Sarkar and Islam, 2020) (Table 13.6).

13.4 CONCLUSION

According to the investigation carried out in this study, we conclude that although the natural factors such as neotectonism and climate change, manifested in numerous study previously done, have played a crucial role in decaying of the river Mathabhanga–Churni, the interventions posed by human activities, mainly the construction of bridge, on-bed agriculture and cutting of soil from bank for brick production, have had an immense significance for altering the system of hydro-morphology of the river and its basin. In addition, the change in LULC (dropping the share of waterbodies and vegetation area and rapid expansion of settled and agricultural area) is also a clear indication of high runoff and sediment yield into river. Moreover, as the river is playing the crucial role in the development of deltaic economy and healthy environment, there is an ample scope of further research in future. As the river helps in enriching the biodiversity of the delta by providing habitat to countless aquatic species, to improve the health of the river all round is a major challenge in the recent time. Moreover, being a unit of the hydrological cycle, the river is controlling the hydrological system and local climate. Therefore, finding the best way to maintain the environmental flow of the river is also a major challenge. Finally, as the river is the source of livelihood to the fishermen residing along the banks of the river Mathabhanga–Churni, sustaining their livelihood on fishing is another pertinent issue before the researchers.

REFERENCES

Abid, I. (2017). *Interaction of pier, contraction, and abutment scour in clear water scour conditions* (Doctoral dissertation, Georgia Institute of Technology).

Adusumilli, N. C., Lacewell, R. D., Rister, M. E., Woodard, J. D., & Sturdivant, A. W. (2010). *Effect of agricultural activity on river water quality: a case study for the Lower Colorado River Basin* (No. 1371-2016-108879).

Allison, M. (1997). Subaqueous delta of the Ganges-Brahmaputra river system. *Marine Geology, 144*, 81–96.

Allison, M. A., & Kepple, E. B. (2001). Modern sediment supply to the lower delta plain of the Ganges-Brahmaputra River in Bangladesh. *Geo-Marine Lettr, 21*, 66–74. doi:10.1007/s003670100069

Al-Shukur, A. H. K., & Obeid, Z. H. (2016). Experimental study of bridge pier shape to minimize local scour. *International Journal of Civil Engineering and Technology, 7*(1), 162–171.

Arnaud, F., Schmitt, L., Johnstone, K., Rollet, A. J., & Piégay, H. (2019). Engineering impacts on the Upper Rhine channel and floodplain over two centuries. *Geomorphology, 330*, 13–27.

Bandyopadhaya, D. K. (2007). *Banglar Nadnadi*. Dey's Publishing, Kolkata, p. 67.

Bhattacharya, P., Jacks, G., Jana, J., Sracek, A., Gustafsson, J. P., & Chatterjee, D. (2001). Geochemistry of the Holocene alluvial sediments of Bengal Delta Plain from West Bengal, India: implications on arsenic contamination in groundwater. *Groundwater Arsenic Contamination in the Bengal Delta Plain of Bangladesh, 3084*, 21–40.

Blanton, P., & Marcus, W. A. (2009). Railroads, roads and lateral disconnection in the river landscapes of the continental United States. *Geomorphology, 112*(3–4), 212–227.

Blanton, P., & Marcus, W. A. (2014). Roads, railroads, and floodplain fragmentation due to transportation infrastructure along rivers. *Annals of the Association of American Geographers, 104*(3), 413–431.

Boix-Fayos, C., Barberá, G. G., López-Bermúdez, F., & Castillo, V. M. (2007). Effects of check dams, reforestation and land-use changes on river channel morphology: case study of the Rogativa catchment (Murcia, Spain). *Geomorphology, 91*(1–2), 103–123.

Botter, M., Burlando, P., & Fatichi, S. (2019). Anthropogenic and catchment characteristic signatures in the water quality of Swiss rivers: a quantitative assessment. *Hydrology and Earth System Sciences, 23*(4), 1885–1904.

Bouska, W. W., Keane, T., & Paukert, C. P. (2010). The effects of road crossings on prairie stream habitat and function. *Journal of Freshwater Ecology, 25*(4), 499–506.

Burchsted, D., Daniels, M., & Wohl, E. E. (2014). Introduction to the special issue on discontinuity of fluvial systems. *Geomorphology, 205*, 1–4.

Charlton, R. (2008). *Fundamentals of Fluvial Geomorphology*. Routledge, London and New York.

Chatterjee, M. (2017). An enquiry into the evolution and impact of human interference on the Churni River of Nadia District, West Bengal. *International Journal of Current Research, 5*(5), 1088–1092.

Chattopadhyay , S. (2007). *Maharajendra Bahadur Krishnachandra*. Dey's Publishing, Kolkata, pp. 52–67.

Chimwanza, B., Mumba, P. P., Moyo, B. H. Z., & Kadewa, W. (2006). The impact of farming on river banks on water quality of the rivers. *International Journal of Environmental Science & Technology, 2*(4), 353–358.

Chin, A., & Gregory, K. J. (2001). Urbanization and adjustment of ephemeral stream channels. *Annals of the Association of American Geographers, 91*(4), 595–608.

Chin, A., Gidley, R., Tyner, L., & Gregory, K. (2017). Adjustment of dryland stream channels over four decades of urbanization. *Anthropocene, 20*, 24–36.

Chorley, R. J., Malm, D. E., & Pogorzelski, H. A. (1957). A new standard for estimating drainage basin shape. *American Journal of Science, 255*(2), 138–141.

Costa, J. E. (1975). Effects of agriculture on erosion and sedimentation in the Piedmont Province, Maryland. *Geological Society of America Bulletin, 86*(9), 1281–1286.

Crickmay, C. H. (1974). *The Work of the River: A Critical Study of the Central Aspects of Geomorphogeny.* Macmillan, London.

Das, G. (2003). *Bangla Bhashar Abhidhan.* Sahitya Samsad, Kolkata. ISBN-81-85 626-08-1.

Datta, D. K., & Subramanian, V. (1997). Texture and mineralogy of sediments from the Ganges-Brahmaputra-Meghna river system in the Bengal Basin, Bangladesh and their environmental implications. *Environmental Geology, 30*(3–4), 181–188.

Downs, P. W., & Piégay, H. (2019). Catchment-scale cumulative impact of human activities on river channels in the late Anthropocene: implications, limitations, prospect. *Geomorphology, 338,* 88–104.

Dunca, A. M. (2018). Water pollution and water quality assessment of major transboundary rivers from Banat (Romania). *Journal of Chemistry, 2018,* 8.

Dutta, S. (1996). *Anjana Nadi Tire.* Dogachi Gram Panchyet, Nadia, p. 44.

Ebisemiju, F. S. (1989). Patterns of stream channel response to urbanization in the humid tropics and their implications for urban land use planning: a case study from southwestern Nigeria. *Applied Geography, 9*(4), 273–286.

Fergusson, J. (1912). On recent changes in the Delta of the Ganges, Bengal Secretariat Press, Calcutta, ed-Biswas K.R. 2001, Rivers of Bengal, p-184, 205. *Quarterly Journal of the Geological Society of London, XIX,* 1863.

Fergusson, J. (1912). On recent changes in the delta of the Ganges. *Quarterly Journal of the Geological Society, 19*(1–2), 321–354.

Fuller, I. C., Macklin, M. G., & Richardson, J. M. (2015). The geography of the Anthropocene in New Zealand: differential river catchment response to human impact. *Geographical Research, 53*(3), 255–269.

Gampathi, G. A. P. (2010). Suitable bridge pier section for a bridge over a natural river. *Engineer: Journal of the Institution of Engineers, Sri Lanka, 43*(3).

Garrett, J. H. E. (1910). *Bengal District Gazetteers: Nadia.* Bengal Secretariat Book Depot, Calcutta.

Garzanti, E., Vezzoli, G., Andò, S., Limonta, M., Borromeo, L., & France-Lanord, C. (2019). Provenance of Bengal shelf sediments: 2. Petrology and geochemistry of sand. *Minerals, 9*(10), 642.

Gergel, S. E., Turner, M. G., Miller, J. R., Melack, J. M., & Stanley, E. H. (2002). Landscape indicators of human impacts to riverine systems. *Aquatic Sciences, 64*(2), 118–128.

Ghorbani, B. (2008). A field study of scour at bridge piers in flood plain rivers. *Turkish Journal of Engineering and Environmental Sciences, 32*(4), 189–199.

Ghosal, U., Sikdar, P. K., & McArthur, J. M. (2015). Palaeosol control of arsenic pollution: the Bengal Basin in West Bengal, India. *Groundwater, 53*(4), 588–599.

Gibling, M. (2018). River systems and the anthropocene: a late pleistocene and holocene timeline for human influence. *Quaternary, 1*(3), 21.

Gilbert, G. K. (1917). *Hydraulic-mining debris in the Sierra Nevada* (No. 105). US Government Printing Office.

Goodbred, S. (2003). Response of the Ganges dispersal system to climate change: a source-to sink-view since the last interstate. *Sedimentary Geology, 162,* 83–104.

Govindasamy, A. V. (2009). *Simplified method for estimating future scour depth at existingbridges.* Texas A&M University.

Gravelius, H. (1914). Grundrifi der gesamten Gewcisserkunde. Band I: Flufikunde. Compendium of Hydrology I. Rivers.

Gregory, K. J. (2006). The human role in changing river channels. *Geomorphology, 79*(3–4), 172–191.

Guchhait, S. K., Islam, A., Ghosh, S., Das, B. C., & Maji, N. K. (2016). Role of hydrological regime and floodplain sediments in channel instability of the Bhagirathi River, Ganga-Brahmaputra Delta, India. *Physical Geography, 37*(6), 476–510.

Guo, C., Jin, Z., Guo, L., Lu, J., Ren, S., & Zhou, Y. (2020). On the cumulative dam impact in the upper Changjiang River: streamflow and sediment load changes. *Catena, 184,* 104250.

Hammer, T. R. (1972). Stream channel enlargement due to urbanization. *Water Resources Research, 8*(6), 1530–1540.

Hans, R. K., Farooq, M., Babu, G. S., Srivastava, S. P., Joshi, P. C., & Viswanathan, P. N. (1999). Agricultural produce in the dry bed of the River Ganga in Kanpur, India–a new source of pesticide contamination in human diets. *Food and Chemical Toxicology, 37*(8), 847–852.

Happ, S. C., Rittenhouse, G., & Dobson, G. C. (1940). *Some principles of accelerated stream and valley sedimentation* (No. 695). US Department of Agriculture.

Hoffmans, G. J., & Verheij, H. J. (1997). Scour Manual (Vol. 96). CRC Press, Rotterdam, Netherlands.

Hooke, J. M. (2006). Human impacts on fluvial systems in the Mediterranean region. *Geomorphology, 79*(3–4), 311–335.

Horton, R. E. (1932). Drainage-basin characteristics. *TrAGU*, *13*(1), 350–361.

Huggett, R. (2007). Fundamentals of Geomorphology. Routledge.

Hunter, W. W. (1877). *A Statistical Account of Bengal* (Vol. 20). Trübner & Company, London.

Islam, A., & Guchhait, S. K. (2017). Analysing the influence of Farakka Barrage Project on channel dynamics and meander geometry of Bhagirathi river of West Bengal, India. *Arabian Journal of Geosciences*, *10*(11), 245.

Islam, A., & Guchhait, S. K. (2018). Analysis of social and psychological terrain of bank erosion victims: a study along the Bhagirathi river, West Bengal, India. *Chinese Geographical Science*, *28*(6), 1009–1026.

Islam, A., Sarkar, B., Das, B. C., & Barman, S. D. (2020). Assessing gully asymmetry based on cross-sectional morphology: a case of Gangani Badland of West Bengal, India. In *Gully Erosion Studies from India and Surrounding Regions* (pp. 69–92). Springer, Cham.

Jacobson, R. B., & Coleman, D. J. (1986). Stratigraphy and recent evolution of Maryland Piedmont flood plains. *American Journal of Science*, *286*(8), 617–637.

Jacobson, R. B., & Pugh, A. L. (1992). Effects of land use and climate shifts on channel instability, Ozark Plateaus, Missouri, USA. In *Proceedings of the Workshop on the Effects of Global Climate Change on Hydrology and Water Resources at the Catchment Scale: Tsukuba, Japan, Japan–US Committee on Hydrology, Water Resources and Global Climate Change* (pp. 423–444).

Jha, V. C., & Bairagya, H. P. (2011). Flood plain evaluation in the Ganga-Brahmaputra Delta: a tectonic review. *Ethiopian Journal of Environmental Studies and Management*, *4*(3), 12–24.

James, L. A., & Marcus, W. A. (2006). The human role in changing fluvial systems: retrospect, inventory and prospect. *Geomorphology*, *79*(3–4), 152–171.

Johnson, P. A., Gleason, G. L., & Hey, R. D. (1999). Rapid assessment of channel stability in vicinity of road crossing. *Journal of Hydraulic Engineering*, *125*(6), 645–651.

Khan, K. A., Muzzammil, M., & Alam, J. (2016). Bridge Pier Scour: a review of mechanism, causes and geotechnical aspects. *Proceedings of Advances in Geotechnical Engineering*, April (pp. 8–9).

Knighton, A. D. (1981). Asymmetry of river channel cross-sections: part I. Quantitative indices. *Earth Surface Processes and Landforms*, *6*, 581–588.

Knox, J. C. (1977). Human impacts on Wisconsin stream channels. *Annals of the Association of American Geographers*, *67*(3), 323–342.

Kong, D., Latrubesse, E. M., Miao, C., & Zhou, R. (2020). Morphological response of the Lower Yellow River to the operation of Xiaolangdi Dam, China. *Geomorphology*, *350*, 106931.

Kothyari, U. C. (2007). Indian practice on estimation of scour around bridge piers–a comment. *Sadhana*, *32*(3), 187–197.

Koenig, C. E. (2011). *Hydrogeochemical site characterization and groundwater flow modeling of the arsenic-contaminated Gotra aquifer, West Bengal, India* (Doctoral dissertation, University of British Columbia).

Kuehl, S. A., Allison, M. A., Goodbred, S. L., & Kudrass, H. E. R. M. A. N. N. (2005). The Ganges-Brahmaputra Delta. Special Publication-SEPM, 83, 413.

Laursen, E. M., & Toch, A. (1956). *Scour Around Bridge Piers and Abutments* (Vol. 4). Iowa Highway Research Board, Ames, IA.

Li, S., Li, Y., Yuan, J., Zhang, W., Chai, Y., & Ren, J. (2018). The impacts of the Three Gorges Dam upon dynamic adjustment mode alterations in the Jingjiang reach of the Yangtze River, China. *Geomorphology*, *318*, 230–239.

Madej, M. A., & Ozaki, V. (1996). Channel response to sediment wave propagation and movement, Redwood Creek, California, USA. *Earth Surface Processes and Landforms*, *21*(10), 911–927.

Mallick, K. (1911). *Nadia Kahini* (in Bengali) (edited by Ray Mohit, 1998, 3rd Ed). Pustak Bipani, Kolkata, p. 300.

Marion, D. A., Phillips, J. D., Yocum, C., & Mehlhope, S. H. (2014). Stream channel responses and soil loss at off-highway vehicle stream crossings in the Ouachita National Forest. *Geomorphology*, *216*, 40–52.

Meade, R. H., & Trimble, S. W. (1974). Changes in sediment loads in rivers of the Atlantic drainage of the United States since 1900. *International Association of Hydrological Sciences Publication*, *113*, 99–104.

Miller, V. C. (1953). A quantitative geomorphologic study of drainage basin characteristics in the clinch mountain area. *Tech Rep*, *3*, 271–300.

Moss, B. (2007). Water pollution by agriculture. *Philosophical Transactions of the Royal Society B: Biological Sciences*, *363*(1491), 659–666.

Mukherjee, A., & Fryar, A. E. (2008). Deeper groundwater chemistry and geochemical modeling of the arsenic affected Western Bengal basin, West Bengal, India. *Applied Geochemistry*, *23*(4), 863–894.

Mukherjee, A., Fryar, A. E., & Thomas, W. A. (2009). Geologic, geomorphic and hydrologic framework and evolution of the Bengal basin, India and Bangladesh. *Journal of Asian Earth Sciences, 34*(3), 227–244.

Mukhopadhaya, S. R. (1805). *Maharaj Krishnachandra Rayasya Charitrang*. Sourindranath Das, Srirampur, pp. 301–303.

Mukhopadhyay, B., Mukherjee, P. K., Bhattacharya, D., & Sengupta, S. (2006). Delineation of arsenic-contaminated zones in Bengal Delta, India: a geographic information system and fractal approach. *Environmental Geology, 49*(7), 1009–1020.

Mutisya, D. K., & Tole, M. (2010). The impact of irrigated agriculture on water quality of rivers Kongoni and Sirimon, Ewaso Ng'iro North Basin, Kenya. *Water, Air, & Soil Pollution, 213*(1–4), 145–149.

Neill, M. (1989). Nitrate concentrations in river waters in the south-east of Ireland and their relationship with agricultural practice. *Water Research, 23*(11), 1339–1355.

Ortega, J. A., Razola, L., & Garzón, G. (2014). Recent human impacts and change in dynamics and morphology of ephemeral rivers. *Natural Hazards and Earth System Sciences, 14*(3), 713–730.

Parua, P. (2009). Erosion problem of Ganga River bank in Malda District. In: *Some aspects about Farakka Barrage Project* (Vol. I) (edited by Parua P. K.). Berhampore: Shilpanagari Publishers.

Parua, P. (2010). *The Ganga: Water Use in the Indian Subcontinent*. Springer Dordrecht Heidelberg London New York.

Pickup, G., & Warner, R. F. (1984). Geomorphology of tropical rivers. II. Channel adjustment to sediment load and discharge in the Fly and lower Purari, Papua New Guinea. *Catena Supplement, 5*, 18–41.

Rahman, M. M., & Haque, M. A. (2003). Local scour estimation at bridge site: modification and application of Lacey formula. *International Journal of Sediment Research, 18*(4), 333–339.

Ray, K. C. (1751). Annada Mangal, Bengali, quoted from Choudhury. J, (2004), Sri Chaitanyadev O Samakalin Nabadwip (in Bengali), Nabadwip Puratatwa Parisad, p. 14.

Ray, K. C. (1875). *Kshitish Vansabali Charita* (edited by Ray Mohit, 1986). Natun Sanskrita Jantra, Kolkata, pp. 69–71.

Ray, K. C. (1986). *Kshitish Banshaboli Charita* (edited by Ray M., 1986). Manjusha, Kolkata, p–57

Reaks, H. G. (1919). Report on the physical and hydraulic characteristics of the rivers of the delta. *Kumud Ranjan Biswas (compiled), Rivers of Bengal, 2*, 27–190.

Richardson, E. V., & Davis, S. R. (1995). *Evaluating scour at bridges* (No. HEC 18). United States. Federal Highway Administration. Office of Technology Applications.

Roy, S. (2013). The effect of road crossing on river morphology and riverine aquatic life: a case study in Kunur River Basin, West Bengal. *Ethiopian Journal of Environmental Studies and Management, 6*(6), 835–845.

Roy, S., & Sahu, A. S. (2017). Potential interaction between transport and stream networks over the lowland rivers in Eastern India. *Journal of Environmental Management, 197*, 316–330.

Rudra, K. (2010). Dynamics of the Ganga in West Bengal, India (1764–2007)—implications for science-policy interaction. *Quaternary International 227*, 161–169.

Rudra, K. (2011). The encroaching Ganga and social conflict: the case of West Bengal, India. Kolkata. http://www. gangawaterway. in/assets/02Rudra. pdf.

Rudra, K. (2014). Changing river courses in the western part of the Ganga Brahmaputra delta. *Geomorphology, 227*, 87–100.

Saha, R., Lee, S., & Hong, S. (2018). A comprehensive method of calculating maximum bridge scour depth. *Water, 10*(11), 1572.

Sahitya Samsad. (2000). *Samsad Bangla Abhidhan*. Kolkata. ISBN-81-86 806-92-X.

Sarkar, B., & Islam, A. (2019). Assessing the suitability of water for irrigation using major physical parameters and ion chemistry: a study of the Churni River, India. *Arabian Journal of Geosciences, 12*(20), 637.

Sarkar, B., & Islam, A. (2020). Drivers of water pollution and evaluating its ecological stress with special reference to macrovertebrates (fish community structure): a case of Churni River, India. *Environmental Monitoring and Assessment, 192*(1), 45.

Schottler, S. P., Ulrich, J., Belmont, P., Moore, R., Lauer, J. W., Engstrom, D. R., & Almendinger, J. E. (2014). Twentieth century agricultural drainage creates more erosive rivers. *Hydrological Processes, 28*(4), 1951–1961.

Schumm, S. A. (1956). Evolution of drainage systems and slopes in badlands at Perth Amboy, New Jersey. *Geological Society of America Bulletin, 67*(5), 597–646.

Surian, N., & Rinaldi, M. (2003). Morphological response to river engineering and management in alluvial channels in Italy. *Geomorphology, 50*(4), 307–326.

Tarolli, P., & Sofia, G. (2016). Human topographic signatures and derived geomorphic processes across landscapes. *Geomorphology, 255*, 140–161.

Toner, P. F. (1986). Impact of agriculture on surface water in Ireland Part I. General. *Environmental Geology and Water Sciences, 9*(1), 3–10.

Trimble, S. W., & Lund, S. W. (1982). *Soil conservation and the reduction of erosion and sedimentation in the Coon Creek Basin, Wisconsin* (No. 1234).

Varnum, N., & Ozaki, V. (1986). *Recent channel adjustments in Redwood Creek, California* (No. 18, pp. 74). National Park Service, Redwood National Park.

Wang, C., Liu, S., Deng, L., Liu, Q., & Yang, J. (2014). Road lateral disconnection and crossing impacts in river landscape of Lancang River Valley in Yunnan Province, China. *Chinese Geographical Science, 24*(1), 28–38.

Wohl, E. (2018). Rivers in the Anthropocene: the US perspective. *Geomorphology*, https://doi.org/10.1016/j.%20geomorph.2018.12.001

Weinman, B., Goodbred Jr, S. L., Zheng, Y., Aziz, Z., Steckler, M., van Geen, A., Singhvi, A. K., & Nagar, Y. C. (2008). Contributions of floodplain stratigraphy and evolution to the spatial patterns of groundwater arsenic in Araihazar, Bangladesh. *Geological Society of America Bulletin, 120*(11–12), 1567–1580.

Wolman, M. G. (1967). A cycle of sedimentation and erosion in urban river channels. *Geografiska Annaler: Series A, Physical Geography, 49*(2–4), 385–395.

Wolman, M. G., & Schick, A. P. (1967). Effects of construction on fluvial sediment, urban and suburban areas of Maryland. *Water Resources Research, 3*(2), 451–464.

Wyzga, B. (1993). River response to channel regulation: case study of the Raba River, Carpathians, Poland. *Earth Surface Processes and Landforms, 18*(6), 541–556.

14 Detecting the Facets of Anthropogenic Interventions on the Palaeochannels of Saraswati and Jamuna

Mehebub Sahana, Mohd Rihan, Samrat Deb,
Priyank Pravin Patel, Wani Suhail Ahmad and Kashif Imdad

CONTENTS

14.1 INTRODUCTION

Almost all civilisations have arisen along rivers, making them an integral part of human history. Changes in river courses have consequently affected settlements (Morozova, 2005) as much as the geomorphology of any region, making any study into their evolutionary history a dual investigation into the physical and cultural mosaic that constitutes a locale (Heckenberger et al., 2007; Schwartzman et al., 2013). Moribund streams, palaeochannels and buried rivers exist as markers of such historical alterations (Dambeck and Thiemeyer, 2002; Gibbard, 1988), arising from tectonic events (Gupta et al., 2014; Page et al., 2009), anthropogenic activities (Chen et al., 1996; Stefani and Vincenzi, 2005) or climatic changes (Maroulis et al., 2007; Page et al., 1996; Vandenberghe, 1995). Examining them and consequently delving into the fluvio-historical evolution of a region is thus an important aspect of river geomorphology, to better decipher the existent morphodynamic conditions (Borgohain et al., 2017; Brooks et al., 2003; Starkel, 2008).

Palaeochannels, also known as palaeodrainages, palaeorivers (Khan, 1987), buried rivers or cut-off rivers (Camporeale et al., 2008), are usually present in a fluvial system, especially in a lowland alluvial setting (Brown, 1999; Latrubesse, 2002; Hesse et al., 2018). They represent those channels which were abandoned or ceased to exist as parts of an active river system (Zankhna and Thakkar, 2014). Palaeochannels are important for land and water management (Syvitski et al., 2005) and have therefore garnered much attention in fluvial geomorphology studies. The existence of palaeochannels is usually confirmed through the field identification of many morphological features such as ox-bow lakes, palaeo-levees and point bars (Kleinhans et al., 2012), with the help of various geological and geophysical survey techniques, e.g. field observations with or without well-logging (Bates et al., 2007), followed by laboratory studies, electromagnetic surveys (Jorgensen et al., 2012) and seismic investigations (Deen et al., 2000), which can be used to determine the sedimentation rates within these features (Ernenwein, 2002), track how they have evolved (Conyers et al., 2008) and gauge their suitability as aquatic habitats (Dieras, 2013). However, sometimes, the locations where they may be present are either not easily accessible or the entire field investigation process becomes cumbersome and too expensive (Salama et al., 1994; Vervoort and Annen, 2006).

In such a scenario, remote sensing techniques can be adequately and efficiently employed for delineating past river courses effectively (Ghose et al., 1979; Srigyan et al., 2019; Wray, 2009), along with the demarcation of the river corridors (cf. Banerji and Patel, 2019) within which the river has flowed and shifted previously or the morphological changes within the channel itself (Patel, 2010) and its adjoining tracts (Patel and Dasgupta, 2009; Chatterjee and Patel, 2016). In this study, a database has been created for the Saraswati and Jamuna palaeochannels of the Hooghly River in southern West Bengal, India, (Figure 14.1).

According to Hindu mythology and popular belief, the main Ganga, Yamuna and Saraswati rivers join together at Triveni Sangam, near the city of Allahabad in Uttar Pradesh. After this confluence, the merged channel is known as *Yuktabeni* (united strands). Further downstream, this single river dissociates and separates into the aforementioned three channels at Tribeni town, where they are known as *Muktabeni* (freed strands). These two channels diverged away from the main Bhagirathi-Hooghly at the ancient town of Tribeni, with the trifurcation being referred to as *Muktabeni* (freeing up of entwined strands – Majumdar, 1941) and had mythological as well as historical bearing on the evolving socio-political and economic importance of the region (Majumdar, 1942). Information on

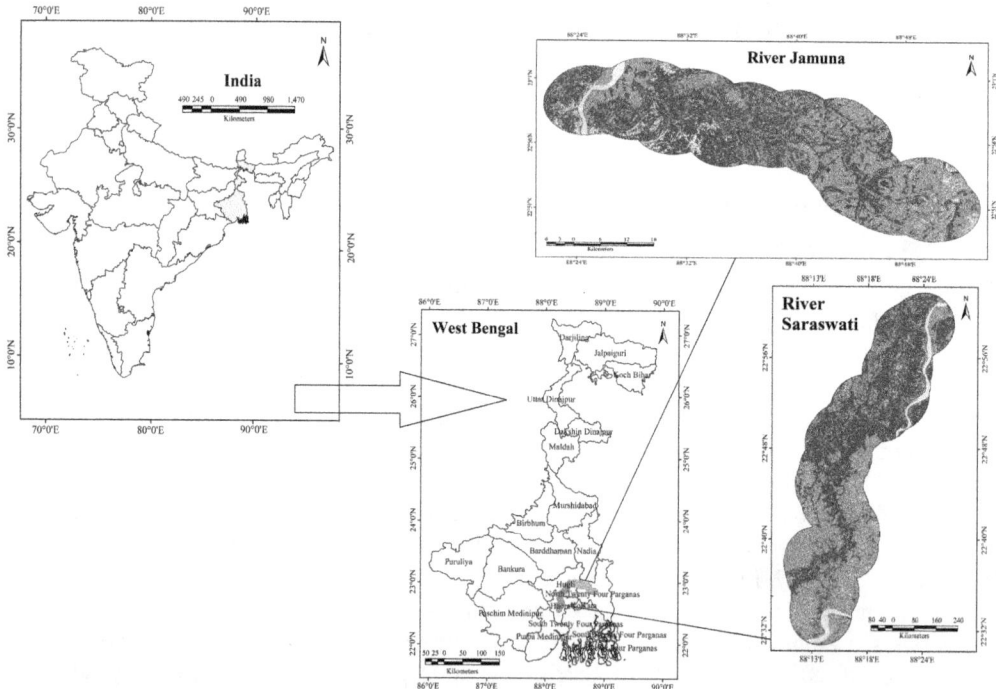

FIGURE 14.1   Location of the studied palaeochannels in the Lower Ganga Delta.

the above two divergent strands was collated and extracted from various remotely sensed geospatial datasets along with detailed accounts of the area from historical maps and documents, since riverine floodplains are composites of the physical-socioeconomic milieu of a region (Dasgupta and Patel, 2017). The obtained facts were further verified through a detailed ground-truthing exercise. The generated database reveals the changes that have occurred along the adjacent riparian tracts of these two palaeochannels and how they have become progressively more and more moribund over time, with the knock-on effect of this on the settlements once situated along them.

To identify and highlight the spatio-temporal changes that have occurred along the riparian tracts of the Saraswati and Jamuna Rivers' meandering palaeochannels in the Lower Bengal Delta (LBD), satellite-based remotely sensed data of Landsat 1 MSS (Spatial Resolution (SR) 79 m, Path/Row 138-45) for November 1973, Landsat TM 5 (SR 30 m, Path/Row 138-45) for November 2000, Landsat 8 OLI (SR PAN 15 m, MSS 30 m, Path/Row 138-45) for October 2017 and SRTM DEM (Shuttle Radar Topographic Mission Digital Elevation Model) (SR 30 m or 1 arc-second) downloaded from USGS EarthExplorer has been utilised (Table 14.1). Image processing software like Erdas Imagine 14 and Arc GIS 10.2 was used to classify and process the above data. The information regarding the other geographical aspects of the region, such as geology, geomorphology, soil cover and communication characteristics, was gleaned and digitised from the respective National Atlas and Thematic Mapping Organisation (NATMO), District Planning Map Series (DPMS) and Geological Survey of India (GSI) District Resource Map (DRM) sheets and Food and Agriculture Organisation (FAO) databases.

To generate the multi-temporal land use and land cover (LULC) database, a maximum-likelihood supervised classification technique was utilised, in which about 200 signature samples for each LULC class were extracted to enhance the classification accuracy. For analysing the surface cross-profile and the overall geomorphology of the river plain, SRTM DEM data was obtained, which can be feasibly used for the extraction of surface profiles across river basins (Patel and Sarkar, 2010; Patel, 2013; Patel and Mondal, 2019). Profile lines were marked along the courses of both the rivers, dividing them into six zones, to better understand the hydro-morphological changes that have occurred and monitor the channel migration. Parameters such as the length of the stretch, its sinuosity, the radius of curvature, meander amplitude, wavelength and channel dimensions of these various cross sections (Figures 14.9 and 14.12) and reaches were also used to gauge the channel morphology alterations. These have been briefly explained as follows:

The radius of curvature (r_c): It denotes the radius of a curve or an arc that clearly depicts the channel curve at that point. To calculate the r_c, the "principle of best fit circular arc" (Bagnold, 1960; Williams, 1986) was used, and it was enumerated using the Arc GIS 10.2 circle tool. In the present study, almost all the examined meanders are irregular in their planform morphology.

Meander amplitude (a_m): The longest or maximum distance between the meander crest and trough, showing the displacement of the river sideways from the meander axis. This meander axis was delineated with the help of the collapse dual line tool in the data management toolbox of Arc GIS 10.2.

TABLE 14.1
Spatial and Non-Spatial Data Used for the Present Work

Data	Data Type	Source	Data Specifications	Time Period
LANDSAT-1 MSS	Spatial	USGS	(59 m resolution) Path/Raw 138, 45	November, 1973
LANDSAT-5 TM	Spatial	USGS	(30 m resolution) Path/Raw 138, 45	November, 2000
LANDSAT-8 OLI	Spatial	USGS	(PAN 15, 30 m resolution) Path/Raw 138, 45	October, 2017
SRTM DEM	Spatial	USGS	30 m spatial and 1 arc-second horizontal resolution	February, 2000

Source: Compiled by the authors.

Distance: It is the straight line distance between two consecutive crests or troughs of a meander bend.

Channel length (C_i): It denotes the curvilinear length between the inflexion points of a medium channel of a specific meander loop. It is double the bend length, therefore,

$$C_l = 2(\text{Bend Length}) \tag{14.1}$$

The rate of migration of the palaeochannels for both the rivers was calculated considering the shifts along the axes of all the meanders within the period of study. Similarly, to analyse the planform changes for these meanders, in each of the demarcated reaches, the Sinuosity Index (SI) was measured by computing the ratio of the mid-channel length to total channel belt length, as measured in a straight line:

$$SI = \text{Actual channel length/Straight length} \tag{14.2}$$

If, for a reach, the SI > 1.05, it indicated that section to be of a sinuous nature. If the SI > 1.5, then it showcased a meandering stretch and a SI > 2.0 indicated a tortuous section (Morisawa, 1985). Besides this, the SI was also used to define and compare the spatial reference of these two rivers, which exhibit significant oscillation along their respective courses from a mean straight line, in terms of their meanders and consequently their flow directions, at different locations along their channel lengths, by analysing their meander loops through eliciting the ratio between the length of the examined channel section and its individual loop wavelength (Langbein and Leopold, 1966).

14.2   THE RIVER AND ITS BASIN

14.2.1   HISTORICAL BACKGROUND OF THE DISTRIBUTARIES

The Saraswati channel in the LBD plain is actually a distributary of the main Bhagirathi River, with its catchment spread across Howrah, Barddhaman (a tiny portion) and Hugli districts of West Bengal (Rudra, 1986) (Figure 14.2a). As per inscriptions found in the present day Burdwan district of West Bengal, the *Vardhaman Bhukti* (an important division of the Sen Region, of which the Saraswati-Adiganaga *Doab* tract was a part) was known as the *Vetadda-Chautaraka*, and comprised of the Saraswati Basin (Ghosh, 1988). Trade and commerce thrived along its banks in the settlements of Janai, Singur, Chanditala, Nasibpur, Domjur and Andul, whereas Saptagram (or *Satgaon*) near Tribeni was till the medieval period, one of Bengal's biggest ports (Mukherjee, 1987). Ptolemy, describing the Ganges' mouths (Rudra, 1981), noted it as the capital of the Gangaridai Kingdom, while Ibn Battuta wrote of its prosperity during his travels in 1345. In 1535, the Portuguese landed here (Sen, 1943) and named it as *Porto Piqueno* (the "little port") (Mukhopadhyay, 1996), establishing one of their Customs Houses in return for aiding the then Bengal Sultan Mahmud Shah to check the advances of Sher Shah's army in the region (Sen, 2019). However, they later decamped and moved their port base to Hugli in 1580 (Sreemani, 2011), as the Saraswati channel slowly became more derelict (Mukherjee, 1987). Thus as the channel silted up, it lost its eminence, and the settlements along it suffered likewise (Rudra, 2018), with only the current remains of Saptagram near the present railway station of Adisaptagram, about 43 km from Kolkata (Seth, 2014), being a marker of its former glorious days. In its place, the newer port of Tamralipta (near present day Haldia) was established in the 3rd century, and was located much further downstream (Mukhopadhyay, 1996). With further decay of the Saraswati, however, both these port cities lost their former prominence (Reaks, 1919), with larger settlements shifting further east across the main Hooghly channel, following the overall trend of migration of the delta's principal streams (Eaton, 1993). A similar situation occurred in the vicinity of the Jamuna channel on the other flank of Hooghly River, with the decline of the once prosperous port of Chandraketugarh (Sen et al., 2015).

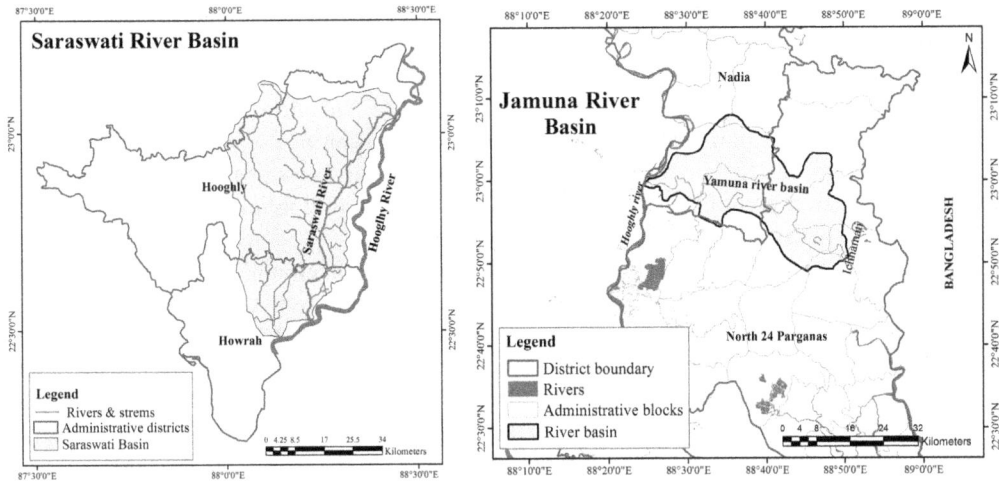

FIGURE 14.2 The individual basins of the Saraswati and Jamuna distributaries.

The town of Tribeni (meaning "three braids" in Sanskrit) is the site where the main Bhagirathi channel, flowing southwards, divides into three separate streams, with the Saraswati flowing out from its right bank, the Hooghly (or Hugli) in the middle and the Jamuna emanating from its left bank (Mukhopadhyay, 1996), and was a bustling trade centre in the 3rd century (Mukherjee, 1995). The River Jamuna is at present virtually dead, and parts of its old pathways can be delineated from historic maps and records (Rudra, 1986) (Figure 14.2b), whereas the Saraswati survives as a shallow, shrinking and at times non-perennial and virtually discontinuous stream across parts of Hooghly and Howrah districts of West Bengal (Seth, 2014). Stream degradation over millennia has thus progressively led to diminishing vessel numbers and capacities (Johnston, 1933), making it lose its former significance in furthering the socioeconomic milieu of the region.

While several antique maps by European travellers such as Gastaldi (1548), Jao De Barros (1615) and Van den Broucke (1660) depict the Saraswati's historic courses (Blochmann, 1873; Raychaudhuri, 1943), the absence of actual field surveys while preparing them negates their cartographic accuracy and only the settlement sites denoted in them with respect to their adjacent streams may be presumed as correct (Mukhopadhyay, 1996). Such maps tally with Abul Fazl's statement in his *Ain-i-Akbari* (1590) of there being three branches at Tribeni, as aforementioned. Blaev's map of 1690 and that by De Barros depicted all three strands to be similar in width, with the Saraswati releasing its discharge at Chouma (modern Chaumuha, situated in the northern portion of Hugli district) and the Jamuna going up to Buram (in the North 24 Parganas district of West Bengal) (Mukhopadhyay, 1996; Seth, 2014). The openings of the Saraswati were depicted differently in Cantle's map (1683), with its flowing independently to the Bay of Bengal, surrounded by clay and nodular limestone beds westwards and altering channel courses eastwards (Mukherjee, 1938). Maps by Rennell (1764–1776, the first authentic surveys of the region) show the Saraswati's upper course (La Touche, 1910) without the nomenclature (e.g. in his *Memoirs of a Map of Hindoostan*, 1783). The decline in the flow of the Saraswati also caused changes in the direction of the Damodar River (Mukherjee, 1938), leading to substantial changes in the hydrology and consequently the socioeconomic landscape of the delta.

A number of pioneering studies have dealt with the history of the ancient Bengal Basin (e.g. Bagchi, 1944), and only a selected few can be considered within the limited scope of this article. According to O'Malley and Chakravarti (1909), the old Saraswati was alluded to as the *Sankrail Khal*, which actually depicts part of its lower course in present day Sankrail and Domjur Police Stations of Howrah District in southern West Bengal which was then (in the early 20th century) navigable up to Andul by large boats, and the present high banks and vestiges of big vessels occasionally excavated along it indicate its former prowess. Mukhopadhyay (1996) pointed out that during Rennell's period, the Saraswati was just a small creek at its exit from the Bhagirathi, and, as per the Addams Report (1920), was a small tidal watercourse that re-entered the River Hugli at the lower end of the Garden Reach area in Sankrail (Seth, 2014). This deterioration had commenced since the middle of the 16th century, partly due to the diversion of the Damodar's inflow from the upper reaches of the Bhagirathi, and in notable works penned post that period, like the *Ain-e-Akbari*, it is the newer port of Hooghly on the Bhagirathi that was highlighted as being superior to older ones on the Saraswati (Mukhopadhyay, 1996; Seth, 2014).

From Tribeni near Hugli and up to the sea, the Saraswati and Bhagirathi ran quite parallel to each other. The Saraswati's lower course obtained a strong tidal flow as well as part of the Bhagirathi's discharge through a narrow east-west link channel, the *Katiganga*, south of Kolkata at Bator, which was dredged and widened by the Dutch in the 18th century, with tolls paid by ships using this course (Mukherjee, 1995). Mukherjee (1986, 1995) reported further such human-aided river captures in the Saraswati and Lower Damodar Basins, with the eventual result that the Saraswati became simultaneously both a distributary and a tributary of the Bhagirathi. Around 1750, Alivardi Khan, the then Nawab of Bengal, again connected both streams by digging a link channel near Uluberia (Sen, 1998) and possibly the Bhagirathi later took over this course as its eastward movement was restricted by the Calcutta–Mymensingh Hinge antiform (Basu, 1992). Many artefacts from old Bengali literature, folk tales and songs also contain viable information regarding these river basins, which are mentioned in *Manasamangal* of Bijoy Gupta (1484–1485), and *Manasabijay* of Bipradas Pipilai (1495–1496), *Chandimangal* (*Abhayamangal*) of Mukundoram (1544) that document various settlements on either bank of these streams and make reference to the affluent status of Tribeni and Satgaon in particular (Sreemani, 2011).

While many different aspects of the Ganges delta and its adjoining parts have been examined previously, there is scant literature on the changing land use pattern along and within the Saraswati and Jamuna Basins. Some studies in this regard have been undertaken by Roy (1972) and Ghosh (1988), on the evaluation of the agricultural land use pattern in a small portion of the upper part of the Saraswati Basin. A detailed study on the pattern of deterioration of the Saraswati and Jamuna Rivers is therefore important.

14.2.2 LANDSCAPE ATTRIBUTES OF THE DISTRIBUTARY BASINS

The Saraswati River courses through the LBD across parts of Barddhaman, Hugli and Howrah districts (Figure 14.2a), and its corridor is bound by the graticules 22° 30′ to 23° 15′ N and 88° 00′ E to 88° 30′ E, covering an area of 1,447.6 km² (Figure 14.3). The river and its tributaries create a low-lying landscape traversed by multiple sluggish channels. Its stretch from Tribeni to Bagdanda (where it meets the Kunti) is demarcated as the Upper or North Saraswati, with the rest of its course being the Lower or South Saraswati. The last 8–10 km of the South Saraswati before its outfall at Sankrail is also known as the Sankrail Khal, which is subjected to tidal flows and is navigable. Except this stretch, it generally remains dry apart from in the monsoonal months when part of the Hooghly's discharge overspills into it along with some portion from the Damodar via the Kana Nadi. Its entire course traverses through the Gangetic alluvial plains at an average slope of about 0.5 m km⁻¹. The slope is maximum near the off-take, and thereby the river flows straight to the south here. The land elevation is naturally maximum near the bank, forming levees, and thus, the altitude is highest along the channel margins and sloping towards either flank of the floodplain, which often form low-lying swamps or marshes, e.g. the eastern Dankuni and Howrah swamps and the Rajapur swamp in the west. Most of these areas that were usually waterlogged annually have been reclaimed and are now cultivated or settled over. The surface lithology comprises of sandy silts and clays primarily, with loams and fine sands towards the interfluves (Figure 14.3a). The entire area is part of the paradeltaic fan surface, arising from the depositional events and changes in the Bengal Basin since the Quaternary (Morgan and McIntire, 1959), with mature and present floodplain tracts along current stream paths (Figure 14.3b). The surface soils of these swampy areas and riverbanks contain clay, whereas the bed is generally sandy, except along silted up stretches which have a mud and clay veneer (WBPCB, 2017). The subsoil layer in most of the area is mainly of clay. Eutric cambisols (Be80-2a) and fluvisols (Je71-2a) and orthic luvisols (Lo48-2a) comprise the main soil groups (Figure 14.3c). These are most suitable for rice cultivation (FAO-UNESCO, 1977). The region is part of the Hooghly's industrial belt and has thus been much altered by growing urbanisation along the right bank of the Hooghly, with numerous towns aligned along the principle rail and road routes (Bose, 1968) (Figure 14.3d).

The Jamuna branches off the Hooghly opposite Tribeni, in Nadia district and courses eastward (Figure 14.2b), eventually flowing across the North 24 Parganas district to the Ichamati River at Tipi in Swarupnagar Block, traversing a total distance of 60.59 km and draining 660.60 km^2, covering the region bounded by 88° 25′30″–88° 55′40″ E and 22° 51′10″–23° 09′50″ N. In its upper reaches, the Jamuna hardly bears any discharge. In the middle reaches, agricultural plots occupy either bank with seasonal cultivation on the dry channel bed in the lean season. The lower

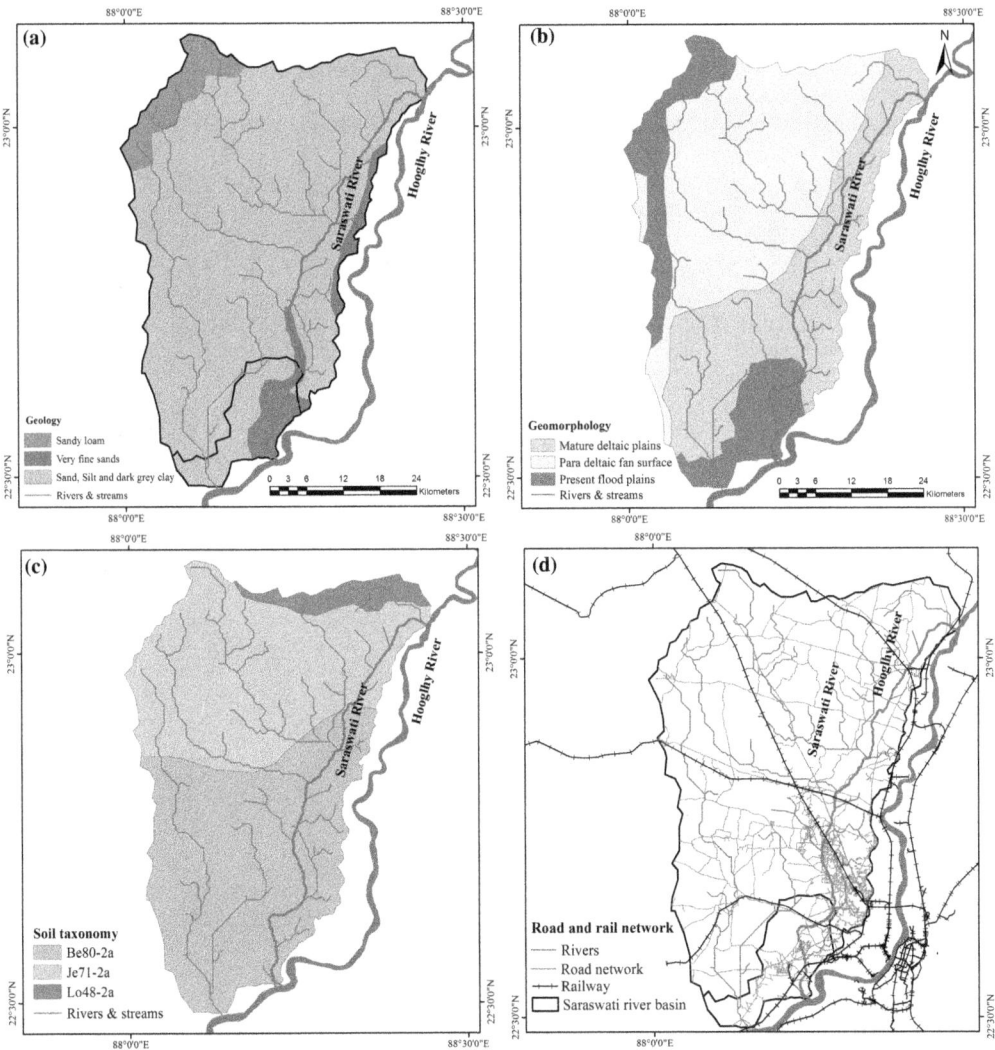

FIGURE 14.3 Geographical aspects of the Saraswati Basin: (a) geology (GSI); (b) geomorphology (GSI); (c) soil cover (FAO soil groups); and (d) communication lines.

reaches near Gobardanga experience tidal flow, as do the other channels in the entire region, which carries in and deposits silt, further choking the channel and making it shallower (Rudra, 2018). Obstructions across the channel for impounding water and catching fish also aid this silt deposition by restricting the natural flow and abetting waterlogging, with embankments being present along most of these stretches. The Jamuna's course has several marked bends, with prominent ox-box lakes, the most famous being the Konkona Boar in Gobordanga, resembling a *Kankan* (a bangle). The Jamuna Basin's lithological groups comprise current floodplains along the Hooghly, made of unconsolidated and oxidised fines (silts, clays and mica sands) that are situated at the lowest elevations (Figure 14.4a). The central basin tracts are covered by oxidised and unoxidised sands, silts and clays (very fine to fine), forming depressions and natural levees. The south-western part is deposited over by very fine sands and silts, giving rise to different bars (mid-channel and point) and meander scrolls. As with the Saraswati, here too the main geomorphic units are floodplains, deltaic plains (mature) and paradeltaic fan surfaces (Figure 14.4b). The former constitutes areas that are presently flooded by the River Bhagirathi. The other two units occupy higher surfaces and above the reach of normal flood levels. Eutric fluvisols (Je71-2a) and calcaric gleysols (Gc9-3a) comprise the main soil groups (Figure 14.4c). These are productive and periodically inundated and suitable for broadcast rice, jute and sugarcane cultivation (FAO-UNESCO, 1977). Communication lines and settlements are more sparsely developed in the Jamuna basin as compared to the Saraswati's (Figure 14.4d).

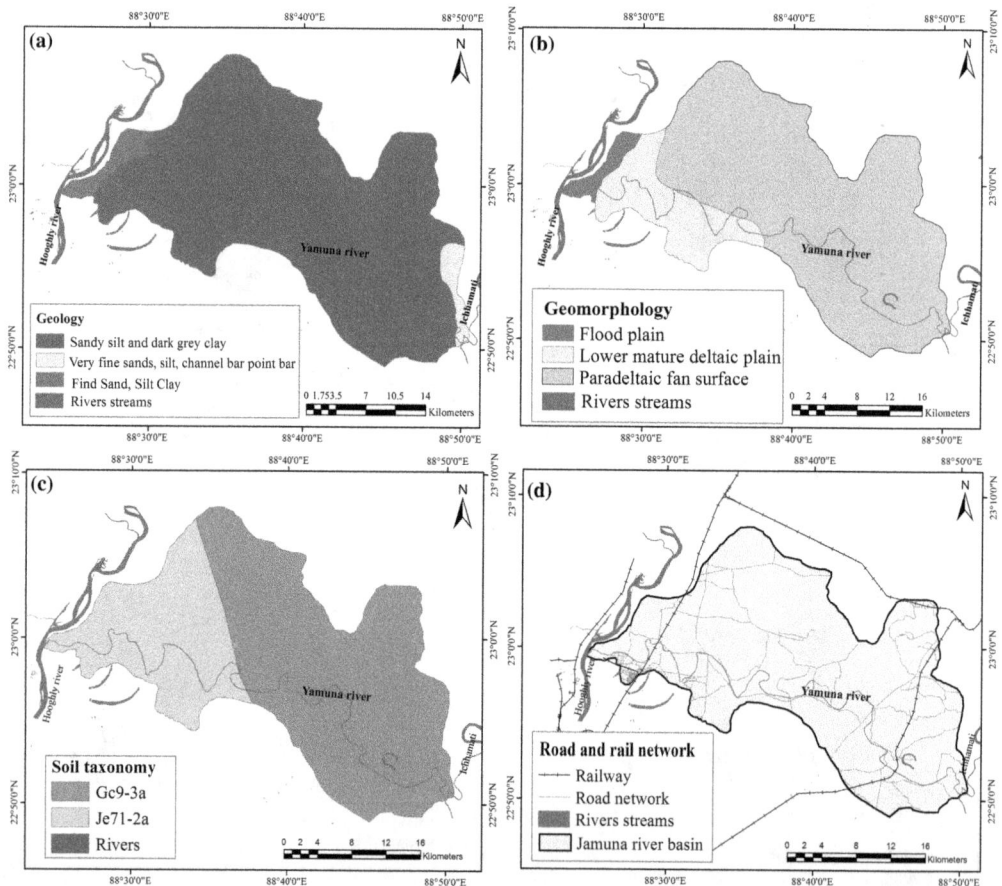

FIGURE 14.4 Geographical aspects of the Jamuna Basin: (a) geology (GSI); (b) geomorphology (GSI); (c) soil cover (UN-FAO soil groups); and (d) communication lines.

14.3 ANTHROPOGENIC SIGNATURES ON FORMS AND PROCESSES

14.3.1 LULC CHANGES

Multi-temporal LULC mosaics play a vital role in understanding historical changes in human occupation and changes in geomorphic processes (Sarkar and Patel, 2017). Four LULC components (water bodies, agricultural land, natural vegetation and built-up area) were extracted and demarcated from the multi-spectral images analysed along both the rivers. Significant changes in these components were noted during the study period of 1973–2017.

14.3.1.1 LULC Changes along the Jamuna River

Table 14.2 displays the recorded LULC changes along the Jamuna's tract from 1973 to 2107 (Figure 14.5). The water body class shows a marked decline, decreasing by about 2,890 ha (a negative change of about 24%). The most probable reasons for this decrease may be attributed to an increase in population and encroachment (particular on wetlands). While climate change/variability may also be an issue, further detailed studies would be needed to ascertain this. Agricultural lands comprise the greatest fraction in 1973, but this too declined by about 2,360 ha from 1973 to 2017. Similarly, natural vegetation (primarily comprising of tree groves) also exhibits a significant decrease in its coverage, from 14,521.3 ha (20.55% of the total area) in 1973 to 11,670.4 ha (16.518% of the total) in 2017, with a negative growth rate of about 20% (less by 2,850.9 ha) in a time period of about 45 years. The only LULC category to increase, and very markedly so, has been the built-up component, with a positive growth rate of about 1,250.3%, from 647.84 ha in 1973 to 8,100.22 ha in 2017. This shows the sustained and increased occupancy of this riparian tract, with the clear transformation of the natural to the artificial (Figure 14.8a). The detailed transformations between the individual components for this riparian corridor are reported in Table 14.3. The maximum

TABLE 14.2
LULC Change along the Jamuna River from 1973 to 2017

Land Use/Land Cover Categories	1973		2000		2017		LULC Change from 1973 to 2017	
	Area (ha)	Area (%)	Area (ha)	Area (%)	Area (ha)	Area (%)	Change (ha)	Change (%)
Water body	12,032.00	17.03	9,030.45	12.78	9,141.38	12.938	−2,890.62	−24.00
Vegetation	14,521.30	20.55	13,767.70	19.49	11,670.40	16.518	−2,850.90	−19.60
Built-up	647.84	0.92	5,639.49	7.98	8,748.06	12.382	8,100.22	1,250.30
Agricultural land	43,452.20	61.50	42,215.70	59.75	41,093.50	58.162	−2,358.70	−5.40

Source: Compiled by the authors on the basis of supervised classification of multi-temporal image datasets.

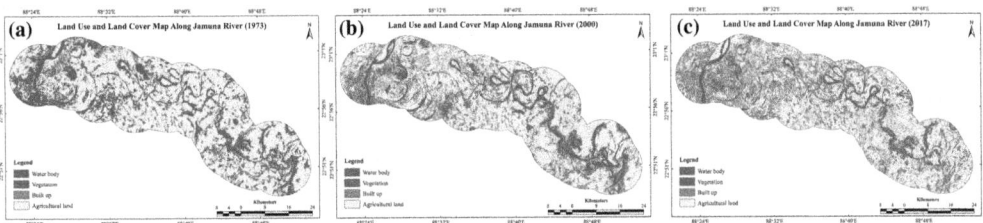

FIGURE 14.5 LULC components along the Jamuna channel in different years: (a) 1973; (b) 2000; and (c) 2017.

TABLE 14.3
LULC Change Matrix along the Jamuna River

Land Use/Land Cover Categories		1973 (in ha)				
		Water Body	Vegetation	Built-up	Agricultural Land	Total (in 2017)
2017 (in ha)	Water body	2,947.32	4,580.00	0.00	1,614.06	9,141.38
	Vegetation	2,716.83	2,224.77	0.00	6,728.80	11,670.40
	Built up	791.89	1,927.83	647.84	5,380.50	8,748.06
	Agricultural land	5,575.96	5,788.70	0.00	29,729.00	41,093.70
	Total (in 1973)	12,032.00	14,521.30	647.84	43,452.36	70,653.50

Source: Compiled by the authors on the basis of supervised classification of multi-temporal image datasets.

encroachment of the built-up areas has obviously been on agricultural and previously naturally vegetated tracts.

14.3.1.2   LULC Changes along the Saraswati River

Table 14.4 displays the recorded LULC changes along the Saraswati's tract from 1973 to 2017 (Figure 14.6). Water bodies covered an area of about 6,569.6 ha (9%) in 1973, which decreased to just 445.6 ha (6.2%) in the year 2000 and further to 434.8 ha (just 6% of the total landscape) in 2017. This clearly depicts that the Saraswati River, along with its adjacent wetlands and linked channels, has suffered a grave shrinkage (as is evident from the photographs of the channel in Figure 14.7, showing how people have encroached and the channel has reduced in dimension) primarily due to the increasing population pressure on the land, as is quite evident from the sharp increase in the built-up component herein, which shows an increase of 529.8% over the same time period. In 1973, this built-up component covered 3,145.9 ha which increased tremendously to 19,813.7 ha in 2017, as it encroached onto the other LULC classes and garnered about 16,667.8 ha to its account. To facilitate this transformation, about 11,679.1 ha agricultural land was converted to built-up, along with 4,222.5 ha of natural vegetation and 776.1 ha of water bodies (Table 14.5). Thus, the natural vegetation shows a negative growth rate of about 67.1%, agricultural land has declined by 9.0%, and most markedly, water bodies have decreased by 33.9%, from 1973 to 2017 (Figure 14.8b). The loss of the wetlands and floodplains along the Saraswati has further choked

TABLE 14.4
LULC Change along the Saraswati River from 1973 to 2017

LULC Categories	1973		2000		2017		LULC Change from 1973 to 2017	
	Area (ha)	Area (%)	Area (ha)	Area (%)	Area (ha)	Area (%)	Change (ha)	Change (%)
Water body	6,569.6	9.1	4,456.9	6.2	4,343.8	6.0	−2,225.7	−33.9
Vegetation	15,243.7	21.2	10,857.7	15.1	5,019.9	7.0	−10,223.9	−67.1
Built-up	3,145.9	4.4	11,093.1	15.4	19,813.7	27.5	16,667.8	529.8
Agricultural land	47,111.1	65.4	45,662.6	63.4	42,892.9	59.5	−4,218.2	−9.0

Source: Compiled by the authors on the basis of supervised classification of multi-temporal image datasets.

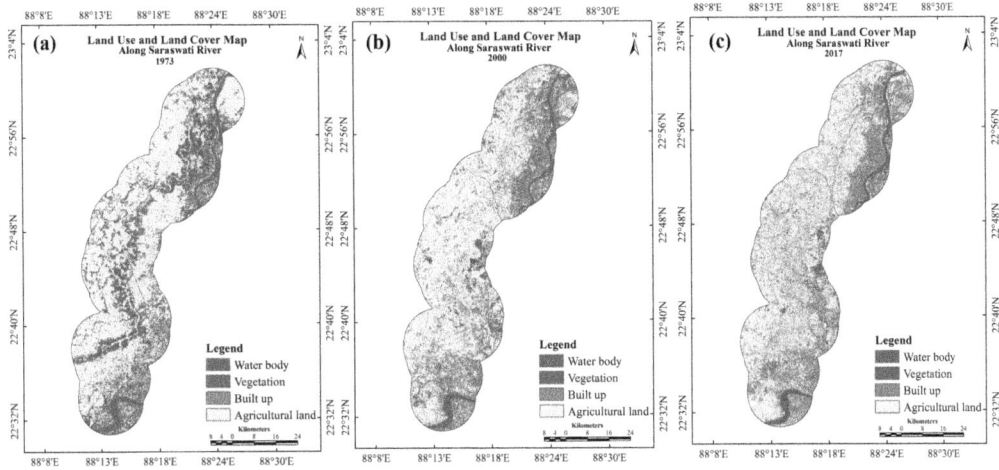

FIGURE 14.6 LULC components along the Saraswati channel in different years: (a) 1973; (b) 2000; and (c) 2017.

FIGURE 14.7 The evident shrinkage in the River Saraswati's dimensions, at (a) Kazidanga and (b) Rajhat, in Hooghly district.

TABLE 14.5
LULC Matrix along the Saraswati River

LULC Categories	1973 (in ha)				
	Water Body	Vegetation	Built-up	Agricultural Land	Total (in 2017)
Water body	2,669.1	546.8	0.0	1,128.0	4,343.8
Vegetation	396.0	1,105.9	0.0	3,518.0	5,019.8
Built-up	766.1	4,222.5	3,145.9	11,679.1	19,813.7
Agricultural land	2,738.4	9,368.6	0.0	30,786.0	42,892.9
Total (in 1973)	6,569.6	15,243.7	3,145.9	47,111.0	72,070.3

Source: Compiled by the authors on the basis of supervised classification of multi-temporal image datasets.

the channel, by taking away its natural replenishment zones and areas over which it could have distributed the silt borne when in spate during the monsoon months. This has led to clogging up of the channel, progressive eutrophication and loss of natural flow and habitats and thereby navigability.

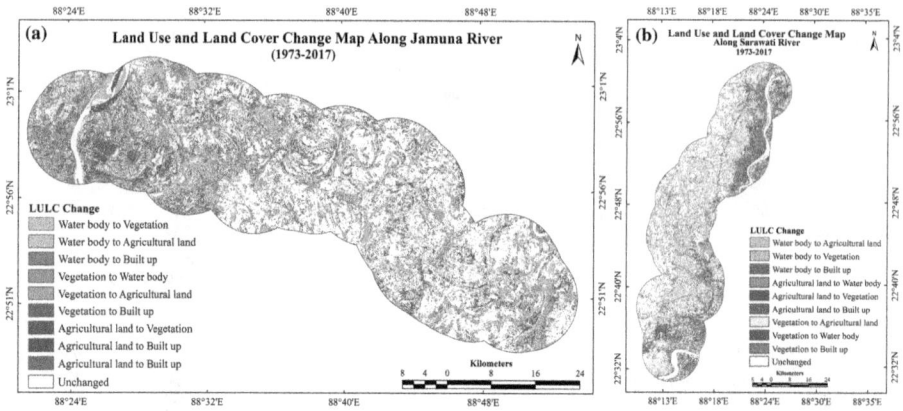

FIGURE 14.8 Transformations of respective LULC components from 1973 to 2017: (a) Jamuna River tract and (b) Saraswati River tract.

14.3.2 LANDSCAPE CHARACTER

To generate the cross sections across the floodplains of the Saraswati and Jamuna channels, SRTM DEM data was used in order to ascertain the nature of the overall topography. While such studies typically demand a finer resolution DEM that are expensive to procure, with the information derived being an artefact (or derivative) of the DEM scale (Das et al., 2016), in the absence of such a dataset, the SRTM data was utilised with prior checks as to its viability for generating basin and reach-level surface profiles (e.g. Patel and Sarkar, 2009; Sarkar and Patel, 2011, 2012; Guha and Patel, 2017). Spot height data for authentication was plotted against the heights extracted from the SRTM DEM. The calculated value ($R^2 = 0.947$) revealed that the SRTM data was useful for the stated purpose and could be feasibly utilised, as the distance surveyed was considerable (across a number of channels and interfluves) and not just from bank to bank of a single channel. The raster calculator utility from the data management toolbox of Arc GIS 10.2 was used to remove the voids in the analysed SRTM DEM. The ascertained profiles provided further insight into the palaeochannels formed and migration done by both these rivers.

14.3.2.1 Jamuna Floodplain Cross-Sections

In the Jamuna's riparian tract, there is a significant change in the shape of the river profile from its off-take to its mouth. Its most visible meandering processes are pointed out in five cross-sections (Figure 14.9). The degree of meandering near its off-take is quite high, and future meander neck cut-offs are probable here, leading to the formation of ox-bow lakes and channel scars (Figure 14.10). However, sections G-H to I-J show minimum deviations from a comparatively straighter course (characterised as such due to its relatively lesser sinuosity). The most intense region of meandering lies between the sections C-D to G-H, where a number of meandering necks and ox-bow lakes have already been formed. From the section A-B to C-D, the intensity of meandering decreases probably since the more sinuous course before this segment has already led to much deposition in the previous section (C-D to G-H) and now it tries to flow in a straighter course. In the section A-B to C-D, the river first meanders towards its right bank and then at mid-section it shifts and bends towards its left bank, repeating this process again. The maximum elevation extracted is found in this reach. Conversely, in the section I-J, the relief is comparatively lower reaching down almost to the base level. Along this line, a number of valley forms are visible (possible older channels and present day meander scars) (Figure 14.11) which were not as prominent in the section A-B. An overall analysis of the cross-sections shows that the elevation decreases from A-B to I-J. For the sections C-D, there is a discernable rise in the surface profile towards the northeast, whereas for the G-H section, this reverses towards the southwest.

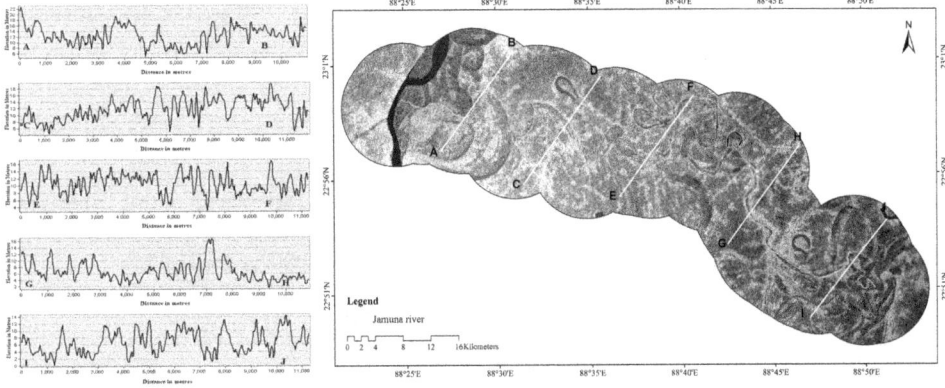

FIGURE 14.9 Surface profile transects across the Jamuna's riparian tract.

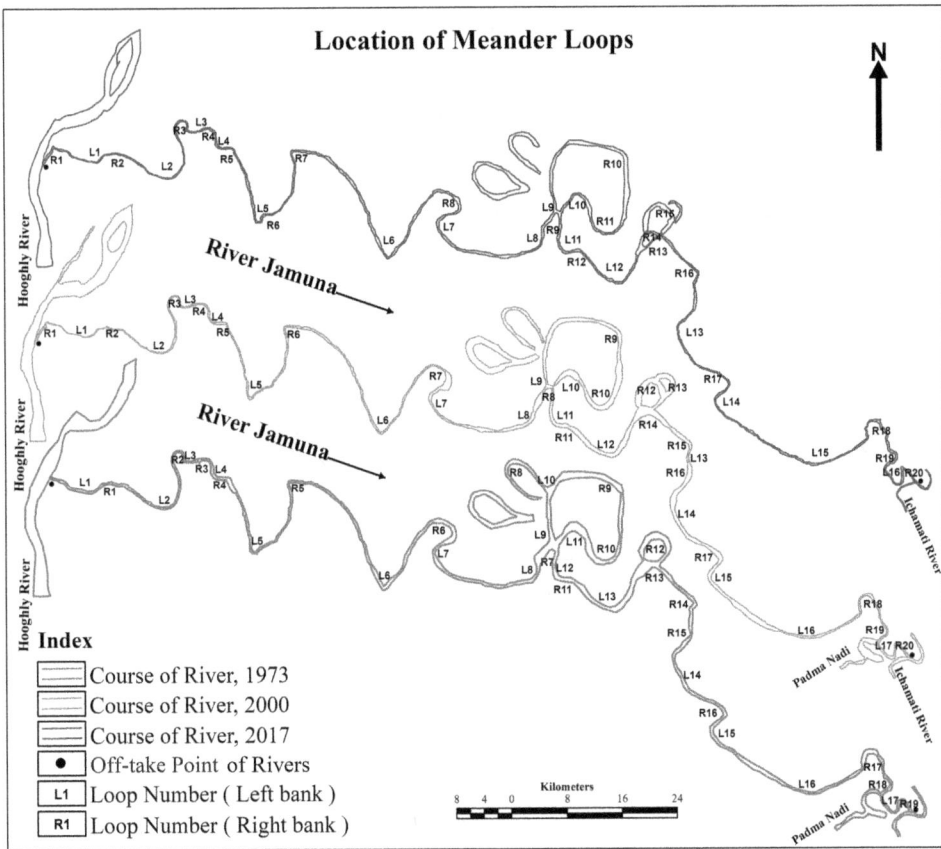

FIGURE 14.10 Changes in meander loops of the Jamuna River from 1973 to 2017.

14.3.2.2 Saraswati Floodplain Cross-Sections

The profiles across the Saraswati's floodplain have been similarly drawn and named (Figure 14.12). From profiles A-B and C-D, two separate stretches of the Saraswati are discernable, and the more eastern branch is more sinuous compared to the western branch. This eastern branch faces a likelihood of developing a number of neck cut-offs and ox-bow lakes in the recent future, as the sections of the stream flow west and then almost double back entirely in loops towards the east (Figure 14.13).

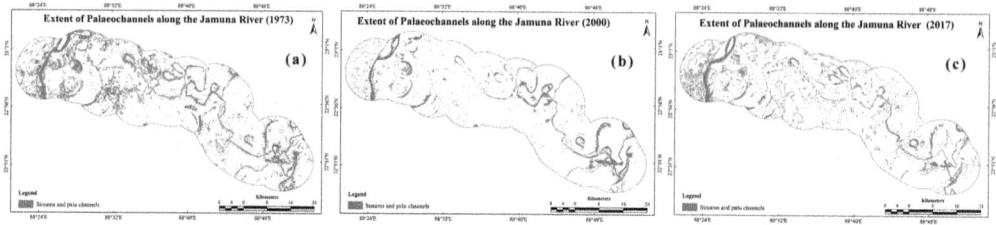

FIGURE 14.11   Palaeochannels along the Jamuna River: (a) 1973; (b) 2000; and (c) 2017.

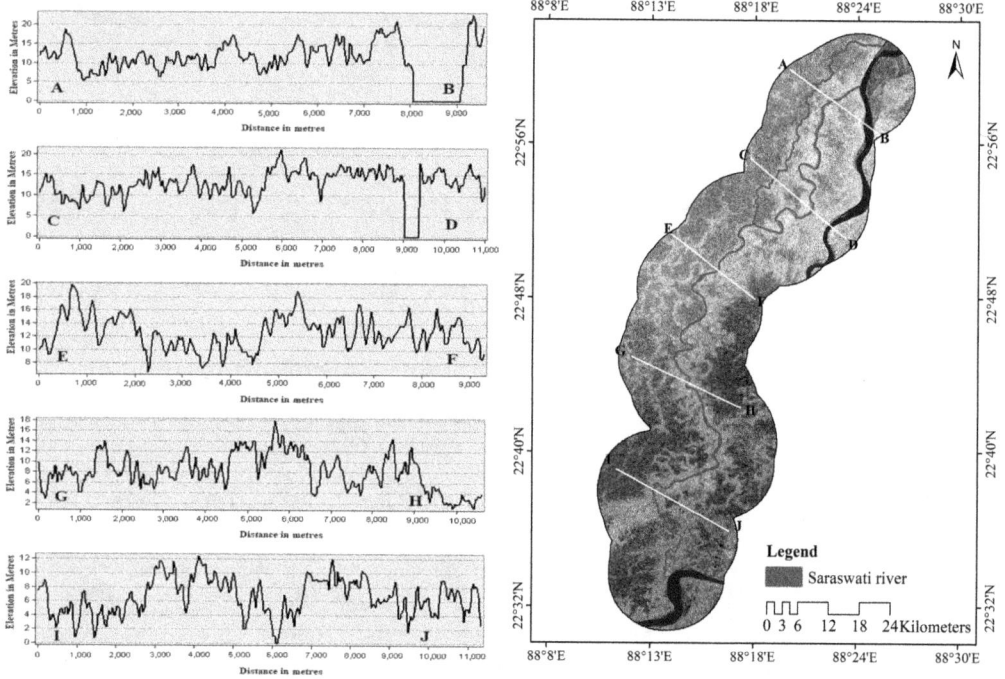

FIGURE 14.12   Surface profile transects across the Saraswati's riparian tract.

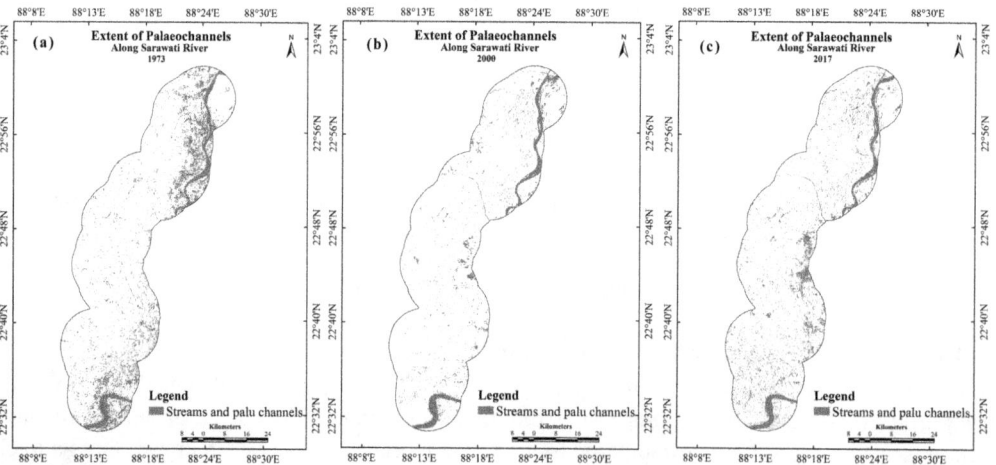

FIGURE 14.13   Palaeochannels along the Saraswati River: (a) 1973; (b) 2000; and (c) 2017.

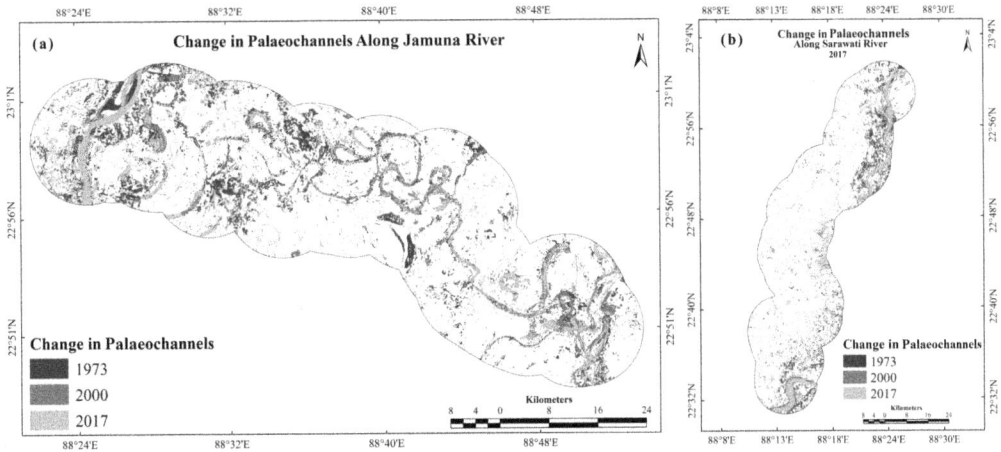

FIGURE 14.14   Changes in the palaeochannels of the two distributaries from 1973 to 2017: (a) Jamuna River tract and (b) Saraswati River tract.

Similarly, along the C-D profile, the stream is quite sinuous, displaying increased meandering and bends, before merging with the western branch. Again in between the sections E-F to G-H, the river is quite sinuous with a number of sharp bends. Downstream of this reach, however, the channel straightens out more and as the river approaches the I-J section, the meandering intensity declines, and it becomes less sinuous. The maximum elevation along the river is about 20 m (embankment height). With each passing section, the river's gradient declines until Section I-J; its valley has the maximum depth and drops down almost to the base level, as seen from the declivity in the drawn profile.

Both the profiles of the Jamuna and the Saraswati are consistent with those obtained from convex floodplains (Bridge, 2003; Bridge and Demico, 2008; Carling and Hargitai, 2014), with numerous old channel scars and cut-offs (Lewin and Ashworth, 2014). There are a number of local slope segments amidst the broader convex profile curvature, which has been created by the depositions within the channel due to the extensive natural levees (later converted to embankments) along their course (Laha, 2015). Such typical surface profile cross-sections are very much historically representative of the convexo-concave nature of the Bengal deltaic floodplain (Wood, 1895), created by overspill silt depositions and naturally built-up levees along streams, which have been further reinforced once settlements flourished. While the positions of the palaeochannels and ox-bow lakes of both these distributaries have remained quite constant, certain tracts have been modified (Figure 14.14), with the resultant loss of wetlands. These have been primarily converted to agricultural lands or built-up areas, as detailed previously.

14.4   CONCLUSION

Since the discharge of the Ganges began to flow primarily through the Padma channel early in the 6th century, the Bhagirathi-Hooghly system began to progressively get starved of upland flow (Rob, 1989). This caused deterioration of the various distributaries, like the Saraswati and Jamuna, emanating from the Hooghly, which itself is essentially a spill-over channel of the main Ganga. The drainage map of West Bengal, particularly that of the lower tracts of the Bhagirathi-Hooghly basin, has thus changed appreciably during known historical times with the constituent streams having markedly altered their courses, or undergone transformative changes in channel character and morphology due to discharge variations, with this impacting severely on the economic aspects of settlements located along them (Sen, 1998). Some have even disappeared altogether or gone dry, or been choked up due to the inability to flush out the deposited sediments. Riparian landscape management in this region has almost totally focused on structural measure such as embankments, as a remnant of its colonial legacy, along with routine engineering measures like dam-building. While parallel and multiple networks of embankments have usually provided security against lower flood peaks in the short term, their prolonged effects have often been negative, leading to river decay and drainage congestion (Das, 2005; Rashid et al., 2013) and often abetting floods, as evidenced from other places in southern Bengal (e.g. Mondal et al., 2015).

The LULC changes in the riparian tracts of both these distributaries have been very marked, particularly so in the case of the Saraswati. Burgeoning urbanisation (DPD-GoWB, 2011; DDM-GoWB, 2017, 2018) has led to the decline of natural vegetation cover and the conversion of agricultural tracts to built-up surfaces. Furthermore, many previous wetland areas (palaeochannels and ponds and ox-bow lakes) have been converted into built-up areas, thereby retarding the amount and rate of surface water abstraction underground and consequently abetting higher runoff, which can eventually lead to waterlogging in these locales, as is symptomatic of similar occurrences across the entire LGD (Bandyopadhyay et al., 2014). During the monsoon months, the Jamuna's confluence with the Ichamati remains waterlogged, damaging standing crops. To remove such drainage congestion, desiltation, dredging and removal of weeds choking these two channels is urgently required. Further resuscitation of the channels can be undertaken by removing low-lying crossing structures that serve as blockages, having restrictions on the dumping of wastes and effluents into these channels, and creating proper riparian buffers on either side of the stream (Mondal and Patel, 2018, 2020), which can enhance the channel flows while also mitigate higher discharges during flood peaks and protect riverbanks. There is also a need for proper analysis of the flood susceptibility of the region due to the continuous choking of such channels, and geospatial data for this purpose can be feasibly employed (e.g. Sahana and Patel, 2019), along with the examination of any further vulnerability to channel migrations that may arise due to its near location to the coastal Sunderbans, which is facing the effects of climate change-induced variations in tides and sea levels (e.g. Sahana and Sajjad, 2019; Sahana et al., 2019), as has occurred in the geological past (Akter et al., 2016; Umitsu, 1993).

NOTES

Details of some of the historical maps consulted for this chapter, and others, can be obtained at the following websites:

1. Nouvelle Carte Du Royaume De Bengale (1750)—https://stock-images.antiqueprints.com/images/sm0064-BengalSchley.jpg
2. A Map of the Mouths of the Ganges in the Bay of Bengal (1757)—http://www.rarefinds.co.in/View/Group/Maps=General
3. A New and Accurate Map of Bengal (1760)—http://www.columbia.edu/itc/mealac/pritchett/00maplinks/colonial/kitchinmaps/bengal1760/bengal1760max.jpg
4. Rennell - Dury Wall Map of Bihar and Bengal, India (1776)—https://upload.wikimedia.org/wikipedia/commons/8/85/1776_Rennell_-_Dury_Wall_Map_of_Bihar_and_Bengal%2C_India_-_Geographicus_-_BaharBengal-dury-1776.jpg
5. Calcutta and Environs (1893)—https://www.indiawaterportal.org/sites/indiawaterportal.org/files/historical_maps_of_india_kolkata_environs_1893_0.jpg
6. Environs of Calcutta (1907–1909)—http://dsal.uchicago.edu/maps/gazetteer/images/gazetteer_V9_pg288.jpg

REFERENCES

Akter, J., Sarker, M.H., Popescu, I. and Roelvink, D. (2016) Evolution of the Bengal Delta and its prevailing processes. *Journal of Coastal Research* 32(5): 1212–1226. doi: 10.2112/JCOASTRES-D-14-00232.1.

Bagchi, K.G. (1944) *The Ganges Delta*. University of Calcutta, Kolkata. Available at https://archive.org/details/in.ernet.dli.2015.21183/page/n5.

Bagnold, R.A. (1960) Some aspects of the shape of river meanders. Geological Survey Professional Paper 282-E, United States Department of the Interior, United States Geological Survey. Available at https://pubs.usgs.gov/pp/0282e/report. pdf.

Bandyopadhyay, S., Kar, N.S., Das, S. and Sen, J. (2014) River systems and water resources of West Bengal: A review. In: Vaidyanadhan, R. (ed.) *Rejuvenation of Surface Water Resources of India: Potential, Problems and Prospects*. Special Publication 3, Geological Society of India, Bangalore, pp. 63–84. doi: 10.17491/cgsi/2014/62893.

Banerji, D. and Patel, P.P. (2019) Morphological aspects of the Bakreshwar River Corridor in Western Fringe of Lower Ganga Basin. In: Das, B., Ghosh, S. and Islam, A. (eds.) *Quaternary Geomorphology in India. Geography of the Physical Environment*. Springer, Cham, pp. 155–189. doi: 10.1007/978-3-319-90427-6_9.

Basu, A.K. (1992) *Ecological and Resource study of the Ganga Delta*. Netaji Institute for Asian Studies, Kolkata.

Bates, M.R., Bates, C.R. and Whittaker, J.E. (2007) Mixed method approaches to the investigation and mapping of buried Quaternary deposits: examples from southern England. *Archaeological Prospection* 14(2): 104–129. doi: 10.1002/arp.303.

Blochmann, H. (1873) *Contribution to the Geography and History of Bengal: Muhammaddan Period-1203 to 1538*. Baptist Mission Press, Kolkata. Available at https://ia801605.us.archive.org/25/items/in.ernet.dli.2015.49139/2015.49139.Geography-And-History-Of-Bengal.pdf.

Borgohain, S., Das, J., Saraf, A.K., Singh, G. and Baral, S.S. (2017) Structural controls on topography and river morphodynamics in Upper Assam Valley, India. *Geodinamica Acta* 29(1): 62–69.doi: 10.1080/09853111.2017.1313090.

Bose, S.C. (1968) *Geography of West Bengal*. National Book Trust, New Delhi. Available at https://ia800603.us.archive.org/26/items/in.ernet.dli.2015.132064/2015.132064.Geography-Of-West-Bengal.pdf.

Bridge, J.S. (2003) *Rivers and Floodplains: Forms, Processes and Sedimentary Record*. Blackwell, Oxford.

Bridge, J.S. and Demico, R. (2008) *Earth Surface Processes, Landforms and Sediment Deposits*. Cambridge University Press, Cambridge. doi: 10.1017/CBO9780511805516.

Brooks, A.P., Brierley, G.J. and Millar, R.G. (2003) The long-term control of vegetation and woody debris on channel and flood-plain evolution: insights from a paired catchment study in southeastern Australia. *Geomorphology* 51(1–3): 7–29. doi: 10.1016/S0169-555X(02)00323-9.

Brown, A.G. (1999) Characterising prehistoric lowland environments using local pollen assemblages. Quaternary Proceedings 7: 585–594. https://doi.org/10.1002/(SICI)1099-1417(199910)14:6%3C585::AID-JQS492%3E3.0.CO;2-C

Camporeale, C., Perucca, E. and Ridolfi, L. (2008) Significance of cutoff in meandering river dynamics. *Journal of Geophysical Research* 113: F01001. doi: 10.1029/2006JF000694.

Carling, P.A. and Hargitai, H. (2014) Floodplain. In: Hargitai, H. and Kereszturi, A. (eds.) *Encyclopedia of Planetary Landforms*. Springer, New York. doi: 10.1007/978-1-4614-9213-9_152-1.

Chatterjee, S. and Patel, P.P. (2016): Quantifying landscape structure and ecological risk analysis in Subarnarekha sub-watershed, Ranchi. In: Mondol, D.K. (ed.) *Application of Geospatial Technology for Sustainable Development*. University of North Bengal, India, North Bengal University Press, Raja Rammohunpur, pp. 54–76.

Chen, W., Qinghai, X., Xuiqing, Z. and Yonghong, M. (1996) Palaeochannels on the North China plain: types and distribution. *Geomorphology* 18(1): 5–14. doi: 10.1016/0169-555X(95)00147-W.

Conyers, L.B., Ernenwein, E.G., Grealy, M. and Lowe, K.M. (2008) Electromagnetic conductivity mapping for site prediction in meandering river floodplains. *Archeaological Prospection* 15: 81–91. doi: 10.1002/arp.326.

Dambeck, R. and Thiemeyer, H. (2002) Fluvial history of the northern Upper Rhine River (southwestern Germany) during the Lateglacial and Holocene times. *Quaternary International* 93–94: 53–63. doi: 10.1016/S1040-6182(02)00006-X.

Das, B.P. (2005) Environmental problem of drainage congestion in Mahanadi Delta, India: case study of a remedial direct cut. *Proceedings of World Water and Environmental Congress on Impacts of Global Climate Change*, Anchorage, Alaska, May 15–20. doi: 10.1061/40792(173)502.

Das, S., Patel, P.P. and Sengupta, S. (2016) Evaluation of different digital elevation models for analyzing drainage morphometric parameters in a mountainous terrain: a case study of the Supin–Upper Tons Basin, Indian Himalayas. *SpringerPlus* 5: 1544. doi: 10.1186/s40064-016-3207-0.

Dasgupta, R. and Patel, P.P. (2017) Examining the physical and human dichotomy in geography: existing divisions and possible mergers in pedagogic outlooks. *Geographical Research* 55(1): 100–120. doi: 10.1111/1745–5871.12220.

DDM-GoWB (Department of Disaster Management, Government of West Bengal) (2017) *Disaster Management Plan 2017–18: Hooghly*. Office of the District Magistrate and Collector, Hooghly. Available at http://www.wbdmd.gov.in/writereaddata/uploaded/DP/Disaster%20Management%20Plan%20of%20Hooghly.pdf.

DDM-GoWB (Department of Disaster Management, Government of West Bengal) (2018) *Disaster Management Plan 2018–19: Hooghly*. Office of the District Magistrate and Collector, Hooghly. Available at http://wbdmd.gov.in/writereaddata/uploaded/DP/DPHooghly%20%20%20%20%20%20%20%20%20%20%20%20%20%20%20%20%20%20%20%20%20%20%20%2064859.pdf.

Deen, T.J., Gohl, K., Leslie, C., Papp, E. and Wake-Dyster, K. (2000) Seismic refraction inversion of a palaeochannel system in the Lachlan Fold Belt, Central New South Wales. *Exploration Geophysics* 31(2): 389–393. doi: 10.1071/EG00389.

Dieras, P.L. (2013) The persistence of oxbow lakes as aquatic habitats: an assessment of rates of change and patterns of alluviation. Unpublished PhD Thesis, Cardiff University, Wales. Available at https://orca.cf.ac.uk/49392/1/Thesis_Pauline_Dieras_final.pdf.

DPD-GoWB (Development and Planning Department, Government of West Bengal) (2001) *District Human Development Report: Hooghly*. Available at https://www.undp.org/content/dam/india/docs/human-development/HDR/DHDR_Hooghly.pdf.

Eaton, R.M. (1993) *The Rise of Islam and the Bengal Frontier, 1204–1760*. University of California Press, Berkeley. Available at http://ark.cdlib.org/ark:/13030/ft067n99v9/.

Ernenwein, E.G. (2002) Establishing a method for locating buried oxbow lake deposits using electrical conductivity, Sacramento Valley, California. Unpublished Master's Thesis, University of Denver.

FAO-UNESCO (Food and Agriculture Organization of the United Nations and United National Educational, Scientific and Cultural Organization) (1977) *Soil Map of the World-1:5000000. Vol. VII: South Asia*. UNESCO, Paris. Available at http://www.fao.org/3/as352e/as352e.pdf accessed on 24.03.2019.

Ghose, B., Kar, A. and Husain, Z. (1979) The lost courses of the Saraswati River in the Great Indian Desert: new evidence from Landsat imagery. *The Geographical Journal* 145(3): 446–451. doi: 10.2307/633213.

Ghosh, J.D. (1988) The decaying Saraswati in Bengal. *Indian Journal of Landscape System and Ecological Studies* 2(2): 81–88.

Gibbard, P.L. (1988) The history of the great Northwest European rivers doing the past three million years. *Philosophical Transactions of the Royal Society B: Biological Sciences* 318(1191): 559–602. doi: 10.1098/rstb.1988.0024.

Guha, S. and Patel, P.P. (2017) Evidence of topographic disequilibrium in the Subarnarekha River Basin, India: a digital elevation model based analysis. *Journal of Earth System Science* 126: 106. doi: 10.1007/s12040-017-0884-1.

Gupta, N., Kleinhans, M.G., Addink, E.A., Atkinson, P.M. and Carling, P.A. (2014) One-dimensional modeling of a recent Ganga avulsion: assessing the potential effect of tectonic subsidence on a large river. Geomorphology 213: 24–37. https://doi.org/10.1016/j.geomorph.2013.12.038

Heckenberger, M.J., Russell, J.C., Toney, J.R. and Schmidt, M.J. (2007) The legacy of cultural landscapes in the Brazilian Amazon: implications for biodiversity. *Philosophical Transactions of the Royal Society B: Biological Sciences* 362(1478): 197–208. doi: 10.1098/rstb.2006.1979.

Hesse, P.P., Williams, R., Ralph, T.J., Fryirs, K.A., Larkin, Z.T., Westaway, K.E. and Farebrother, W. (2018) Palaeohydrology of lowland rivers in the Murray-Darling Basin, Australia. *Quaternary Science Reviews* 200: 85–105. doi: 10.1016/j.quascirev.2018.09.035.

Johnston, J. (1933) *Inland Navigation on the Gangetic Rivers*. Thacker, Spink and Co., Kolkata. Available at https://ia801603.us.archive.org/22/items/in.ernet.dli.2015.206394/2015.206394.Inland-Navigation.pdf.

Jorgensen, F., Scheer, W., Thomsen, S., Sonnenborg, T.O., Hinsby, K., Wiederhold, H., Schamper, C., Burschil, T., Roth, B., Kirsch, R. and Auken, E. (2012) Transboundary geophysical mapping of geological elements and salinity distribution critical for the assessment of future sea water intrusion in response to sea level rise. *Hydrology and Earth System Sciences* 16: 1845–1862. doi: 10.5194/hess-16-1845-2012.

Khan, Z.A. (1987) Paleodrainage and paleochannel morphology of a Barakar river (early permian) in the Rajmahal Gondwana Basin, Bihar, India. *Palaeogeography, Palaeoclimatology, Palaeoecology* 158(3–4): 235–247. doi: 10.1016/0031-0182(87)90063-0.

Kleinhans, M.G., Ferguson, R.I., Lane, S.N. and Hardy, R.J. (2012) Splitting rivers at their seams: bifurcations and avulsions. *Earth Surface Processes and Landforms* 38(1): 47–61. doi: 10.1002/esp.3268.

La Touche, T.H.D. (1910) (ed.) *The Journals of Major James Rennel: First Surveyor-General of India- Written for the Information of the Governors of Bengal During his Surveys of the Ganges and Brahmaputra Rivers 1764 to 1767*. The Asiatic Society, Kolkata. Available at https://ia800903.us.archive.org/0/items/journalsofmajorj00renn/journalsofmajorj00renn.pdf.

Laha, C. (2015) Oscillation of meandering Bhagirathi on the alluvial flood plain of Bengal Basin, India; as controlled by the palaeo-geomorphic architecture. *International Journal of Geomatics and Geosciences* 5(4): 564–572. http://www.ipublishing.co.in/jggsarticles/volfive/EIJGGS5049.pdf.

Langbein, W.B. and Leopold, L.B. (1966) River meanders: theory of minimum variance. United States Geological Survey Professional Paper 422-H, USGS. https://pubs.usgs.gov/pp/0422h/report.pdf.

Latrubesse, E.M. (2002) Evidence of Quaternary palaeohydrological changes in middle Amazonia: the Aripuana-Roosevelt and Jiparana fans. *Zeitschrift fur Geomorphologie Supplementband* 129: 61–72.

Lewin, J. and Ashworth, P.J. (2014) The negative relief of large river floodplains. *Earth-Science Reviews* 129: 1–23. doi: 10.1016/j.earscirev.2013.10.014.

Majumdar, S.C. (1941) *Rivers of the Bengal Delta*. Bengal Government Press, Alipore. Available at https://ia801904.us.archive.org/15/items/in.ernet.dli.2015.51632/2015.51632.Rivers-Of-The-Bengal-Delta-1941.pdf.

Majumdar, S.C. (1942) *Rivers of the Bengal Delta*. Calcutta University Readership Lectures, University of Calcutta, Kolkata. Available at https://ia801605.us.archive.org/34/items/in.ernet.dli.2015.511780/2015.511780.Rivers-Of.pdf.

Maroulis, J.C., Nanson, G.C., Price, D.M. and Pietsch, T. (2007) Aeolian-fluvial interaction and climate change: source-bordering dune development over the past ~100 ka on Cooper Creek, central Australia. *Quaternary Science Reviews* 26(3–4): 386–404. doi: 10.1016/j.quascirev.2006.08.010.

Mondal, S. and Patel, P.P. (2018) Examining the utility of river restoration approaches for flood mitigation and channel stability enhancement: a recent review. *Environmental Earth Sciences* 77: 195. doi: 10.1007/s12665-018-7381-y.

Mondal, S. and Patel, P.P. (2020) Implementing Vetiver grass-based riverbank protection programmes in rural West Bengal, India. Natural Hazards, https://doi.org/10.1007/s11069-020-04025-5

Mondal, S., Sarkar, A. and Patel, P.P. (2016) Causes of drainage congestion in Moyna Block, Purba Medinipur, West Bengal. In: Mondal, D.K. (ed.) *Application of Geospatial Technology for Sustainable Development*. North Bengal University Press, North Bengal University, Raja Rammohunpur, pp. 1–9.

Morgan, J.P. and McIntire, W.G. (1959) Quaternary geology of the Bengal Basin, East Pakistan and India. *Geological Society of America Bulletin* 70(3): 319–341. doi: 10.1130/0016-7606(1959)70[319:QGOTB B]2.0.CO;2.

Morisawa, M. (1985) *Rivers: Form and Process*. Longman, New York.

Morozova, G.S. (2005) A review of Holocene avulsions of the Tigris and Euphrates rivers and possible effects on the evolution of civilizations in lower Mesopotamia. *Geoarchaeology: An International Journal* 20(4): 401–423. doi: 10.1002/gea.20057.

Mukherjee, B.N. (1987) The Territory of the Gangaridai. *Indian Journal of Landscape Systems and Ecological Studies* 10(2): 65–70.

Mukherjee, K.N. (1986) Genesis of monsoon. *Indian Journal of Landscape Systems and Ecological Studies* 9(2): 58–65.

Mukherjee, K.N. (1995) *Rivers of Bengal and Their Impact On History*. S.K Saraswati Memorial Lecture, S.H.C.s, Kolkata.

Mukherjee, R. (1938) *The changing face of Bengal*. Calcutta University Press, Kolkata. Available at https://archive.org/details/in.ernet.dli.2015.125962.

Mukhopadhyay, A.K. (1996) Changing land use pattern in the Saraswati basin West Bengal. Unpublished PhD Thesis, University of Calcutta, Kolkata. Available at http://shodhganga.inflibnet.ac.in/handle/10603/164558.

O'Malley, L.S.S. and Chakravarti, M. (1909) *Bengal District Gazetteer: Howrah*. Bengal Secretariat Book Depot, Kolkata. Available at https://archive.org/details/in.ernet.dli.2015.206872.

Page, K., Nanson, G.C. and Price, D.M. (1996) Chronology of Murrumbidgee River palaeochannels on the Riverine Plain, southeastern Australia. *Journal of Quaternary Science* 11(4): 311–326. doi: 10.1002/(SICI)1099-1417(199607/08)11:4<311::AID-JQS256>3.0.CO;2-1.

Page, K.J., Kemp, J. and Nanson, G.C. (2009) Late Quaternary evolution of Riverine Plain paleochannels, southeastern Australia. Australian Journal of Earth Sciences, 56: S99–S33. https://doi.org/10.1080/08120090902870772.

Patel, P.P. (2010) The shifting 'Charlands' of the Godavari at Rajahmundry. *Indian Journal of Spatial Science*, 1(1): 1–17. Article 2.

Patel, P.P. (2013) GIS Techniques for landscape analysis – case study of the Chel River Basin, West Bengal. *Proceedings of State Level Seminar on Geographical Methods in the Appraisal of Landscape*, held at Dept. of Geography, Dum Dum Motijheel Mahavidyalaya, Kolkata, on 20th March, 2012, pp. 1–14.

Patel, P.P. and Dasgupta, R. (2009) Flood induced land use change in the Dulung River Valley, West Bengal. In: Singh, R.B., Roy, S.D.D., Samuel, H.D.D.K., Singh, V.D. and Biji, G.D. (eds.) *Geoinformatics for Monitoring and Modelling Land-Use, Bio-diversity and Climate Change - Contribution Towards International Year of Planet Earth*, Vol. 1. NMCC Publication, Marthandam, pp. 103–123.

Patel, P.P. and Mondal, S. (2019) Terrain – landuse relation in Garbeta-I Block, Paschim Medinipur District, West Bengal. In: Mukherjee, S. (ed.) *Importance and Utilities of GIS*. Avenel Press, Burdwan, pp. 82–101.

Patel, P.P. and Sarkar, A. (2009) Application of SRTM Data in evaluating the morphometric attributes: a case study of the Dulung River Basin. *Practicing Geographer* 13(2): 249–265.

Patel, P.P. and Sarkar, A. (2010) Terrain characterization using SRTM data. *Journal of the Indian Society of Remote Sensing* 38(1): 11–24. doi: 10.1007/s12524-010-0008-8.

Rashid, M.B., Mahmud, A., Ahsan, M.K., Khasru, M.H. and Islam, M.I. (2013) Drainage congestion and its impact on the environment in the south-western coastal part of Bangladesh. *Bangladesh Journal of Geology* 26: 359–371.

Raychaudhuri, H.C. (1943) Chapter I: Physical and historical geography. In: Majumdar, R.C. (ed.) *The History of Bengal: Volume I- Hindu Period*. The University of Dacca, Dhaka, pp. 1–34. Available at https://ia801900.us.archive.org/24/items/in.ernet.dli.2015.69821/2015.69821.History-Of-Bengal-Vol1.pdf.

Reaks, H.G. (1919) Report on the physical and hydraulic characteristics of the rivers of the delta (Appendix-II). In: Stevenson-Moore, C.J., Ryder, C.H.D., Nandi, M.C., Law, R.C., Haydem, H.H., Campbell, J., Murray, A.R., Addams-Williams, C. and Constable, E.A. (eds.) *Report on the Hooghly River and its Head-waters- Vol-I*. The Bengal Secretariat Book Depot, Kolkata, pp. 23–139. Available at https://ia800700.us.archive.org/1/items/dli.bengal.10689.23789/10689.23789.pdf.

Rob, M.A. (1989) Fluvial geomorphology of the Gangetic Delta. Unpublished PhD Thesis, Aligarh Muslim University, Aligarh. Available at https://core.ac.uk/download/pdf/144525175.pdf.

Roy, A. (1972) Land use and major agricultural characteristics of Damodar-Saraswati Doab. *Geographical Review of India* 34(1): 28–35.

Rudra, K. (1981) Identification of the ancient mouths of the Ganga as described by Ptolemy. *Geographical Review of India* 43(2): 97–104.

Rudra, K. (1986) The history of development of the Bhagirathi Hooghly river system and some connected considerations. Unpublished PhD Thesis, University of Calcutta, Kolkata Available at http://shodhganga.inflibnet.ac.in/handle/10603/165248.

Rudra, K. (2018) *Rivers of the Ganga-Brahmaputra-Meghna Delta: A Fluvial Account of Bengal*. Springer International Publishing, Cham. doi: 10.1007/978-3-319-76544-0.

Sahana, M., Hong, H., Ahmed, R., Patel, P.P., Bhakat, P., Sajjad, H. (2019) Assessing coastal island vulnerability in the Sundarban biosphere reserve, India, using geospatial technology. *Environmental Earth Sciences* 78: 304. doi: 10.1007/s12665-019-8293-1.

Sahana, M. and Patel, P.P. (2019) A comparison of frequency ratio and fuzzy logic models for flood susceptibility assessment of the lower Kosi River Basin in India. *Environmental Earth Sciences* 78: 289. doi: 10.1007/s12665-019-8285-1.

Sahana, M. and Sajjad, H. (2019) Assessing influence of erosion and accretion on landscape diversity in Sundarban biosphere reserve, Lower Ganga Basin: a geospatial approach. In: Das, B., Ghosh, S. and Islam, A. (eds.) *Quaternary Geomorphology in India. Geography of the Physical Environment*. Springer, Cham, pp. 155–189. doi: 10.1007/978-3-319-90427-6_10.

Salama, R.B., Tapley, I., Ishii, T. and Hawkes, G. (1994) Identification of areas of recharge and discharge using Landsat-TM satellite imagery and aerial photography mapping techniques. *Journal of Hydrology* 162(1–2): 119–141. doi: 10.1016/0022-1694(94)90007-8.

Sarkar, A. and Patel, P.P. (2011) Topographic analysis of the Dulung R. Basin. *Indian Journal of Spatial Science* 2(1): 2.

Sarkar, A. and Patel, P.P. (2012) Terrain classification of the Dulung drainage basin. *Indian Journal of Spatial Science* 3(1): 1–8. Article 6.

Sarkar, A. and Patel, P.P. (2017) Land use – terrain correlations in the piedmont tract of Eastern India: a case study of the Dulung River Basin. In: Santra, A. and Mitra, S.S. (eds.) *Remote Sensing Techniques and GIS Applications in Earth and Environmental Studies*. IGI Global, Hershey, pp. 147–192. doi: 10.4018/978-1-5225-1814-3.ch008.

Schwartzman, S., Boas, A.V., Ono, K.Y., Fonseca, M.G., Doblas, J., Zimmerman, B., Junqueira, P., Jerozolimski, A., Salazar, M., Junqueira, R.P. and Torres, M. (2013) The natural and social history of the indigenous lands and protected areas of the Xingu River Basin. *Philosophical Transactions of the Royal Society B: Biological Sciences* 368(1619): 20120164. doi: 10.1098/rstb.2012.0164.

Sen, A. (1998) The decay of the river Hugli and its impact on the port of Calcutta a geographical appraisal. Unpublished PhD Thesis, University of Calcutta, Kolkata. Available at http://shodhganga.inflibnet. ac.in/bitstream/10603/164560/8/08_chapter%202.pdf.

Sen, J., Basu, A., Sengupta, P. and Mukherjee, A. (2015) *Chandraketugarh- rediscovering a missing link in Indian history (Project Codes AIB and GTC).* SandHI and Indian Institute of Technology, Kharagpur. Available at http://www.iitkgpsandhi.org/Chandraketugarh_Report.pdf.

Sen, S. (2019) *Ganga: The Many Pasts of a River.* Penguin Random House, Gurgaon.

Sen, S.N. (1943) The Portuguese in Bengal. In: Sarkar, J. (ed.) *The History of Bengal: Volume II- The Muslim Period.* B.R. Publishing Corporation, New Delhi, pp. 351–370. Available at https://ia801603.us.archive. org/12/items/in.ernet.dli.2015.24396/2015.24396.History-Of-Bengal--Vol-2.pdf.

Seth, M. (2014) Hydro-geomorphic analysis restoration and sustainable management of Saraswati river corridor in West Bengal. Unpublished PhD Thesis, University of Calcutta, Kolkata. Available at http://shodhganga.inflibnet.ac.in/handle/10603/163728.

Sreemani, S. (2011) *The Gangetic West Bengal [c. Seventeenth-Nineteenth Century]: A Survey.* New Central Book Agency (P) Ltd., London.

Srigyan, M., Basu, A., Mukherjee, A., Sengupta, P. and Sen, J. (2019) Identification of paleochannels in and around "Chandraketugarh", Ganges Delta through remote sensing techniques using fuzzy inference system. *Archaeological and Anthropological Sciences* 11(3): 839–852. doi: 10.1007/s12520-017-0577-3.

Starkel, L. (2008) Palaeohydrology: the past as a basis for understanding the present and predicting the future. In: Harper, D., Zalewski, M. and Pacini, N. (eds.) *Ecohydrology: Processes, Models and Case Studies.* CAB International, Wallingford, pp. 276–302. doi: 10.1079/9781845930028.0276.

Stefani, M. and Vincenzi, S. (2005) The interplay of eustasy, climate and human activity in the late Quaternary depositional evolution and sedimentary architecture of the Po Delta system. *Marine Geology* 222–223: 19–48. doi: 10.1016/j. margeo.2005.06.029.

Syvitski, J.P.M., Kettner, A.J., Correggiari, A. and Nelson, B.W. (2005) Distributary channels and their impact on sediment dispersal. *Marine Geology* 222–223: 75–94. doi: 10.1016/j.margeo.2005.06.030.

Umitsu, M. (1993) Late Quaternary sedimentary environments and land forms in the Ganga delta. *Sedimentary Geology* 83: 177–186. doi: 10.1016/0037-0738(93)90011-S.

Vandenberghe, J. (1995) Timescales, climate and river development. *Quaternary Science Reviews* 14(6): 631–638. doi: 10.1016/0277-3791(95)00043-O.

Vervoort, R.W. and Annen, Y.L. (2006) Palæochannels in Northern New South Wales: Inversion of electromagnetic induction data to infer hydrologically relevant stratigraphy. *Australian Journal of Soil Research* 44(1): 35–45. doi: 10.1071/SR05037.

WBPCB (West Bengal Pollution Control Board) (2017) *State of Environment Report: West Bengal-2016.* West Bengal Pollution Control Board, Government of West Bengal, Kolkata. Available at https://www.wbpcb. gov.in/writereaddata/files/SOE_Report_2016.pdf.

Williams, G.P. (1986) River meanders and channel size. *Journal of Hydrology* 88: 47–164. doi: org/10. 1016/0022-1694(86)90202-7.

Wood, W.H.A. (1895) *A Short Geography of Bengal.* George Bell and Sons, London. Available at https:// ia802700.us.archive.org/7/items/ashortgeography00woodgoog/ashortgeography00woodgoog. Pdf.

Wray, R.A.L. (2009) Palaeochannels of the Namoi River Floodplain, New South Wales, Australia: the use of multispectral Landsat imagery to highlight a Late Quaternary change in fluvial regime. *Australian Geographer* 40(1): 29–49. doi: 10.1080/00049180802656952.

Zankhna, S. and Thakkar, M.G. (2014) Palaeochannel investigations and geo hydrological significance of Saraswati River of Mainland Gujarat, India: using remote sensing and GIS techniques. *Journal of Environmental Research and Development* 9(2): 472–479. Available at http://www.jerad.org/ppapers/ dnload.php?vl=9andis=2andst=472.

15 Facets of Anthropogenic Encroachment within the Palaeo-Fluvial Regime of *Kana*–Ghia–Kunti System, Damodar Fan Region

Arghyadip Sen, Ujwal Deep Saha,
Arijit Majumder and Nilanjana Biswas

CONTENTS

15.1 INTRODUCTION

World's largest inhabited river delta, the Ganga–Brahmaputra Delta (GBD), and its surrounding alluvial tracts started being occupied by inhabitants more than 2,100 years ago (Alam, 2015; Chattopadhyay and Roy Choudhury, 2017; Siddiq and Habib, 2017). The interfluvial part of lower Damodar and Bhagirathi-Hooghly is being intensively cultivated and settled by farming communities since the beginning of the Zamindari period during the late 18th century (Hunter, 1876; Bhattacharyya, 2011). The dwellers have constantly modified their livelihood with the hazards generated by frequent monsoonal floods of Damodar, and with time, they have changed the natural riparian landscape to a great extent (Choudhury, 1990; Lahiri-Dutt, 2003).

Palaeochannels, a channel which is no more a part of the active channel system and may or may not carry water (Borisova et al., 2006; Wray, 2009; Sümeghy and Kiss, 2012), often experience intense rate of human encroachment in order to utilise its relatively higher moisture content for agricultural practices. This has been a common scenario for most of the abandoned channels of the rivers in the GBD (Bandyopadhyay, 1996; Bandyopadhyay, 2007b; Rudra, 2006, 2010, 2014). The human encroachment within the abandoned fluvial environment is responsible for the changing ecological characteristics of the riparian tract, diminishing geomorphic efficiency higher than the natural rate. These cumulative effects enhance the process of channel decaying at a much higher rate (Jacobson et. al., 1995; Kondolf, 1997; Charlton, 2008). Studies carried out regarding channel decay on the GBD under anthropogenic pressure by the scientist of the research community haven't showcased the detail analysis of the human encroachment (Bhattacharyya, 2011). The facets of direct and indirect anthropogenic impacts will control the intensity of channel decay. This will in return diminish the geomorphic role of the fluvial body. Organised and regulated interactions between the human society and palaeochannel system usually help to maintain the functioning of the palaeochannels to carry its local geomorphic, ecological and economic significance (Singh, 1996; Brown, 2002; Sinha et al. 2005; Wray, 2009).

This study has been aimed towards studying the facets of anthropogenic encroachment within and around the palaeo-fluvial corridor of *Kana*–Ghia and the Kunti system over the Damodar fan-deltaic surface. The autogenic response over the palaeo-fluvial environment is needed to be studied in order to regulate the restoration strategies for the natural system (Fischer and Cyffka, 2014). The direct and indirect impoundments and over utilisation of the natural resources related to the palaeo-fluvial environment studied with a holistic approach are required to prepare any draft of institutional regulation strategy in order to understand the cost-benefit profile or the loops that will drive the strategic frame. The palaeo-fluvial system of *Kana*–Ghia and Kunti rivers has been studied with the main thrust put on the demarcation of the palaeosegments within the study and spatial clustering of the land use categories to quantify the human dependence on the natural system along with the present geomorphic scenario of the palaeo-fluvial corridor.

Multiple sensor-borne satellite images, as well as digital datasets, were used to fulfil different purposes in this study, after carefully pre-processing of the images. Modified Normalised Differential Wetness Index (MNDWI), the proposed method of Gao, 1996, was used for assessing the surface wetness. Multispectral images of Landsat 7 during the pre-monsoon season of 2001 and Landsat 8 multispectral images of 2019 have been used for LULC mapping and associated quantification of LULC categories (Table 15.1). A post-flood period image of 2000 captured by Landsat 7 was used to demarcate the palaeocourses within the *Kana*–Ghia–Kunti system in order to utilise the ground moisture which remains higher within the palaeocourses. The fine resolution of Sentinel 2 images captured in April 2019, Orb view 3 images of 2003 captured on different dates and Google Earth image (dated 20 January 2019) has been utilised for spatial verification of demarcated palaeocourses and other geomorphic as well as land use attributes. All the satellite images used in this study were collected free of cost from the web-based platform of USGS (http://glovis.usgs.gov). In the case of Landsat images utilised for LULC mapping, the season was kept the same to wipe out the seasonal bias in terms of agricultural practices and the amount of soil moisture present.

15.1.1 BUFFER DELINEATION

Anthropogenic activities over the floodplain often obliterate the palaeo-fluvial indicators by modifying the spatial extent. Continuous exposure of the palaeo-fluvial markers often disappears making it extremely difficult to demarcate the spatial extent. To resolve this difficulty, the levee present around the channels has been considered to demarcate the exact corridor of the palaeo-channels, although, in various areas, it became almost impossible to locate the same. A sample

TABLE 15.1
Data Used

Satellite	Sensor	P/R or Coordinate	Date of Acquisition	Bands	Resolution (m)
Landsat 7	ETM+	138/44	25.10.2000	MS	30
Landsat 7	ETM+	138/44	26.04.2001	MS	30
Landsat 8	OLI/TIRS	138/44	20.04.2019	MS	30
Sentinel 2	MSI	T45QXF N02.07	25.04.2019	MS	10
Orb view 3	Orb view 3	22.86820000 88.36550000	16.11.2003	PAN	1
Orb view 3	Orb view 3	22.86850000 88.29910000	05.12.2003	PAN	1
Orb view 3	Orb view 3	22.87080000 88.03400000	08.12.2003	PAN	1
Survey of India Topographical sheet	79A/4, 79A/8, 79B/1, 79B/5; Surveyed in 1968–1969; R.F – 1:50000				

survey has been used to extract the realistic nature of the human response to the palaeo-fluvial corridor. Nineteen locations were chosen randomly based on their accessibility where topographic study, interview of the local farmers and knowledgeable persons regarding means of the utilisation of the channel as well as the nature of crop being cultivated were done in 2019 (Table 15.2). Buffer region of 950 m on either side of the palaeochannel has been considered while providing significant emphasis on an average width of the palaeo-corridor, palaeolevee and area in which the channel water serves for irrigation.

15.1.2 GEOMORPHIC AND LAND USE INDICATORS

Palaeo-fluvial environments are usually associated with several surface geomorphic tracers in the form of cut-offs, levees, distinct palaeo-bank lines and connected channels (Bandyopadhyay, 2007a; Sinha, 2008; Nandini et al., 2013; Saha and Bhattacharya, 2016). The geomorphic tracers in the study area were utilised to demarcate the palaeo-fluvial corridors of the *Kana*–Ghia–Kunti system (Table 15.2). Locations of the levee, abandoned channels and cut-offs were confirmed and traced from high-resolution satellite images of recent date, and ground survey records of palaeo-bank lines were merged to extract the valid location of the palace corridors. An extensive field survey was conducted in 2019 using Garmin Map 66s GPS with ±2 m of vertical accuracy to locate the significant fluvio-geomorphic markers at 19 locations.

Palaeochannels are the preferential locations of groundwater storage (Revil et al., 2005; Samaddar et al., 2011; Nandini et al., 2013). Groundwater level remains at a much lower depth within the palace corridors because of silt-rich sediment clasts forming suitable grounds for digging ponds and tanks. Abandoned channels are also impounded with earthen sediments to separate individual plots and for aquaculture practices. Location of ponds and other water bodies in a more or less continuous-linear fashion had been utilised to trace the palaeochannels from the works of Nandini et al. (2013) and Saha and Bhattacharya (2019).

15.1.3 LULC MAPPING AND QUANTIFICATION

Feature space plotting of satellite images was applied to generate LULC maps for2001 and 2019 classified into six categories, within the 950 m buffer on either side of the major palaeocourses. LULC of the entire *Kana*–Ghia–Kunti system was carried out for 2019. Accuracy of the LULC maps was

TABLE 15.2
On-Field Observations of the Palaeocourses at 19 Selected Locations

Locations	Active Channel	Palaeo-Corridor	Levee	Old Floodplain<500m from the Active Channel	Old Floodplain<1,000m from the Active Channel	Old Floodplain<1,500 m from the Active Channel
Betragarh	Aquaculture; *active channel width 17 m*	Agriculture	Settlement, ponds and orchard	Comes under levee on either side	Agriculture on both the sides (double cropping) *Use river for source of irrigation*	Agriculture (single crop)
Gohaldaha	*Channel width 21 m*	Agriculture	Settlement, ponds and orchard	Comes under levee on either side	Agriculture on both the sides (single cropping)	Agriculture (single crop)
Madhusudanpur	Channel width 20 m	Swamps	Settlement, ponds, orchard Agriculture (double cropping)	Comes under levee on either side	Agriculture on both the sides (double cropping) *Use river for source of irrigation*	Agriculture (double crop)
Byaspur	Channel width 18 m	Agriculture (double cropping) *Road crossing*	Settlement, orchard	Agriculture (double crop) *Use river for source of irrigation*	Agriculture on both the sides (double cropping)	Agriculture (double crop)
Maheshpur	Channel width 13 m	Agriculture (double cropping) *Road crossing*	Settlement, ponds and orchard	Agriculture (double crop) *Use river for source of irrigation*	Agriculture on both the sides (single cropping)	Agriculture (single crop)
Amgachi	Channel width 13 m	Agriculture (double cropping) *Road crossing*	Settlement and ponds	Wetland along the right bank; agriculture (double crop) *Use river for source of irrigation*	Agriculture on both the sides (single cropping)	Agriculture (single crop)
Hamirgachi	Channel width 18.5 m	Ponds and marshes	Settlement and ponds	Agriculture (double crop) *Use river for source of irrigation*	Agriculture on both the sides (single cropping)	Agriculture (single crop)
Choalpara (Nalikul)	Channel width 22.5 m	Settlement ponds and agricultural field	Settlement and metaled road	Agriculture on the left bank side and railway track on the right bank side	Agriculture on both the sides (double cropping)	Agriculture (single crop)

(Continued)

TABLE 15.2 (*Continued*)
On-Field Observations of the Palaeocourses at 19 Selected Locations

Locations	Active Channel	Palaeo-Corridor	Levee	Old Floodplain<500m from the Active Channel	Old Floodplain<1,000m from the Active Channel	Old Floodplain<1,500 m from the Active Channel
Gopalnagar	Channel width 21m	Settlement and ponds	Settlement and orchard	Agriculture (double crop) *Use river for source of irrigation*	Fallow land (due to land acquisition for industrial setup)	Fallow land (due to land acquisition for industrial setup) on the left bank side and agricultural field(single crop) on the right bank side
Tegachi	No continuous flow	Settlement, ponds and	Settlement, ponds and orchard	Agriculture (double crop)	Agriculture on both the sides (double cropping)	Agriculture (single crop)
Kanuibanka	Dried up beds	Swamps and ponds	Settlement, ponds and orchard	Agriculture (double crop)	Agriculture (single crop)	Agriculture (single crop)
Kankuria	Channel width 9m	Swamps and ponds	Settlement, ponds and orchard	Agriculture (doublecrop)	Agriculture (single crop)	Agriculture (single crop)
Bhangabari	Dried up river beds	Swamps *Road crossing*	Settlement and orchard	Agriculture on the left bank side and on the right bank side it comes under the levee	Agriculture (doublecrop)	Agriculture (single crop)
Berabari	Channel width 9m	Ponds and agriculture (multi-cropping)	Settlement and orchard	Agriculture on the left bank side and on the right bank side it comes under the levee	Agriculture (doublecrop)	Agriculture (single crop)
Palasi	Channel width 20.5m	Swamps	Settlement and orchard	Agriculture (double crop)	Agriculture (single crop)	Agriculture (single crop)
Payraura	Channel width 40.5 m	Agriculture *Road crossing*	Settlement and orchard	Agriculture (doublecrop) and settlement	Agriculture (single crop)	Agriculture (single crop)
Rajhat	Channel width 36m	Agriculture and marsh *Road crossing*	Settlement and vegetation	Vegetation and ponds	Vegetation and ponds	Vegetation and ponds
Notungram (Mogra)	Channel width 26m	Marsh *Road crossing*	Settlement, pond and vegetation	Settlement	Settlement	Settlement and marsh
Kuntighat	Channel width 39.4m	Marsh *Road crossing and railway bridge*	Settlement, pond and vegetation	Settlement	Settlement	Settlement and marsh

measured from the error matrix prepared during the classification of images in Edras Imagine (v 14). Overall accuracy and Khat value was considered as the level of accuracy for the classified images of different durations. Overall accuracy was calculated by dividing the sum of diagonal cells of the error matrix by sum of row totals (Σt), and expected accuracy is the product of the sum of diagonal cells of the error matrix and Σt that value is calculated using the following formula:

$$\left[\left(\text{Observed accuracy} - \text{Expected accuracy} \right) \big/ \left(\Sigma t \right)^2 - \text{Expected accuracy} \right]$$

15.1.4 Spatial Quantification of LULC Categories

Inter-class conversion of LULC categories and its nature of integration are quite useful while identifying the actual facets of human encroachment within the fluvial regime (Yousefi et al., 2017; Debnath et al., 2017). In the case of anthropogenic encroachment, a general experience-based framework can be drawn where the conversion of moist grounds into the agricultural field and fragmentation of continuous water bodies are likely to have occurred. Thus, encroachment within the fluvial regime will showcase a picture of fragmentation likely to be of moist ground and clustering of cultivable land (James and Lecce, 2013; Grabowski et al., 2014).

Spatial occupation by each LULC class has been generated by summing up the area of pixels grouped into each class. The class to the class conversion of the LULC attribute was calculated to probe any changes in the nature of land use practices within the palaeo-fluvial environment. It was performed using the combining module in Arc GIS10.3.1. The integration or disintegration of spatial segments or patches can be utilised to analyse the temporal change over the investigated spatial units (Deng et al., 2009; Castillo et al., 2011; Wu et al., 2018). Spatial clustering, as well as the fractal dimension of the LULC attributes of 2001 and 2019, was calculated using the Patch Analysis toolsets in Arc GIS and LISA (Local Indicator of Spatial Association) module in GeoDa. The fragmentation statistics were performed at a class level where each LULC attributes were placed in individual vectors. Each vector was calculated based on a union performed between the vectors and grid net of 200 m^2 as it proved to be the most suitable while integrating the minimum possible spatial extent. The cultivated lands, as well as the wet fallow lands, are spread in a continuous trend with larger patches when compared to the settlements and orchards.

15.1.5 Spatial Association

The evolution of spatial features is well understood from judging its spatial association (Sokal and Oden, 1978; Weeks et al., 2004). In order to study the spatial clustering of LULC attributes based on the experienced-based framework of anthropogenic encroachment over the considered temporal duration LISA analysis was performed. LISA was considered suitable because of its ability to analyse the neighbourhood of the spatial features (Anselin, 1995; Castillo et al., 2011). Value of Moran's I has been calculated for settlement and orchard, cultivable land along with the wet fallow for the respective years using GeoDa. A 95% significance level has been used to test the spatial allocation of the clustering of the selected LULC attributes.

15.2 THE RIVER AND ITS BASIN

The Damodar fan delta unit is associated with several palaeo-systems of Damodar which were once active in the past as the major passageways of monsoonal floodwater (Bandyopadhyay, 2007a; Rudra, 2006). Based on the flow direction and connectivity between the palaeochannels along with considerations of Digital Elevation Model (DEM)-derived watershed boundaries, six individual palaeo-fluvial systems were delineated within the fan delta unit (Figure15.1). The Ghia–Kunti sub-system had developed within the transitional zone of the Memari fan and Tarakeswar fan. Both the

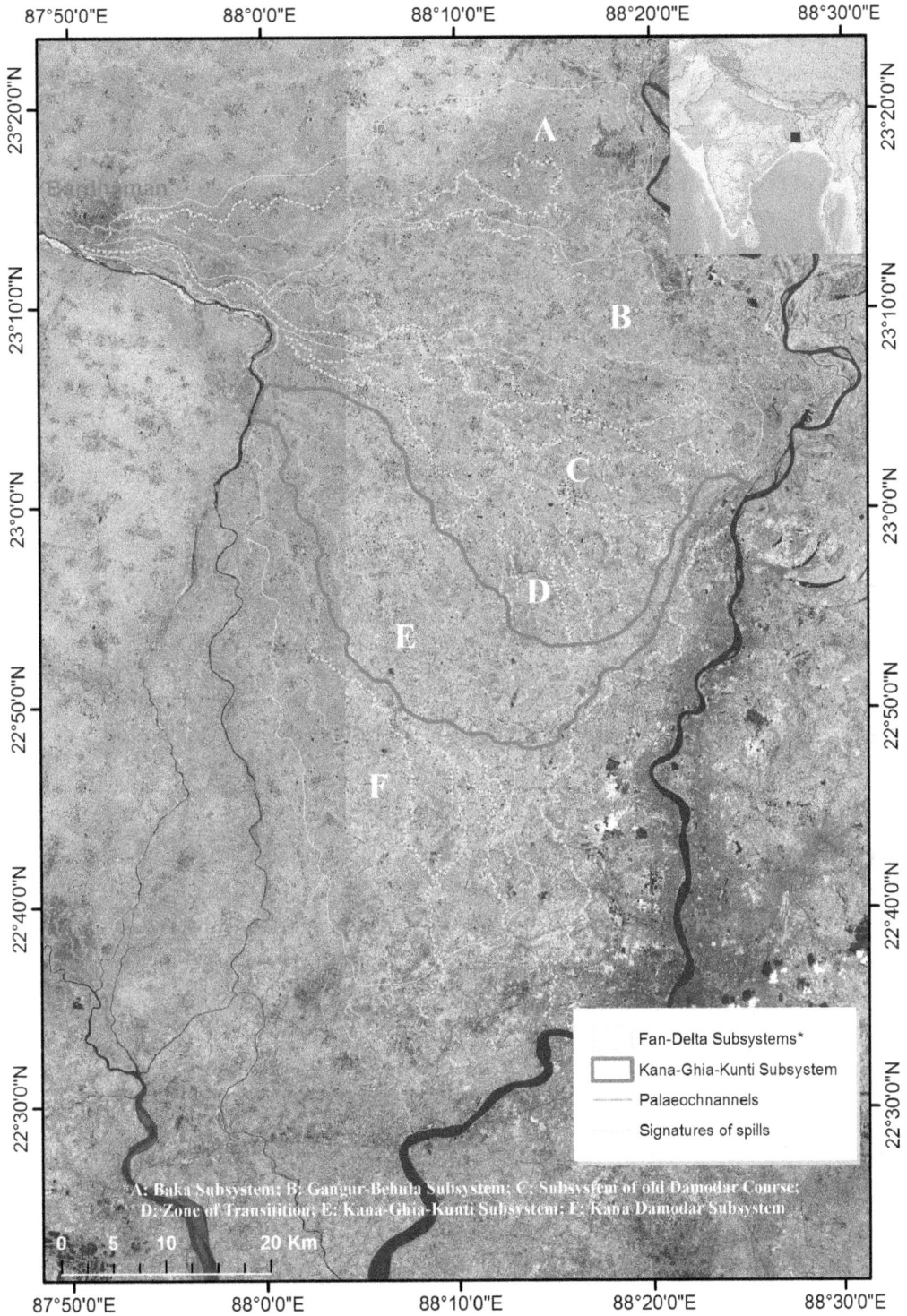

FIGURE 15.1 Palaeochannel systems over the Damodar fan delta. The catchment system highlighted in bold black is the studied system of *Kana*–Ghia–Kunti rivers.

channels flow from the north-west to south-east direction and join near Bainchipota at the eastern margin of the fandelta. The unified channel flowed downstream with a new name "Kunti" that took an unusual bend towards north-east following a parallel drainage lineament. The channel gradients of all the palaeochannels within the Kana–Ghia–Kunti system are quite low but have varied in an inverted manner. In most cases, the relatively lower gradient is associated with higher values downstream possibly due to in-channel sedimentation (Table 15.3).

Shreds of evidence from historical maps, literature and satellite images suggest that the ancient course of Damodar used to flow eastward through the course of the Gangur River to meet river Hooghly near Kalna (Rudra, 2010; Bhattacharyya, 2011; Ghosh and Jana, 2019). The present course of Damodar during the earlier period was probably a minor distributary flowing towards the south. Later, the downstream segment of the main flow started rotating clockwise after it approached Damodar fault line (Singh et al., 1998; Roy and Sahu, 2016; Ghosh and Islam, 2016) and gradually shifted its mouth 128 km southward to meet river Hooghly at Garchumuk (Rudra, 2010; Majumder and Sivaramakrishnan, 2014). This gradual process of shifting of the main course had distributed the quaternary sediment over the interfluvial region to form the fan-delta and had given rise to several flood passages of Damodar. Still, the process of channel abandonment on the depositional unit with a very low channel gradient hasn't followed a clockwise manner throughout the past (Ghosh and Mistri, 2012). Different channels became active in the course of time, due to avulsion following a major flood (Sen, 1978; Rudra, 2006).

During the late 18th century, the major flow of Damodar shifted more towards the south followed by the abandonment of the river Gangur, and *Kana* Damodar emerged as the main branch (Bhattacharyya, 2011). Ghia–Kunti system and *Kana* River were active during that time. *Kana* and Ghia–Kunti had an arch-like path which used to flow towards SE and then diverted towards NE (Figure 15.3). Although in Rennell's survey map was prepared during 1767–1777 (Rudra, 2016), the present course of Damodar or Amta Channel had been evident to be more active along with Kunti River forming an important branch (Figure 15.2).

TABLE 15.3
General Characteristics of the Palaeochannels Extracted from Satellite Images and Averaged Based on Survey Results at 19 Locations

River	Distance from Source (km)	Gradient (mkm⁻¹)	Average Channel Width (m)	Average Corridor Width (m)	Depth in m (Bankfull)	Directions
Kana	0–10	0.21	11.38	567	3.2	SSE
	10–20	0.22	26	310	2.92	SSE
	20–30	0.30	29.2	303	2.84	SE
	30–40	0.31	28.10	210	3.4	ESE
	40–50	0.16	24.76	185	3.56	E
	50–59.3	0.16	24.56	71	3.5	ENE
Ghia	0–10	0.505	8.30	37	-	ESE
	10–20	0.26	6.78	25	-	SSE
	20–30	0.22	9.98	43	1.9	SE
	30–40	0.14	10.81	86	2.2	SSE
	40–51.5	0.31	27.50	66	2.89	EES
Kunti	0–10	0.22	20.30	69	3.85	NE
	10–20	0.131	12.10	54	4.2	NNE
	20–29.1	0.42	27.4	47	4.62	NE
Jhimki	0–10	0.21	7.04	25	-	ESE
	10–20	0.30	20.75	52	-	SE
	20–30	0.41	16.83	30	1.88	ESE

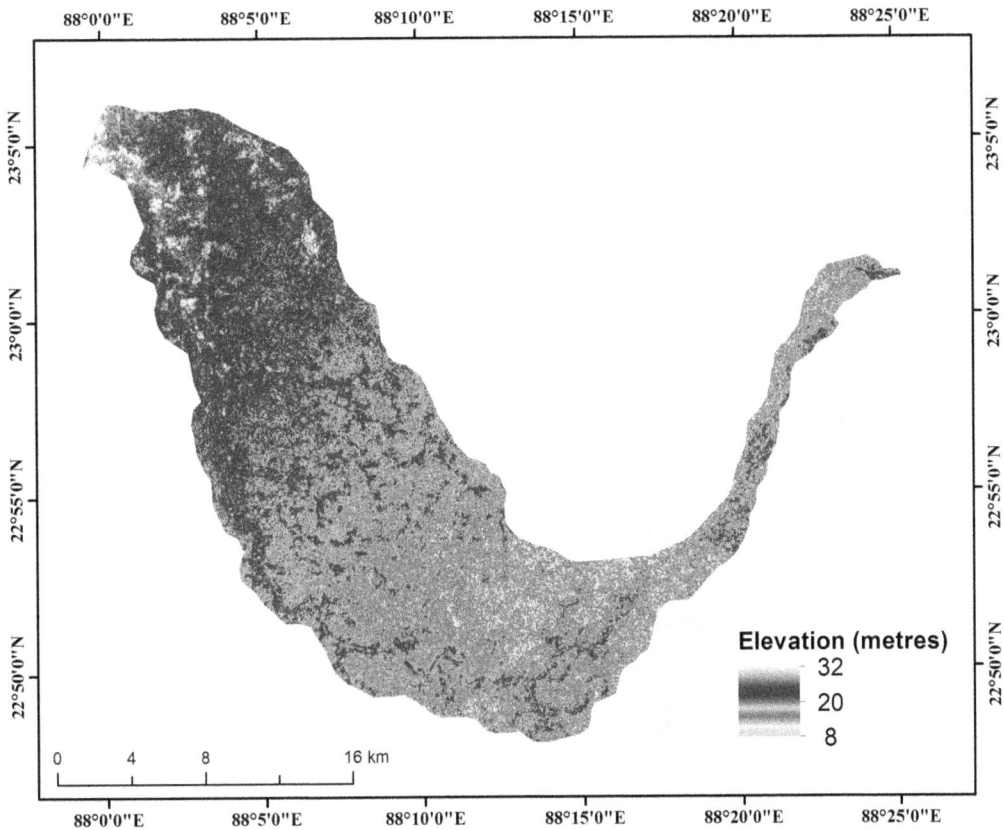

FIGURE 15.2   Elevation characteristics of the *Kana*–Ghia–Kunti system.

Later on, almost all the maps prepared during the post-Rennell period (19th century) had featured *Kana*–Kunti as an important distributary of Damodar (Majumder, 1941; Rudra, 2010). In the south, the Amta channel continued to be the main branch since the 18th century, whereas during the occurrence of high floods, Damodar reopened and chose the easterly channels to pass the overflow (Ghosh and Mistri, 2012). The map published by Laurie and Whittle in 1804 had shown *Kana* Damodar as a major distributary with no reference to *Kana* and Kunti, which implies a temporary shift of flow to the former channel. Later during 1848, the Sidney Morse's map had demarcated a resurrection of the Gangur system possibly followed by a major flood event when the *Kana* and Kunti channels received a considerable amount of discharge. With the construction of DVC dams, the actual flow of Damodar was regulated and almost all the distributaries including *Kana*, Ghia and Kunti were delinked from the main course of Damodar. The development of the *Kana*–Ghia–Kunti course denotes an intermediate phase of shifting of the main flow of Damodar from east to south. These emerged courses have had spread its channels within the boundaries of the eastern as well as the southern flowing channels.

15.2.1   DEGRADATION OF THE CHANNELS

The migration of Damodar river had developed a deltaic fan region with remnants of several abandoned courses that used to carry a fair share of discharge in the past and even a few of those got reoccupied after its abandonment. Over the quaternary depositional surface, palaeochannels have been radiated eastward from the left bank of the Damodar towards Bhagirathi-Hooghly River. In previous studies of Majumdar (1941), Rudra (2010) and Bhattacharyya (2011), river training since

FIGURE 15.3 Condition of Damodar River and its distributaries during different periods; (a) map prepared by Van-Den Brouke in 1658–1664; (b) map prepared by James Rennell in 1776. (c) map prepared by Rennell, Lauri and Whittl, *1794*; (d) map prepared by Morse and Sidney in 1848.

the colonial period and construction of the earthen embankments by the landlords had been portrayed as reasons for delinking the eastward distributaries from the main channel, including those of *Kana* and Ghia. After the construction of DVC dams in the upper catchment of Damodar in the late 1950s, both the velocity and magnitude of average annual flow decreased drastically, thereby causing a higher rate of sedimentation in the beds of fan-delta channels as well as the main channel of Damodar (Sen, 1985). The flow became much irregular and seasonal in the channels like *Kana* and Ghia, although Kunti remains relatively more active due to its contact with river Hooghly, through which tidal water keeps invading regularly. Finally, the formation of embankments and the DVC canal on the left bank of Damodar completely delinked *Kana* and Ghia from main Damodar, and they started receiving water supply only from the DVC canal, which itself has very less capacity. Since then, the degradation of the *Kana–Ghia–Kunti* sub-system was accelerated, which became intense with increasing population pressure and resultant anthropogenic encroachment in the form of expansion of agronomical activities on the channel bed, the formation of new transport links and settlements in the latter part of the century (Bhattacharyya, 2011). In recent dates, the land use practices might be affecting the palaeo-fluvial corridors of Damodar River. Throughout the past, the channels of fan-delta have supplied sizeable amount of fresh silt, which contributed to the development of this geomorphological unit (Mukherjee, 2008). With the increasing pressure of population, expanding agricultural activities, river training works that have delinked the flow and resultant siltation has changed the flow behaviour and geomorphic efficiency to a great extent. The intensive use of the low-lying floodplains, the relatively higher grounds of the palaeolevee and the gradual transformation of palaeochannels into irrigation canals claiming its water from a highly regulated river Damodar has been a threat to the geomorphological performance as well as discontinuity of the palaeochannels. Such human responses and their dreadful consequences usually dragged down the ecological value of the palaeo-corridors, channel connectivity as well as its efficiency to transfer rainwater (Siddique and Mukherjee, 2017).

15.2.2 PHYSICAL SETUP OF THE STUDY AREA

A fan-delta can be termed as a fan-shaped depositional surface which has been deposited, partly or entirely, at the interface between the active deltaic system and the standing fluvial body due to a sudden change in surface gradient (Nemec and Steel, 1988). The Damodar fan-delta, part of the adjacent GBD has been formed by the conjugated efforts of its palaeocourses of river Damodar. Damodar fan-delta, with signs of numerous avulsions, spills and palaeo-corridors, is relatively older than the GBD and not active anymore in terms of lateral expansion (Bagchi, 1944). The palaeo-corridor or palaeo-fluvial corridor has been derived from the concept of the fluvial corridor proposed by Piegay et al. (2005). It signifies the concentration of the channels within a spatial extent incorporating its once active floodplain and levee.

The study area includes the *Kana*, Ghia and Kunti channel system that comprises the central part of Damodar fan-delta. The Damodar fan-delta was formed due to sedimentation envisaged by the old distributaries of Damodar during late quaternary flood episodes (Rudra, 2010; Ghosh, 2017). Sediments eroded from the exposed quartz-rich Archaean gneiss and Gondwana sandstone of Chhotanagpur plateau fringe at the headwaters and middle stretch of Damodar River, respectively, were deposited by its old courses flowing eastward to form the fan within the Bengal basin (Bhattacharyya, 2011). The eastern margin of this quaternary alluvial unit has been truncated by Bhagirathi-Hooghly River and coated with relatively newer alluvium on the top (Majumder and Shivaramakrishnan, 2014; Ghosh, 2017). Several minor faults are present in the underlying bedrock, along with a major fault line called the Damodar fault lying in an SSW-NNE direction. The structural deformations and confinements have influenced the southward shifting of the main distributary of Damodar to a great extent (Singh, et. al., 1998; Roy, 2019; Mahata and Maiti, 2019). Physiographically, this fan-delta is a deltaic floodplain, consisting of two parts. The Memari fan forms the eastern part, while the Tarakeswar fan consists of south-east and south of the fan-delta (Mallick and Niyogi, 1972;

Acharyya and Shah, 2006). The entire fan-delta has showcased numerous geomorphic signatures in the form of abandoned channels, levees and scars which are evidence of frequent avulsions. In this deltaic floodplain, the average slope is extremely low (less than 1°), which provides a suitable ground for frequent channel shifting. The orientation of two fans and the history of avulsion had established the scenario of the fan-delta building process which possibly had taken place from the take-off of one of the old courses of Damodar near *Kana* up to the present off-take located 128 km southward near Garchumuk. The *Kana*–Ghia–Kunti system has an average elevation of 22 m, ranging from 32 m on the west and gradually lowering to 10 m towards the south (Figure 15.2). The gradient of these channels mostly resembles a dead flat terrain downstream (Table 15.3).

15.3 ANTHROPOGENIC SIGNATURES ON FORMS AND PROCESSES

15.3.1 PALAEOCHANNELS

The NDWI prepared from the post-flood period imagery of 2000 was of great help while identifying the palaeocourses of the *Kana*–Ghia–Kunti system. The high moisture containing channel of *Kana* which is presently disconnected from Damodar, due to the DVC canal, continues to flow in a similar manner as before. The process of channel decay and continuous sedimentation had compelled the flow to be concentrated into a feeble channel. Presently, *Kana* River has been found as a channel that is a misfit in character with respect to its palaeo-corridor. Possibly, a spill channel of the river *Kana*, locally named Jhimki, was found with an abandoned head fragmented due to the construction of ponds and settlements near Tegachi (Figures 15.4 and 15.5). The course of Jhimki

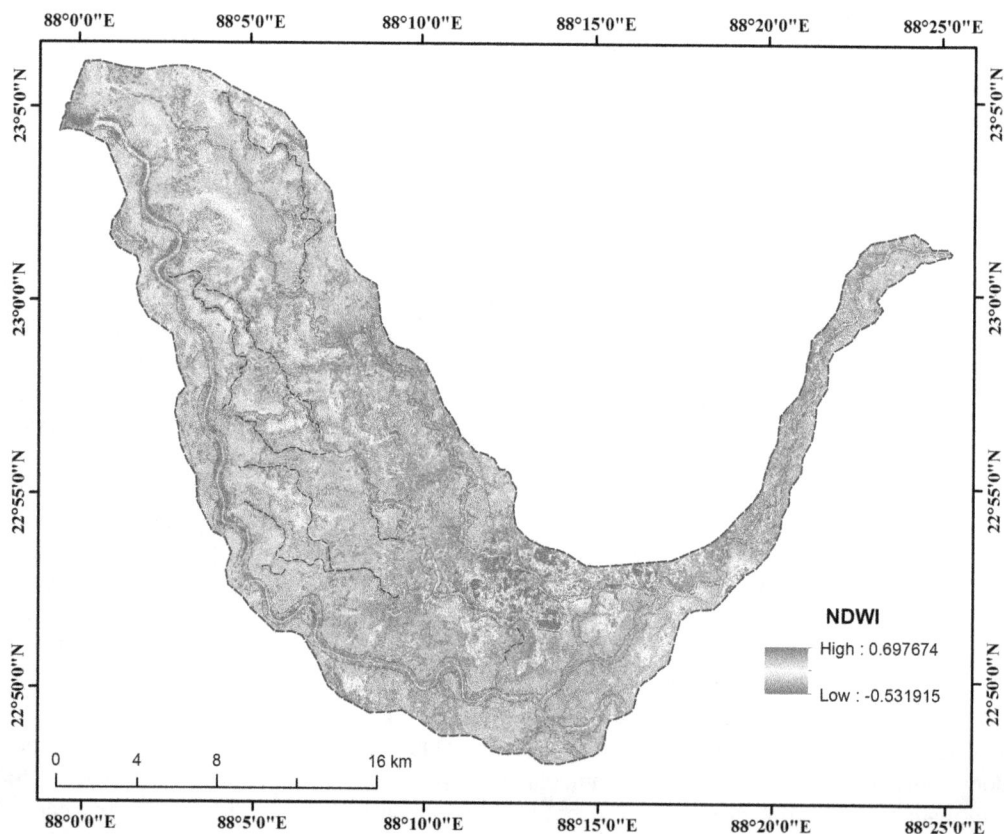

FIGURE 15.4 NDWI of the *Kana*–Ghia–Kunti system. Demarcated palaeo channels have been superimposed on the NDWI map.

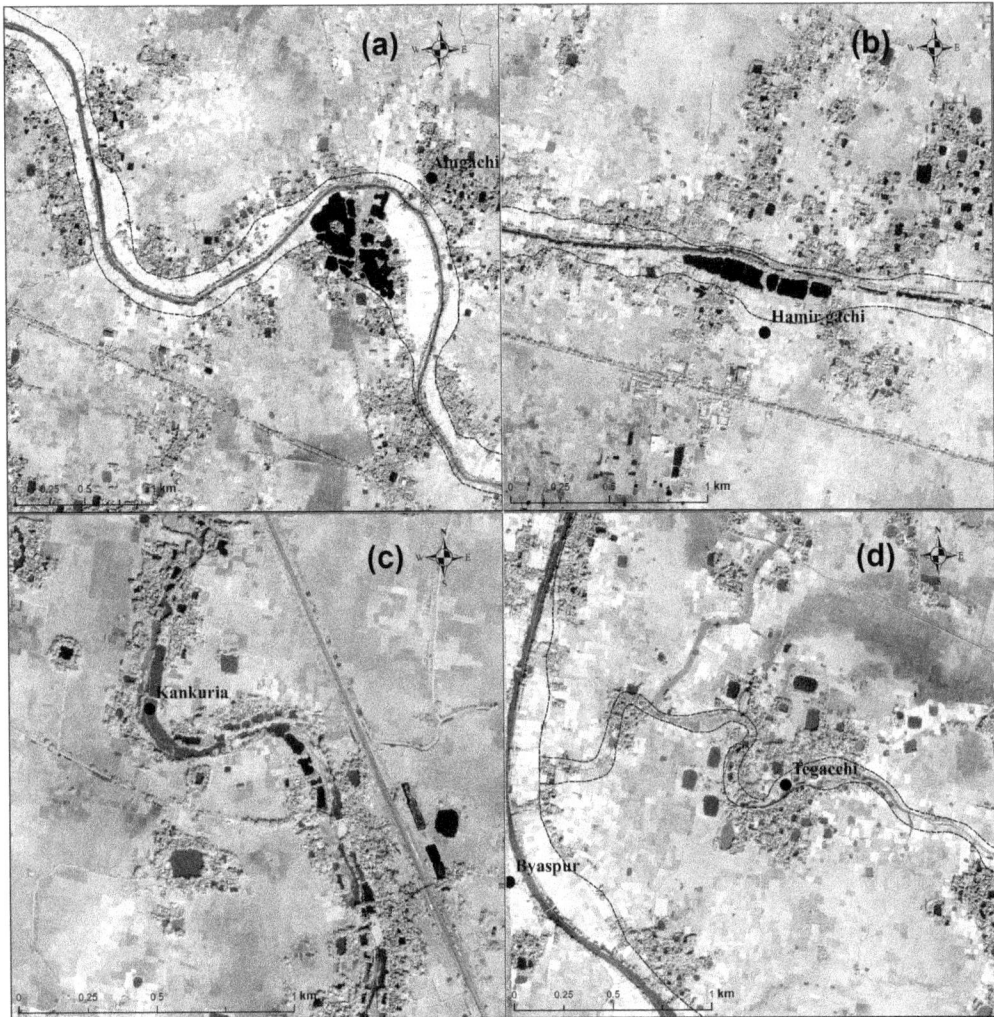

FIGURE 15.5 Spatial windows showing demarcated palaeo-corridors including the palaeochannels and palaeo bank lines:(a) near Amgachhi (22°52′0.79″ N/88°6′40.98″E); (b) near Hamirgachi (22°50′29.58″N/88°8′32.99″E); (c) near Kankuria (22°53′6.46″N/88°10′17.97″ E); (d) near Tegachi (22°56′57.99″N/88°4′29.93″ E). Fragmentation of the palaeochannels and conversion of moist grounds within the corridor into agricultural plots are well evident in these windows.

in the downstream segment beyond its fragmented head flows almost continuously meeting *Kana* River near Payraura (Figure 15.4). Alike *Kana* River, the course of Ghia was also traceable for its continuity downstream till it meets *Kana* near Payraura. Ghia River is also disconnected from the Damodar because of the DVC canal, but the upstream segment of Ghia is difficult to identify as the region is intensively cultivated. A linear string of wetness was traced in this concerned region which has had helped to demarcate the upstream segment of Ghia near Betragarh (Figure 15.4). The palaeocourses have been identified after locating the palaeolevee, higher grounds with settlement and orchards along with a series of ponds that appear in a linear manner. Still, at many places, the galloping encroachment had gulped the levee in order to pass the channel water during early summer for irrigation.

15.3.2 ACCURACY OF THE LULC MAPS

The prepared LULC maps using the feature space plotting method had gained scientifically acceptable accuracy level. The overall accuracy of 2001 classified LULC maps was 98.06%, and Kappa value was 97.34% (see Table 15.4). The overall accuracy of 2019 classified LULC map was 97.68%, and Kappa value was 96.79% (see Table 15.5).

15.3.3 CHANGES IN LULC WITHIN THE BUFFER

The LULC map of the entire sub-system (2019) has depicted that the relatively low lands of the study area lying between two corresponding channels have been engaged in agricultural practices (Figure15.6). Areas lying near the palaeochannels where the usual moisture level is comparatively higher are cultivated more than once a year. The LULC maps of the buffers prepared for 2001 and 2019 (Figure 15.7) have illustrated that the amount of the cultivable fallow increases as the distance from the palaeocourses increases (Table 15.2). Within the delineated buffer, the concentration of settlement has increased by 19.82% over the last 18 years at a rate of 0.56 km^2 year^{-1}. The maximum conversion had occurred from fallow with moisture to settlement by a percentage of 57% of total except for the static settlement patches (Tables 15.6 and 15.7). The moist grounds that were

TABLE 15.4
Kappa Statistics for the LULC Classification of 2001

Classified Data	Water Bodies	Cultivated Land	Vegetation	Cultivable Dry Fallow	Settlement	Fallow with Moisture	Row Total
Water bodies	182	1	0	0	0	3	186
Cultivated field	12	3,555	0	0	0	4	3,571
Vegetation	0	11	563	0	0	0	574
Cultivable dry fallow	0	0	0	2,236	0	13	2,249
Settlement	0	0	0	39	349	42	430
Fallow with moisture	18	0	0	3	32	2,155	2,208
Column total	212	3,567	563	2,278	381	2,217	9,218

Overall accuracy = 98.06%; Kappa value = 97.34%.

TABLE 15.5
Kappa Statistics for the LULC Classification of 2019

Classified Data	Water Bodies	Cultivated Land	Vegetation	Cultivable Dry Fallow	Settlement	Fallow with Moisture	Row Total
Water bodies	216	1	3	0	0	1	221
Cultivated field	4	3,910	3	0	0	1	3,917
Vegetation	2	9	446	0	0	0	447
Cultivable dry fallow	0	1	0	2,236	0	82	2,319
Settlement	8	0	3	0	481	68	560
Fallow with moisture	1	34	0	19	0	2,856	2,910
Column total	231	3,955	475	19	481	3,007	10,404

Overall accuracy = 97.68%; Kappa value = 96.79%.

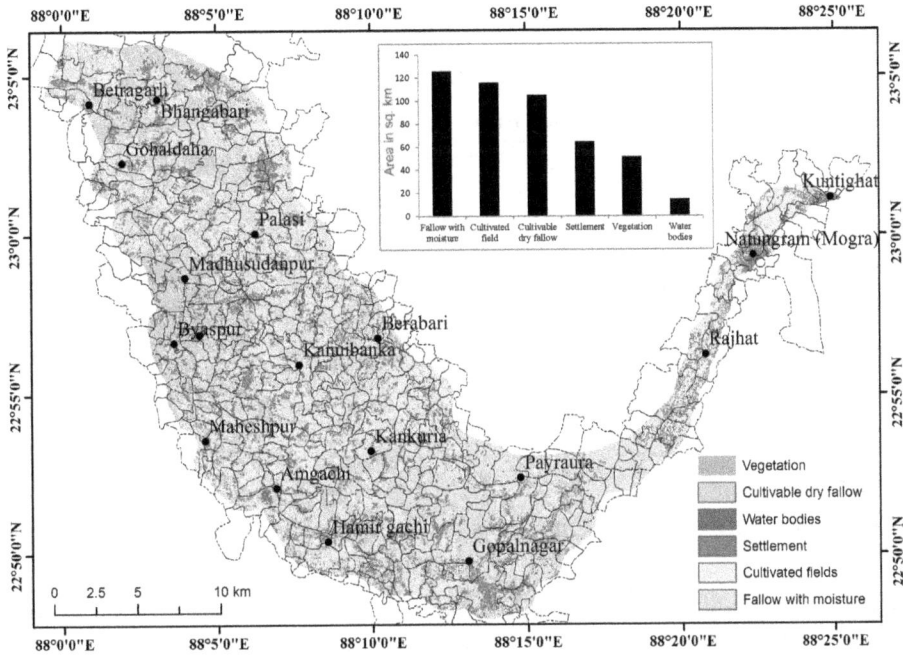

FIGURE 15.6   Prepared LULC map for the entire sub-system of *Kana*–Ghia–Kunti from Landsat 8 dataset of 2019.

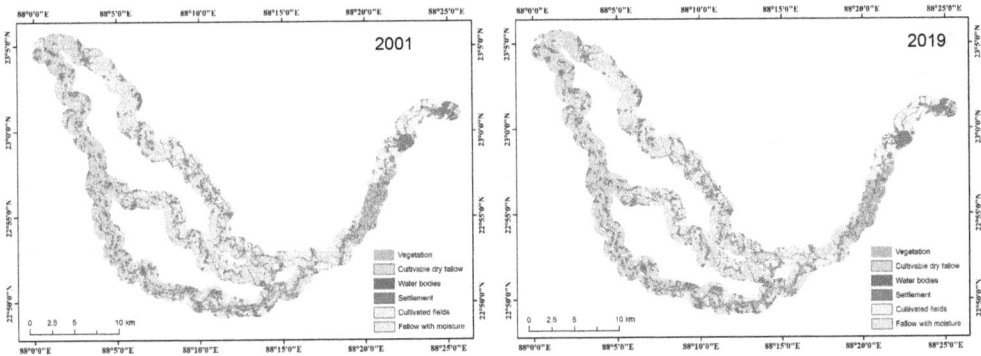

FIGURE 15.7   Prepared LULC of the buffer region of the major palaeochannels of the sub-system for 2001 and 2019.

converted into settlement were mainly associated along the palaeochannels. A sharp decrease in vegetation cover was observed within the concerned time interval; 23.8% of it was converted into a settlement (3.74 km²), and 16.17% was converted into the cultivated field along with 21.52% that was converted into fallow with moisture. A sharp increase in water bodies has also been observed with a percentage of 66.82 at an annual rate of 0.55 km².

Although the channels in 2019 were present with a lesser amount of water, the encroached palaeosegments, as well as the areas just associated along the levee, had been well fragmented into farm ponds. It was mainly done in order to get water for cultivation during early summer because of the decreasing amount of water that is being gushed out by DVC into the major canals. Besides, the increase of settlement within the buffer region another phenomenon that supports galloping encroachment was decreased in fallow with moisture. The decrease was nearly 17%

TABLE 15.6

Changes in the LULC Categories between 2001 and 2019

LULC	Area in km²				
	2001	2019	Change	% Change	Rate of Change
Vegetation	15.17	6.15	−9.02	−146.67	−0.47
Cultivable dry fallow	54.88	56.51	1.63	2.88	0.09
Water bodies	5.18	15.61	10.43	66.82	0.55
Settlement	42.99	53.62	10.63	19.82	0.56
Cultivated fields	77.50	78.28	0.78	1.00	0.04
Fallow with moisture	102.68	88.23	−14.45	−16.38	−0.76
Total	298.40	298.40			

TABLE 15.7

Areal Conversion between LULC Categories between 2001 and 2019

Final State (Area in km²)	Initial State (Area in km²)					
	Vegetation	Cultivable Dry Fallow	Water Bodies	Settlement	Cultivated Fields	Fallow with Moisture
Vegetation	4.16	0.01	0.19	0.63	0.20	0.98
Cultivable dry fallow	0.55	25.49	0.13	0.95	13.11	16.29
Water bodies	0.82	0.72	2.72	0.95	2.10	4.25
Settlement	3.74	3.04	0.29	25.49	4.18	16.19
Cultivated fields	2.54	10.82	0.89	5.70	35.19	23.16
Fallow with moisture	3.38	14.80	0.97	5.24	22.73	41.11

from 2001 to 2019 (Table 15.7). Thirty-two percent of the total area under moist fallow in 2001 was converted into a cultivable fallow and settlement in 2019. The problem associated with land acquisition in Singur during 2008–2009 was partly responsible for a steady conversion of cultivable fallow and cultivated field into fallow with moisture. However, there were no such significant changes in the areas under cultivated field and cultivable dry fallow. Nine out of 19 surveyed places were found with agricultural plots within the palace corridors, and four out of those nine sites were present with agricultural plots being cultivated more than once a year (Table 15.2).

15.3.4   Spatial Clustering of the LULC Attributes

The LISA analysis of fractal dimension (FD) of each LULC attributes of both 2001 and 2019 has generated spatial cluster maps of the corresponding categories (Figures 15.8 and 15.9). The lower the FD, the higher the clustering. The changes in cluster maps of each corresponding LULC categories have also reflected the nature of the associated changes in areal occupancy of the LULC categories during 2001 and 2019. The low-low cluster reflects higher incidences of spatial agglomeration, whereas the high-high clusters reflect incidences of the highest fragmentation of the patches. Vegetation, cultivable dry fallow and fallow with moisture had experienced a decrease in low-low clustering, whereas the other three categories had experienced an increase in low-low clustering (Table 15.8). The increasing fragmentation of the cultivable fallow was possibly due to excavation of ponds and growth of settlement, whereas the fragmentation of the moist fallow was due to transformation of it into cultivable land units.

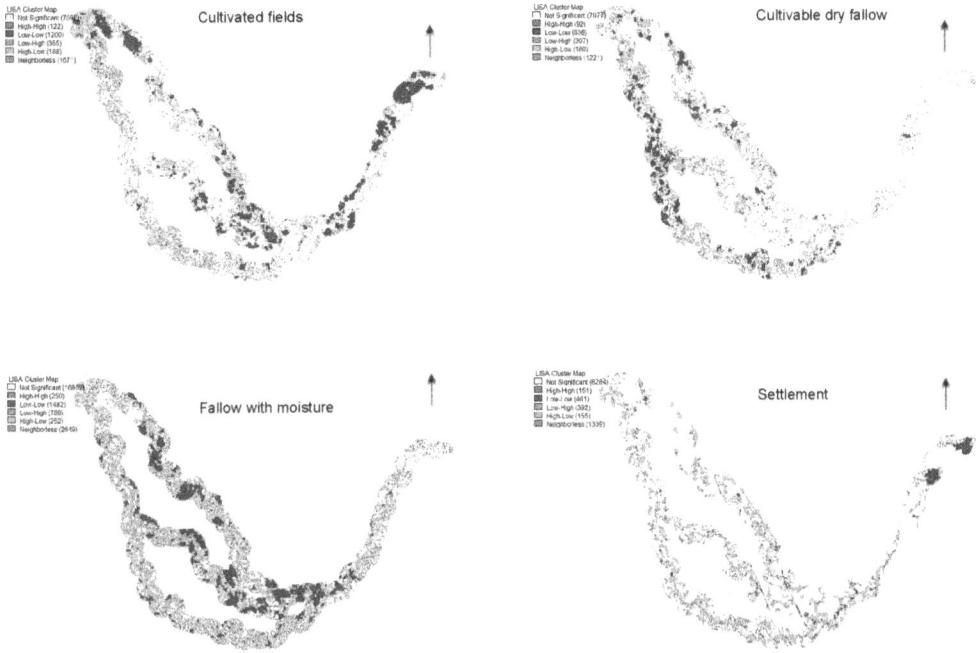

FIGURE 15.8 LISA cluster for the four LULC categories in 2001. The four LULC categories were considered from the theoretical framework considered to bring out the nature of human encroachment.

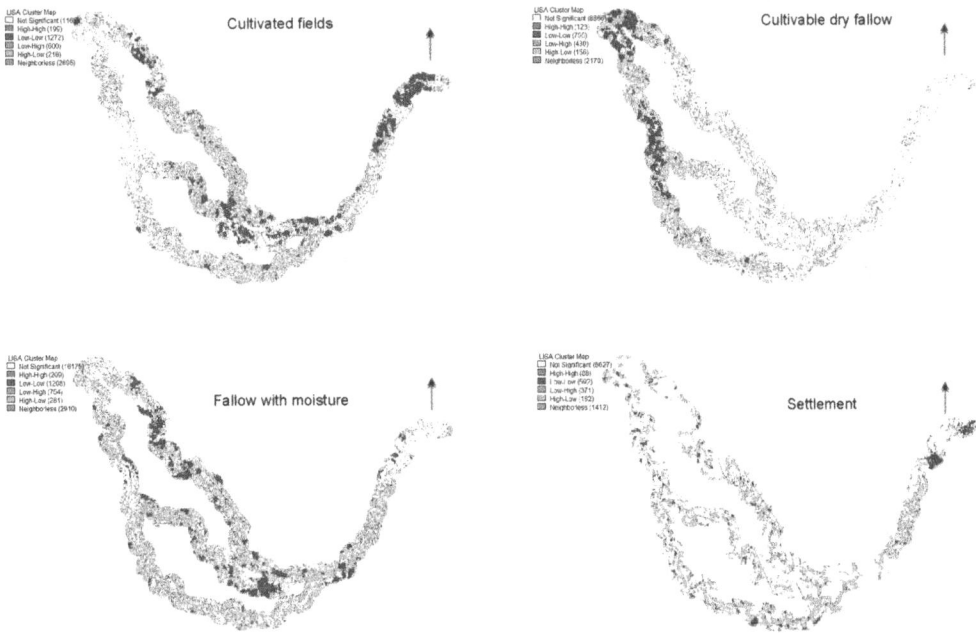

FIGURE 15.9 LISA cluster for the four LULC categories in 2019. The four LULC categories were considered from the theoretical framework considered to bring out the nature of human encroachment.

TABLE 15.8

Spatial Clustering of Fractal Dimension (FD) of LULC Attributes during 2001 and 2019 Using LISA

LULC	Low-Low Cluster		Low-High Cluster		High-High Cluster		High-Low Cluster		Sum	
	2001	2019	2001	2019	2001	2019	2001	2019	2001	2019
Vegetation	271	113	149	91	43	21	36	23	499	248
Cultivable dry fallow	838	755	307	430	92	123	160	156	1,397	1,464
Water bodies	59	272	89	402	26	89	26	114	200	877
Settlement	461	592	392	371	151	88	155	182	1,159	1,233
Cultivated fields	1,206	1,272	355	600	122	199	188	218	1,871	2,289
Fallow with moisture	1,482	1,208	789	754	250	209	252	281	2,773	2,452

The Moran's I value for vegetation has reflected a negative auto-correlation due to the higher fragmentation of it. The dynamicity of the area underwater bodies was quite significant as all kinds of clustering features of it have shown a significant rise during 2019 from that of in 2001 (Table 15.9). Even the number of patches significant at 95% level of confidence was also characteristic of a significant increase between 2001 and 2019, but it was negatively auto-correlated in 2019 (Table 15.10). It was possibly because of the agglomeration of small patches reflecting individual farm ponds positioned at spatially distant locations from each other. With time, the palaeochannels had lost its flow efficiency as well as continuation due to further fragmentation into several parts. Stretches present with water in 2001 had been converted mostly into moist fallow. In the case of the settlement, there was a sharp increase in low-low clustering as well as high-low clustering which is evident in a higher degree of spatial agglomeration. The significance of clustering has also increased for settlement at 0.05 and 0.01 level of significance (Figures 15.10 and 15.11). Although in both the years it was positively correlated in terms of its spatial distribution (Table 15.10), the coefficient value had been lifted up due to less variability in spatial association in 2019 compared to 2001.

TABLE 15.9

Significance Testing of Spatial Clustering of FD

LULC	Not Significant		P = 0.05		P = 0.01		P = 0.001		Sum	
	2001	2019	2001	2019	2001	2019	2001	2019	2001	2019
Vegetation	2,155	1,321	190	109	173	52	136	87	499	248
Cultivable dry fallow	7,077	8,866	746	686	453	372	198	406	1,397	1,464
Water bodies	1,599	5,955	94	458	40	175	66	244	200	877
Settlement	8,284	8,627	683	724	257	304	219	205	1,159	1,233
Cultivated fields	7,598	11,658	751	896	537	658	583	735	1,871	2,289
Fallow with moisture	16,863	16,175	1,227	1,234	755	615	791	603	2,773	2,452

TABLE 15.10

Global Auto-Correlation Value (Moran's I) for 2001 and 2019 for LULC Attributes

Moran's I	Vegetation	Cultivable Dry Fallow	Water Bodies	Settlement	Cultivated Fields	Fallow with Moisture
2001	0.014	0.024	−0.001	0.004	0.058	0.016
2019	−0.048	−0.002	−0.047	0.012	0.029	−0.013

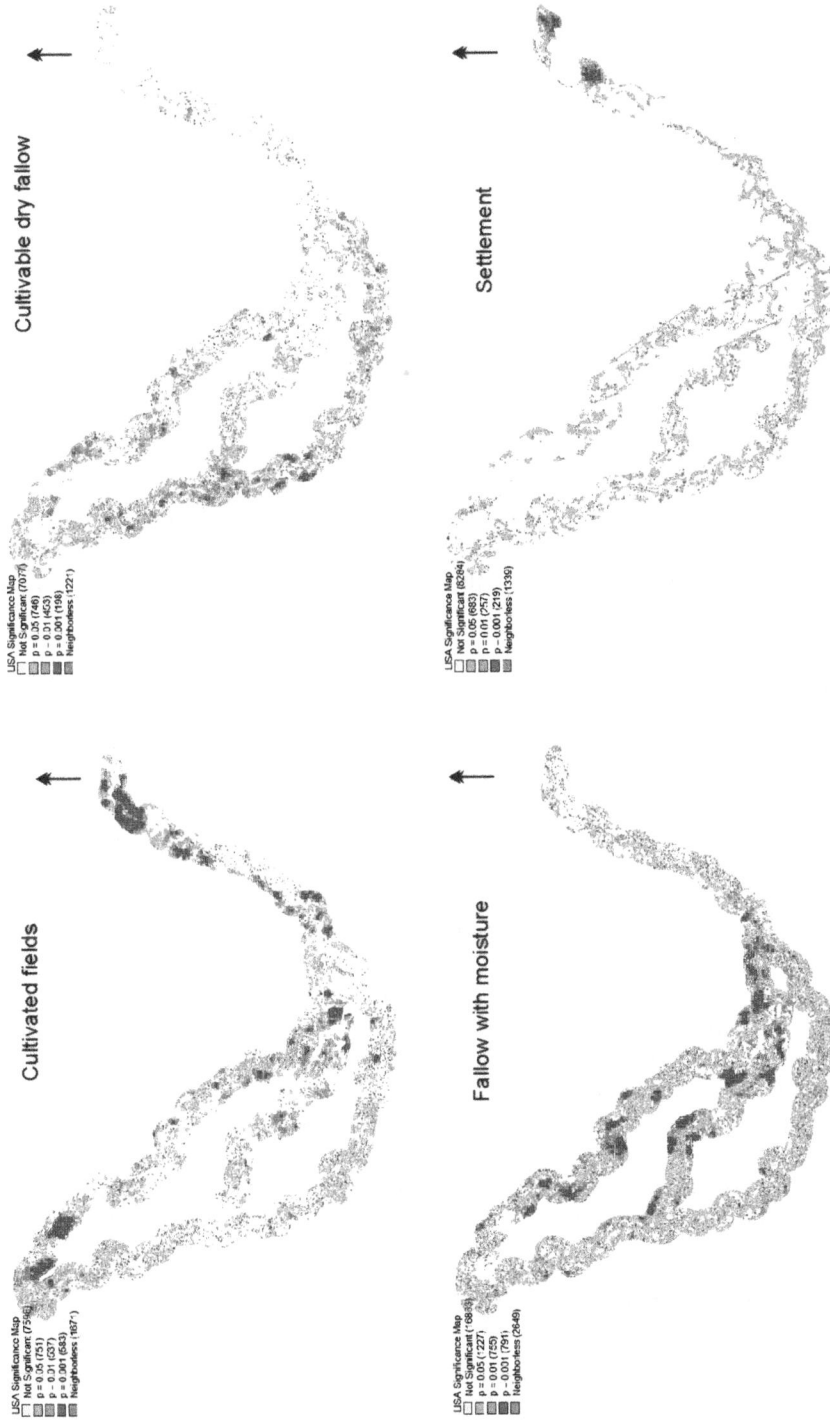

FIGURE 15.10 Statistical significance testing of the clusters analysis of the four considered LULC categories of 2001.

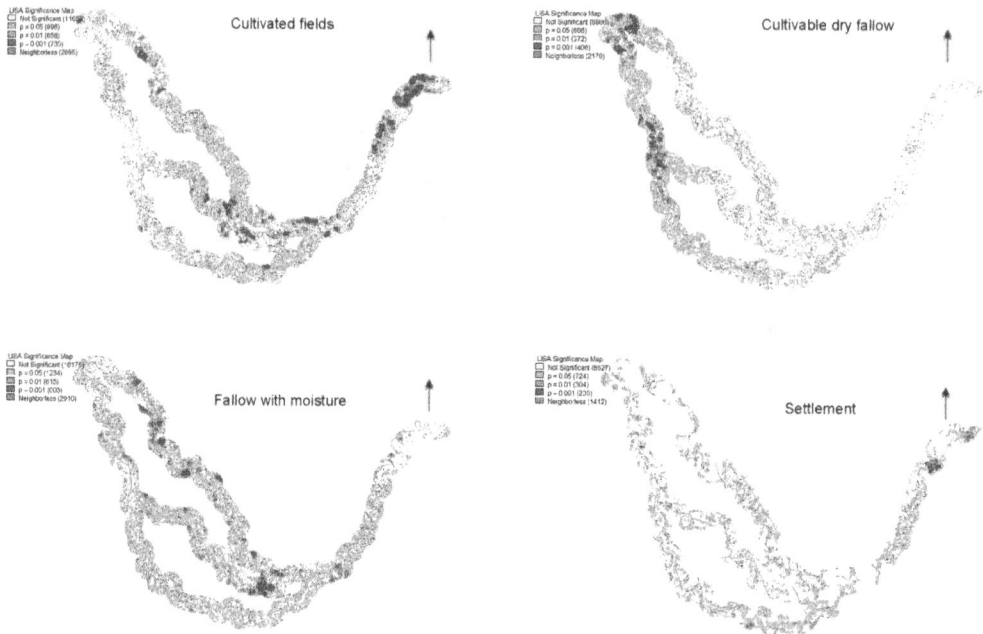

FIGURE 15.11 Statistical significance testing of the clusters analysis of the four considered LULC categories of 2019.

15.3.5 Nature of Channel Impoundments

Encroachment through clogging the palaeochannels after converting it into agricultural plots or ponds has been found abundantly within the studied segment. In this study, fragmenting the corridor into agricultural plots, ponds and construction of bridge piers has been considered as the means of invading the corridor (Figure 15.12). These locations were marked using OrbView-3images and Google Earth. Fifty-eight locations were found with linear strings of the ponds within the corridor in 2019. It has been averaged to 6.58 km per location of the ponds within the corridor for *Kana* River, 2.71 km for Ghia River, 4.83 km for Kunti River and 1.25 km for Jhimki. Besides the avulsion off-take of Jhimki from *Kana*, a total of 23 locations within the corridor, either the channel or within the floodplain of Jhimki, have been found to be fragmented into ponds. A total of 129 locations were found within the corridors of all four channels where agriculture is being practiced at least once a year. The highest occupancy of agricultural plots was found for the Jhimki River with 39 locations placed on an average distance of 0.76 km (Table 15.11). One hundred and thirteen bridges and culverts were found within the study area constructed across the palaeochannels. And the highest amount was on the course of *Kana* with 47 bridges constructed across the channel. This leads to a reduction in channel width and changes of floodplain topography, where agricultural terraces were introduced and ponding effects at the immediate downstream of the bridge piers were observed as the resultant effects of channel impoundments.

15.3.6 Channel Geomorphic Characteristics

Reduction in the width of the flow passage and the nature of confinement of the present channel within the palaeo-corridor has been considered to study the present geomorphic scenario of the palaeochannels. Width ratio of the active passageway to traceable corridor width and width contraction by the bridge piers as well as culverts were calculated from Google Earth and was also verified during the field survey. The respective widths were calculated at a 2 km interval starting from the off-takes. The present channel width ratio to the palaeo-corridor width as well as the relative variation of it was

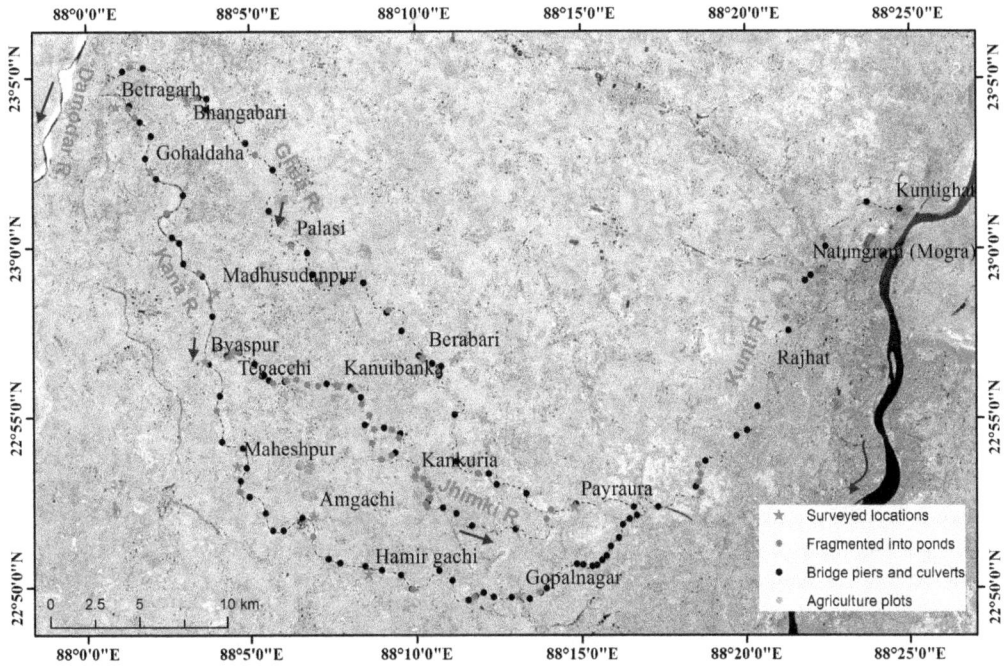

FIGURE 15.12 Occurrences of direct impoundments within the palaeo-corridor; three major phenomenon were recorded; fragmentation of the palaeochannels into ponds, construction of bridges and culverts on the channel and transforming palaeo-corridor into agricultural plots.

highest for *Kana* River. Kana River has registered the ratio value of 8.82 followed by Ghia (6.50), Jhimki (6.07) and Kunti (5.55) (Table 15.3). *Kana* had been a major passageway of Damodar River for some time between the 16th and 18th centuries (Bandyopadhyay, 2007a; Bhattacharyya, 2011; Ghosh, 2017). It was the reason that Kana was found with 274.33 m of average width of the palaeo-corridor, highest among the other concerned palaeochannels. Among the four palaeochannels, maximum width reduction of the present passage way was found for Kana with a value of 36.80 followed by Jhimki with 17.01, Ghia with 15.97 and Kunti with 10.35 (Table 15.12). While characterising stability in terms of width reduction between present passageway and palaeo-corridor, Kana has been found least stable whereas Kunti has registered the lowest variation in width reduction due to the modifications of the channels along its entire stretch. Kunti was the only channel with continuous flow present almost round the year. Either side of its bank has not been found with settlement congestion except at Magra and near its outfall in Bhagirathi River.

TABLE 15.11

Reduction in Channel Width Downstream due to Culverts and Bridge Piers Construction

Channel	Number of Culvert and Bridges			Locations with Farm Ponds within the Corridor			Culvert or Bridge Per Unit Length (km)			Reduction in Width at Culvert or Bridge (m) in 2019		
	1959	2000	2019	1959	2000	2019	1959	2000	2019	Max	Min	Avg
Kana	29	46	50	6	21	22	2.04	1.28	1.18	62.1	0.9	17.94
Ghia	10	23	27	3	17	17	5.15	2.23	1.83	27.38	0.81	8.47
Kunti	8	12	15	3	7	8	3.63	2.42	1.94	23.09	0	7.04
Jhimki	7	18	21	6	15	26	4.28	1.66	1.5	16.93	1	7.35
Total	54	99	113	18	60	73						

TABLE 15.12

Width of the Wetted Perimeter and Channel Width and Corridor Width Ratio for the Major Palaeochannels in 2019

Parameters	Active Channel Width (m)				Present Channel and Palaeo Corridor Width Ratio			
	Kana R.	Jhimki R.	Ghia R.	Kunti R.	Kana R.	Jhimki R.	Ghia R.	Kunti R.
Max	43.3	22.6	29.7	42.3	36.80	17.01	15.97	10.35
Min	13	7.11	2.46	6.86	3.20	1.89	2.93	1.42
Avg	26.31	14.58	13.26	17.27	8.82	6.07	6.50	5.55
SD	8.59	5.60	5.74	8.78	7.10	4.04	3.75	2.37
CV	32.67	38.42	43.34	50.87	80.57	66.50	57.67	42.73

Construction of bridges and culverts at 113 locations across the palaeochannels has rigorously reduced active flow passage for the channels. The average width reduction for *Kana* River was 17.94 m; for Jhimki, it was 8.47 m; for Ghia, it was 7.04 m; and for Kunti, it was 7.35 m (Figure 15.13). Compared to the average width of the corresponding channels, the average reduction in width due to bridge and culvert construction across the active flow was found highest for *Kana* River followed by Jhimki River, Ghia River and Kunti River with corresponding values of 68%, 58%, 53% and 42%, respectively (see Table 15.11).

15.3.7 Discussion

The close integration of the fluvial landscape with the increasing anthropogenic burden has transformed the fluvial landscape into a constricted and controlled system to act accordingly. The facets of anthropogenic intervention have been analysed only at an interval of 18 years which has provided the recent most picture of it. The area that has been studied has emerged during the quaternary system where the history of growing human society on it is thousands of years old. Thus, the encroachment within the fluvial corridor has been a continuous process carried out through hundreds of decades.

The linear stretches of the levee have been thoroughly utilised for the growth of settlement, while the low-lying floodplains have been engaged in cultivation (Figures 15.14a and 15.15). Monsoonal rainfall in the floodplain is being used for cultivation and with increasing distance from the palaeochannels frequency of cultivation gets reduced. The moisture content of the palaeocourses and its surrounding areas is being utilised for the cultivation of winter crops. During early summer, water supplied by the DVC through its irrigation canals and water present in the palaeocourses is utilised for irrigation in the areas lying along with the palaeocourses and canals. Impoundments of the palaeocourses were quite significant in this region. It has helped to aggravate the decaying process of the palaeochannels.

Entwistle et al. (2019) have discussed the how did floodplain functioning decrease due to intensive utilisation of it through agricultural practices in England. The increasing intensity of the human encroachment had successfully reduced the geomorphic efficiency of the floodplains as well as the channel in England, where both the natural units were used to create a geomorphic as well as hydrological symphony between each other in terms of efficiency and functioning (Macklin et al., 2010). The present study had also revealed that the different means of human encroachment that had confined the flow continuity within the palaeocourses has also reduced the channel efficiency at a rapid rate. Clark and Wilcock (2000), Stammel et al. (2012, 2018) and Yousefi et al. (2017) had shown the rigorous impact of floodplain land use on channel efficiency where the conversion of the natural tract had shrunk the riparian vegetation and also made the channel confined within the corridor.

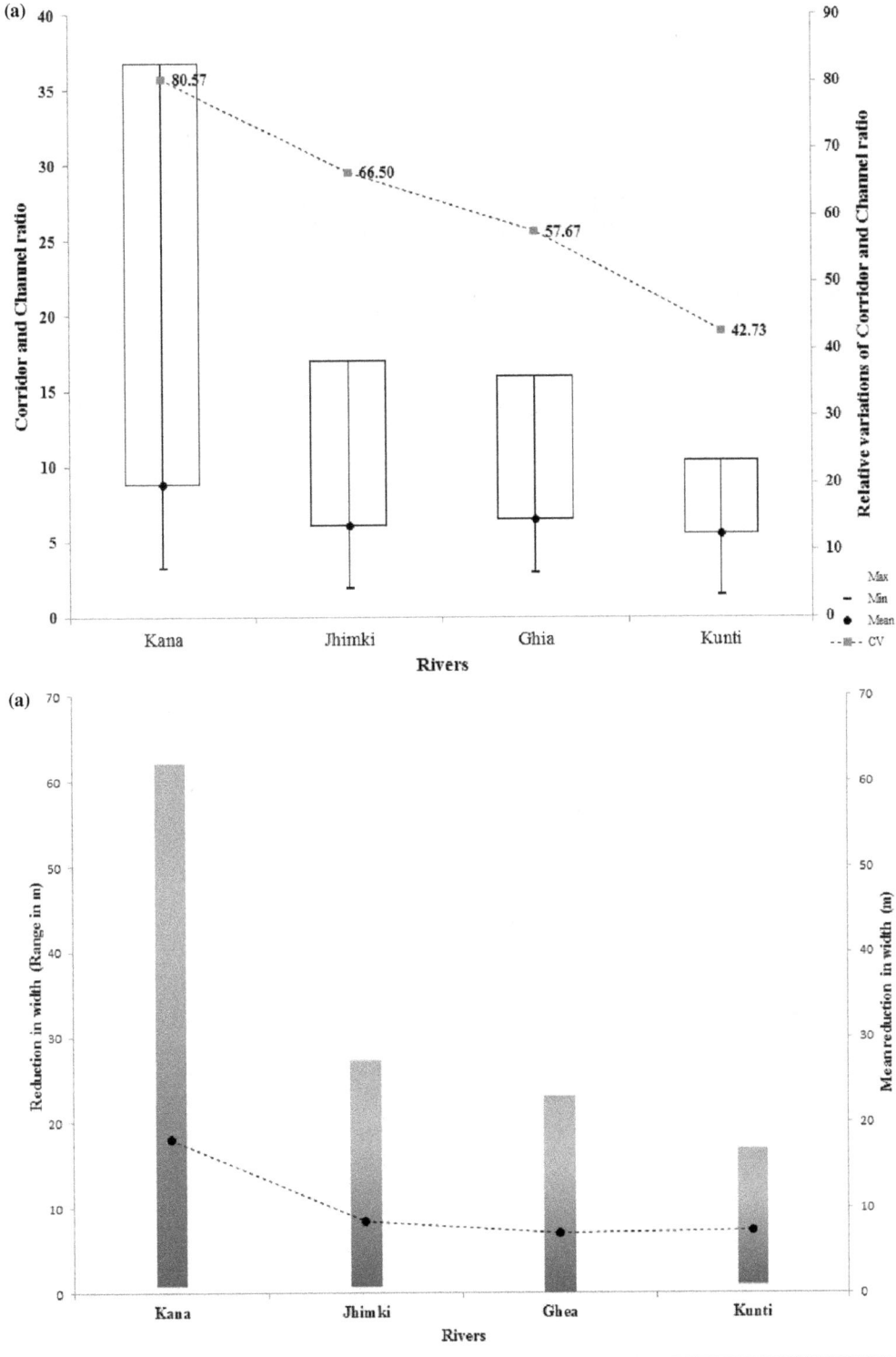

FIGURE 15.13   The human encroachment has a resultant effect on the changing geometric character of the palaeochannels. (a) Ratio of corridor width and existing channel width. (b) Reduction in channel width immediately downstream of the bridges and culverts.

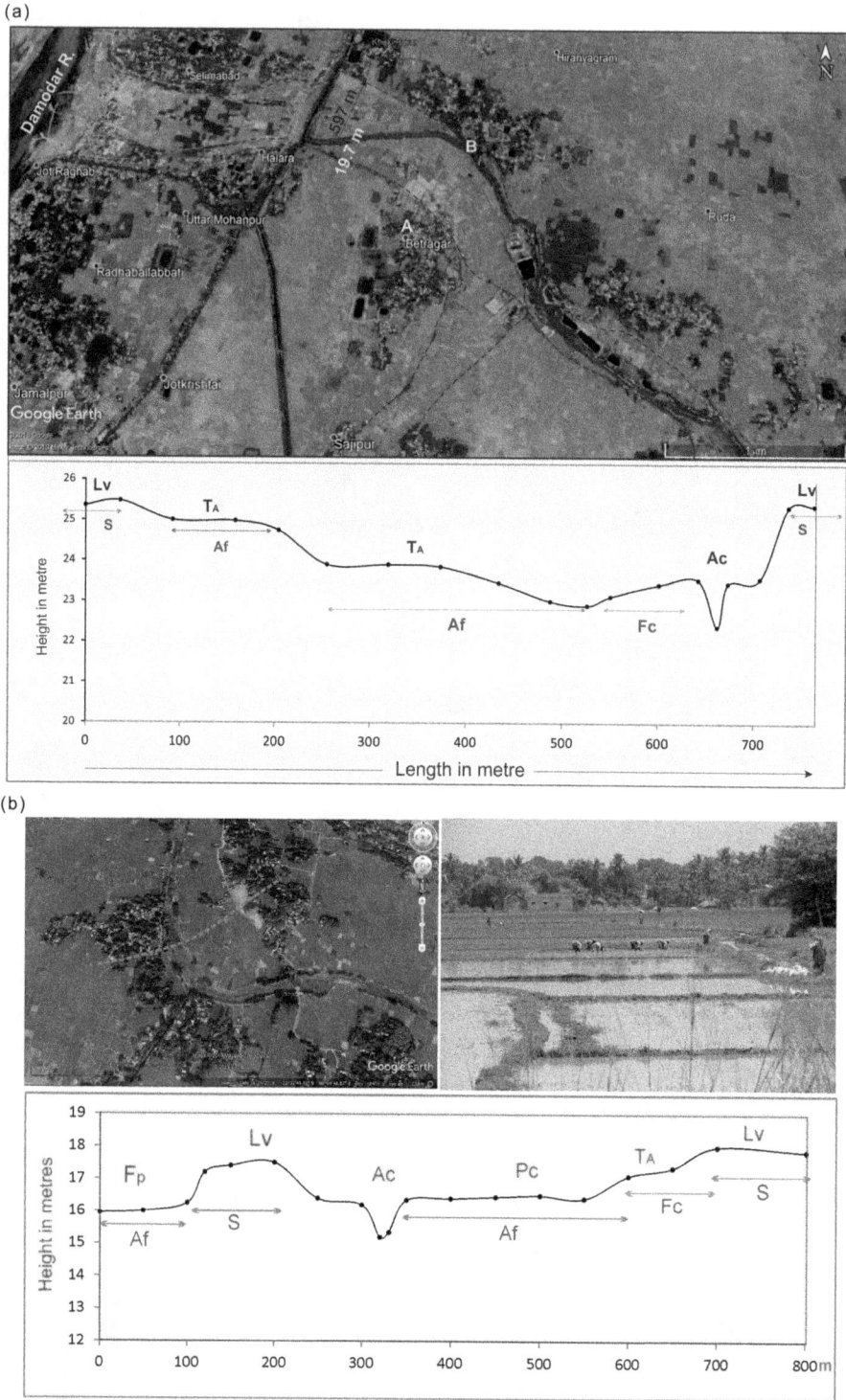

FIGURE 15.14   (a) Utilisation of palaeo-corridors near the head of *Kana* River. Humans have not only encroached the fluvial regime; it has altered the natural association of the landscape, i.e., agricultural terraces within the floodplain., (b) Utilisation of palaeochannel of *Kana* near Maheshpur. Cross section extended from 22°52′43.69″N/88°4′31.61″E on the left to 22°52′56.82″ N/88°4′59.24″E.

FIGURE 15.15 Utilisation of palaeo-corridors of *Kana* River near Hamirgachi. (a) Diagonal view of the stretch and along its bottom to top the cross section was prepared. (b) Relict condition of the palaeochannel of Ghia near Kankuria. (c) Agricultural practices during the winter on the plots situated near the palaeochannels. (d) Palaeochannel fragmented into pond. (e) Restricted course of the palaeochannel into a shallow canal. (f) Converted stretch of Ghia into canal. (g) Dried up beds due to unavailability of water that used to be supplied from DVP canal.

The increasing population pressure (Figure 15.16) of occupying the levee and its surroundings to avoid the menace of floods had often seemed to reduce the natural vegetation and the riparian tract within the palaeo-corridors of *Kana*–Ghia and Kunti sub-system (Figures 15.14a and 15.15). The development of roadways and railways across the channels had been accompanied by the construction of Bridges and culverts which has resulted in consequent stream impoundments leading to an incorrigible amount of flow congestion along with the reduction of the channel width. The fragmentation of palaeocourses into ponds or aquaculture sites had developed continuous breaks, thus resulting in a sluggish amount of flow within the palaeocourses as evidenced by this work.

Apart from the decreasing efficiency of the physical unit, the encroachment has also created other geomorphic problems which have resulted into social menace (see Table 15.13). Waterlogged condition both in the channel and nearby floodplain after a prolonged rainfall tolling the agricultural loss each year became a common phenomenon. Entrapment of silt within the channel and unavailable irrigational water is also making farmers to concentrate only on Kharif cropping if they don't have plots near to the channels.

In order to avoid further encroachment and decaying, the channels of *Kana*–Ghia and Kunti are in a grave need for proper restoration planning and its implementation. Rudra (2010) has discussed that the dwellers living in this area, who become a victim of inundation and waterlogging in the monsoon and late monsoon, and the farmers who face the loss of agricultural production in the dry pre-monsoon due to lack of available irrigation, feel the need of rejuvenation of these channels. Due to less availability of water coupled with the increasing irregularity of monsoon, agriculture in these areas depends on groundwater to a large extent, which is also evidential by the interviews of the farmers. Serious extraction of groundwater has dragged most of the CD blocks composing the study area to medium groundwater level zone and among them, marked Polba, Chinsura and Singur block as "semi-critical" in terms of groundwater availability (Patra et al., 2015). The works of Chen et al. (1996), Magee (2009), Wang et al. (2012) and Roy and Sahu (2015) have indicated that palaeochannels often appear as either a way of groundwater recharge or a favourable zone of shallow aquifer due to the presence of unconsolidated materials. Restoration of the channels will not only rejuvenate the rivers but also find a way to recharging the groundwater. Acharyya and Shah (2006), Majumder and Shivaramakrishnan (2014) and Ghosh (2017) found an impermeable clay layer as bed material in most of the palaeochannels. We also found that increasing concentration of shrubs and practicing agriculture is trapping finer sediments and reducing flow efficiency. Restoration of the channels will deepen the channel for relatively regular flow and also opens a way of semi-artificial groundwater recharge as it will expose the layer of permeable quaternary clayey sand and sand exists lying beneath.

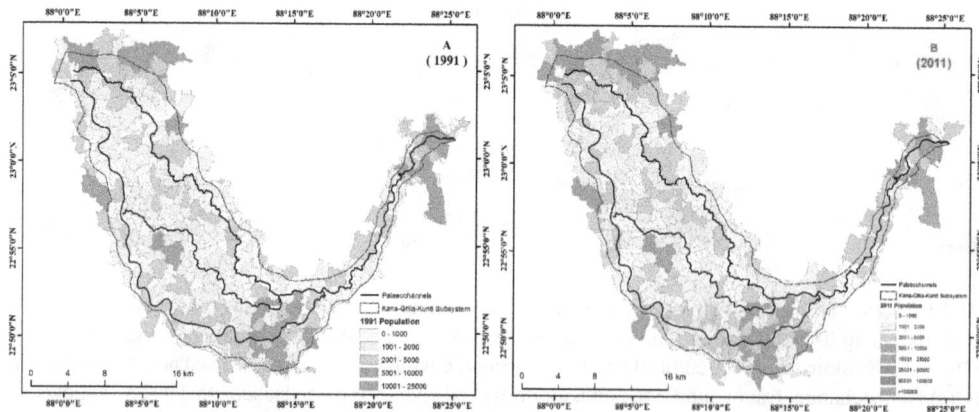

FIGURE 15.16 Population concentrations in 1991 and 2011. The number of increase in population has been accommodated on the relatively higher grounds of the palaeolevee.

TABLE 15.13

Problems Associated with Decaying of Palaeochannels

Locations	Channel Waterlogging during Monsoon	Flood	Stagnancy of Flood Water into Nearby Floodplain	Growth of Marshes within the Channel	Availability of Irrigation Water during Lean Period	Loss of Fertility	Obstructed Channel Flow	Dry Bed during Pre-Monsoon
Betragarh		√			√	√		
Gohaldaha		√			√			
Madhusudanpur		√	√		√	√		
Byaspur	√	√	√		√	√		
Maheshpur	√	√	√	√	√			
Amgachi	√	√	√	√			√	
Hamirgachi	√	√	√	√		√	√	√
Choalpara (Nalikul)	√	√	√	√			√	√
Gopalnagar	√	√	√	√		√	√	√
Tegachi	√	√	√	√		√	√	√
Kanuibanka				√		√	√	√
Kankuria				√		√	√	√
Bhangabari		√					√	
Berabari		√	√	√		√	√	
Palasi		√	√					
Payraura	√	√			√			
Rajhat				√	√		√	
Notungram (Mogra)					√			
Kuntighat					√			

15.4 CONCLUSION

The ongoing transformation of the palaeo-fluvial corridor gave rise to a conflict between human society and the geomorphic processes. It has been evident that the process of channel decay has shot up way faster than the natural rate. The major findings of this study have portrayed how did the human manifestation of its modes of usage of the palaeo-corridor has helped the fluvial system to choke faster and how well the geomorphic benefits of the palaeo-fluvial regime they have incorporated into their forms of land utilisation.

i. The higher grounds, mainly the levee, have been utilised for settling the inhabitants in order to prevent flood damage. On the other hand, the lower grounds have been utilised for cultivation as these grounds can hold rainwater available for a longer period of time. A decrease in spatial clustering of agricultural fallow along the channel was possibly due to increased settlement.

ii. With increasing distance from the palaeocourses, the frequency of cultivation decreases. Agricultural fields present closest to the palaeochannels have been cultivated more than once a year. The Rabi crops, mostly potatoes, and few vegetables were cultivated using the ground moisture present during the winter in the lands associated close to the channels.

Means of direct impoundments in the forms of culverts and bridges along with the agricultural practices within the channels had played a great role in choking the flow condition. As a resultant effect, it has started trapping sediments due to increasing channel resistance. The increasing nature of spatial clustering has also proved the characteristics of the ongoing impoundments and fragmentation of palaeochannels. The connection of these major palaeocourses with the trunk channel had been abolished by constructing irrigation canals. Presently, the courses of *Kana* and Ghia are used as irrigation channels which were currently found with shallow flow passage and filled with bushes and field crops at certain locations. Installing culverts, agricultural practices within the channel have fragmented the courses into several parts. It has resulted into ponding effect that has decreased the flow efficiency of the courses carrying runoff water during the monsoon.

REFERENCES

Acharyya, S. K., & Shah, B. A. (2006). Arsenic-contaminated groundwater from parts of Damodar fan-delta and west of Bhagirathi River, West Bengal, India: influence of fluvial geomorphology and Quaternary morphostratigraphy. *Environmental Geology, 52*, 489–501.

Alam, A. (2015). Geographical factors in cultural aspects of early Bengal (5th to 13th centuries CE). *Journal of Bengal Art, 20*, 309–322.

Anselin, L. (1995). Local indicators of spatial association – LISA. *Geographical Analysis, 27*, 93–115.

Bagchi, K. (1944). *The Ganges Delta.* Calcutta: University of Calcutta.

Bandyopadhyay, D. K. (2007a). *Banglar Nad-Nadi (Rivers of Bengal).* Kolkata: Dey's Publishing.

Bandyopadhyay, S. (1996). Location of the Adi Ganga Palaeochannel, South 23-Parganas, West Bengal: a review. *Geographical Review of India, 58* (2), 93–109.

Bandyopadhyay, S. (2007b). Evolution of Ganga Brahmaputra Delta: a review. *Geographical Review of India, 69* (3), 235–268.

Bhattacharyya, K. (2011). *The Lower Damodar River, India: Understanding the Human Role in Changing Fluvial Environment.* New York: Springer.

Borisova, O., Sidorchuk, A., & Panin, A. (2006). Palaeohydrology of the Seim River basin, Mid-Russian Upland, based on palaeochannel morphology and palynological data. *Catena, 66*, 53–73.

Brown, A. G. (2002). Learning from the past: palaeohydrology and palaeoecology. *Freshwater Biology, 47*, 817–829.

Castillo, K. C., Korbl, B., Stewart, A., Gonzalez, J. F., & Ponce, F. (2011). Application of spatial analysis to the examination of dengue fever in Guayaquil, Ecuador. *Procedia Environmental Sciences, 7* (2011), 188–193.

Chander, G., & Markham, B. (2003). Revised Landsat-5 TM radiometric calibration procedures and postcalibration dynamic ranges. *IEEE Transactions on Geoscience and Remote Sensing, 41* (11), 2674–2677.

Chandra, S. (2003). *India: Flood Management – Damodar River Basin.* New Delhi: WMO/GWP Associated Programme on Flood Management.

Charlton, R. (2008). *Fundamentals of Fluvial Geomorphology.* New York: Routledge.

Chattopadhyay, S., & Roy Chowdhury, A. (2017). Southern Bengal Delta –ahub of ancient civilization and cultural assimilation: a case study of Chandraketugarh and allied sites. *Heritage: Journal of Multidisciplinary Studies in Archaeology, 5*, 283–299.

Chen, W., Xuanqing, Z., Naihua, H., & Yonghong, M. (1996) Compiling the map of shallow-buried palaeochannels on the North China Plain. *Geomorphology, 18*, 47–52.

Choudhury, J. (1990). *Vardhaman: Itihash O Samskriti (The history and culture of Barddhaman district, West Bengal)* (Vol. 1). Calcutta: Pustak Bipani.

Clark, J. J., & Wilcock, P. R. (2000). Effects of land-use change on channel morphology in northeastern Puerto Rico. *Geological Society of America Bulletin, 112* (12), 1763–1777.

Debnath, J., Das, N., Ahmed, I., & Bhowmik, M. (2017). Channel migration and its impact on land use/land cover using RS and GIS: a study on Khowai River of Tripura, North-East India. *The Egyptian Journal of Remote Sensing and Space Sciences, 20*, 197–210.

Deng, J. S., Wang, K., Hong, Y., & Qui, J. G. (2009). Spatio-temporal dynamics and evolution of land use change and landscape pattern in response to rapid urbanization. *Landscape and Urban Planning, 92*, 187–198.

Entwistle, N. S., Heritage, G. L., Schofield, L. A., & Williamson, R. J. (2019). Recent changes to floodplain character and functionality in England. *Catena, 174*, 490–498.

Fischer, P., & Cyffka, B. (2014). Floodplain Restoration on the Upper Danubee by Re-establishing back water dynamics: first results of the hydrological monitoring. *Erdkunde, 68* (1), 3–18.

Ghosh, P. K., & Jana, N. C. (2019). Historical evidences in the identification of palaeochannels of Damodar River in Western Ganga-Brahmaputra Delta. *Quaternary Geomorphology in India, Geography of the Physical Environment*, 127–138.

Ghosh, S. (2017). Palaeochannels as groundwater repository: a sustainable way to manage water resource in Damodar Fan Delta, West Bengal. *Problems and Sustainability of Surface and Groundwater Resources in Deltaic West Bengal*, 95–119.

Ghosh, S., & Islam, A. (2016). Quaternary alluvial stratigraphy and palaeoclimatic reconstruction in the Damodar River Basin of West Bengal. *Neo-Thinking on Ganges-Brahmaputra Basin Geomorphology*, 1–18.

Ghosh, S., & Mistri, B. (2012). Reconstructing the phases of channel shifting through identification of palaeochannels and historical accounts of extreme floods of Damodar River in West Bengal. *Indian Journal of Geomorphology, 17* (2), 65–80.

Grabowski, R. C., Surian, N., & Gurnell, A. M. (2014). Characterizing geomorphological change to support sustainable river restoration and management. *WIREs Water, 1*, 483–512.

Hunter, W. W. (1876). *A Statistical Account of Bengal* (Vols. III-IV). London, UK: Trubner & Co.

Jacobson, P. J., Jacobson, K. M., & Seely, M. K. (1995). *Ephemeral Rivers and their Catchments, Sustaining People and Development in Western Namibia*. Windhoek, Namibia: Desert Research Foundation of Namibia.

James, L. A., & Lecce, S. A. (2013). Impacts of land-use and land-cover change on river systems. In J. F. Shroder(Ed.), *Treatise on Geomorphology* (Vol. 9, pp. 768–793). San Diago: Academic Press.

Kondolf, G. M. (1997). Hungry water: effects of dams and gravel mining on river channels. *Environmental management, 21*, 533–551.

Fluvial Systems: Scientific Approaches, Analyses, and Tools (Vol. 194, pp. 29–43). Barkeley, CA: American Geophysical Union, paet of American Geophysical Union.

Lahiri-Dutt, K. (2003). People, power and rivers: experiences from the Damodar River, India. *Water Nepal, 9* (1), 251–267.

Macklin, M. G., Jones, A. F., & Lewin, J. (2010). River response to rapid Holocene environmental change: evidence and explanation in British catchments. *Quaternary Science Reviews, 29*, 1555–1576.

Magee, J. (2009). *Palaeovalley Groundwater Resources in Arid and Semi-Arid Australia - A Literature Review*. Canberra: Geoscience Australia, National Water Commission, Australian Government.

Mahata, H. K., & Maiti, R. (2019). Evolution of Damodar fan delta in the Western Bengal Basin, West Bengal. *Journal of the Geological Society of India, 93* (6), 645–656.

Majumdar, S. C. (1941). *River of The Bengal Delta*. Kolkata: Bengal Government Press.

Majumder, A., & Shivaramakrishnan, L. (2014). Groundwater budgeting in alluvial Damodar fan delta: a study in semi-critical Pandua block of West Bengal. *International Journal of Geology, Earth & Environmental Sciences, 3* (4), 23–37.

Mallick, S., & Niyogi, D. (1972). Application of geomorphology in groundwater prospecting in the alluvial plains around Burdwan, West Bengal. *India. Geohydrology*, 8, 86–98.

Mukherjee, R. (2008). *The Changing Face of Bengal: A Study in Riverine Economy*. Kolkata: University of Calcutta.

Nandini, C. V., Sanjeevi, S., & Bhaskar, A. S. (2013). An integrated approach to map certain palaeochannels of South India using remote sensing, geophysics and sedimentological techniques. *International Journal of Remote Sensing, 34* (19), 6507–6528.

Nemec, W., & Steel, R. J. (1988). What is a fan delta and how do we recognize it? *Fan Deltas: Sedimentology and Tectonic Settings, Blackie and Sons*, 3–13.

Patra, S., Mahapatra, S.C., Mishra, P., and Sahoo, S. (2015). A quantitative analysis of groundwater status of Hooghly district in West Bengal, India. *International Journal of Current Research, 7*(07), 18810–18818.

Piégay, H., Darby, S. E., Mosselman, E., & Surian, N. (2005). A review of techniques available for delimiting the Erodible River corridor: a sustainable approach to managing bank erosion. *River Research and Applications, 21*, 773–789.

Rennell, J. (2016). *A Bengal Atlas/Major James Rennell (F.R.S.)* (Vols. I-II, K. Rudra, Ed.) Kolkata: Shishu Sahitya Samsad Pvt. Ltd.

Revil, A., Cary, L., Fan, Q., Finizola, A., & Trolard, F. (2005). Self-potential signals associated with preferential ground water flow pathways in a buried palaeo-channel. *Geophysical Research Letter, 32* (7), L07401.

Roy, S. (2019). Influence of faulting on the extra-channel geomorphology of the Ajay-Damodar interfluve in Lower Ganga Basin. *Quaternary Geomorphology in India, Geography of the Physical Environment*, 79–87.

Roy, S., & Sahu, A. S. (2015). Investigation for potential groundwater recharge area over the Kunur river basin, Eastern India: an integrated approach with geosciences. *Journal of Geomatics, 9* (2), 165–177

Roy, S., & Sahu, A. S. (2016). Palaeo-path investigation of the lower Ajay River (India) using archaeological evidence and applied remote sensing. *Geocarto International, 31* (9), 966–984.

Rudra, K. (2006). *The Palaeochannels and Wetlands of Southern Deltaic West Bengal. Landforms, Processes and Environmental Management*. Kolkata: ACB Publication.

Rudra, K. (2010). *'Banglar Nadikotha' (A Tale of Rivers of Bengal)*. Kolkata: Sahitya Samsad.

Rudra, K. (2014). Changing river courses in the Western part of the Ganga-Brahmaputa delta. *Geomorphology, 227*, 87–100.

Saha, U. D., & Bhattacharya, S. (2016). A study of river induced major hydrogeomorphic issues and associated problems in Torsa basin, West Bengal. *Geographical Review of India, 78* (2), 132–145.

Saha, U. D., & Bhattacharya, S. (2019). Reconstructing the channel shifting pattern of the Torsa River on the Himalayan Foreland Basin over the last 250 years. *Bulletin of Geography. Physical Geography Series, 16*, 99–114.

Samaddar, R. K., Kumar, S., & Gupta, R. P. (2011). Palaeochannels and their potentials for artificial groundwater recharge in the western Ganga plains. *Journal of Hydrology, 400* (1), 154–164.

Sen, P. K. (1978). *Evaluation of the Hydro-Geomorphological Analysis of the Bhagirathi-Hooghly and Damodar Interfluve*. Burdwan: UGC Research Project, Department of Geography, The University of Burdwan.

Sen, P. K. (1985). Environmental changes and degradation in the Damodar basin: impact of development on environment. *Geographical Society of India, 1*, 6–12.

Siddiq, A. B., & Habib, A. (2017). The formation of bengal civilization: a glimpse on the socio-cultural assimilations through political progressions socio-cultural assimilations through political progressions. *Artuklu Human and Social Science Journal, 2* (2), 1–12.

Siddique, G., & Mukherjee, N. (2017). Transformation of agricultural land for urbanisation, infrastructural development and question of future food security: cases from parts of Hugli District, West Bengal. *Space and Culture, 5* (2), 47–68.

Singh, K.P. (1996). *Significance of Palaeochannels for Hydrogeological Studies – A case Study from Alluvial Plains of Punjab & Haryana State India. Sub surface – water Hydrology*. Dordrecht: Kluwer Academic Publishers, 245–249.

Singh, L. P., Parkasha, B., & Singhvi, A. K. (1998). Evolution of the lower Gangetic plain landforms and soils in West Bengal, India. *Catena, 33* (2), 75–104.

Sinha, A.K., Raghav, K.S., & Sharma, A. (2005). Palaeochannels and their recharge as drought proofing measure: study and experiences from Rajasthan, Western India. *Proceedings of the International Conference Drought mitigation and prevention of land desertification*, Bled, Slovenia.

Sinha, C. P. (2008, November 15). Management of floods in Bihar. *Economic & Political Weekly, 43*. 40–42.

Sokal, R. R., & Oden, N. L. (1978). Spatial autocorrelation in Biology. *Biological Journal of the Linnean Society, 10* (2), 199–228.

Stammel, B., Amtmann, M., Gelhaus, M., & Cyffka, B. (2018). Change of regulating ecosystem services in the Danube floodplain over the past 150 years induced by land use change and human infrastructure. *Die Erde, Journal of the Geographical Society of Berlin, 149* (2), 145–156.

Stammel, B., Cyffka, B., Geist, J., Müller, M., Pander, J., & Blasch, G. (2012). Floodplain restoration on the Upper Danube (Germany) by re-establishing water and sediment dynamics: a scientific monitoring as part of the implementation. *River Systems, 20* (1), 55–70.

Sümeghy, B., & Kiss T. (2012). Morphological and hydrological characteristics of palaeochannels on the alluvial fan of the Maros River, Hungary. *Journal of Environmental Geography, 5* (1–2), 11–19.

Wang, X., Guo, Z., Wu, L., Zhu, C., & He, H. (2012). Extraction of palaeochannel information from remote sensing imagery in the east of Chaohu Lake, China. *Frontiers of Earth Science, 6* (1), 75–82.

Weeks, J. R., Getis, A., Hill, A. G., Gadalla, M. S., & Rashed, T. (2004). The fertility transition in Egypt intraurban patterns in Cairo. *Annals of the Association of American Geographers, 94* (1), 74–93.

Wray, R. A. L., (2009). Palaeochannels of the Namoi River Floodplain, New South Wales, Australia: the use of multispectral Landsat imagery to highlight a Late Quaternary change in fluvial regime. *Australian Geographer, 40* (1), 29–49.

Wu, L., Xu, Y., Yuan, J., Wang, Q., Xu, X., & Wen, H. (2018, May). Impacts of land use change on river systems for a river network plain. *Water, 10* (5), 609.

Yousefi, S., Moradi, H. R., Pourghasemi, H. R., & Khatami, R. (2017). Assessment of floodplain landuse and channel morphology within meandering reach of the Talar River in Iran using GIS and aerial photographs. *Geocarto International, 33* (12), 1367–1380.

16 The Anjana
A Journey from River to Canal

Balai Chandra Das and Debabrata Das

CONTENTS

16.1 INTRODUCTION

"Water is life". But neither the water itself is "life" nor the water has "life". Yet the statement is true in two senses. First, life began to evolve within the water from water. Second, life cannot thrive without water. W.H. Auden (1907–1973) aptly said that "Thousands have lived without love, not one without water". That is why from the prehistoric period (Peking Man 750,000 BCE) man preferred to live near a source of water, and generally, most of those water sources are rivers. During historic period, Harappa (6,000 BCE on Indus river), Mohenjo Daro (2,500 BCE on Indus river), Mesopotamia (3,500–150 BCE on Tigris and Euphrates rivers), Egypt (3,100 BCE on Nile river), Greek (1,200 BCE–600 AD on Ilisos River), etc., all the civilisations were flourished on the river banks. The river provided all the fortunate environments for the great deeds of human beings and helped in making the kingly history of the world. Time is like river (**Marcus Aurelius, 121-180),** and river is material of time (past, present, and future) and medium of expression of time (history). And in times, river expunges that history by destroying mighty deeds of man.

Like large rivers and water bodies, little rivers do have a great role in the native rural life, economy and culture. They also materialise the plan of present living and plot of past history. The river Anjana is such a small river which has a great role in the economic, social, political and cultural life of the Nadia district of West Bengal, India. It witnessed tales of numerous local histories of joy and sorrow. Apart from being material and methods of history, it has its own tales also.

The river Anjana completed the mesh of waterlines connecting relatively two large rivers Jalangi and Churni. It was flowing through the densely populated agrarian belt of the Moribund Deltaic West Bengal (MDWB). It took part in building this lower delta in collaboration with rivers Bhagirathi, Jalangi and Churni; it moistened those lands to supply water to plants and crops; irrigated plots during dry months; saved man and other life from being inundated during floods by hasty pass of flood volume through its deep wide and efficient channel; nourished the dwellers on its banks supplying fishes to their diet; made waterways to transport loads and bulks; offered life to drink; and provided a watery neighbour to be in day-to-day life with dwellers on its banks. But days are gone. It has been transformed into a stagnant segmented shallow and narrow ditch in rural places and mere sewage in urban areas. No way it can serve its basin as did in bygone days.

Man reworked on the basin and channel of the river Anjana reshaping its channel form and the overall landscape of the basin. He intervened with its flow. His active effort multiplied the natural processes of river channel deterioration. He has highly encroached the river Anjana, dumped its bed with solid waste, blocked its flow by the earthen dam across the channel, made its water filthy and polluted to the poisonous level pouring sewage of settlements (News18, 19.08.2019; Telegraph India, 22.06.2019). The river has lost its capacity to carry flood volumes. It lost its value and lowered the values of men on its banks.

In these contexts, the present chapter will recall the VIBGIYOR (colourful and glorious) past of the river and its basin, and tell the tales of anthropogenic efforts of reshaping the channel and basin forms and deteriorating the river to a sewage drain via canal (ABP, 26.08.2019; http://riverbangla.com/author/tapas/).

16.2   THE RIVER AND ITS BASIN

Like Jalangi River, Anjana River is also a distributary as well as a tributary in nature in the Bhagirathi-Hooghly River System in the province of West Bengal of India (see Chapter 1). It receives from the river Jalangi at Krishnagar and debouches in the river Churni at Byaspur (although the river is bifurcated into two branches and the minor branch, known as Heler Khal, falls in the river Churni at Hanskhali) (Figure 16.1). The basin is located between 23°16′10″ to 25′02″ N latitudes and 88°27′39″–88°36′46″ E longitudes.

16.2.1   THE BIRTH

When and how Anjana, the distributary of the river Jalangi and tributary of the river Churni, appeared in the scenario of the MDWB is not so clear. But the general history of the evolution of MDWB and intuition reveals that the river Anjana must have come into the scene after the origin of the river Jalangi because the river Anjana would feed by the river Jalangi. As Jalangi River took off from the river Padma, it is obvious that both Jalangi and Anjana originated far later the origin or flourishing of Padma. The river Jalangi took off its supply from the river Padma near Char Madhubona village, north-east to Karimpur, at the border of the districts of Nadia and Murshidabad. During 1918, Jalangi and Mathabhanga would take off from the river Padma at the village Bausmari. Fergusson

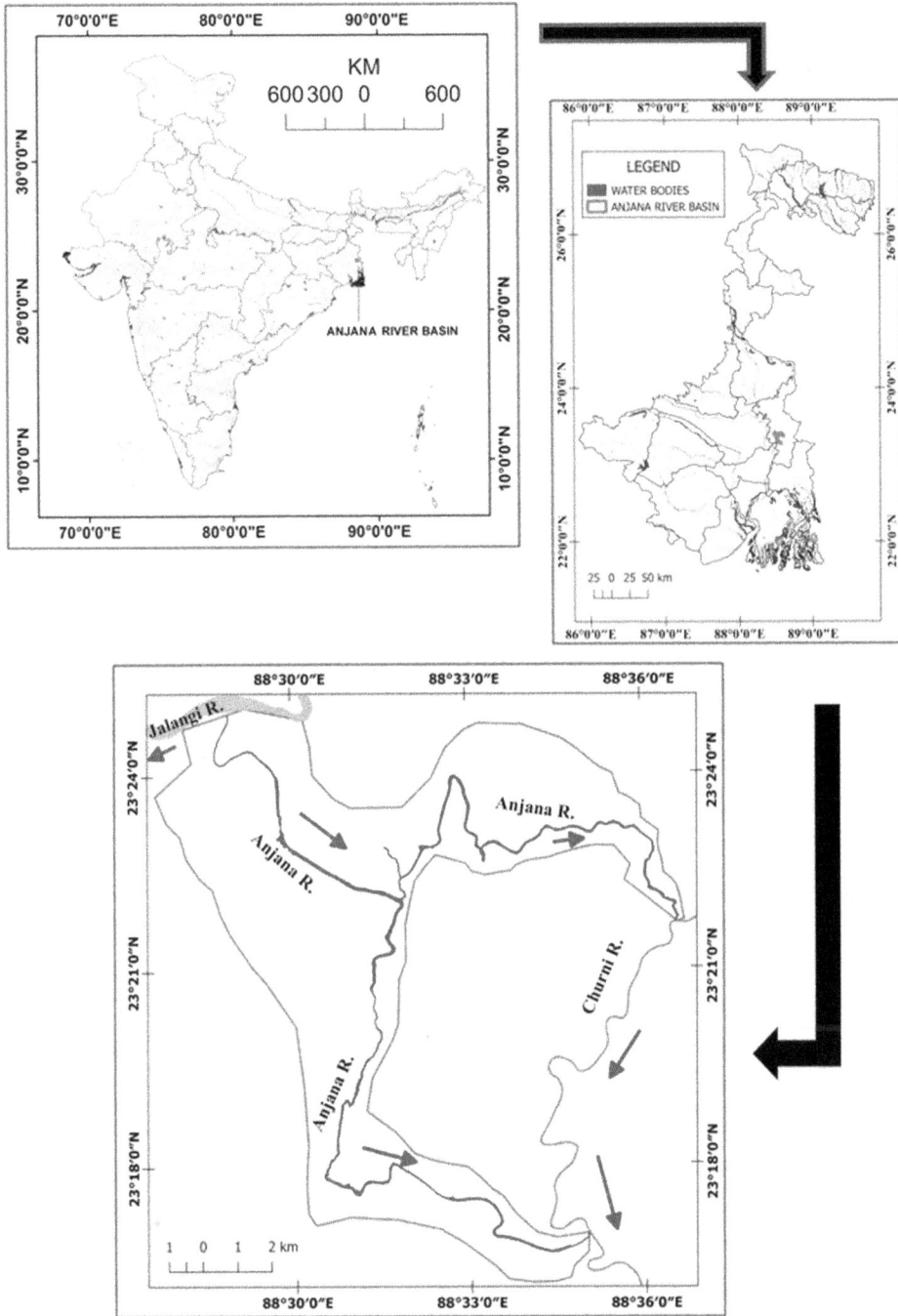

FIGURE 16.1 Location of Anjana River basin.

(1912) opined that river Jalangi was originated in between the period 15th and 16th century. Because during that period, the main flow of the Ganges shifted from the river Bhagirathi to the river Padma (Sherwill, 1858; Reaks, 1919; Mukherjee, 1938; Majumder, 1941; Bagchi, 1978). So, consequent to this great shifting of the course of the river Ganges, at first Jalangi and then Anjana came into the scenario of the region during the period between the 15th and 16th centuries.

Fergusson (1912) and Majumder (1978) argued that the river Padma (and subsequently Jalangi) was opened up from the river Bhagirathi about 300–400 years ago. On the other hand, Moor (1919), Bose (1972) and Chatterjee (1972) argued that the river Padma (and subsequently Jalangi) must have originated during the 17th or 18th century because during that period Padma River became the prime channel of the river Ganges. But the river Jalangi was in existence even during the 15th century (Das, 2013). Therefore, if the river Jalangi was originated in the 15th century or earlier, then the Anjana, the daughter river of Jalangi, also has every possibility to take birth during that period also.

But how the river Anjana initiated its course from Jalangi to debouch into the river Churni is an issue. Philosophy of uniformitarianism and catastrophism for geological and geomorphological events are suggested by different earth scientists (Knighton, 1998). Catastrophism suggests a sudden event of larger magnitude and brings lasting change on the earth's surface. On the other hand, uniformitarianism (Hutton, 1788; Playfair, 1802) advocates orderliness (cyclic and gradual) of nature where gradual evolutionary processes are of relatively much lesser magnitude (Gilbert, 1877; Wolman and Millar, 1960; Hack, 1960). Those gradual processes of moderate frequency and magnitude make a significant change in the earth's surface but in the long run (Wolman and Millar, 1960). Another tenet refers to the doctrine that the gradual processes we see on earth have been supplemented by huge natural catastrophes and called "Neocatastrophism" (Shatskiy, 1937). However, regarding the origin of the river Anjana, it is thought that gradual processes of erosion and deposition of moderate frequency and magnitude had prepared the background for the initiation of a new channel in the deltaic plain. Gradual deposition of eroded sediment within the bed of the river Jalangi raised its bed. At this point of prepared background (gradualism), perhaps during the 15th or 16th century, when the load-volume equilibrium of the river Jalangi crossed threshold point, a huge flow of flood overtopped banks and the river Anjana found its spill-way through the breached left bank (catastrophism) at Krishnagar, the present-day headquarter of the district of Nadia. Anjana was born.

16.2.2 The Course

River Anjana passes through the C.D. Blocks of Krishnagar-I, Hanskhali and Ranaghat-I (Figure 16.1). Although it is rather impossible than difficult to the precise delineation of the drainage basin of a deltaic river like Anjana, an attempt of approximation in this regard was taken. For the purpose, Google earth image of the inundated area of 2000 flood, contour map (Bagchi and Mukherjee, 1978), flood slope map (Biswas, 2001, plate-19) and transport networks were taken into consideration. Calculated drainage area is 89.54 km².

After birth, the river Anjana flowed for more than 400 years in the middle part of the landscape of the district of Nadia. Anjana is simultaneously a distributary (Mallik, 1911, Dutta, 1996; Bandyopadhaya, 2007; Das, 2013) and tributary (Das, 2013) channel in the drainage system of the world's largest Ganges–Brahmaputra delta. Taking off from the river Jalangi (Figure 16.2) at Krishnagar (23°25′7.5″ N and 88°29′17.5″ E), Anjana runs south through Manik Para on the right bank and Nagendranagar on the left bank (Figure 16.2). But this straight reach (from feeder river Jalangi to the culvert under Krishnagar–Nabadwip road) of the river Anjana raised a debate whether it is an artificial canal or a natural river bed.

Some aged knowledgeable persons of Krishnagar like Sri Devashis Mandal, retired teacher of Badkulla United Academy, and Sri Nirmal Syanal, retired AI of School, Nadia, said that the original off-take of the river Anjana was near present-day Kadamtala Ghat (Figure 16.3) at coordinate 23°25′1.7″ N and 88°29′36″ E. During past, the river Anjana would follow a south-westerly path through Nagendranagar (Figures 16.3 and 16.4). But another opinion is that the palaeochannel from Kadamtala Ghat through Nagendranagar is of Jalangi, not of Anjana. If we accept the second opinion as truth, several objections come from a micro-relief pattern of the region. First, the magnitude of the width and depth of the river Jalangi no way matches the probable magnitude of this palaeochannel at its largest estimate.

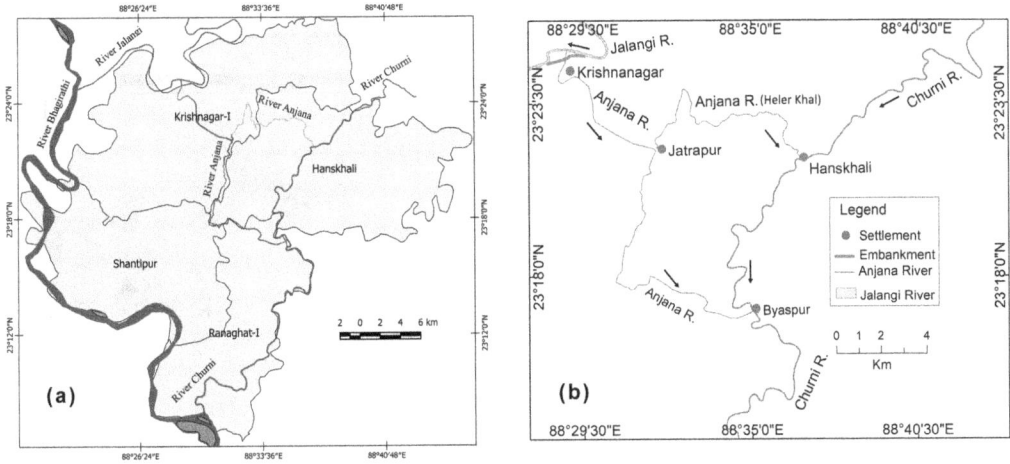

FIGURE 16.2 (a) The river Anjana through four C.D. Blocks Krishnagar-I, Hanskhali, Shantipur and Ranaghat-I of the district of Nadia (prepared by Sri Debabrata Das). (b) Path of the river Anjana linking the river Jalangi and Churni.

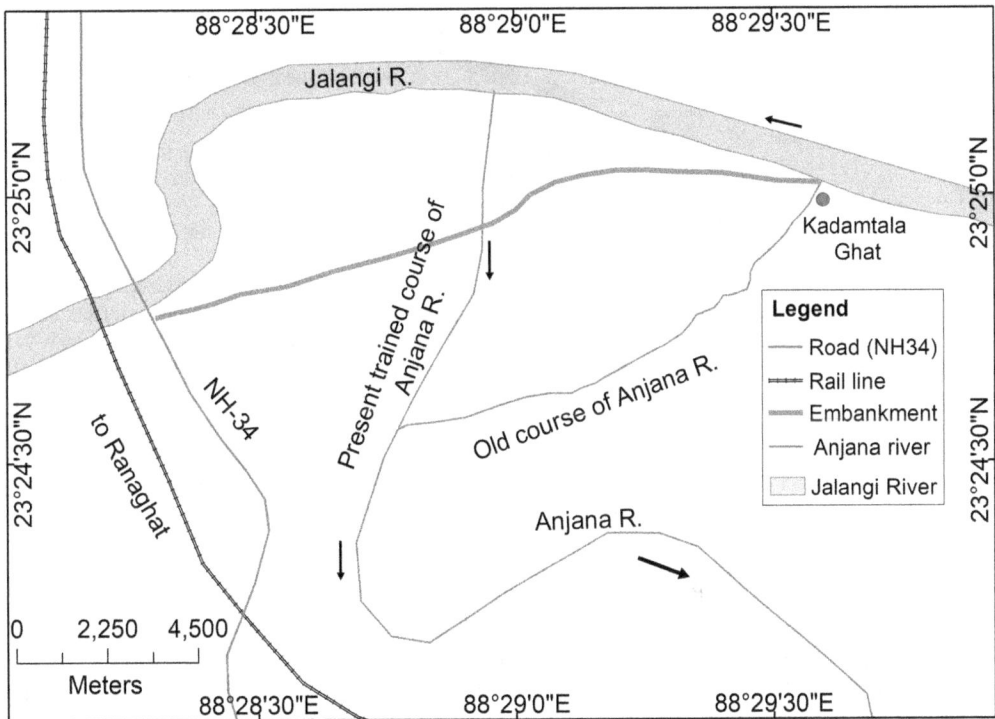

FIGURE 16.3 Old off-take of the river Anjana at Kadamtala Ghat.

Second, view opines that the dead course from Kadamtala Ghat to the north of Circuit House via ditch behind Old and New Hindu Hostel of Krishnagar Government College is the palaeochannel of the river Jalangi, not of Anjana (Figure 16.4). If it is true, then "where was the course of the river Jalangi towards downstream next to Circuit House?" However, sedimentological study and dating may bring the truth unfold.

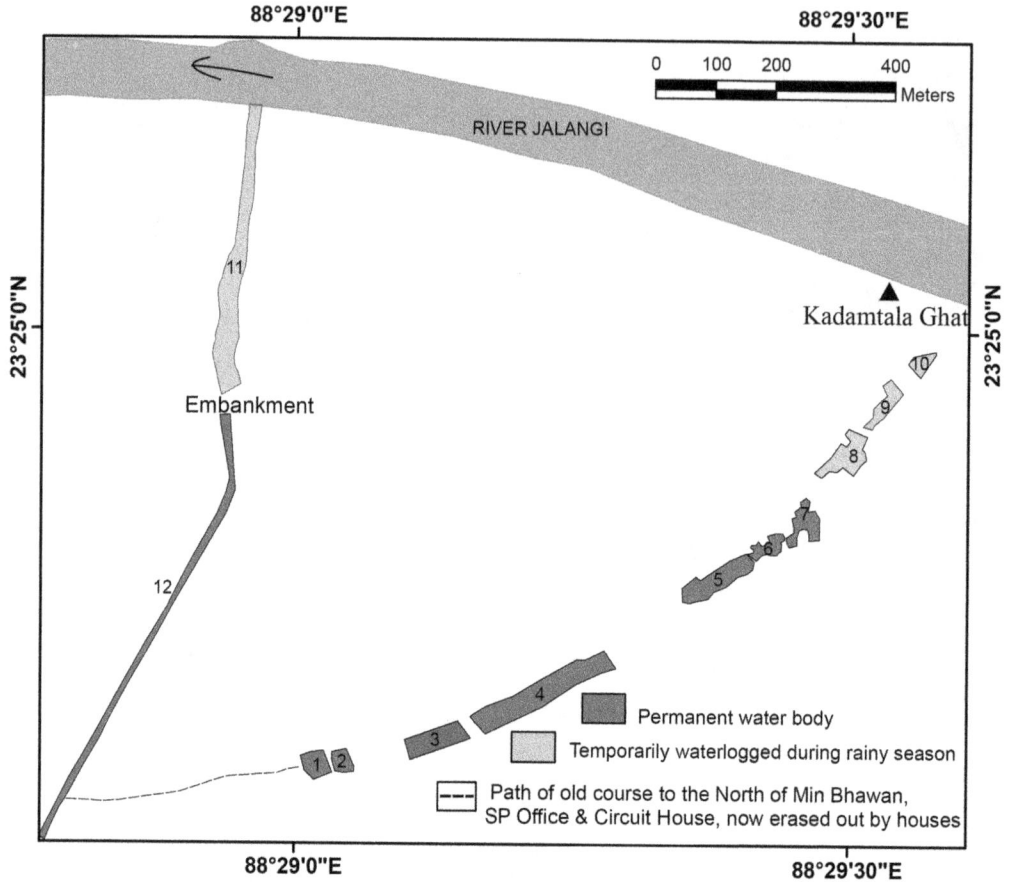

FIGURE 16.4 The linear arrangement of ponds from north of the old Circuit House to the Kadamtala Ghat is the segmented palaeochannel of the river Anjana. 1. Pond 1 to the west of Kanai Chakraborty Dhal (slope). 2. Pond 2 to the west of Kanai Chakraborty Dhal (slope). 3. Pond to the north of Jalangi Bhawan, I&WWD, Govt. of West Bengal. 4. Pond to the north of Old Hindu Hostel, Krishnagar Govt. College. 5. Waterbody to the east of BD Mukherjee Road, north-west of Malo Para. 6. Waterbody 1 to the north of Sri Chaitanya Gaudiya Math. 7. Waterbody 2 to the north of Sri Chaitanya Gaudiya Math. 8. Waterbody to the north of Dafadar Para, West of Hemanta Sarkar Lane. 9. Waterbody to the south-west of Golapatti Sub Post Office. 10. Waterbody to East of Golapatti Sub Post Office, North of Rakhal Raj Biswas Road. 11 and 12 show the present reshaped path of the river Anjana. (Drawn by Miss Sanchita Saha.)

After culvert under Krishnagar–Nabadwip road, the river flows due south for a few hundred meters until it turns to north-east from south-west end of the Banashree Para. Thence, the river runs eastwards keeping Baghadanga, Chandsarak Para, Krishnagar Cathedral, and Bejikhali to its right bank and joins the north-east corner of the moat of Rajbari, the royal palace of the king of Nadia. The western part of the moat is actually the artificially modified Anjana River. The river passes through eight wards of the Krishnagar Municipality (from Ward No. 24 to the north, where its closed off-take is) to Ward No. 10 to the south (from where the river leaves the city of Krishnagar) via Ward Nos. 17, 18, 15, 13, 12 and 11 (as in 2020). From Rajbari to the Nadia District Hospital at Shaktinagar, the river pursues a southerly course wherefrom it turns east up to Jatrapur village with *Jurisdiction List Number* (J.L. No.) 111 and *Permanent Location Code Number* (PLCN) 01550800 (*District Census Handbook, Nadia*, 1981 and 2001; Ray, 1875). After passing Kshirpuli village, the river bifurcates at Jatrapur village at coordinate 23°22′4.7″ N and 88°31′52″ E. The right and main branch passes by Kshirpuli (J.L. No. 112; PLCN. 01550900), Jalalkhali (J.L. No. 115; PLCN. 01551200),

Hant Boyalia (J.L. No. 114; PLCN. 01551100), Khamar Simulia (J.L. No. 123; PLCN. 01552000), Dharmadaha (J.L. No. 124; PLCN. 01552100), Gopalpur (J.L. No. 125; PLCN. 01552200), Patuli (J.L. No. 43; PLCN. 01563100), Badkulla (J.L. No. 44; PLCN. 01563200), Nawpukuria (J.L. No. 41; PLCN. 01562900) and Chandandaha (J.L. No. 31) and finally debouches into the river Churni at Byaspur (J.L. No. 34) at coordinate 23°16′58″ N and 88°35′1.5″ E.

The left branch after Jatrapur village flows towards north passing by Mahishnengra village (J.L. No. 108; PLCN. 01550500) wherefrom the river turns southward making a hairpin bend and runs by Beraberia (J.L. No. 109; PLCN. 01550600), Jaypur (J.L. No. 39; PLCN. 01562700), Purba Khamar Simulia (J.L. No. 40; PLCN. 01562800), Mandab Ghat (J.L. No. 51; PLCN. 01563900) and Mamjoyan (J.L. No. 60; PLCN. 01564800) and joins the river Churni at 23°16′56″ N and 88°36′40″ E, north-east of Hanskhali (J.L. No. 53; PLCN. 01564100).

From its present-day off-take at Jalangi River to the point of bifurcation at Jatrapur, the length of the river Anjana is 11 km. From Jatrapur, the length of the right branch of the river is 18 km, and the length of the left branch of the river is 13 km. Therefore, the total length of the river is 42 km. But the river was brought under 100 days' work scheme under Mahatma Gandhi National Rural Employment Guarantee Act-2005 (MGNREGA) in 2013, and it was mentioned there that total length of the river is 29 km (Dutta, 2013). Actually, the left branch (known as Heler Khal) of the river after Jatrapur was excluded from that project because that branch is almost erased out in places and not traceable – therefore insignificant to consider.

Taking the right branch of the river into consideration, the total length of the river from its off-take from Jalangi at Krishnagar to its confluence with Churni at Byaspur is 29 km and straight length from off-take to confluence is 18.51 km. Therefore, the sinuosity index of the river is 1.57.

16.2.3 Spills

Anjana is a very small river in the district of Nadia. No such distinct namable spill channels are there contributing the river. It drains low lying areas or beels along its course. Yet, following spills and their drained areas are worthy to be mentioned (Table 16.1).

All these spills would spread silt-laden flood water on to low lying beels where sediments are silted. During the lean period when the level of water in the Anjana went down, reverse flow of sediment-free clean water recharged the volume of the river and would maintain its channel through this self-maintenance mechanism. These processes repeated in hundreds of years and raised the level of the delta.

Anjana River basin covers an area of 89.54 km². The average annual rainfall of the basin is 1,310.4 mm (Census of India, 2001). Therefore, the total input of rain in the basin is 117,215,280,000,000 cm³ or 117,215,280 m³.

TABLE 16.1
Spills of the River Anjana

Sl. No.	The Place Where the Spill Joins the River Anjana	Location	Drained Area/Locality
1	North of Gopalpur	23°19′00″ N and 88°30′52″ E	Right-bank spill drains low lying area east to the Dharmadah–Gopalpur village road
2	East of Gopalpur	23°18′14.5″ N and 88°30′44″ E	Left-bank spill drains low lying area (beel) between Badkulla and Gopalpur
3	Mugrail	23°17′27″ N and 88°32′40″ E	Right-bank spill drains low lying area (beel) between Bhaduri and Mugrail
4	Gagrakhali	23°16′40.5″ N and 88°34′10.5″ E	Right-bank spill drains low lying area (beel) between Gagrakhali and Hatpukuria

Average population density of the district of Nadia is 1,172.6 person km^2 (Census of India, 2001). In the town of Krishnagar, the headquarters of the district, the population density is 8,694.4 person km^2 (Census of India, 2001). Anjana River flows through the municipality of Krishnagar, and this is the most densely populated area of the Anjana basin.

16.2.4 Major Landforms

The river passes through the monotonous deltaic plain of the Bhagirathi-Hooghly River system. Whatever the minor and micro variations in the landforms of the basin are either of the river channel itself and its relict scars or of the anthropogenic origin. The general slope of the basin is towards the south and south-east. The highest relief of >16 m is found mainly at the Krishnagar Municipality area and other settled areas of the basin. Made grounds are common in areas having a height of ~16 m. The lowest relief of <9 m is found at back-swamps and beels (Figure 16.5).

The largest area (32.27%) of the basin falls under a height class of 9–12 m, and the smallest area (7.58%) of the basin has the lowest relief of <9 m (Table 16.2). The highest relief of >16 m covers 21.13% area of the basin (Figure 16.5) and found mostly at Krishnagar and Badkulla, the first and second largest settlements and the most anthropogenically modified landforms (made ground) within the basin. The highest slope of >4° is also found in settled areas (12.09% of the basin area).

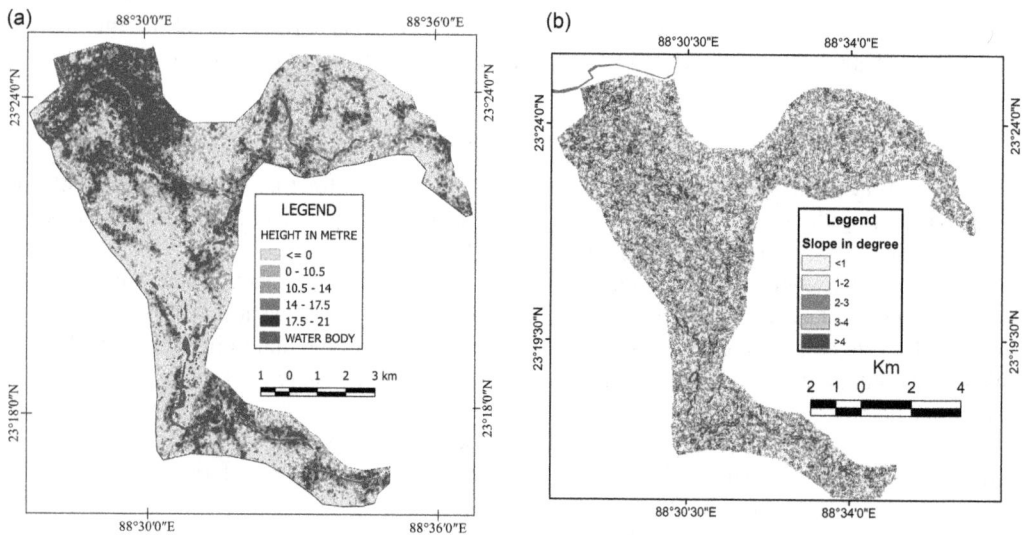

FIGURE 16.5 (a) Relief and (b) slope of the Anjana Basin.

TABLE 16.2
Area under Different Relief and Slope Classes of the Anjana Basin

Slope (°)	Area in km^2	% of Area	Relief (m)	Area in km^2	% of Area
<1	1,157.03	12.92	<9	678.29	7.58
1–2	2,908.34	32.48	9–12	2,889.53	32.27
2–3	2,342.42	26.16	12–14	2,048.34	22.84
3–4	1,464.16	16.35	14–16	1,449.56	16.18
>4	1,082.16	12.09	>16	1,892.41	21.13
Total	8,954.12	100.00	Total	8,958.12	100.00

Average reduced level (RL) of the lowest bed level or thalweg level (Figure 16.6; Table 16.3) of the river Anjana is 3.92 m with a maximum of 6.08 m and a minimum of 0.750 m. The total length of the river is 29 km, and the RLs of thalweg level at off-take from Jalangi and at Churni confluence are 3.625 and 2.145 m, respectively, with a relative fall of 1.480 m. Therefore, the average bed slope of the channel is 1:20000 or 0.00005. It is clear from the long profile of the river that the lowest thalweg level (and maximum depth) of 0.750 m is recorded in urban reach at Rajar Dighi (the moat) which is surely the result of artificial excavation of the moat within the river channel. The maximum RL of thalweg level of 6.080 m is found in rural reach. The average width (w) to depth (d) ratio of the channel is 46.92 with a maximum of 162.07 and a minimum of 4.76 at the moat of the royal palace (Figure 16.6).

Reduced level (RL) of left bank, right bank and thalweg level (lowest river bed) shows a significant irregularity all along its course (Figure 16.7). A sharp break in thalweg level is noticed after bifurcation of the channel at village Jatrapur. At a distance of 2.65 km from off-take, RL of lowest bed level was 5.670 m, highest in the urban reach. At moat of royal palace and downstream to Nadia District Hospital, Shaktinagar, depth of the river is maximum. After Jatrapur, depth again reduced. The lowest bed level at offtake (2.935 m) from river Jalangi is higher than the lowest bed level at Anjana-Churni confluence (2.145 m).

16.2.5 THE GLORIOUS PAST

Before spotting anthropogenic signatures over basin and channel geomorphology which has been killing and erasing the dying river from the scenario of landscape of the district of Nadia, let us recall the glorious and fabulous past of the river. During the 16th and 17th centuries, the river Anjana was

TABLE 16.3
Reach-Wise Comparison of RLs of Thalweg Level of the River Anjana

Reach Name	Maximum	Minimum	Average	SD	CV
Total reach	6.080	0.750	3.920	1.352	0.345
Urban reach	5.860	0.750	2.682	1.172	0.437
Rural reach	6.080	2.145	4.621	0.856	0.185

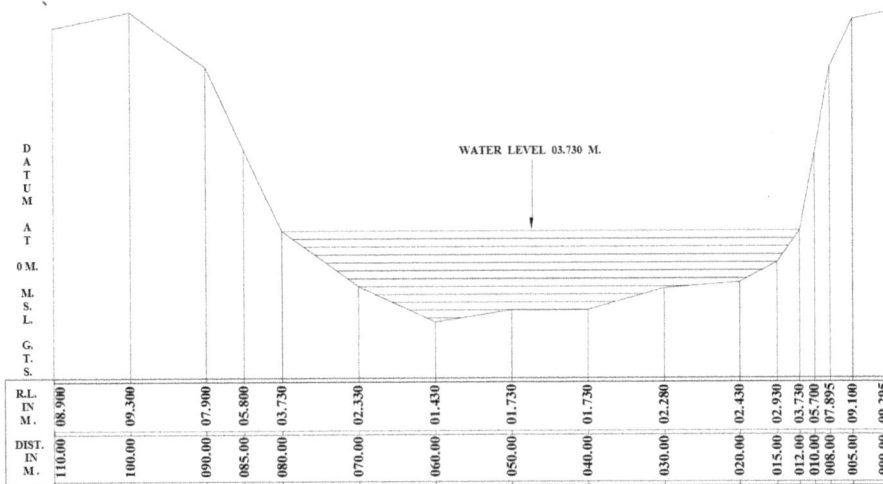

FIGURE 16.6 Anjana bed was deepened to make the moat difficult to pass. The lowest w/d ratio 4.76 is found here, at Rajar Dighi.

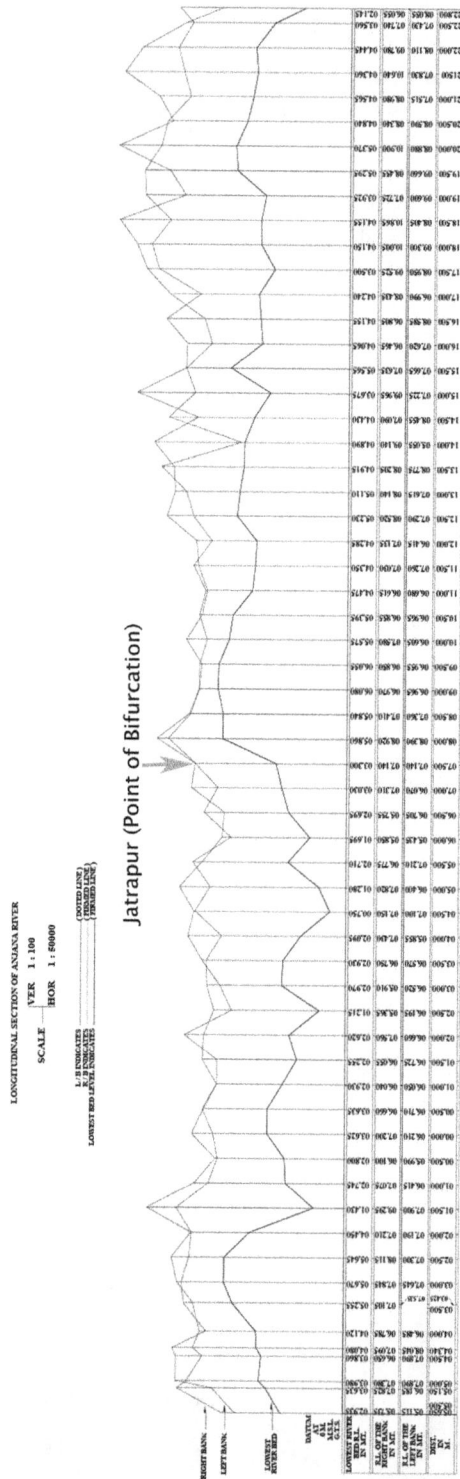

FIGURE 16.7   Long profile of the river showing bank heights and the lowest bed level.

of such a significant river that the royal family of Nadia found the most suitable place for their capital and the royal palace on the bank of the river. In 1606, when Mughal Emperor Jahangir was in the throne of Delhi, ruler of Bengal Pratapaditya (1,561–1,611 CE; Garrett, 1910; Ray, 1875) revolt and fought against the Mughal Empire. Jahangir sent his commander Senapati Raja Man Singh to Bengal to defeat Pratapaditya. At the end of Khagraghat battle, the Mughals were offered a truce in spite of a marginal win, but Pratapaditya was put in chains and kept confined at Dhaka and his kingdom was annexed. After Pratapaditya's death, Bhavananda Majumdar, who had been in the service of Pratapaditya, was given the throne by Raja Man Singh, and he subsequently became the founder of the Nadia Raj family (Bhattacharya, 1896; Mallick, 1911). Capital of Bhavananda Majumder (1606–1611) was at Matiari. Raghab Majumder, grandson of Bhavananda Majumder, shifted his capital at Reui village (Mallick, 1911; Rarhi, 1927; Chattopadhyay, 2007), on the bank of the river Anjana. He was a worshiper of Lord Krishna, and he renamed Reui as Krishnanagar. From this historic preference of Nadia Raj family to shift their palace on the bank of the river Anjana, it could be concluded that the then Anjana was a river of significant importance in transport and navigation, irrigation and agriculture, fishing and health. However, since the establishment of the royal palace at Krishnanagar in 1648 (Sanyal, 2014), the river Anjana also passed its royal phase of volume and velocity up to 1743, when Maharaja Krishnachandra Ray shifted his capital on the bank of a mightier river Churni.

In 1676, Raja Rudra Ray, great grandfather of Maharaja Krishna Chandra Ray, faced an attack by the naval force of Nawab from Murshidabad coming through the river Anjana (Mallick, 1911; Dutta, 1996). This fact indicates to the vigorous and voluminous river Anjana because the boat of the navy force could pass through the river.

Fr. Thomas Zubiburu, a priest from Spain, purchased a huge plot of land on 18.11.1845 (Lazaro, 1996) on the bank of the river Anjana where Krishnagar Cathedral got its foundation on 26.01.1846 on a very beautiful healthy place on that pleasant river bank. This choice of the priest of Christ missionary approves the glorious and healthy past of the river Anjana. From 1855 to 1860, Fr. Limana purchased more land on the bank of the river Anjana to establish a Christian colony for a community living (Lazaro, 1996). He felt the necessity of sisters for spiritual teaching among girls and women of the area, and he did accordingly. On 7 February 1860, four sisters from Milan sailed for Krishnanagar, and they reached Kolkata on 11 March 1860. They are Sr. Agostina Baruffini, Sr. Benedetta Danielli, Sr. Lucia Viero and Sr. Antonia Ferrari. From Kolkata, they started by boat on 14 March 1860 and reached the cathedral on 17 March 1860. In the memory of that auspicious sailing through the Anjana River, a stone (Figure 16.8) was set on the spot encrypting:

> Sister of Charity. The first four Italian missionaries of the sister of charity of Sr. Bartolomea Capitanio and Vincenza Gerosa, sailed through this Anjana canal and set foot on this hallowed spot on 17th March 1860 to spread the message of God's Love among his people in India.

So, in the 19th century also, the river Anjana was navigable.

After 15 years of this event, Sr. Brigida Zanella of Jessore was in a fever of Kalajar and she wished to breathe her last at Krishnanagar Cathedral. So, she being accompanied by Sr. Nazarena

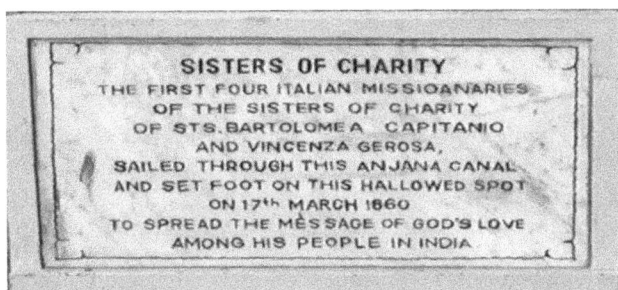

FIGURE 16.8 Marble plate on the bank of the river Anjana encrypting memories of gone days.

Cavalotti sailed on a boat through the river Bhairab–Churni–Anjana for Krishnanagar Cathedral. This event also indicates towards the voluminous state of the river Anjana.

Before the construction of the Ranaghat–Lalgola railway in 1905, rivers were the principal route for transport. Gagan Chandra Biswas (1850–1936), a civil engineer from Shibpur Bengal Engineering College, Howrah, founded his zamindari palace on the left bank of the river Anjana at Badkulla. Members of the zamindar family used to sail through Anjana River for Ranaghat, Krishnagar, Shibnibas and Krishnaganj. Indira Biswas, wife of the grandson of Gagan Chandra Biswas, said that their zamindari was extended near Krishnagar and farmers used to carry crops, fodders, giant earthen pots and other household items in boats through this Anjana River. The manager of the zamindar used to go to Krishnagar and Ranaghat for official and business purposes through the Anjana riverway. Zamindar family had a motor-driven speed boat. The author of this chapter also witnessed (1980s) the large bricked bathing *ghat* of zamindar palace on the river Anjana. All this testifies the VIBGYOR (glorious/colourful) past of the river.

Cultivation of Indigofera in the 18th century spread out over the Bengal. "Nadia alone produced over 8000 maunds of indigo a year or approximately one-fifth of the outturn of the old province of Bengal" (Pringle and Kemm, 1928). In 1860, Krishnagar was the headquarter of indigo planters (Majumder, 1978). For the manufacturing of indigo from the plant, a plentiful supply of water was required. Moreover, for carrying of indigo plants from fields to the factory and finished product to the export centres, conveyance of factory owners or indigo planters, workers and Lathials (force armed with bamboo), transport of machinery, etc., waterways through the rivers were the principal mode of transport. Anjana basin was not an exception. There were more than a dozen of the indigo factories (Table 16.4) on the banks of the river Anjana. A list is provided in Table 16.4.

TABLE 16.4
Location of Indigo Factories on the Bank of the River Anjana

Sl. No.	Present Status of the Indigo Factory	Location	Source of Information
1	Office of the District Inspector of Schools	At Krishnagar, on the left bank of the old course of the river Anjana	Mallick (1911)
2	Holy Family Girls High School	At Krishnagar, on the left bank of the river Anjana	Mallick (1911)
3	The old building of the Government Girls High School	At Krishnagar, on the right bank of the river Anjana	Mallick (1911)
4	Judge Bungalow (residence of district judge)	At Krishnagar, on the right bank of the river Anjana	Mallick (1911)
5	Bungalow of the District Magistrate	At Krishnagar, on the left bank of the river Anjana	Dutta (1996)
6	Office of the Superintendent of Police	At Krishnagar, on the left bank of the river Anjana	Mallick (1911)
7	House of Samir Basak	Sanko Ghat Para, Dogachi, on the right bank of the river Anjana	Dutta (1996)
8	Ruins at Hatboalia Ghat	Kagmari, Hatboalia, on the right bank of the river Anjana	Dutta (1996)
9	Ruins at Teghari	Dogachi, on the right bank of the river Anjana	Dutta (1996)
10	Ruins at Joypur	Jaypur, on the right bank of the river Anjana	Dutta (1996)
11	Ruins at Hazrapota		
12	Ruins at Chandandaha	Chandandaha, on the right bank of the river Anjana	Biswas (2002)
13	Ruins at House of Kalipada De	Ballavpur, P.S. Shantipur, on the right bank of the river Anjana	Biswas (2002)

The array of all these indigo factories along the course of the river indicates good past of the river Anjana with sufficient width (w), depth (d), velocity (v) and discharge (Q).

Many celebrated persons have uttered the glorious presence of the river Anjana even during the late 19th century. Bharatchandra Ray (1752), the author of *Annada Mangal Kavya*, Rabindranath Tagore (Tagore, 1900, 1979), Michal Madhusudan Dutta, Haridas Goswami, Kaji Nazrul Islam, Ramtanu Lahiri (Dutta, 1996) have either mentioned the river in their writings or had sailed on a boat through this river. These events also tell us the glorious bygone days of the river Anjana.

16.3 ANTHROPOGENIC SIGNATURES ON FORMS AND PROCESSES

16.3.1 Cross-Profiles of the River

For understanding the present-day channel geomorphology of long and cross profile, 64 cross-profiles (18 in urban reach + 46 in rural reach) at an interval of 0.5 km from off-take to confluence were collected (18 in urban reach + 46 in rural reach) from Irrigation and Waterways Directorate, Government of West Bengal, S.D.O., I & P Subdivision, (C) No. II, Shantipur, Nadia. The average width of the river is 81.66 m (Table 16.5) with a maximum of 155 m at the south of Purba Badkulla (22 km downstream from off-take) and a minimum of 4.75 m at 50 m downstream from the off-take from the river Jalangi. The standard deviation of width is 45.50, and the coefficient of variation is 0.56. Out of 64 cross-sections, 16 are within the urban area of Krishnagar Municipality and 46 are in rural areas. The average width of urban reach is 30.04 m, while it is 101.86 m in rural reach.

TABLE 16.5

Comparative Statistics on Channel Morphology of the River in Urban Reach and Rural Reach Shows the Level of the Anthropogenic Signature on Channel Geomorphology

Reach Name	Statistics	Width (w)	Cross-Sectional Area (A)	Average Depth (d)	Maximum Depth	RL of the thalweg point of the Cross Section
Total reach	Max	155.00	541.06	5.41	7.02	6.75
	Min	4.75	1.81	0.38	0.76	0.75
	Average	81.66	175.42	1.96	3.58	3.97
	STDEV	45.50	135.64	0.88	1.37	1.29
	CV	0.56	0.77	0.45	0.38	0.33
Upper (Urban) Reach of the River Anjana from Off-Take to Culvert on Krishnagar Railway Station to Shaktinagar District Hospital						
Urban reach	Max	100.00	541.06	5.41	7.02	5.67
	Min	4.75	1.81	0.38	0.76	2.28
	Average	30.04	71.06	1.76	2.91	3.98
	STDEV	27.94	124.89	1.02	1.26	1.00
	CV	0.93	1.76	0.58	0.43	0.25
Lower (Rural) Reach of the River Anjana from Culvert on Krishnagar Railway Station to Shaktinagar District Hospital to Confluence with the River Churni						
Rural reach	Max	155.00	490.13	3.96	6.35	6.75
	Min	11.50	15.42	0.61	0.89	0.75
	Average	101.86	216.25	2.04	3.85	3.97
	STDEV	34.15	119.21	0.83	1.35	1.41
	CV	0.34	0.55	0.41	0.35	0.36

Out of 18 cross-sections on the reach (a reach of 5 km from off-take) within the city of Krishnagar, 17 (94.44%) have $w < \bar{w}$. In rural reach (a reach of 23 km), 10 (21.74%) have $w < \bar{w}$. Only one cross section having $w > \bar{w}$ is located within the municipality of Krishnagar at Rajar Dighi, widened artificially and maintained by Nadia Raj family (Table 16.6). On the other hand, in the reach of the rural areas, less affected by man, 36 (78.26%) cross sections have $w > \bar{w}$ (Table 16.6). In the city of Krishnagar, 14 (77.78%) cross sections are abnormally narrow and have $w < (\bar{w} - 1\sigma)$ and 4 (22.22%) have $w > (\bar{w} - 1\sigma)$. In rural reach, only 4 (8.70%) have $w < (\bar{w} - 1\sigma)$ and 42 (91.30%) cross sections have $w > (\bar{w} - 1\sigma)$. Of urban reach, 18 (100%) cross sections have $w < (\bar{w} + 1\sigma)$, whereas 36 (78.26%) are in rural reach. On the other hand, there are 10 (21.44%) cross sections have $w > (\bar{w} + 1\sigma)$ in rural areas. These sharp differences in the width of the channel between urban and rural reaches are because of the human hand and which is well visualized by different mean curve in Figure 16.9.

It is clear from the location of these cross sections that as the price of land for housing in Krishnagar city is five to ten times higher than that in rural areas. So, for housing purpose, encroachment/poaching of the river bed has narrowed down the width of the river in urban areas. Like the closing of tanks and ponds in urban areas, the river bed is also partially closed from both the banks. On the other side, the price of water bodies for pisciculture in rural areas is higher than cropping lands. So, the fishermen/(or others) did not encroach the river to keep a wide fishing area in the river bed.

Average bankfull depth d (hereafter depth) of 64 cross sections were derived by dividing the cross-sectional area by width (A/w). The maximum depth of 5.41 m was recorded at 4,150 m downstream from off-take (at Rajar Dighi) which was excavated and maintained by Nadia Raj family (Mallik, 1911, Dutta, 1996; Bandyopadhaya, 2007). The minimum depth of 0.38 m was recorded at 50 m downstream from the off-take at Parmedia (PLCN 01541600) from the river Jalangi. The average depth (d) of the river is 1.96 m. The average depth in urban reach was 1.76 and 2.04 m in rural reach (Table 16.6a).

Out of 64 cross sections (18 in urban reach + 46 in rural reach), 35 cross sections have $d < \bar{d}$ (15 in the 5 km reach in Krishnagar Municipality area and 20 in 23 km reach in rural areas). Only 3 out of 18 cross sections have $d > \bar{d}$ within the urban reach, whereas 26 (56.52%) out of 46 cross sections in rural areas have $d > \bar{d}$. Numbers of cross sections having $d > (\bar{d} + \sigma)$ are five in the reach of rural areas and one in the reach of the urban area of Krishnagar (Table 16.6b). This sharp contrast in the average depth (Figure 16.10) of the river also indicates strong control of urban man on the channel morphology of the river Anjana (http://riverbangla.com/author/tapas/).

Like width (w) and average depth (d), maximum depth (d_{max}) of the river Anjana is also much lower in urban reach of Krishnagar Municipality area than that of rural reach. The highest maximum depth of 6.35 m was also recorded at Rajar Dighi, 4.5 km downstream from off-take. The lowest d_{max} of 0.76 m was recorded at 50 m downstream from off-take. The average d_{max} is 3.58 m with $\sigma = 1.37$ m. For urban reach, average d_{max} is 2.91 m, while it is 3.85 m for rural reach (Figure 16.11).

Out of 18 cross-sections on the urban reach (a reach of 5 km from off-take) within the city of Krishnagar, 16 (88.89%) have $d_{max} < \bar{d}_{max}$. In rural reach (a reach of 23 km), 18 (39.13%) have $d_{max} < \bar{d}_{max}$. Only two cross sections having $d_{max} > \bar{d}_{max}$ are located within the municipality of Krishnagar at Rajar Dighi, deepened artificially and maintained by Nadia Raj family. On the other hand, in the reach of the rural areas, 28 (60.87%) cross sections have $d_{max} > \bar{d}_{max}$ (Table 16.6a). In the city of Krishnagar, 4 (22.22%) cross sections have $d_{max} < (\bar{d}_{max} - 1\sigma)$ and 14 (77.78%) have $d_{max} > (\bar{d}_{max} - 1\sigma)$. In rural reach, less affected by man, only 6 (13.04%) have $d_{max} < (\bar{d}_{max} - 1\sigma)$ and 40 (86.96%) cross sections have $d_{max} > (\bar{d}_{max} - 1\sigma)$. Of rural reach in the city of Krishnagar, 17 (94.44%) cross sections have $d_{max} < (\bar{d}_{max} + 1\sigma)$, and only one cross section has $d_{max} > (\bar{d}_{max} + 1\sigma)$. The figures are 33 (71.74%) and 13 (28.26), respectively, for rural reach.

TABLE 16.6

Descriptive Statistics of Width, Depth, Cross-Sectional Area and Maximum Depth of Two Distinct Reaches in Urban and Rural Areas of the River Anjana

	$w < \bar{w}$		$w > \bar{w}$		$w < (\bar{w} - 1\sigma)$		$w > (\bar{w} - 1\sigma)$		$w < (\bar{w} + 1\sigma)$		$w > (\bar{w} + 1\sigma)$	
Reach	Number	%	Number	%	Number	%	Number	%	Number	%	Number	%
A. Width (w)												
In urban areas	17.00	94.44	1.00	5.56	14.00	77.78	4.00	22.22	18.00	100.00	0.00	0.00
In rural areas	10.00	21.74	36.00	78.26	4.00	8.70	42.00	91.30	36.00	78.26	10.00	21.74
Total	27.00	42.19	37.00	57.81	18.00	28.13	46.00	71.88	54.00	84.38	10.00	15.62

	$d < \bar{d}$		$d > \bar{d}$		$d < (\bar{d} - 1\sigma)$		$d > (\bar{d} - 1\sigma)$		$d < (\bar{d} + 1\sigma)$		$d > (\bar{d} + 1\sigma)$	
Reach	Number	%	Number	%	Number	%	Number	%	Number	%	Number	%
B. Average depth (d)												
In urban areas	15.00	83.33	3.00	16.67	3.00	16.67	15.00	83.33	17.00	94.44	1.00	5.56
In rural areas	20.00	43.48	26.00	56.52	7.00	15.22	39.00	84.78	41.00	89.13	5.00	10.87
Total	35.00	54.69	29.00	45.31	10.00	15.63	54.00	84.38	58.00	90.63	6.00	9.38

(Continued)

TABLE 16.6 (Continued)

Descriptive Statistics of A. Width, B. Depth, C. Cross-Sectional Area and D. Maximum Depth of Two Distinct Reaches in Urban and Rural Areas of the River Anjana

	Reach	$A < \bar{A}$		$A > \bar{A}$		$A < (\bar{A} - 1\sigma)$		$A > (\bar{A} - 1\sigma)$		$A < (\bar{A} + 1\sigma)$		$A > (\bar{A} + 1\sigma)$	
		Number	%	Number	%	Number	%	Number	%	Number	%	Number	%
C. Cross-sectional area (w)	In urban areas	17.00	94.44	1.00	5.56	12.00	66.67	6.00	33.33	17.00	94.44	1.00	5.56
	In rural areas	18.00	39.13	28.00	60.87	4.00	8.70	42.00	91.30	36.00	78.26	10.00	21.74
	Total	35.00	54.69	29.00	45.31	16.00	25.00	48.00	75.00	53.00	82.81	11.00	17.19

	Reach	$d_{max} < \bar{d}_{max}$		$d_{max} > \bar{d}_{max}$		$d_{max} < (\bar{d}_{max} - 1\sigma)$		$d_{max} > (\bar{d}_{max} - 1\sigma)$		$d_{max} < (\bar{d}_{max} + 1\sigma)$		$d_{max} > (\bar{d}_{max} + 1\sigma)$	
		Number	%	Number	%	Number	%	Number	%	Number	%	Number	%
D. Maximum depth (d_{max})	In urban areas	16.00	88.89	2.00	11.11	4.00	22.22	14.00	77.78	17.00	94.44	1.00	5.56
	In rural areas	18.00	39.13	28.00	60.87	6.00	13.04	40.00	86.96	33.00	71.74	13.00	28.26
	Total	34.00	53.12	30.00	46.88	10.00	15.63	54.00	84.38	50.00	78.12	14.00	21.88

Source: Computed by the author from data of I&WWD, Nadia, 2012/13.

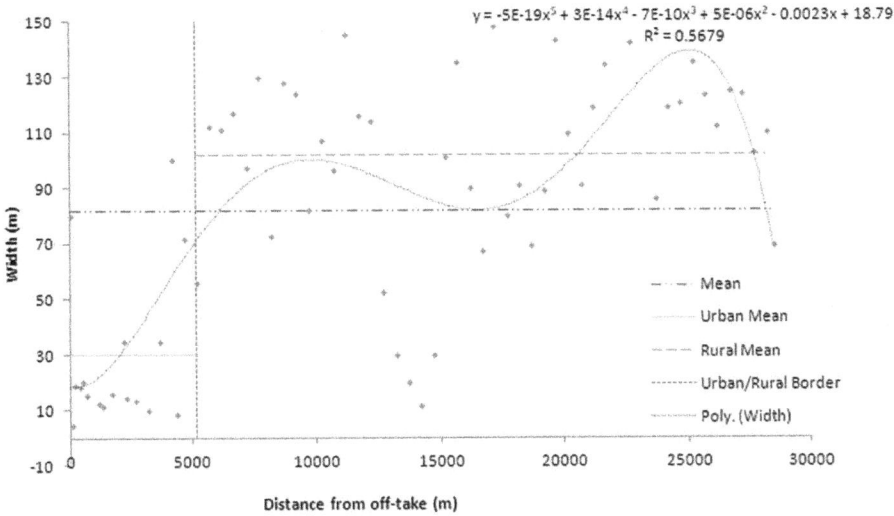

FIGURE 16.9 Downstream variation of channel width and sharp difference between urban and rural mean of channel width from total mean.

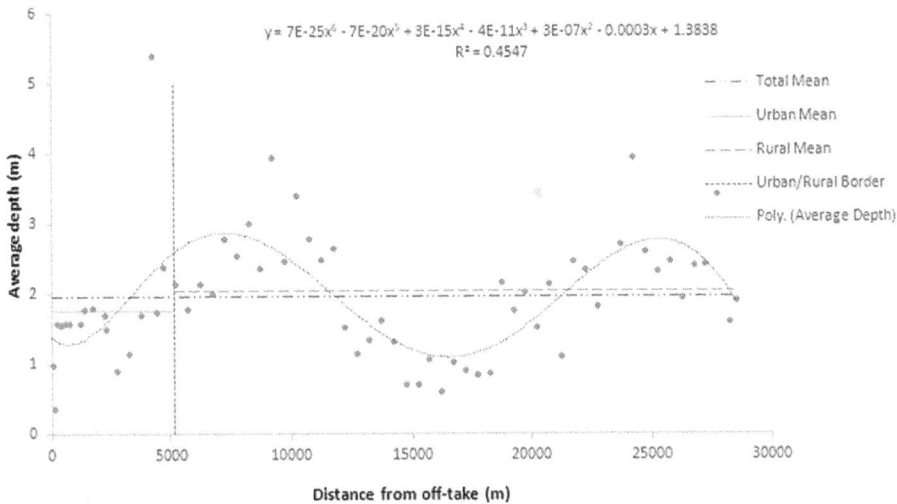

FIGURE 16.10 Downstream variation of channel average depth and sharp difference between urban and rural mean of channel average depth from total mean.

Width/depth ratio is another indicator of channel shape as well as human intervention on the channel morphology (Figure 16.12). Because of encroachment, the ratio is significantly low in urban reach (mean curve is shown as solid line) than that of rural reach (mean curve is shown as dashed line).

The cross-sectional area (A) of 64 stations is plotted against distance from off-take (D) in Figure 16.8, which shows a gradual increase in the cross-sectional area towards downstream. The highest cross-sectional area of $541.06\,m^2$ was found in Rajar Dighi. It is the minimum $(1.81\,m^2)$ near the off-take. The general trend represents the downstream increase in magnitude of channel cross-section (Figure 16.13). However, although this is true for all the rivers, yet in the case of the river Anjana, this is mainly because of anthropogenic interventions within the domain of the river. The first 5–6 km reach of the river is within Krishnagar Municipality where channel dimensions are

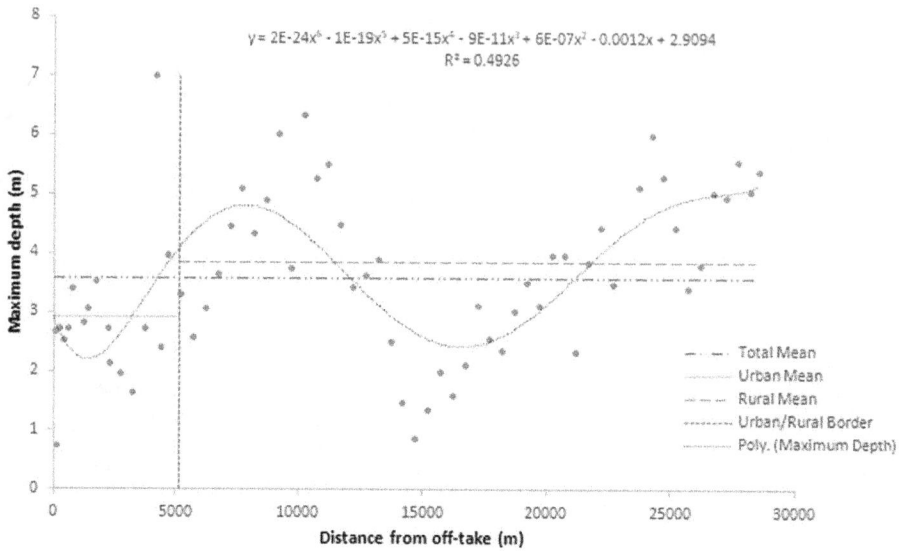

FIGURE 16.11   Downstream variation of channel maximum depth and sharp difference between urban and rural mean of channel maximum depth from total mean.

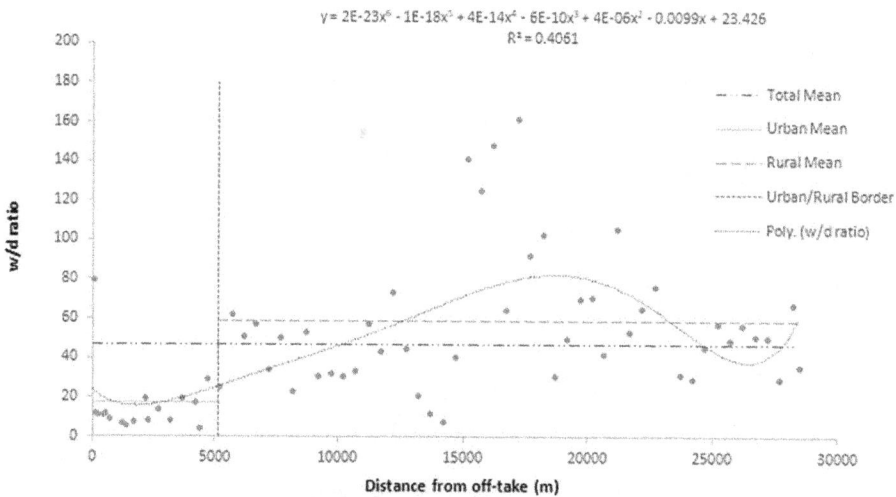

FIGURE 16.12   Downstream variation of channel width/depth ratio and sharp difference between urban and rural mean of channel width/ depth ratio from total mean.

reduced by human effort. But towards downstream, in rural areas, the channel remained wider and deeper because of lesser encroachment by man. All these signify the human impact on the channel morphology of the river Anjana.

Anthropogenic contribution in earth-forms (A^L) in the Anjana channel was derived as mean difference of magnitude of natural landforms ($\bar{x}$) from the magnitude of artificially modified observed landforms (x).

$$A^L = \frac{x - \bar{x}}{n}$$

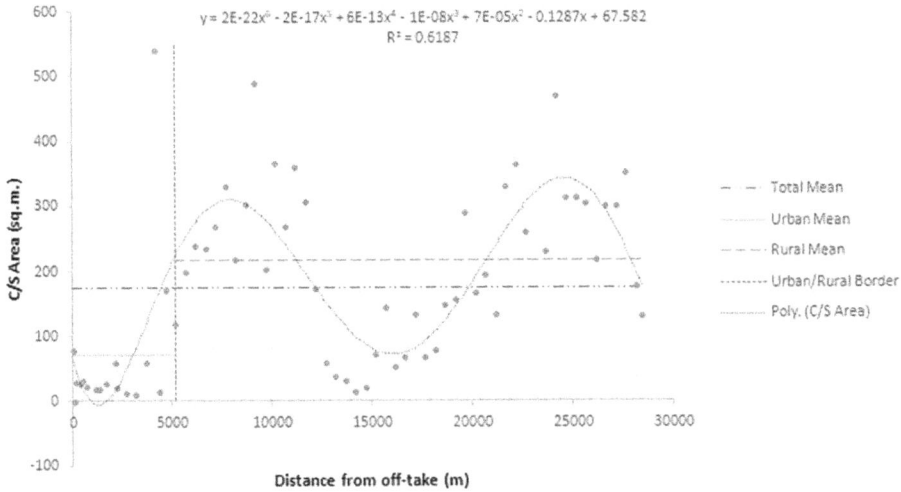

FIGURE 16.13 Downstream variation of channel cross-sectional area and sharp difference between urban and rural mean of channel cross-sectional area from total mean.

Average value of 64 cross-sections was taken as natural magnitude of dimension of landforms ($\overline{x}$) and n is the reach-wise number of observations.

If the value is positive, then the anthropogenic effort has increased the magnitude of the dimension of landforms and if the value is negative, then the anthropogenic effort has reduced the magnitude of the dimension of landforms.

Anthropogenic contribution in channel width, $\left(A^{cw}\right) = \dfrac{x - \overline{x}}{n}$

A^{cw} was −51.62 m for urban reach and 20.20 m for rural reach.

Anthropogenic contribution in channel depth, $\left(A^{cd}\right) = \dfrac{x - \overline{x}}{n}$

A^{cd} was −0.02 m for urban reach and 0.08 m for rural reach.

Anthropogenic contribution in channel cross-sectional area, $\left(A^{ca}\right) = \dfrac{x - \overline{x}}{n}$

A^{ca} was −104.36 m² for urban reach and 40.83 m² for rural reach.

Anthropogenic contribution in channel maximum depth, $\left(A^{cmax}\right) = \dfrac{x - \overline{x}}{n}$

A^{cmax} was −0.67 m for urban reach and 0.27 m for rural reach.

The volume of earth materials put into the river/removed from the river is crudely estimated as the product of the mean difference between the average cross-sectional area ($\overline{x}$) and observed cross-sectional area (x) and the reach length (l). For urban reach, $(x - \overline{x}) \times l =$ 104.36 m² × 5150 m = 537454 m³ of earth materials (soil and solid waste) was put into the river to poach into its bed.

16.3.1.1 Cross-Profile: Krishnagar Rajbari (Royal Palace)

The first human effort that impacted the channel morphology of the river Anjana during the historical time was perhaps the conversion of the river channel into the moat of the royal palace of Nadia Raj family. Raghab Majumder, the grandson of Bhavananda Majumder, shifted the capital from Matiari to Krishnagar in 1648. To ensure adequate depth of water in the ditch of the moat, the site of the royal palace was chosen on the bank of the river Anjana. River Anjana was deepened (maximum depth 7.02 m at bankfull stage) to convert it into the moat to the west of the palace. Moats were excavated to all the four sides of the royal palace leaving an entrance to the east and were

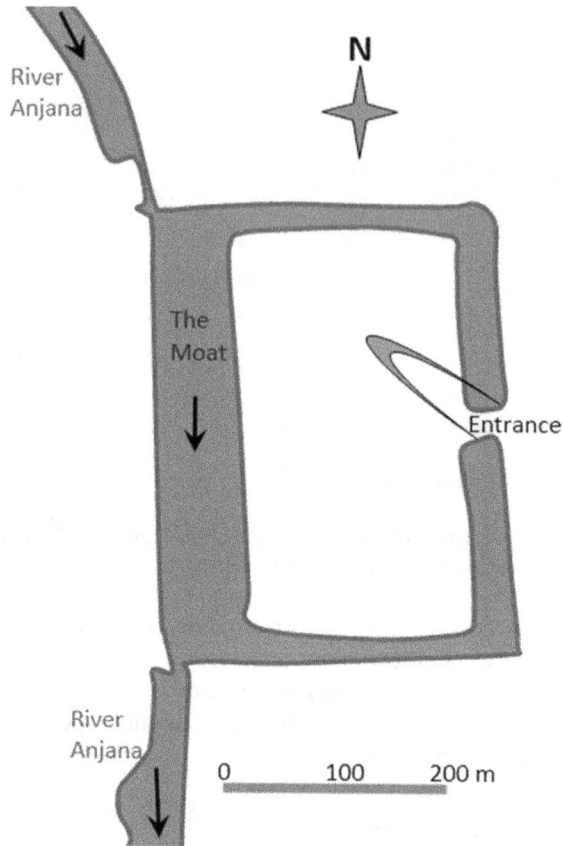

FIGURE 16.14   At Rajbari, the river was converted into the moat around the palace. Width, depth and shape of the channel of the river have been reshaped by royal interference.

joined with the river Anjana (Figure 16.14). The flow of the river Anjana would maintain the depth of water in the moat and safety of the palace. From off-take to Shaktinagar District Hospital, the average width of the river is 30.04 m, while it is 100 m at the moat to the west of the palace called Rajar Dighi. The average bankfull depth of the urban reach is 1.76 m, while it is 5.41 m at Rajar Dighi. This reshaping of the channel morphology of the river Anjana at a local scale was executed by the Nadia Raj towards the interest of the royal family. The anthropogeomorphologic transformation index $\left(R_{ag} = \dfrac{Va}{Vn} \right)$ as formulated by Lóczy and Pirkhoffer (2009) as the ratio between earth material translocated by anthropogenic processes (Va) to the earth material translocated by natural processes (Vn) can be applied to the moat around royal palace in terms of width, depth and channel cross-sectional area (A).

The anthropogeomorphologic transformation index for channel width $\left(R_{ag}^{w} \right) = \dfrac{Vw}{Vn}$ is given by

$$R_{ag}^{w} = \frac{100 \text{ m}}{30.04 \text{ m}} = 3.33$$

And for channel's average depth $\left(R_{ag}^{d} \right) = \dfrac{Vd}{Vn}$

$$R_{ag}^{d} = \frac{5.41\,m}{1.76\,m} = 3.07$$

The anthropogeomorphologic transformation index for channel cross-sectional area (A) combining both depth (d) and width (w) of the channel is given by

$$R_{ag}^{d} = 541.06\,m^{2}/71.06\,m^{2} = 7.61$$

16.3.1.2 Cross-Profile: Nadia Raj Family

The most impacted anthropogenic activity that caused the river Anjana to be deteriorated and left the most significant and lasting signature on the morphology of the river came in 1676 from Raja Rudra Ray, great grandfather of Maharaja Krishna Chandra Ray during monsoon months (Mallick, 1911). In that year, some Muslim naval forces from Murshidabad were passing by a navy vessel through the river. They made quarrel and conflict with the Maharaja's force. Next year, Maharaja Rudra Ray closed the off-take of the river Anjana from the river Jalangi so that any vessel cannot proceed towards "Rajbari" through waterways of Jalangi–Anjana route (Dutta, 1996). *Bargis* was a group of Maratha soldiers who indulged in large-scale plundering of the countryside of the western part of Bengal for about ten years (1741–1751) during the Maratha expeditions in Bengal. To escape *bargi* invasion, Maharaja Krishna Chandra shifted his capital in 1743 from Krishnagar to Shibnibas, a safer site on the bank of the river Churni. Shibnibas was encircled by a dilapidated ox-bow lake which was excavated to convert into the moat of the royal palace and named as *Kankana*. To prevent *bargi* invasion, adequate water depth in the moat was ensured by connecting the moat with the river Churni to the east and with the confluence of the river Anjana to the west (Mallick, 1911; Dutta, 1996; Chattopadhyay, 2007). So, Raja Rudra Ray and Maharajendra Bahadur Krishna Chandra kept their significant signature on the channel and basin geomorphology of the river Anjana.

16.3.1.3 Cross-Profile: Indigo Planters

For uninterrupted smooth navigation through rivers ways and for assured supply of adequate water for indigo processing, indigo planters were very much concerned about the river Anjana, and they made a significant change in the channel forms. The second branch of the river Anjana which is known as Heler Khal (canal), named after an indigo planter, was excavated in between Jaypur and Ichapur villages (Dutta, 1996). Peoples in Nadia district who were active participants of indigo revolt in the 19th century Bengal used to move through the river Anjana at midnight. But the indigo planters were informed about their movement through the Anjana. So, to stop their movement at midnight, indigo planters made earthen *bāndh* from both sides of the banks for partial closure of the river course. Then, central part of the river was fenced by iron grille mesh at Joypur indigo factory (Dutta, 1996). All the indigo factories on the bank of the river Anjana (Table 16.4) had their significant signature on channel geomorphology in the form of excavation of canal, dredging of the channel to improve navigability locally, making barriers across the channel, making *ghat*, etc.

16.3.1.4 Cross-Profile: Urbanisation and Constructions within Anjana

In the editorial introduction of the first volume of the book *Rivers of Bengal*, Biswas (2001) commented that "Modern man is intoxicated by the power he has acquired through his scientific inventions and technological skills and he viewed the world as a mere object of his enjoyment". Man's attitude towards rivers is nothing different from the statement of Biswas (2001). He exploited the river to his level best and never thought about rivers' health. As a result, in urban areas, like all other rivers, the Anjana is deadly exploited to change it as sewerage. There are only two urban centres on the course of the river Anjana. One is Krishnagar Municipality, and the other is Baruihuda Census Town. In 2001, total area under Krishnagar Municipality was 15.96 km². The total population of the

Krishnagar Municipality was 121,110 in 1991 which has been 153,062 in 2011. Population of the town was 121,110 in 1991; 139,110 in 2001; and 153,062 in 2011. The decadal growth rate of the population of Krishnagar Municipality from 1991 to 2001 was 14.9% and from 2001 to 2011 was 10.03%. Baruihuda is a small census town having a geographical area of 1.9 km², and the population in 2001 and 2011 was 9,599 and 11,474, respectively. The decadal growth rate of Baruihuda was 19.53%.

To keep pace with the rapid growth of the population in the town, vertical expansion of the town started in 2003, when Mr. Paresh Kundu, the then proprietor of Popular Pharmacy, first introduced flat "Sriniketan" for sale in Chunaripara of Krishnagar. During the 2010s, the town started sprawling both horizontally and vertically to the west of the NH-34. Yet, horizontal expansion of the town around its old centre is also continuing. Obviously, the pressure is felt on the river, and it is being encroached significantly.

The old course of the river Anjana in Ward No. 24 is almost closed and erased under constructions of the built-up area except some ponds aligned along the erased course (Figures 16.3 and 16.4). There is a remnant of palaeochannels of the river Anjana at present-day Nagendranagar 4th and 5th lanes, north of the *Old Circuit House, Meen Bhawan* (Fishery Department, Govt. of West Bengal), *Jalangi Bhawan* (Irrigation and Waterways Department, Govt. of West Bengal) and Krishnagar Government College Hostels. The present-day artificially straightened channel from off-take to the road crossing at Government Girls High School is choked by urban solid waste dumped within the river bed. This section of the reach was re-excavated during 2011 and 2013 but remained unchanged.

Construction within the river Anjana bed is a common phenomenon within the urban area of Krishnagar Municipality (Figure 16.16a and b). At the end of Bejikhali Lane and at A.C. Patra Road near Rajar Dighi and North Garh (Figure 16.15a), construction by poaching the river bed is at such an extent that it is difficult for a stranger to trace the last sign of the river Anjana. About 32 houses (Table 16.7) are constructed within the river bed where Manashadaha Lane crosses the river Jalangi (Figure 16.16). This scenario is common all along the course of the river within the city of Krishnagar (Figures 16.17 and 16.18).

The reach of the river from road crossing on Krishnagar–Nabadwip road to Bejikhali road culvert is highly encroached and poached due to urbanisation (Figure 16.19). Several houses have been constructed partly or fully within the river channel. The most surprising matter is that all these house owners have their legal papers for illegal occupation of the natural channel of the river Anjana. As they claimed, they have their purchase deeds, and they have sanctioned (by the authority of Krishnagar Municipality) house plan but were unwilling to show those papers! In answering

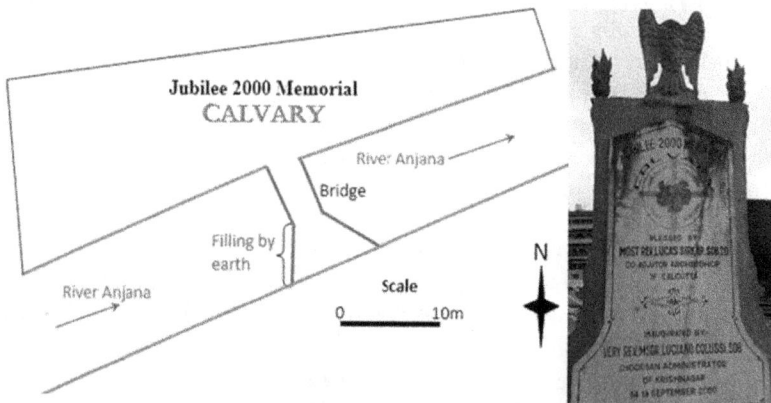

FIGURE 16.15 Jubilee 2000 Memorial Calvary was constructed by Krishnagar Cathedral authority in 2000. A bridge was constructed over the river Anjana, filling about half of the river's width by the earth which was really a nail on the dying river.

TABLE 16.7

Encroachment within the River Anjana

Sl. No.	Locality	Nature of Construction	The Built-up Area within Anjana River (m²)	Number of Houses or Buildings
1.	Culvert on S.K. Basu Road Near Banasree Para to Baghadanga Culvert	Commercial	1,347	17
2.	Baghadanga culvert to Cathedral culvert	Residential + commercial	1,299	8
3.	Culvert in front of Cathedral to Bejikhali culvert	Residential + commercial	19,056	75
4.	Bejikhali culvert to Nediar Para culvert	Residential + commercial	2,236	28
5.	Nediar Para culvert to Culver on A.C. Patra Road near Harijan Para	Residential + commercial	9,110	44
6.	A.C. Patra Road near Harijan Para to South end of Rajmatar Dighi	No encroachment	0	0
7.	South end of Rajmatar Dighi to Manasadaha Lane Culvert	Residential + commercial	6,875	31
8.	Manasadaha Lane Culvert to Raja Road Culvert	Residential + commercial	6,472	10
9.	Raja Road Culvert to Shaktinagar Hospital Culvert on Kabiguru Road	Residential + commercial	7,259	34
10.	Shaktinagar Hospital Culvert south	commercial	3,496	9

FIGURE 16.16 (a) Encroachment to the north of Raj Matar Dighi shown by encircling the area. (b) Constructions within Anjana River bed at the culvert on Manashadaha Road have been legalised by Krishnagar Municipality (by sanctioning plans) and by L & LRO (by giving purchase deeds).

questions (2008–2012) by the present author, Sri Prabhat Saha, vice chairman of the municipality of Krishnagar said that no housing plan encroaching Anjana River is being passed during their term. He accepted the fact of encroachment of the river but was unwilling to remove those encroachers.

In front of Krishnagar Cathedral, several houses are built up within the Anjana bed narrowing its width alarmingly. In this reach of 615 m, the largest width of the channel at present is 25 m which is only 4–5 m at places. The width of the channel is encroached from both banks by constructions of residential or commercial buildings and even by cathedral authority narrowing down its

FIGURE 16.17 Severe encroachment of the river Anjana within the urban reach at Bejikhali Mor, Krishnagar. The arrow shows the course of the river.

FIGURE 16.18 (a) Construction within river bed and fragmentation into ponds of the river Anjana at the culvert on Station to Shaktinagar Hospital approach road. (b) Silent poaching of the river bed by dumping solid waste. The thin dotted line marks in filled ground made by putting plastic and other solid waste within the river.

FIGURE 16.19 Construction within the River Anjana (a) at Bejikhali Para Krishnagar Municipality (maximum and minimum widths of the channel is 50.5m and 4.0 m, respectively) and (b) at RC Para, Krishnagar Municipality (maximum and minimum widths of the channel is 25m and 4.0 m, respectively). Places labelled as smaller alphabets A = Dr. Bidhanchandra Ray Statue, B = Swami Vivekananda Statue, C = Sadar Hospital Mor and D = Krishnagar Cathedral.

width (Figure 16.14). It appears from the microterrain of the area that the pond to the north of the Cathedral's Jubilee 2000 Memorial Calvary was once the channel of the river Anjana.

16.3.1.5 Cross-Profile: Fishing

Widening of river cross-profile, opposite the situation of urban areas, by anthropogenic effort in the river Anjana is exclusively confined to pisciculture and obviously in rural areas. Next to Nadia District Hospital at Shaktinagar, the width of the river is widened or at least original width is maintained for pisciculture by clearing hyacinths and other aquatic weeds and waste. Immediate downstream of the district hospital, the width of the channel is 95 m, and including banks, it is 140 m. At Dogachi and around, the river is intensively used for pisciculture and its channel width is maintained by cutting of banks and clearing of weeds and wastes by fishermen.

Actually, downstream to district hospital up to Jalalkhali, the width of the river including banks ranges from 100 to 200 m and nowhere is less than 100 m (Figure 16.20).

FIGURE 16.20　Along Dogachi gram panchayat, the river width is retained for pisciculture.

Channel geomorphology of the river Anjana became more and more artificial during the mid-twentieth century when the government leased the river to fishermen or their cooperative society. The river was brought under *Bengal Tank Improvement Act 1945* and was fragmented into many ponds of private ownership, such as Ghosh Pukur, BosePukur and Rajmata Dighi (Dutta, 1996).

Zamindar Haralal Bandyopadhyay, Kishorilal Bandyopadhyay and Bhuban Mohan Bandyopadhyay made several *bāndhs* across the river and fragmented it into different ponds (Dutta, 1996). *Raj Matar Dighi, Manosa Pukur, Ghosh Pukur, Bose Pukur,* etc. are the example of such ponds (Figure 16.21). At the Dogachi gram panchayat, such ponds are now being used for pisciculture by the Fishermen Cooperative Society. During 1945–1946, Sarat Chandra Biswas, a refugee from Faridpur of present-day Bangladesh, got lease a reach of the river from Baruihuda to Jalalkhali and started fish cultivation (Dutta, 1996). Significant change of the river was made by widening its width, dredging its bed and fragmenting its length into ponds (Table 16.8). In 1945, fishermen's cooperative took ownership of the river through lease and started pisciculture. The number of the cooperative societies of fishermen increased by 1951, and the river was fragmented by earthen *bāndh* into many pieces. A part of Anjana River at and near Krishnagar came under Bengal Tank Improvement Act (1961–1962), and Anjana was segmented into ten ponds for pisciculture (Dutta, 1996). This practice not only fastened the decaying and dying processes of the river but also fragmented river sections were captured by some powerful men, and the un-organised poor fishermen were deprived of catching their livelihood from the river.

16.3.1.6　Cross-Profile: Road and Rail Crossings

There are 18 major road crossings/culverts on the river Jalangi from its off-take to the culvert at district hospital at Shaktinagar, within the city of Krishnagar. From the culvert near the district hospital to Jalalkhali, there are seven road crossings on the river. From Jalalkhali to the confluence with the river Churni at Byaspur, there are six road crossings on the river Anjana. The Krishnagar–Ranaghat railway line crossed the river Anjana twice. Therefore, on the main course of the river, the total number of road crossings is 33. Two railway crossings and two road crossings at Jalalkhali and Badkulla are quite open, and wide passage is given for the river to pass. Except these four, all the 29 road crossings are designed and made in such a way that the river is almost closed by earthen cross-*bāndh* leaving only a narrow passage for the river (Figure 16.22).

16.3.2　Long Profile of the River: Cultivation of Crops within the River Bed

The reach of the river from Jalalkhali to Gopalpur is decayed to such a level that the RL of the river bed is almost to the level of the surrounding ground. As a result, except for one or two

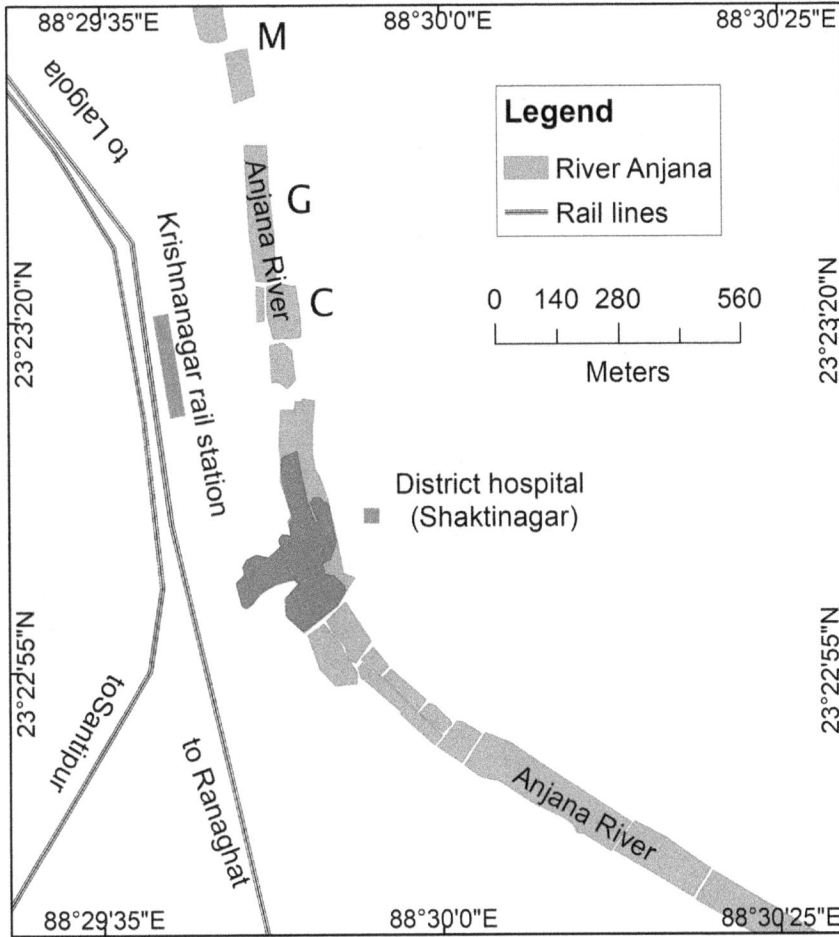

FIGURE 16.21   The river is fragmented into several small ponds. The northmost one M = Manosa Daha or Pukur; G = Ghosh Pukur; C = Chowdhury Pukur.

TABLE 16.8

List of Ponds Created by Fragmented Anjana from Krishnagar Cathedral to Kabiguru Road, Dogachi

Sl. No.	Fragmented Reach Name	Extension	
		From	**To**
1	Joler Para	Culvert in front of Cathedral Church	Bejikhali Culvert
2	Bejikhali	Bejikhali culvert	H.C. Sarkar Road Culvert (At Harijan Para)
3	Rajmatar Dighi	H.C. Sarkar Road Culvert (At Harijan Para)	End of the moat of Rajbari
4	Manasa Daha	Manasadaha Lane Culvert	Raja Road Culvert
5	Ghosh Pukur	Raja Road Culvert	Haritaki Tala Culvert on Choudhury Para approaching road
6	Choudhury Pukur	Haritaki Tala Culvert, Choudhury Para approaching road	Culvert of Railway Station to Shaktinagar Hospital approaching road
7-14	Tank Nos. 1–8 (Figure 16.21)	Shaktinagar Hospital	Kabiguru Road, Dogachi

FIGURE 16.22 A part of the river along Dogachi Gram Panchayat is fragmented into tanks (Nos. 1–8) under "The Bengal Tanks Improvement (West Bengal Amendment) Act, XXIV 1948".

months of the monsoon period, the river remains dry, and peasants cultivate paddy, jute and other crops within the river bed and alter and rework on channel geomorphology. They reshape the original gradually slopping channel into a stepped banked channel. Fishermen also flattened the river bed (Figure 16.23) for better utilisation of the water resources for their purpose, reshaping channel geometry.

16.3.3 Flood Checking Embankment and Road Network

To protect the city of Krishnagar from flood water of the river Jalangi, an earthen embankment of 6.32 m height and 1,700 m length (total length of the embankment is 2,300 m out of which 1,700 m is earthen and 600 m is of concrete wall) was constructed in 1972 along the northern boundary of the municipality from Kadamtala Ghat to the *Dwijendra Setu* of NH-34 on the river Jalangi under the regency of Sri Kashikanta Maitra, MLA and Food Minister of the Government of West Bengal. The embankment is along the left bank of the river Jalangi and across the river Anjana closing its off-take from Jalangi. It saved the city from being inundated by flood, killing the dying river Anjana. The average cross-sectional area of the earthen part of the embankment is 141.458 m² (Figure 16.24), and the total volume of earth translocated was 240,478.6 m³.

This embankment was designed to the objective of arresting floodwater from the river Jalangi which would inundate frequently the town of Krishnagar. This obviously met the objective but at the cost of the river Anjana. Before this embankment, during monsoon months, the river Anjana would get water from its mother channel Jalangi. Embankment stopped it permanently.

Except for this flood-checking embankment, numerous embankments within the basin have significantly reshaped the landforms. Some of these embankments are converted into metaled roads to facilitate road transport such as Krishnagar–Hanskhali road and Krishnagar–Badkulla road. Keeping pace with the urbanisation, transport network of the basin has developed significantly (Table 16.9). During 1917–1921, the Krishnagar–Badkulla road was unmetalled. Road density was also very low (Figures 16.25 and 16.26).

FIGURE 16.23 (a) Blocked-road crossing leaving a pipe passage of few centimetre diameters. (b) Private culverts (leaving very little passage for the river) for to and from houses in front of Krishnagar Cathedral.

But with the passage of time, road density of the basin increased to 0.97 km/km^2 (Figure 16.27) in 2019. The total length of the roads of the Anjana basin, as derived from Google, was 86.89 km. Particularly, in the municipality area of Krishnagar, the road density was increased many folds.

16.3.4 BRICKFIELDS

Brickfields play a significant role in reshaping the earth forms by cutting soil and making ditch at one place (negative change), and dumping of soil makes a positive change of relief increasing its RL temporarily at another place. There are four brickfields on the bank of the main channel of the river and one kiln on the bank of the left branch of the river Anjana.

The purpose of excavating ponds on river banks by brickfields is to trap silt in those ponds deposited during floods (Das, 2014). But the river Anjana is almost a stagnant waterbody and not a flowing river. So, silt trapping opportunity is nil. They make ponds on banks mainly to ensure an adequate supply of water for brick making. This kind of human intervention widens and deepens the channel reshaping channel geometry and bringing changes in the channel-line direction locally. There are 11 brickfields within the basin area out of which five are located on the bank of the river Anjana (Figure 16.28 and 16.29 and Table 16.10).

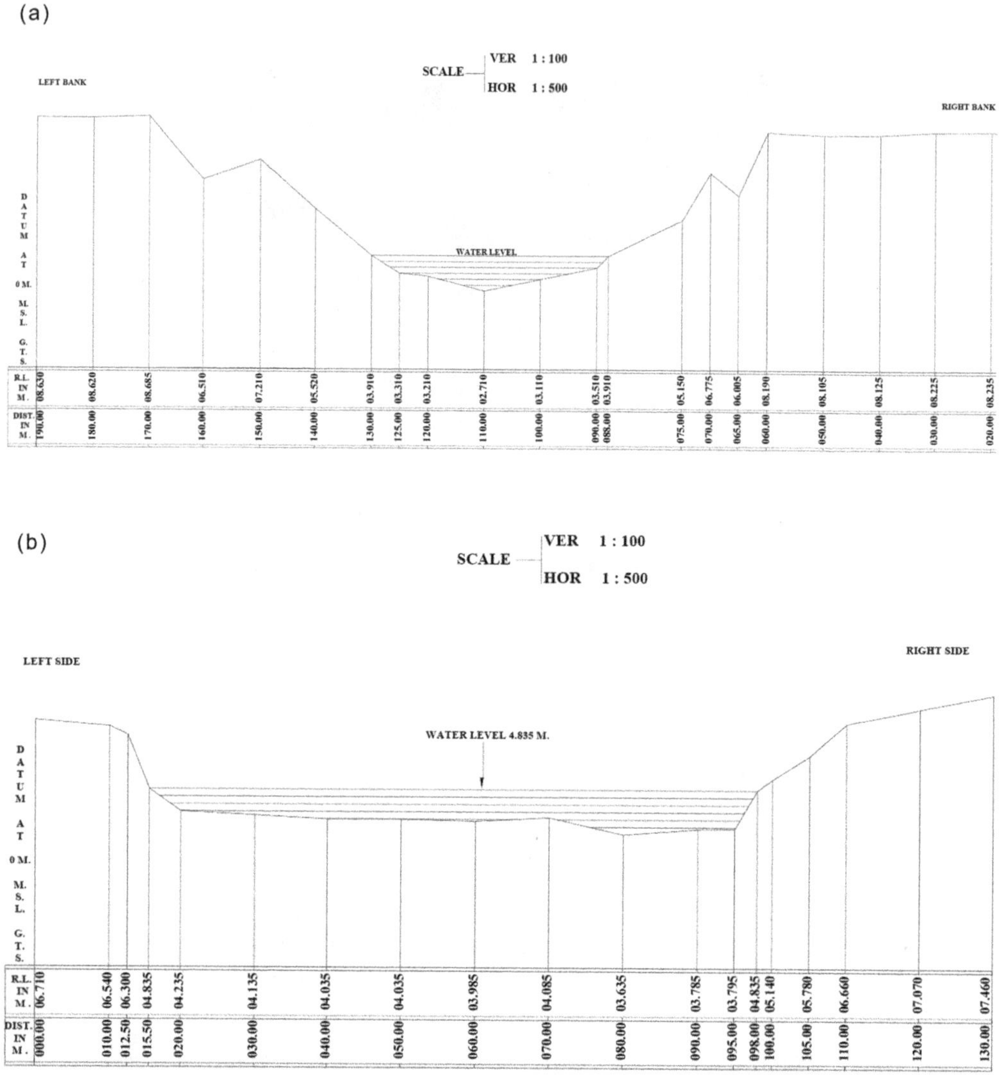

(a)

(b)

FIGURE 16.24 Fishermen reshape the channel geometry at Dogachi. (a) Gradually slopping arc-shaped natural channel and (b) flatbed reshaped by fishermen.

TABLE 16.9

Man's Effort in Reshaping the Earth in the Form of Roads and Embankments

Basin Area (m²)	Basin Area (km²)	Basin Perimeter (m)	Road Length in the Basin (m)	Road Density (km/km²)	The Volume of Earth Translocated (m³)
89,541,176	89.54	70,719.54	86,893	0.97	1,737,860

During the first decade of the 21st century, soil cut per brickfields was much less than now. D.L. & L.R.O., Nadia, Government of West Bengal (2009) record revealed that average soil cut by each brickfield is 133,110 ft³. Therefore, the total volume of soil translocated by brickfields annually in the basin of the river Anjana, at the lowest estimate, is >1,464,210 ft³. As the river Anjana is a dead spill and has no flow even in monsoon months (except days of rains when overland flow finds

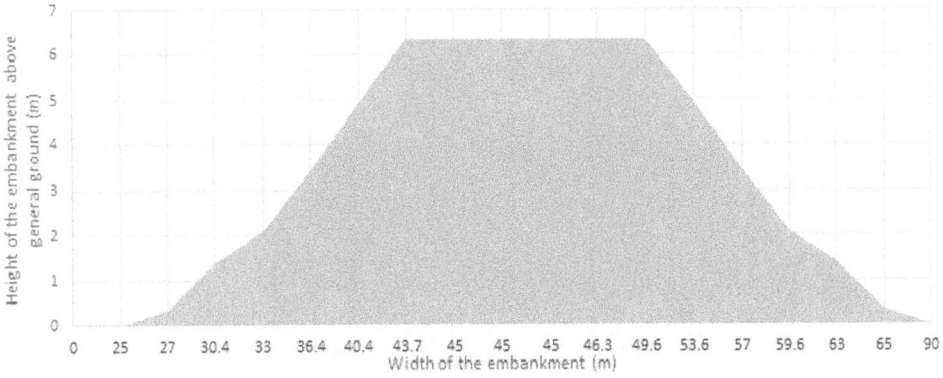

FIGURE 16.25 Cross section of the earthen embankment along the left bank of the river Jalangi and closing the entrance of the river Anjana to protect the town Krishnagar from being inundated by floodwater.

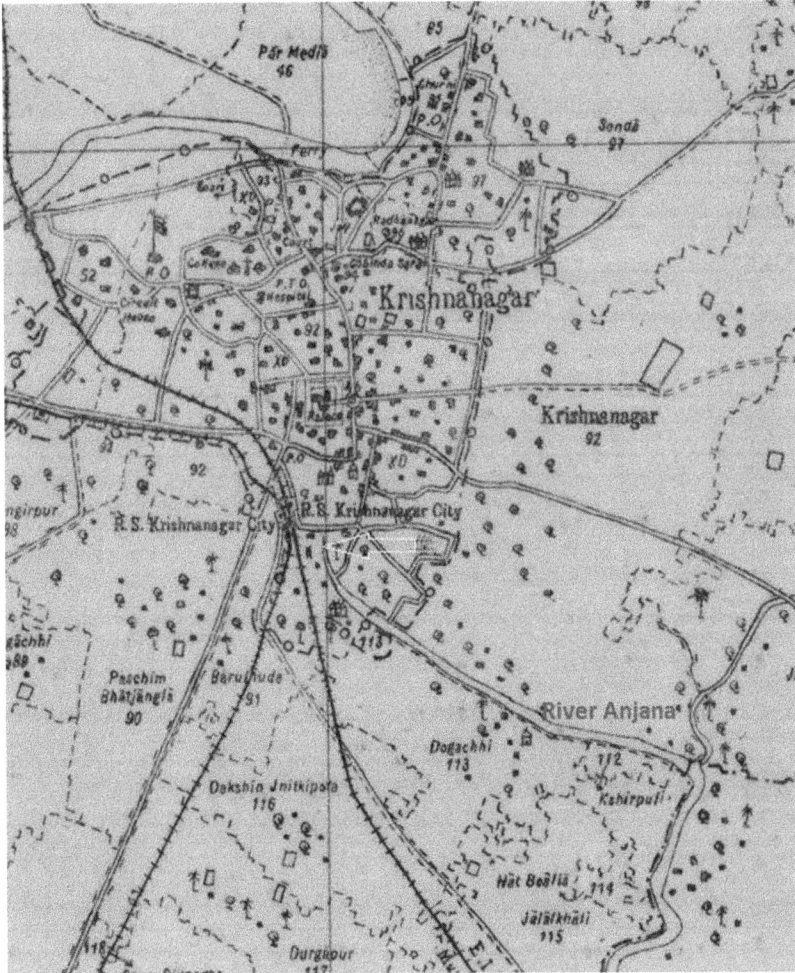

FIGURE 16.26 The river Anjana is absent within the city of Krishnagar in this Police Station Map (surveyed in 1917–1921) of Krishnagar (the end of the river is marked by the arrow). During that period, the Krishnagar-Badkulla road was unmetalled.

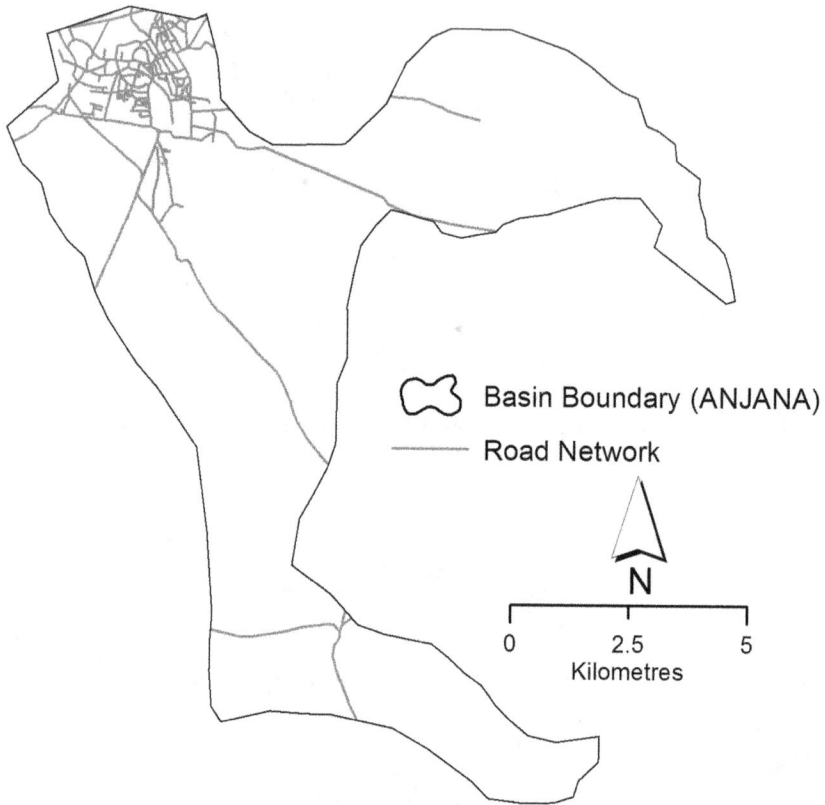

FIGURE 16.27   Road network in urban and rural areas of the basin. (Prepared by Dr. Suvendu Roy.)

FIGURE 16.28   Hatboyalia Brick Fields on the right bank of the main branch of the Anjana.

its path through Anjana to the river Churni), sediment carried annually by the river Anjana to the river Churni and no way supersedes the estimated volume of soil translocated by brickfields of the basin. Therefore, anthropogeomorphic process is playing the leading role over natural geomorphic processes in reshaping and reworking the landforms of the basin.

FIGURE 16.29 (a) Dogachi Samabay Brick Field has reshaped the bank and (b) brickfields at Bhutpara on the left bank of the Heler Khal widened the river channel and modified channel morphology significantly. Reshaped channel and river banks as *Khadans* are marked by arrows.

Anthropogeomorphologic transformation index of brickfields in Anjana River basin is

$$\left(R_{ag}\right) = \frac{\text{Soil moved by brickfields}}{\text{Soil moved by Anjana River system}}$$

$$R_{ag} = \frac{> 1464210 \text{ ft}^3}{< 1464210 \text{ ft}^3} => 1.0$$

16.3.5 Land Use Land Cover (LULC) Change as the Anthropogenic Signature on Landforms

For detecting the changing pattern of land use, satellite images (Landsat 5 TM C1 Level-1 images with path/row 138/44) of 1990 and 2011 were collected from USGS Earth Explorer. In the pre-processing stage of image classification, all the layers have been stacked which have become the subject to supervised maximum likelihood classification (Figure 16.31). In this classification technique, a number of training signatures have been identified and finally, the image has been classified into four categories (water body, green coverage, built-up area and others). The accuracy of the classification has been estimated using two indices and taking help from the topographical map

TABLE 16.10

List of the Brickfields on the Bank of the River Anjana

Sl. No.	Distance from Off-Take	Name of the Brickfields	Locality	Channel Width (m)	Channel Width at Kiln (m)	Channel Depth (m) from Bankfull Stage (at 500 m Upstream) & RL at Thalweg	Channel Depth at Kiln (m) from Bankfull Stage) & RL at Thalweg	*Khadan* (Pond) Size Perimeter (m)	Area (m²)
1	200	SONA	Parmedia (PLCN 01541600), off-take at Jalangi river 23°25′07″ N 88°28′53″ E	4.75	4.75	2.585 (3.625)	2.600 (3.625)	Nil	Nil
2	8,534	Dogachi Samabay Brick Field (DSBF)	Dogachi 23°22′10″ N 88°31′10″ E	70	80	2.53 (5.860)	1.52 (5.840)	1,292	19,686
3	11,471	SHAKTI	Hatboyalia 23°21′18″ N 88°31′39″ E	63	98	2.140 (4.475)	2.68 (4.350)	NA	NA
4	21,650	STAR	Badkulla 23°17′50″ N 88°31′43″ E	45	21	2.95 (4.565)	3.47 (4.360)	NA	NA
5		JYOTI	Bhutpara, on the bank of Heler Khal 23°22′55″ N 88°32′09″ E	39	150	1.47 (4.485)	1.88 (4.975)	628	14,969

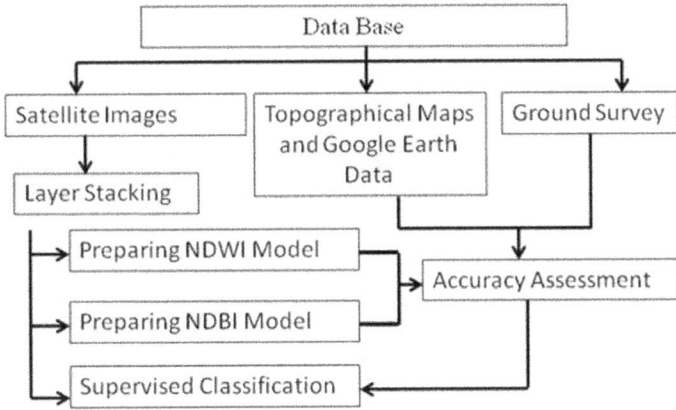

FIGURE 16.30 Methods applied for detecting decadal LULC change.

(NF 45-3 and NF 45-4 within series U502, scale 1:250000), google earth image and ground survey (Figure 16.31). Output pixel values were compared in GIS environment (NDWI & NDBI) for further accuracy/justification. To minimize error, before image classification, we did the NDWI, and the NDBI model for identifying the pixel value to prepare the LULC map. Due to lack of ground data, we did not measure statistical accuracy of our findings.

For extracting waterlogged areas, normalized difference water index (NDWI) has been calculated (McFeeters, 1996) using the following formula:

$$NDWI = (Green - NIR)/(Green + NIR)$$

where Green is the reflectance of green band and NIR is that of the near-infrared band. The index value of NDWI ranges from –1 to +1 where the higher negative value indicates the bright surface having almost no vegetation and water content, whereas the higher positive value is the indication of the presence of water bodies.

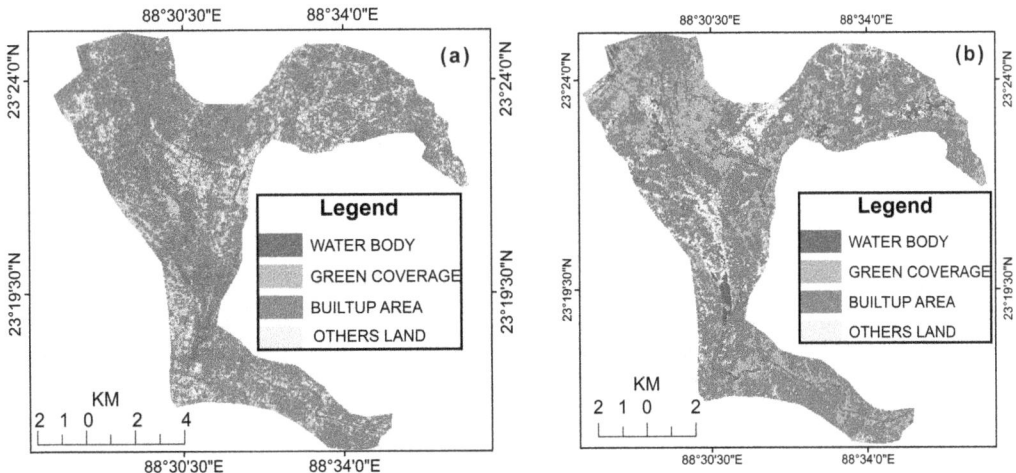

FIGURE 16.31 LULC of Anjana Basin in (a) 1990 and (b) 2011.

TABLE 16.11

LULC Changes in the Anjana River Basin from 1990 to 2011

LULC Categories	1990		2011		LULC Change from 1990 to 2011	Percentage Change to Individual Category	Percentage Change to Total Area
	Area (ha)	Area (%)	Area (ha)	Area (%)	Changes (ha)		
Waterbody	1,159.59	12.95	609.18	6.8	−550.42	−47%	−6.15
Green coverage	1,701.85	19.01	2,207.79	24.66	505.94	30%	5.65
Built-up	5,250.76	58.64	5,553.95	62.03	303.19	6%	3.39
Other lands	841.92	9.4	583.2	6.51	−258.72	−31%	−2.89
Total	8,954.12	100	8,954.12	100	−0.01		

Similarly, for extracting built-up areas, normalised difference built-up index (NDBI) has been calculated (Zha et al., 2003) using the following algorithm:

$$NDBI = (SWRI - NIR)/(SWRI + NIR)$$

where SWRI is the reflectance of the short-wave near-infrared band, and NIR is that of the near-infrared band. NDBI value ranges from −1 to +1 where the higher negative value represents the water bodies, whereas the higher positive value is the indication of the built-up area.

In 1990, the waterbody was 12.95% of the basin area, whereas in 2011, it was only 6.80% (Table 16.11). A total of 550.42 ha areas of water bodies were erased out which was 47% of total water bodies and 6.15% to the total basin area. This indicates the substantial decrease in water bodies by the expansion of built-up areas (most significant sector of man-altered geomorphology) on *made ground* and agricultural land. The green coverage area was 1,701.85 ha (19.01%) in 1990 and 2,207.79 ha (24.66%) in 2011. The total increase was 505.94 ha which was 30% to green coverage and 5.65% to the total basin area.

The built-up area was 5,250.76 ha (58.64%) in 1990 and 5,553.95 ha (62.03%) in 2011. The total 303.19 ha (6% to the total built-up area) lands increased under this category was 3.39% of the total basin area. This increase in the *made ground* was possible at the cost of water bodies and other types of lands.

16.3.6 Dumping and Debouching of Solid and Liquid Wastes and Chocking and Filthening of the River

The reach of the river Anjana in the Krishnagar Municipality area is simultaneously a combination of sewerage and dumping ground in the true sense (https://www.youtube.com/watch?v=_pPtNL-guR8I retrieved in December 2019). All the drains of wards 24, 18, 17, 15, 13, 12, 11 and 10 of the municipality are released into the river. As a result, the flow within the city is only of sewage of households and commercial places. Flow is extremely filthy, and no aquatic animals including fish can thrive in it. Moreover, not only the inhabitants on banks throw garbage into the river bed but also the municipality use the river for their dumping ground (Figure 16.32). These practices not only choked the course but also degraded the riverine environment to such a level that it became the birthplace of mosquito and other insects and microorganisms very harmful to human health. The filthy smell of water spread into the air causing air pollution.

FIGURE 16.32 Dumping plastic and other solid waste within the river Anjana at Bejikhali Culvert not only by local vendors but also by sweepers/drain-cleaners of the municipality.

16.3.7 PROJECT "SUJALA NADIA"

The natural processes of decaying and dying of deltaic rivers are fastened by human activities in the present-day world. Anjana is one of such rivers. It is dead now. But not ought to. Because the highest flood level (HFL) and lowest water level (LWL) of the river Jalangi found at the gauge station at *Dwijendra Setu* from 1978 to 2013 (36 years) are 9.13 and 4.40 m, respectively. So, the mean water level of the river Jalangi is $\left(\dfrac{9.13+4.40}{2}\right) = 6.77$ m (I&WD 2013). Due to the construction of Farakka Barrage, the water level of the river Bhagirathi has been increased considerably in comparison with the pre-Farakka period. As a result, the backpressure of water from Bhagirathi has risen the level of water of the river Jalangi. The water level of Jalangi on 25 April 2012 was 2.645 m, and the lowest bed level of the river Anjana at off-take was 2.935 m. Therefore, except for a few days in the rainy season, water from the river Jalangi does not enter the river Anjana. But hope is that the water level (1.620 m) of the river Churni on the same day was 1.025 m (2.645–1.620 m) lower than the water level of the river Jalangi, and the bed level of the river Anjana at confluence was 4.255 m. Therefore, if excavated sufficiently, the water from the river Jalangi will move through Anjana to the river Churni. In November 2013, the re-excavation under the "SUJALA NADIA" project started from the Anjana–Churni confluence at Byaspur to the off-take of the river from the river Jalangi at Krishnagar. The main aims of this project were

 i. To renovate the river Anjana for hasty pass of floodwater
 ii. To link the river Jalangi with the river Churni.

The estimated cost of re-excavation was Rs. 3,20,46,000 for 2,373 m (ABP 04.12.2013, http://archives.anandabazar.com/archive/1131204/4mur1.html). The total length of the river Anjana is 29 km. So, the total cost will be Rs. 39,16,28,320. According to the executive engineer of the I&W Directorate of District of Nadia, the cost of re-excavation was Rs. 74, 04,000 per 850 m. Therefore, the total cost for a total length of 29 km was Rs. 25,26,07,060. But some experts on Anjana expressed their doubts about the prospect of the project (ABP 04.12.2013). However, the project stopped.

16.3.7.1 Past Experience of Re-excavation

During 1971–1972, the off-take of Anjana was absolutely detached from the Jalangi by constructing flood checking embankment (*bāndh*), after 100 years of constitution of Krishnagar Municipality in 1864. In May of 1985, Sri Sadhan Chatterjee, Ex Member of Legislative Assembly (MLA) of West Bengal from Krishnagar and Sri Sukumar Mondal, Ex MLA from Hanskhali, proposed for the re-excavation of Anjana according to 185 amendments. It was the first endeavour of the re-excavation of this river (Dutta, 1996). Later on, from 2006 to 2011, Anjana was re-excavated from Gopalpur (near Badkulla) to Nagendranagar (in Krishnagar) funded by MLA fund of Sri Subinay Ghosh, Ex MLA from Krishnagar. But within a span of two years, the channel was jammed by the huge growth of shrubs, hyacinth and dumping of polyethylene and other solid waste.

16.3.7.2 Will She Flow Again?

There are several dimensions of the question "Will she flow again?". First, the water level of the river Churni in the lean period was 1.025 m lower than the water level of the river Jalangi, maintaining a very gentle slope of 1.025/29000 or 1:28293 or 0.002°. Therefore, water may come into the river Churni from the river Jalangi provided the river Anjana is properly and adequately dredged for the purpose. Second, if the course within the municipality remains very narrow (minimum 4.75 m only), then during floods very few cusecs of water will enter through this reach which will never be able to scour the loose silt deposited annually from the wider reach of downstream. Because the concentrated flow of water passing through narrow off-take will spread over a wider channel downstream resulting in much lower velocity. Thus, lower velocity will not have enough potency to scour and carry loose silt deposited annually within the channel downstream. And within a few years, the river will be decayed again making the "SUJALA NADIA" project meaningless (ABP 4.12.2013, 29.05.2014, 18.05.2015, 16.06.2015, 14.07.2016; Telegraph, 22.06.2019; News18, 19.08.2019). Therefore, the first and foremost task of the project should be widening of the reach within the city to give the project sustainability, and it is only possible by removing the occupants from the Anjana bed. But the executing body might have faced extreme political restrictions when they moved for the end. Whatever be the reason, the project stopped.

16.4   CONCLUSION

Bengali newspaper *Anandabazar Patrika* on 29 May 2014 reported that even Krishnagar Municipality mentioned the river Anjana as a *khal* as if an artificial canal. The acute decaying of the river is a serious unsolved anthropogeomorphological issue of the district of Nadia. Form the above study, it can be concluded that although the thalweg levels of the river and its feeder river Jalangi are in such a condition that renovation of the river Anjana is theoretically possible, but urbanisation has encroached the river to such an extent that all the attempt to renovate the river may fail to be materialised until and unless fair political will and iron determined administration is employed to the end. And if not, floods like 2000 will take revenge again and again against limit-breaking anthropogenic interference with the landforms and processes shaping those landforms in the Anjana River basin.

NOTE

Debabrata Das, the second author of this chapter generated data and drew Figures 16.1, 16.2, 16.5, 16.30 and 16.31. He converted all other figures from JPEG to TIFF format. He also extracted data on width, cross-sectional area and the maximum depth of 64 cross-sections from diagrams by I&WD, Nadia, collected in 2014.

REFERENCES

ABP (04.12.2013) http://archives.anandabazar.com/archive/1131204/4mur1.html.

ABP (29.05.2014) https://www.anandabazar.com/district/nadia-murshidabad/%E0%A6%AA%E0%A6%B0-%E0%A6%9A%E0%A7%9F-%E0%A6%B9-%E0%A6%B0-%E0%A7%9F-%E0%A6%AA-%E0%A6%B0%E0%A6%B8%E0%A6%AD-%E0%A6%B0-%E0%A6%A8-%E0%A6%9F-%E0%A6%B8-%E0%A6%93-%E0%A6%85%E0%A6%9E-%E0%A6%9C%E0%A6%A8-%E0%A6%8F%E0%A6%96%E0%A6%A8-%E0%A6%96-%E0%A6%B2-1.35801.

ABP (16.06.2015) https://www.anandabazar.com/district/nadia-murshidabad/jalangi-river-and-anjana-canal-blocked-due-to-forcible-occupation-1.160686.

ABP (18.05.2015) https://www.anandabazar.com/state/anjana-river-will-vanish-in-coming-future-1.147150.

ABP (14.07.2016) https://www.anandabazar.com/district/nadia-murshidabad/krishnanagar-municipality-at-a-heavy-start-on-mamata-s-yes-1.434171.

ABP (14.01.2020) https://www.anandabazar.com/district/nadia-murshidabad/krishnanagar-people-want-anjana-river-back-in-their-life-1.1036453.

ABP (14.01.2020) https://www.anandabazar.com/state/anjana-river-will-vanish-in-coming-future-1.147150.

Bagchi K and Mukherjee KN (1978). *Diagnostic Survey of Deltaic West Bengal. A Research and Development Project*, Department of Geography, Calcutta University, Calcutta, p. 17.

Bandyopadhaya DK (2007). *Banglar Nad Nadi*. Dey's Publishing, Kolkata, p. 67.

Bharatchandra R (1752). *Annada Mangal*, Calcutta.

Bhattacharya JN (1896). *Hindu Castes and Sects*. Thacker, Spink and Co., Calcutta. Retrieved 6 February 2015.

Biswas KR (2001). *Rivers of Bengal*, Vol-I, Government of West Bengal, pp. xviii, xxix, 87, plate-18 and 19.

Biswas SK (2002). *Baadkulla Parichay* (in Bengali). Sahitya Sanshad, Badkulla, p. 38.

Bose NK (1972). The Bhagirathi-Hooghly – a few remarks. In Bagchi, K.G. (Ed.). *Proceedings of The Interdisciplinary Symposium*. University of Calcutta, Calcutta, pp. viii–xii.

Census of India, West Bengal, Nadia (2001), https://www.census2011.co.in/census/district/10-nadia.html.

Chatterjee SP (1972). The Bhagirathi-Hooghly Basin. In Bagchi, K.G. (Ed.). *Proceedings of The Interdisciplinary Symposium*. University of Calcutta, Calcutta, pp. XIX–XXIV.

Chattopadhyay S (2007). *Maharajendra Bahadur Krishnachandra* (in Bengali). Dey's Publishing, Kolkata, pp. 33–35, 52–55.

Das BC (2013). Changes and deterioration of the course of river Jalangi and its impact on the people living on its banks, West Bengal, Ph.D. Thesis, University of Calcutta, pp. 15, 104–120.

Das BC (2014) Impact of in-bed and on-bank soil cutting by brick fields on moribund deltaic rivers: a study of Nadia River in West Bengal. *The NEHU Journal*, ISSN. 0972-8406, Vol. XII, No. 2, July–December 2014, pp. 101–111.

Das T http://riverbangla.com/author/tapas/.

District Census Handbook, Nadia (1981). Directorate of census operations West Bengal, Village & Town Directory.

District Census Handbook, Nadia (2001). http://lsi.gov.in:8081/jspui/bitstream/123456789/5381/1/39718_2001_NAD.pdf.

Dutta J (2013). *Anjana River a Report*. Government of West Bengal. Executive Engineer. Nadia Irrigation Division, Krishnagar, Nadia, p. 2.

Dutta S (1996). *Anjana Nadi Tire* (in Bengali). Dogachi Gram Panchayet, pp. 25, 26, 30, 45, 47, 52–55.

Fergusson J (1912). *On Recent Changes in The Delta of the Ganges*. Bengal Secretariat Press, Calcutta. Reprinted from the Quaternary Journal of the Geological Society of London. Vol. XIX, in Biswas KR (2001). *Rivers of Bengal*, pp. 184–205.

Garrett JHE (1910). *Bengal District Gazetteer*. Bengal Secretariat Book Depot, Nadia. Reprinted in 2001. Published by Biswas KR (2001), pp. 1–28.

Gilbert GK (1877). *Report on the Geology of the Henry Mountains.* United States Geological Survey, Washington, DC. Rocky Mountain Region.

Hack JT (1960). Interpretation of erosional topography in humid temperate region. *American Journal of Science*, 258A, 80–97.

Hutton J (1788). Theory of the earth; or an investigation of the laws observable in the composition, dissolution, and restoration of land upon the globe. *Transactions of the Royal Society of Edinburgh*, 1(Part 2), 209–304.

Knighton D (1998). *Fluvial Forms and Processes, a New Perspective.* Arnold, New York. pp. 279.

Lazaro FV (1996). *Krishnagar Yajan Kshetrer Sankhipta Itihas* (in Bengali). Sadhu Josef Press, Krishnagar, pp. 20, 21, 23.

Lóczy D and Pirkhoffer E (2009). Mapping direct human impact on the topography of Hungary. *Zeitschrift für Geomorphologie*, 53(3), 145–155.

Majumder D (1978), *District Gazetteer.* Government of West Bengal, Nadia, p. 7.

Majumder SC (1941). *Rivers of the Bengal Delta.* Edited by Biswas K.R. (2001) *Rivers of Bengal*, Vol: I. Department of Higher Education, Govt. of West Bengal, pp. 17, 18, 54.

Mallick K (1911). *Nadia Kahini* (in Bengali). Ranaghat. Edited by Ray Mohit (1998). Kumudnath Mallik, Nadia Kahini. Pustak Bipani, Kolkata, pp. 32–34, 19,21, 121, 232-233, 229, 299, 322, 338.

McFeeters SK (1996). The use of the normalized difference water index (NDWI) in the delineation of open water features. *International Journal of Remote Sensing*, 17(7), 1425–1432. https://doi.org/https://doi.org/10.1080/01431169608948714

Moor Committee's Report (1919). *Report on the Hooghly River and Its Headwaters.* Edited by Biswas KR (2001), *Rivers of Bengal*, Vol-II, Government of West Bengal, pp. 1–15.

Mukherjee RK (1938). *The Changing Face of Bengal a Study in Riverine Economy.* University of Calcutta, Calcutta. Reprinted in 2009, pp. 1–12.

News18 (19.08.2019) https://bengali.news18.com/videos/south-bengal/news18-exclusive-on-why-anjana-river-has-lost-its-flow-rm-358621.html

News18 (19.08.2019) https://bengali.news18.com/videos/south-bengal/news18-exclusive-on-why-anjana-river-has-lost-its-flow-rm-358621.html

Playfair J (1802). *Illustrations of the Huttonian Theory of the Earth.* William Creech, Edinburgh.

Pringle JM and Kemm AH (1928). *Final Report on Survey and Settlement Operation in the District of Nadia, 1918–1926.* Government of Bengal, Calcutta.

Rarhi KC (1927). *Nabadwip Mohima.* Edited by Choudhuri J. (2004). Nabadwip Puratattva Parishad, p. 177.

Ray KC (1875). *Kshitish Banshavali Charita* (in Bengali). Natun Sanskrita Yantra. Edited by Ray Mohit (2003). Kshitish Banshavali Charita. Balaka, Kolkata, pp. 33, 54.

Reaks HG (1919). Report on the physical and hydraulic characteristics of the rivers of the delta. In *Report on the Hooghly River and Its Head-Waters.* Vol-I, The Bengal Secretariat Book Depot, Calcutta, In Biswas (2001), *Rivers of Bengal*, Vol-II, Government of West Bengal, pp. 87, 107.

Sanyal N (2014). Krishnanagar Saharer Patton O Bibartan (in Bengali). In Saha AK, Saha RN and Saha S (Eds.) *Raja Raghab Rayer Krishnanagar (Bengali).* Published by Debabrata Sarkar, p. 77–86.

Shatskiy NS (1937). On neocatastrophism: a contribution to the problem of organic phases and on folding. *Probl. Sov. Geol.*, 7(7), 532–551, 1937a. (Referenced in: AI Suvorov, "Nikolay Syergyeyevich Shatskiy and the present time (in commemoration of his centennial birthday)", Geotectonics, July-August 1995, English Translation in *Geotectonics*, Vol.29 No.4 Feb 1996.

Sherwill WS (1858). *Report on the Rivers of Bengal*, 29. G. A. Savielle Printing and Publishing Co. Ltd, Calcutta, p. 39.

Tagore R (1900). *Ek Gnaye.* Kshanika, Sanchayita. (1969). Visva Bharati Grantha Bivag. Kolkata, p. 418.

Tagore R (1979). *Aagamani* (in Bengali). Edited by Visva Bharati. Sahaj Path. Dwitiya Bhag. Rabindra Rachanavali. (Sulav Sanskaran), p. 467, 468.

Telegraph India (22.06.2019) https://www.telegraphindia.com/states/west-bengal/on-the-trail-of-the-vanishing-waterways-of-bengal/cid/1692964

Wolman MG and Millar JP (1960). Magnitude and frequency of forces in geomorphic processes. *Journal of Geology*, 68, 54–74.

Zha Y, Gao J and Ni S (2003). Use of normalized difference built-up index in automatically mapping urban areas from TM imagery. *International Journal of Remote Sensing*, 24(3), 583–594.

https://www.youtube.com/watch?v=_pPtNLguR8I.

Index

Note: **Bold** page numbers refer to tables and *Italic* page numbers refer to figures.